# Study Guide and Student Solutions Manual

# Fundamentals of Organic Chemistry

**SEVENTH EDITION**

**John McMurry**



Prepared by

**Susan McMurry**

BROOKS/COLE
CENGAGE Learning

Australia • Brazil • Japan • Korea • Mexico • Singapore • Spain • United Kingdom • United States

BROOKS/COLE
CENGAGE Learning



For product information and technology assistance, contact us at **Cengage Learning Customer & Sales Support, 1-800-354-9706**

For permission to use material from this text or product, submit all requests online at **www.cengage.com/permissions**
Further permissions questions can be emailed to **permissionrequest@cengage.com**

ISBN-13: 978-0-538-73676-3
ISBN-10: 0-538-73676-3

**Brooks/Cole**
20 Davis Drive
Belmont, CA 94002-3098
USA

Cengage Learning is a leading provider of customized learning solutions with office locations around the globe, including Singapore, the United Kingdom, Australia, Mexico, Brazil, and Japan. Locate your local office at: **www.cengage.com/global**

Cengage Learning products are represented in Canada by Nelson Education, Ltd.

To learn more about Brooks/Cole, visit **www.cengage.com/brookscole**

Purchase any of our products at your local college store or at our preferred online store **www.CengageBrain.com**

Printed in the United States of America
1 2 3 4 5 6 7 14 13 12 11 10

# Contents

## Solutions to Problems:



## Appendices:

# Preface

If you're in a typical organic chemistry course, you attend lecture and take notes, read the text, work some problems, and take tests. For many of you, performing this ritual is all that's necessary to succeed in organic chemistry. Many others of you, however, follow all the correct steps but still feel bewildered by the course. In such a situation, supplementary material is needed.

This study guide has two functions. First, it gives an overview of the course, both in individual chapters and in appendices. Second, it furnishes solutions to the problems presented in the text. The first of these functions might be described as "the big picture," and the second as "the details." Understanding both the big picture and the details is necessary if organic chemistry is to be more than the memorization of assorted unrelated facts.

### How to use this book

This *Study Guide and Solutions Manual* can't perform miracles if you don't read *Fundamentals of Organic Chemistry*. In all cases, step one is to go to class, take notes, and read the text. Then, turn to the Study Guide.

Study the Chapter Outline. The outline should help you to see how the topics in the chapter are related. Many people are able to learn the facts in each chapter, but are unable to recognize the underlying principles.The outline will clarify both the relationships between reactions and how these reactions are related to larger concepts.

Solve the problems in the text. At first, don't use the *Solutions Manual* to help you with the problems. After completing the problems, consult the *Solutions Manual* to see if your answer is correct and if your method of solution is logical and systematic. If you're totally confused by a problem, carefully read the solution; then try to solve similar problems on your own. If possible, use molecular models to make the three-dimensional nature of organic compounds more clear.

Before a test or final exam, consult the appendices in the *Study Guide*. Many of these appendices summarize or tabulate information that has been presented over several chapters. Especially helpful before an exam are the following sections: Glossary, Reagents Used in Organic Synthesis, Summary of Functional Group Preparations, and Summary of General Reaction Mechanisms. Other appendices present interesting chemical facts and tables.

Review quizzes are located at the end of the book. Each quiz covers material from a few chapters and allows you check your knowledge of more than one chapter at a time.

For most people, understanding organic chemistry takes a long time—sometimes longer than the duration of an organic chemistry course. I hope that the combination of *Fundamentals of Organic Chemistry* plus this *Study Guide and Solutions Manual* makes the study of organic chemistry easier and more rewarding for you.

## Acknowledgments

I would like to thank the following people for their assistance with this book: John McMurry (of course), Bette Kreuz, and Elizabeth Woods.

# Chapter 1 – Structure and Bonding; Acids and Bases

## Chapter Outline

I. Atomic structure (Sections 1.1–1.2).
   A. Introduction to atomic structure (Section 1.1).
      1. Atoms consist of a dense, positively charged nucleus surrounded by negatively charged electrons.
         a. The nucleus is made up of positively charged protons and uncharged neutrons.
         b. The nucleus contains most of the mass of the atom.
         c. Electrons move about the nucleus at a distance of about $2 \times 10^{-10}$ m (200 pm).
      2. The atomic number *(Z)* gives the number of protons in the nucleus.
      3. The mass number *(A)* gives the total number of protons and neutrons.
         a. Isotopes of an element have the same number of protons and electrons but differ in the number of neutrons.
      4. All atoms of a given element have the same value of Z.
   B. Orbitals.
      1. The distribution of electrons in an atom can be described by a wave equation.
         a. The solution to a wave equation is an orbital, represented by $\Psi$.
         b. $\Psi^2$ predicts the volume of space in which an electron is likely to be found.
      2. There are four different kinds of orbitals (*s*, *p*, *d*, *f*).
         a. The *s* orbitals are spherical.
         b. The *p* orbitals are dumbbell-shaped.
         c. Four of the five *d* orbitals are cloverleaf-shaped.
      3. An atom's orbitals are organized into shells.
         a. The shells differ in the numbers and kinds of orbitals they have.
         b. Electrons in different orbitals have different energies.
         c. Each orbital can hold two electrons.
         d. The two lowest-energy electrons are in the 1*s* orbital.
   C. Electronic configuration of atoms (Section 1.2).
      1. The ground-state electron configuration of an atom is a listing of the orbitals occupied by the electrons of the atom.
      2. Rules for predicting the ground-state electron configuration of an atom:
         a. Orbitals with the lowest energy levels are filled first.
            i. The order of filling is 1*s*, 2*s*, 2*p*, 3*s*, 3*p*, 4*s*, 3*d*.
         b. Only two electrons can occupy each orbital, and they must be of opposite spin.
         c. If two or more orbitals have the same energy, one electron occupies each orbital until all are half-full (Hund's rule). Only then does a second electron occupy one of the orbitals.
            i. All of the electrons in a half-filled shell have the same spin.

II. Chemical bonds (Sections 1.3–1.5).
   A. Development of chemical bonding theory (Section 1.3).
      1. Kekulé and Couper proposed that carbon is tetravalent.
      2. Kekulé also suggested that carbon can form double bonds, triple bonds and rings.
      3. Van't Hoff proposed that the 4 atoms to which carbon forms bonds sit at the corners of a regular tetrahedron.
      4. In a drawing of a tetrahedral carbon, a wedged line represents a bond pointing toward the viewer, a dashed line points behind the plane of the page, and a solid line lies in the plane of the page.

B. Chemical bonds (Section 1.4).
  1. Atoms bond together because the resulting compound is more stable than the individual atoms.
    a. Atoms tend to achieve the electron configuration of the nearest noble gas.
    b. Atoms either lose electrons (groups 1A, 2A) or gain electrons (group 7A) to form ionic compounds.
    c. Atoms in the middle of the periodic table share electrons by forming covalent bonds.
  2. The number of covalent bonds formed by an atom depends on the number of valence electrons it has and on the number it needs to achieve an octet.
  3. Covalent bonds can be represented two ways.
    a. In electron-dot structures, bonds are represented as pairs of dots.
    b. In line-bond structures, bonds are represented as lines drawn between two atoms.
  4. Valence electrons not used for bonding are called lone-pair electrons.
    a. Lone-pair electrons are represented as dots.

C. Formation of covalent bonds: valence bond theory (Section 1.5).
  1. Covalent bonds are formed by the overlap of two atomic orbitals, each of which contains one electron. The two electrons have opposite spins.
  2. Bond strength is the measure of the amount of energy needed to break a bond.
  3. Bond length is the optimum distance between nuclei.
  4. Every bond has a characteristic bond length and bond strength.

III. Hybridization (Sections 1.6–1.8).

A. $sp^3$ Orbitals (Sections 1.6, 1.7).
  1. Structure of methane (Section 1.6).
    a. When carbon forms 4 bonds with hydrogen, one 2*s* orbital and three 2*p* orbitals combine to form four equivalent atomic orbitals ($sp^3$ hybrid orbitals).
    b. These orbitals are tetrahedrally oriented.
    c. Because these orbitals are unsymmetrical about the nucleus, they can form stronger bonds than unhybridized orbitals can.
    d. Each C–H bond has an energy of 439 kJ/mol.
    e. These bonds have a specific geometry and a bond angle of 109.5°.
  2. Ethane has the same type of hybridization as occurs in methane (Section 1.7).

B. $sp^2$ Orbitals (Section 1.8).
  1. If one carbon 2*s* orbital combines with two carbon 2*p* orbitals, three hybrid $sp^2$ orbitals are formed, and one *p* orbital remains unchanged.
  2. The three $sp^2$ orbitals lie in a plane at angles of 120°, and the *p* orbital is perpendicular to them.
  3. Two different types of bond form between two carbons.
    a. A $\sigma$ bond forms from the overlap of two $sp^2$ orbitals.
    b. A $\pi$ bond forms by sideways overlap of two *p* orbitals above and below the internuclear axis.
    c. This combination is known as a carbon–carbon double bond.

C. *sp* Orbitals.
  1. If one carbon 2*s* orbital combines with one carbon 2*p* orbital, two hybrid *sp* orbitals are formed, and two *p* orbitals are unchanged.
  2. The two *sp* orbitals are 180° apart, and the two *p* orbitals are perpendicular to them and to each other.
  3. Two different types of bonds form.
    a. A $\sigma$ bond forms from the overlap of two *sp* orbitals.
    b. Two $\pi$ bonds form by sideways overlap of four *p* orbitals.
    c. This combination is known as a carbon–carbon triple bond.

IV. Polar covalent bonds: electronegativity (Section 1.9).
  A. Although some bonds are totally ionic and some are totally covalent, most chemical bonds are polar covalent bonds.
    1. In these bonds, electrons are attracted to one atom more than to the other atom.
  B. Bond polarity is due to differences in electronegativity (EN).
    1. Elements on the right side of the periodic table are more electronegative than elements on the left side.
  C. The difference in EN between two elements can be used to predict the polarity of a bond.
    1. If $\Delta EN < 0.4$, a bond is nonpolar covalent.
    2. If $\Delta EN$ is between 0.4 and 2.0, a bond is polar covalent.
    3. If $\Delta EN > 2.0$, a bond is ionic.
    4. The symbols $\delta+$ and $\delta-$ are used to indicate partial charges.
    5. A crossed arrow is used to indicate bond polarity.
      a. The arrow points in the direction of the negative charge.
  D. An inductive effect is an atom's ability to polarize a bond.
V. Acids and bases (Sections 1.10–1.12).
  A. Brønsted–Lowry definition (Section 1.10).
    1. A Brønsted–Lowry acid donates an $H^+$ ion; a Brønsted–Lowry base accepts an $H^+$ ion.
    2. The product that results when a base gains $H^+$ is the conjugate acid of the base; the product that results when an acid loses $H^+$ is the conjugate base of the acid.
  B. Acid and base strength.
    1. A strong acid reacts almost completely with water.
    2. The strength of an acid in water is indicated by $K_a$, the acidity constant.
    3. Strong acids have large acidity constants, and weaker acids have smaller acidity constants.
    4. The $pK_a$ is normally used to express acid strength.
      a. $pK_a = -\log K_a$.
      b. A strong acid has a small $pK_a$, and a weak acid has a large $pK_a$.
      c. The conjugate base of a strong acid is a weak base, and the conjugate base of a weak acid is a strong base.
    5. Predicting acid–base reactions from $pK_a$.
      a. An acid with a low $pK_a$(strong acid) reacts with the conjugate base of an acid with a high $pK_a$(strong base).
      b. A proton always goes from the stronger acid to the stronger base in an acid–base reaction.
      b. In other words, the products of an acid–base reaction are more stable than the reactants, and the reaction favors formation of the weaker acid/base pair.
  C. Organic acids and organic bases (Section 1.11).
    1. There are two main types of organic acids:
      a. Acids that contain hydrogen bonded to oxygen.
      b. Acids that have hydrogen bonded to the carbon next to a C=O group.
    2. The main type of organic base contains a nitrogen atom with a lone electron pair.
    3. Amino acids contain both organic acid and organic base functional groups in the same molecule.
  C. Lewis acids and bases (Section 1.12).
    1. A Lewis acid accepts an electron pair.
      a. A Lewis acid may have either a vacant low-energy orbital or a polar bond to hydrogen.
      b. Examples include metal cations, halogen acids, Group 3 compounds and transition-metal compounds.
    2. A Lewis base has a pair of nonbonding electrons.
      a. Most oxygen- and nitrogen-containing organic compounds are Lewis bases.

b. Many organic Lewis bases have more than one basic site.

3. A curved arrow shows the movement of electrons from a Lewis base to a Lewis acid.
   a. A pair of electrons moves from the atom at the tail of the arrow to the atom at the head of the arrow.

## Solutions to Problems

**1.1** The elements of the periodic table are organized into groups that are based on the number of outer-shell electrons each element has. For example, an element in group 1A has one outer-shell electron, and an element in group 5A has five outer-shell electrons. To find the number of outer-shell electrons for a given element, use the periodic table to locate its group.

(a) Potassium is a member of group 1A and thus has one outer-shell electron.
(b) Calcium (group 2A) has two outer-shell electrons.
(c) Aluminum (group 3A) has three outer-shell electrons.

**1.2** (a) To find the ground-state electron configuration of an element, first locate its atomic number. For boron, the atomic number is 5; boron thus has 5 protons and 5 electrons. Next, assign the electrons to the proper energy levels, starting with the lowest level. Fill each level *completely* before assigning electrons to a higher energy level.

Boron

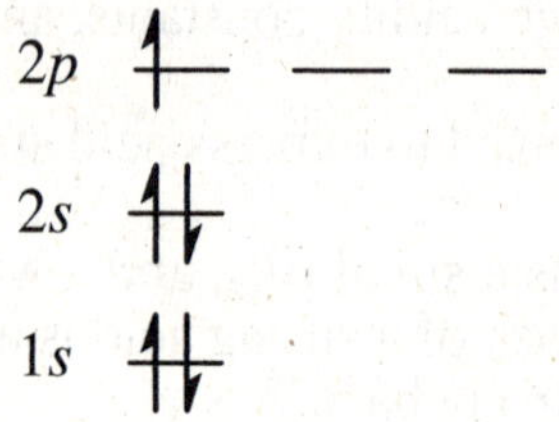

Remember that only two electrons can occupy the same orbital, and that they must be of opposite spin.

A different way to represent the ground-state electron configuration is to simply write down the occupied orbitals and to indicate the number of electrons in each orbital. For example, the electron configuration for boron is $1s^2\ 2s^2\ 2p^1$.

(b) Let's consider an element with many electrons. Phosphorus, with an atomic number of 15, has 15 electrons. Assigning these to energy levels:

Phosphorus

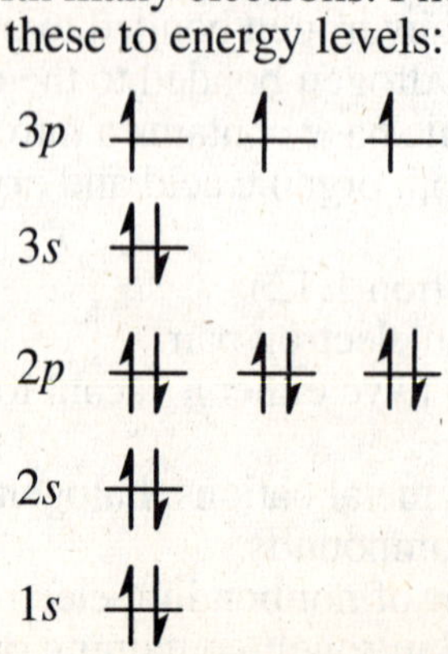

Notice that the 3*p* electrons are all in different orbitals; we must place one electron into each orbital of the same energy level until all orbitals are half-filled.

The more concise way to represent ground-state electron configuration for phosphorus: $1s^2\ 2s^2\ 2p^6\ 3s^2\ 3p^3$.

(c) Oxygen (atomic number 8)

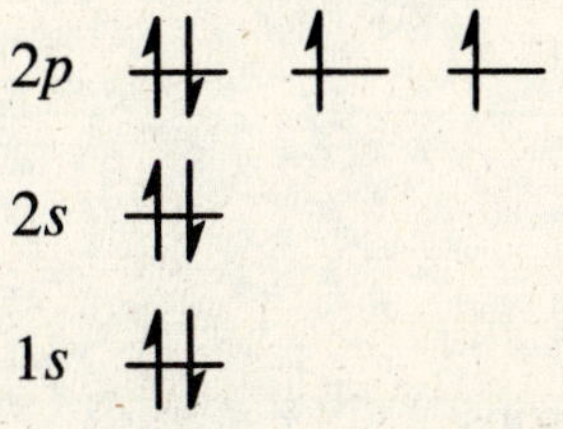


$1s^2\ 2s^2\ 2p^4$

(d) Argon (atomic number 18)

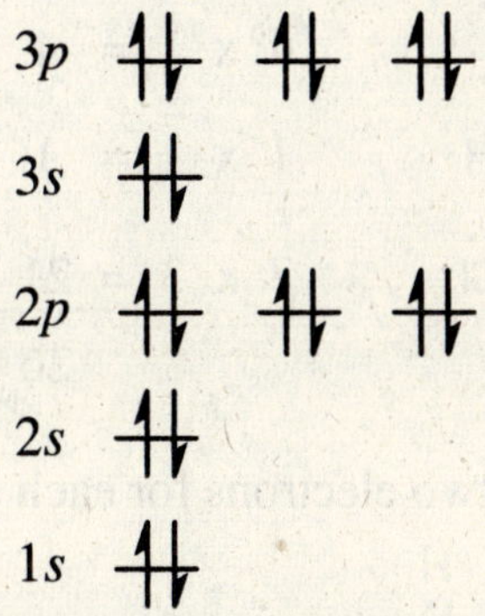


$1s^2\ 2s^2\ 2p^6\ 3s^2\ 3p^6$

**1.3** A solid line represents a bond lying in the plane of the page, a wedged bond represents a bond pointing out of the plane of the page toward the viewer, and a dashed bond represents a bond pointing behind the plane of the page.

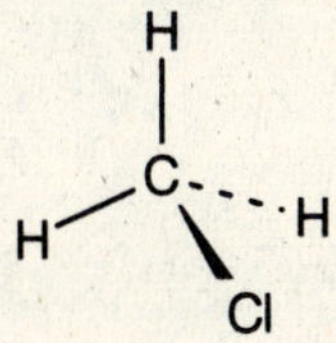


Chloromethane

**1.4**

Ethane

**1.5** Identify the group of the central element to predict the number of covalent bonds the element can form.

(a) Carbon (group 4A) has four electrons in its valence shell and forms four bonds to achieve the noble-gas configuration of neon. Thus, a likely formula is $CCl_4$.

| *Element* | *Group* | *Likely Formula* |
|---|---|---|
| (b) Al | 3A | $AlH_3$ |
| (c) C | 4A | $CH_2Cl_2$ |
| (d) Si | 4A | $SiF_4$ |

**1.6** Follow these three steps for drawing the electron-dot structure of a molecule.

(1) Determine the number of valence, or outer-shell, electrons for each atom in the molecule. For chloroform, we know that carbon has four valence electrons, hydrogen has one valence electron, and each chlorine has seven valence electrons.

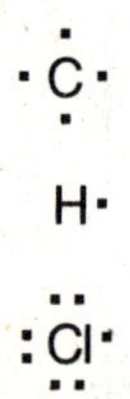

$4 \times 1 = 4$

$1 \times 1 = 1$

$7 \times 3 = 21$

26 total valence electrons

(2) Next, use two electrons for each single bond.

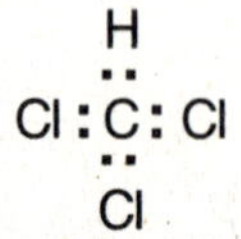

(3) Finally, use the remaining electrons to achieve an noble gas configuration for all atoms.

| | *Molecule* | *Electron-dot structure* | *Line-bond structure* |
|---|---|---|---|
| (a) | $CHCl_3$ | 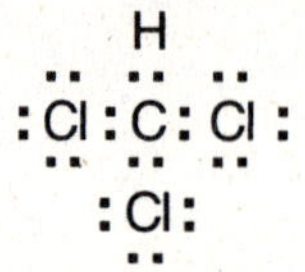 | 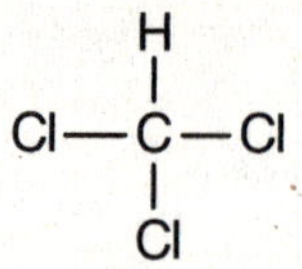 |
| (b) | $H_2S$<br>8 valence electrons | 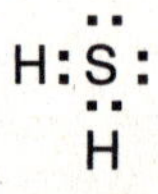 | 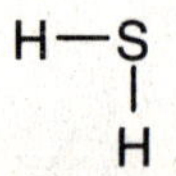 |
| (c) | $CH_3NH_2$<br>14 valence electrons | 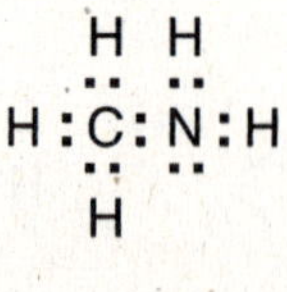 | 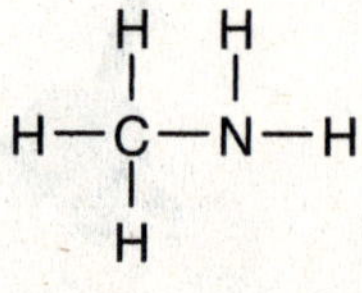 |

**1.7** Each of the two carbons has 4 valence electrons. Two electrons are used to form the carbon–carbon bond, and the 6 electrons that remain can form bonds with a maximum of 6 hydrogens. Thus, the formula $C_2H_7$ is not possible.

**1.8**

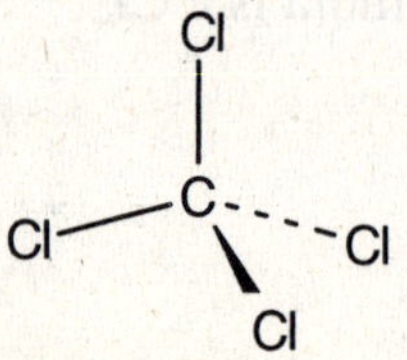

Tetrachloromethane

**1.9** An electron in the second shell is farther from the nucleus than an electron in the first shell. Thus, a bond formed by the overlap of an $sp^3$ orbital of carbon (second shell) and a 1*s* orbital of hydrogen (first shell) is longer than a bond formed from two 1*s* orbitals (H–H bond).

**1.10**

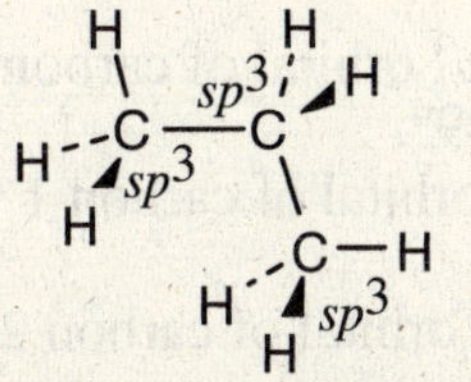

Propane

The geometry around all carbon atoms is tetrahedral, and all bond angles are approximately 109°.

**1.11** The atoms of acetaldehyde contribute 18 valence electrons. Ten electrons are used in the 5 single bonds, 4 electrons are involved in the carbon–oxygen double bond, and 4 electrons form the 2 lone pairs of electrons on oxygen.

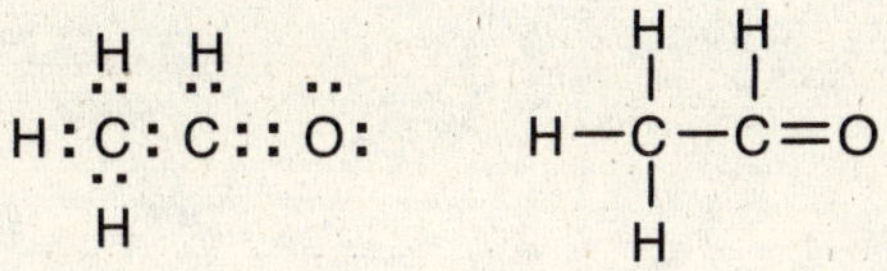

Acetaldehyde

**1.12**

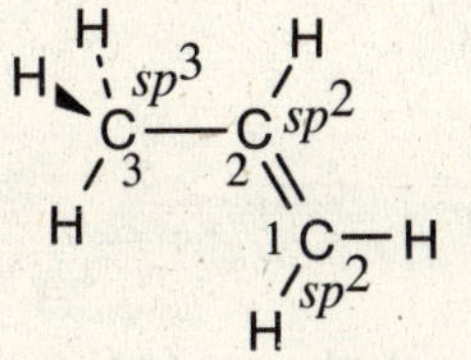

Propene

The C3–H bonds are $\sigma$ bonds formed by overlap of an $sp^3$ orbital of carbon 3 with an *s* orbital of hydrogen. Bond angles at C3 are approximately 109°.

The C2–H and C1–H bonds are $\sigma$ bonds formed by overlap of an $sp^2$ orbital of carbon with an *s* orbital of hydrogen.

The C2–C3 bond is a $\sigma$ bond formed by overlap of an $sp^3$ orbital of carbon 3 with an $sp^2$ orbital of carbon 2.

There are two C1–C2 bonds. One is a $\sigma$ bond formed by overlap of an $sp^2$ orbital of carbon 1 with an $sp^2$ orbital of carbon 2. The other bond is a $\pi$ bond formed by overlap of a 2*p* orbital of carbon 1 with a 2*p* orbital of carbon 2. All four atoms connected to the carbon–carbon double bond lie in the same plane, and all bond angles between these atoms are 120°.

**1.13**

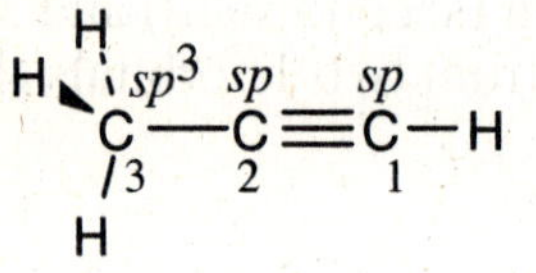


Propyne

The C3–H bonds are $\sigma$ bonds formed by overlap of an $sp^3$ orbital of carbon 3 with an $s$ orbital of hydrogen. Bond angles at C3 are approximately 109°.

The C1–H bond is a $\sigma$ bond formed by overlap of an $sp$ orbital of carbon 1 with an $s$ orbital of hydrogen.

The C2–C3 bond is a $\sigma$ bond formed by overlap of an $sp$ orbital of carbon 2 with an $sp^3$ orbital of carbon 3.

There are three C1–C2 bonds. One is a $\sigma$ bond formed by overlap of an $sp$ orbital of carbon 1 with an $sp$ orbital of carbon 2. The other two bonds are $\pi$ bonds formed by overlap of two $2p$ orbitals of carbon 1 with two $2p$ orbitals of carbon 2.

The three carbon atoms of propyne lie on a straight line, with a bond angle of 180°. The H–$C_1$–$C_2$ bond angle is also 180°.

**1.14**

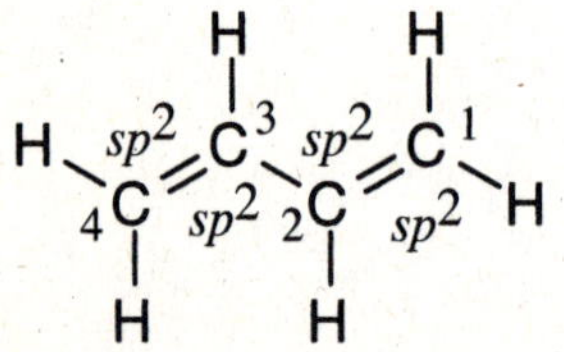


Buta-1,3-diene

All atoms lie in the same plane, and all bond angles are approximately 120°.

**1.15**

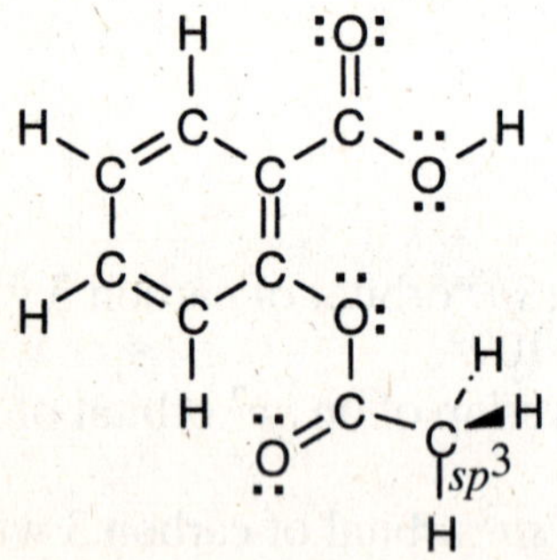


Aspirin

All carbons are $sp^2$ hybridized, with the exception of the indicated carbon. All oxygen atoms have two lone pairs of electrons.

**1.16** Use Figure 1.14 to answer this problem. The larger the EN, the more electronegative the element.

| | *More electronegative* | | *Less electronegative* | |
|---|---|---|---|---|
| (a) | H | (2.1) | Li | (1.0) |
| (b) | Br | (2.8) | Be | (1.6) |
| (c) | Cl | (3.0) | I | (2.5) |

**1.17** As in Problem 1.16, use Figure 1.14.

(a) $\overset{\delta+}{H_3C}-\overset{\delta-}{Br}$

(b) $\overset{\delta+}{H_3C}-\overset{\delta-}{NH_2}$

(c) $\overset{\delta-}{H_2N}-\overset{\delta+}{H}$

(d) $H_3C-SH$

Carbon and sulfur have identical electronegativies.

(e) $\overset{\delta-}{H_3C}-\overset{\delta+}{MgBr}$

(f) $\overset{\delta+}{H_3C}-\overset{\delta-}{F}$

**1.18** Use Figure 1.14 to locate the EN of each element. The larger the difference in EN, the more ionic the bond.

| $CCl_4$ | $MgCl_2$ | $TiCl_3$ | $Cl_2O$ |
|---|---|---|---|
| Cl : EN = 3.0 | Cl : EN = 3.0 | Cl : EN = 3.0 | Cl : EN = 3.0 |
| C : EN = 2.5 | Mg : EN = 1.2 | Ti : EN = 1.5 | O : EN = 3.5 |
| Δ EN = 0.5 | Δ EN = 1.8 | Δ EN = 1.5 | Δ EN = 0.5 |

*Least ionic* —> *Most ionic*

$CCl_4$ and $Cl_2O$, $TiCl_3$, $MgCl_2$

**1.19** In an electrostatic potential map, the color red indicates the areas of a molecule that are most electron-rich. The map shows that chlorine is the most electronegative atom in chloromethane, and the direction of polarity of the C–Cl bond is:

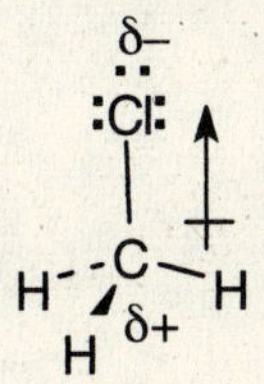


Chloromethane

**1.20** (a) On a calculator, enter 3.75, and give it a negative sign, using the +/- key. Use INV, then LOG (or use ANTILOG), and convert the answer to scientific notation. $K_a$ of formic acid is 1.8 x $10^{-4}$, and $K_a$ of picric acid is 0.42.

(b) Since a lower p$K_a$ value indicates a stronger acid, picric acid is stronger than formic acid.

**1.21** HO–H is a stronger acid than $H_2N$–H. Since $H_2N^-$ is a stronger base than $HO^-$, the conjugate acid of $H_2N^-$ ($H_2N$–H) is a weaker acid than the conjugate acid of $HO^-$ (HO–H).

**1.22**

(a)

$$H{-}CN + CH_3CO_2^-\,Na^+ \xrightarrow{?} Na^+\,^-CN + CH_3CO_2H$$

p$K_a$ = 9.3 (HCN), Weaker acid; p$K_a$ = 4.8 ($CH_3CO_2H$), Stronger acid

Remember that the lower the p$K_a$, the stronger the acid. Thus $CH_3CO_2H$, not HCN, is the stronger acid, and the above reaction will not take place in the direction written.

(b)

$$CH_3CH_2O{-}H + Na^+ \ {}^-CN \xrightarrow{?} CH_3CH_2O^- \ Na^+ + HCN$$

$pK_a = 16$ Weaker acid — $pK_a = 9.3$ Stronger acid

Using the same reasoning as in part (a), we can see that the above reaction will not take place.

**1.23** A Lewis base has a nonbonding electron pair to share. A Lewis acid has a vacant orbital to accept an electron pair. Look for a lone electron pair when identifying a Lewis base.

Lewis acids: $MgBr_2$, $B(CH_3)_3$, ${}^+CH_3$

Lewis bases: $CH_3\ddot{N}HCH_3$, $CH_3\ddot{P}CH_3$ (with $CH_3$ on P)

Both: $CH_3CH_2\ddot{O}{-}H$

**1.24** Locate the electron pair(s) of the Lewis base and draw a curved arrow from the electron pair to the Lewis acid. The electron pair moves from the atom at the tail of the arrow (Lewis base) to the atom at the point of the arrow (Lewis acid).

(Note: electron dots have been omitted from Cl⁻ to reduce clutter.)

(a)

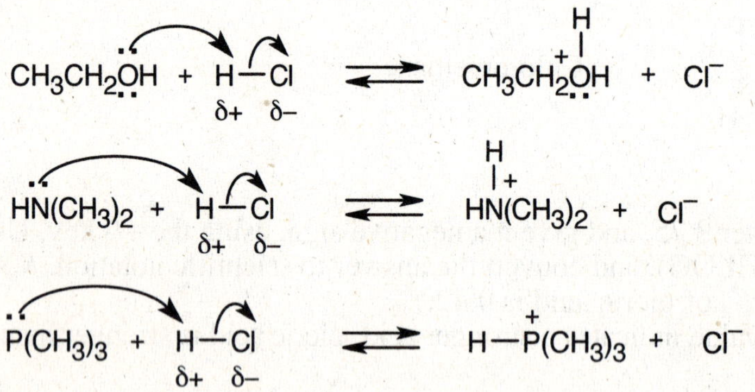


(b)

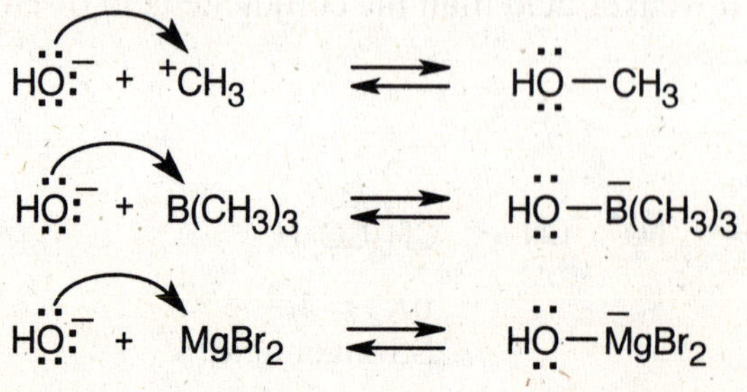

**1.25** According to the electrostatic potential map, the starred hydrogen is most acidic (blue), and the starred nitrogen is most basic (red ).

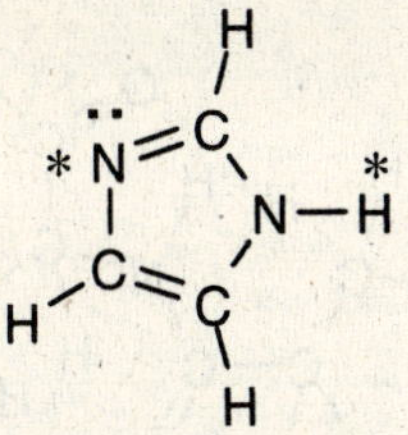

Imidazole

## Visualizing Chemistry

**1.26** Citric acid ($C_6H_8O_7$) contains seven oxygen atoms, each of which has two electron lone pairs. Three of the oxygens form double bonds with carbon.

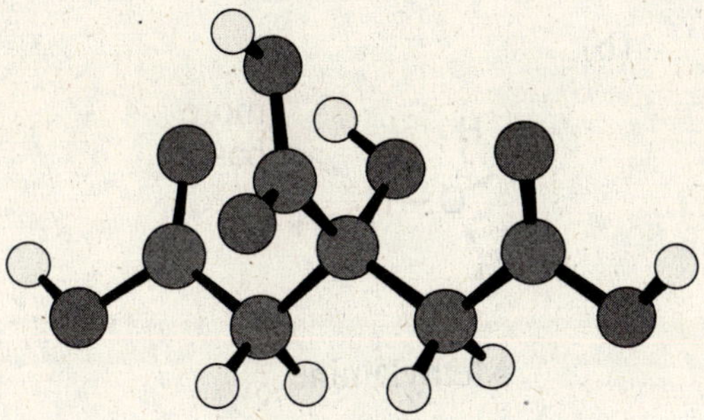

**Citric acid**

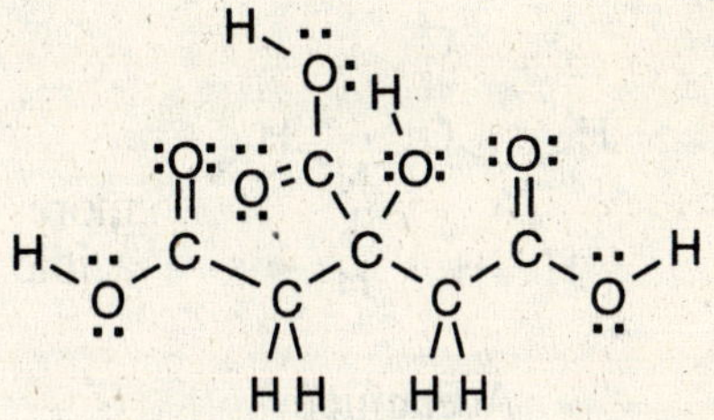

**1.27**

(a)

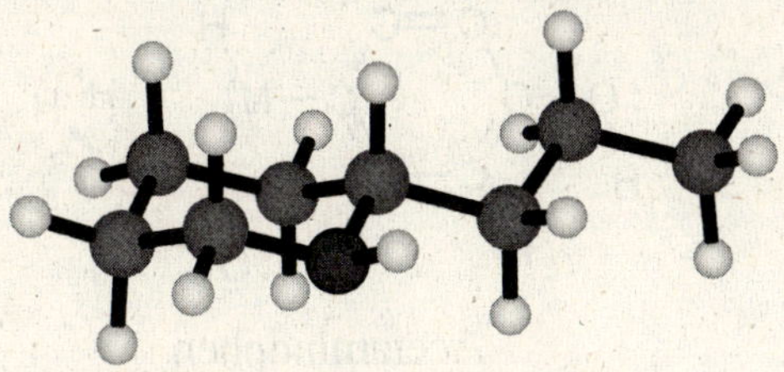

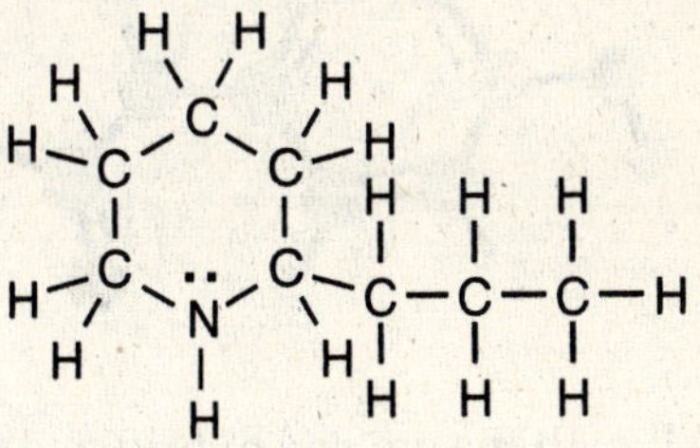

$C_8H_{17}N$

(b)

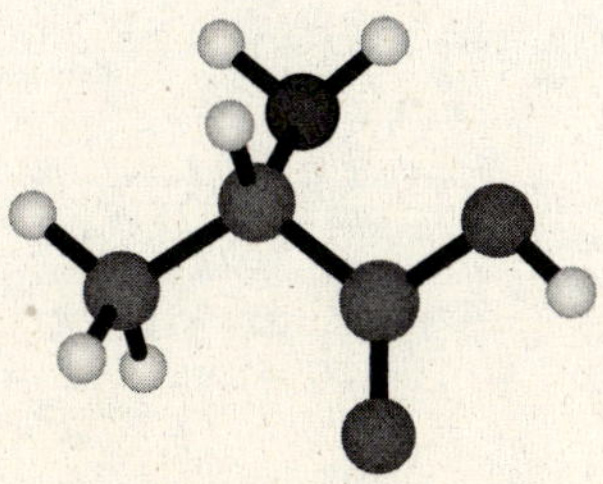

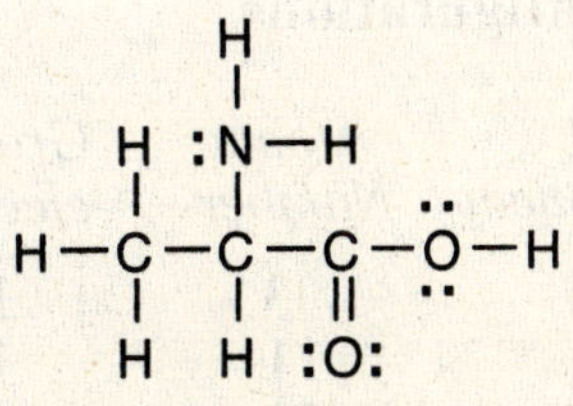

$C_3H_7NO_2$

**1.28**

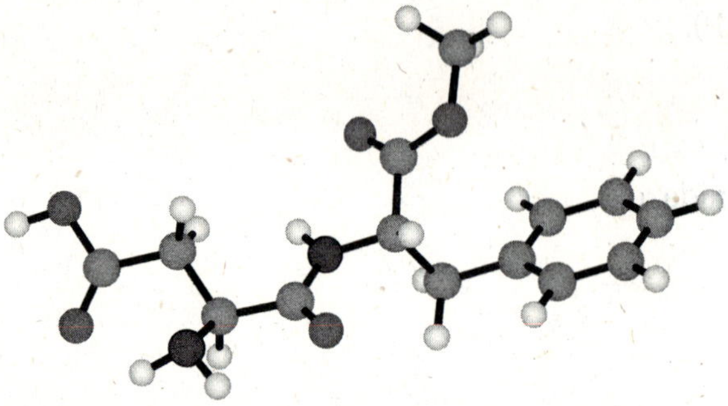

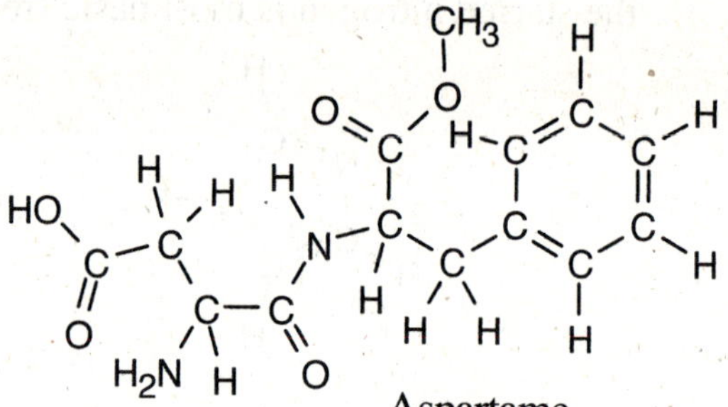

Aspartame

**1.29** Electrostatic potential maps show that the most electron-rich atom in (a) is oxygen, not nitrogen. In (b), nitrogen is electron-rich and is thus more basic than the nitrogen in (a). In (a), the hydrogen atoms bonded to nitrogen are more acidic (electron-poor) than all other hydrogens in both molecules. Remember that electron-rich regions are shown in red, and electron-poor regions are shown in blue.

(a)

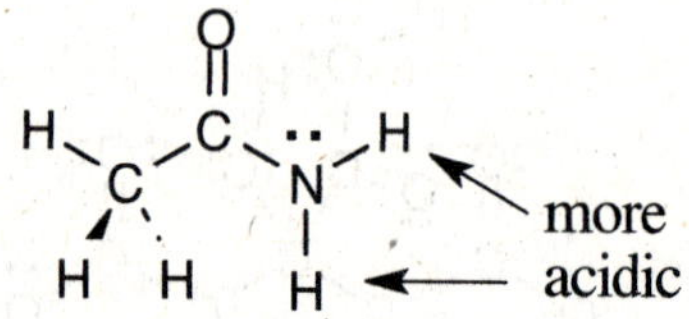


Acetamide

(b)

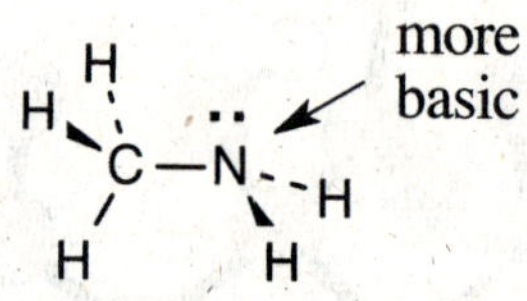


Methylamine

**1.30**

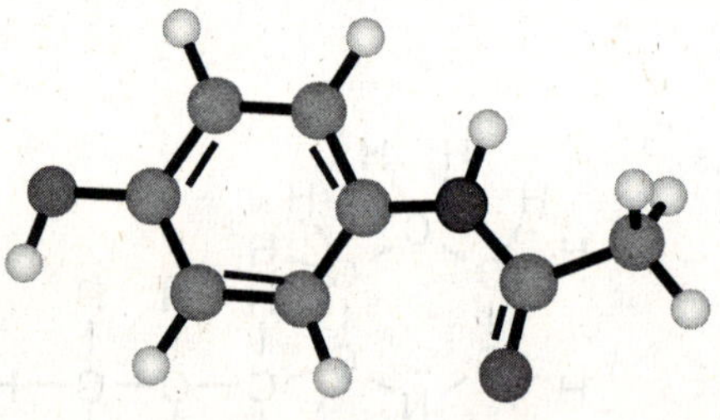

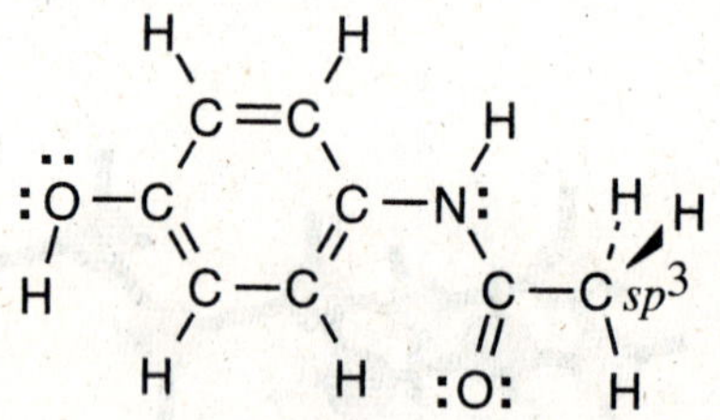


Acetaminophen

All carbons are $sp^2$ hybridized, except for the carbon indicated as $sp^3$. The two oxygen atoms and the nitrogen atom have lone pair electrons, as shown.

## Additional Problems

### Electron Configurations

**1.31**

| Element | Atomic Number | Ground-state electron configuration |
|---|---|---|
| (a) Li | 3 | $1s^2\ 2s^1$ |
| (b) Na | 11 | $1s^2\ 2s^2\ 2p^6\ 3s^1$ |
| (c) Al | 13 | $1s^2\ 2s^2\ 2p^6\ 3s^2\ 3p^1$ |
| (d) S | 16 | $1s^2\ 2s^2\ 2p^6\ 3s^2\ 3p^4$ |

**1.32** See Problem 1.1 if you need help.

| *Element* | *Group* | *Number of outer shell electrons* |
|---|---|---|
| (a) Oxygen | 6A | 6 |
| (b) Magnesium | 2A | 2 |
| (c) Fluorine | 7A | 7 |

## Electron-dot Structures

**1.33**

(a)

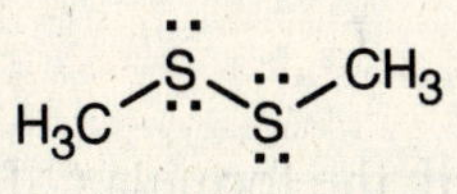

(b)

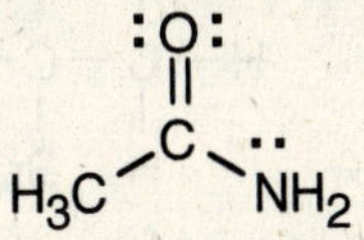

(c)

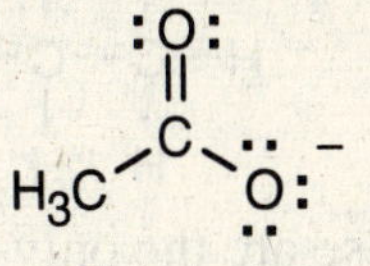

**1.34** (a) The 4 valence electrons of carbon can form bonds with a maximum of 4 hydrogens. Thus, it is not possible for the compound $CH_5$ to exist.

(b) If you try to draw a molecule with the formula $C_2H_6N$, you will see that it is impossible for both carbons and nitrogen to have a complete octet of electrons. Therefore, $C_2H_6N$ is unlikely to exist.

(c) A compound with the formula $C_3H_5Br_2$ doesn't have filled outer shells for all atoms and is thus unlikely to exist.

**1.35** (a) $AlCl_3$ (b) $CF_2Cl_2$ (c) $NI_3$ (d) $CH_2O$

**1.36**

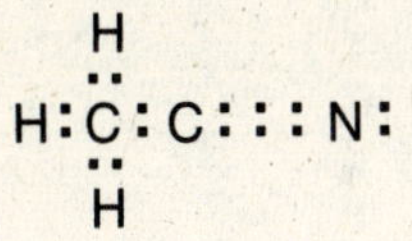

Acetonitrile

In acetonitrile, nitrogen has 8 electrons in its valence shell. Six are used in the carbon–nitrogen triple bond, and two are a nonbonding electron pair.

## Structural Formulas

**1.37** To solve a problem of this sort, draw all possible structures that have the correct number of bonds. You must systematically consider all possible arrangements of atoms, including those that have branches, rings and multiple bonds.

(a)

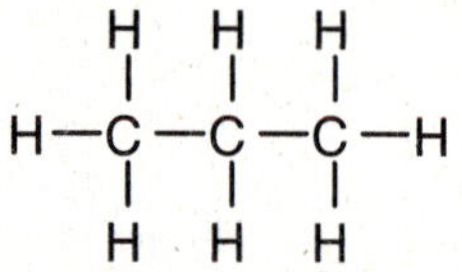

(b)

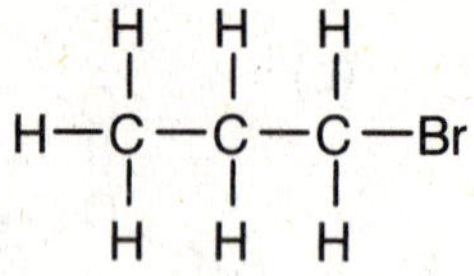 *and* 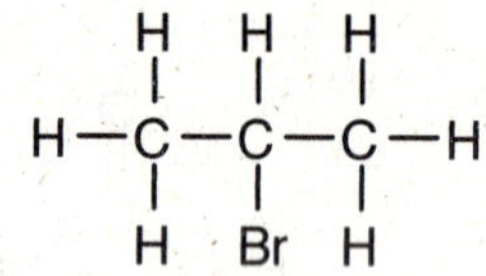

These are the only two possible structures for compounds with the formula $C_3H_7Br$.

(c)

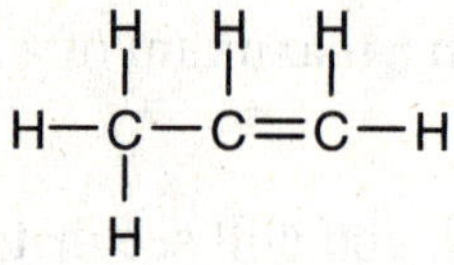 *and* 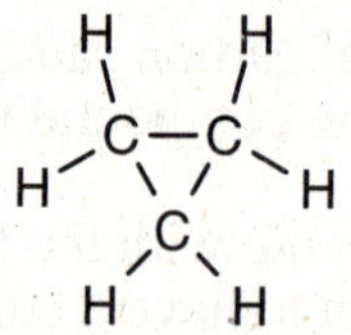

(d)

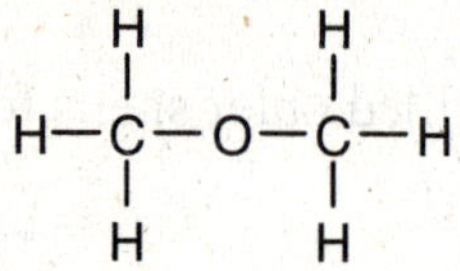 *and* 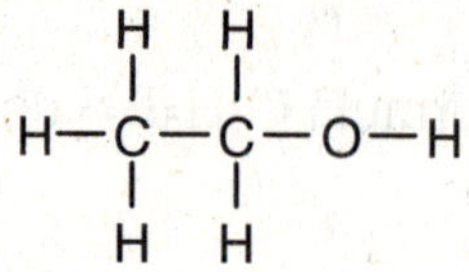

**1.38** In molecular formulas of organic molecules, carbon is listed first, followed by hydrogen. All other elements are listed in alphabetical order.

| *Compound* | *Molecular Formula* |
|---|---|
| (a) Vitamin C | $C_6H_8O_6$ |
| (b) Nicotine | $C_{10}H_{14}N_2$ |
| (c) Glucose | $C_6H_{12}O_6$ |

**1.39**

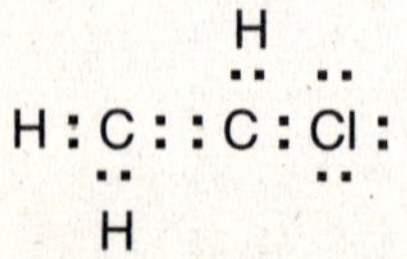 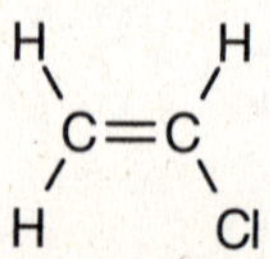 Vinyl chloride

**1.40**

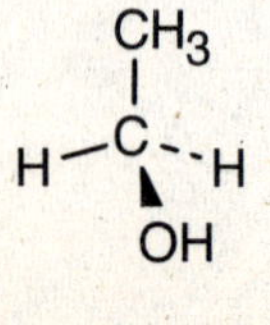 Ethanol

**1.41**

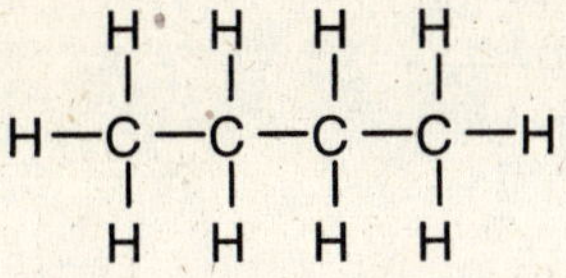

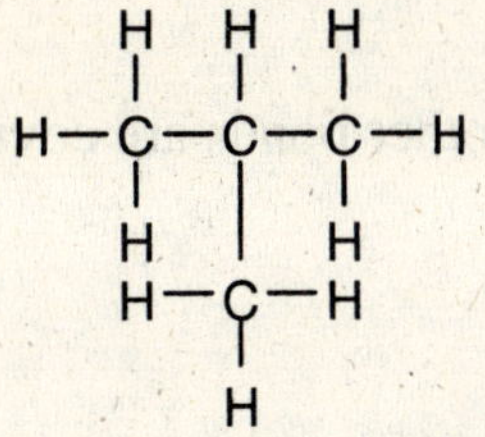

The two structures differ in the connection of the carbon atoms.

**1.42** Three structures are possible.

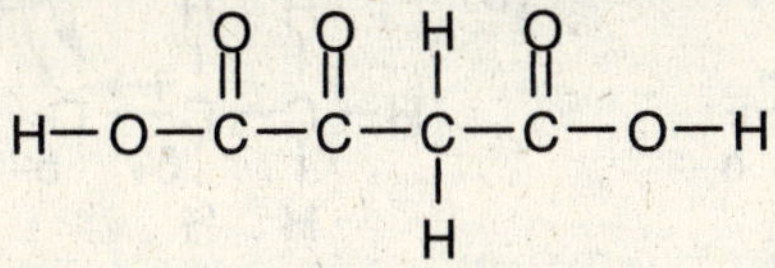

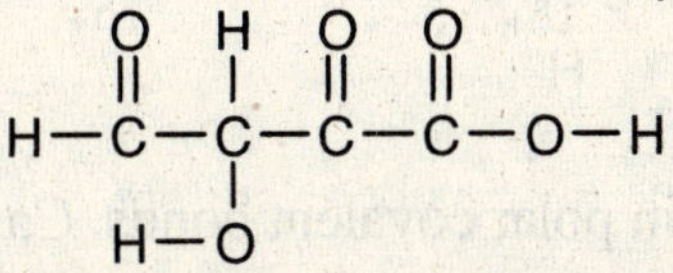

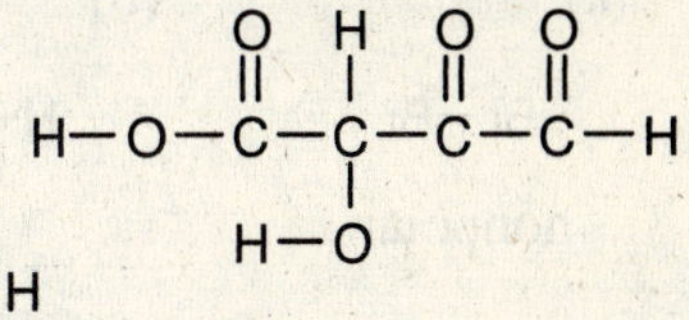

**1.43**

(a)

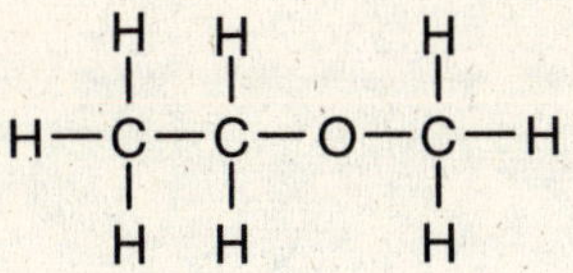

Ethyl methyl ether

(b)

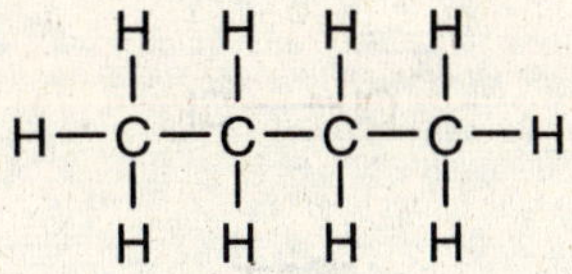

Butane

(c)

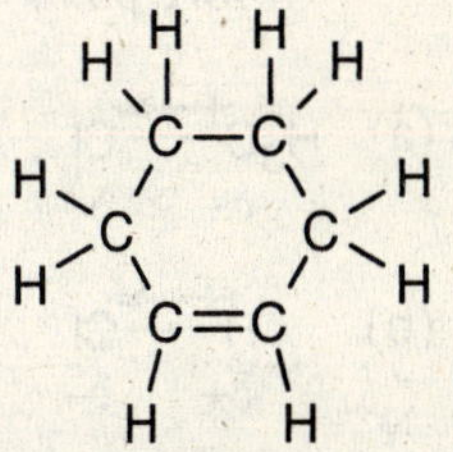

Cyclohexene

## Electronegativity

**1.44** Use Figure 1.14 if you need help. The most electronegative element is starred.

(a) $CH_2\overset{*}{F}Cl$ (b) $\overset{*}{F}CH_2CH_2CH_2Br$ (c) $H\overset{*}{O}CH_2CH_2NH_2$ (d) $CH_3\overset{*}{O}CH_2Li$

**1.45** Carbon has the most positive character when it is bonded to the most electronegative atom.

*Most negative carbon* ———————> *Most positive carbon.*

$CH_3Li$, $CH_3–CH_3$, $CH_3–I$, $CH_3–NH_2$, $CH_3–OH$, $CH_3–F$

**1.46**

(a)

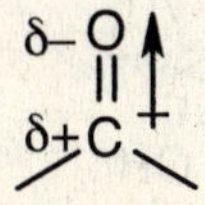

ketone

(b)

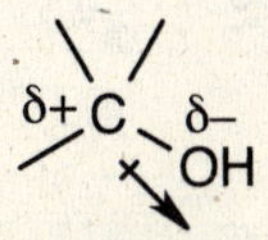

alcohol

(c)

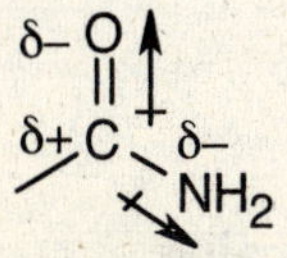

amide

(d)

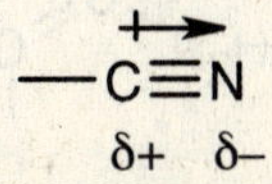

nitrile

**1.47**

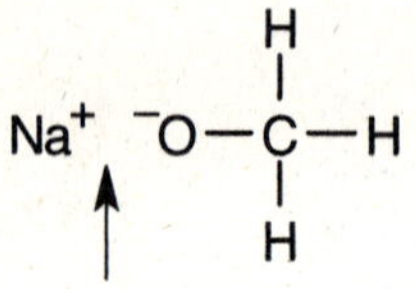

ionic

All other bonds are covalent.

**1.48** Ionic bonds: $BeF_2$ ($\Delta$ EN = 2.4)
Covalent bonds: $SiH_4$ ($\Delta$ EN = 0.3), $CBr_4$ ($\Delta$ EN = 0.3)

**1.49**

(a) Br–Br

nonpolar

(b)

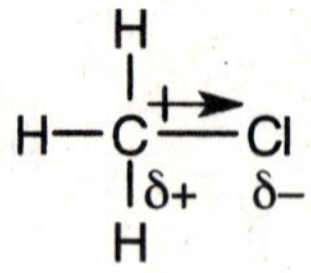

(c)

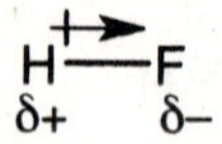

(d)

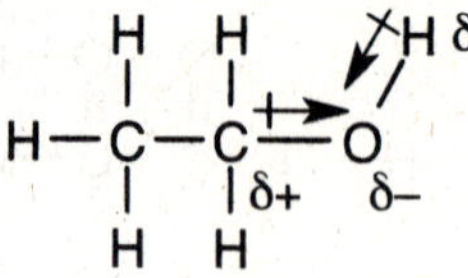

Molecules (b)–(d) contain polar covalent bonds. Carbon–hydrogen bonds are only slightly polar.

**1.50**

*More polar* / *Less polar*

(a)

(b)

(c)

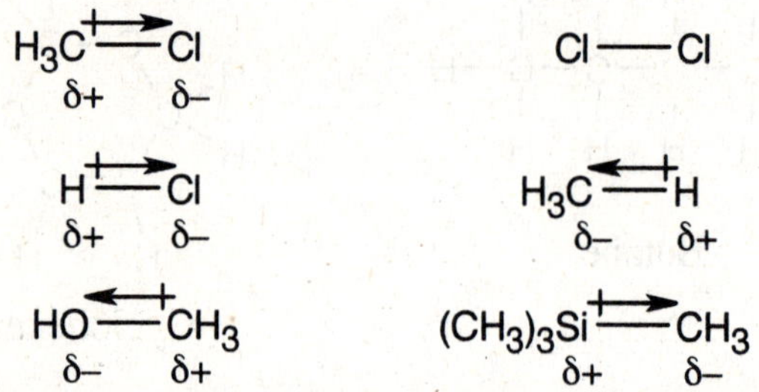

## Hybridization

**1.51**

(a) $CH_3CH_2CH{=}CH_2$ (b) $H_2C{=}CHCH{=}CH_2$ (c) $H_2C{=}CHC{\equiv}CH$

**1.52**

(a)

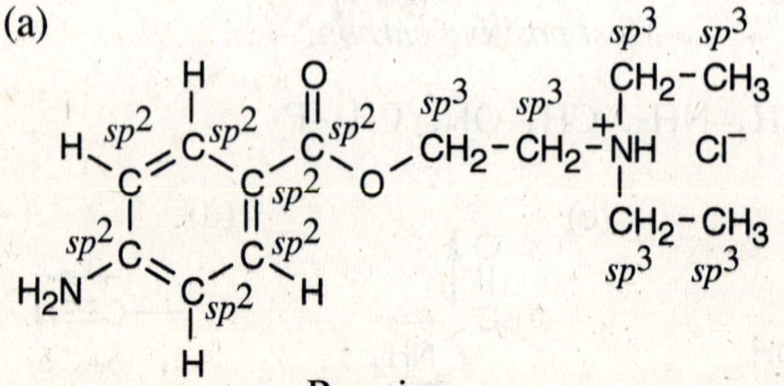

Procaine

(b)

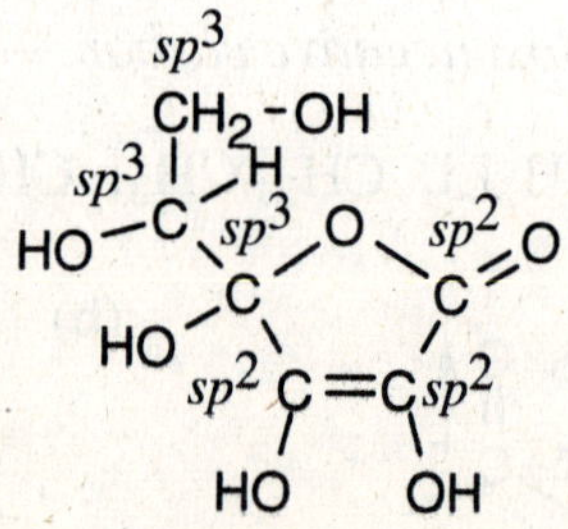

Vitamin C

**1.53**

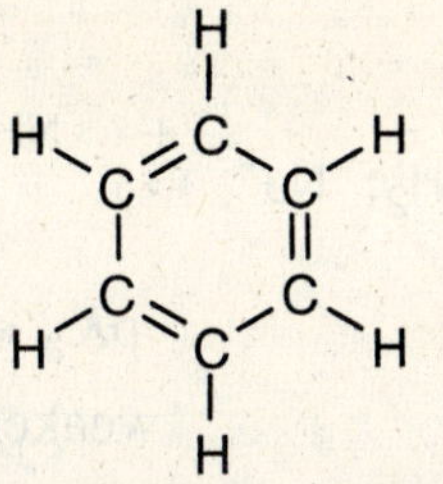


Benzene

All carbon atoms of benzene are $sp^2$-hybridized, and all bond angles of benzene are 120°. Benzene is a planar molecule.

**1.54**

(a)

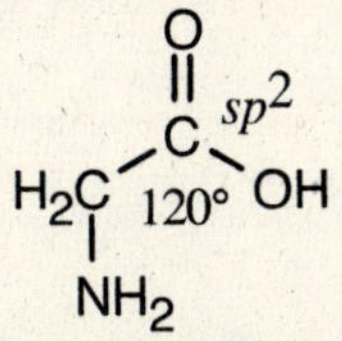


Glycine

(b)

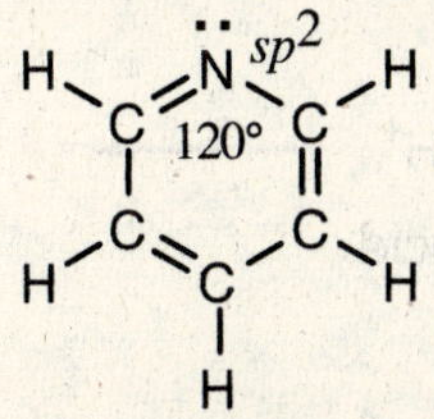


Pyridine

(c)

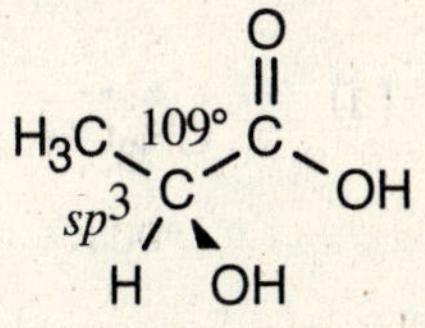


Lactic acid

**1.55** Refer to Problem 1.36 for the structure of acetonitrile. The $H_3C$– carbon is $sp^3$ hybridized, and the –CN carbon is *sp* hybridized.

## Acids and Bases

**1.56** The substances with the largest values of p$K_a$ are the least acidic.

*Least acidic* ⟶ *Most acidic*

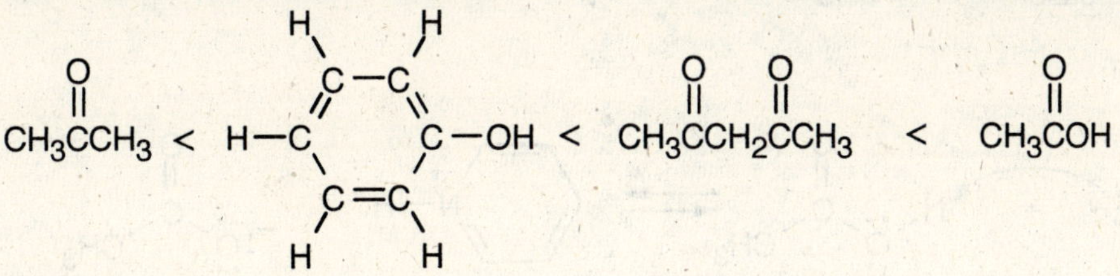


p$K_a$ = 19.2 p$K_a$ = 9.9 p$K_a$ = 9 p$K_a$ = 4.76

**1.57** To react completely with NaOH, an acid must have a p$K_a$ considerably lower than the p$K_a$ of $H_2O$. Thus, all substances in the previous problem except acetone react completely with NaOH.

**1.58** Lewis acids: $AlBr_3$, HF

Lewis bases: $CH_3CH_2\ddot{N}H_2$, $CH_3\ddot{S}CH_3$

**1.59** The stronger the acid (lower p$K_a$), the weaker its conjugate base. Since $NH_4^+$ is a stronger acid than $CH_3NH_3^+$, $CH_3NH_2$ is a stronger base than $NH_3$.

**1.60** The reaction between methanol and bicarbonate does not take place in the indicated direction because methanol (p$K_a$ = 15.5) is a weaker acid than $H_2CO_3$ (p$K_a$ = 6.4).

**1.61**

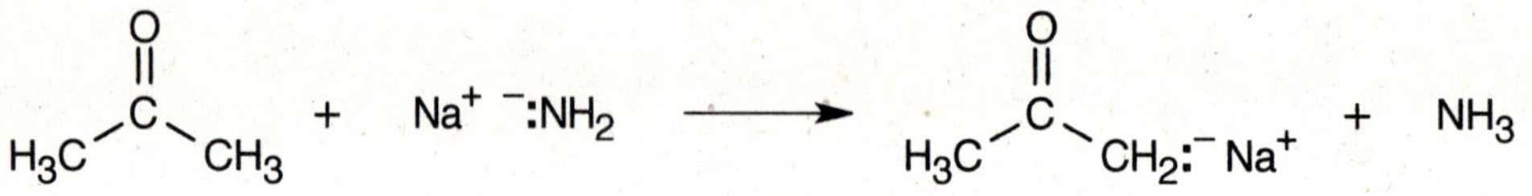

$pK_a \approx 19$ stronger acid

$pK_a \approx 36$ weaker acid

The above reaction will take place as written because acetone is a stronger acid than ammonia.

**1.62**

(a) $CH_3OH$ (base) + $H^+$ (acid) $\longrightarrow$ $CH_3OH_2^+$

(b) $CH_3OH$ (acid) + $^-NH_2$ (base) $\longrightarrow$ $CH_3O^-$ + $NH_3$

(c)

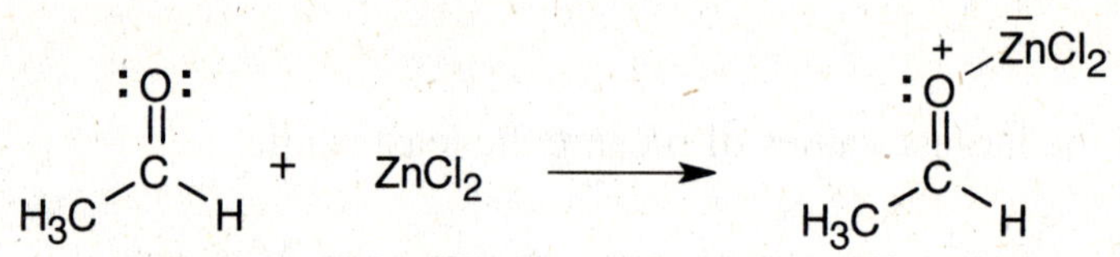

base acid

**1.63**

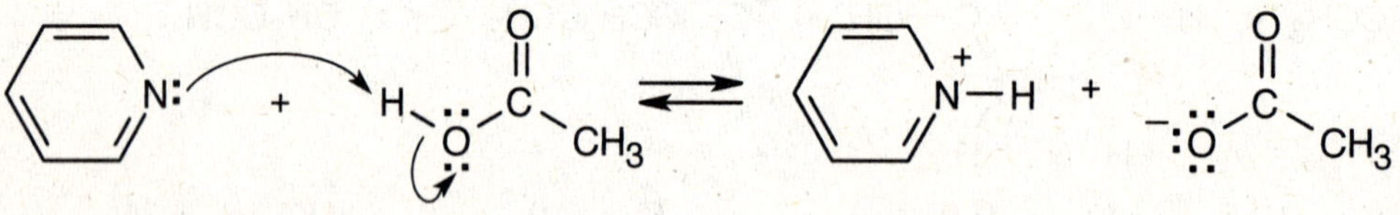

## General Problems

**1.64** In a compound containing a carbon-carbon triple bond, atoms bonded to the *sp*-hybridized carbons must lie in a straight line. It is not possible to form a five-membered ring if four carbons must have a linear relationship.

**1.65**

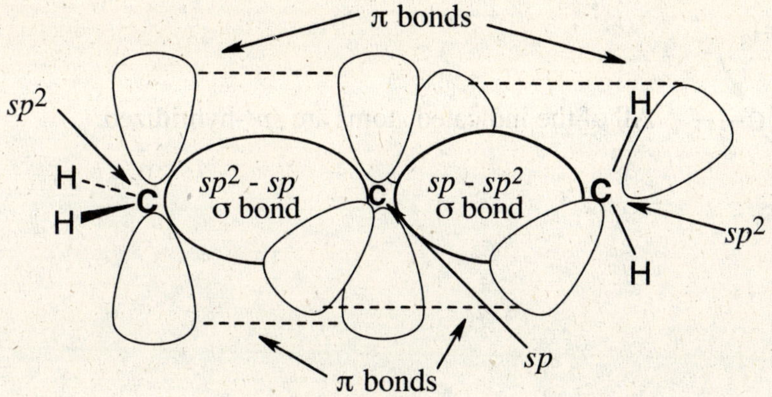


The central carbon of allene forms two σ bonds and two π bonds. The central carbon is *sp*-hybridized, and the two terminal carbons are $sp^2$-hybridized. The bond angle formed by the three carbons is 180°, indicating linear geometry for the carbons of allene.

**1.66** The carbon atom of $CO_2$ is *sp*-hybridized. Allene and $CO_2$ are both linear molecules.

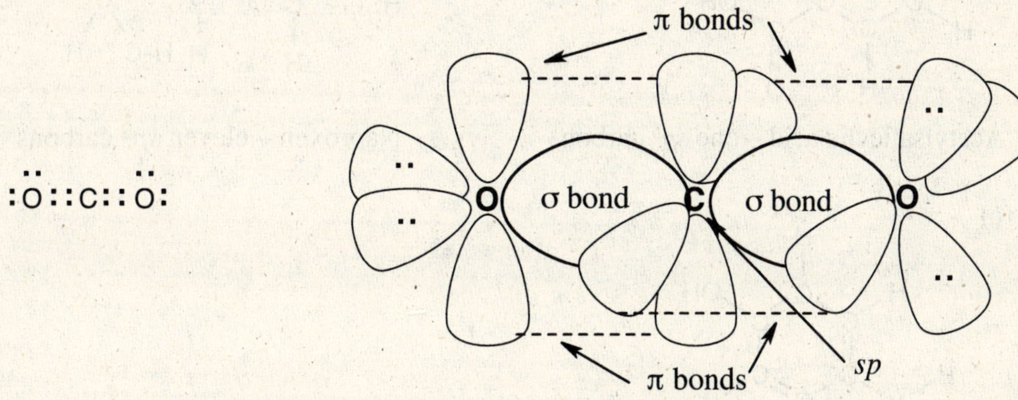


**1.67** The carbon atom, which has six valence shell electrons, is $sp^2$-hybridized. A carbocation is planar and is *isoelectronic* with (has the same number of electrons as) a trivalent boron compound.

**1.68**

Ammonium ion

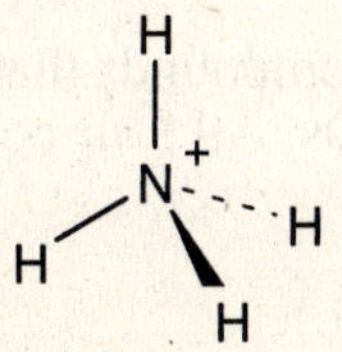

The nitrogen atom of the tetrahedral ammonium ion is $sp^3$-hybridized because, like the carbon atom of methane, nitrogen forms bonds to four different hydrogen atoms.

**1.69**

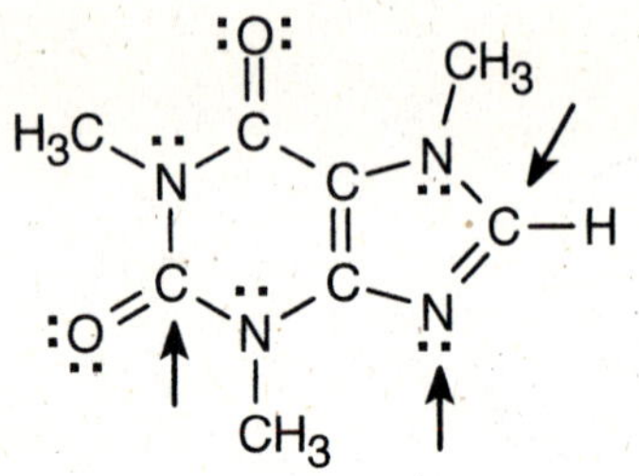

All of the indicated atoms are $sp^2$-hybridized.

## In the Medicine Cabinet

**1.70**

(a)

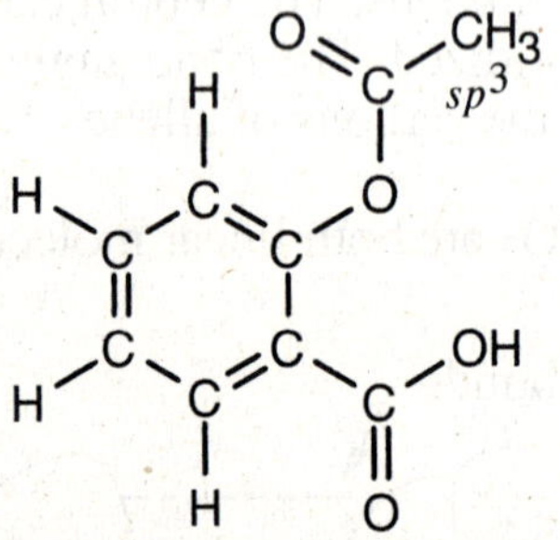

Acetylsalicylic acid – one $sp^3$ carbon

(b)

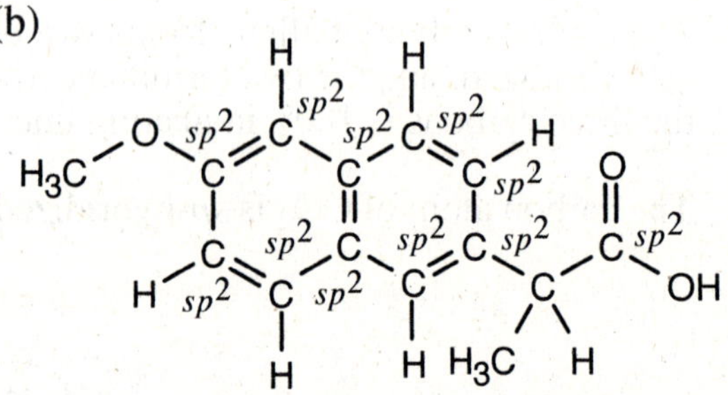

Naproxen – eleven $sp^2$ carbons

(c)

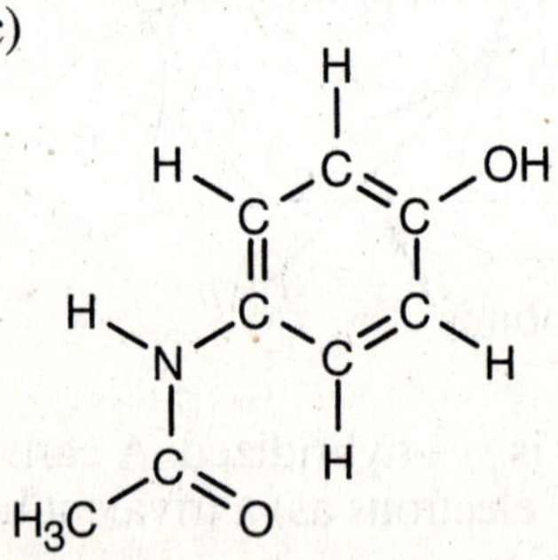

Acetaminophen $C_8H_9NO_2$

(d) Cyclooxygenase, apparently, must require compounds that interact with it to have the structural features common to these four NSAIDs. All four compounds have at least one benzene ring (Problem 1.54) and a C=O group.

## In the Field

**1.71**

(a)

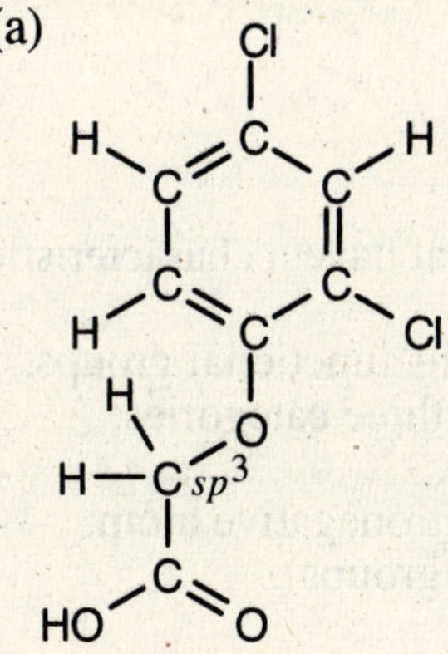

2,4-D
one $sp^3$ carbon

(b)

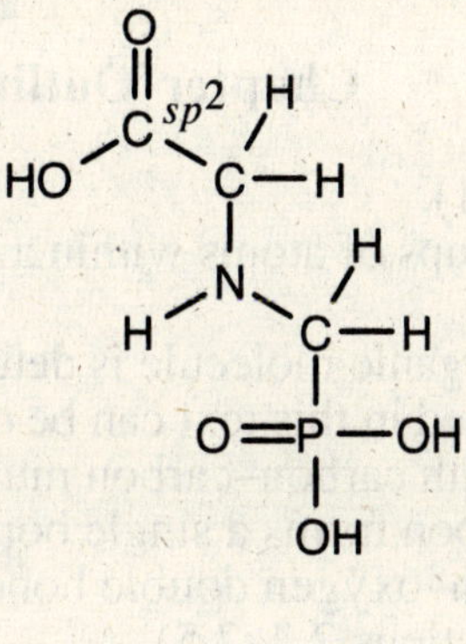

Round Up
one $sp^2$ carbon

(c)

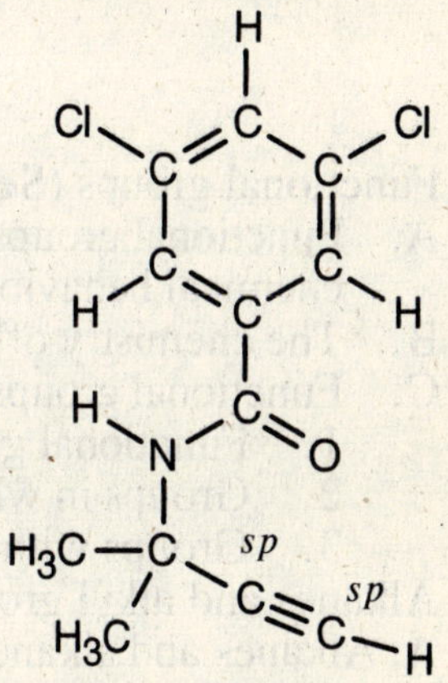

Pronamide
two *sp* carbons

(d)

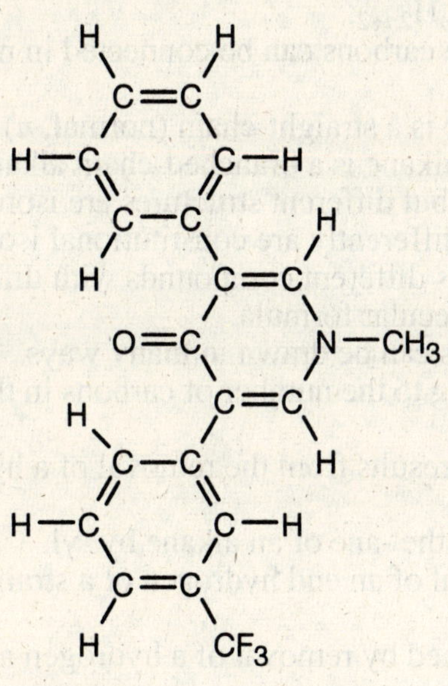

Fluridone $C_{19}H_{14}F_3NO$

# Chapter 2 – Alkanes: The Nature of Organic Compounds

## Chapter Outline

I. Functional groups (Section 2.1).
   A. Functional groups are groups of atoms within a molecule that have a characteristic chemical behavior.
   B. The chemistry of every organic molecule is determined by its functional groups.
   C. Functional groups described in this text can be divided into three categories:
      1. Functional groups with carbon–carbon multiple bonds.
      2. Groups in which carbon forms a single bond to an electronegative atom.
      3. Groups with a carbon–oxygen double bond (carbonyl groups).

II. Alkanes and alkyl groups (Sections 2.2–2.5).
   A. Alkanes and alkane isomers (Section 2.2).
      1. Alkanes are formed by overlap of carbon $sp^3$ orbitals.
      2. Alkanes are described as saturated hydrocarbons.
         a. They are hydrocarbons because they contain only carbon and hydrogen.
         b. They are saturated because all bonds are single bonds.
         c. The general formula for alkanes is $C_nH_{2n+2}$.
      3. For alkanes with four or more carbons, the carbons can be connected in more than one way.
         a. If the carbons are in a row, the alkane is a straight-chain (normal, *n*) alkane.
         b. If the carbon chain has a branch, the alkane is a branched-chain alkane.
      4. Alkanes with the same molecular formula but different structures are isomers.
         a. Isomers whose atoms are connected differently are constitutional isomers.
            i. Constitutional isomers are always different compounds with different properties but with the same molecular formula.
         b. Alkanes with more than three carbons can be drawn in many ways.
      5. Straight-chain alkanes are named according to the number of carbons in their chain.
   B. Alkyl groups.
      1. An alkyl group is the partial structure that results from the removal of a hydrogen atom from an alkane.
         a. Alkyl groups are named by replacing the -ane of an alkane by -yl.
         b. *n*-Alkyl groups are formed by removal of an end hydrogen of a straight-chain alkane.
         c. Branched-chain alkyl groups are formed by removal of a hydrogen atom from an internal carbon.
            i. The prefixes *sec*- and *tert*- refer to the degree of substitution at the branching carbon atom.
      2. There are four possible degrees of alkyl substitution for carbon.
         a. A primary carbon is bonded to one other carbon.
         b. A secondary carbon is bonded to two other carbons.
         c. A tertiary carbon is bonded to three other carbons.
         d. A quaternary carbon is bonded to four other carbons.
         e. The symbol **R** refers to the rest of the molecule.
   C. Naming branched-chain alkanes (Section 2.3).
      1. The system of naming used in this book is the IUPAC system.
         a. In this system, a chemical name has a prefix, a parent, a locant, and a suffix.
            i. The prefix shows the location of functional groups.
            ii. The parent shows the number of carbons in the principal chain.
            iii. The locant gives the location of the primary functional group.

iv. The suffix identifies the primary functional group family.

2. Naming an alkane:
   a. Find the parent hydrocarbon.
      i. Find the longest continuous chain of carbons, and use its name as the parent name.
      ii. If two chains have the same number of carbons, choose the one with more branch points.
   b. Number the atoms in the parent chain.
      i. Start numbering at the end nearer the first branch point.
      ii. If branching occurs an equal distance from both ends, begin numbering at the end nearer the second branch point.
   c. Identify and number the substituents.
      i. Give each substituent a number that corresponds to its position on the parent chain.
      ii. Two substituents on the same carbon receive the same number, using as many numbers as there are substituents.
   d. Write the name as a single word.
      i. Use hyphens to separate prefixes and commas to separate numbers.
      ii. Use the prefixes, *di-*, *tri-*, *tetra-* if necessary, for multiples of the same substituent, but don't use them for alphabetizing.

D. Properties of alkanes (Section 2.4).
   1. Alkanes are prepared by fractional distillation of petroleum.
   2. Alkanes are chemically inert to most laboratory reagents.
      a. Alkanes react with $O_2$ and $Cl_2$.

E. Conformations of ethane (Section 2.5).
   1. Rotation is possible around carbon–carbon single bonds.
      a. The different arrangements of atoms that result from rotation are conformers.
   2. Conformations can be represented in two ways.
      a. Sawhorse representations of alkanes view the C–C bond from an oblique angle.
      b. Newman projections represent the two carbon atoms by a circle and view the C–C bonds end-on.
   3. There is a slight energy barrier to rotation around a C–C bond.
      a. In a staggered conformation, hydrogens are as far apart as possible.
      b. In an eclipsed conformation, hydrogens are as close together as possible.
      c. For most compounds, a staggered conformation is more stable than an eclipsed conformation.

III. Drawing chemical structures (Section 2.6).

A. Skeletal structures are a shorthand way of drawing organic compounds.
   1. Carbon atoms are assumed to be at the intersection of two lines.
   2. Hydrogen atoms bonded to carbon aren't shown.
   3. All other atoms are shown.

IV. Cycloalkanes (Sections 2.7–2.11).

A. The general formula for cycloalkanes is $C_nH_{2n}$.

B. Naming substituted cycloalkanes (Section 2.7).
   1. Count the number of carbon atoms in the ring, and add the prefix *cyclo-* to the name of the corresponding alkane.
   2. Start at a point of attachment, and number around the ring.
      a. For two or more substituents, begin numbering at the group with alphabetical priority, and give the other substituents the lowest possible number.

C. Cis–trans isomerism in cycloalkanes (Section 2.8).
   1. Unlike open-chain alkanes, cycloalkanes have much less rotational freedom.
      a. Very small rings are rigid.
      b. Large rings have more rotational freedom.

2. Cycloalkanes have a "top" side and a "bottom" side.
   a. If two substituents are on the same side of a ring, the ring is cis-disubstituted.
   b. If two substituents are on opposite sides of a ring, the ring is trans-disubstituted.
3. Substituents in the two types of disubstituted cycloalkanes are connected in the same order but differ in spatial orientation.
   a. These cycloalkenes are a class of stereoisomers known as cis–trans isomers.
   b. Cis–trans isomers are stable compounds that can't be interconverted.

C. Conformations of cycloalkanes (Sections 2.9–2.11).
1. Ring conformations (Section 2.9).
   a. Cyclopropane is a flat molecule with bond angles of 60°.
   b. Cyclobutane and cyclopentane are slightly puckered.
   c. Cyclohexane exists as a strain-free puckered ring known as a chair conformer .
      i. The chair conformation of cyclohexane is strain-free.
      ii. In a standard drawing of cyclohexane, the lower bond is in front.
2. Axial and equatorial bonds in cyclohexane (Section 2.10).
   a. There are two kinds of positions on a cyclohexane ring.
      i. Six axial hydrogens are perpendicular to the plane of the ring.
      ii. Six equatorial hydrogens are roughly in the plane of the ring.
   b. Each carbon has one axial hydrogen and one equatorial hydrogen.
   c. Each side of the ring has alternating axial and equatorial hydrogens.
   d. All hydrogens on the same side of the ring are cis.
3. Conformational mobility of cyclohexanes (Section 2.11).
   a. Different chair conformations of cyclohexanes interconvert by a ring-flip.
   b. After a ring-flip, an axial bond becomes an equatorial bond, and vice versa.
   c. Both conformations aren't equally stable at room temperature.
      i. In methylcyclohexane, 95% of molecules have the methyl group in the equatorial position.

## Solutions to Problems

**2.1**

(a)

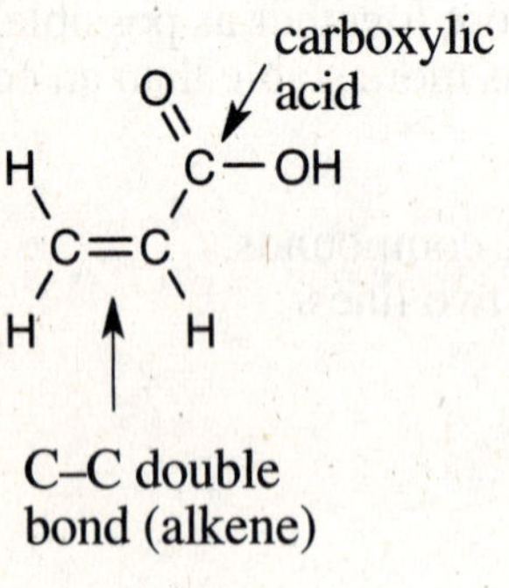


Acrylic acid

(b)

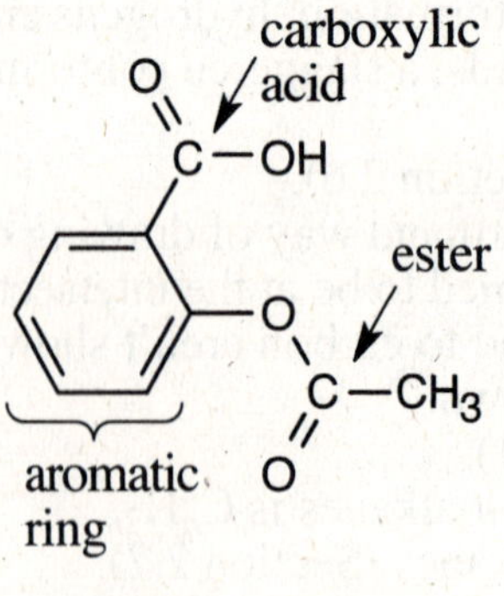


Aspirin

(c)

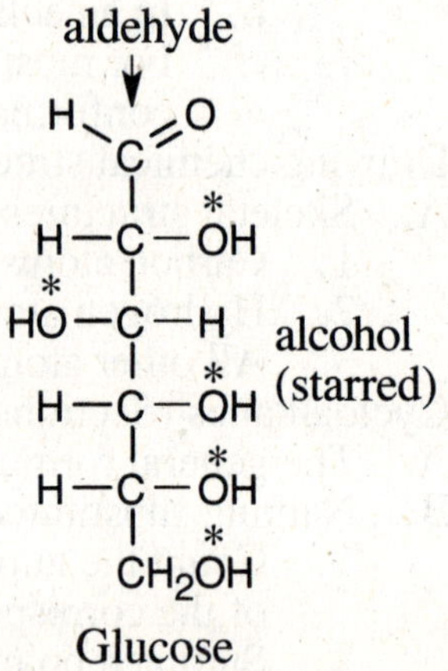


Glucose

**2.2**

(a) $CH_3OH$

Methanol

(b)

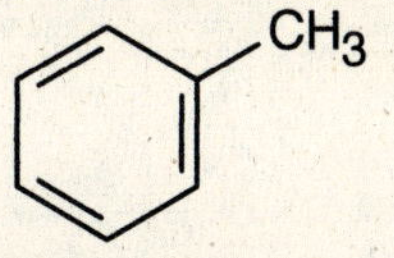

Toluene

(c)

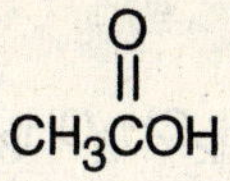

Acetic acid

(d) $CH_3NH_2$

Methylamine

(e)

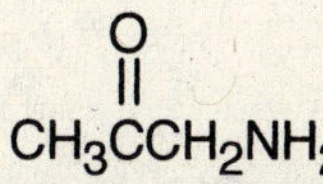

Aminoacetone

(f) $H_2C{=}CHCH{=}CH_2$

Buta-1,3-diene

Many other compounds containing these functional groups can be drawn.

**2.3**

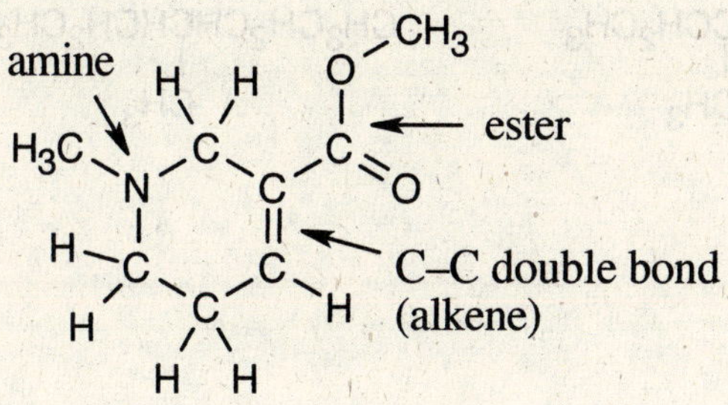


Arecoline

**2.4** We know that carbon forms four bonds and hydrogen forms one bond. Thus, if you draw all possible six-carbon skeletons and add hydrogens so that all carbons have four bonds, you will arrive at the following formulas.

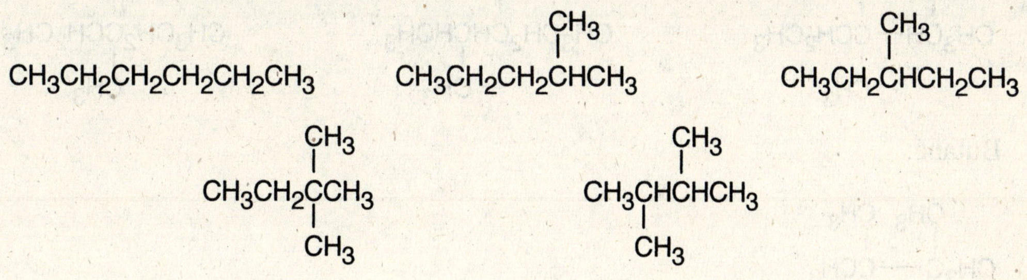

**2.5** (a) There are 18 isomers with the formula $C_8H_{18}$.

Octane:

$CH_3CH_2CH_2CH_2CH_2CH_2CH_2CH_3$

Heptanes:

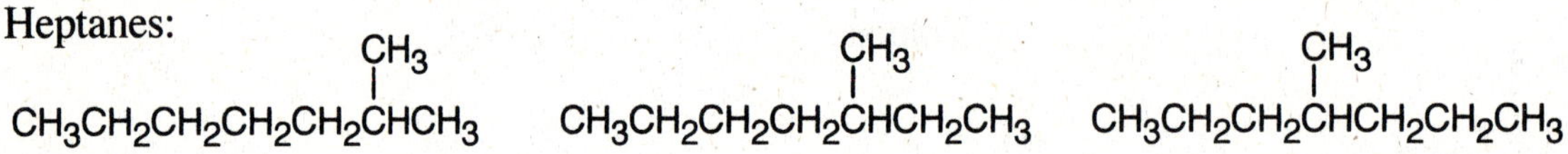

Hexanes:

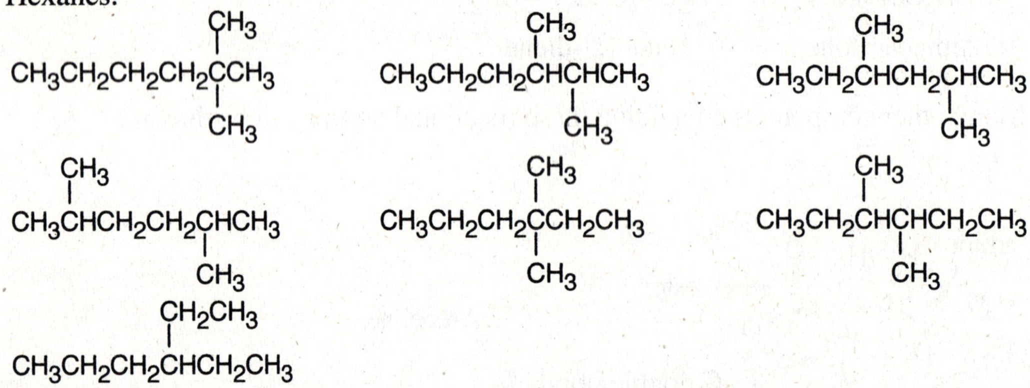

Pentanes:

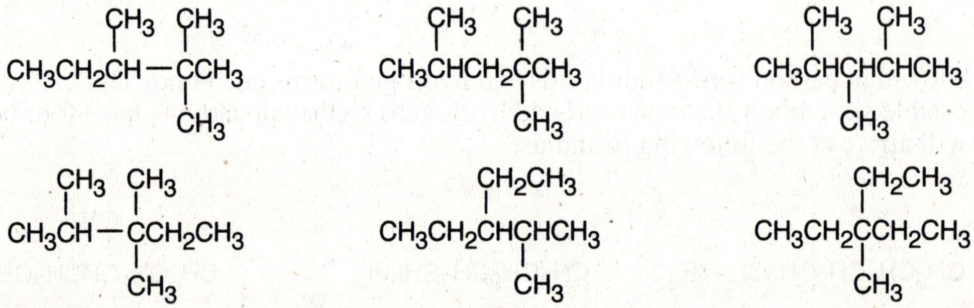

Butane:

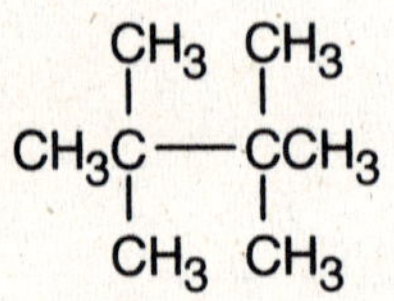

(b) Many isomers of the formula $C_4H_8O_2$ containing different functional groups can be drawn. Here are three examples:

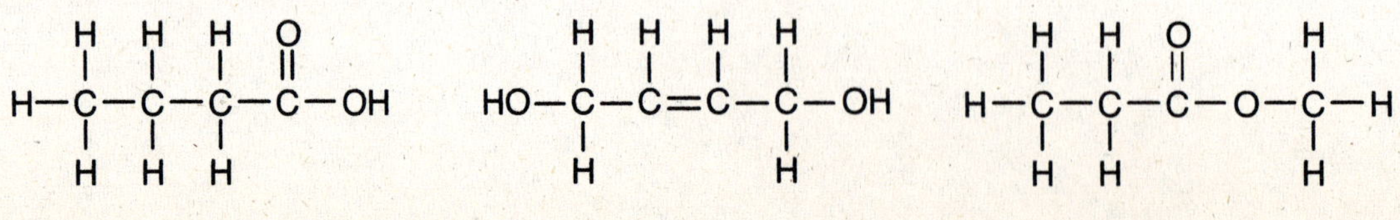

**2.6**

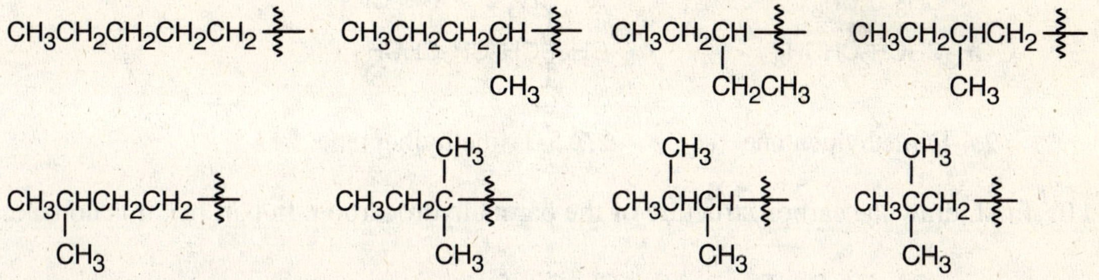

**2.7** Many other answers to these problems are acceptable.

(a)

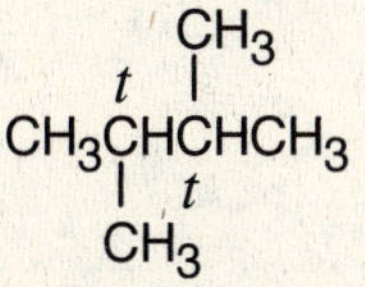

(b)

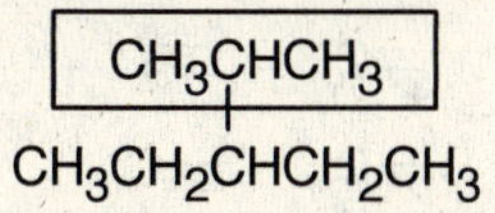

(c)

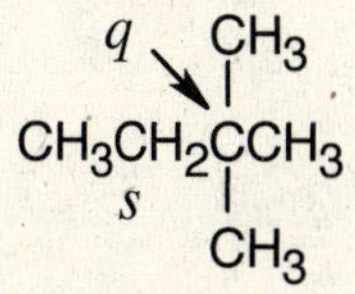

*p* = primary; *s* = secondary; *t* = tertiary; *q* = quaternary

**2.8**

(a)

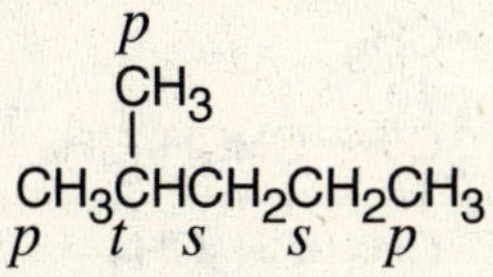

(b)

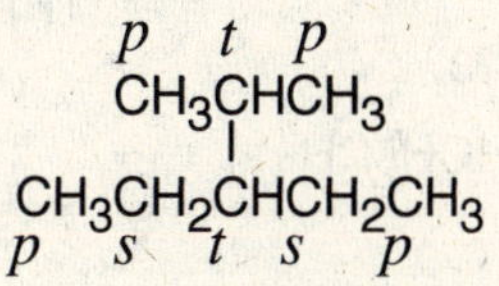

(c)

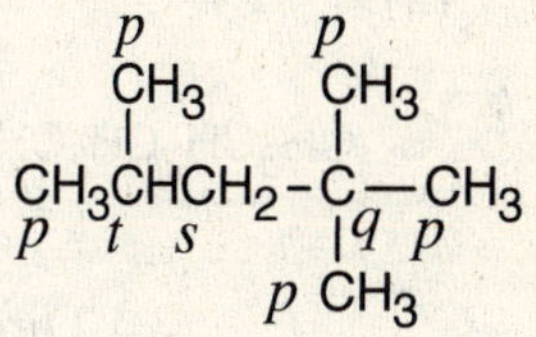

*p* = primary; *s* = secondary; *t* = tertiary; *q* = quaternary

**2.9**

(a)

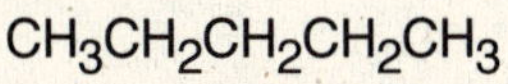

Pentane

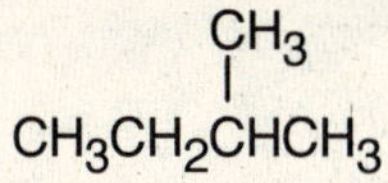

2-Methylbutane

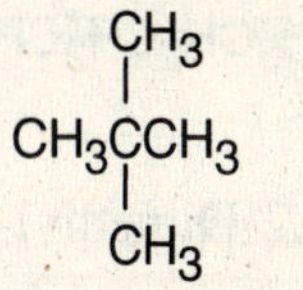

2,2-Dimethylpropane

(b)

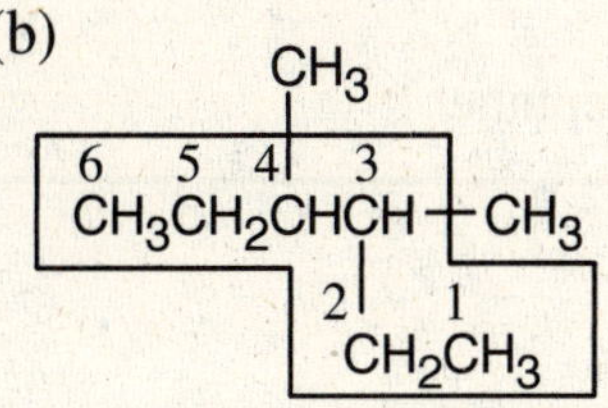

3,4-Dimethylhexane

*Step 1*: Find the longest continuous carbon chain and use it as the parent name. In (b), the longest chain is a hexane (boxed).
*Step 2*: Identify the substituents. In (b), both substituents are methyl groups.
*Step 3*: Number the substituents. In (b), numbering can start from either end of the carbon chain, and the methyl groups are in the 3- and 4- positions.
*Step 4*: Name the compound. Remember that the prefix *di-* must be used when two substituents are the same. The IUPAC name is 3,4-dimethylhexane.

(c)

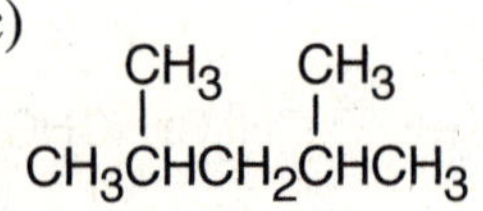

2,4-Dimethylpentane

(d)

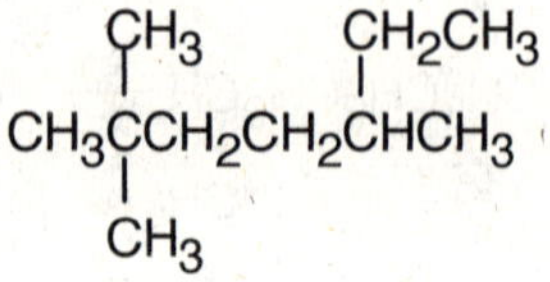

2,2,5-Trimethylheptane

**2.10** First, draw the carbon structure of the parent hydrocarbon. In part (a), it is nonane.

C—C—C—C—C—C—C—C—C

Then, add the substituents—methyl groups at C3 and C4.

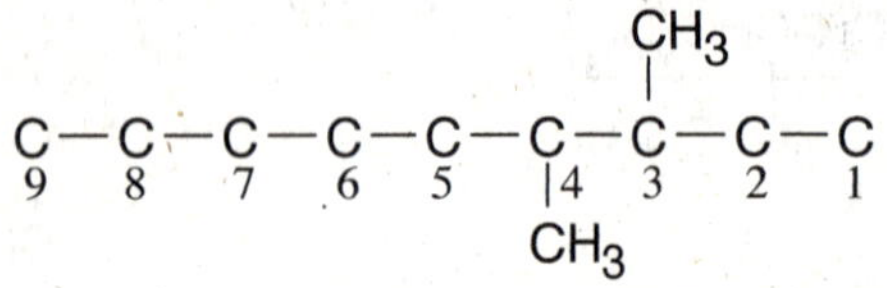

Finally, add hydrogens to complete the structure.

(a)

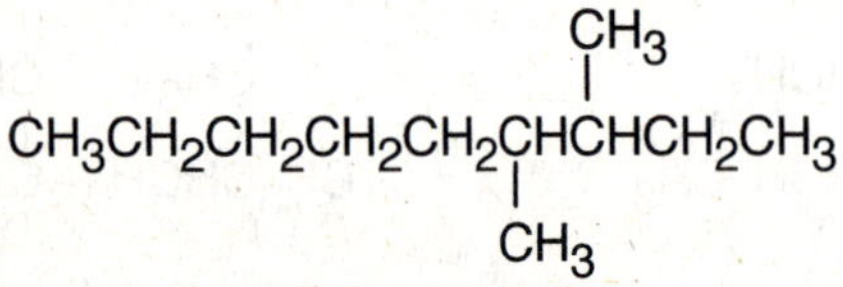

3,4-Dimethylnonane

(b)

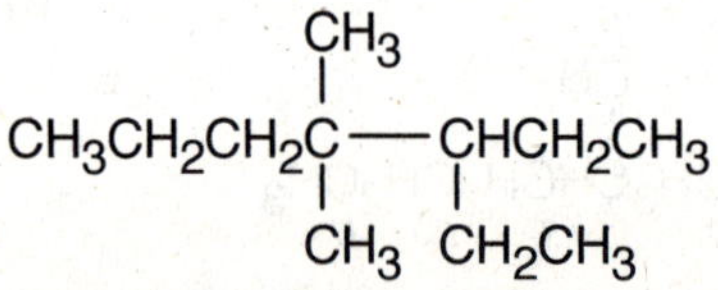

3-Ethyl-4,4-dimethylheptane

(c)

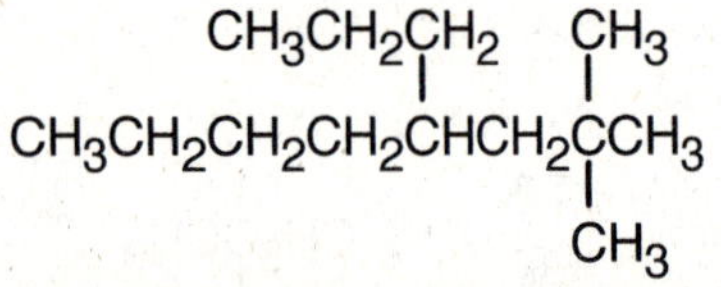

2,2-Dimethyl-4-propyloctane

(d)

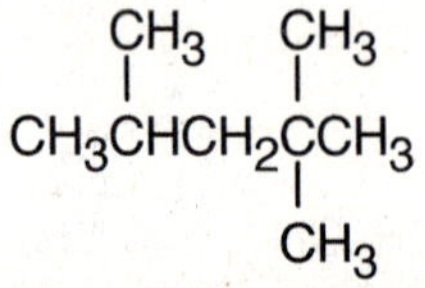

2,2,4-Trimethylpentane

**2.11**

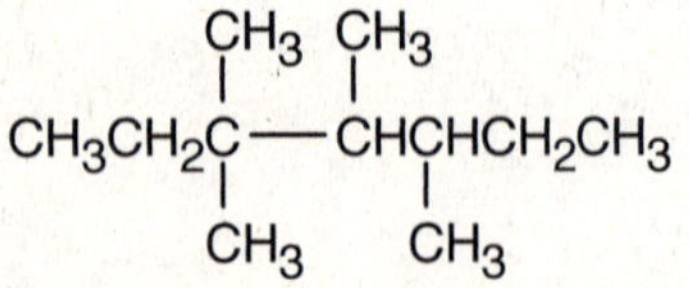

3,3,4,5-Tetramethylheptane

**2.12**

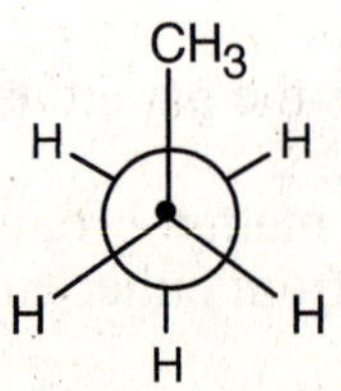

Most stable conformation (staggered)

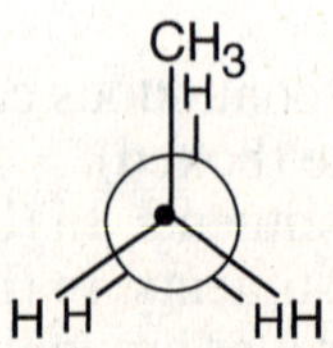

Least stable conformation (eclipsed)

**2.13**

Staggered butane — Eclipsed butane

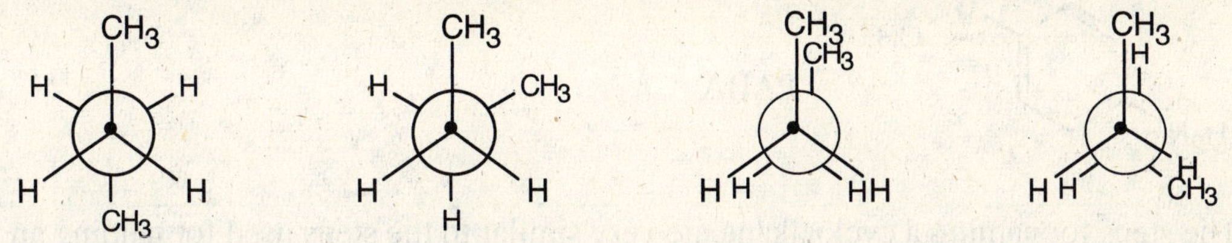

**2.14** The first staggered conformation of butane (pictured above) is the most stable, because the relatively large methyl groups are as far apart as possible.

**2.15**

(a)

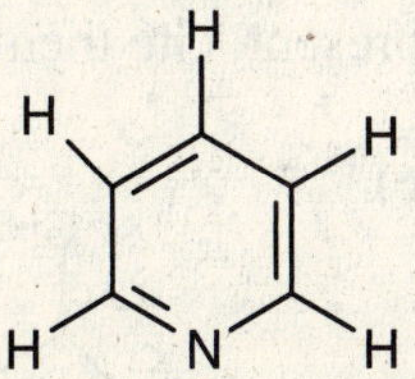

Pyridine

$C_5H_5N$

One hydrogen is bonded to each carbon.

(b)

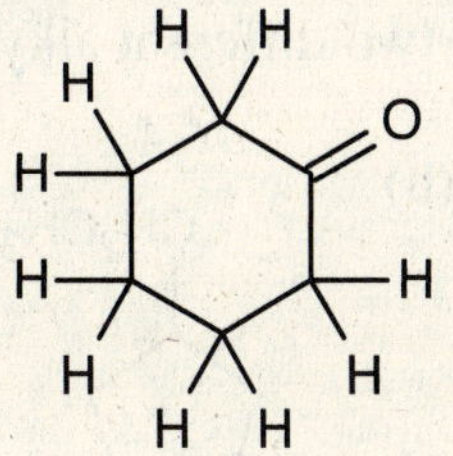

Cyclohexanone

$C_6H_{10}O$

Two hydrogens are bonded to each carbon (except for the carbon bonded to oxygen)

(c)

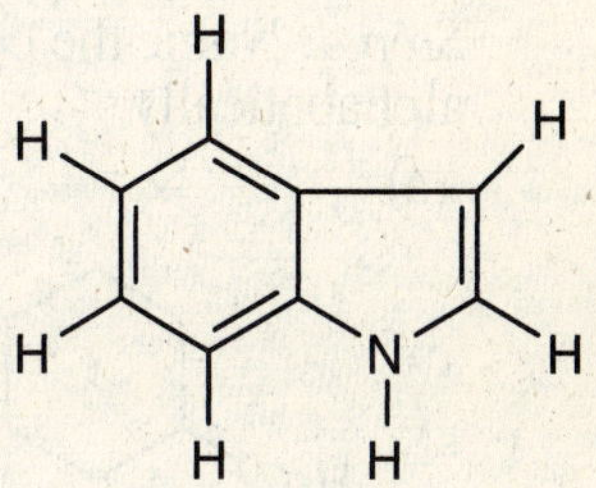

Indole

$C_8H_7N$

One hydrogen is bonded to each carbon (except for the two carbons at the junction of the two rings)

Although a particular shorthand structure has only one molecular formula, a particular molecular formula may represent many structures.

**2.16** Several skeletal structures can satisfy each molecular formula.

(a)

$C_4H_8$

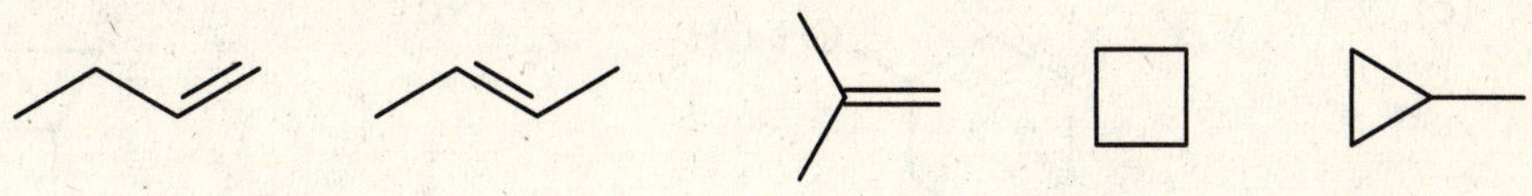

(b)

$C_3H_6O$

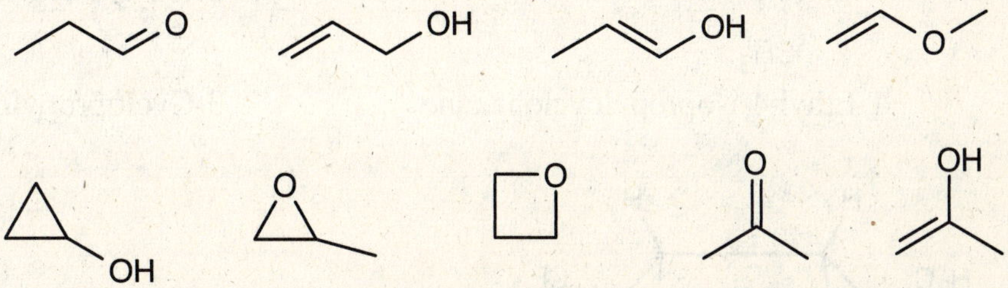

(c)

$C_4H_9Cl$

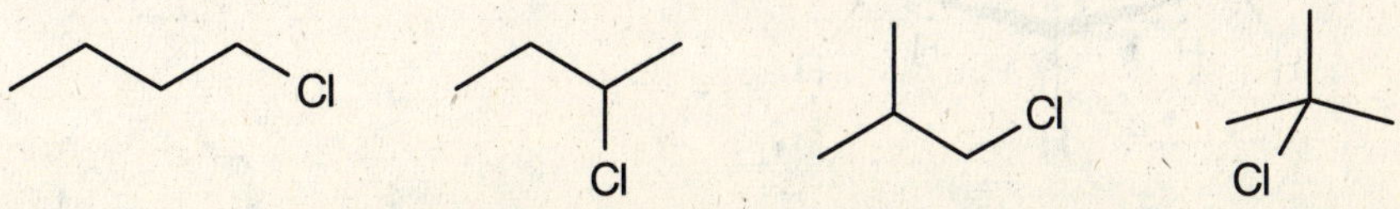

**2.17**

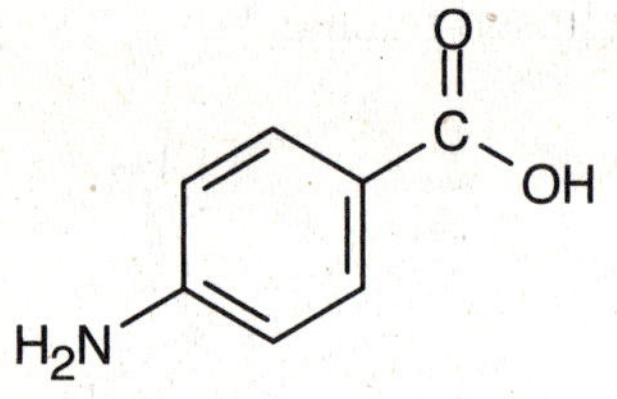


PABA

**2.18** The steps for naming a cycloalkane are very similar to the steps used for naming an open-chain hydrocarbon.
*Step 1*: Name the parent cycloalkane. In (a), the parent is cyclohexane.
*Step 2*: Identify the substituents. In (a), both substituents are methyl groups.
*Step 3*: Number the substituents so that the second substituent receives the lowest possible number. In (a), the substituents are in the 1- and 4- positions.
*Step 4*: Name the compound. If two different alkyl groups are present, cite them alphabetically.

(a)

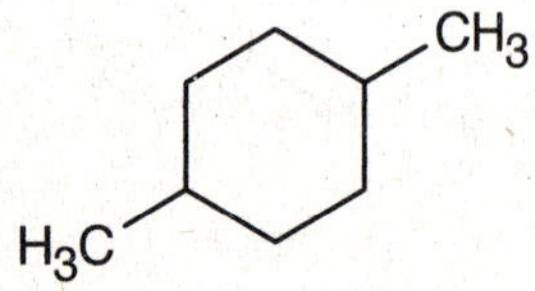


1,4-Dimethylcyclohexane

(b)

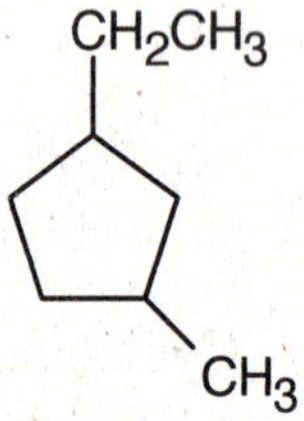


1-Ethyl-3-methylcyclopentane

(c)

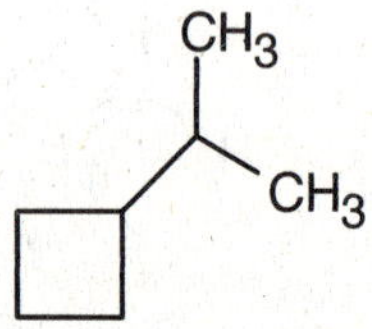


Isopropylcyclobutaane

**2.19**

(a)

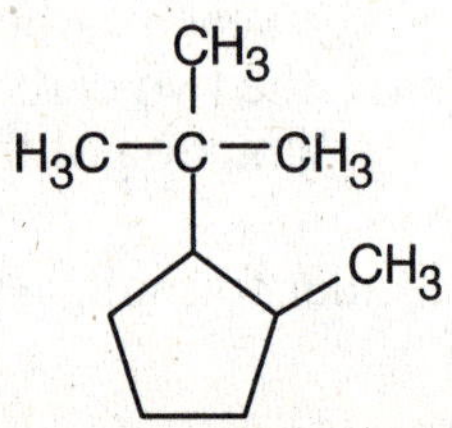


1-*tert*-Butyl-2-methylcyclopentane

(b)

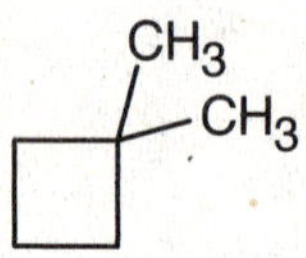


1,1-Dimethylcyclobutane

(c)

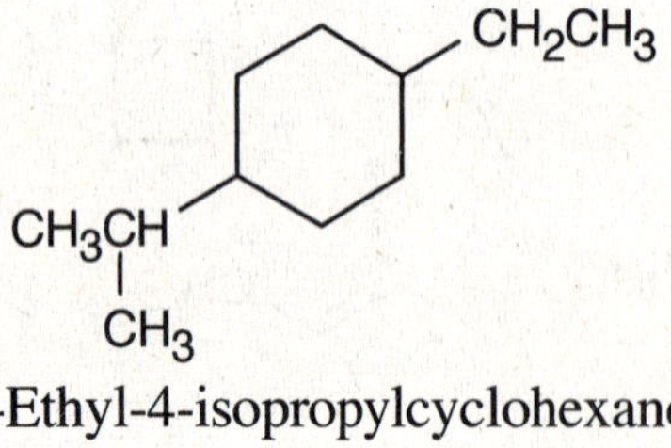


1-Ethyl-4-isopropylcyclohexane

(d)

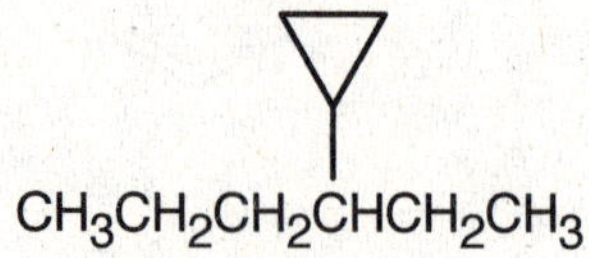


3-Cyclopropylhexane

**2.20**

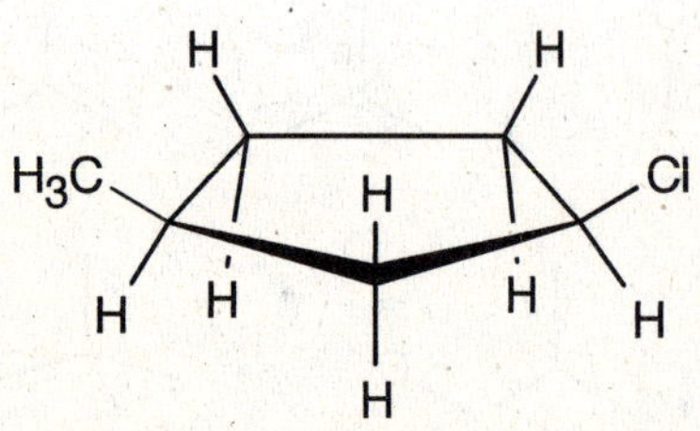


*cis*-1-Chloro-3-methylcyclopentane

**2.21**

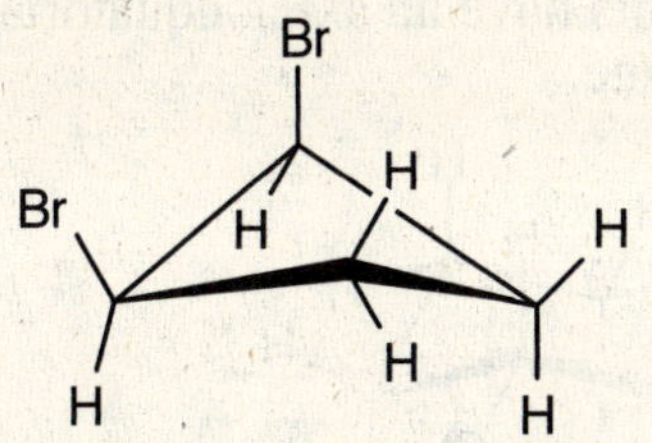

*cis*-1,2-dibromocyclobutane

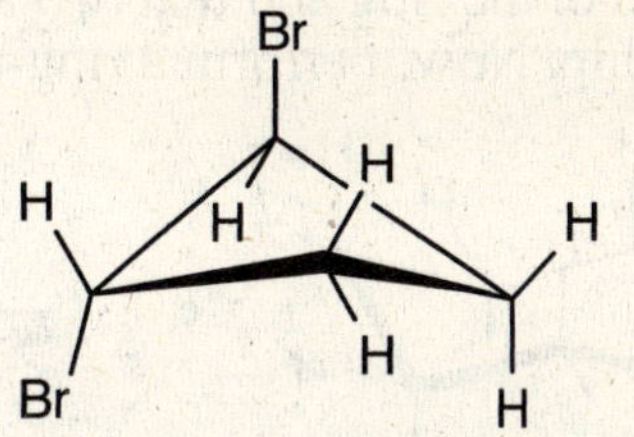

*trans*-1,2-dibromocyclobutane

**2.22**

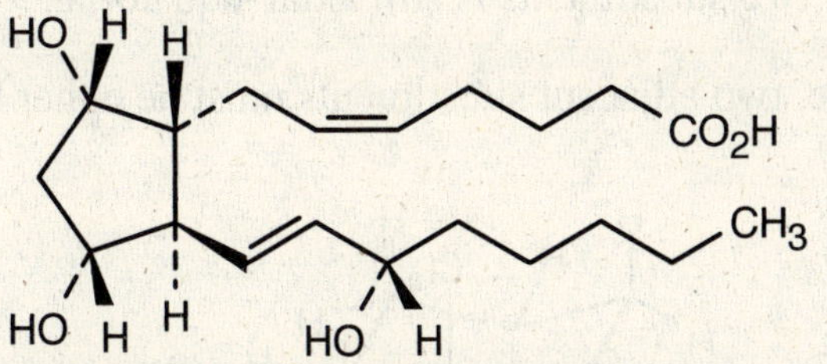

Prostaglandin $F_{2\alpha}$

The two hydroxyl groups are cis because they both point behind the plane of the page, as shown by the dashed lines. The carbon chains have a trans relationship (one is dashed and the other is wedged).

**2.23**

(a)

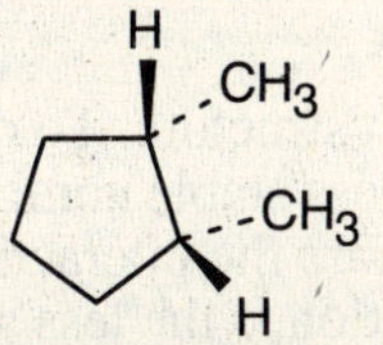

*cis*-1,2-Dimethylcyclopentane

(b)

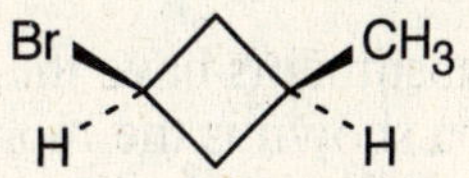

*cis*-1-Bromo-3-methylcyclobutane

**2.24** An axial group is parallel to the axis of the ring, and an equatorial group lies in a band around the ring equator

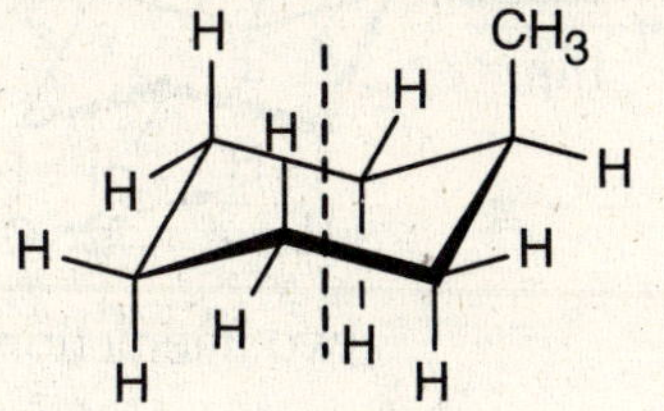

axial methyl group

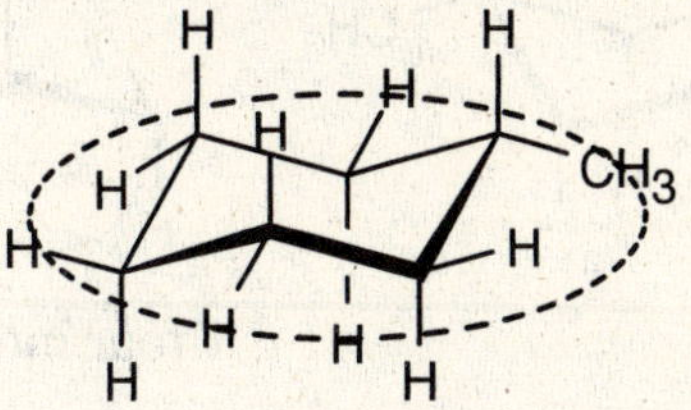

equatorial methyl group

**2.25**

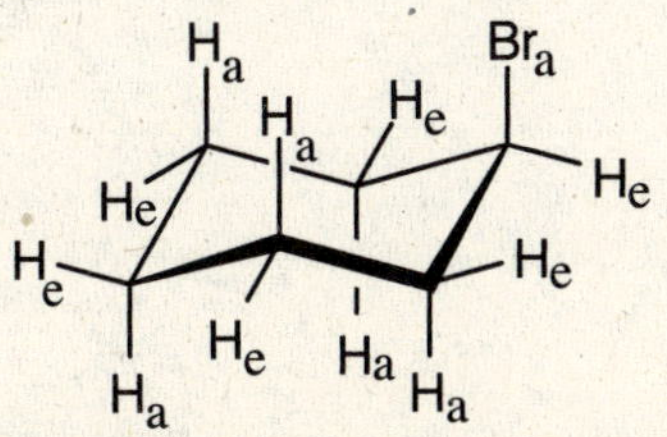

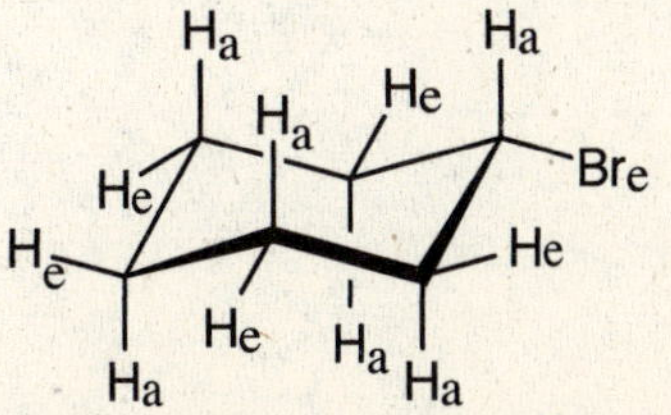

The conformation having bromine in the equatorial position is more stable.

**2.26** Make a model of *cis*-1,2-dichlorocyclohexane. Notice that both cis substituents are on the same side of the ring and that two adjacent cis substituents have an axial–equatorial relationship. Now, perform a ring-flip on the cyclohexane.

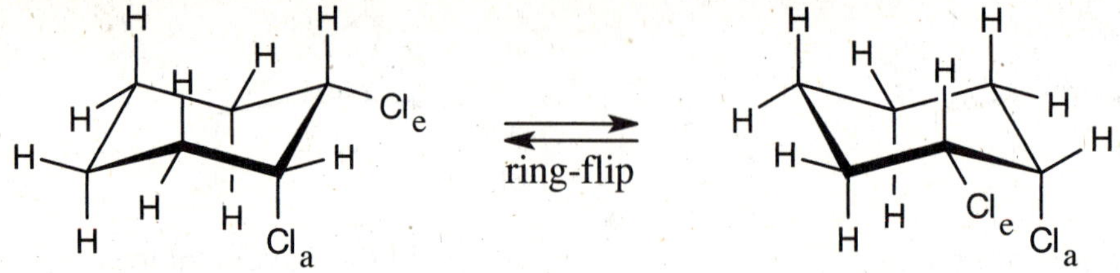


After the ring-flip, the relationship of the two substituents is still axial–equatorial.

**2.27** For a *trans*-1,2-disubstituted cyclohexane, two adjacent substituents must be either both axial or both equatorial.

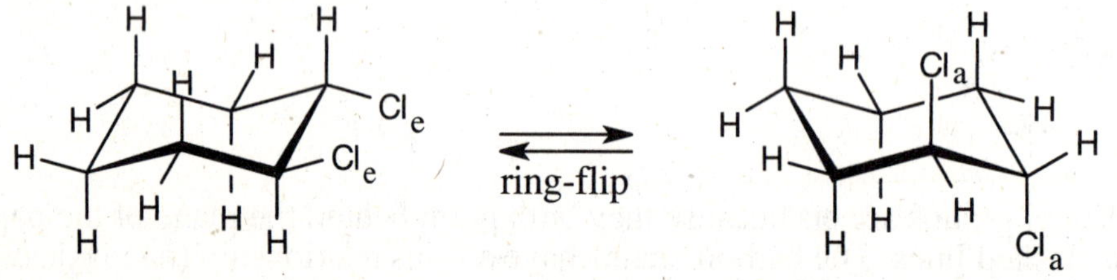


A ring-flip converts two adjacent axial substituents into equatorial substituents, and *vice versa*.

**2.28** The three substituents have the orientations shown in the first structure. To decide if the conformation shown is the more stable conformation or the less stable conformation, perform a ring-flip on the illustrated conformation, and compare the interactions among substituents of the two conformers. The illustrated conformation is the less stable chair form because it has more interactions.

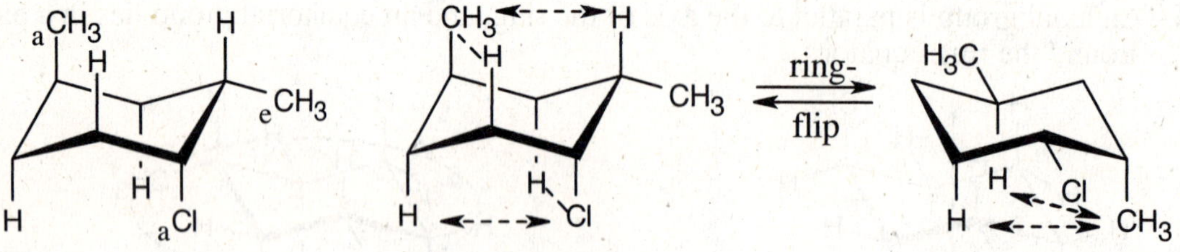


four interactions

two interactions

## Visualizing Chemistry

**2.29**

(a)

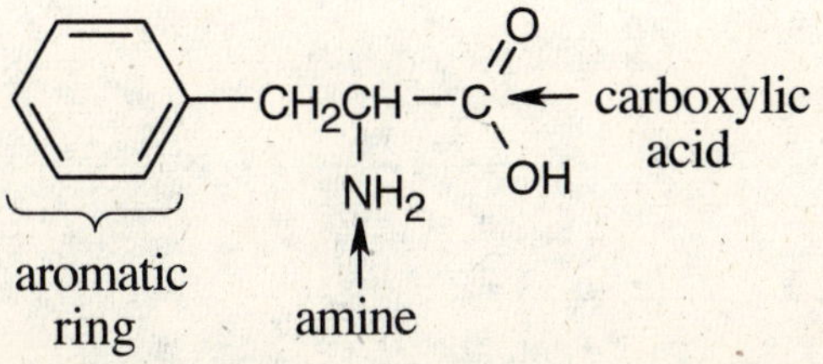


Phenylalanine $C_9H_{11}NO_2$

(b)

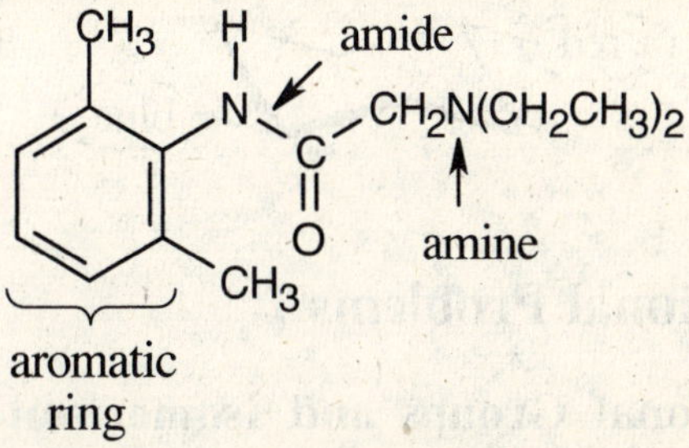


Lidocaine $C_{14}H_{22}N_2O$

**2.30**

(a)

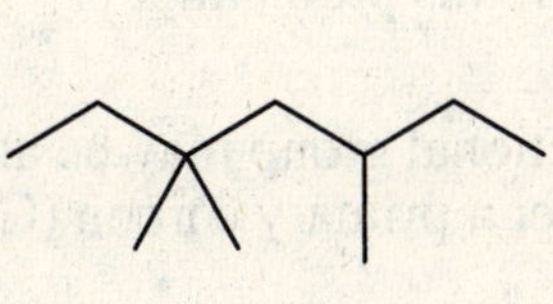

3,3,5-Trimethylheptane

(b)

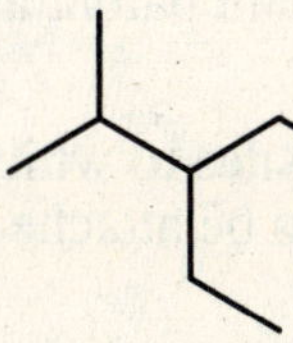

3-Ethyl-2-methylpentane

(c)

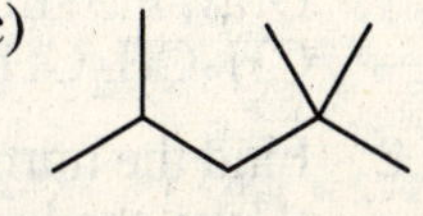

2,2,4-Trimethylpentane

(d)

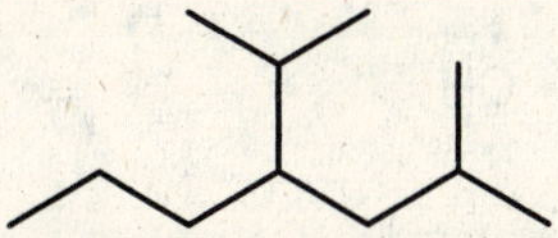

4-Isopropyl-2-methylheptane

**2.31**

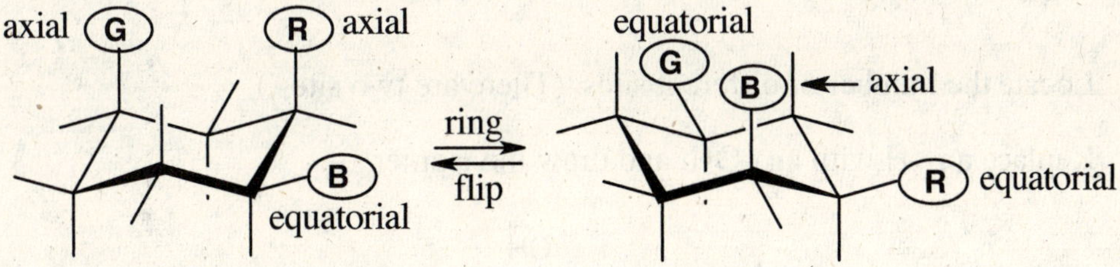


**2.32**

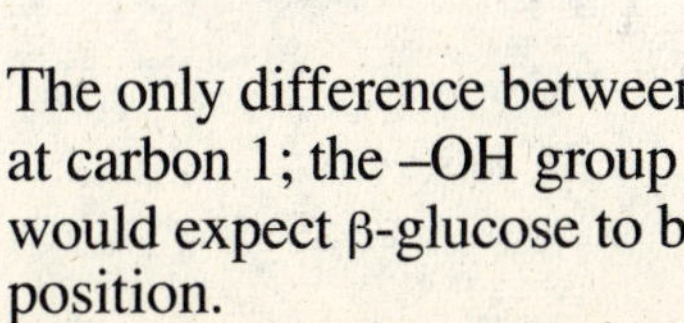


α-Glucose

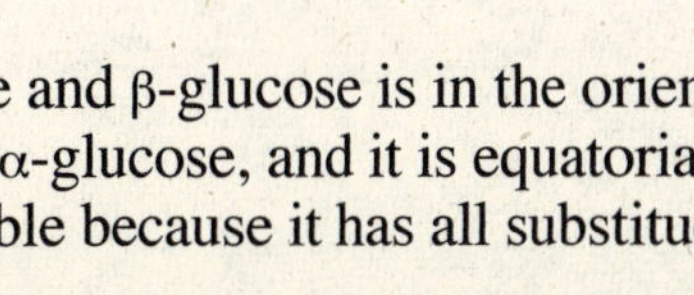


β-Glucose

The only difference between α-glucose and β-glucose is in the orientation of the –OH group at carbon 1; the –OH group is axial in α-glucose, and it is equatorial in β-glucose. You would expect β-glucose to be more stable because it has all substituents in the equatorial position.

**2.33** The green substituent is axial, and the red and blue substituents are equatorial.

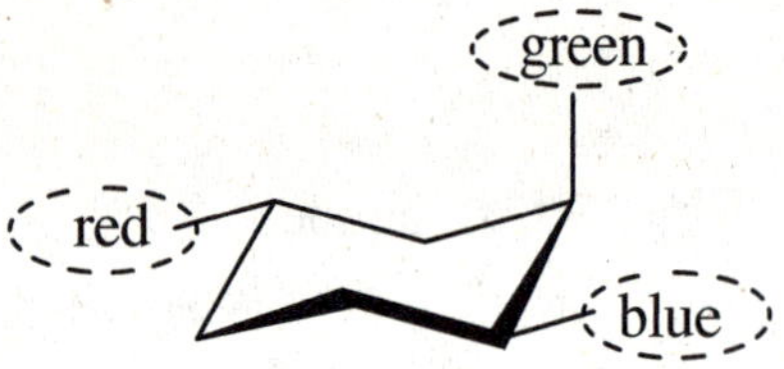


cis relationship: blue–green
trans relationship: green–red, blue–red

## Additional Problems

### Functional Groups and Isomerism

**2.34** (a) To solve this problem, you must check all possibilities in a systematic way. The following procedure may be helpful.

1. Draw the simplest straight-chain parent alkane. (In this problem, it is $CH_3CH_2CH_2CH_3$).
2. Find the number of different sites to which a functional group may be attached. (Here, the hydroxyl group can be attached to either a primary carbon ($CH_3$–) or to a secondary carbon (–$CH_2$–).
3. At each different site, replace an –H with an –OH, and draw the structure.

$CH_3CH_2CH_2CH_2—OH$ and 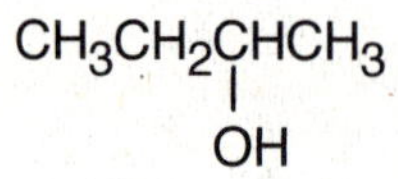

4. Draw the simplest branched alkane.

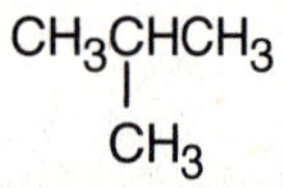

5. Locate the number of different sites. (There are two sites.)
6. Replace an –H with an –OH, and draw the isomer.

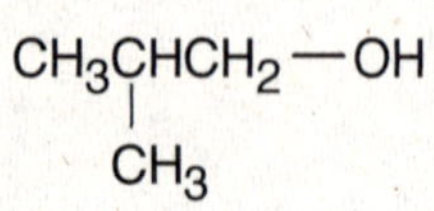 and 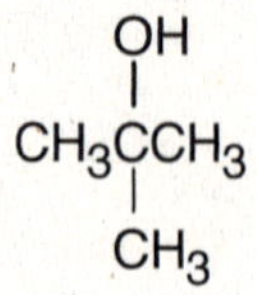

7. Repeat steps 4–6 with the next simplest branched alkane. In this problem, we have already drawn all isomers.

(b) There are 17 isomers of $C_5H_{13}N$. Nitrogen can be bonded to one, two or three alkyl groups.

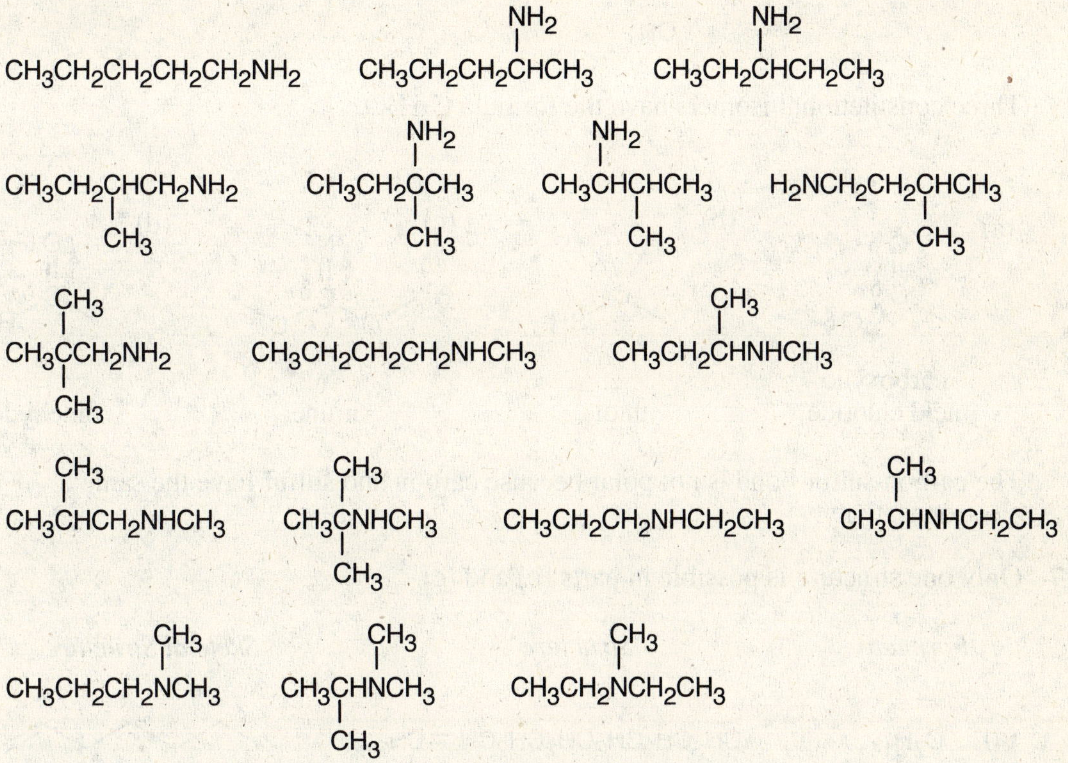

(c) There are 3 ketone isomers with the formula $C_5H_{10}O$.

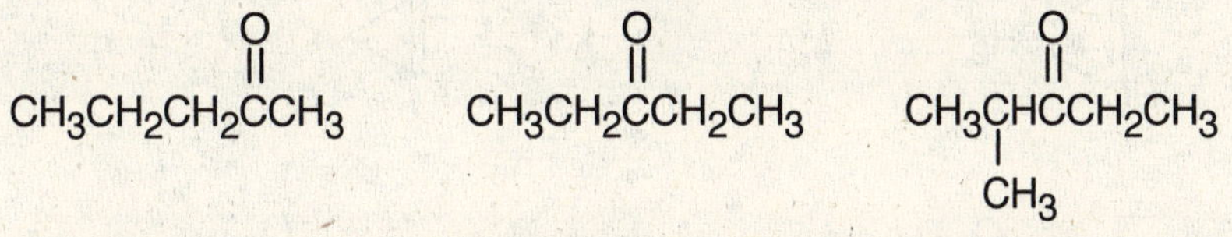

(d) There are 4 isomeric aldehydes with the formula $C_5H_{10}O$. Remember that the aldehyde functional group can occur only at the end of a chain.

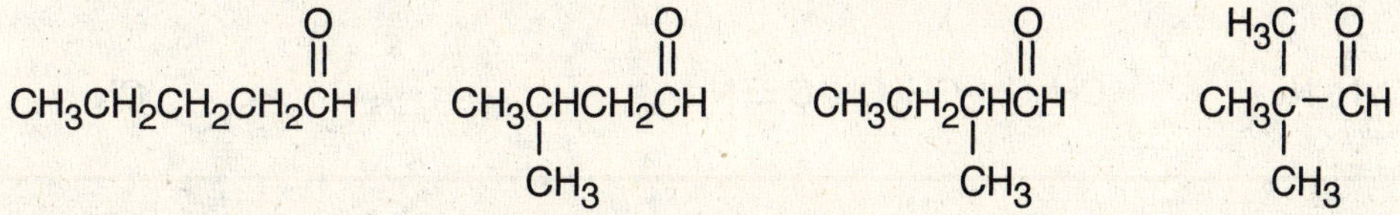

(e) There are 3 ethers with the formula $C_4H_{10}O$.

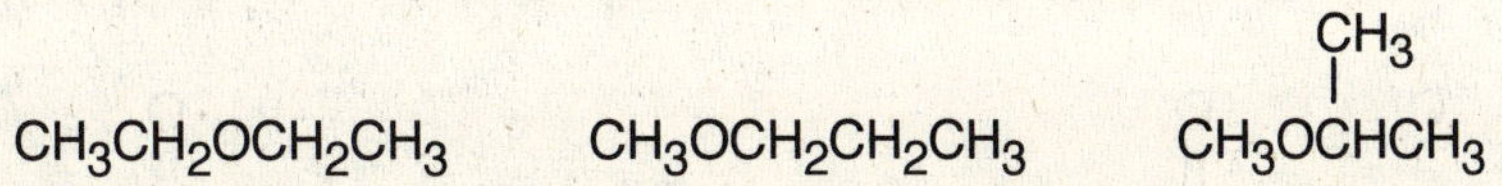

(f) There are 4 esters with the formula $C_4H_8O_2$.

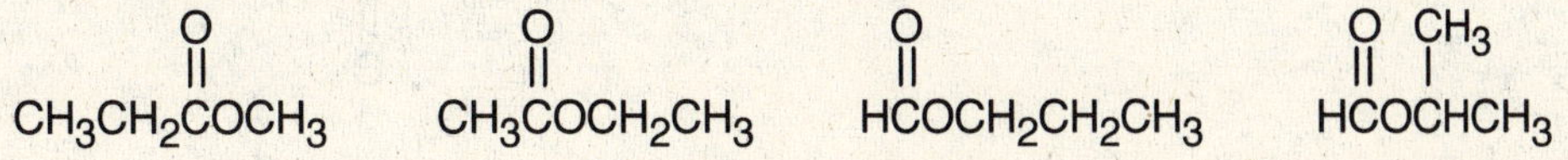

**2.35**

$CH_3CH_2CH_2OH$ $\qquad$ $CH_3CHCH_3$ (with OH on the middle carbon) $\qquad$ $CH_3OCH_2CH_3$

Three constitutional isomers have the formula $C_3H_8O$.

**2.36**

(a) 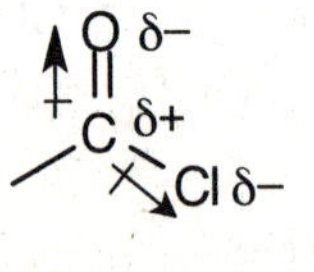
carboxylic acid chloride

(b) 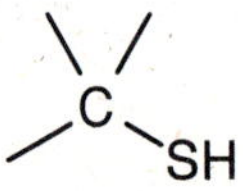
thiol

(c) 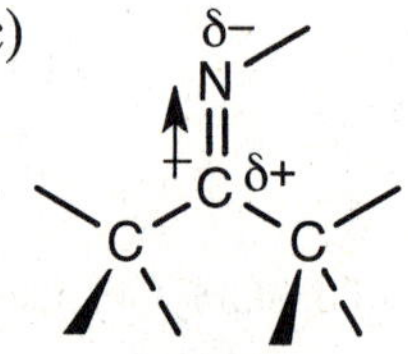
imine

(d) 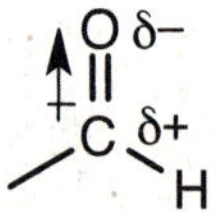
aldehyde

The carbon-sulfur bond is not polar because carbon and sulfur have the same electronegativity.

**2.37** Only one structure is possible in parts (b) and (c).

| | *Formula* | *Structure* | *Skeletal Structure* |
|---|---|---|---|
| (a) | $C_7H_{14}$ | $CH_3CH_2CH_2CH_2CH_2CH{=}CH_2$ |  |
| (b) | $C_3H_4$ | 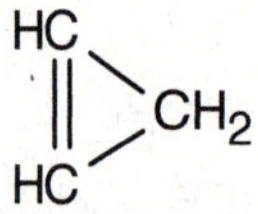 |  |
| (c) | $C_4H_8O$ | 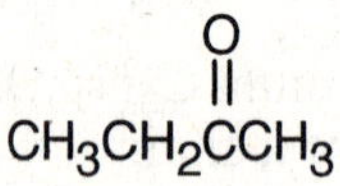 | 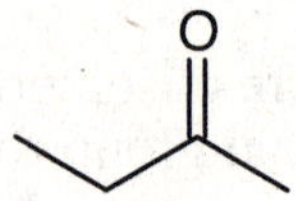 |
| (d) | $C_5H_9N$ | $CH_3CH_2CH_2CH_2C{\equiv}N$ | 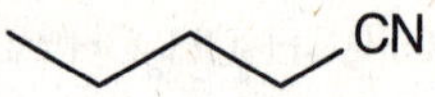 |
| (e) | $C_5H_8$ | $CH_3CH{=}CHCH{=}CH_2$ | 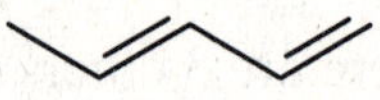 |
| (f) | $C_4H_6O_2$ | 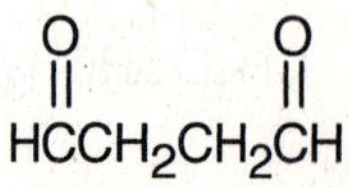 | 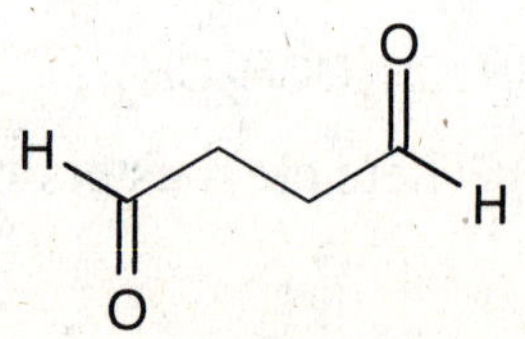 |

**2.38** There are many acceptable answers to each part of this problem. Correct answers include:

*Skeletal structure*

(a) $CH_3CH_2CH_2CH_2CH{=}CH_2$

(b)

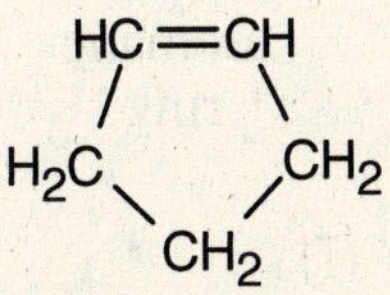


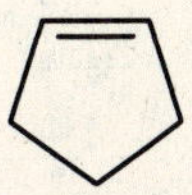

(c) $CH_3CH_2\overset{O}{\overset{\|}{C}}CH_2CH_3$

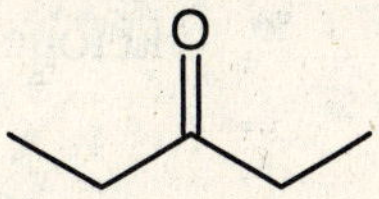


(d) $CH_3CH_2\overset{O}{\overset{\|}{C}}NHCH_3$

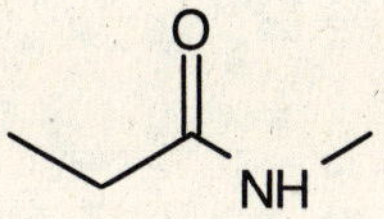


(e) $CH_3\overset{O}{\overset{\|}{C}}OCH_2CH_2CH_3$

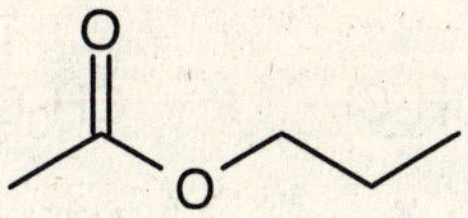


(f)

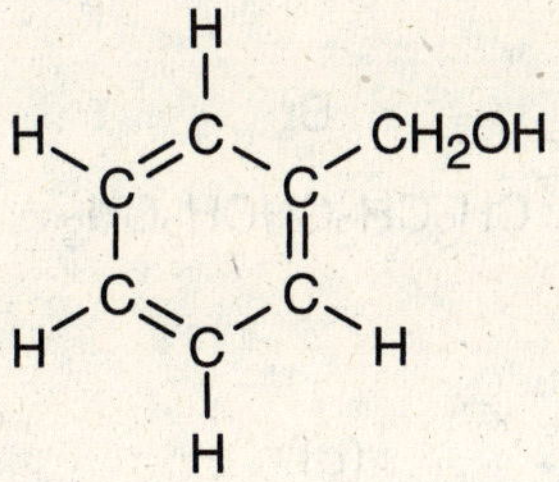


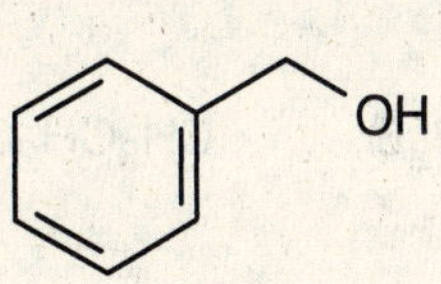


**2.39** Many other structures can be drawn.

(a)

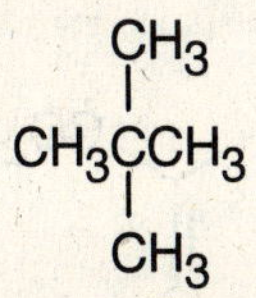


(b)

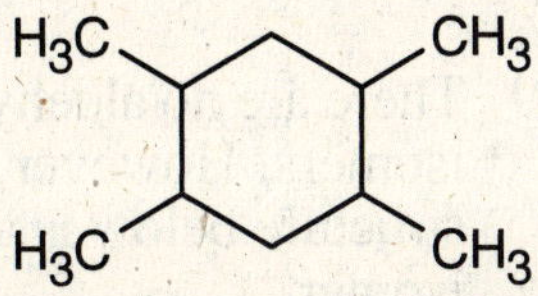


(c)

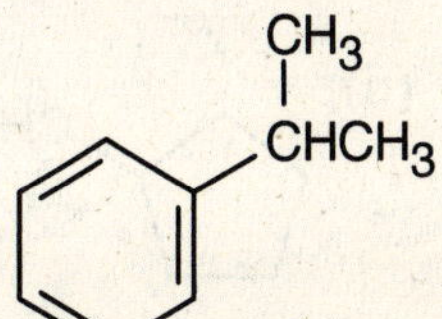


(d)

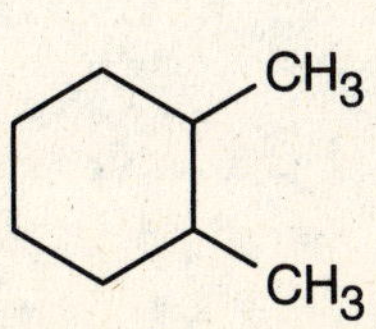


(e) $CH_3CH_2\overset{NH_2}{\overset{|}{C}}HCH_3$

**2.40**

(a)

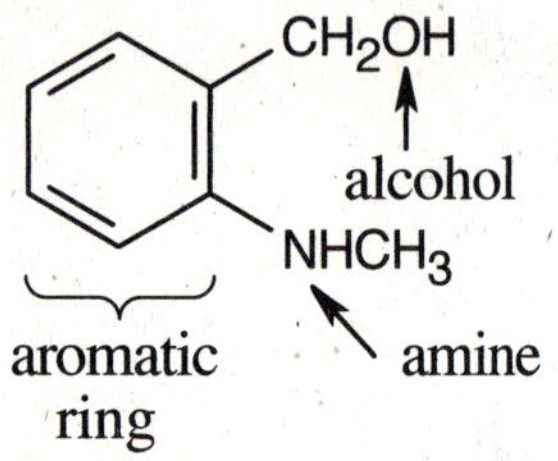


(b)

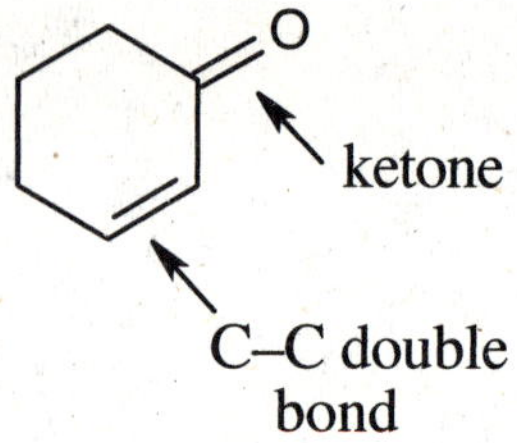


(c)

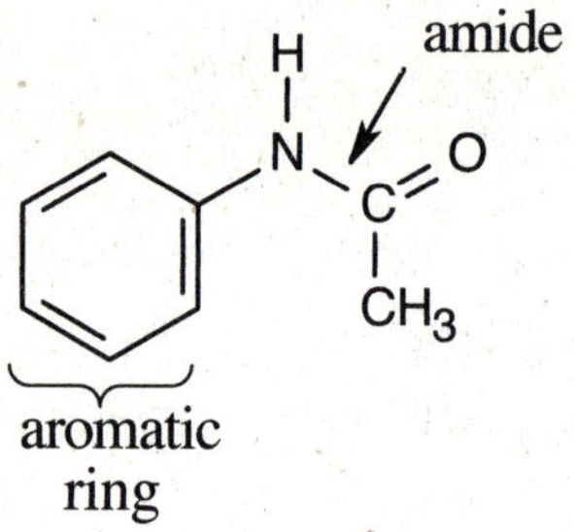


(d)

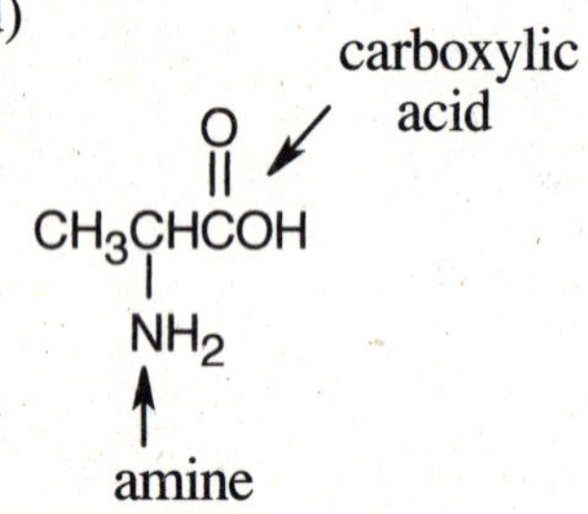


(e)

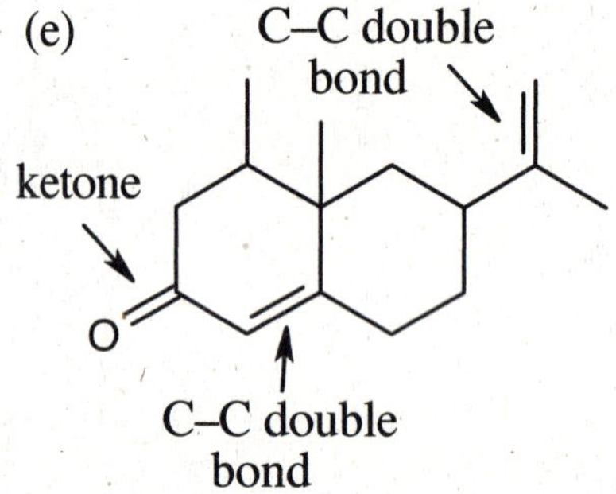


(f)

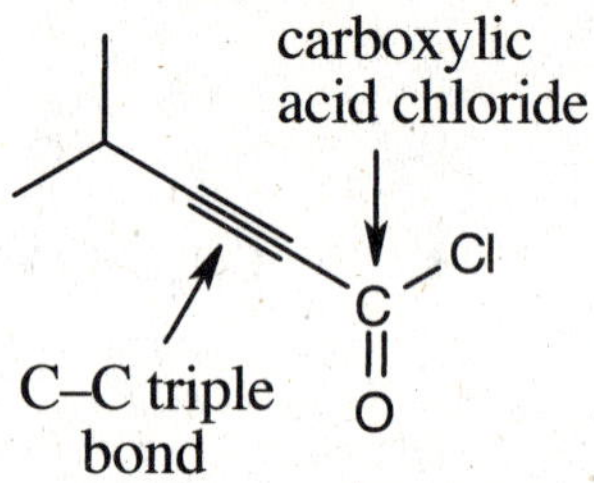


**2.41**

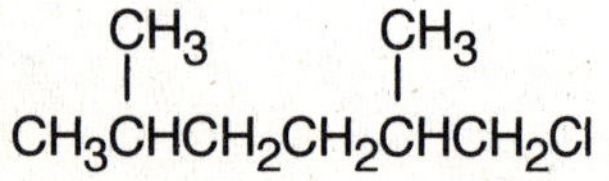

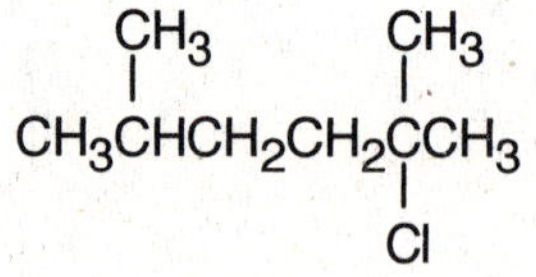

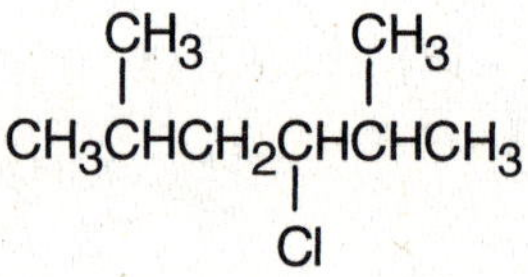

**2.42**

$CH_3CH_2CH_2CH_2CH_2Br$

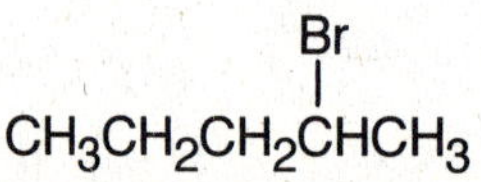

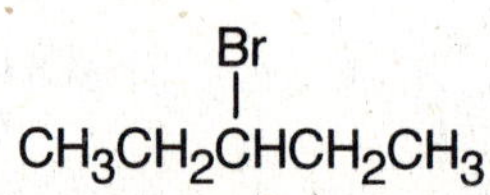

**2.43**

(a) $CH_3CH_2CH_2CH_2CH_2Br$

(b)

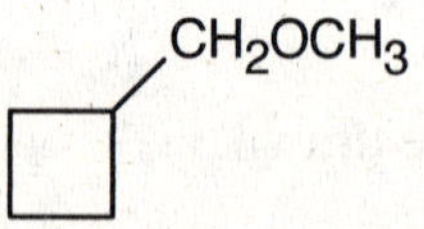

(c)

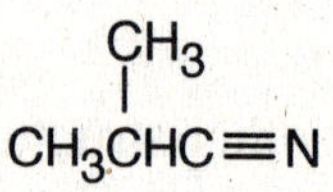

(d)

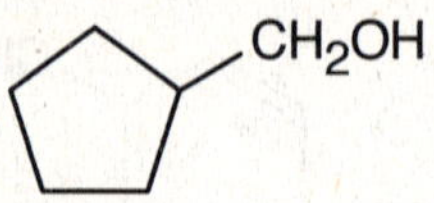

(e) There are no aldehyde isomers. However, the structure below is a ketone isomer.

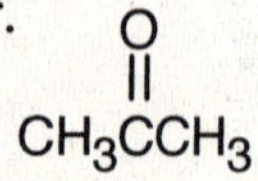

(f)

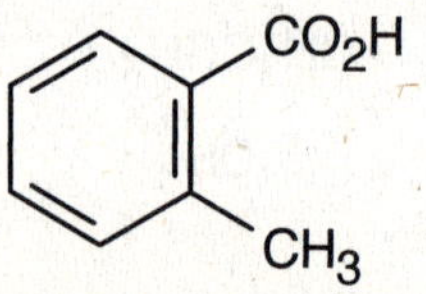

**2.44**

(a)

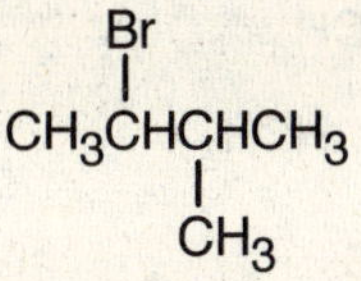

same

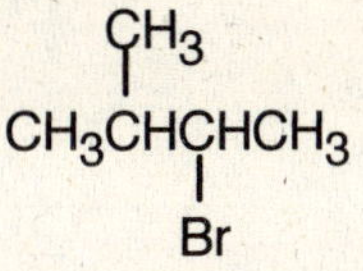

same

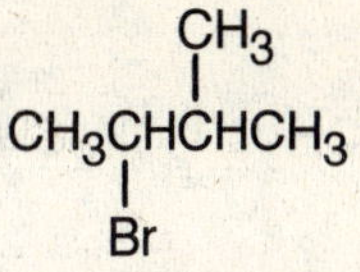

same

(b)

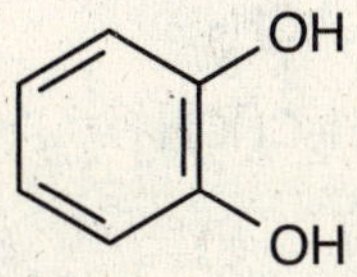

same

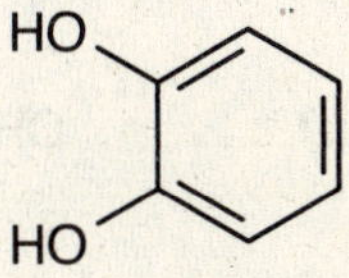

same

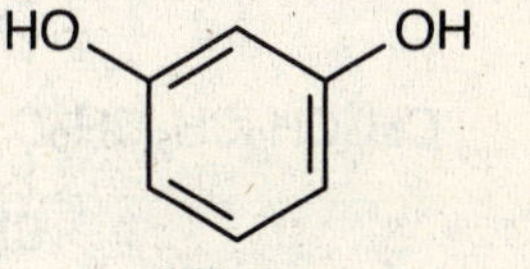

different

(c)

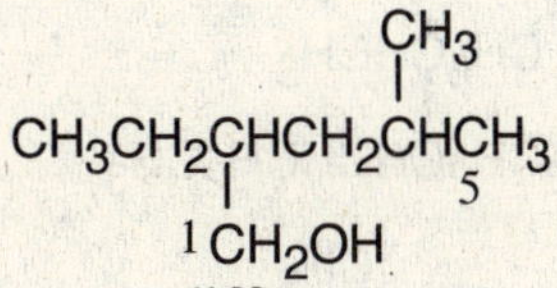

different

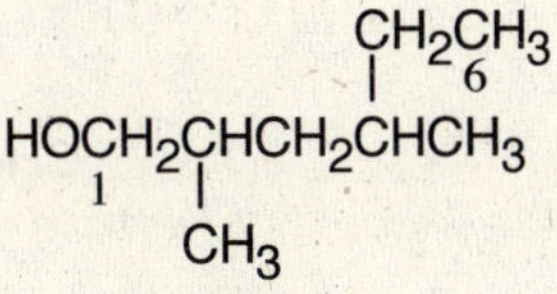

same

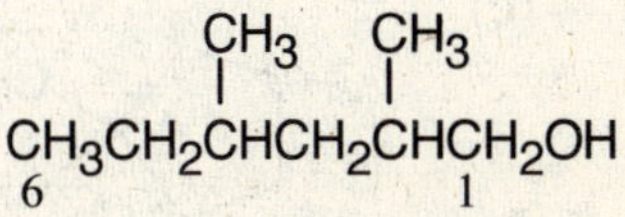

same

Give the number "1" to the carbon bonded to –OH, and count to find the longest chain containing the –OH group.

## Naming and Drawing Chemical Structures

**2.45**

(a)

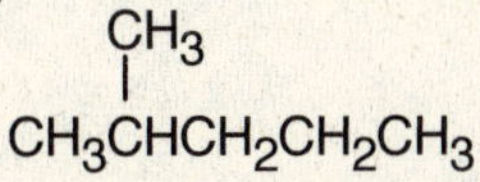

2-Methylpentane

(b)

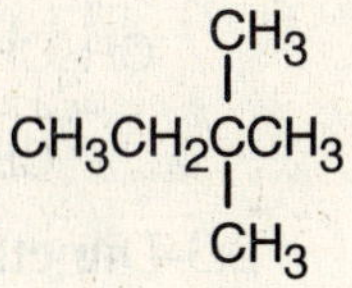

2,2-Dimethylbutane

(c)

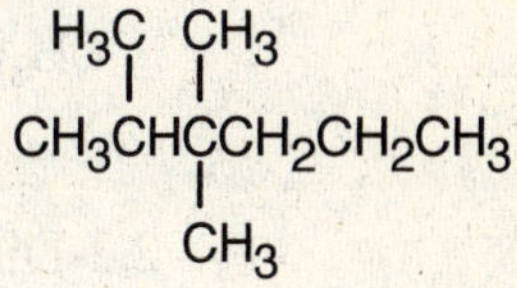

2,3,3-Trimethylhexane

(d)

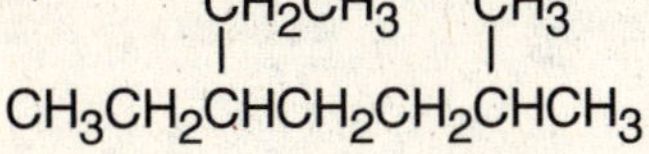

5-Ethyl-2-methylheptane

(e)

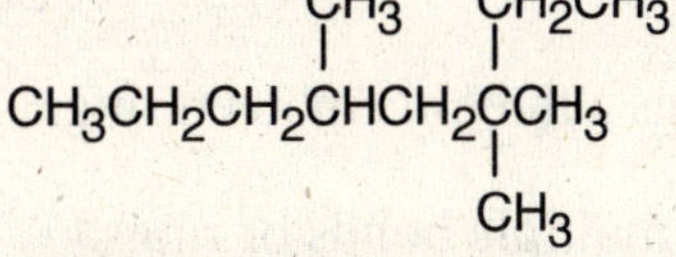

3,3,5-Trimethyloctane

(f)

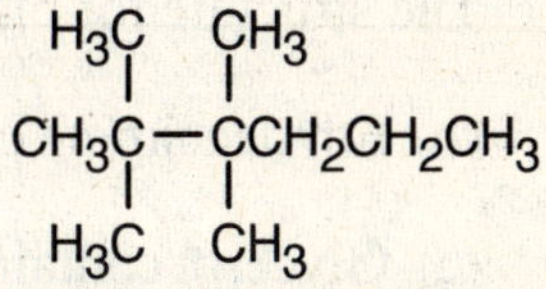

2,2,3,3-Tetramethylhexane

**2.46**

(a)

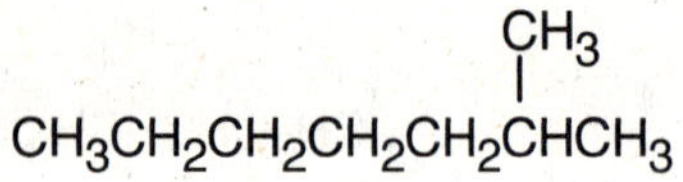

2-Methylheptane

(b)

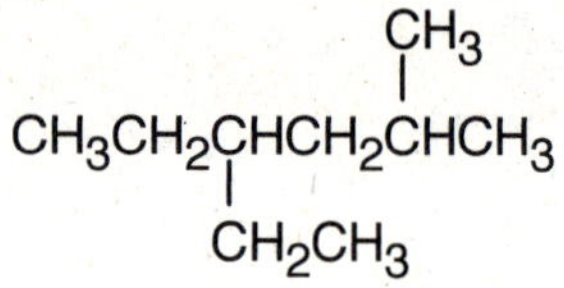

4-Ethyl-2-methylhexane

(c)

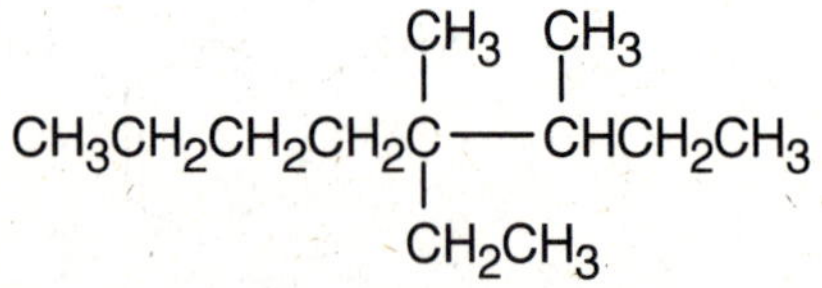

4-Ethyl-3,4-dimethyloctane

(d)

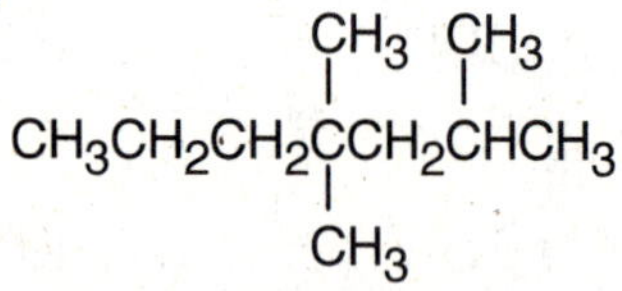

2,4,4-Trimethylheptane

(e)

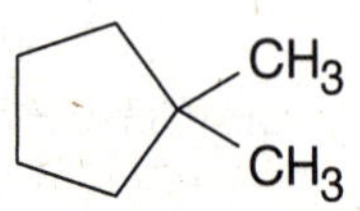

1,1-Dimethylcyclopentane

(f)

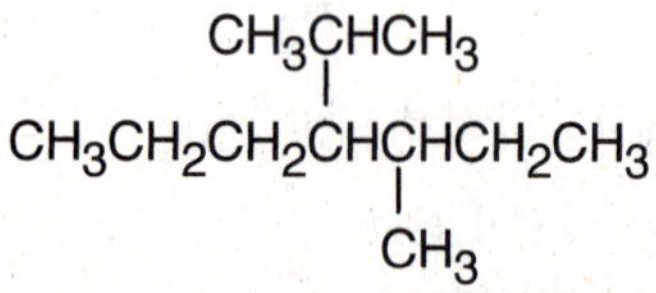

4-Isopropyl-3-methylheptane

**2.47**

$CH_3CH_2CH_2CH_2CH_2CH_3$

Hexane

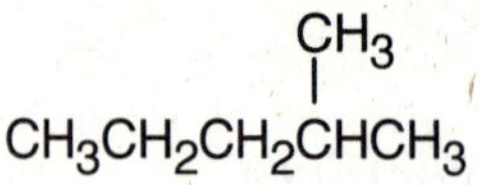

2-Methylpentane

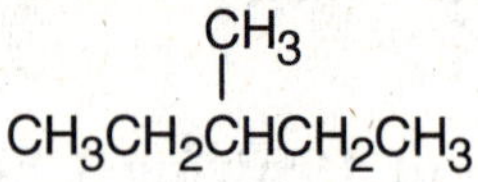

3-Methylpentane

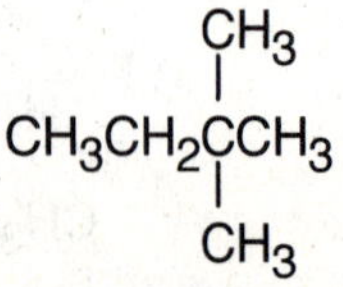

2,2-Dimethylbutane

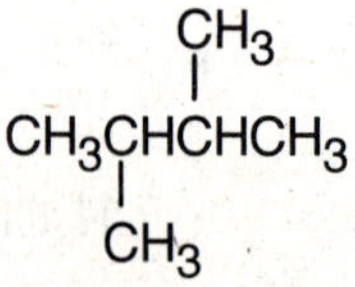

2,3-Dimethylbutane

**2.48** (a) It is stated that biacetyl contains no rings or carbon–carbon double bonds. However, it is obvious from the formula for biacetyl that some sort of multiple bond must be present. The structure for biacetyl contains two carbon-oxygen double bonds.

(b) Ethyleneimine contains a three-membered ring.

(c) Glycerol contains no multiple bonds or rings.

(a)

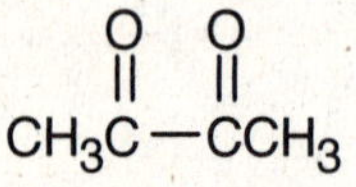

Biacetyl

(b)

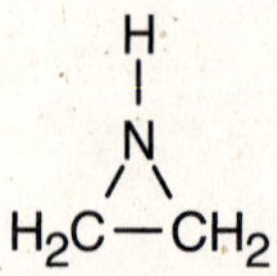

Ethyleneimine

(c)

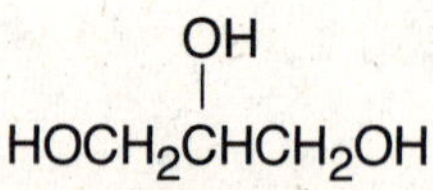

Glycerol

**2.49**

(a)

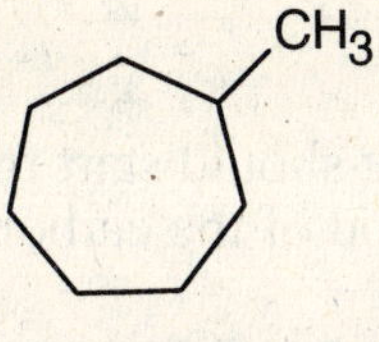

Methylcycloheptane

(b)

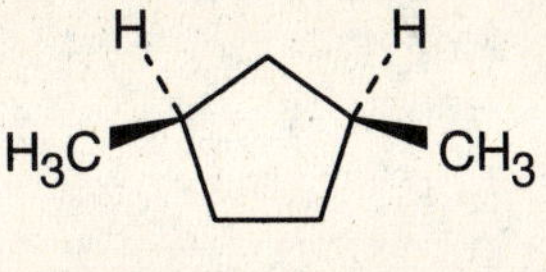

*cis*-1,3-Dimethylcyclopentane

(c)

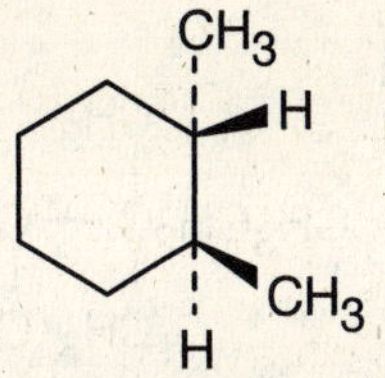

*trans*-1,2-Dimethylcyclohexane

(d)

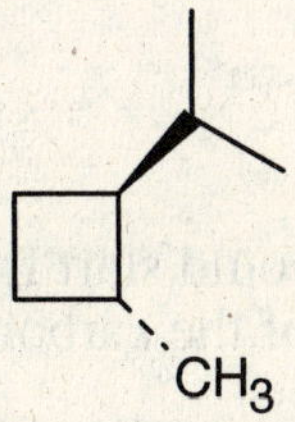

*trans*-1-Isopropyl-2-methylcyclobutane

(e)

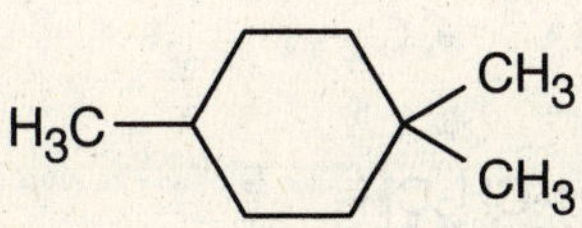

1,1,4-Trimethylcyclohexane

**2.50** Other structures are possible.

(a)

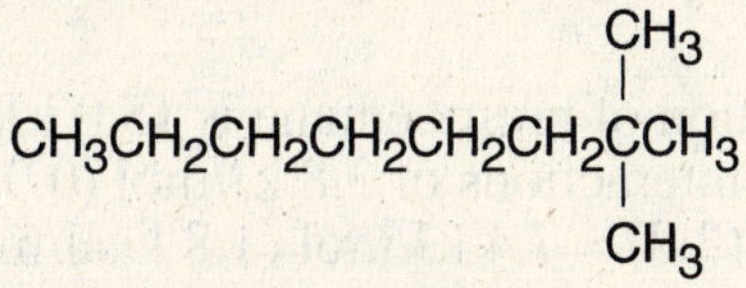

2,2-Dimethyloctane

(b)

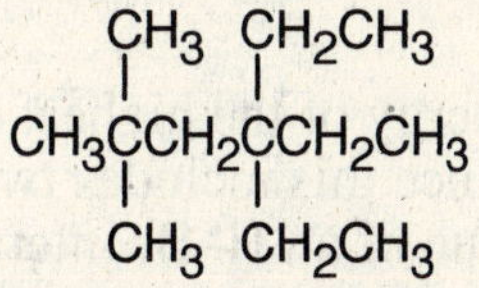

4,4-Diethyl-2,2-dimethylhexane

(c)

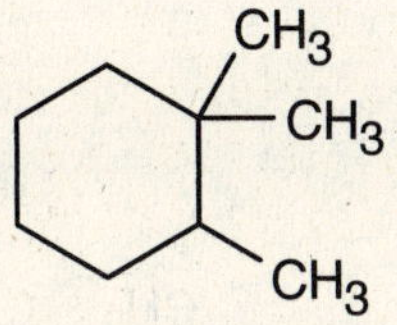

1,1,2-Trimethylcyclohexane

**2.51** *Structure and Correct Name* — *Error*

(a)

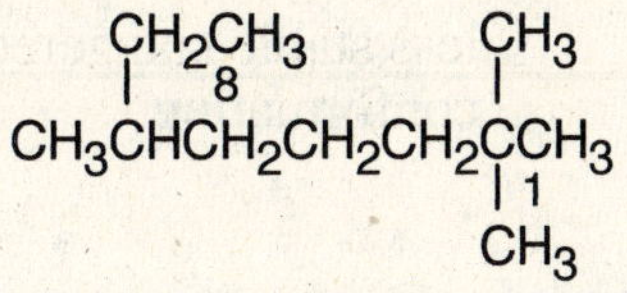

Correct name: 2,2,6-Trimethyloctane

The longest chain is an octane.

(b)

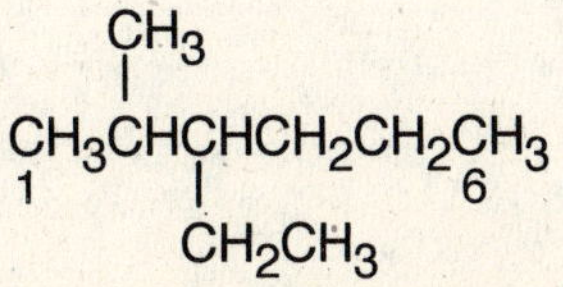

Correct name: 3-Ethyl-2-methylhexane

The longest chain is a hexane. Numbering should start from the opposite end of the carbon chain, nearer the first branch.

| *Structure and Correct Name* | *Error* |
|---|---|
| (c) 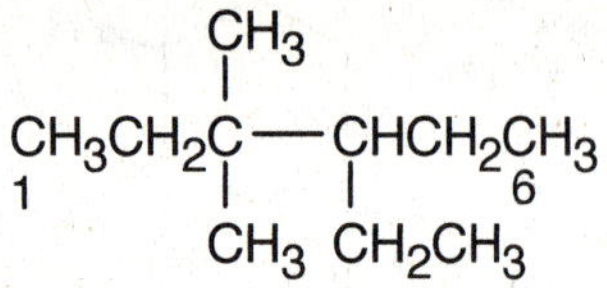 Correct name: 4-Ethyl-3,3-dimethylhexane | Numbering should start from the opposite end of the carbon chain. |
| (d) 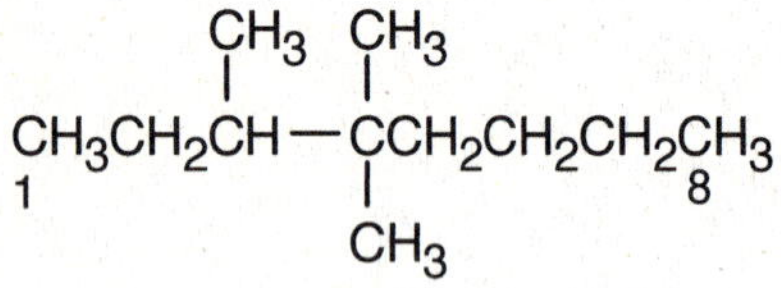 Correct name: 3,4,4-Trimethyloctane | Numbering should start from the opposite end of the carbon chain. |

## Conformations and Cis–Trans Isomerism

**2.52** The strain energy of the highest energy conformation of bromoethane is 15.0 kJ/mol (3.6 kcal/mol). Since this includes two H–H eclipsing interactions of 3.8 kJ/mol (0.9 kcal/mol) each, the value of an H–Br interaction is 15.0 – 2(3.8) = 7.4 kJ/mol (1.8 kcal/mol).

**2.53** Since *cis*-1-*tert*-butyl-4-methylcyclohexane exists in the conformation shown, a *tert*-butyl group must be much larger than a methyl group.

**2.54**

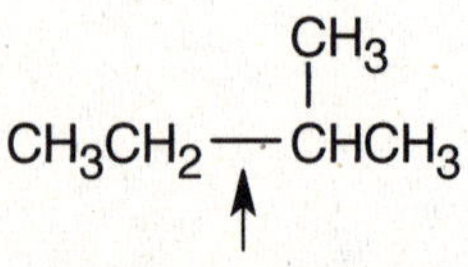

C2–C3 bond

2-Methylbutane

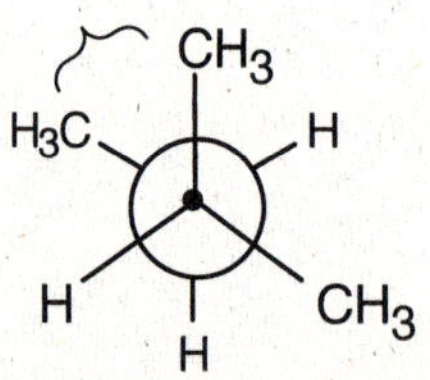

more stable staggered conformation

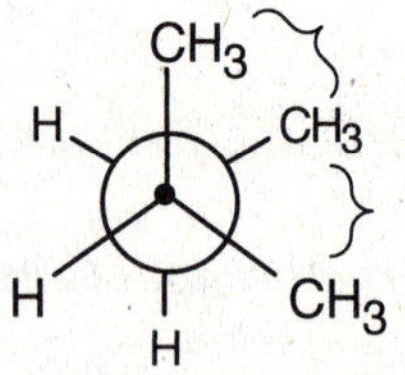

less stable staggered conformation

Brackets indicate steric interactions

In the less stable conformation, there are two interactions between methyl groups that are 60° apart. Only one of these interactions occurs in the more stable isomer.

**2.55**

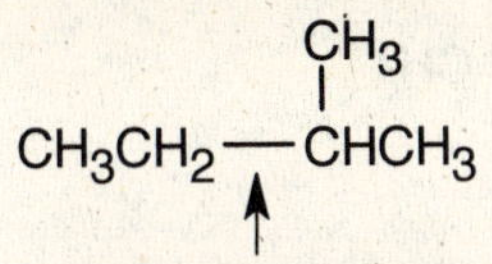


C2–C3 bond

2-Methylbutane

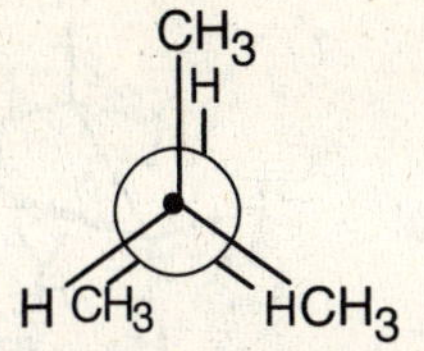


lower energy eclipsed conformation

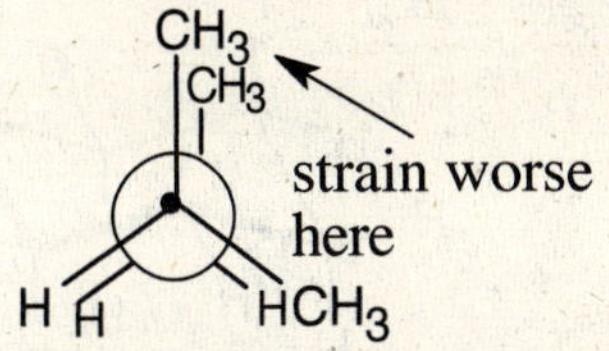


higher energy eclipsed conformation

In the higher energy conformation, the two methyl groups that are eclipsed produce more strain than occurs in the other isomer.

**2.56**

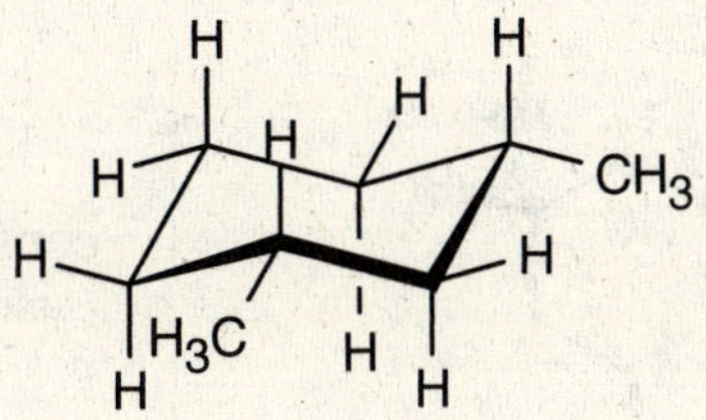


*cis*-1,3-Dimethylcyclohexane

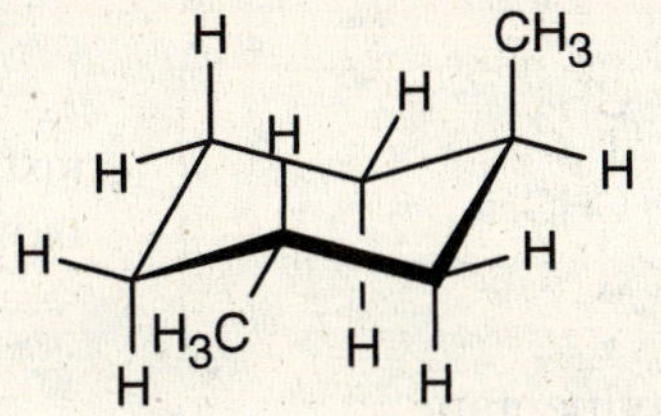


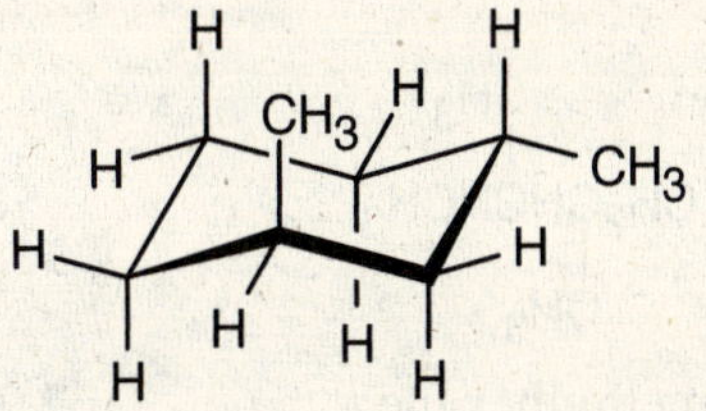


*trans*-1,3-Dimethylcyclohexane

The lowest energy conformations of both 1,3-dimethylcyclohexanes are drawn. *cis*-1,3-Dimethylhexane is the more stable isomer because both methyl groups are equatorial in the most stable conformation. For *trans*-1,3-dimethylhexane, one methyl group must always be in the higher energy axial orientation. (A high energy diaxial conformation of *cis*-1,3-dimethylcyclohexane can also be drawn.)

**2.57**

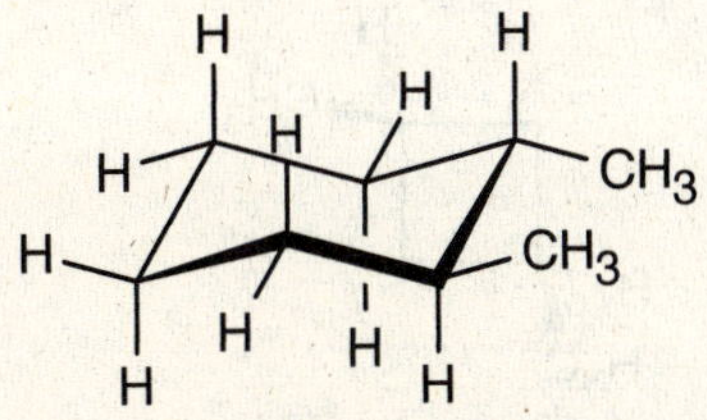


*trans*-1,2-Dimethylcyclohexane

The methyl groups are equatorial in the more stable chair conformation of *trans*-1,2-dimethylcyclohexane.

**2.58**

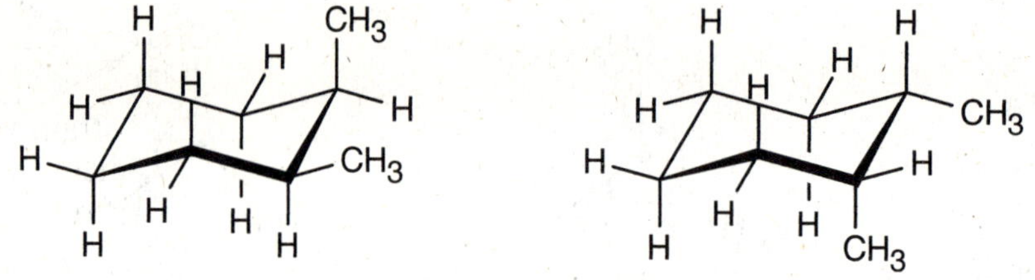

*cis*-1,2-Dimethylcyclohexane

There are two chair conformations of *cis*-1,2-dimethylcyclohexane that are of equal stability. In each conformation, one methyl group is axial and one is equatorial. Both cis conformations are less stable than the more stable conformation of *trans*-1,2-dimethylcyclohexane because of steric strain caused by the axial methyl group.

**2.59**

(a)

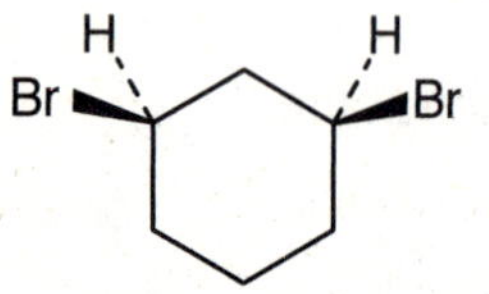

*cis*-1,3-Dibromocyclohexane

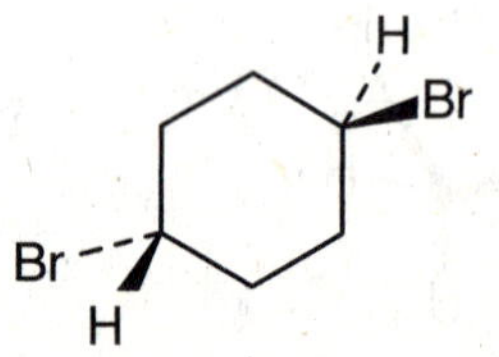

*trans*-1,4-Dibromocyclohexane

constitutional isomers

(b)

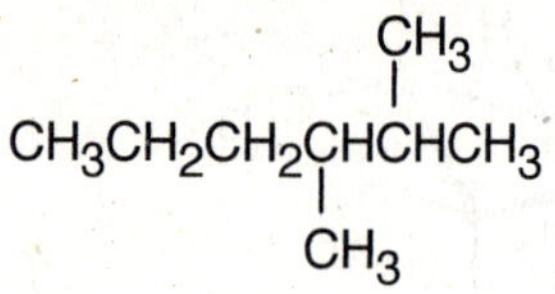

2,3-Dimethylhexane

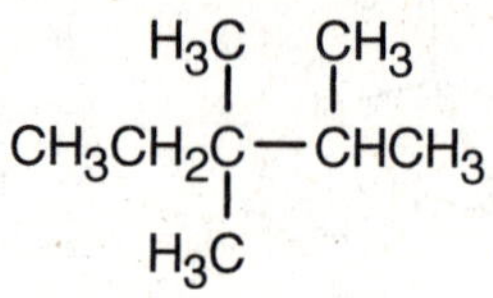

2,3,3-Trimethylpentane

constitutional isomers

(c)

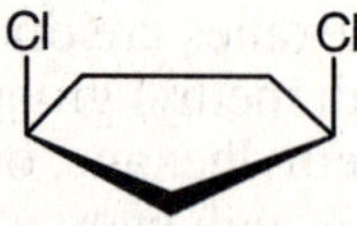

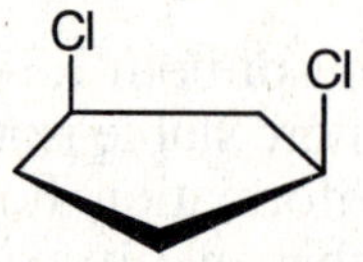

identical

**2.60**

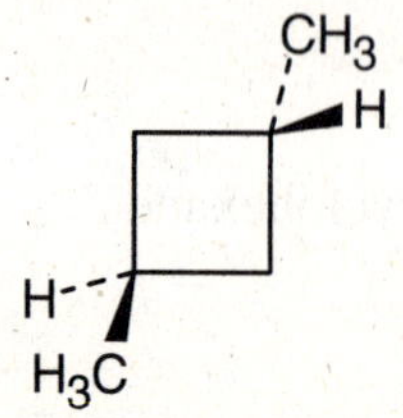

*trans*-1,3-Dimethylcyclobutane

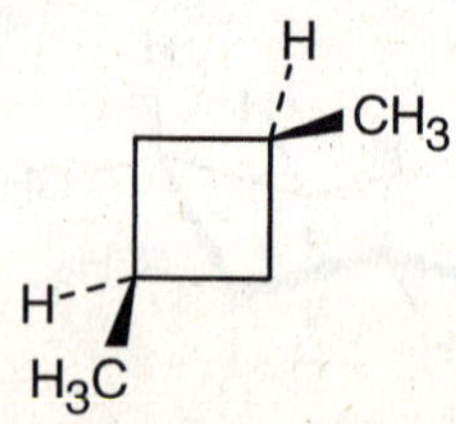

*cis*-1,3-Dimethylcyclobutane

**2.61**

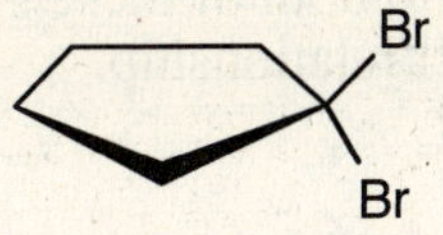


1,1-Dibromocyclopentane

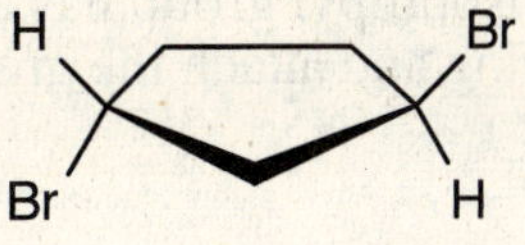


*trans*-1,3-Dibromocyclopentane

Many other constitutional isomers can also be drawn, including *cis*-1,3-dibromocyclopentane.

## General Problems

**2.62**

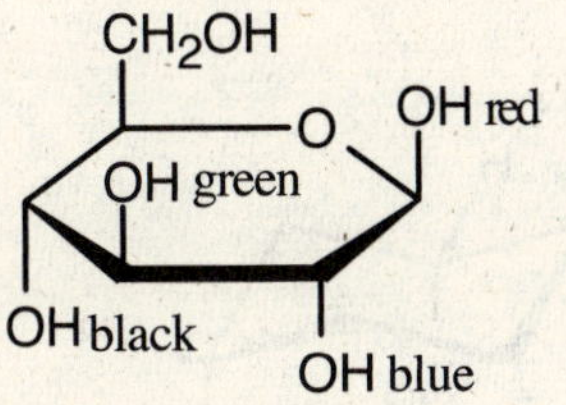


cis relationship: red–green, blue–black

trans relationship: red–blue, green–blue
red–black, green–black

**2.63** (a) Because malic acid has two $-CO_2H$ groups, the formula for the rest of the molecule is $C_2H_4O$. Possible structures for malic acid are:

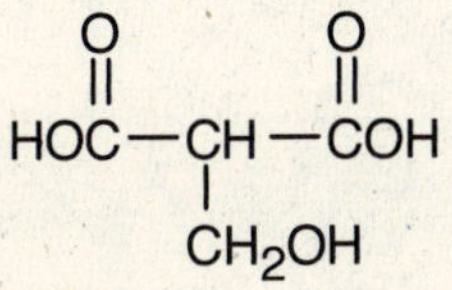

primary alcohol

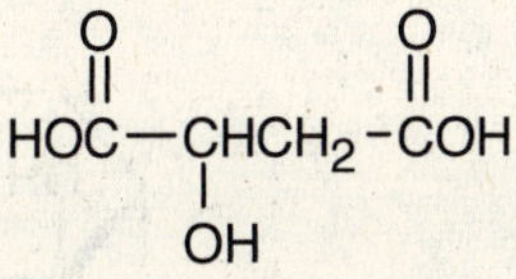

secondary alcohol

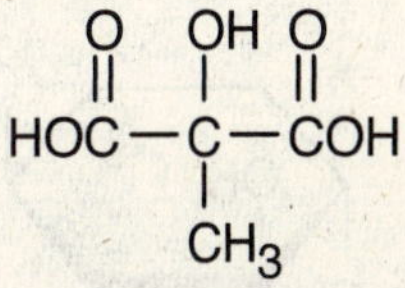

tertiary alcohol

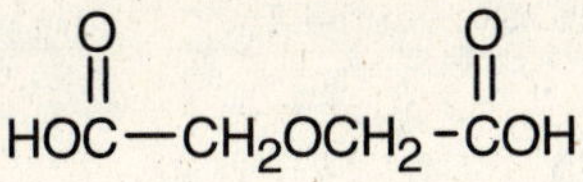

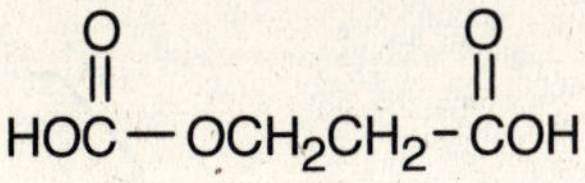

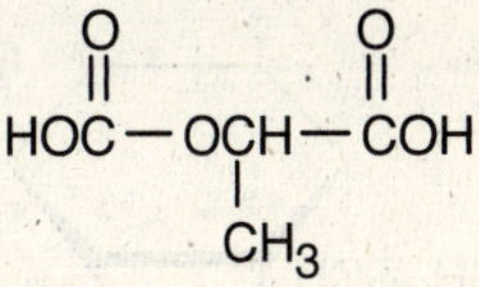

(b) Because only one of these compounds (the second one) is also a secondary alcohol, it must be malic acid.

**2.64** Since the methyl group of *N*-methylpiperidine prefers an equatorial conformation, the steric requirements of a methyl group must be greater than those of an electron lone pair.

**2.65** From the flat-ring drawing you can see that the methyl group and the –OH group have a cis relationship, and the isopropyl group has a trans relationship to both of these groups. Draw a chair cyclohexane ring and attach the groups with the correct relationship.

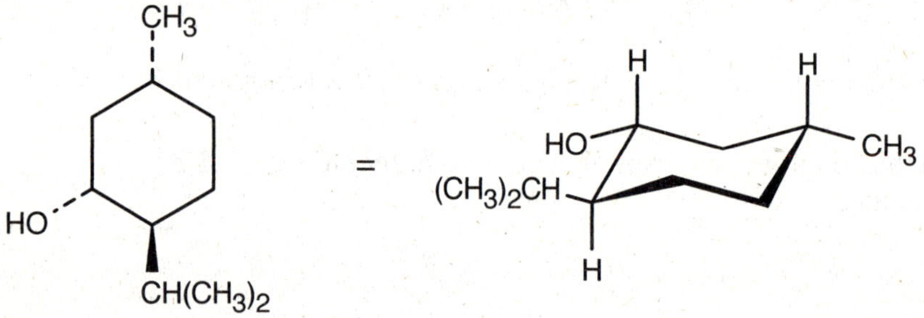


In this conformation, all substituents are equatorial. Now, perform a ring flip.

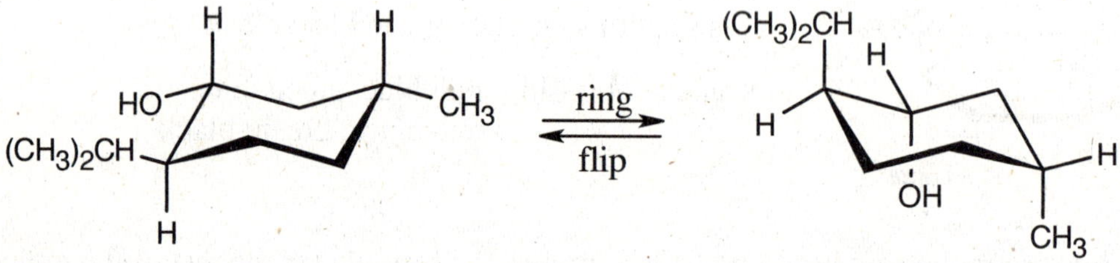


The second conformation is less stable because all substituents are axial.

**2.66**

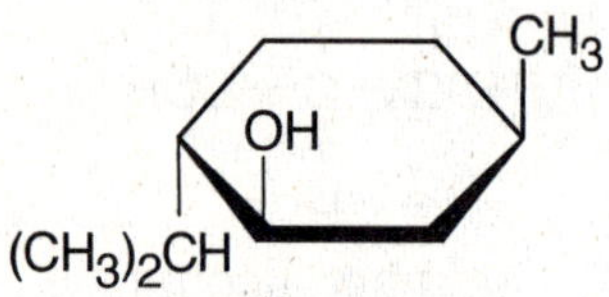


Menthol

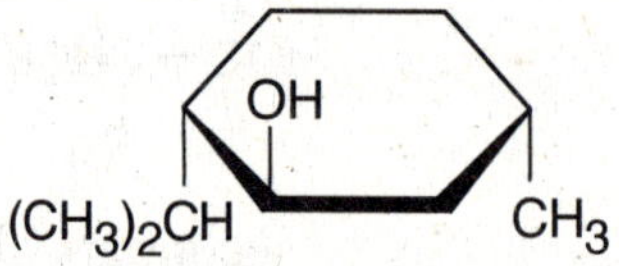


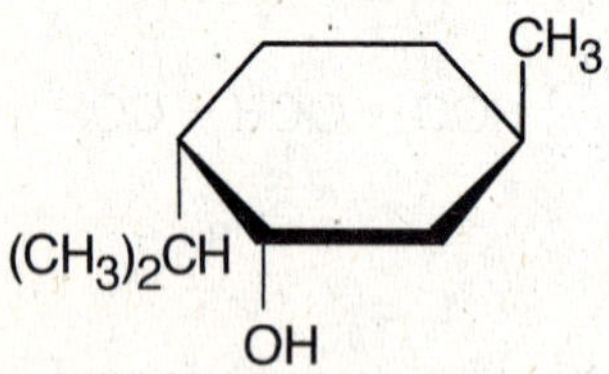


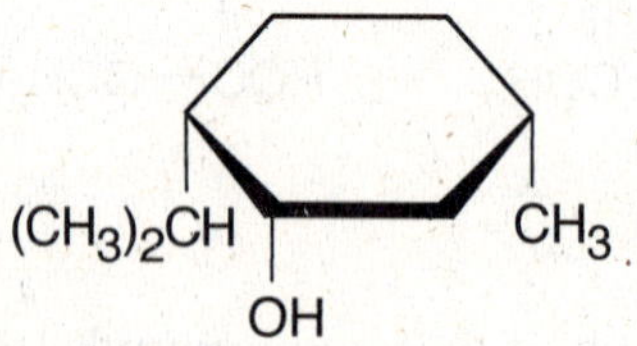


**2.67**

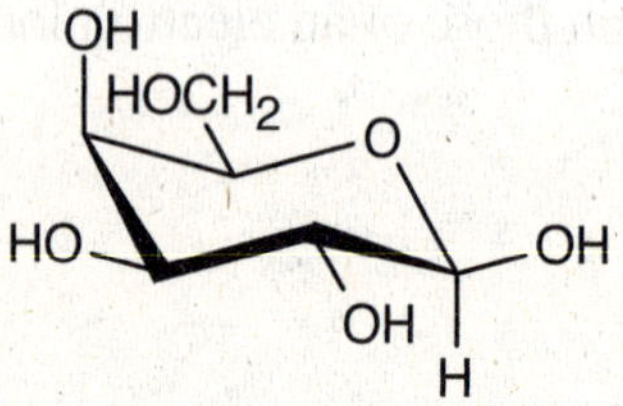


Galactose

In this conformation, all substituents, except for one hydroxyl group, are equatorial.

**2.68**

The two *trans*-1,2-dimethylcyclopentanes are mirror images.

**2.69**

Two cis–trans isomers of 1,3,5-trimethylcyclohexane are possible. In one isomer (**A**), all methyl groups are cis; in **B**, one methyl group is trans to the other two.

**2.70**

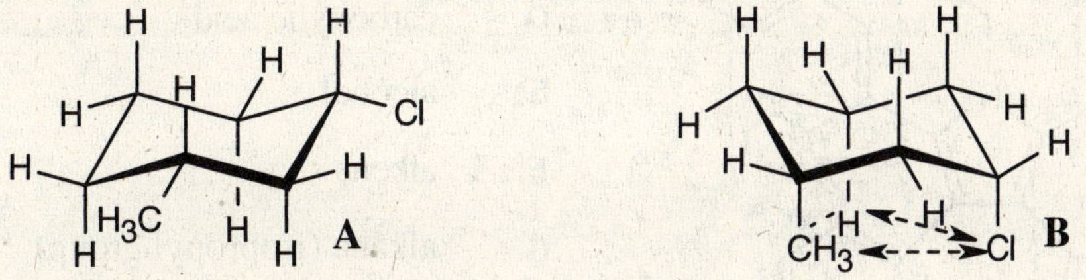


Conformation **A** of *cis*-1-chloro-3-methylcyclohexane has no 1,3-diaxial interactions and is the more stable conformation. Steric strain in **B** is due to one $CH_3 \leftrightarrow H$ interaction, one $Cl \leftrightarrow H$ interaction and one $CH_3 \leftrightarrow Cl$ interaction.

## In the Medicine Cabinet

**2.71**

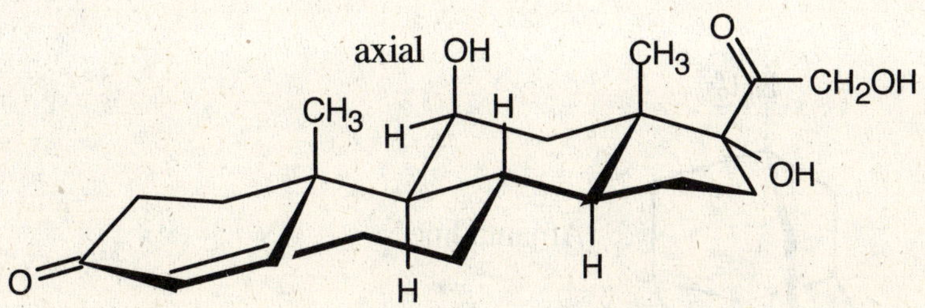

**2.72**

(a)

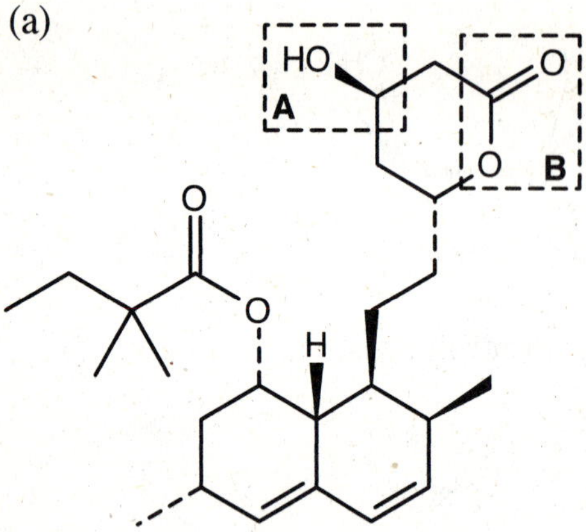


Zocor $C_{25}H_{38}O_5$

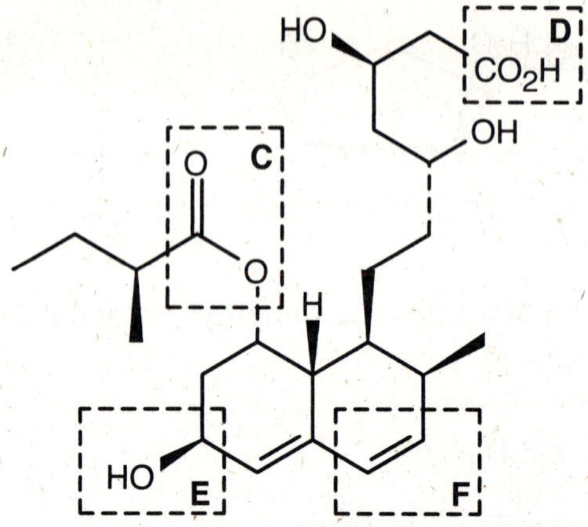


Pravachol $C_{23}H_{36}O_7$

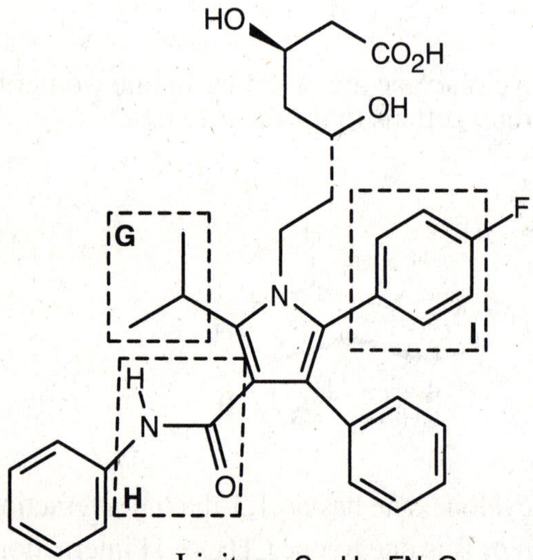


Lipitor $C_{33}H_{35}FN_2O_5$

| | |
|---|---|
| **A** | alcohol |
| **B** | ester |
| **C** | ester |
| **D** | carboxylic acid |
| **E** | alcohol |
| **F** | alkene |
| **G** | alkane (isopropyl group) |
| **H** | amide |
| **I** | arene (aromatic ring) |

(b) **C** and **E** have a trans relationship.

(c) Groups **G**, **H**, and **I** are bonded to $sp^2$ carbons of a planar ring and lie in the same plane.

**2.73**

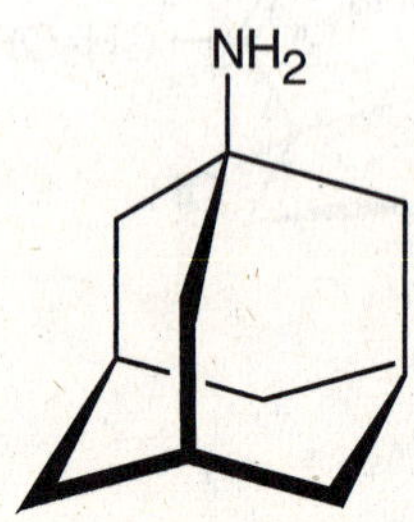


Amantadine

## In the Field

**2.74**

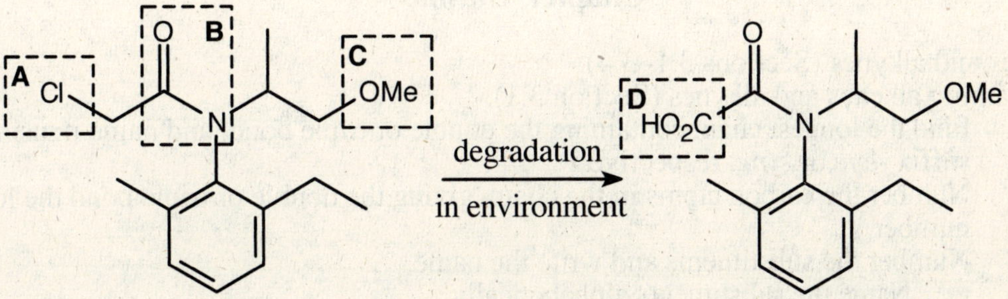


**A** chloride **B** amide **C** ether **D** carboxylic acid

After degradation, the chloride (**A**) is replaced by a carboxylic acid.

# Chapter 3 – Alkenes and Alkynes: The Nature of Organic Reactions

## Chapter Outline

I. Alkenes and alkynes (Sections 3.1–3.4).
  A. Naming alkenes and alkynes (Section 3.1).
    1. Find the longest chain containing the double or triple bond, and name it, using the suffix *-ene* or *-yne*, respectively.
    2. Number the carbon atoms in the chain, giving the double or triple bond the lowest number.
    3. Number the substituents and write the name.
      a. Name the substituents alphabetically.
      b. Indicate the position of the double or triple bond.
        i. Place the locant immediately before the suffix.
      c. Use the suffixes *-diene*, *-diyne*, etc. if more than one double or triple bond is present.
    4. Cycloalkenes are named in a similar way, but the double bond is placed between C1 and C2, and the substituents receive the lowest possible numbers.
  B. Electronic structure of alkenes (Section 3.2).
    1. Carbon atoms in a double bond are $sp^2$-hybridized.
    2. The two carbons in a double bond form one $\sigma$ bond and one $\pi$ bond.
    3. Free rotation doesn't occur around double bonds.
    4. 350 kJ/mol of energy are required to break a $\pi$ bond.
  C. Isomerism in alkenes (Sections 3.3–3.4).
    1. Cis–trans isomerism (Section 3.3).
      a. A disubstituted alkene can have substituents either on the same side of the double bond (cis) or on opposite sides (trans).
      b. These isomers don't interconvert spontaneously because free rotation about a double bond isn't possible.
      c. Cis–trans isomerism doesn't occur if one carbon in the double bond is bonded to identical substituents.
      d. Cis alkenes are less stable than their trans isomers because of steric strain.
    2. *E,Z* isomerism (Section 3.4).
      a. The *E,Z* system is used to describe the arrangement of substituents around a double bond that can't be described by the cis–trans system.
      b. Sequence rules for *E,Z* isomers:
        i. For each double bond carbon, rank the substituents by atomic number; an atom with a high atomic number receives a higher priority than an atom with a lower atomic number.
        ii. If a decision can't be reached after looking at the first atom, look at the second or third atom until a difference is found.
        iii. Multiple-bonded atoms are equivalent to the same number of single-bonded atoms.
      c. In a *Z* isomer, the two groups of higher priority are on the same side of the alkene double bond.
      d. In an *E* isomer, the two groups of higher priority are on opposite sides of the alkene double bond.

II. Organic reactions (Sections 3.5–3.10).
  A. Kinds of organic reactions (Section 3.5).
    1. Addition reactions occur when two reactants add to form one product, with no atoms left over.
    2. Elimination reactions occur when a single reactant splits into two products.

3. Substitution reactions occur when two reactants exchange parts to yield two new products.
4. Rearrangement reactions occur when a single product undergoes a rearrangement of bonds to yield an isomeric product.

B. How reactions occur (Sections 3.6–3.7).
   1. Mechanisms (Section 3.6).
      a. A reaction mechanism describes the bonds broken and formed in a chemical reaction and accounts for all reactants and products.
      b. Bond breaking and formation in chemical reactions.
         i. Bond breaking is symmetrical (heterolytic) if one electron remains with each fragment.
         ii. Bond breaking is unsymmetrical (homolytic} if both electrons remain with one fragment and the other fragment has a vacant orbital.
         iii. Movement of electrons is shown by curved arrows.
            (a). A double-headed arrow indicates movement of an electron pair.
            (b). A fishhook arrow represents a single electron.
         iv. Bond formation can be either symmetrical or unsymmetrical.
      c. Radical reactions are symmetrical and involve the movement of single electrons.
      d. In polar reactions, electrons move in pairs.
   2. Mechanisms of polar reactions.
      a. Polar reactions occur as a result of partial positive and negative charges within molecules.
      b. These charge differences are usually due to electronegativity differences between atoms.
      c. In polar reactions, electron-rich sites in one molecule react with electron-poor sites in another molecule.
      d. The reacting species:
         i. A nucleophile is a species with an electron-rich atom and can be either negatively charged or neutral.
         ii. An electrophile is a species with an electron-poor atom and can be either positively charged or neutral.
         iii. Some molecules can behave as both nucleophiles and as electrophiles.
         iv. Many polar reactions can be explained in terms of acid–base reactions.
   3. An example of a polar reaction: addition of HCl to ethylene (Section 3.7).
      a. This reaction is known as an electrophilic addition.
      b. The $\pi$ electrons in ethylene behave as a nucleophile.
      c. The reaction begins by the addition of the $\pi$ electrons of the double bond to $H^+$.
      d. The resulting intermediate carbocation reacts with $Cl^-$ to form chloroethane.

C. Describing a reaction (Sections 3.8–3.9).
   1. Energy diagrams and transition states (Section 3.8).
      a. Reaction energy diagrams show the energy changes that occur during a reaction.
         i. The vertical axis represents energy changes, and the horizontal axis represents the progress of a reaction.
      b. The transition state is the highest-energy species in the reaction.
         i. It is possible for a reaction to have more than one transition state.
      c. The difference in energy between the reactants and the transition state is the energy of activation $E_{act}$.
         i. Values of $E_{act}$ range from 40–125 kJ/mol.

d. After reaching the transition state, a molecule can go on to form products or can revert to starting material.
e. In a reaction of at least two steps, an intermediate is the species that lies at the energy minimum between two transition states.
f. Even though an intermediate lies at an energy minimum between two transition states, it is a high-energy species and usually can't be isolated.
g. Although each individual step of a reaction has its own energy change, the overall energy change of a reaction is the energy difference between reactants and products.

2. Catalysis (Section 3.9).
   a. The rate of a reaction can't be predicted by the difference in energy between reactants and products.
   b. A catalyst increases the rate of a reaction by providing an alternate mechanism, but undergoes no net change.
   c. A catalyst can change the rate of a reaction but can't change the energy difference between reactant and product.

## Solutions to Problems

**3.1**

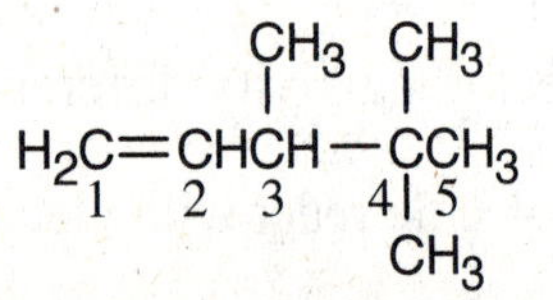

1. Find the longest carbon chain containing the double bond, and name the parent compound. In this problem, the longest chain contains five carbons, and the compound is a *pentene*.

2. Number the carbon atoms, giving to the double bond the lowest possible number.

3. Identify the substituents and number them. In this example, there are three methyl groups – one bonded to carbon 3 and two bonded to carbon 4.

4. Name the compound, listing the substituents alphabetically, giving the position of each, and identifying the position of the double bond with the locant preceding the *–ene* ending.The name is 3,4,4-trimethylpent-1-ene.

(a)

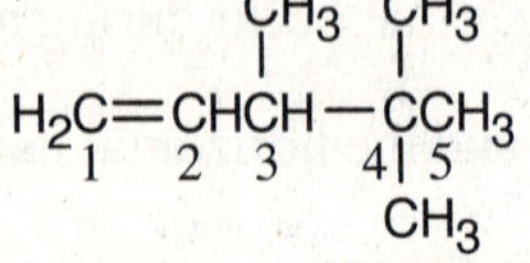

3,4,4-Trimethylpent-1-ene

(b)

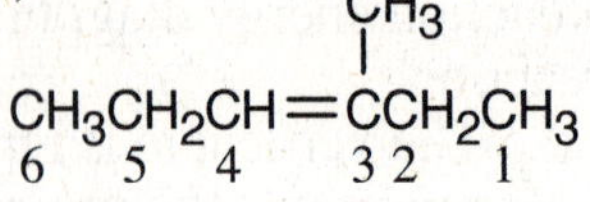

3-Methylhex-3-ene

(c)

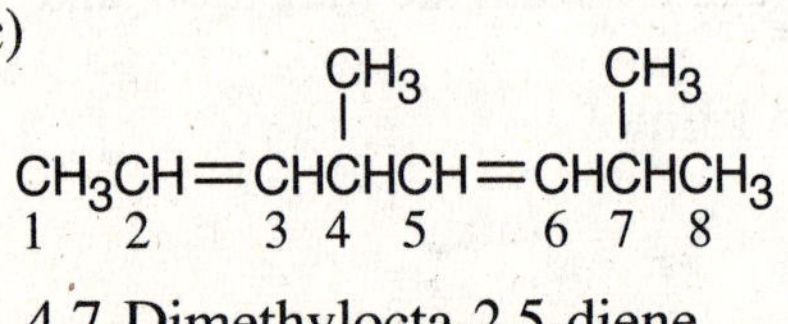

4,7-Dimethylocta-2,5-diene

(d)

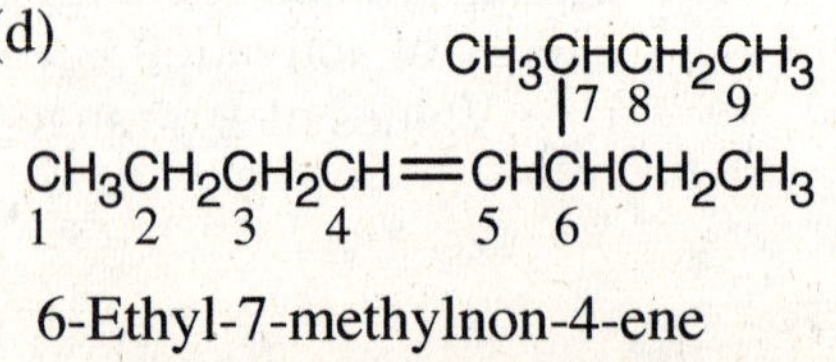

6-Ethyl-7-methylnon-4-ene

**3.2** The double bond is always between carbons 1 and 2 and doesn't need a locant.

(a)

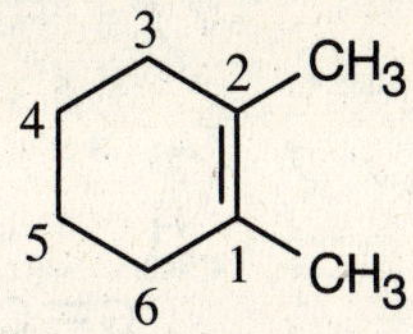

1,2-Dimethylcyclohexene

(b)

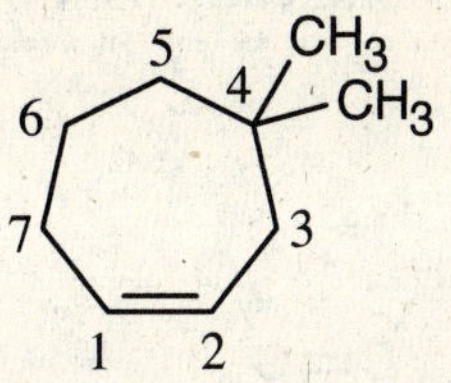

4,4-Dimethylcycloheptene

(c)

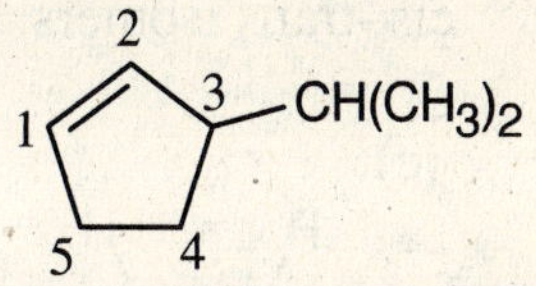

3-Isopropylcyclopentene

**3.3**

(a)

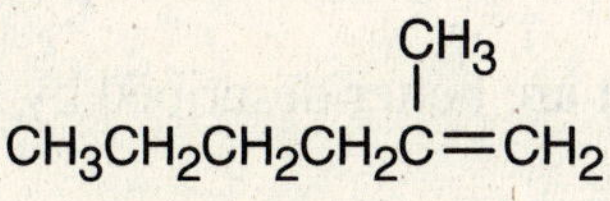

2-Methylhex-1-ene

(b)

$(CH_3)_3CC\equiv CCH_3$

4,4-Dimethylpent-2-yne

(c)

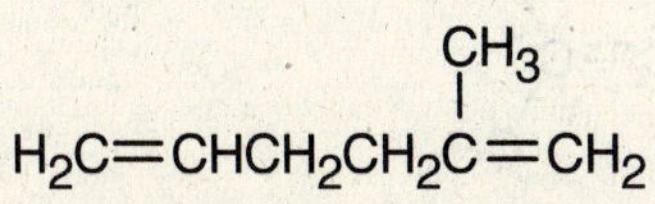

2-Methylhexa-1,5-diene

(d)

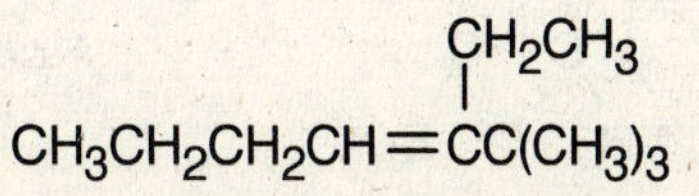

3-Ethyl-2,2-dimethylhept-3-ene

**3.4** Alkynes are named by the same rules as alkenes.

(a)

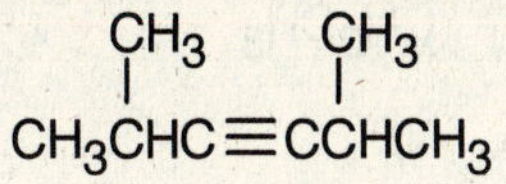

2,5-Dimethylhex-3-yne

(b)

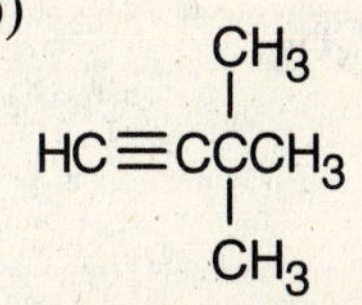

3,3-Dimethylbut-1-yne

(c)

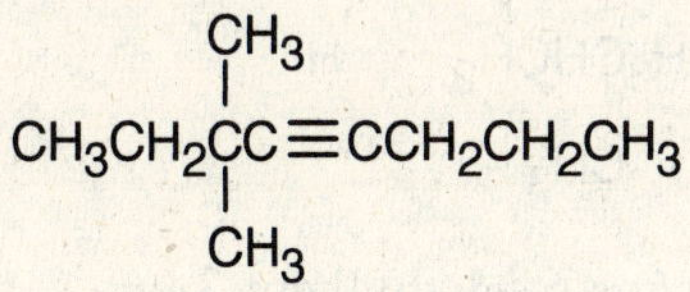

3,3-Dimethyloct-4-yne

(d)

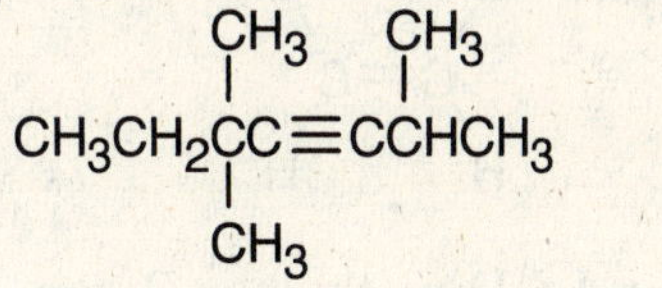

2,5,5-Trimethylhept-3-yne

**3.5**

(a)

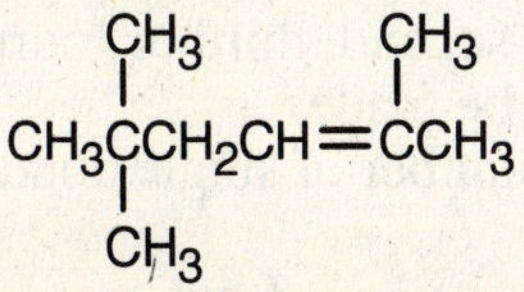

2,5,5-Trimethylhex-2-ene

(b)

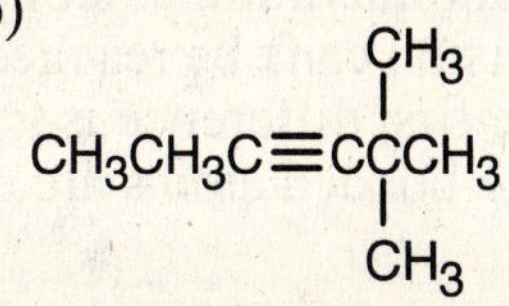

2,2-Dimethylhex-3-yne

(c)

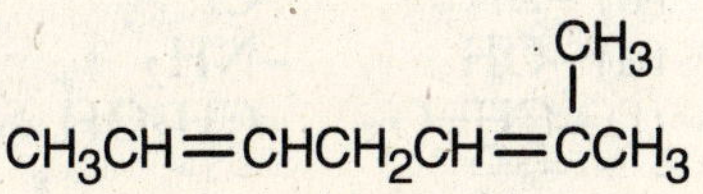

2-Methylhepta-2,5-diene

(d)

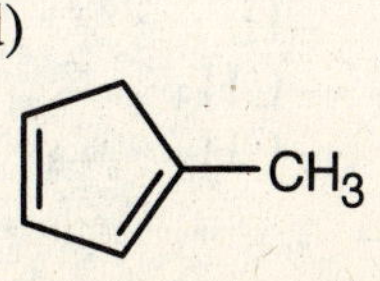

1-Methylcyclopenta-1,3-diene

**3.6** Compounds (a) and (b) don't have cis–trans isomers because two identical groups are bonded to one of the double bond carbons. Compounds (c) and (d) can exist as pairs of cis–trans isomers.

(c)

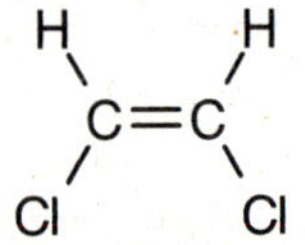

cis

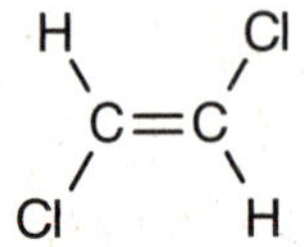

trans

(d)

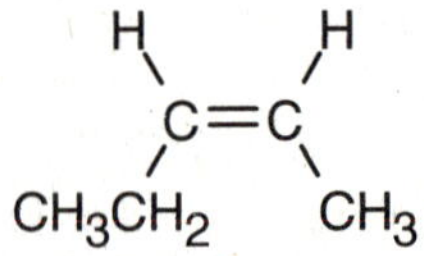

cis

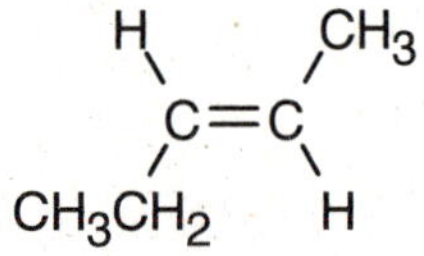

trans

Compounds (e) and (f) can exist as pairs of isomers that are better described by *E*, *Z* designations, which are explained in Section 3.4.

(e)

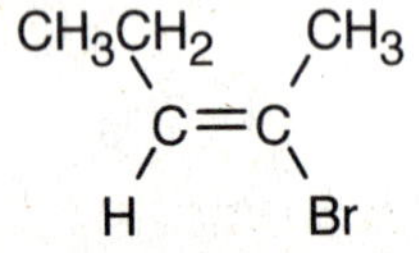

*and*

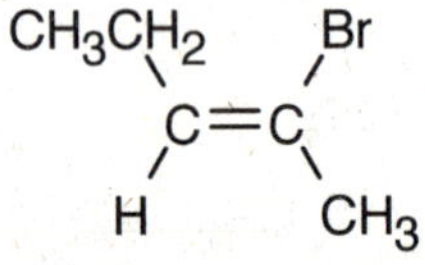

(f)

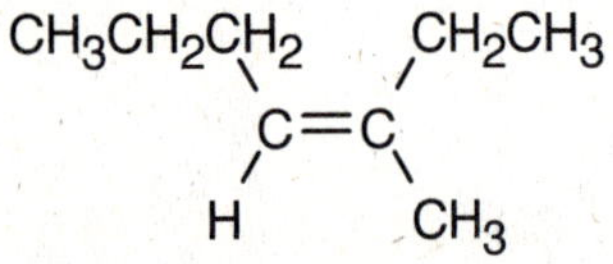

*and*

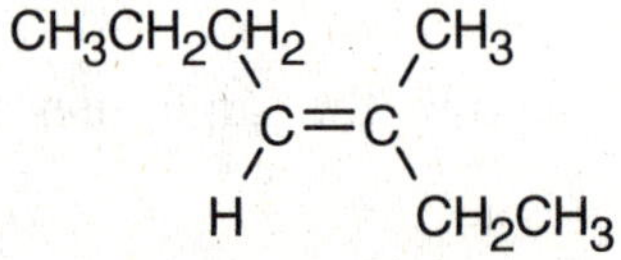

**3.7**

(a)

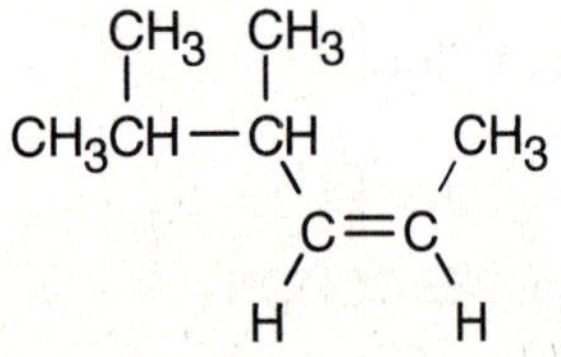

*cis*-4,5-Dimethylhex-2-ene

(b)

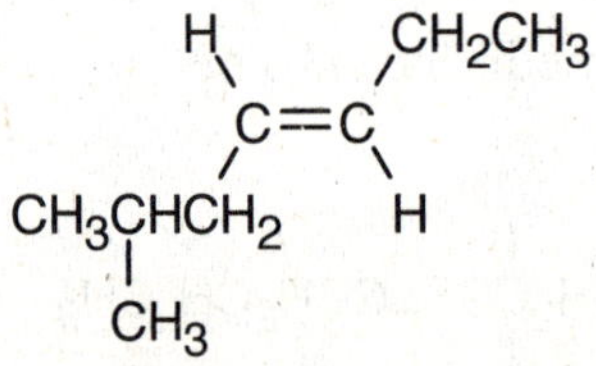

*trans*-6-Methylhept-3-ene

**3.8** Review the sequence rules of Section 3.4. In summary:

*Rule 1*: A high-atomic-number atom ranks higher than a low-atomic-number atom.
*Rule 2*: If a decision can't be reached by Rule 1, look at the second, third or fourth atom out until the first difference is found and a decision can be made.
*Rule 3*: Multiple-bonded atoms are equivalent to the same number of single bonded atoms.

| *High* | *Low* | *Rule* | *High* | *Low* | *Rule* |
|---|---|---|---|---|---|
| (a) –Br | –H | 1 | (b) –Br | –Cl | 1 |
| (c) $–CH_2CH_3$ | $–CH_3$ | 2 | (d) –OH | $–NH_2$ | 1 |
| (e) $–CH_2OH$ | $–CH_3$ | 2 | (f) –CH=O | $–CH_2OH$ | 3 |

**3.9**

(a)

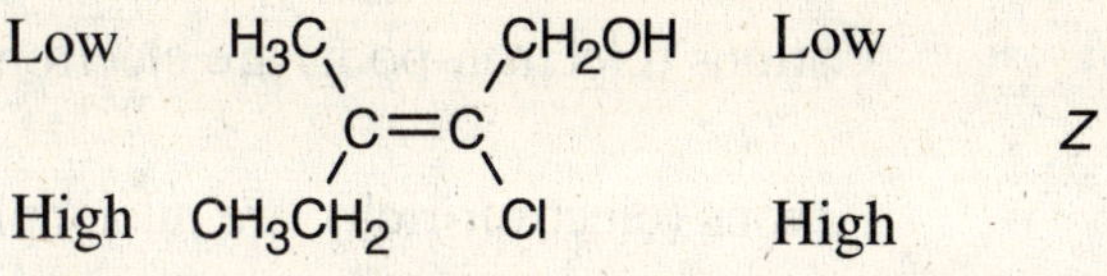


First, consider the substituents on the right side of the double bond. –Cl ranks higher than $–CH_2OH$ by Rule 1 of the Cahn–Ingold–Prelog sequence rules. On the left side of the double bond, $–CH_2CH_3$ ranks higher than $–CH_3$

(b)

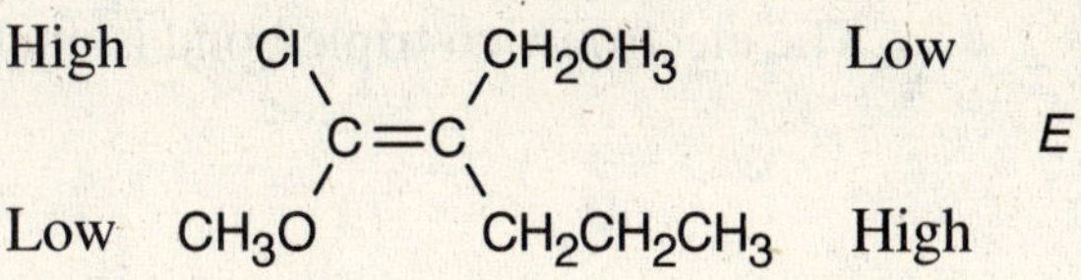


(c)

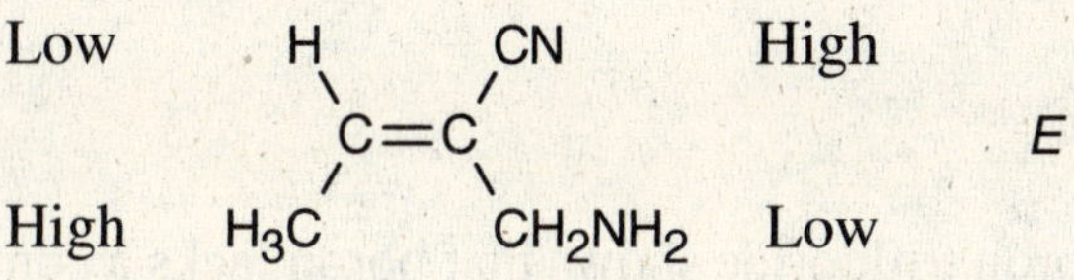


**3.10**

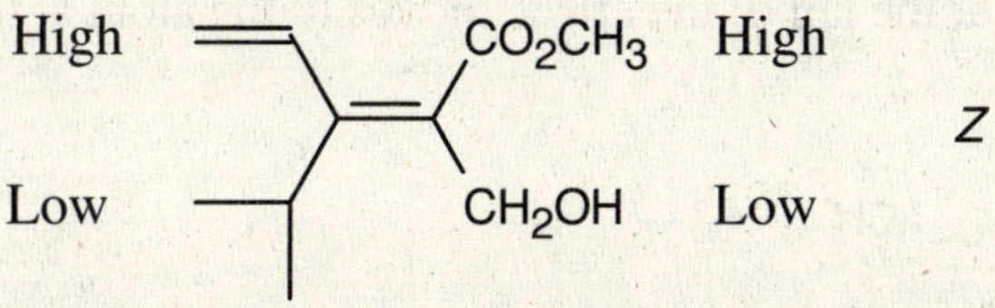


**3.11**

(a) $CH_3Br + KOH \longrightarrow CH_3OH + KBr$ substitution

(b) $CH_3CH_2Cl + NaOH \longrightarrow H_2C{=}CH_2 + NaCl + H_2O$ elimination

(c) $H_2C{=}CH_2 + H_2 \longrightarrow CH_3CH_3$ addition

**3.12** Use Figure 1.14 to identify the more electronegative element, and draw a δ– next to it. Draw a δ+ next to the less electronegative element, and show the direction of polarity with the arrow pointing to the more electronegative atom.

(a)

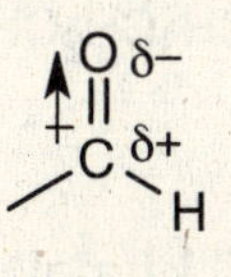


aldehyde

(b)

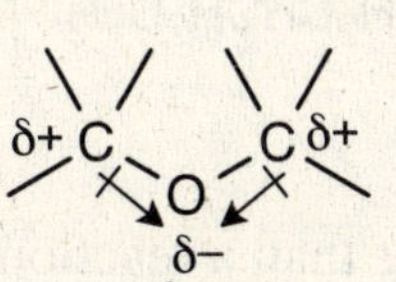


ether

(c)

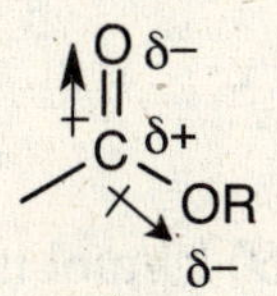


ester

(d)

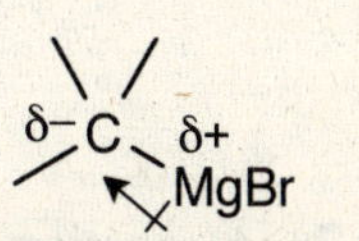


alkylmagnesium bromide

**3.13** *Electrophile or nucleophile?* *Reason*

(a) $NH_4^+$ is an electrophile. Cations (electron-poor) are electrophiles.

(b) $CN^-$ is a nucleophile. Anions (electron-rich) are nucleophiles.

(c) $Br^+$ is an electrophile. Cations (electron-poor) are nucleophiles.

(d) $CH_3NH_2$ is a nucleophile. Compounds with lone-pair electrons are usually nucleophiles.

(e) HC≡CH is a nucleophile. The electron-rich triple bond is nucleophilic.

**3.14**

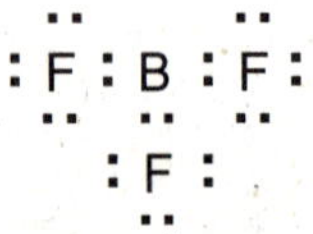

Boron trifluoride

The electron-dot structure for boron trifluoride shows that boron lacks a complete electron octet and is electron-poor. Boron trifluoride is thus likely to be an electrophile.

**3.15** Remember that an electron pair at the tail of the arrow moves to the atom at the head of the arrow.

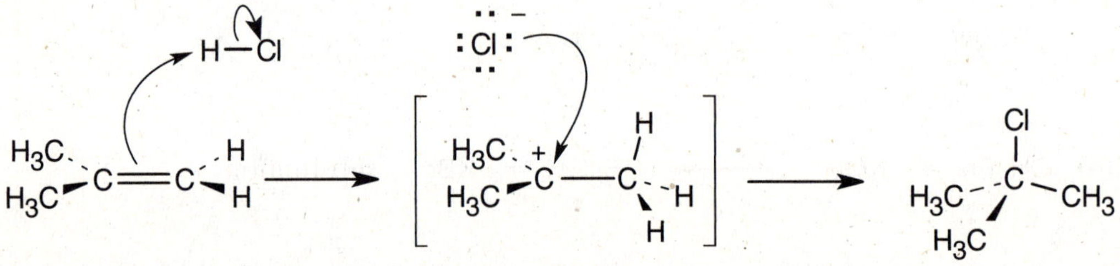

carbocation

**3.16**

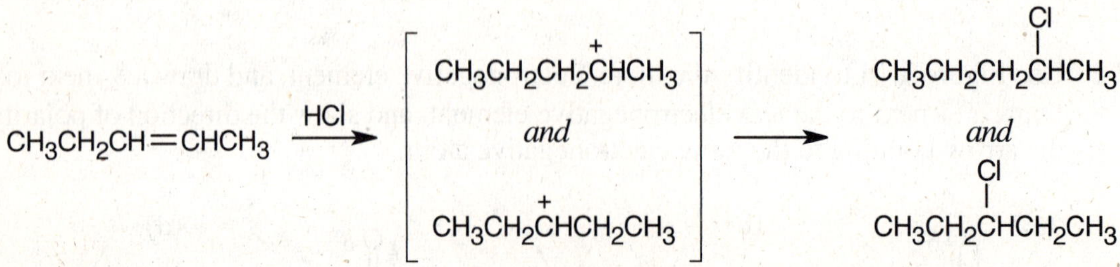


**3.17** A reaction with $E_{act}$ = 60 kJ/mol is faster than a reaction with $E_{act}$ = 80 kJ/mol. (Remember: the larger the $E_{act}$, the slower the reaction).

**3.18**

(a) One-step reaction that releases energy:

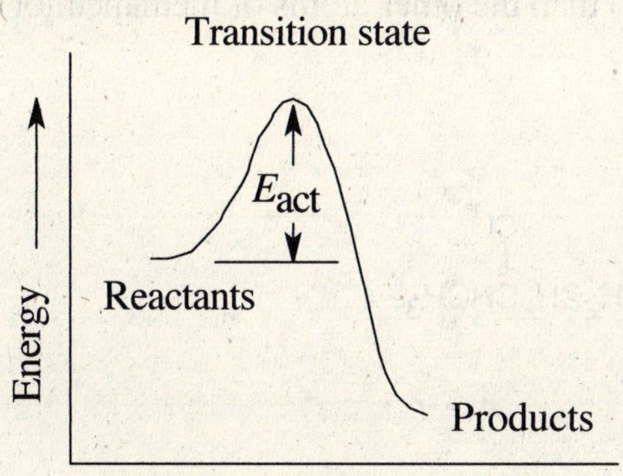


(b) One-step reaction that absorbs energy:

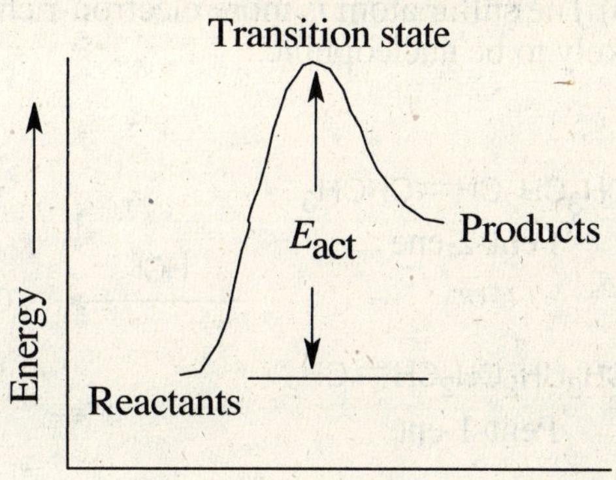


**3.19**

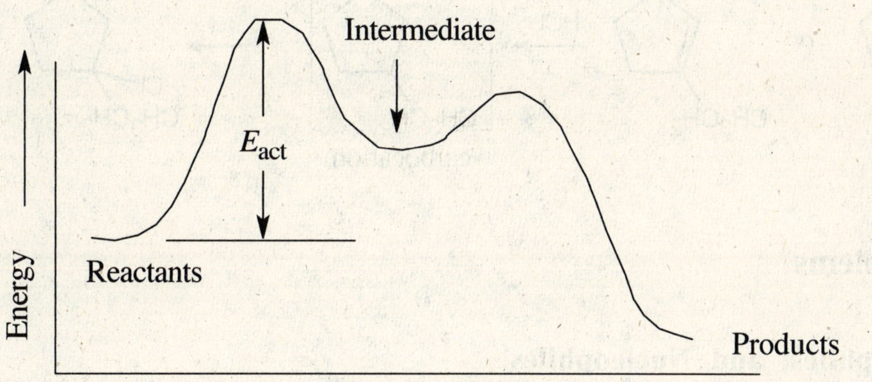


## Visualizing Chemistry

**3.20**

(a)

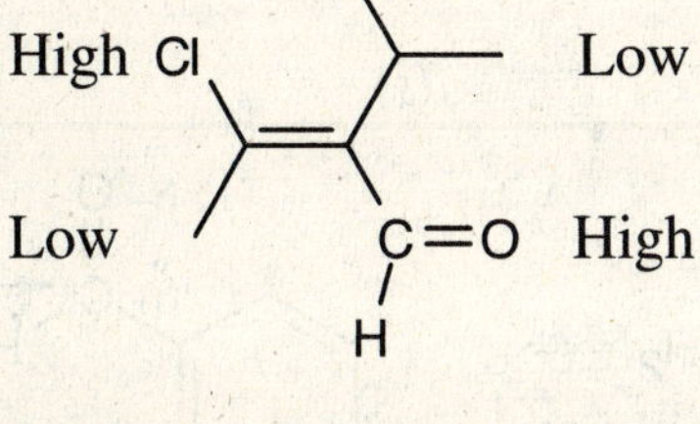


*E*

(b)

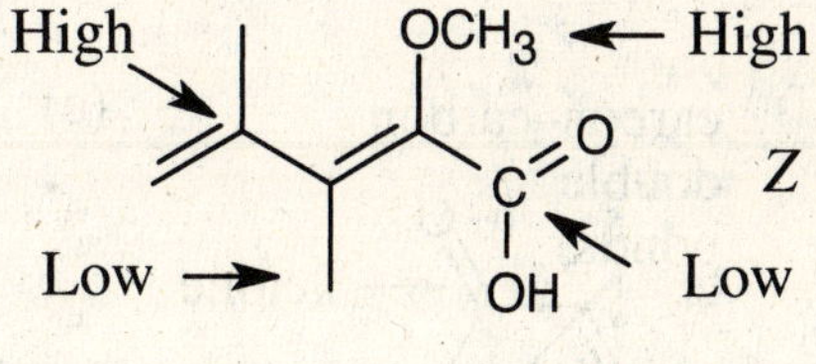


*Z*

**3.21**

(a)

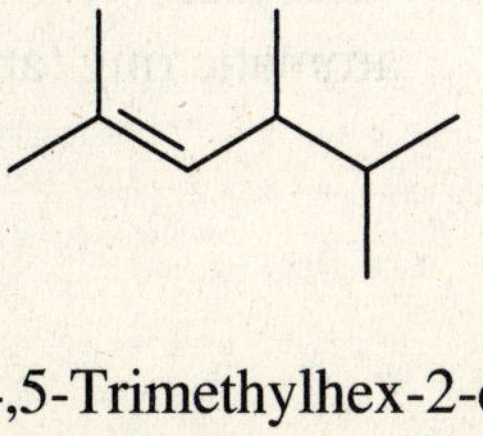

2,4,5-Trimethylhex-2-ene

(b)

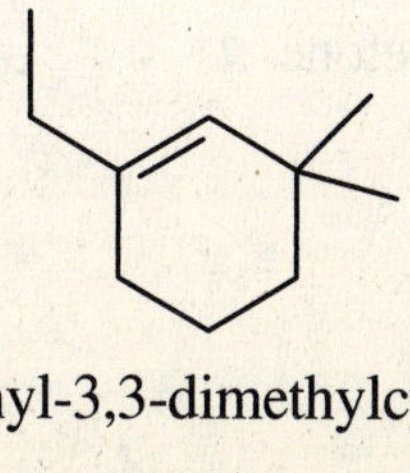

1-Ethyl-3,3-dimethylcyclohexene

**3.22** (a) The electrostatic potential map shows that the formaldehyde oxygen is electron-rich (red), and the carbon–oxygen bond is polarized. The carbon atom is thus relatively electron-poor (blue) and is likely to be electrophilic.
(b) The sulfur atom is more electron-rich (red) than the other atoms of methanethiol and is likely to be nucleophilic.

**3.23**

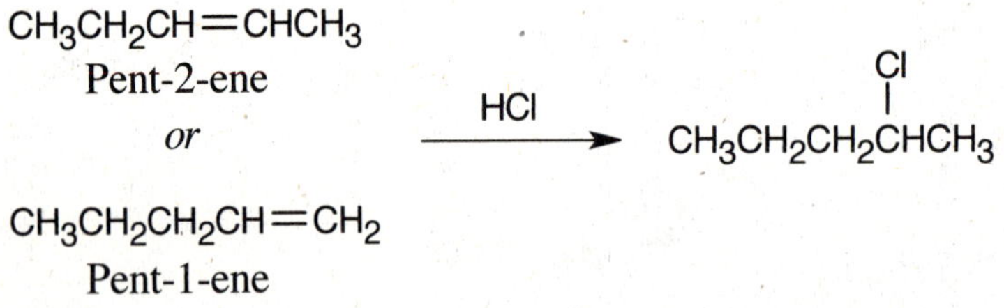


**3.24**

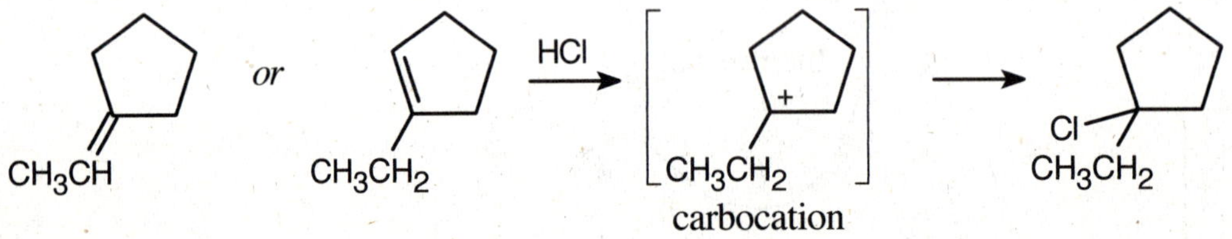


## Additional Problems

### Polarity, Electrophiles, and Nucleophiles

**3.25**

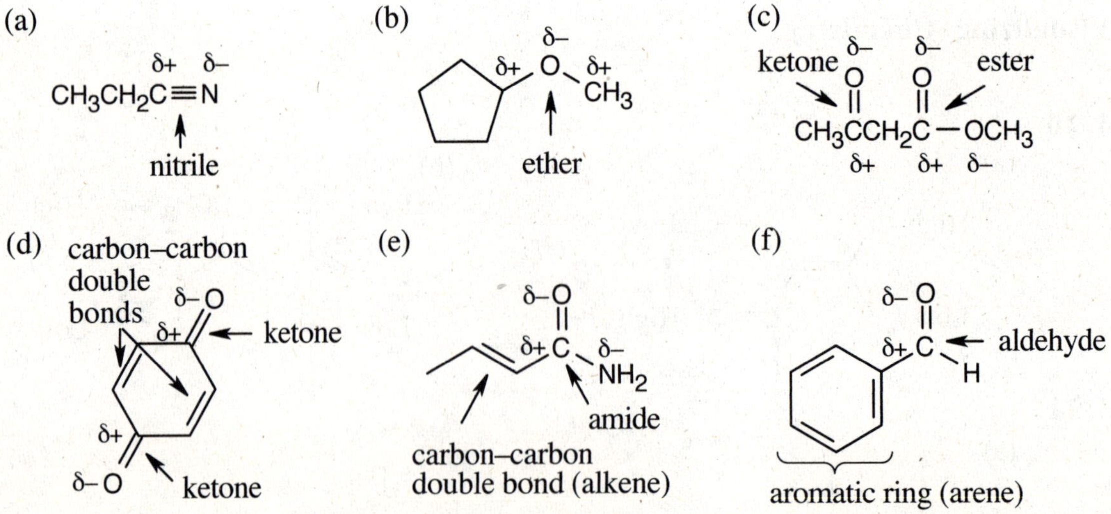

**3.26**

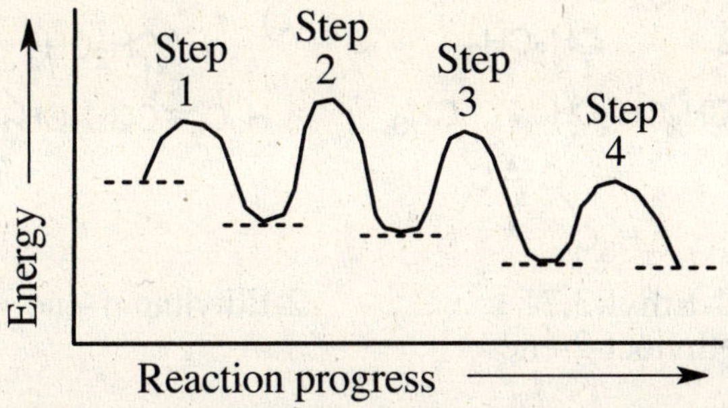


(a) The reaction involves four steps, noted above.
(b) Step 1 is fastest because it has the smallest $E_{act}$. Step 2 is slowest because it has the largest value of $E_{act}$.

**3.27** Nucleophiles: $Cl^-$, $N(CH_3)_3$, $CH_3S^-$
Electrophiles: $Hg^{2+}$, $CH_3^+$

**3.28** e = electrophilic site n = nucleophilic site

(a)

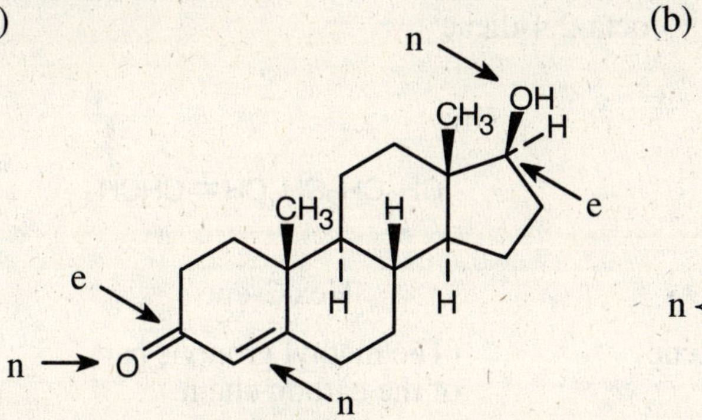


Testosterone

(b)

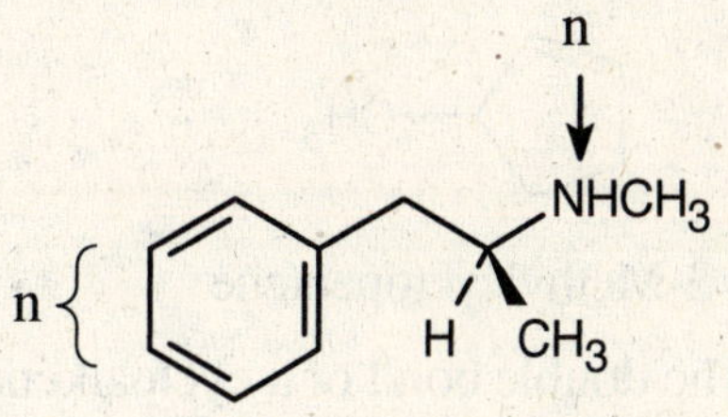


Amphetamine

## Naming Alkenes and Alkynes

**3.29**

(a)

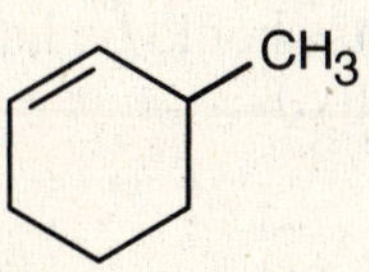


3-Methylcyclohexene

(b)

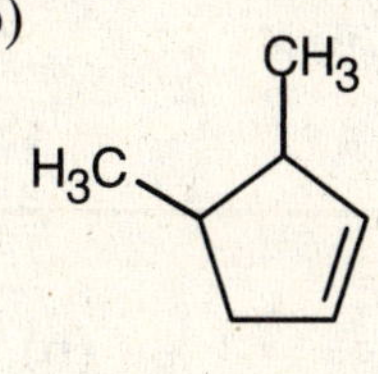


2,3-Dimethylcyclopentene

(c)

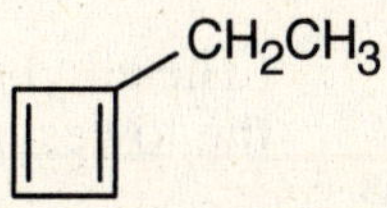


Ethylcyclobutadiene

(d)

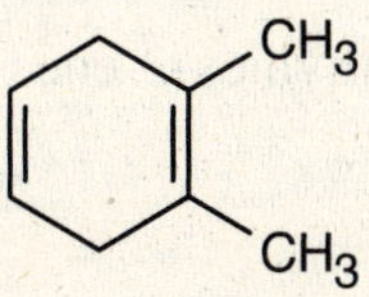


1,2-Dimethylcyclohexa-1,4-diene

In a ring with more than one double bond, the bonds must be numbered.

**3.30**

(a)

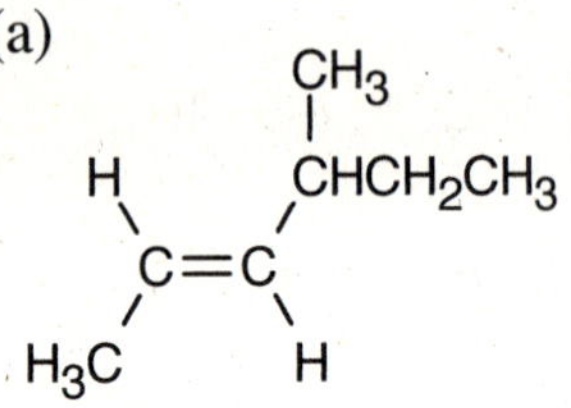

(*E*)-4-Methylhex-2-ene

(b)

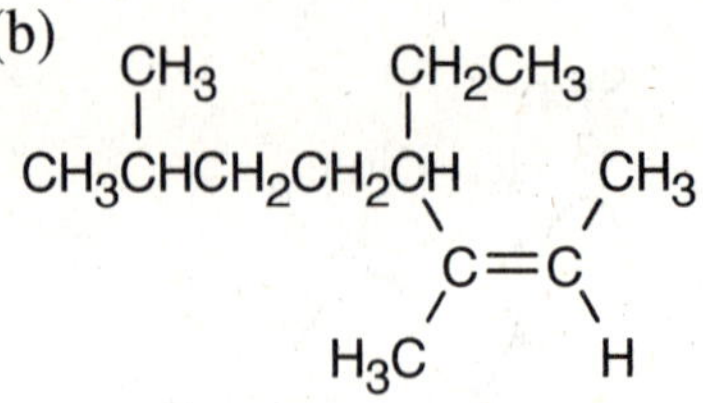

(*Z*)-4-Ethyl-3,7-dimethyloct-2-ene

(c)

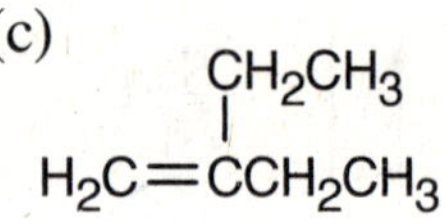

2-Ethylbut-1-ene

(d)

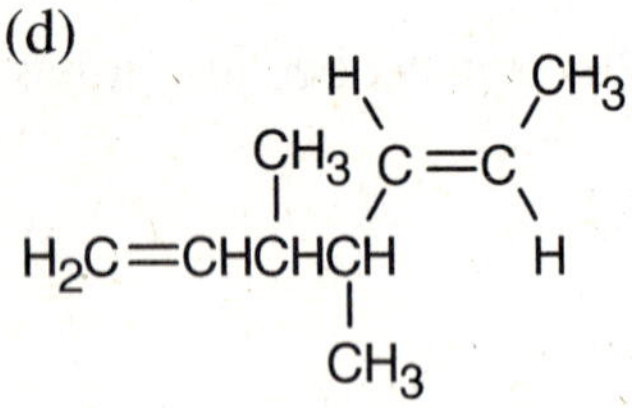

(5*E*)-3,4-Dimethylhepta-1,5-diene

(e)

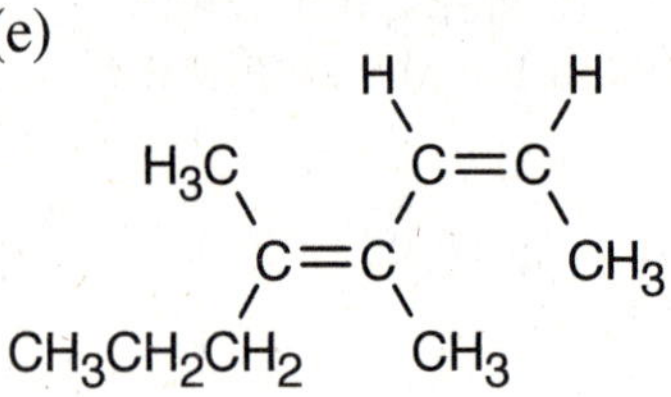

(2*Z*,4*E*)-4,5-Dimethyl-octa-2,4-diene

(f)

$H_2C{=}C{=}CHCH_3$

Buta-1,2-diene

**3.31**

(a)

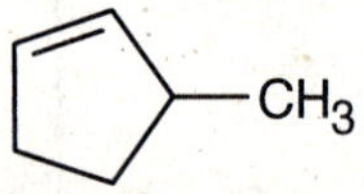

3-Methylcyclopentene

(The double bond of a cycloalkene occurs between C1 and C2)

(b)

$CH_3CH_2CH_2CH{=}CHCH_3$

Hex-2-ene

(The methyl group is part of the carbon chain.)

(c)

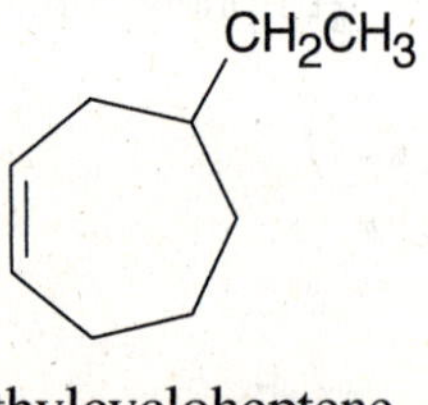

4-Ethylcycloheptene

(The ethyl group receives the lowest possible number.)

(d)

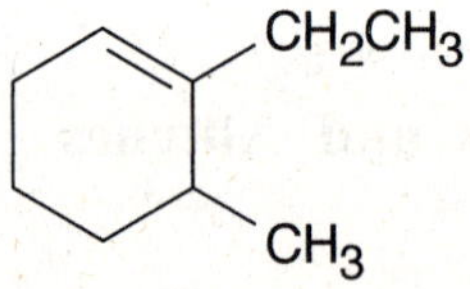

2-Ethyl-3-methylcyclohexene

(Substituents should be listed alphabetically.)

**3.32**

(a)

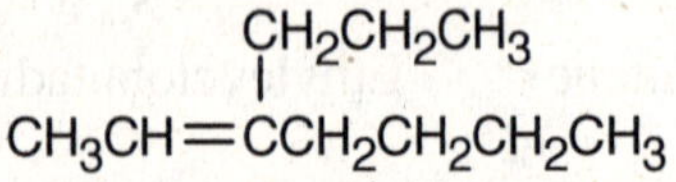

3-Propylhept-2-ene

(b)

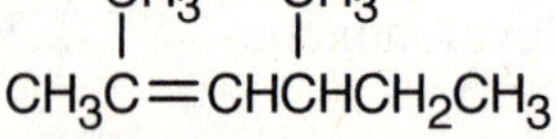

2,4-Dimethylhex-2-ene

(c)

$CH_3CH_2CH{=}CHCH_2CH_2CH{=}CH_2$

Octa-1,5-diene

(d)

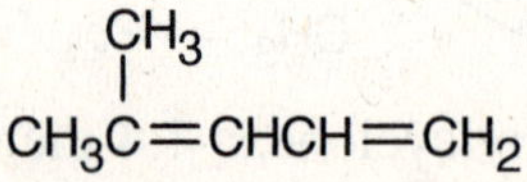

4-Methylpenta-1,3-diene

(e)

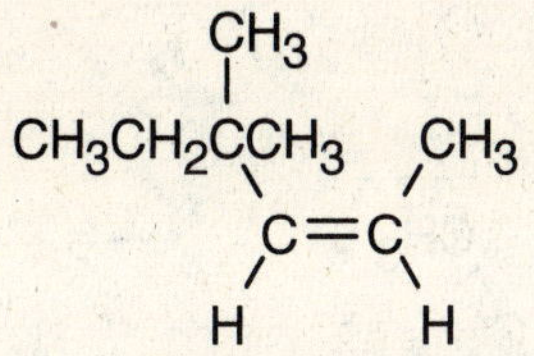

*cis*-4,4-Dimethylhex-2-ene

(f)

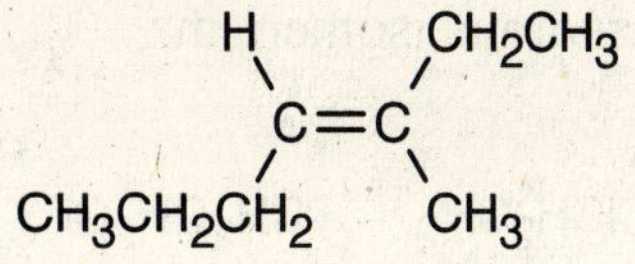

(*E*)-3-Methylhept-3-ene

**3.33**

(a)

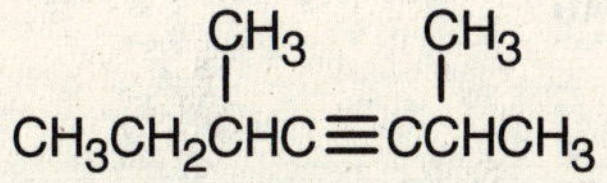

2,5-dimethylhex-3-yne

(b)

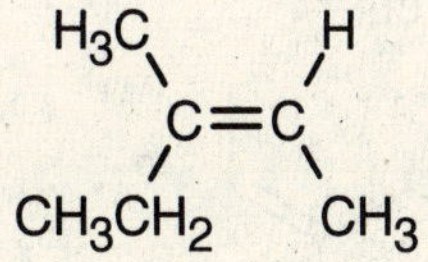

(*Z*)-3-Methylpent-2-ene

**3.34**

(a)

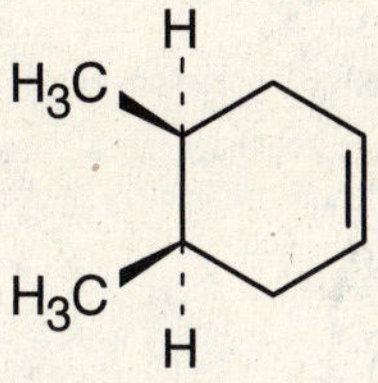

*cis*-4,5-Dimethylcyclohexene

(b)

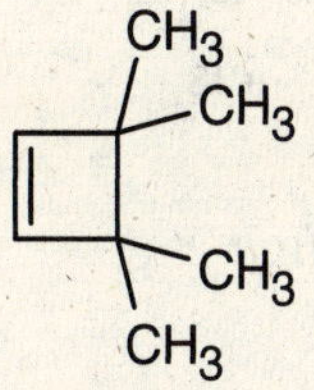

3,3,4,4-Tetramethylcyclobutene

## Cis–Trans Isomers

**3.35**

$CH_3CH_2CH{=}CH_2$

But-1-ene

$CH_3CH{=}CHCH_3$

But-2-ene

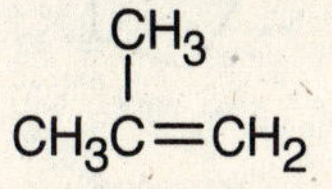

2-Methylpropene

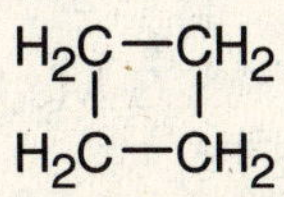

Cyclobutane

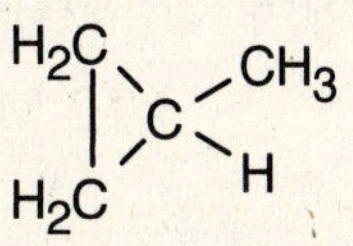

Methylcyclopropane

**3.36** Of the above structures, only but-2-ene shows cis–trans isomerism.

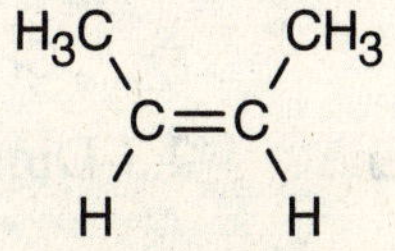

*cis*-But-2-ene

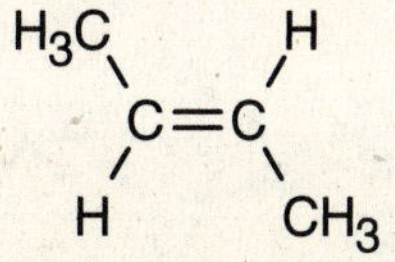

*trans*-But-2-ene

**3.37** *Highest rank ———> Lowest rank*

(a) –I, –Br, $–CH_3$, –H
(b) $–OCH_3$, –OH, $–CO_2H$, –H
(c) $–CO_2H$, –CHO, $–CH_2OH$, $–CH_3$
(d) $–CH{=}CH_2$, $–CH(CH_3)_2$, $–CH_2CH_3$, $–CH_3$

**3.38**

(a) No cis-trans isomerism.

(b)

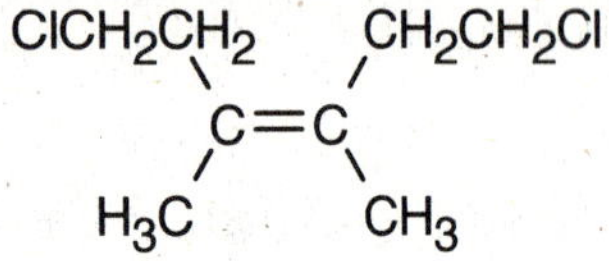

*Z*

cis

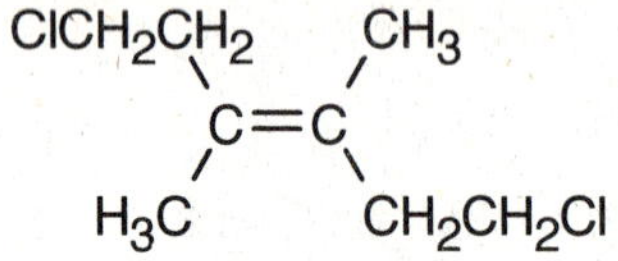

*E*

trans

(c)

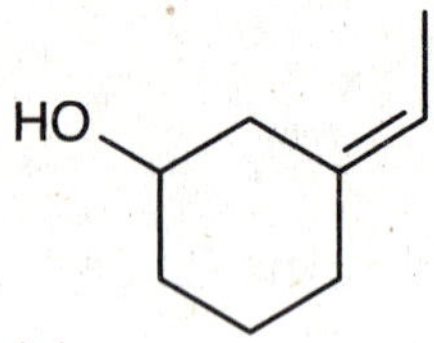

*Z*

cis

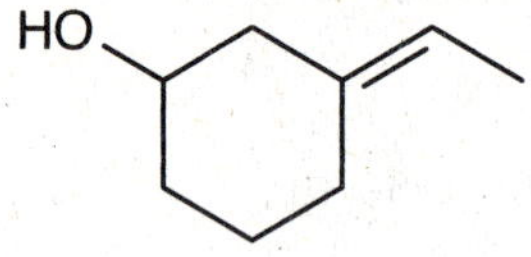

*E*

trans

**3.39**

| | *High* | *Low* | | *High* | *Low* |
|---|---|---|---|---|---|
| (a) | $—CH_2CH_2CH_3$ | $—CH_2CH_3$ | (b) | 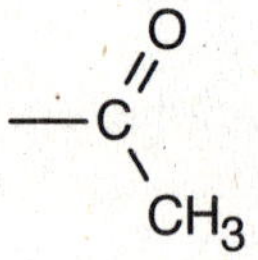 | $—CH_2OH$ |
| (c) | 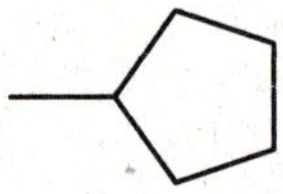 | $—CH(CH_3)_2$ | (d) | 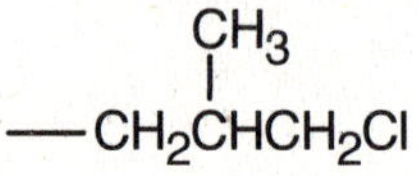 | $—CH_2CH_2CH_2Br$ |

**3.40**

(a)

$CH_3CH_2CH_2CH_2CH{=}CH_2$

Hex-1-ene

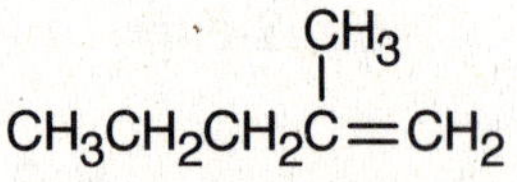

2-Methylpent-1-ene

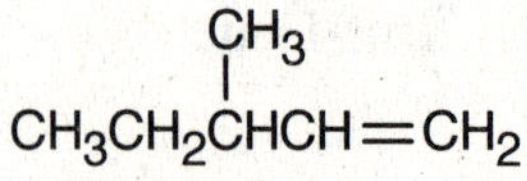

3-Methylpent-1-ene

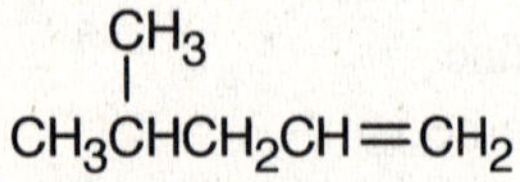

4-Methylpent-1-ene

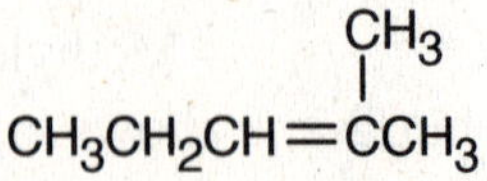

2-Methylpent-2-ene

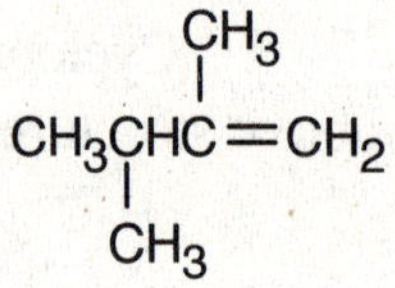

2,3-Dimethylbut-1-ene

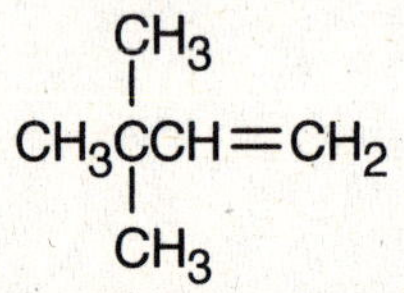

3,3-Dimethylbut-1-ene

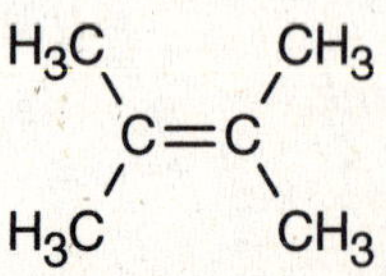

2,3-Dimethylbut-1-ene

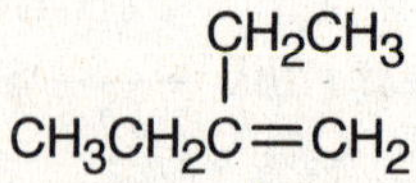

2-Ethylbut-1-ene

(b)

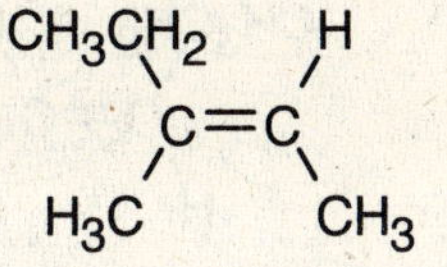

(E)-3-Methylpent-2-ene

(c)

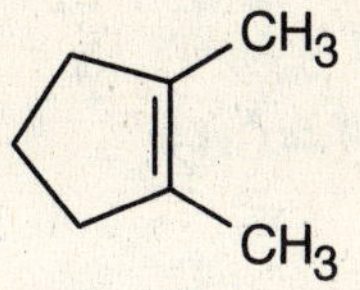

1,2-Dimethylcyclopentene

**3.41** A six membered ring is too small to contain a trans double bond without causing severe ring strain. A model shows that a ten-membered ring is flexible enough to include either a cis or a trans double bond, although the cis isomer has less ring strain than the trans isomer.

**3.42**

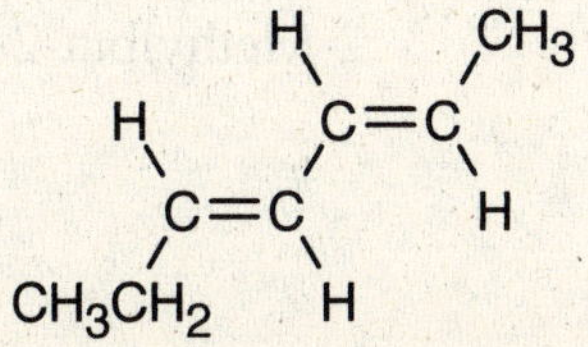

(2*E*,4*E*)-Hepta-2,4-diene

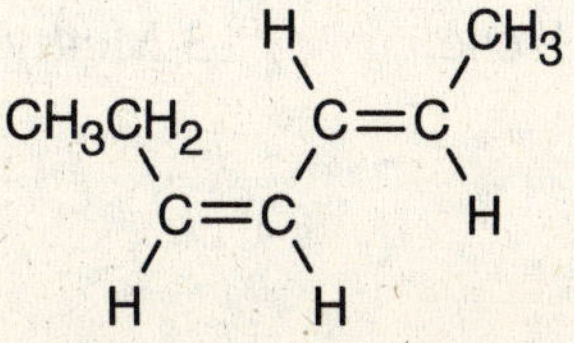

(2*E*,4*Z*)-Hepta-2,4-diene

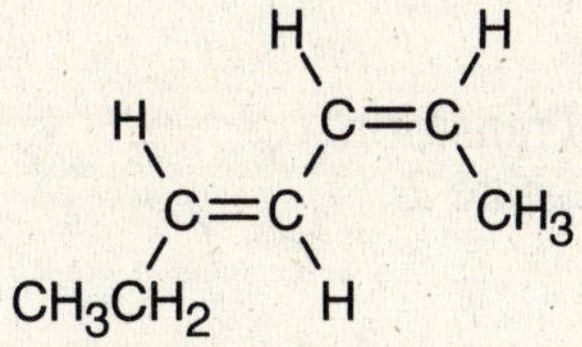

(2*Z*,4*E*)-Hepta-2,4-diene

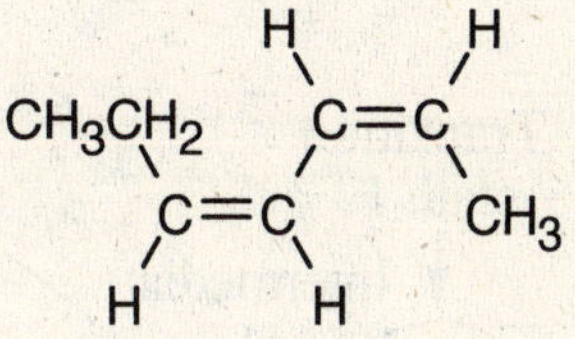

(2*Z*,4*Z*)-Hepta-2,4-diene

**3.43**

(a)

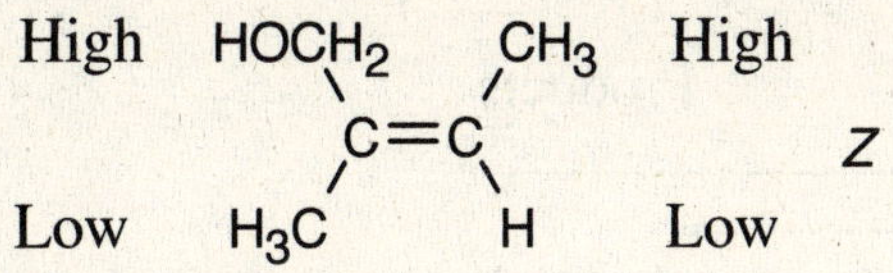


(b)

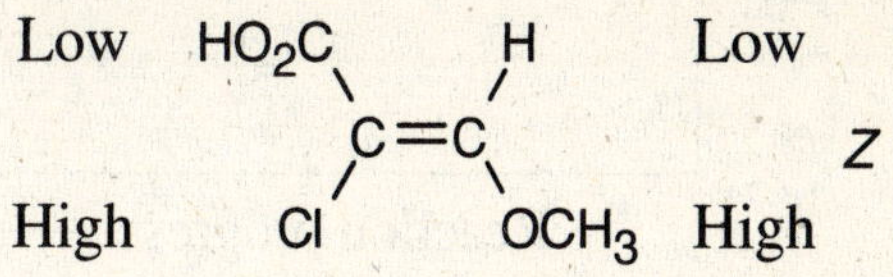


(c)

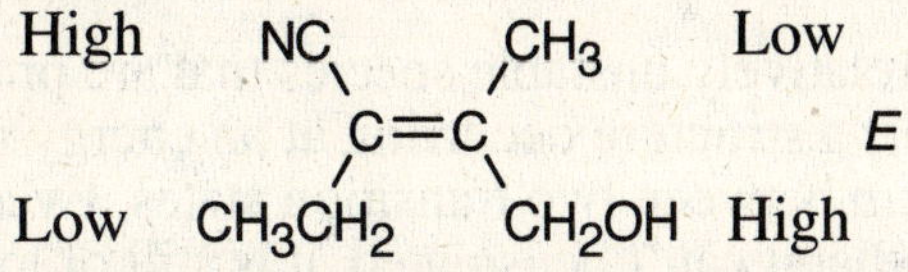


(d)

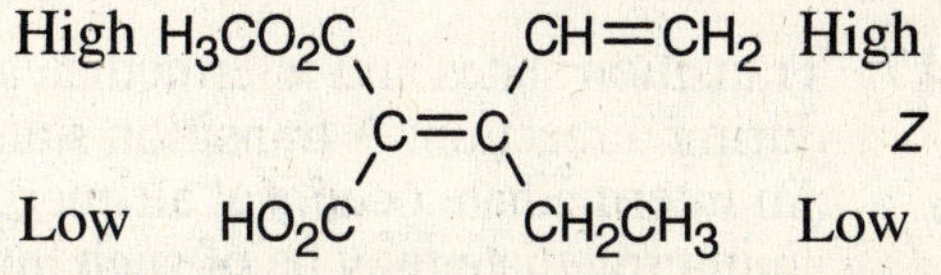


**3.44**

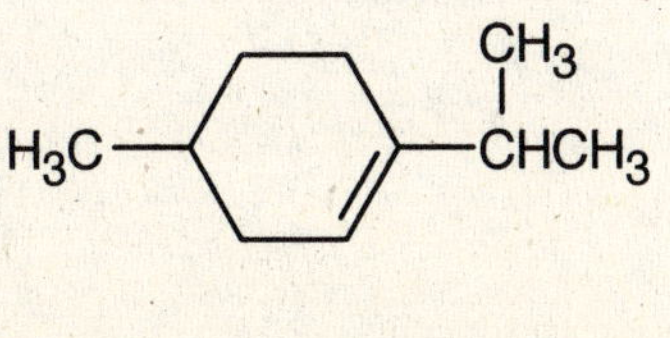

Menthene (1-Isopropyl-4-methylcyclohexene)

**3.45**

$CH_3CH_2CH_2CH{=}CH_2$

Pent-1-ene

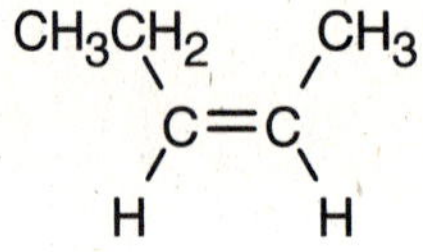

(*Z*)-Pent-2-ene

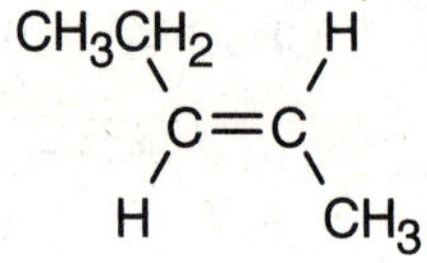

(*E*)-Pent-2-ene

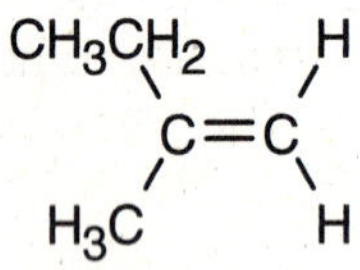

2-Methylbut-1-ene

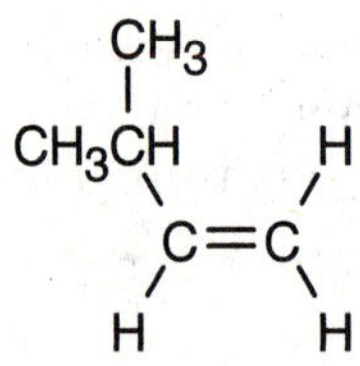

3-Methylbut-1-ene

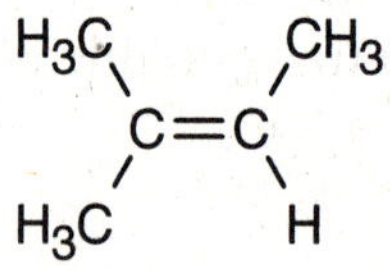

2-Methylbut-2-ene

## Energy Diagrams

**3.46**

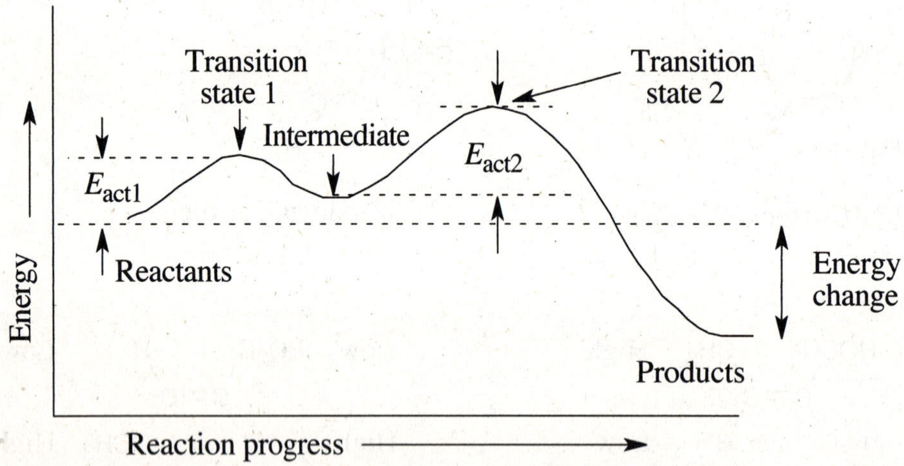


The first step is faster because $E_{act1}$ is smaller than $E_{act2}$.

**3.47** Transition states and intermediates are both relatively unstable species that are produced during a reaction. A transition state represents a structure occurring at an energy maximum. An intermediate occurs at an energy minimum between two transition states. Even though an intermediate may be of such high energy that it can't be isolated, it is still of lower energy than a transition state.

**3.48**

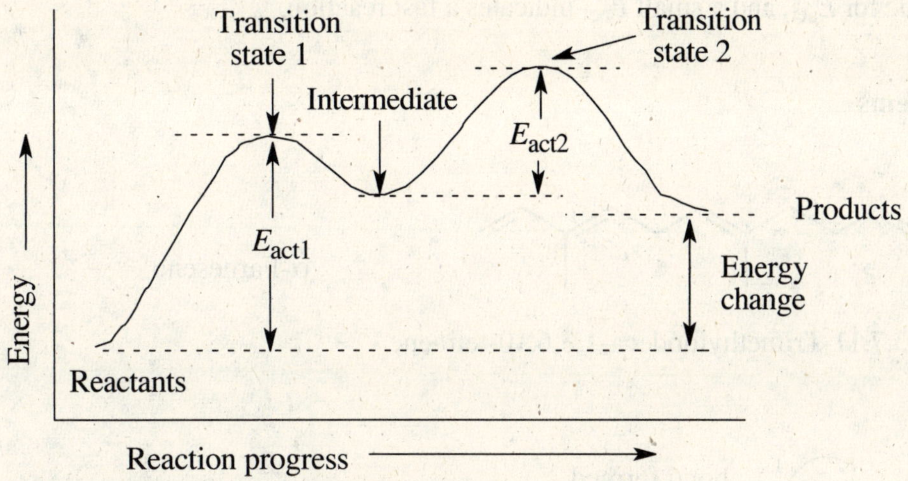


(a) The energy change is positive (energy is absorbed).
(b) There are two steps in the reaction.
(c) The second step is faster because $E_{act2}$ is smaller than $E_{act1}$.
(d) There are two transition states.

**3.49**

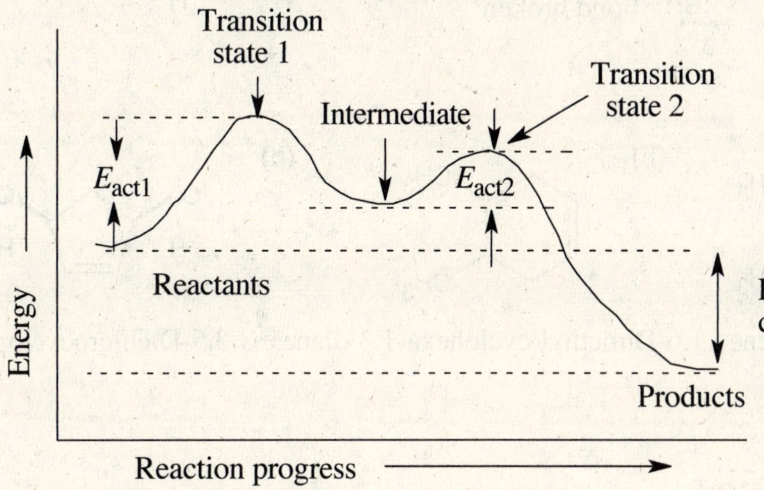


In this reaction, the second step is faster because $E_{act2}$ is smaller than $E_{act1}$.

**3.50**

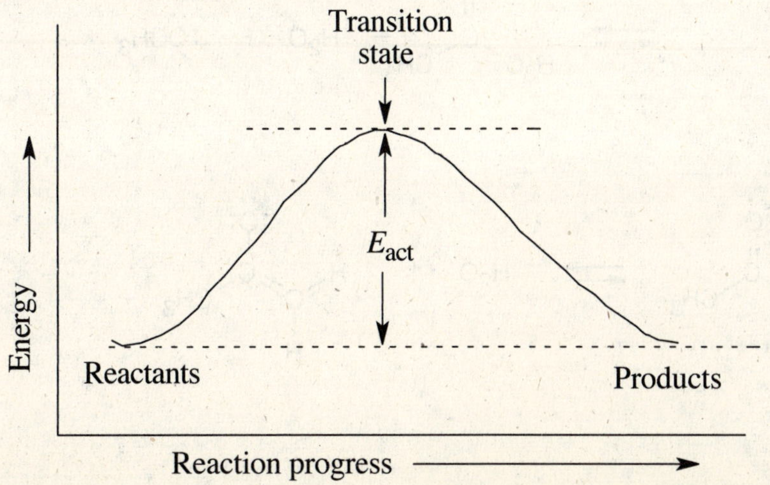

**3.51** A reaction with $E_{act}$ = 15 kJ/mol at room temperature is likely to be fast since 15 kJ/mol is a small value for $E_{act}$, and a small $E_{act}$ indicates a fast reaction.

## General Problems

**3.52–3.53**

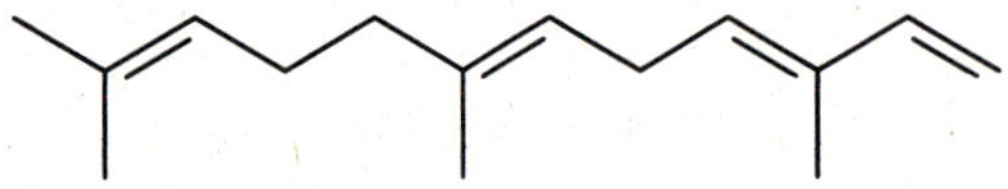

α-Farnesene

(3*E*,6*E*)-3,7,11-Trimethyldodeca-1,3,6,10-tetraene

**3.54**

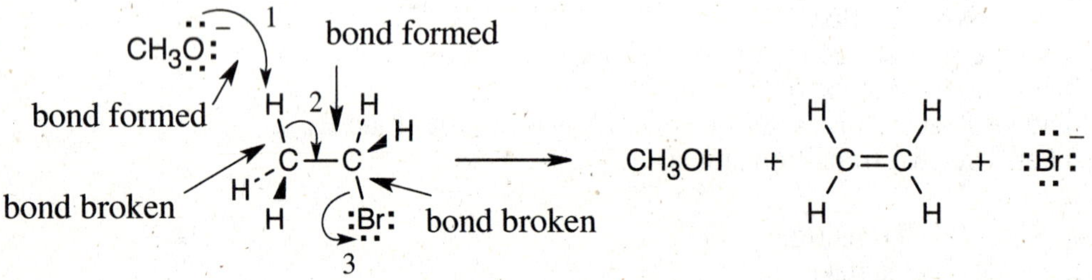


**3.55**

(a)

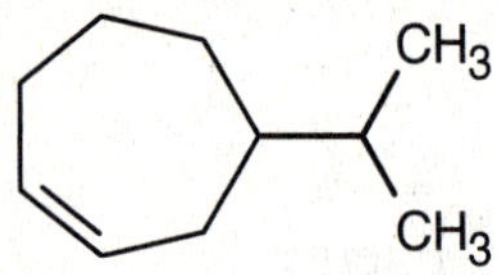

(b)

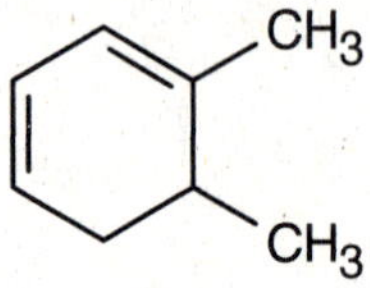

(c)

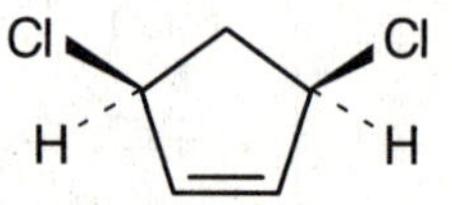

4-Isopropylcycloheptene 1,6-Dimethyl-cyclohexa-1,3-diene *cis*-3,5-Dichlorocyclopentene

**3.56**

(a)

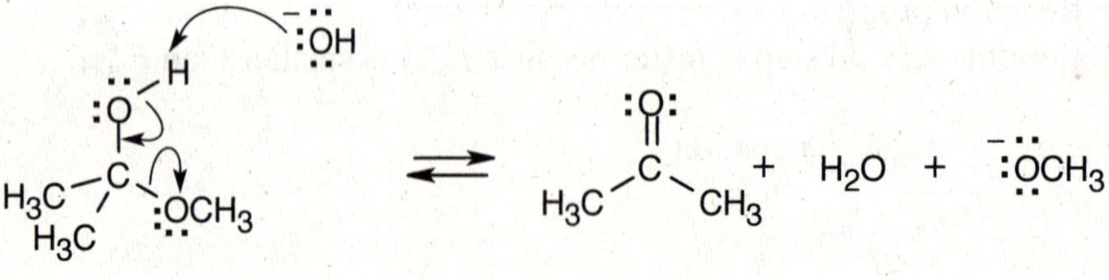

(b)

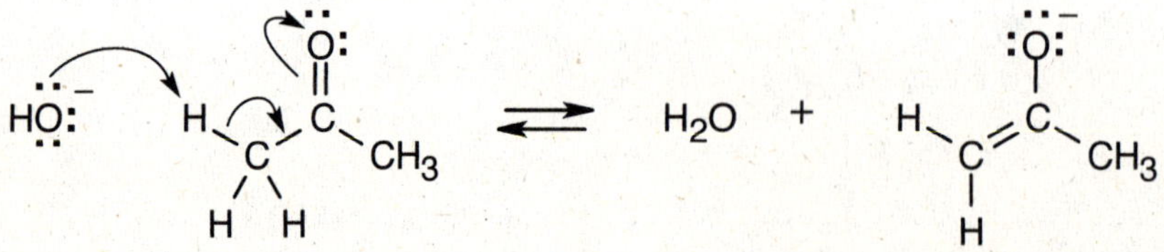

**3.57**

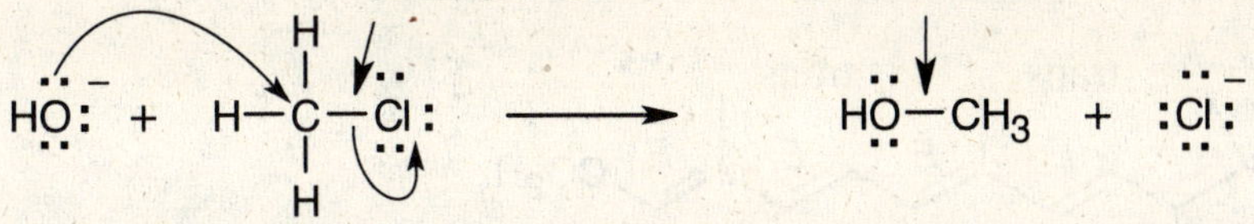


A bond has formed between the hydroxide oxygen and the carbon of chloromethane. The bond between chlorine and carbon has been broken.

**3.58**

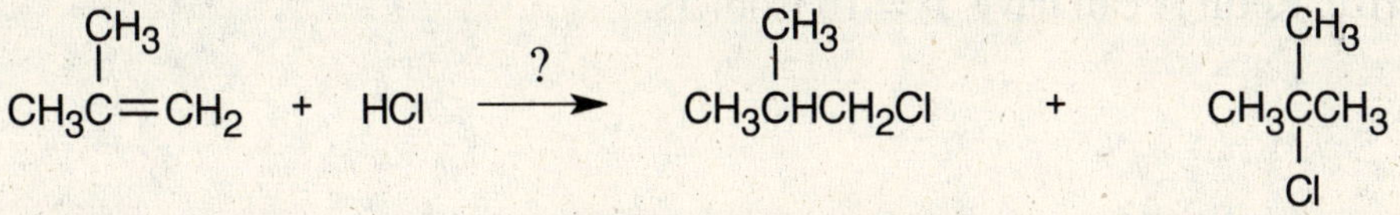


2-Methylpropene 1-Chloro-2-methylpropane 2-Chloro-2-methylpropane

(2-Chloro-2-methylpropane is the observed product.)

**3.59**

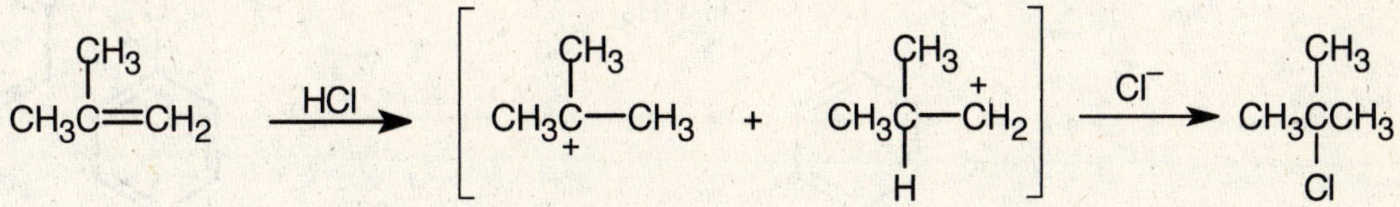


The first carbocation is more stable because more alkyl groups are bonded to the positively charged carbon. The expected product is 2-chloro-2-methylpropane.

**3.60** (a) The reaction is a polar rearrangement.

(b)

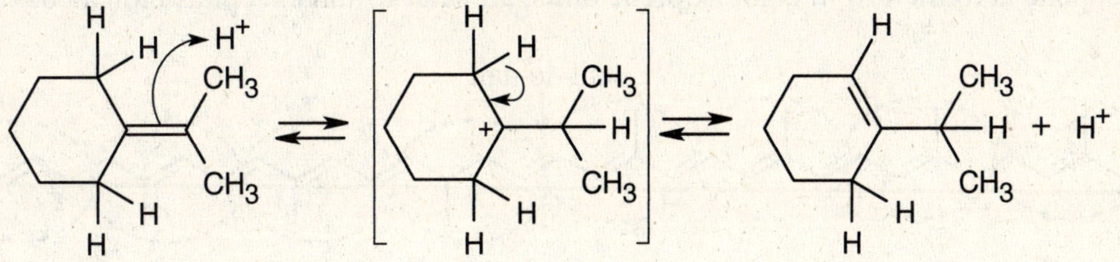

## In the Medicine Cabinet

**3.61**

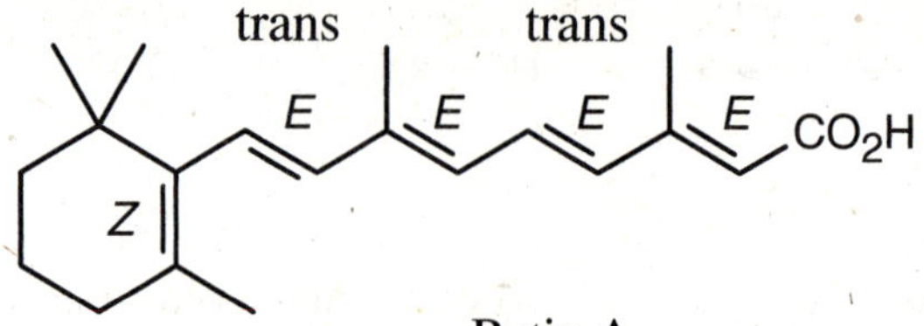


Retin A

Retin A contains five carbon–carbon double bonds. Since the ring double bond can't isomerize, Retin A can have $2^4 = 16$ isomers.

**3.62**

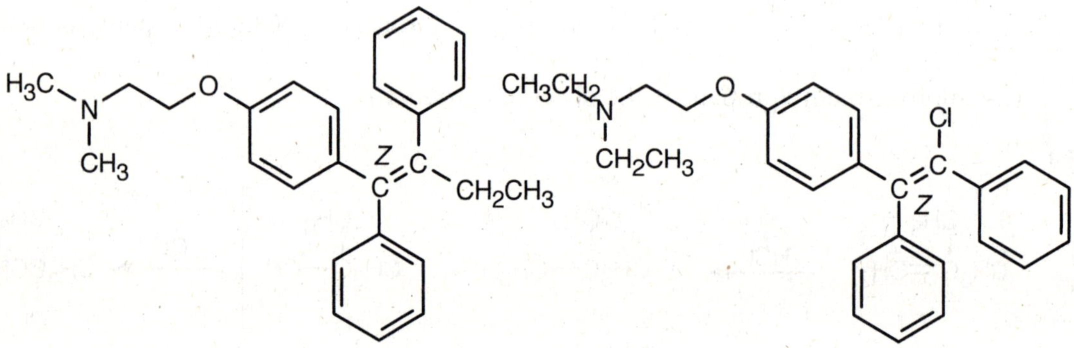


Tamoxifen

Clomiphene

## In the Field

**3.63** Lycopene is formed from eight isoprene units, all head-to-tail except as indicated.

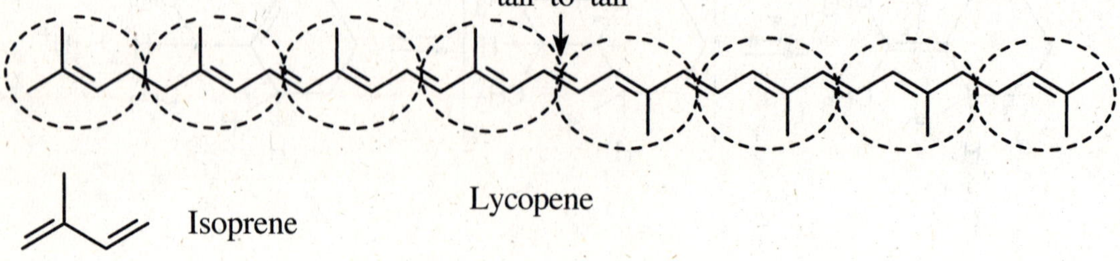


Isoprene

Lycopene

# Chapter 4 – Reactions of Alkenes and Alkynes

## Chapter Outline

I. Alkenes (Sections 4.1–4.6).
  A. Reactions of alkenes (Sections 4.1–4.6).
    1. Addition of HX – a regiospecific reaction (Sections 4.1–4.2).
      a. Orientation of addition – Markovnikov's rule (Section 4.1).
        i. In the addition of HX to a double bond, H attaches to the carbon with fewer substituents, and X attaches to the carbon with more substituents.
        ii. If the carbons have the same number of substituents, a mixture of products results.
      b. Carbocation structure and stability (Section 4.2).
        i. Carbocations are planar; the unoccupied *p* orbital extends above and below the plane.
        ii. The stability of carbocations increases with increasing substitution.
    2. Addition of water to alkenes – hydration (Section 4.3).
      a. Water adds to alkenes to yield alcohols in the presence of a strong acid catalyst, following Markovnikov's rule.
      b. Although this reaction is important industrially, reaction conditions are too severe for most molecules.
    3. Addition of halogens to alkenes – halogenation (Section 4.4).
      a. $Br_2$ and $Cl_2$ react with alkenes to yield 1,2-dihaloalkanes.
      b. Reaction occurs with anti stereochemistry – both bromines come from opposite faces of the double bond.
      c. The reaction intermediate is a cyclic bromonium intermediate that is formed in a single step by interaction of an alkene with $Br^+$.
    4. Reduction of alkenes – hydrogenation (Section 4.5).
      a. Catalytic hydrogenation reduces alkenes to saturated hydrocarbons.
      b. Catalytic hydrogenation is a heterogeneous process that takes place on the surface of a solid catalyst.
      c. Hydrogenation occurs with syn stereochemistry.
    5. Oxidation of alkenes (Section 4.6).
      a. Reaction of an alkene with a peroxyacid forms an epoxide.
        i. Acid-catalyzed opening of the epoxide (hydroxylation) yields a diol.
        ii. Protonation of the epoxide is followed by nucleophilic addition of water to yield a *trans* -1,2-diol.
      b. Reaction of an alkene with basic $KMnO_4$ also produces a diol.
        i. The reaction occurs in a single step with syn stereochemistry and yields a *cis*-1,2-diol.
      c. Cleavage occurs when an alkene is treated with acidic $KMnO_4$.
        i. The products may be ketones, carboxylic acids or $CO_2$.
  B. Alkene polymers (Section 4.7).
    1. Many types of polymers can be formed by radical polymerization of alkene monomers.
    2. There are 3 steps in a radical polymerization reaction.
      a. Initiation involves homolytic cleavage of a weak bond to form a radical
        i. The radical adds to an alkene to generate an alkyl radical.
      b. The alkyl radical adds to another alkene molecule to yield a second radical.
        i. This step is repeated many, many times (up to 100,000 times for some polymers).
      c. Termination occurs when two radical fragments combine.

II. Conjugated dienes and resonance (Sections 4.8–4.10).
  A. Conjugated dienes (Section 4.8–4.9).
    1. In conjugated dienes, there is an electronic interaction due to overlap of the $\pi$ orbitals of the double bonds across the central single bond.
    2. This overlap leads to formation of both 1,2 and 1,4 addition products.
      a. An allylic carbocation is involved as an intermediate.
    3. Stability of allylic cations: resonance (Section 4.9).
      a. An allylic cation is a resonance hybrid of two different forms.
      b. In general, the more resonance forms that can be drawn for a structure, the greater the stability.
  B. Drawing and interpreting resonance structures (Section 4.10).
    1. Resonance forms are imaginary.
    2. Resonance forms differ only in the placement of $\pi$ or nonbonding electrons.
    3. Different resonance forms of a structure don't have to be equivalent.
    4. Resonance forms must be valid electron-dot structures and must obey normal rules of valency.
    5. Resonance leads to stability.

III. Alkynes (Section 4.11).
  A. Naming of alkynes.
    1. The rules for naming alkynes are like the rules for alkenes (Sec. 3.1), with a few exceptions.
      a. The suffix *-yne* is used for an alkyne.
      b. Compounds with both double bonds and triple bonds are *enynes*.
      c. When there is a choice in numbering, double bonds receive lower numbers than triple bonds.
  B. Reactions of alkynes.
    1. Addition of $H_2$ uses a Lindlar catalyst to yield a cis alkene.
    2. Addition of HX yields a vinylic halide with Markovnikov regiochemistry.
    3. Addition of $X_2$ gives a dihalide with anti regiochemistry.
    4. Hydration of alkynes.
      a. The –OH group adds to the more substituted carbon to give Markovnikov product.
      b. The intermediate enol product tautomerizes to a ketone.
      c. A mixture of products is formed from an internal alkyne, but a terminal alkyne yields a methyl ketone.
    5. Formation of acetylide anions.
      a. Terminal alkynes are weakly acidic (p$K_a$ = 25).
      b. Very strong bases ($^-NH_2$) can deprotonate terminal alkynes.
      c. Acetylide anions can react with haloalkanes to form substitution products.
        i. The nucleophilic acetylide anion attacks the electrophilic carbon of a haloalkane to produce a new alkyne.
        ii. This reaction is called an alkylation reaction.
      d. Acetylide alkylations are limited to primary alkyl bromides and iodides.

## Solutions to Problems

**4.1**

(a)

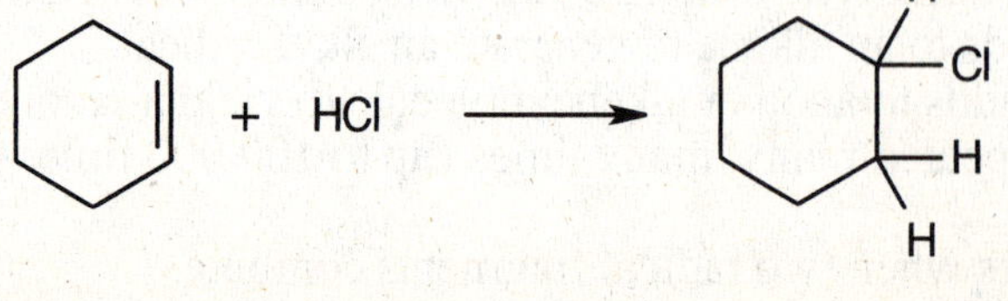


Chlorocyclohexane

(b)

$$(CH_3)_2C{=}CHCH_2CH_3 + HBr \longrightarrow (CH_3)_2C(Br)CH_2CH_2CH_3$$

2-Bromo-2-methylpentane

In accordance with Markovnikov's rule, –H forms a bond to the carbon with fewer substituents, and –Br forms a bond to the carbon with more substituents.

(c)

$$CH_3CH(CH_3)CH_2CH{=}CH_2 \xrightarrow[H_2SO_4]{H_2O} CH_3CH(CH_3)CH_2CH(OH)CH_3$$

(d)

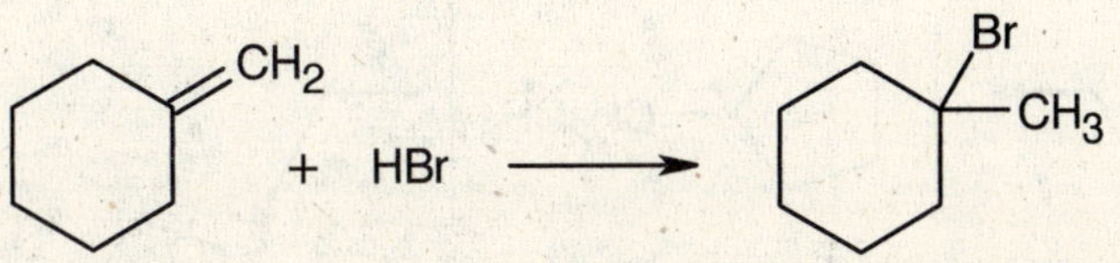

1-Bromo-1-methylcyclohexane

**4.2**

(a)

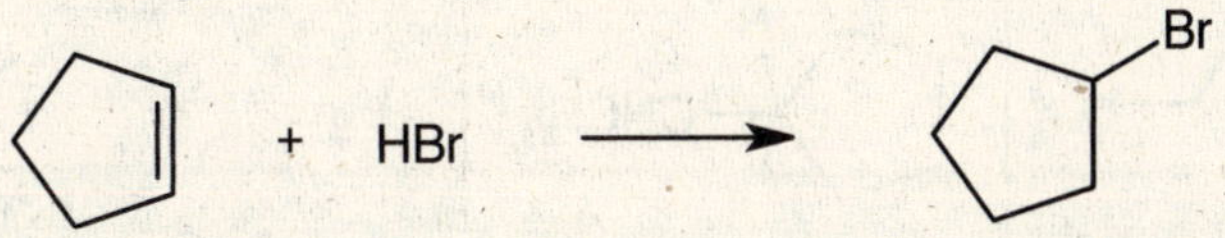

Cyclopentene Bromocyclopentane

(b)

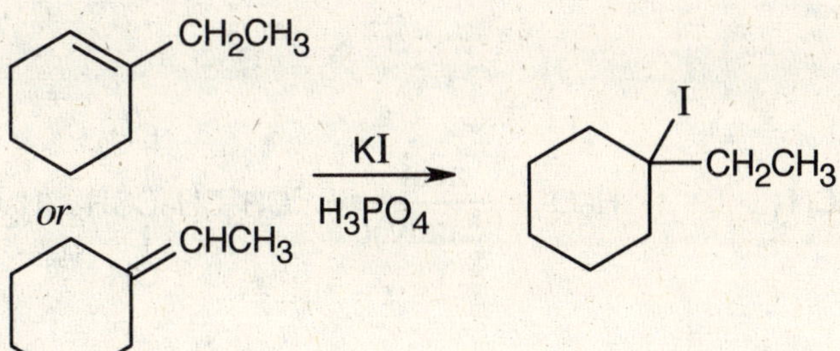

(c)

$$CH_3CH_2CH{=}CHCH_2CH_3 + HBr \longrightarrow CH_3CH_2CH_2CH(Br)CH_2CH_3$$

Hex-3-ene

If hex-2-ene were used, a mixture of products would be formed.

(d)

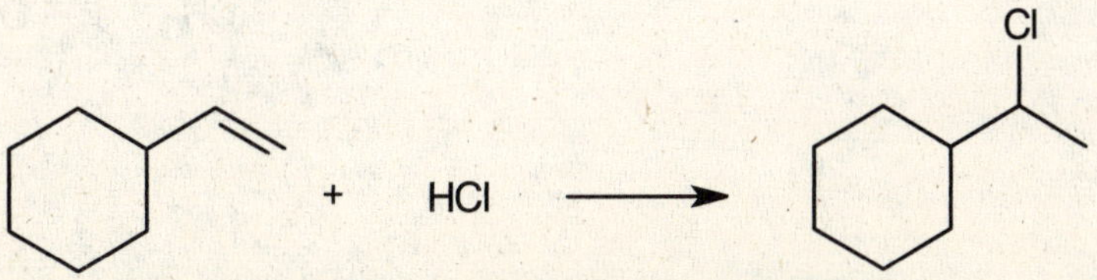

This is the only alkene that yields the desired product.

**4.3**

(a)

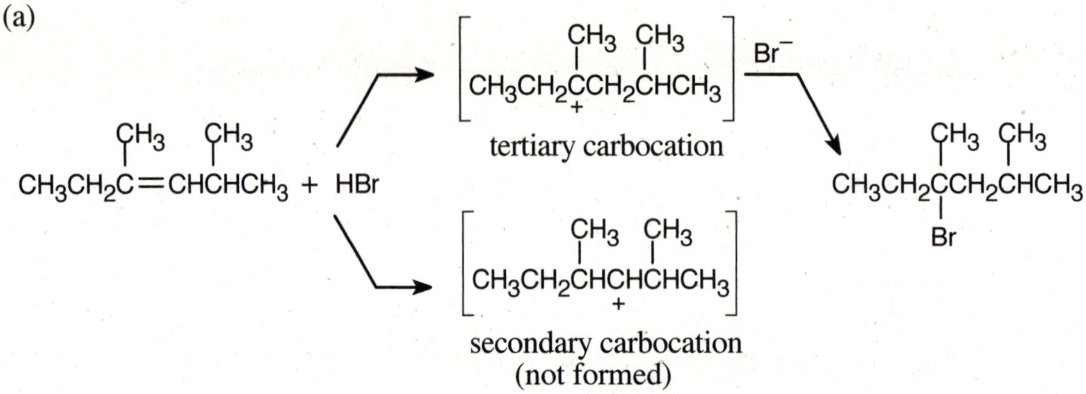
tertiary carbocation
secondary carbocation
(not formed)

(b)

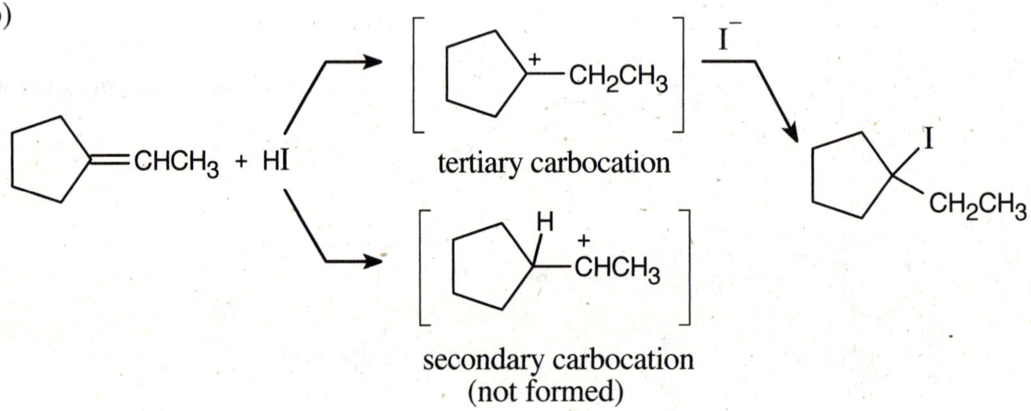
tertiary carbocation
secondary carbocation
(not formed)

**4.4**

(a)

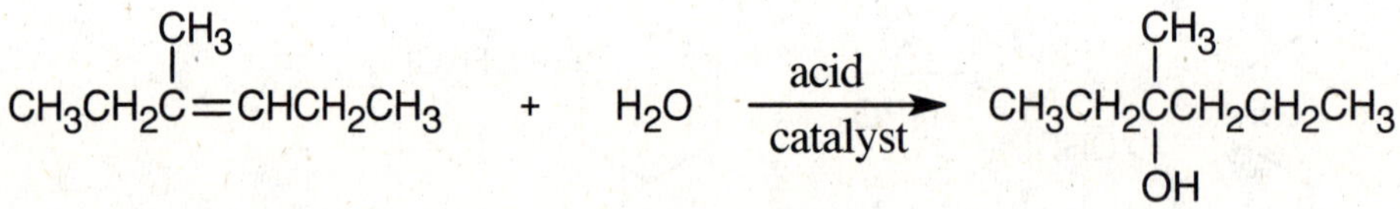
acid
catalyst

(b)

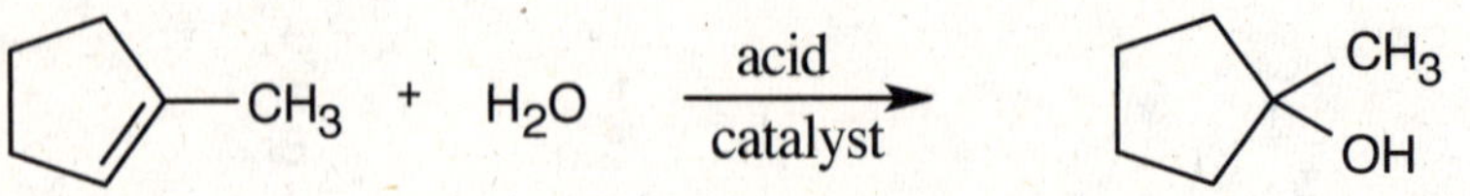
acid
catalyst

(c)

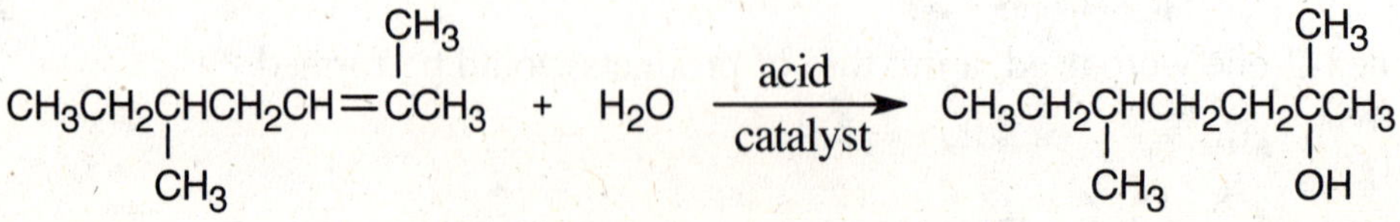
acid
catalyst

**4.5**

(a)

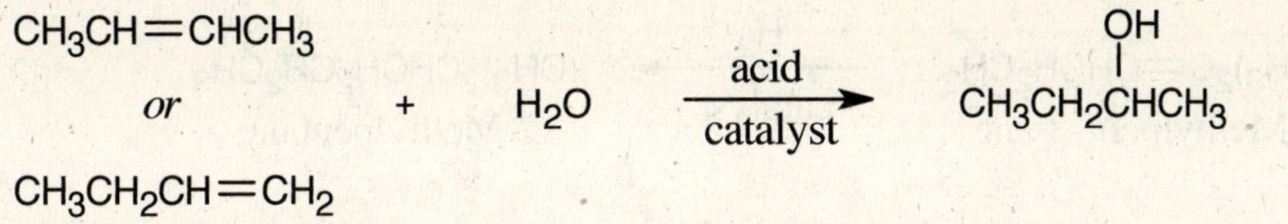


(b)

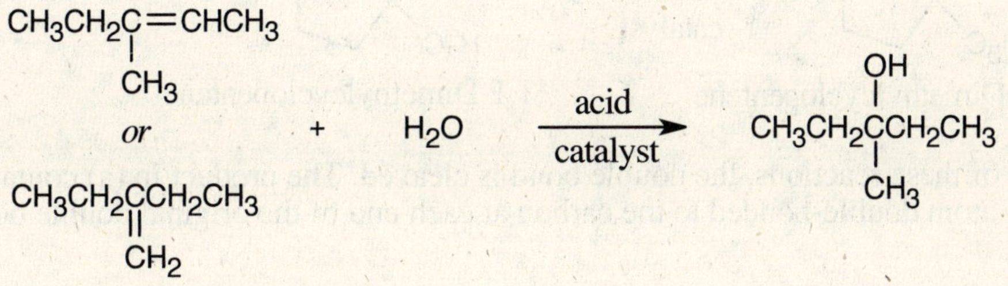


(c)

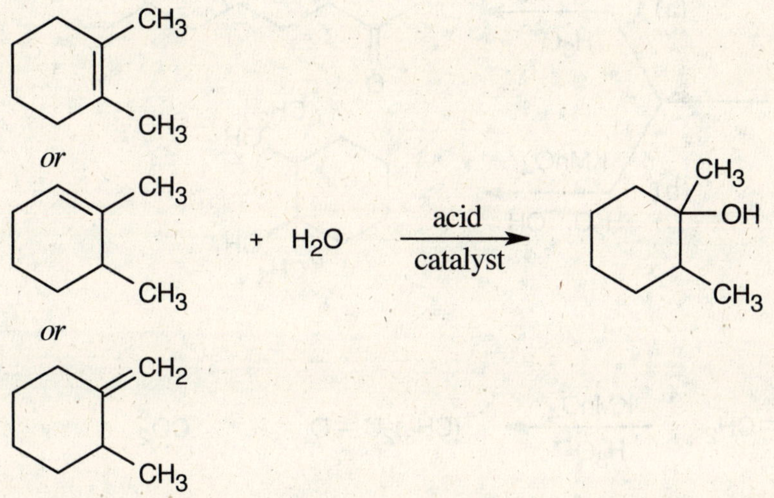


**4.6–4.7**

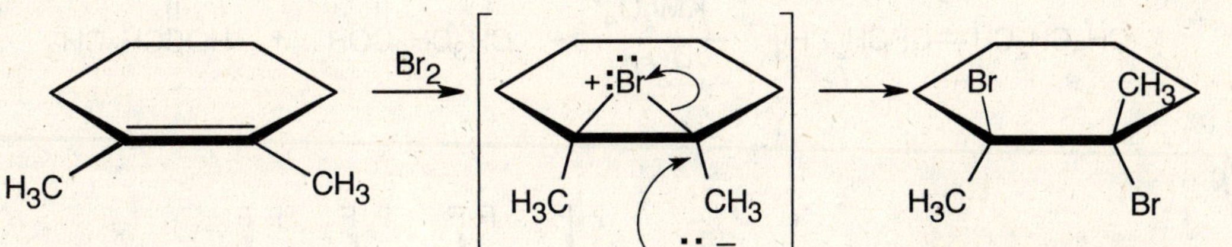


1,2-Dimethylcyclohexene

Bromonium ion intermediate

*trans*-1,2-Dibromo-1,2-dimethylcyclohexane

**4.8**

(a)

$$(CH_3)_2C{=}CHCH_2CH_3 \xrightarrow[\text{catalyst}]{H_2} (CH_3)_2CHCH_2CH_2CH_3$$

2-Methylpent-2-ene → 2-Methylpentane

(b)

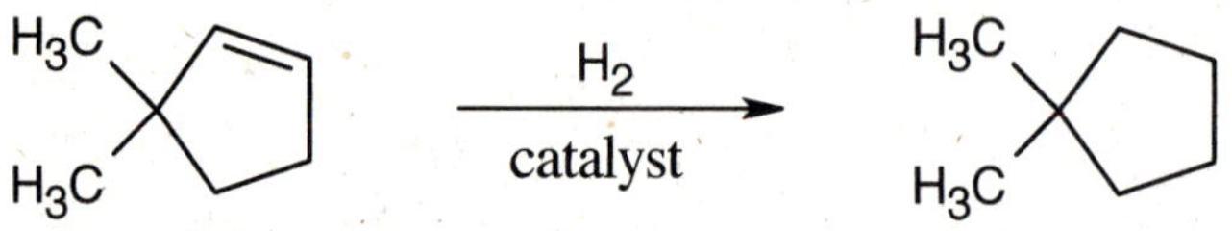

3,3-Dimethylcyclopentene → 1,1-Dimethylcyclopentane

**4.9** In both of these reactions, the double bond is cleaved. The product in (a) contains an oxygen atom double-bonded to the carbon at each end of the original double bond.

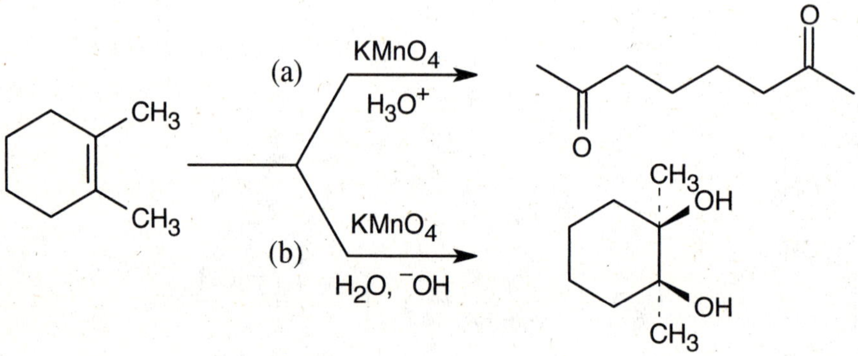

**4.10**

(a)

$$(CH_3)_2C{=}CH_2 \xrightarrow[H_3O^+]{KMnO_4} (CH_3)_2C{=}O \quad + \quad CO_2$$

(b)

$$CH_3CH_2CH{=}CHCH_2CH_3 \xrightarrow[H_3O^+]{KMnO_4} CH_3CH_2\overset{O}{\overset{\|}{C}}OH \quad + \quad HO\overset{O}{\overset{\|}{C}}CH_2CH_3$$

**4.11**

many $F_2C{=}CF_2$ ⟶

Tetrafluoroethylene

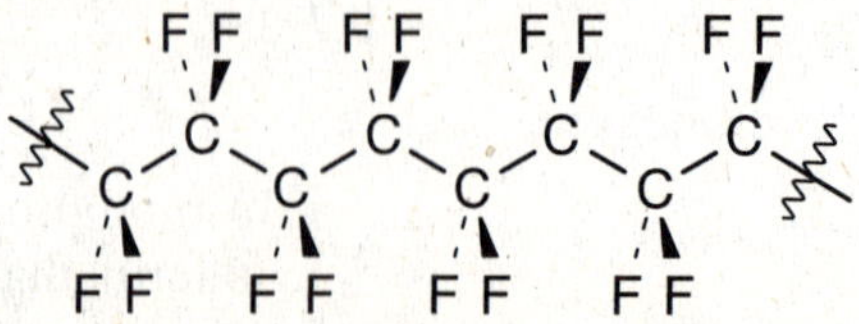

**4.12**

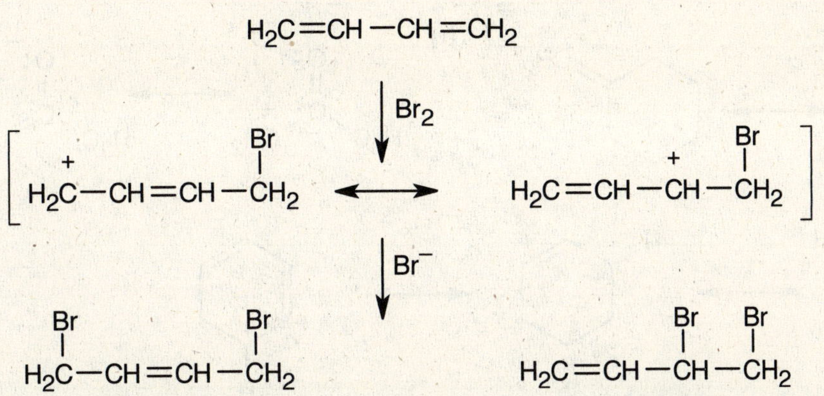

1,4 addition — 1,2 addition

**4.13** Only the products that result from allylic carbocation intermediates are shown.

$CH_3CH{=}CHCH{=}CH_2$ Penta-1,3-diene

| *Product* | *Name* | *Results from:* |
|---|---|---|
| $CH_3CH{=}CHCH(Cl)CH_3$ | 4-Chloropent-2-ene | 1,2 addition<br>1,4 addition |
| $CH_3CH_2CH(Cl)CH{=}CH_2$ | 3-Chloropent-1-ene | 1,2 addition |
| $CH_3CH_2CH{=}CHCH_2Cl$ | 1-Chloropent-2-ene | 1,4 addition |

**4.14**

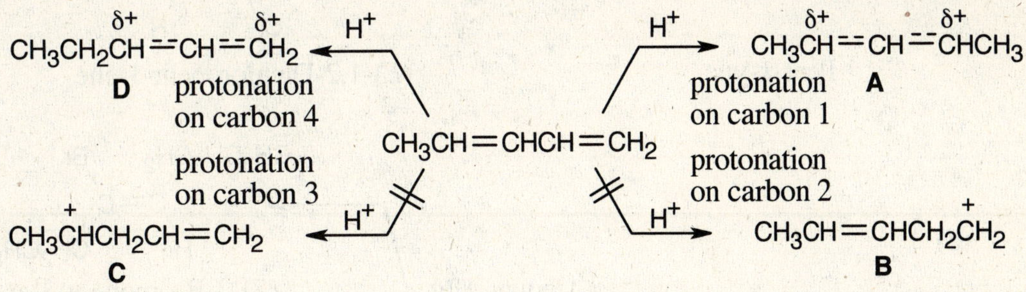


**A** and **D**, which are resonance-stabilized, are formed instead of **B** and **C**, which are not. The positive charge of allylic carbocation **A** is delocalized over two secondary carbons, while the positive charge of carbocation **D** is delocalized over one secondary and one primary carbon. We therefore predict that carbocation **A** is the major intermediate formed, and that 4-chloropent-2-ene predominates. Note that this product results from both 1,2– and 1,4–addition.

**4.15**

(a)

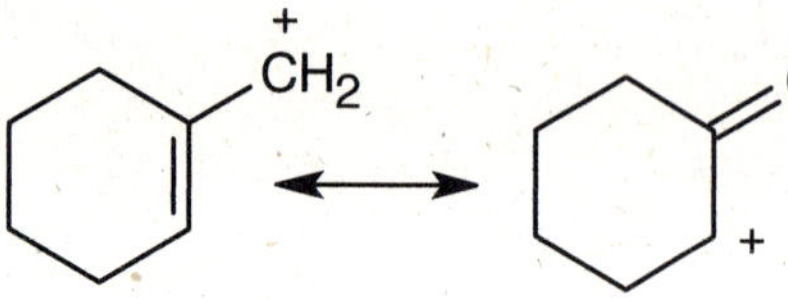

(b)

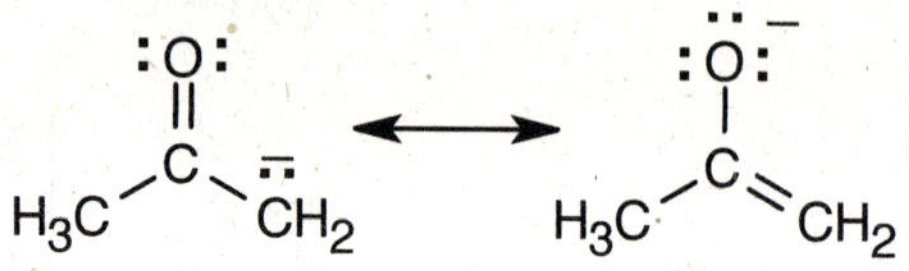

(c)

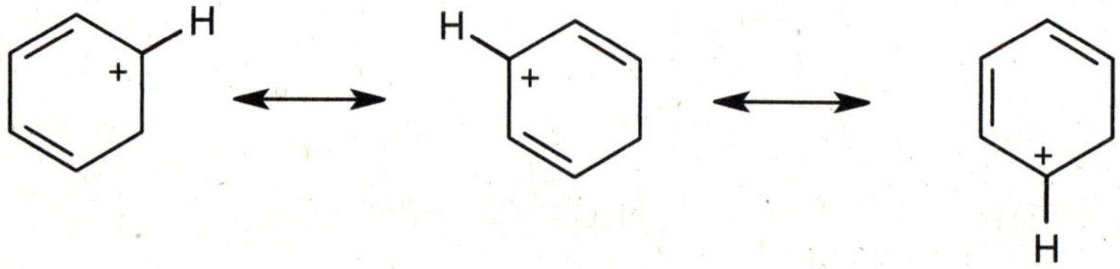

**4.16**

(a)

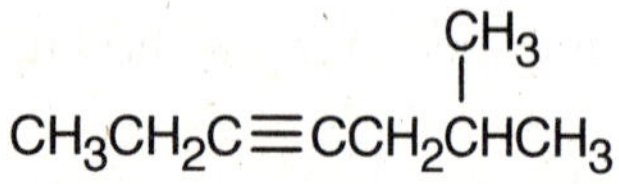

6-Methylhept-3-yne

(b)

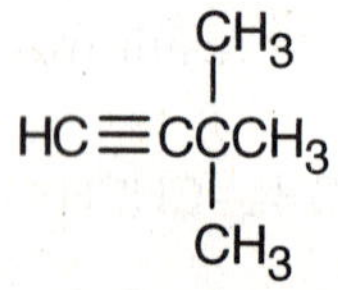

3,3-Dimethylbut-1-yne

(c)

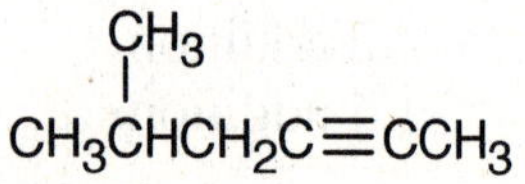

5-Methylhex-2-yne

(d)

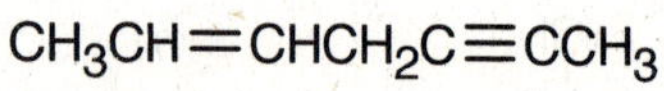

Hept-2-en-5-yne

**4.17**

(a)

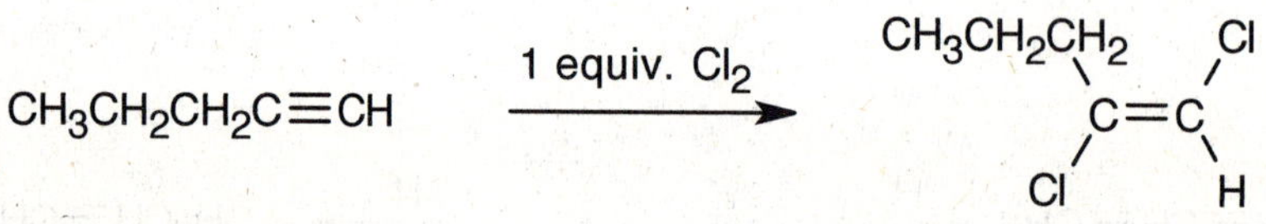

Pent-1-yne

(*E*)-1,2-Dichloropent-1-ene

(b)

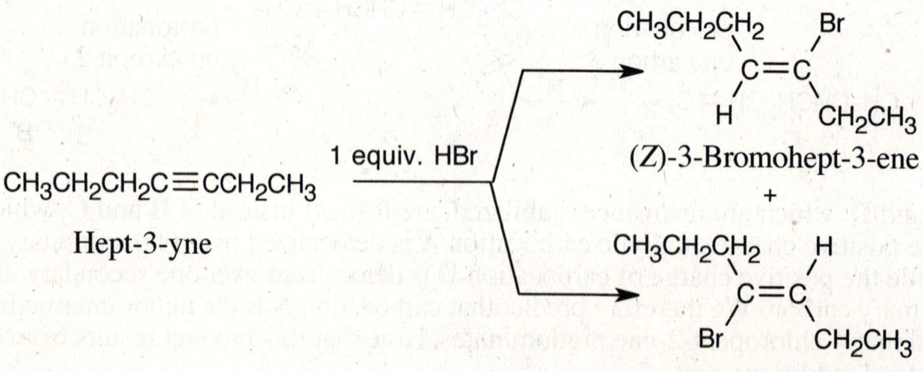

Hept-3-yne

(*Z*)-3-Bromohept-3-ene

\+

(*Z*)-4-Bromohept-3-ene

(c)

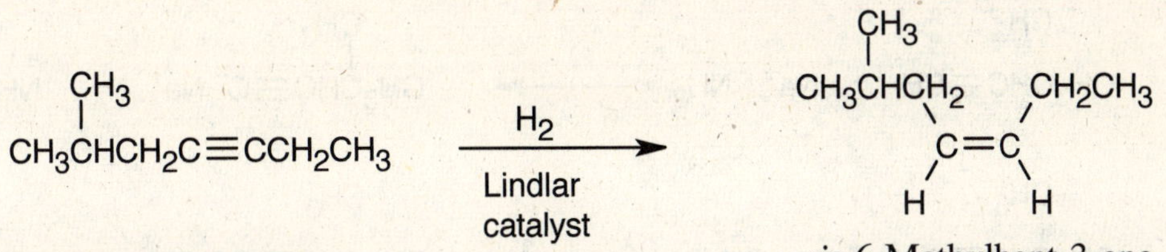

6-Methylhept-3-yne cis-6-Methylhept-3-ene

A Lindlar catalyst yields cis-alkene, not alkane, products.

**4.18**

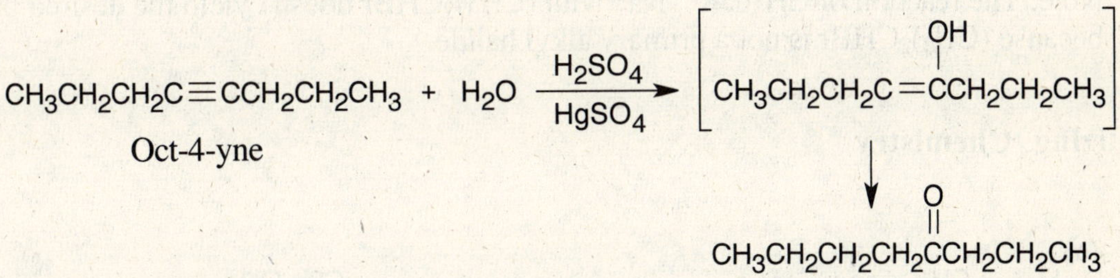


Only one product is observed because of the symmetry of the alkyne.

**4.19**

(a)

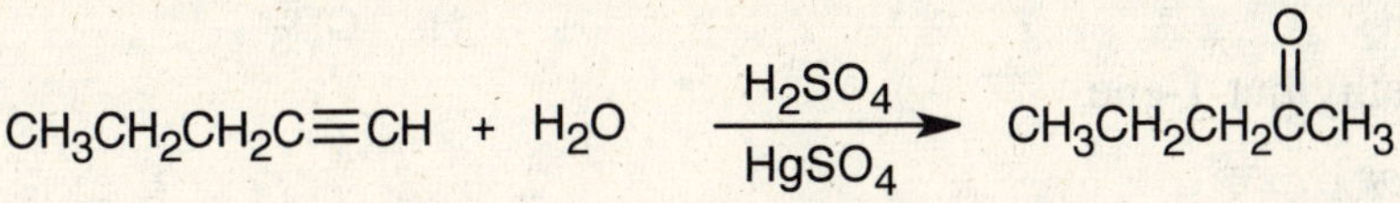

(b)

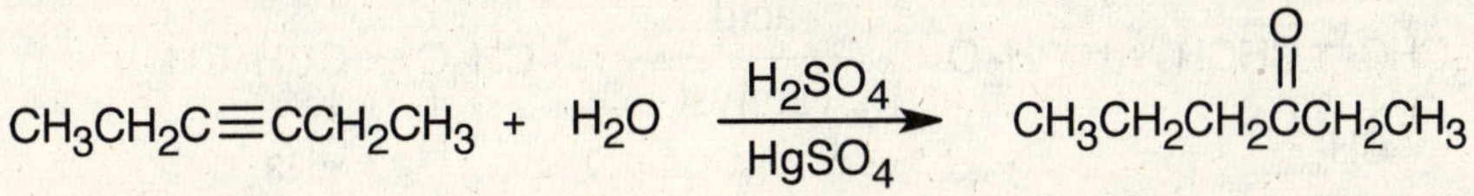

**4.20**

(a)

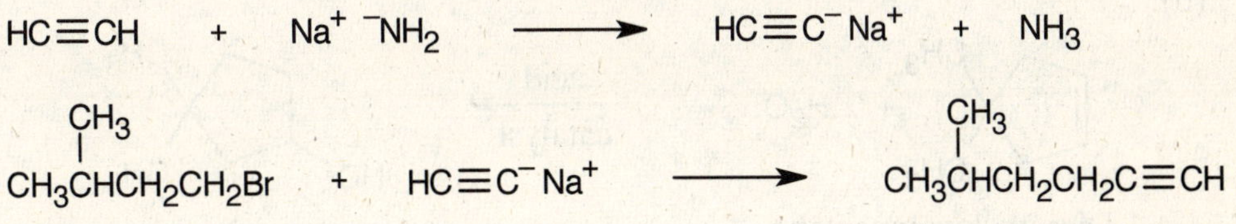

5-Methylhex-1-yne

(b)

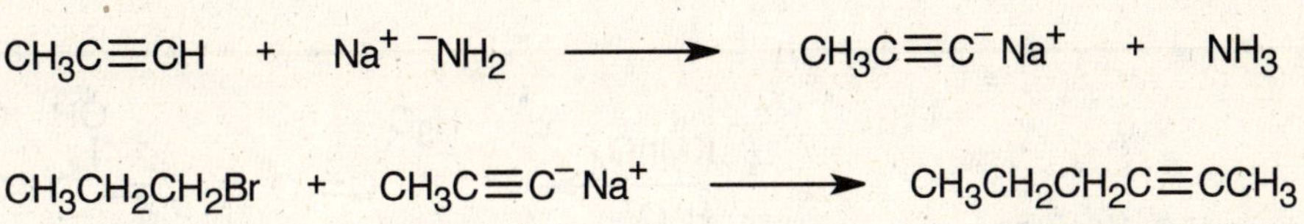

Hex-2-yne

*or*

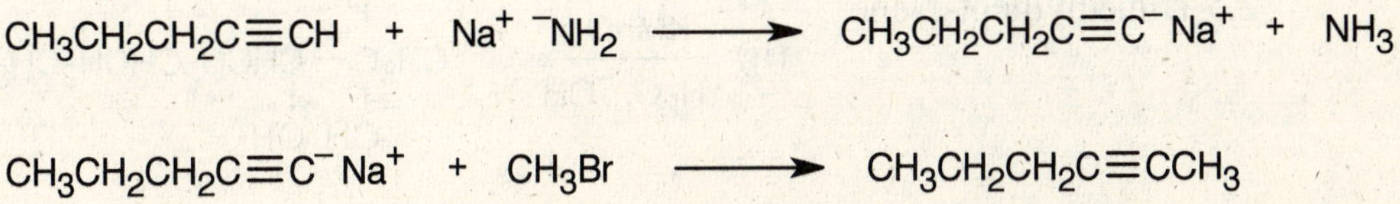

Hex-2-yne

(c)

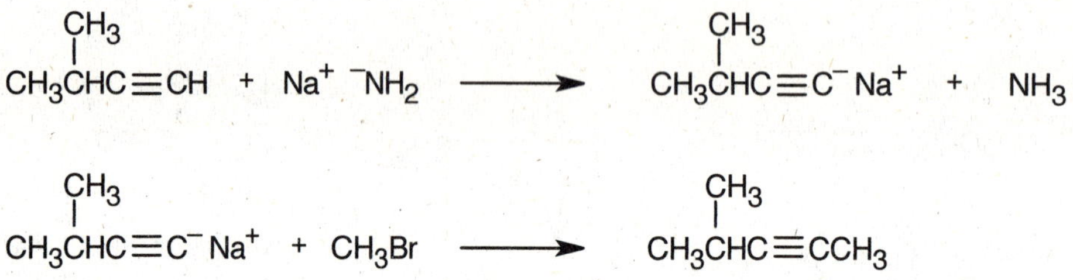

4-Methylpent-2-yne

Note: The reaction of $CH_3C{\equiv}C^- \, Na^+$ with $(CH_3)_2CHBr$ doesn't yield the desired product because $(CH_3)_2CHBr$ is not a primary alkyl halide.

## Visualizing Chemistry

**4.21**

(a)

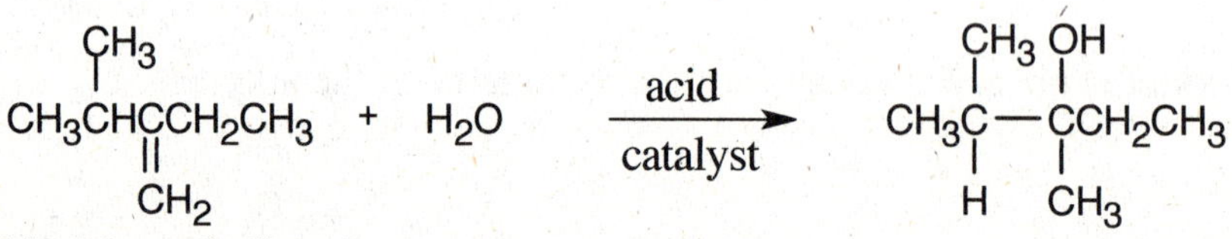

2-Ethyl-3-methylbut-1-ene

*or*

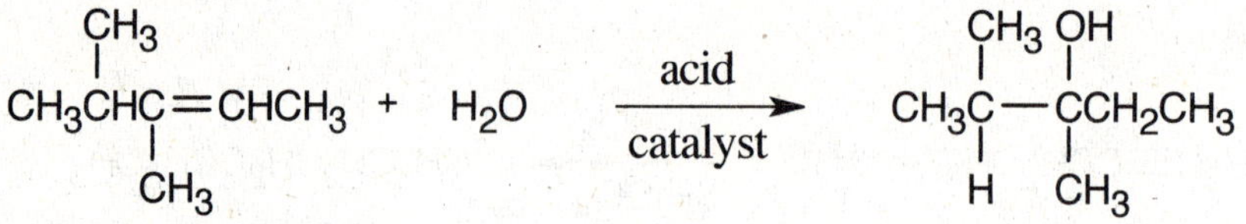

3,4-Dimethylpent-2-ene

(b)

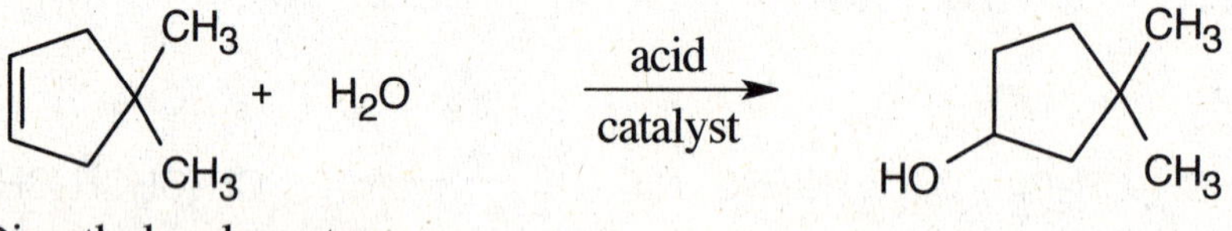

4,4-Dimethylcyclopentene

**4.22**

(a)

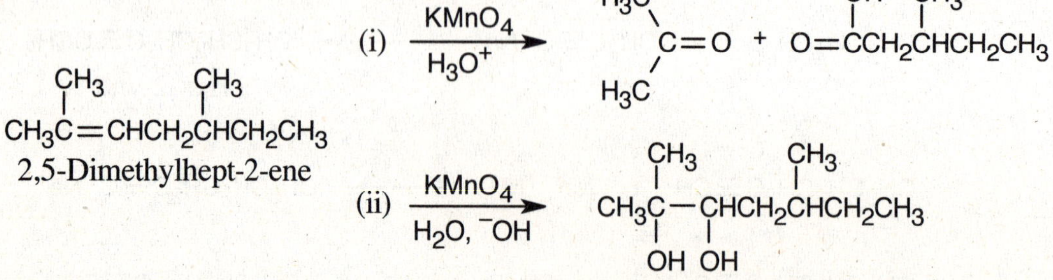

2,5-Dimethylhept-2-ene

(b)

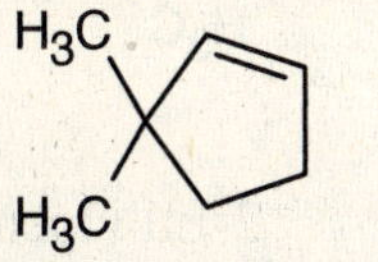

3,3-Dimethylcyclopentene

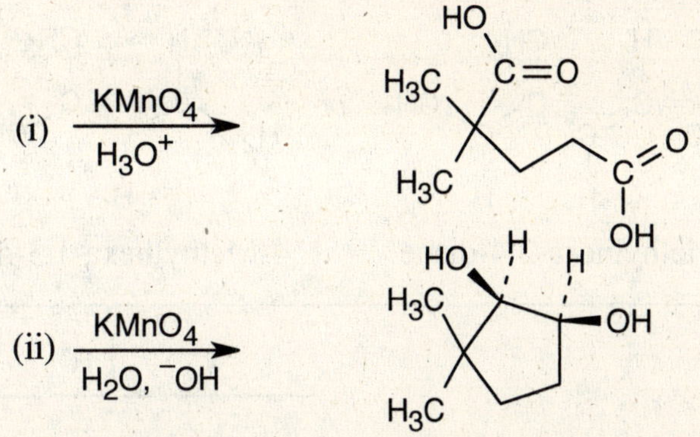

**4.23**

(a)

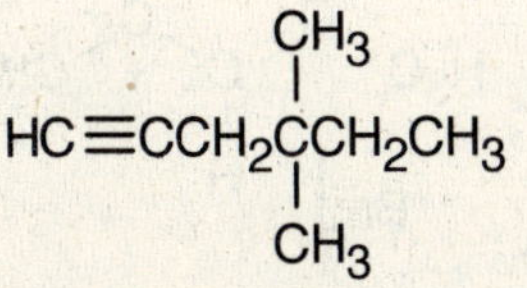

4,4-Dimethylhex-1-yne

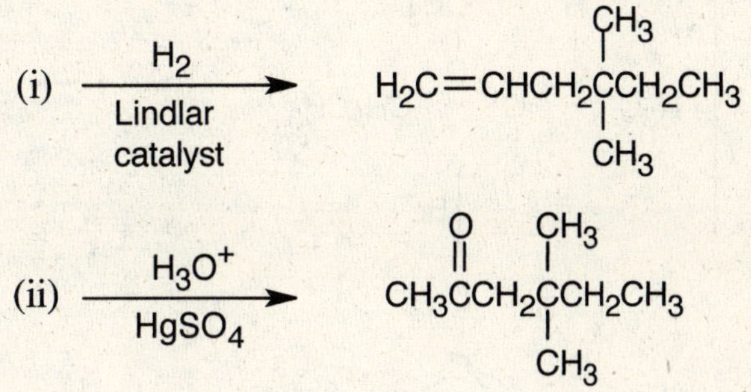

(b)

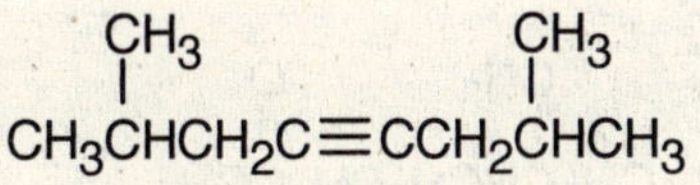

2,7-Dimethyloct-4-yne

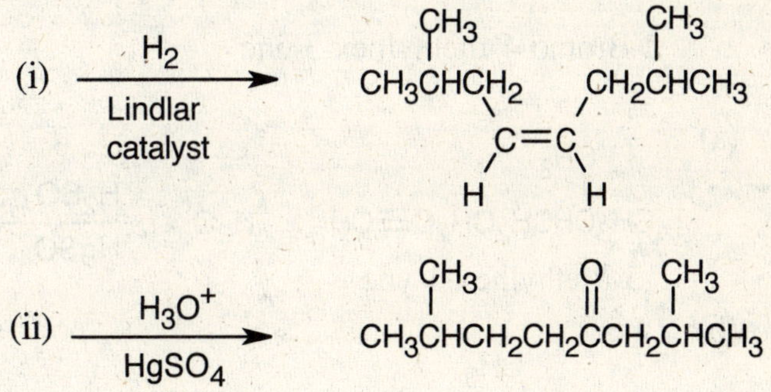

**4.24**

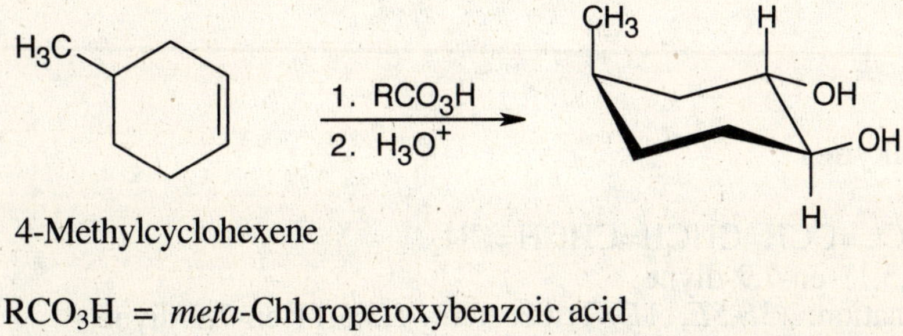

trans

4-Methylcyclohexene

$RCO_3H$ = *meta*-Chloroperoxybenzoic acid

Since the hydroxyl groups in the diol product have a trans relationship, the product can only be formed by epoxide hydrolysis. (Treatment of the alkene with $KMnO_4$ yields a product in which the two –OH groups have a cis relationship.)

**4.25**

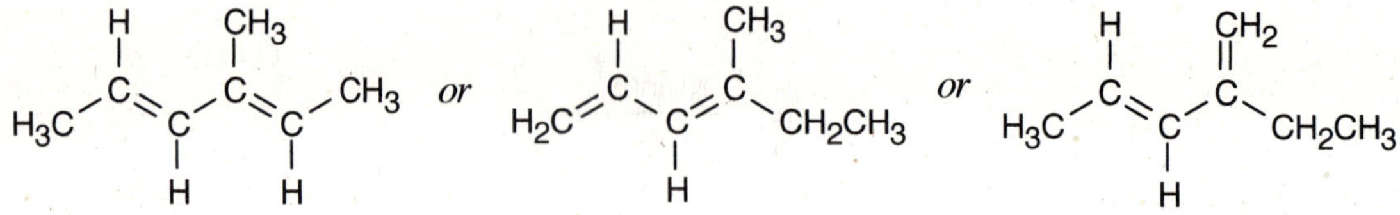

3-Methylhexa-2,4-diene *or* 4-Methylhexa-1,3-diene *or* 2-Ethylpenta-1,3-diene

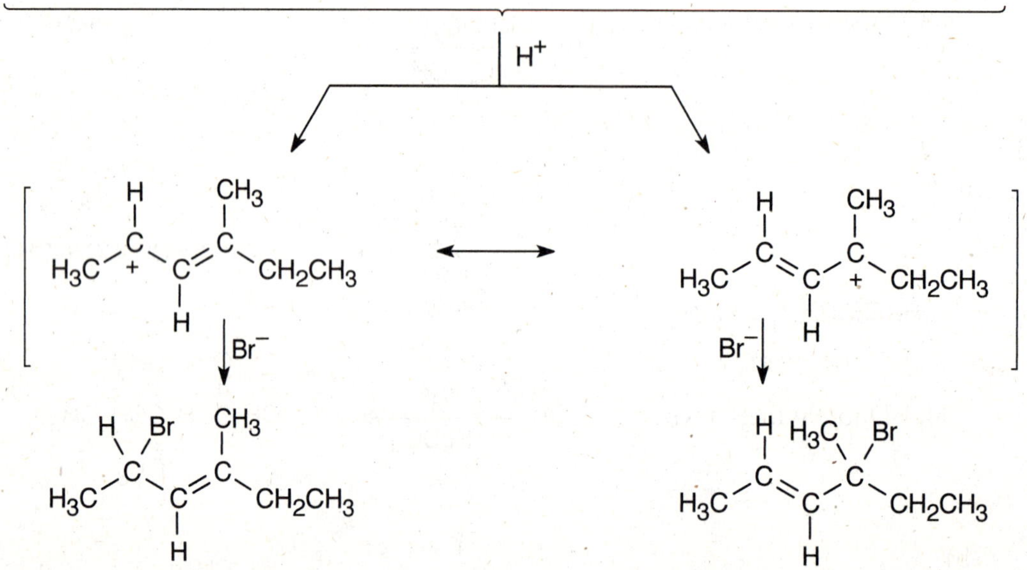

2-Bromo-4-methylhex-3-ene 4-Bromo-4-methylhex-2-ene

**4.26**

(a)

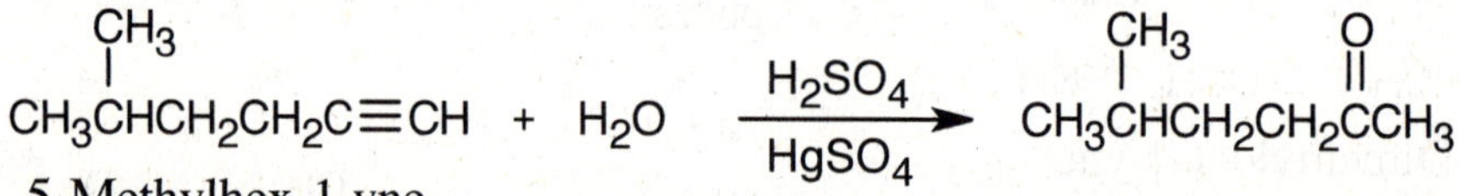

5-Methylhex-1-yne

(b)

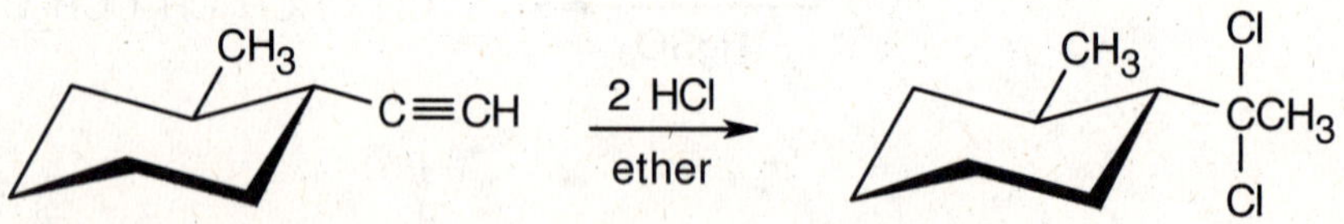

## Additional Problems

### Naming Alkenes and Alkynes

**4.27** (a) $CH_3CH{=}CHC{\equiv}CC{\equiv}CCH{=}CHCH{=}CHCH{=}CH_2$
Tridecatetra-1,3,5,11-en-7,9-diyne
Using *E*-*Z* designation: (3*E*,5*E*,11*E*)-Tridecatetra-1,3,5,11-en-7,9-diyne
The parent alkane of this hydrocarbon is tridecane.

(b) $CH_3C{\equiv}CC{\equiv}CC{\equiv}CC{\equiv}CC{\equiv}CCH{=}CH_2$
Tridec-1-en-3,5,7,9,11-pentayne
This hydrocarbon is also in the tridecane family.

**4.28**

(a) $C_6H_8$:

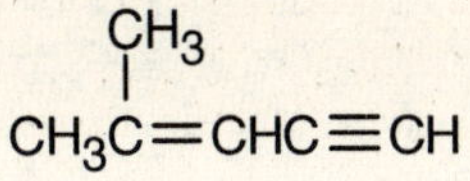

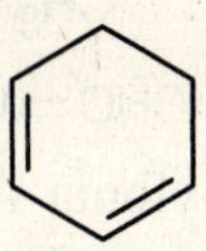

$H_2C{=}CHCH{=}CHCH{=}CH_2$

(b) $C_6H_8O$:

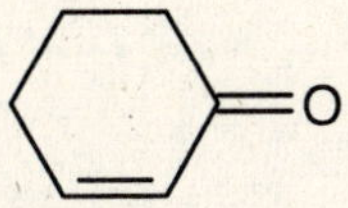

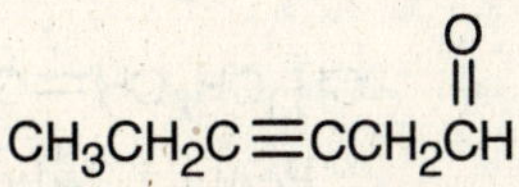

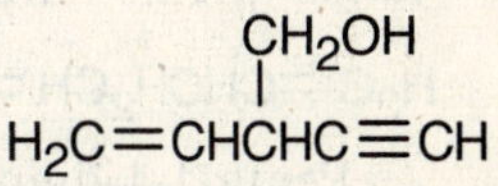

**4.29**

(a)

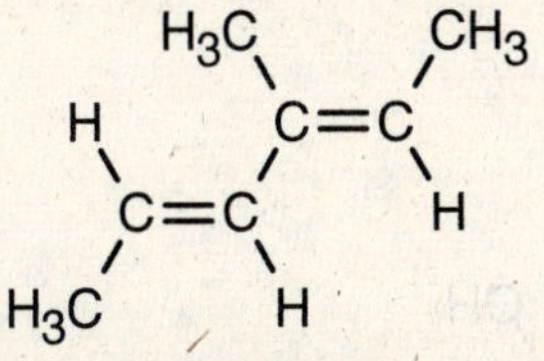

(2*Z*,4*E*)-3-Methylhexa-2,4-diene

(b)

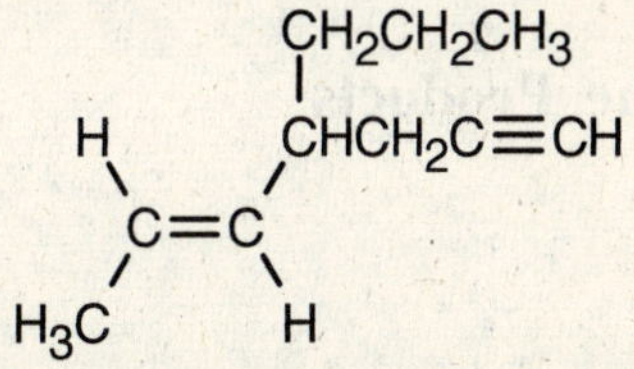

(5*E*)-4-Propylhept-5-en-1-yne

(c)

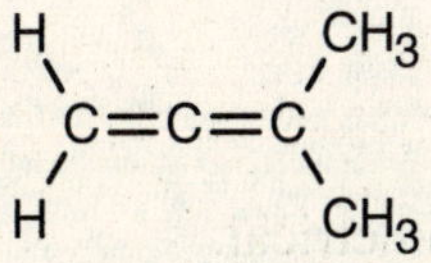

3-Methylbuta-1,2-diene

(d)

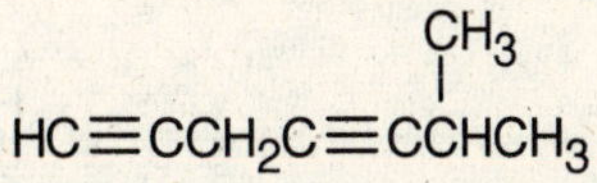

6-Methylhepta-1,4-diyne

**4.30**

(a)

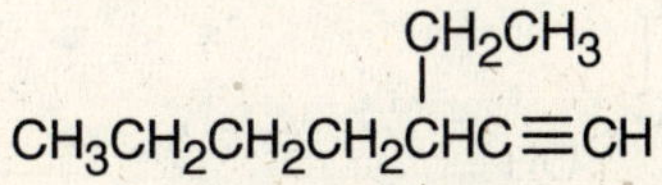

3-Ethylhept-1-yne

(b)

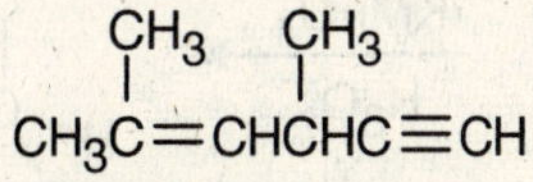

3,5-Dimethylhex-4-en-1-yne

(c)

$CH_3C{\equiv}CCH_2CH_2C{\equiv}CH$

Hepta-1,5-diyne

(d)

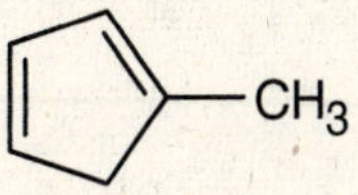

1-Methylcyclopenta-1,3-diene

**4.31**

$CH_3CH_2CH_2C{\equiv}CH$

Pent-1-yne

$CH_3CH_2C{\equiv}CCH_3$

Pent-2-yne

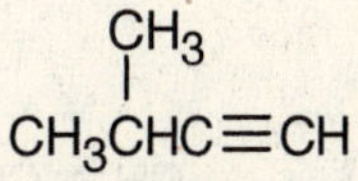

3-Methylbut-1-yne

**4.32**

*Conjugated dienes:*

$CH_3CH{=}CHCH{=}CH_2$
Penta-1,3-diene

$H_2C{=}CHC(CH_3){=}CH_2$
2-Methylbuta-1,3-diene

*Nonconjugated dienes:*

$H_2C{=}CHCH_2CH{=}CH_2$
Penta-1,4-diene

$CH_3CH_2CH{=}C{=}CH_2$
Penta-1,2-diene

$CH_3CH{=}C{=}CHCH_3$
Penta-2,3-diene

$H_2C{=}C{=}C(CH_3)_2$
3-Methylbuta-1,2-diene

## Predict the Products

**4.33**

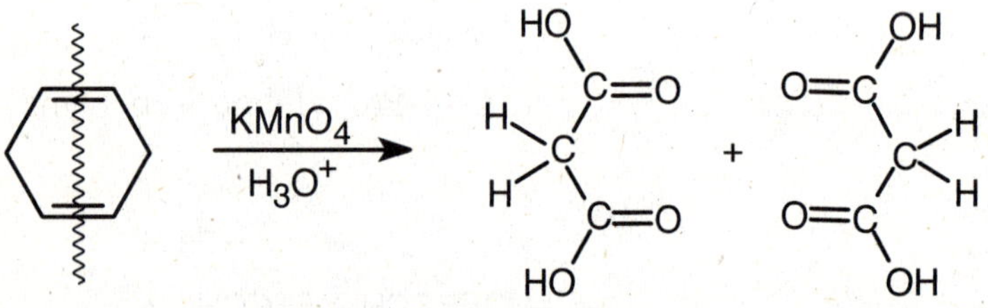


This cyclohexadiene yields only one product because it is symmetrical.

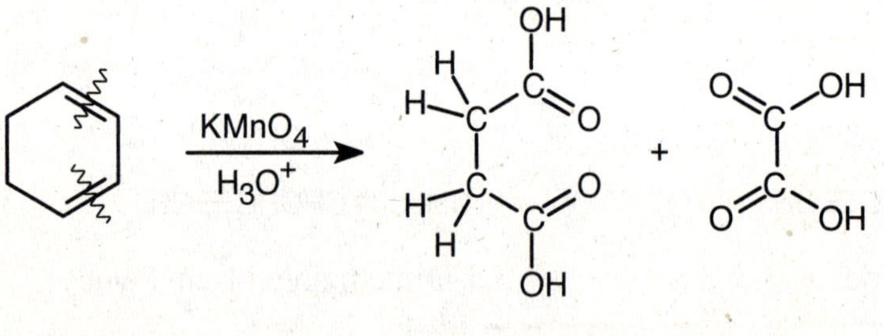


Two different products result from oxidative cleavage of this diene.

**4.34**

(a)

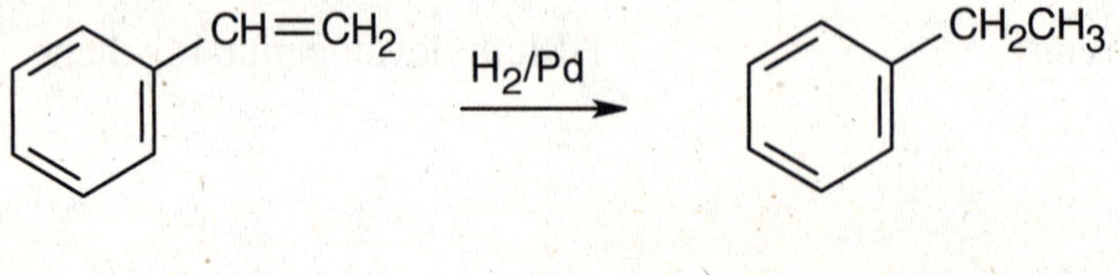


(b)

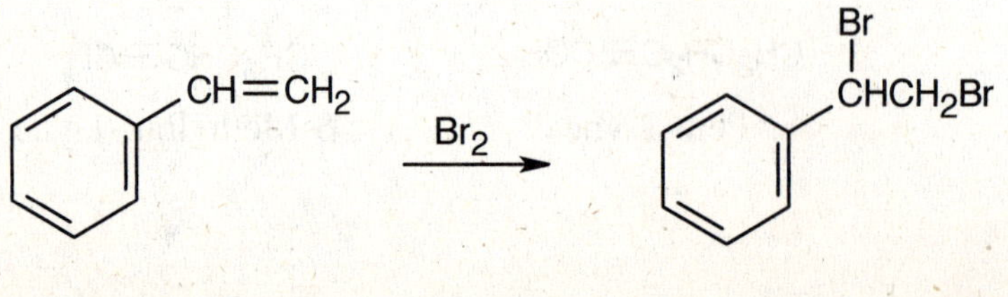

(c)

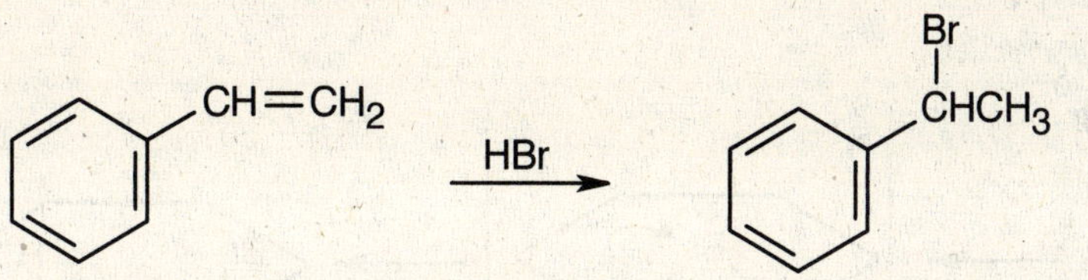

(d)

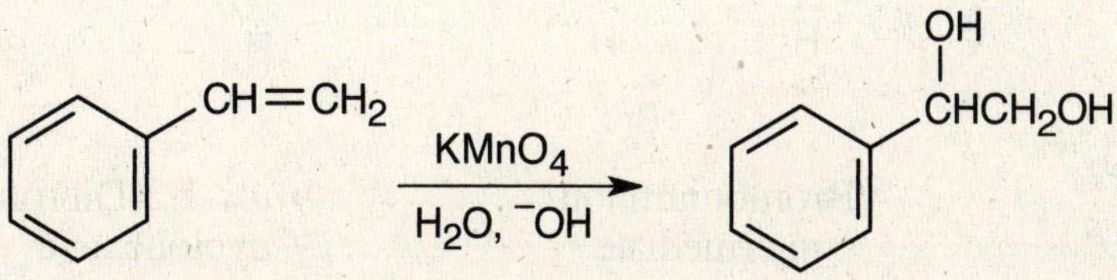

**4.35**

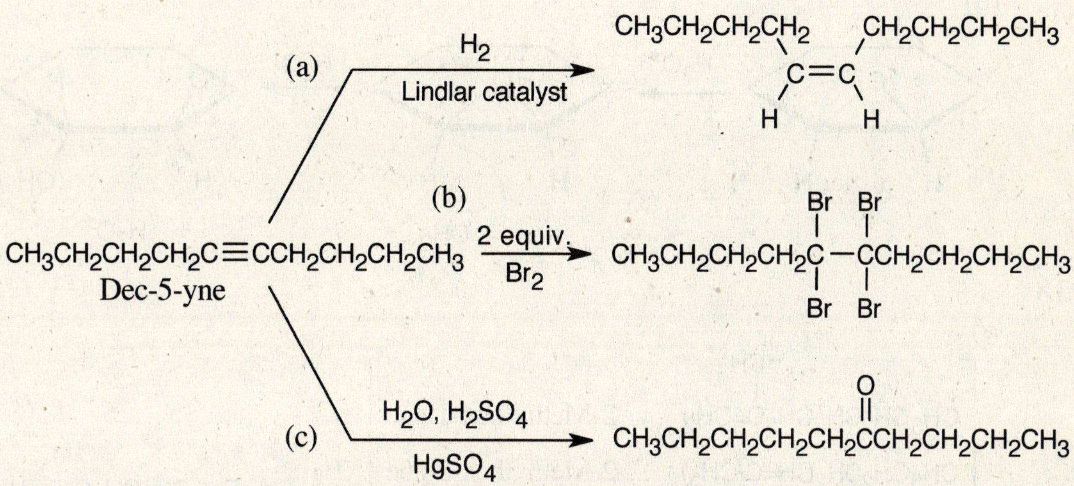

**4.36**

(a)

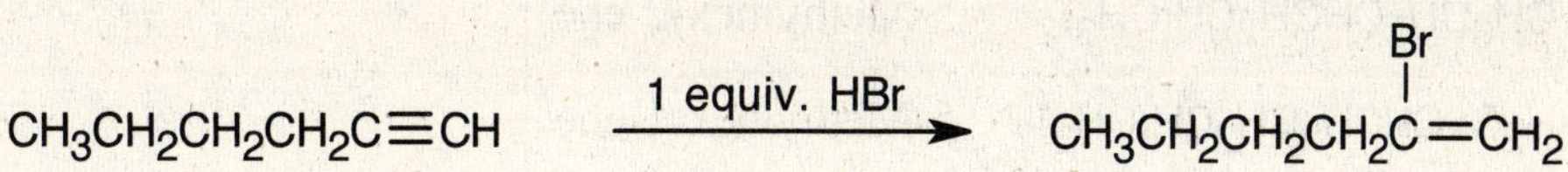

(b)

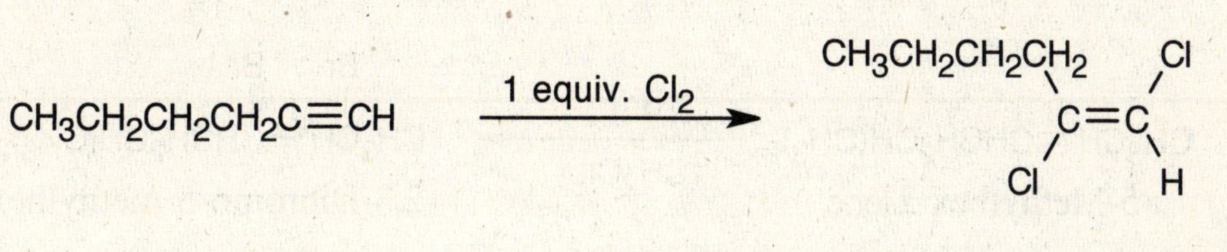

(c)

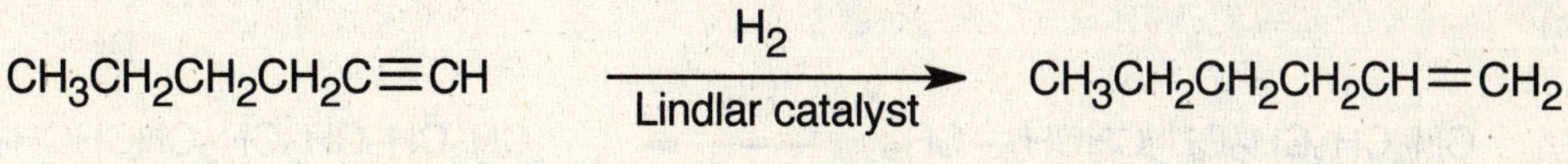

**4.37**

(a)

(b)

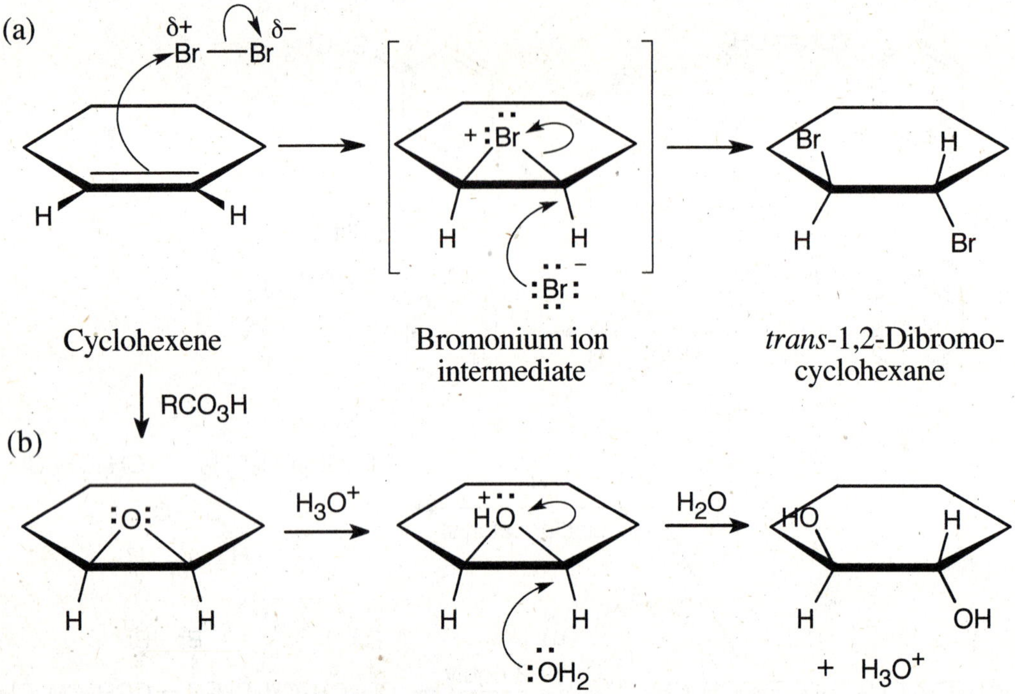


**4.38**

(a)

$$\left[\begin{array}{ll} CH_3CH_2CH_2CH_2C(CH_3){=}CH_2 & \text{2-Methylhex-1-ene} \\ CH_3CH_2CH_2CH{=}C(CH_3)_2 & \text{2-Methylhex-2-ene} \\ CH_3CH_2CH{=}CHCH(CH_3)_2 & \text{2-Methylhex-3-ene} \\ CH_3CH{=}CHCH_2CH(CH_3)_2 & \text{5-Methylhex-2-ene} \\ H_2C{=}CHCH_2CH_2CH(CH_3)_2 & \text{5-Methylhex-1-ene} \end{array}\right] \xrightarrow[\text{Pd catalyst}]{H_2} \underset{\text{2-Methylhexane}}{CH_3CH_2CH_2CH_2CH(CH_3)_2}$$

(b)

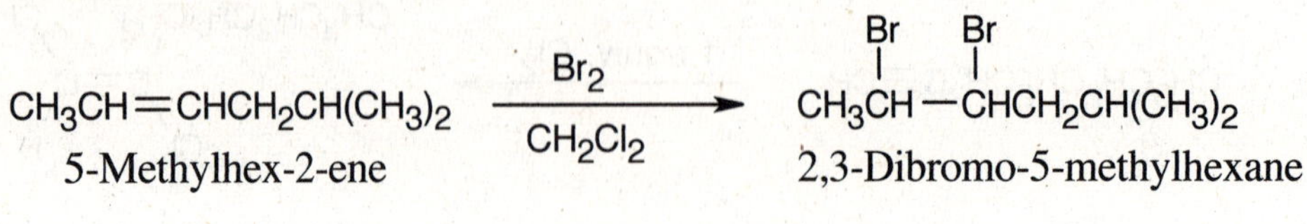


(c)

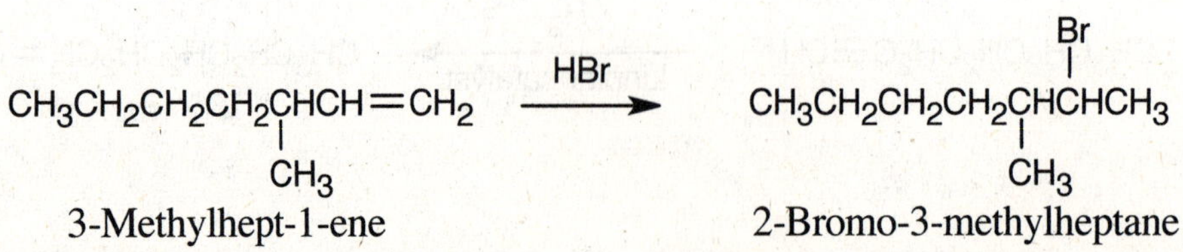

(d)

$$\underset{\text{6-Methylhept-3-ene}}{CH_3\overset{\overset{CH_3}{|}}{C}HCH_2CH{=}CHCH_2CH_3} \xrightarrow[H_2O,\ ^{-}OH]{KMnO_4} CH_3\overset{\overset{CH_3}{|}}{C}HCH_2\overset{\overset{OH}{|}}{C}H{-}\overset{\overset{OH}{|}}{C}HCH_2CH_3$$

**4.39**

(a)

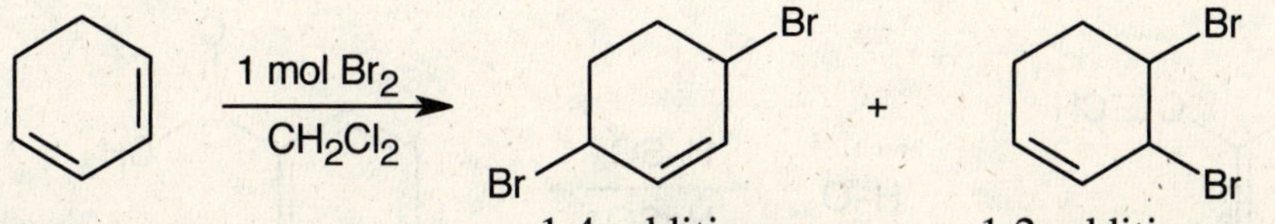

1,4-addition 1,2-addition

(b)

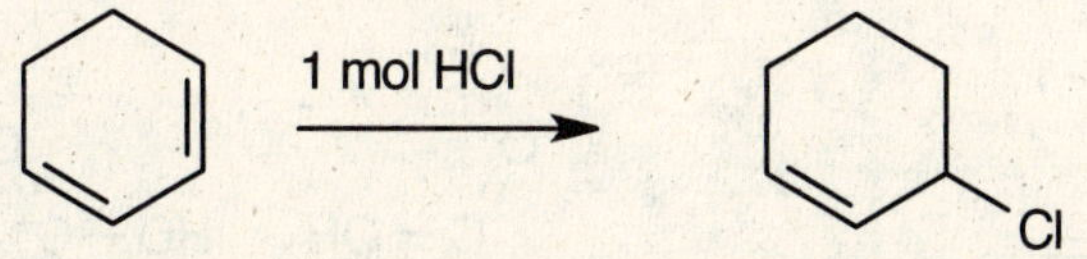

(c)

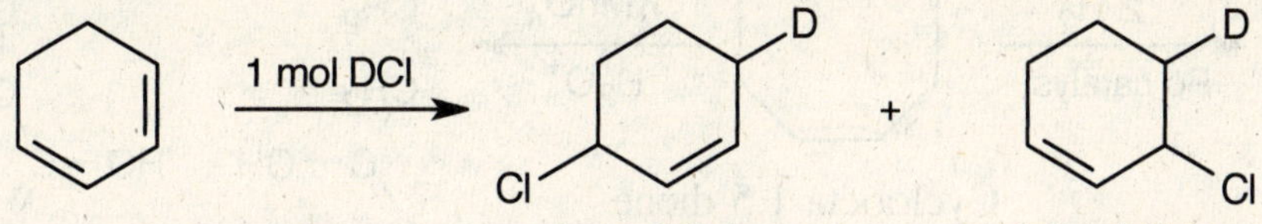

1,4-addition 1,2-addition

(d)

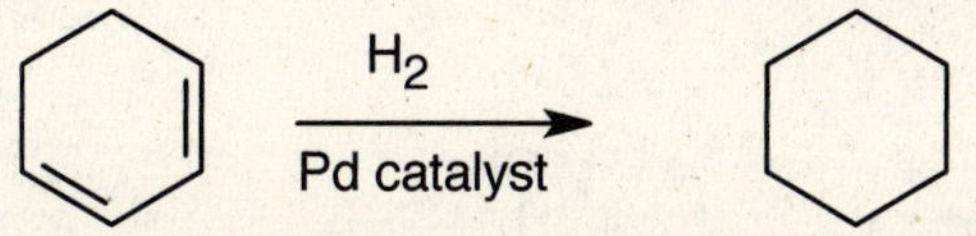

## Predict the Reactants

**4.40**

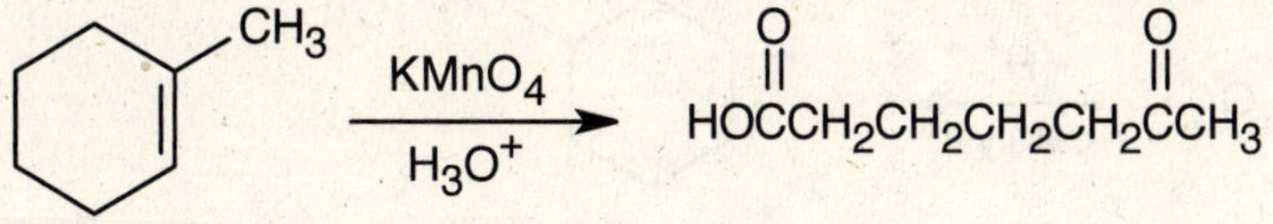

**4.41**

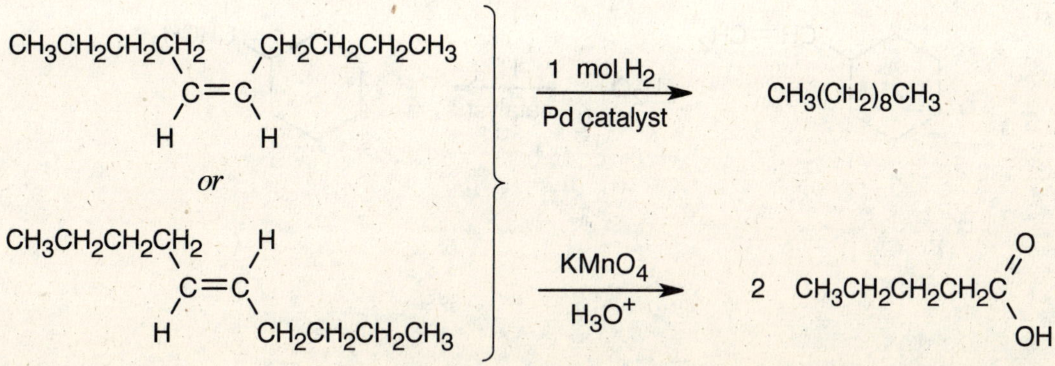

**4.42**

(a)

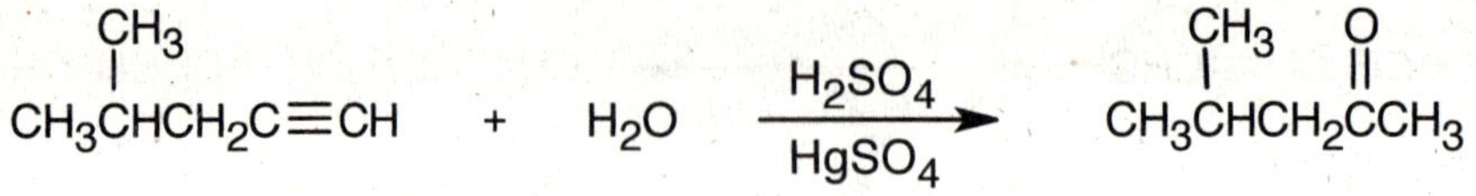

(b)

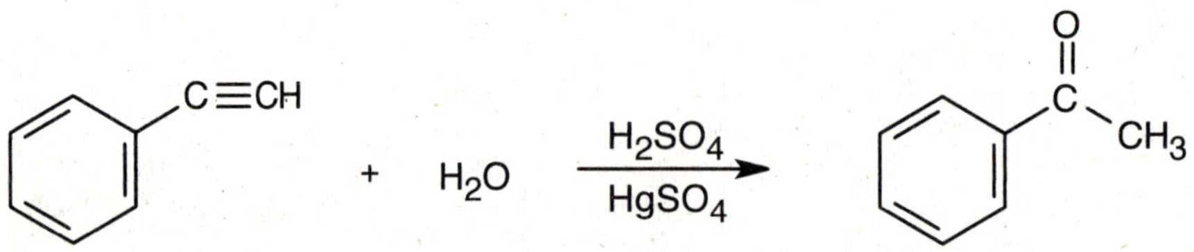

**4.43**

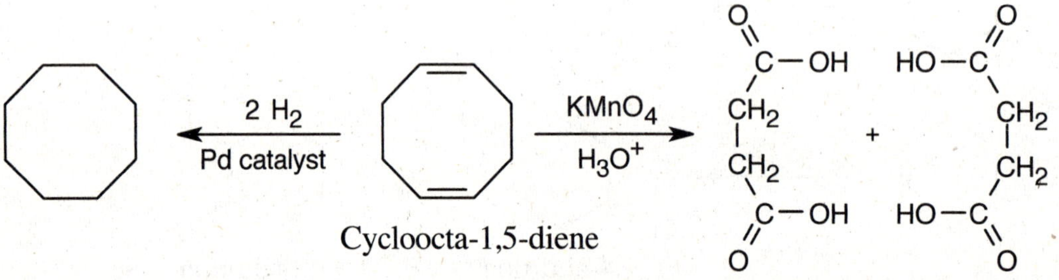


**4.44**

(a)

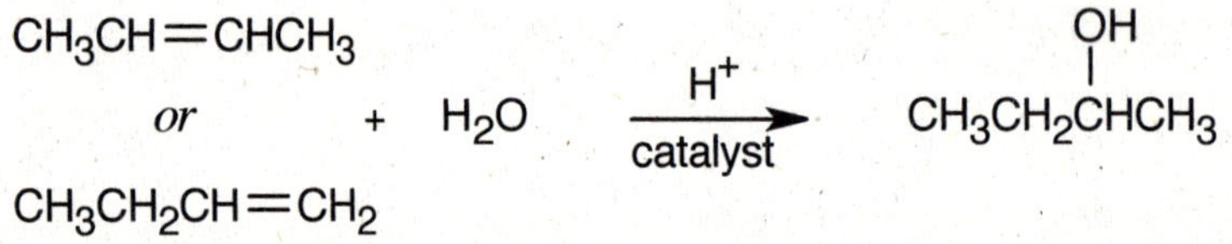

(b)

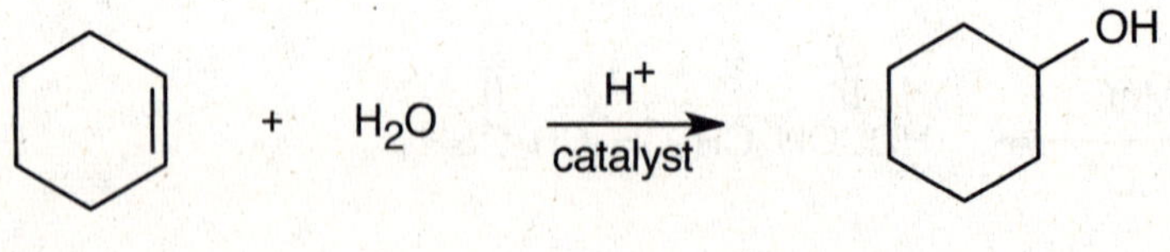

(c)

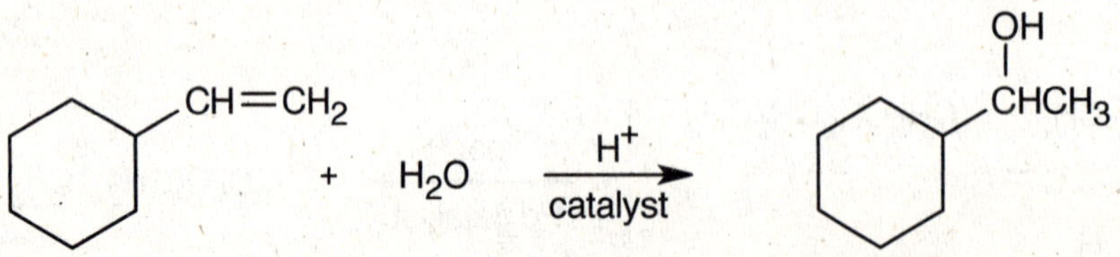

## Synthesis

**4.45**

(a)

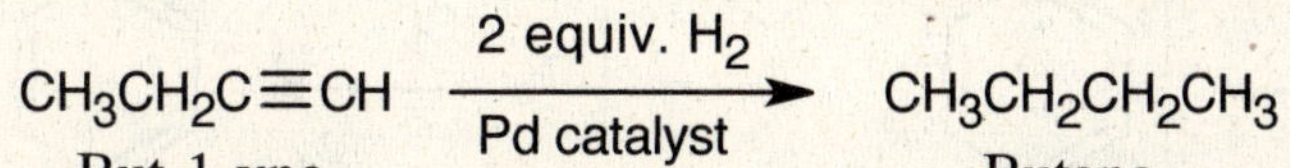


But-1-yne Butane

(b)

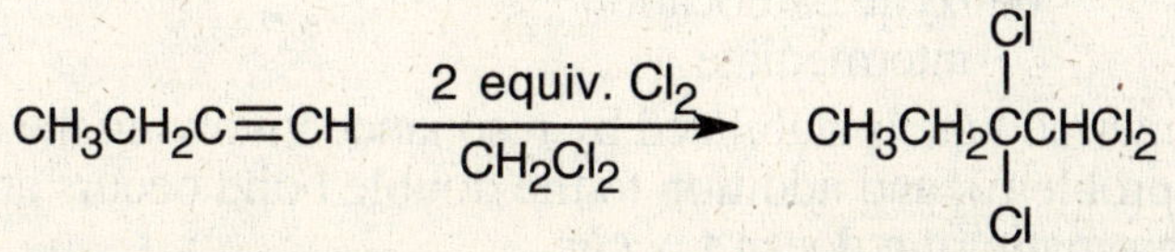


1,1,2,2-Tetrachlorobutane

(c)

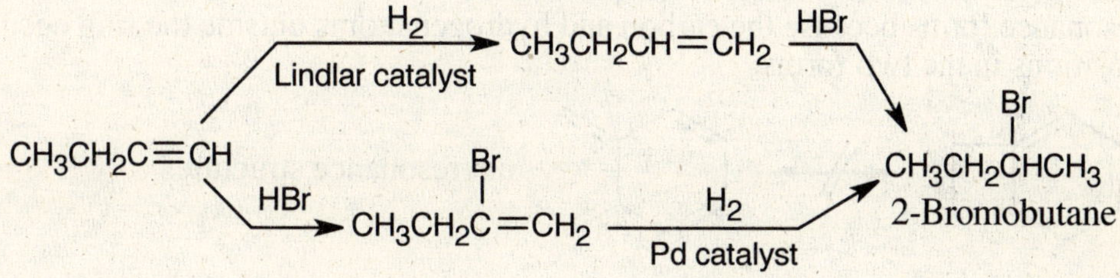


2-Bromobutane

(d)

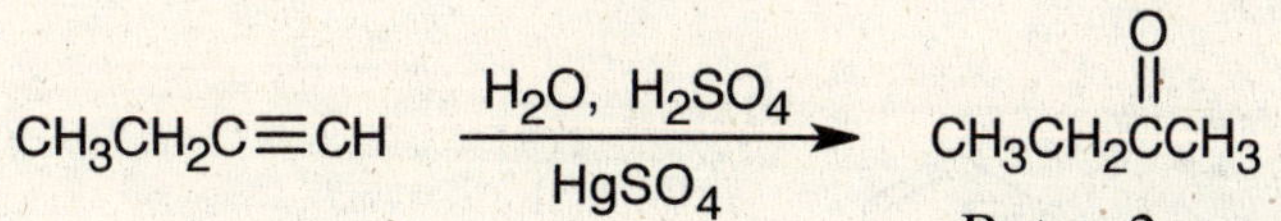


Butan-2-one

**4.46** (a) Reaction of 2-methylbut-2-ene with HBr gives a product in which bromine is bonded to the *more* substituted carbon.
(b) Hydroxylation of double bonds with $KMnO_4$ produces cis, not trans, diols.
(c) Hydration of a terminal alkyne produces a ketone, not an aldehyde.

**4.47**

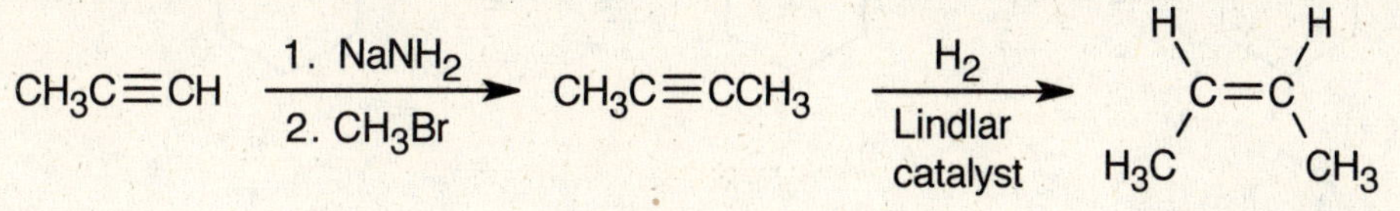


*cis*-But-2-ene

## Resonance

**4.48**

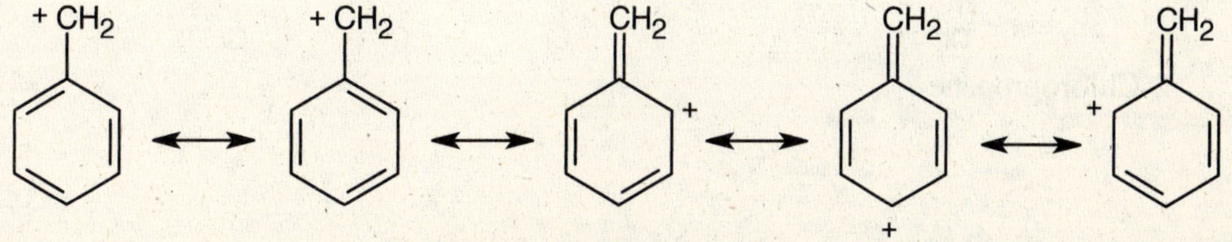

**4.49**

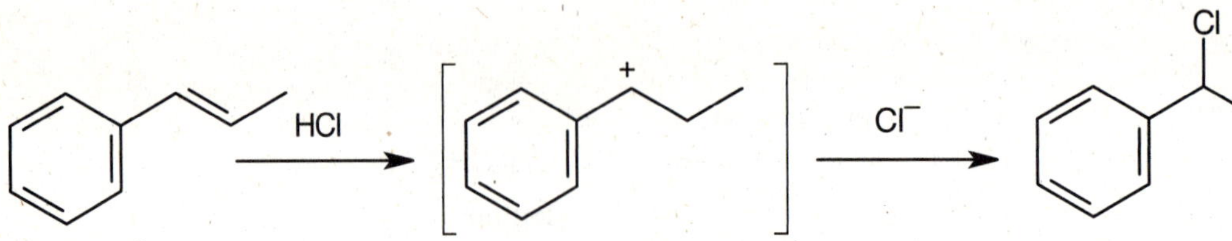

benzylic carbocation
intermediate

The benzylic carbocation intermediate is stabilized by resonance involving the aromatic ring (as shown in the previous problem), and addition to the double bond occurs at the position indicated. Addition to the aromatic ring doesn't occur.

**4.50** Resonance forms can't differ in the position of nuclei. The two structures in (a) are not resonance forms because the carbon and hydrogen atoms outside the ring occupy different positions in the two forms.

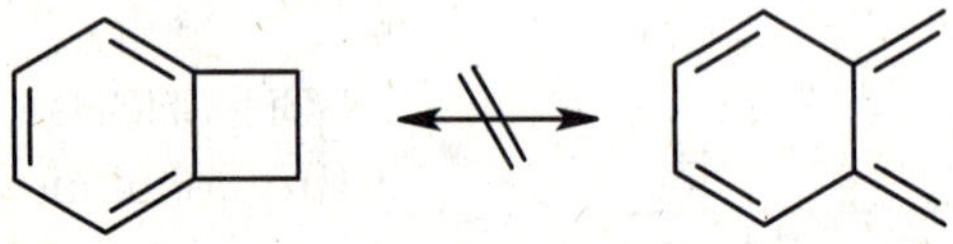

not resonance structures

**4.51**

(a)

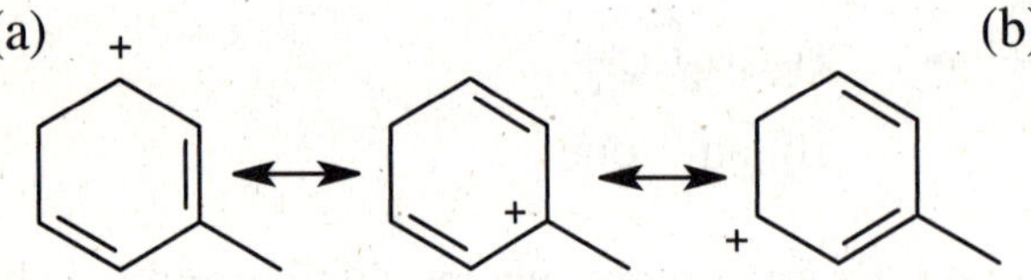

(b)

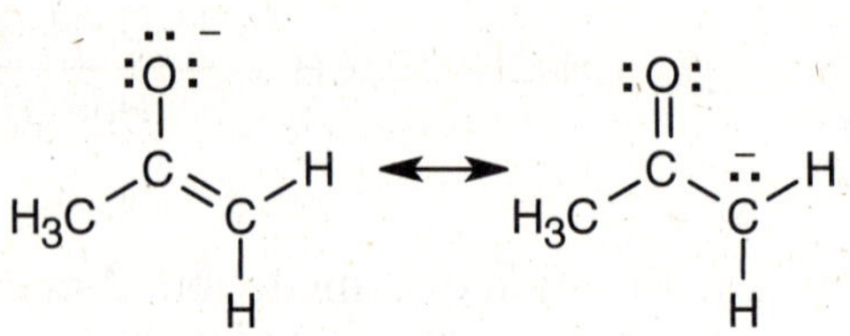

(c)

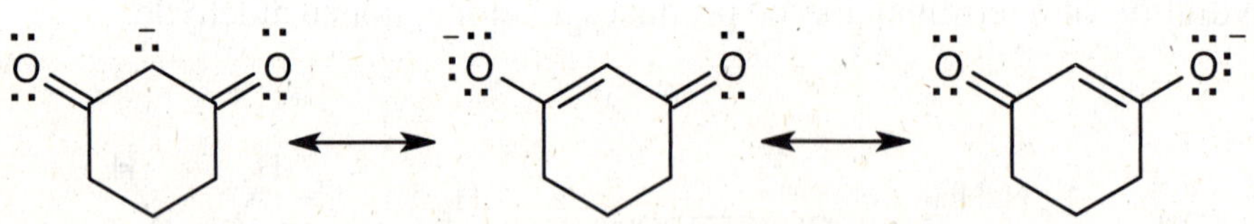

## Polymers

**4.52**

many

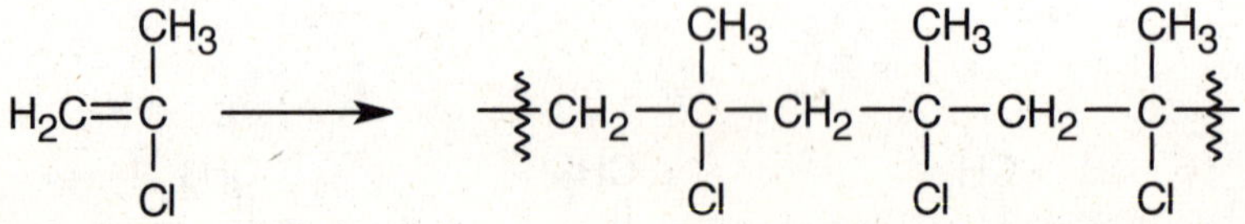

2-Chloropropene

**4.53**

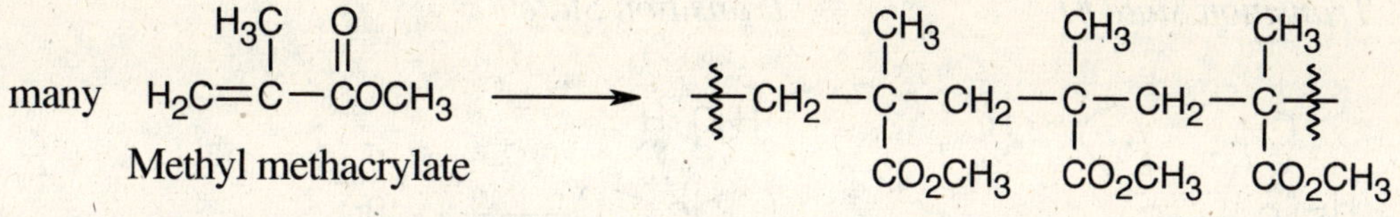


**4.54**

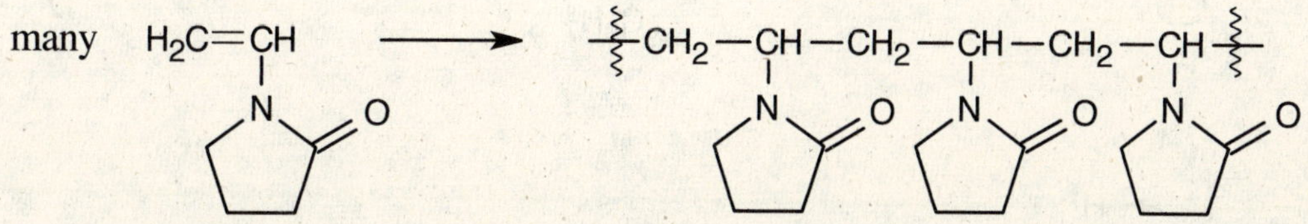


*N*-Vinylpyrrolidone

Poly(vinyl pyrrolidone)

## General Problems

**4.55**

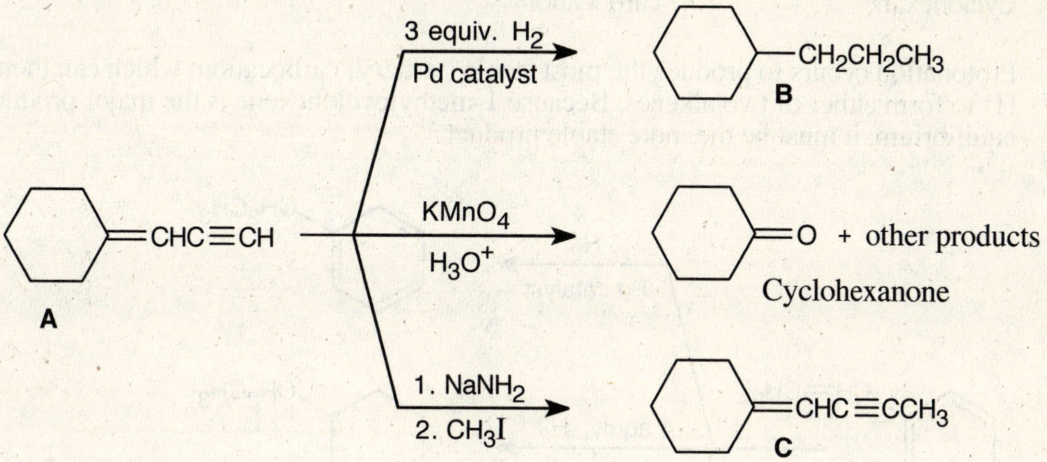


**4.56**

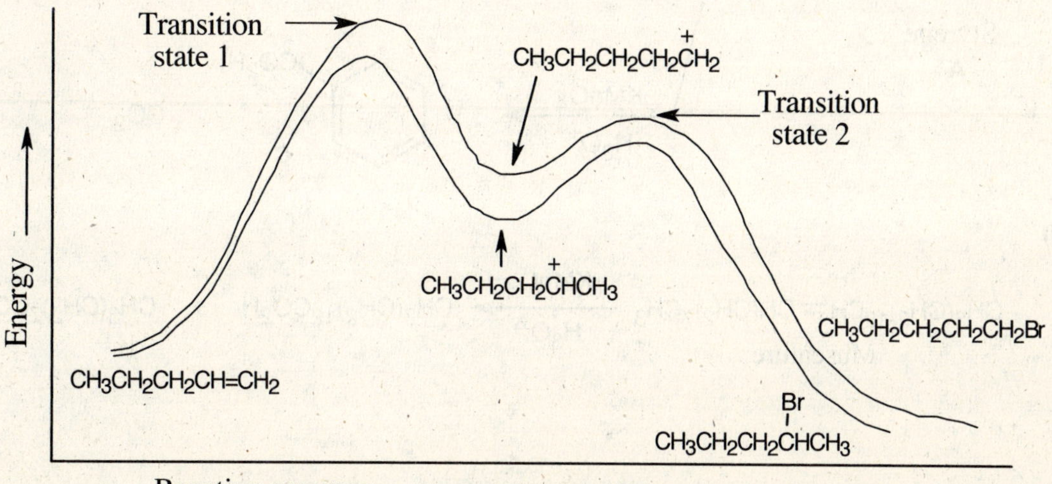

**4.57**

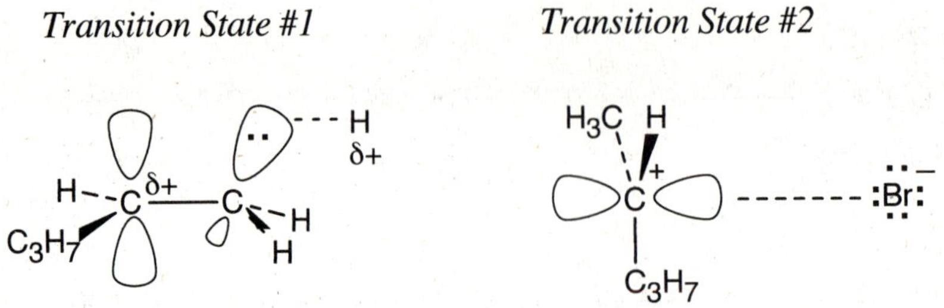


**4.58**

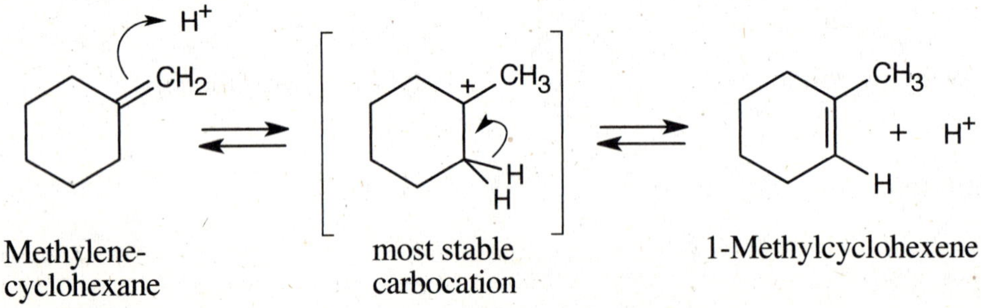


Protonation occurs to produce the most stable (tertiary) carbocation, which can then lose $H^+$ to form either of two alkenes. Because 1-methylcyclohexene is the major product of this equilibrium, it must be the more stable product.

**4.59**

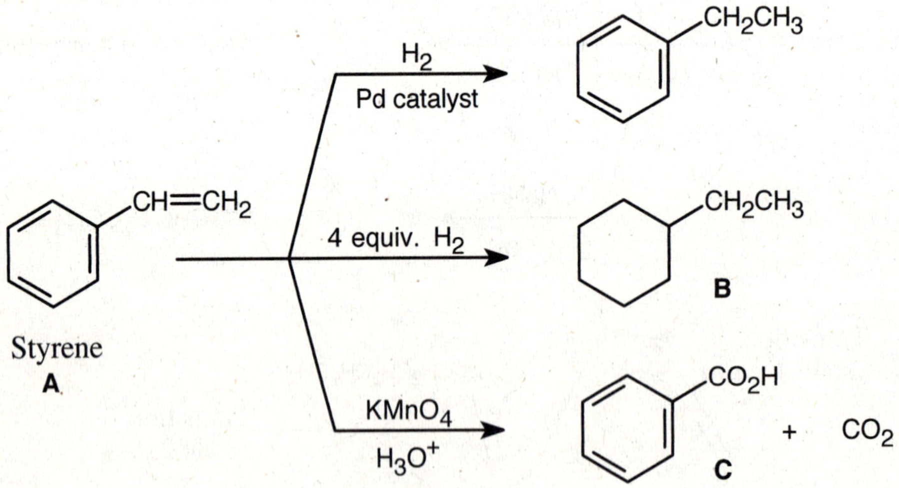


**4.60**

$$CH_3(CH_2)_{12}CH{=}CH(CH_2)_7CH_3 \xrightarrow[H_3O^+]{KMnO_4} CH_3(CH_2)_{12}CO_2H + CH_3(CH_2)_7CO_2H$$

Muscalure

**4.61**

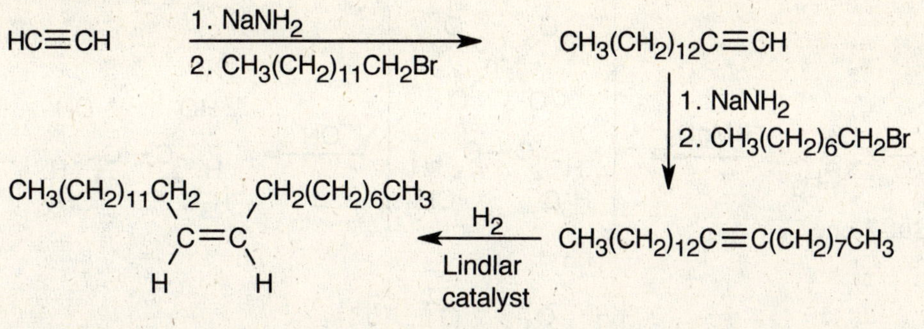


**4.62**

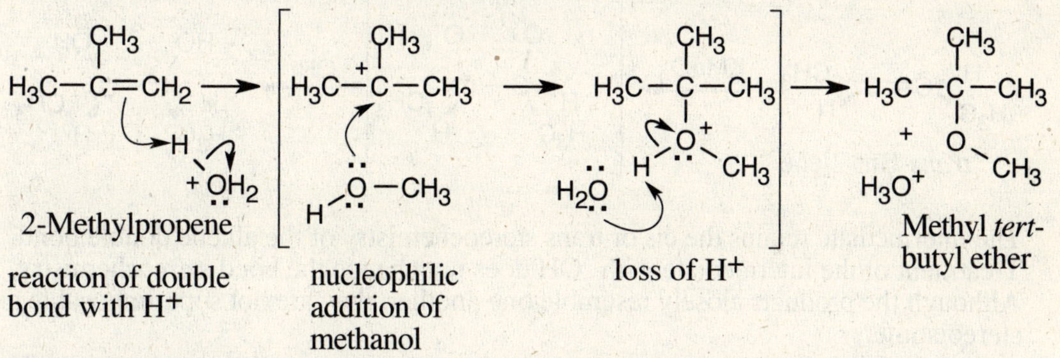


This mechanism differs from the mechanism of acid-catalyzed hydration in only one respect: Methanol substitutes for water in the second step.

**4.63**

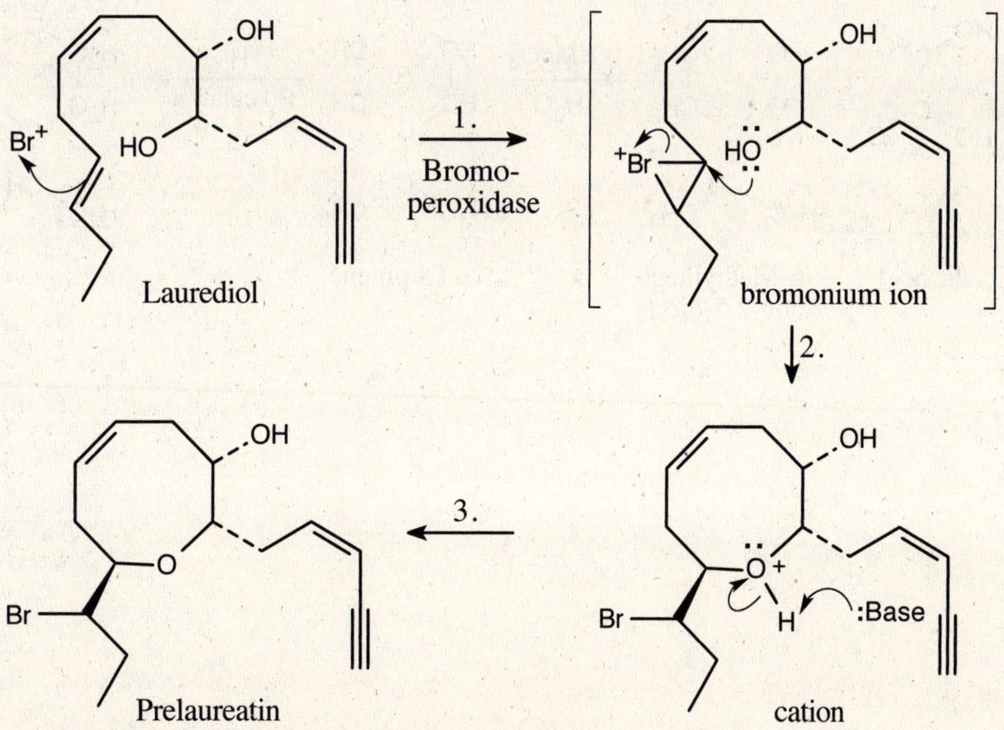


**Step 1:** Formation of a cyclic bromonium ion.
**Step 2:** Opening of bromonium ring by –OH.
**Step 3:** Removal of proton.

**4.64**

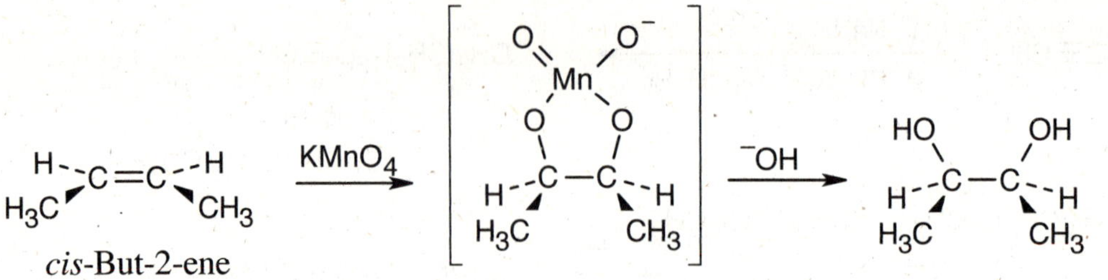

*cis*-But-2-ene

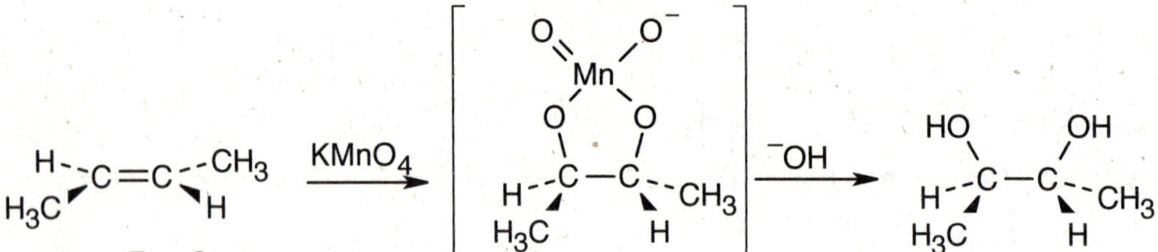

*trans*-But-2-ene

The intermediate retains the cis or trans stereochemistry of the alkene double bond. Treatment of the intermediate with $^{-}$OH does not change the bond stereochemistry. Although the products closely resemble one another, they are not superimposable and are stereoisomers.

**4.65**

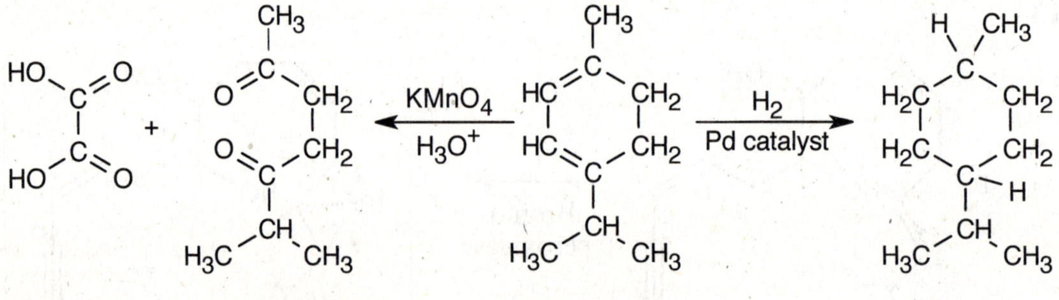

Oxalic acid    6-Methylheptane-2,5-dione    α-Terpinene

## In the Medicine Cabinet

**4.66**

(a)

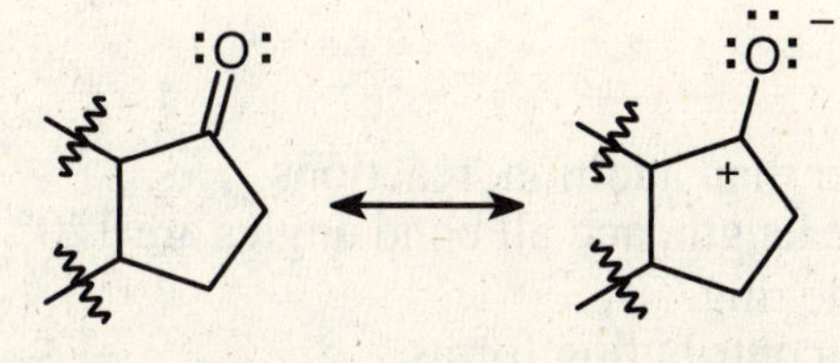

(b)

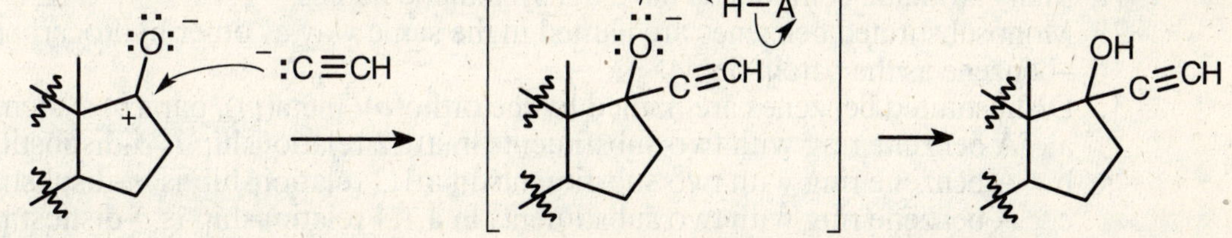


The negatively charged acetylide anion adds to the positively polarized carbon.

## In the Field

**4.67**

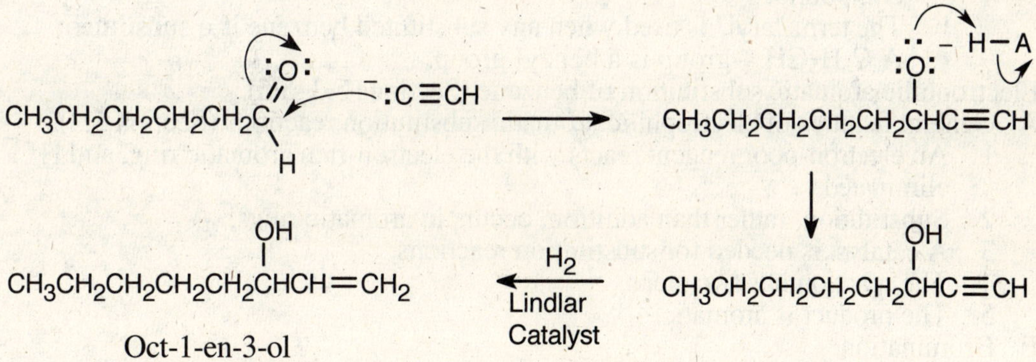


Oct-1-en-3-ol

The addition of acetylide occurs by the same route as shown in Problem 4.66.

**4.68**

One repeat unit of synthetic rubber:

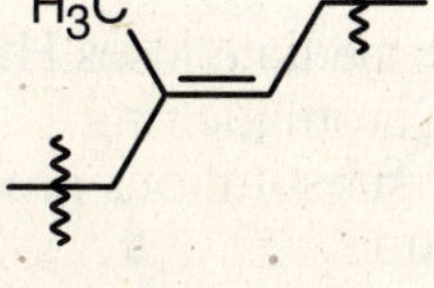


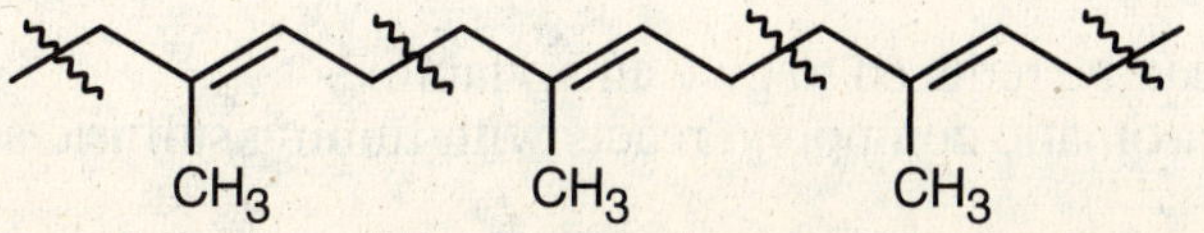


Synthetic Rubber

# Chapter 5 – Aromatic Compounds

## Chapter Outline

I. Aromatic compounds (Sections 5.1–5.2).
   A. Structure of benzene (Sections 5.1).
      1. Benzene undergoes substitution, rather than addition, reactions.
      2. All carbon–carbon bonds are the same length, and all bond angles are 120°.
      3. The $\pi$ electrons are delocalized over the ring.
      4. Benzene is a resonance hybrid of two contributing forms.
   B. Naming aromatic compounds (Section 5.2).
      1. Many aromatic compounds have nonsystematic names.
      2. Monosubstituted benzenes are named in the same way as other hydrocarbons, with –benzene as the parent name.
      3. Disubstituted benzenes are named by the ortho(*o*), meta(*m*), para(*p*) system.
         a. A benzene ring with two substituents in a 1,2 relationship is *o*-disubstituted.
         b. A benzene ring with two substituents in a 1,3 relationship is *m*-disubstituted.
         c. A benzene ring with two substituents in a 1,4 relationship is *p*-disubstituted.
      4. Benzenes with more than two substituents are named by numbering the position of each substituent.
         a. Number so that the lowest possible combination of numbers is used.
         b. Substituents are listed alphabetically.
      5. Aromatic groups.
         a. The term "phenyl" is used when a benzene ring is a substituent of another compound.
         b. The term "aryl" is used when any substituted benzene is a substituent.
         c. A $C_6H_5CH_2$– group is a benzyl group.

II. Electrophilic aromatic substitution of benzene (Sections 5.3–5.5).
   A. General features of electrophilic aromatic substitution reactions (Section 5.3).
      1. An electron-poor reagent reacts with the electron-rich aromatic ring, and $H^+$ is eliminated.
      2. Substitution, rather than addition, occurs in aromatic rings.
      3. A catalyst is needed for substitution reactions.
      4. The reaction has two steps.
      5. The product is aromatic.
   B. Bromination.
      1. Mechanism of bromination.
         a. $Br_2$ complexes with $FeBr_3$.
         b. The polarized electrophile is attacked by the $\pi$ electrons of the ring in a slow, rate-limiting step.
         c. The cation intermediate is doubly allylic but is much less stable than the starting aromatic compound.
         d. The carbocation intermediate loses $H^+$ from the bromine-bearing carbon in a fast step to regenerate an aromatic ring.
   C. Other electrophilic aromatic substitutions proceed by similar mechanisms (Section 5.4).
      1. Chlorination and iodination.
      2. Nitration results when an aromatic compound is treated with a mixture of $HNO_3$ and $H_2SO_4$.
         a. The product of nitration can be reduced to give an arylamine.
      3. Sulfonation results when an aromatic compound reacts with fuming sulfuric acid.

D. Friedel–Crafts alkylation and acylation (Section 5.5).
  1. The Friedel–Crafts alkylation reaction introduces an alkyl group onto an aromatic ring.
    a. An alkyl chloride, plus an $AlCl_3$ catalyst, produces an electrophilic carbocation that loses $H^+$ to yield an alkylated aromatic ring.
    b. Only alkyl halides – not aryl or vinylic halides – can be used.
    c. Friedel–Crafts reactions don't succeed on rings that have amino substituents or deactivating groups.
  2. Friedel–Crafts acylation occurs when an aromatic ring reacts with a carboxylic acid chloride in the presence of $AlCl_3$.

III. Substituent effects in electrophilic aromatic substitution (Sections 5.6–5.7).
  A. General effects (Section 5.6).
    1. Substituents affect the reactivity of an aromatic ring.
    2. Substituents affect the orientation of further substitution.
    3. Substituents can be arranged in three groups:
      a. Ortho/para-directing activators.
      b. Ortho/para-directing deactivators.
      c. Meta-directing deactivators.
  B. Explanation of effects (Section 5.7).
    1. Activation and deactivation.
      a. All activating groups donate electrons to an aromatic ring.
      b. All deactivating groups withdraw electrons from a ring.
      c. Electron donation or withdrawal may be due to inductive or resonance effects.
    2. Orienting effects.
      a. Ortho, para directors.
        i. The intermediates from *o,p* attack are more stable because they can be stabilized by resonance donation by the substituent.
      b. Any substituent with a positively polarized atom bonded to the ring is a meta director.

IV. Oxidation and reduction of aromatic compounds (Section 5.8).
  A. Side-chain oxidation occurs at the benzylic position when an aromatic ring is treated with $KMnO_4$.
    1. Substituted benzoic acids are the products.
  B. Aromatic rings can be reduced by catalytic hydrogenation using a powerful catalyst.

V. Polycyclic aromatic hydrocarbons consist of two or more fused benzene rings (Section 5.9).
    1. Compounds such as pyridine and pyrrole, both heterocycles, are also aromatic.

VI. Organic synthesis (Section 5.10).
  A. To synthesize substituted benzenes, it is important to introduce groups so that they have the proper orienting effects.
  B. To plan a synthesis, work backward from the product (retrosynthetic analysis).

## Solutions to Problems

**5.1** According to resonance theory, *o*-dibromobenzene is not described properly by either of the two structures shown in Problem 5.1, but is a resonance hybrid of the two.

**5.2** Number the positions on the ring. Give number 1 to one substituent and the lowest possible number to the other substituent. Substituents at positions 1 and 2 have an ortho relationship; substituents at positions 1 and 3 have a meta relationship; substituents at positions 1 and 4 have a para relationship.

(a)

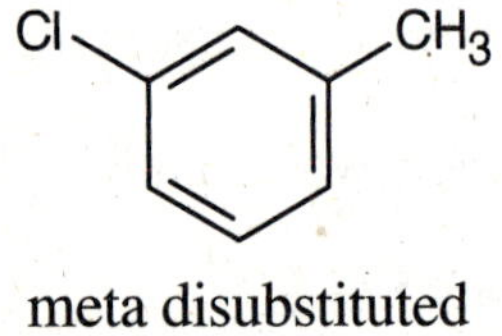


meta disubstituted

(b)

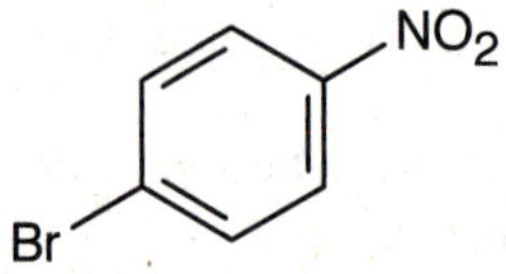


para disubstituted

(c)

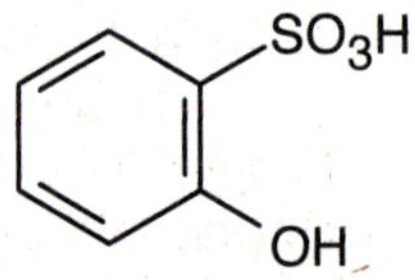


ortho disubstituted

**5.3** For compounds with two substituents, determine the ortho, meta, para relationship of the substituents, and cite them in alphabetical order.

(a)

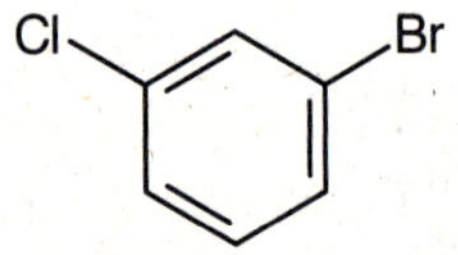


*m*-Bromochlorobenzene

(b)

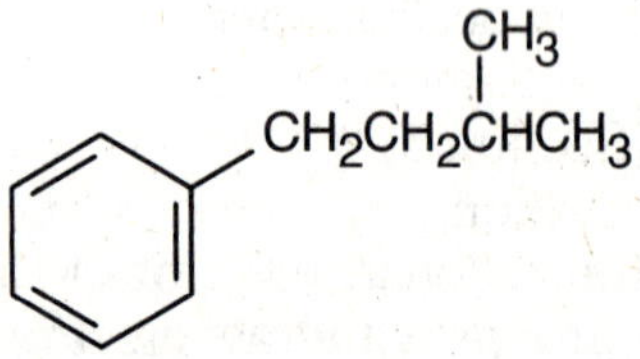


(3-Methylbutyl)benzene

(c)

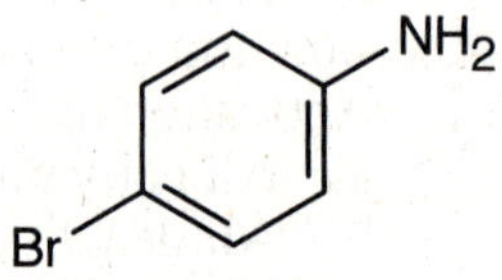


*p*-Bromoaniline

(d)

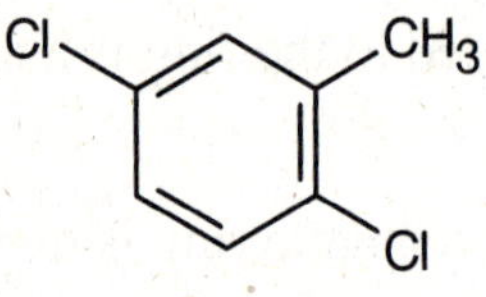


2,5-Dichlorotoluene

(e)

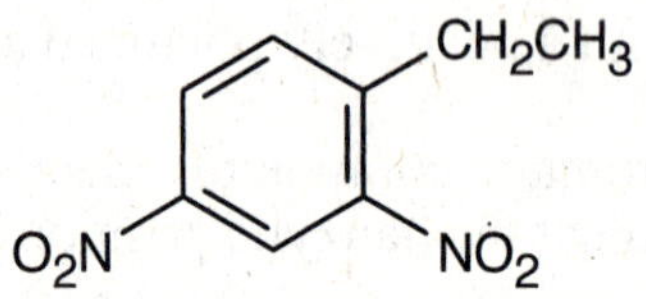


1-Ethyl-2,4-dinitrobenzene

(f)

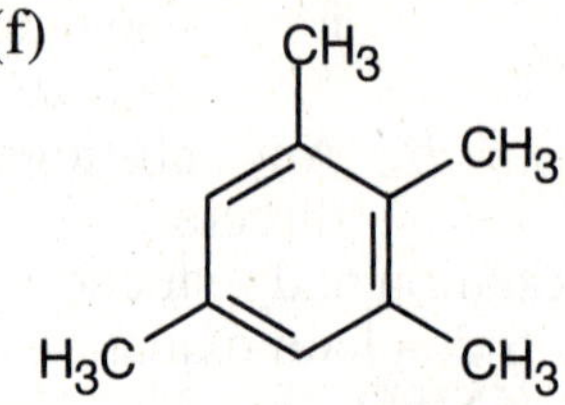


1,2,3,5-Tetramethylbenzene

**5.4**

(a)

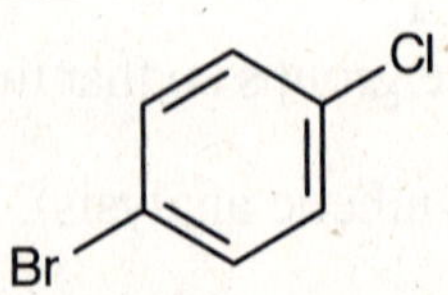


*p*-Bromochlorobenzene

(b)

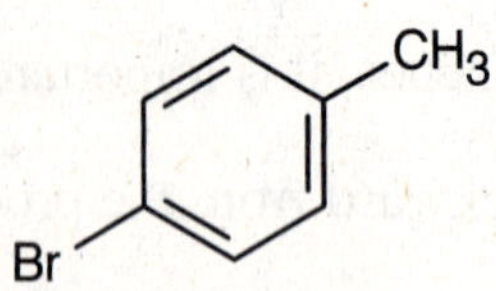


*p*-Bromotoluene

(c)

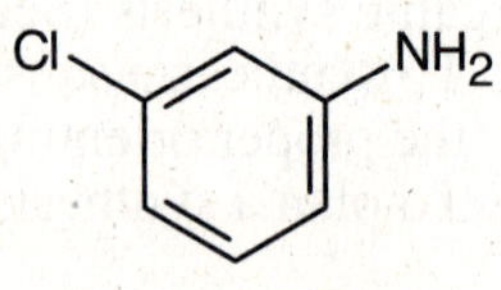


*m*-Chloroaniline

(d)

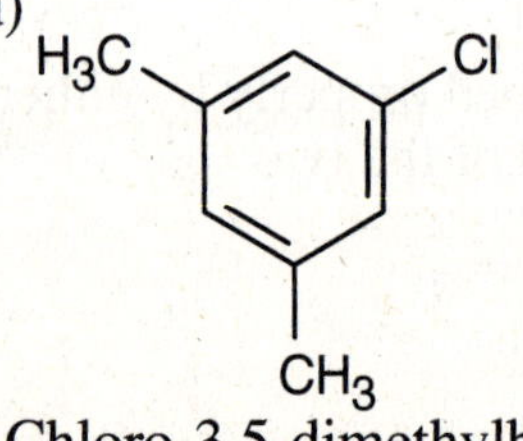


1-Chloro-3,5-dimethylbenzene

**5.5**

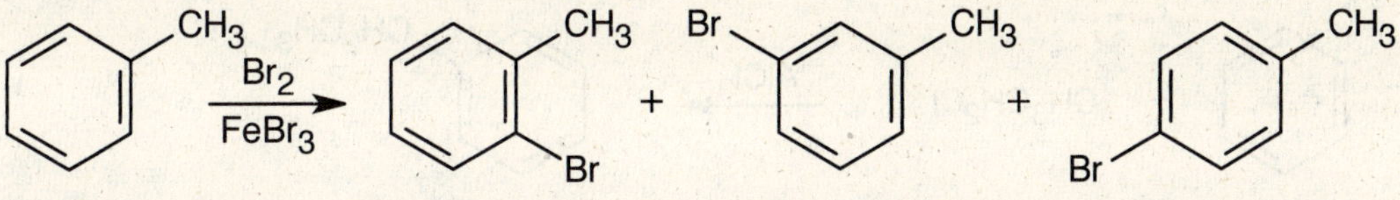

Toluene $\quad$ *o*-Bromotoluene $\quad$ *m*-Bromotoluene $\quad$ *p*-Bromotoluene

**5.6** The mechanism of nitration is the same as the mechanism of other electrophilic aromatic substitution mechanisms we have studied. The nitronium ion is formed by treating nitric acid with sulfuric acid

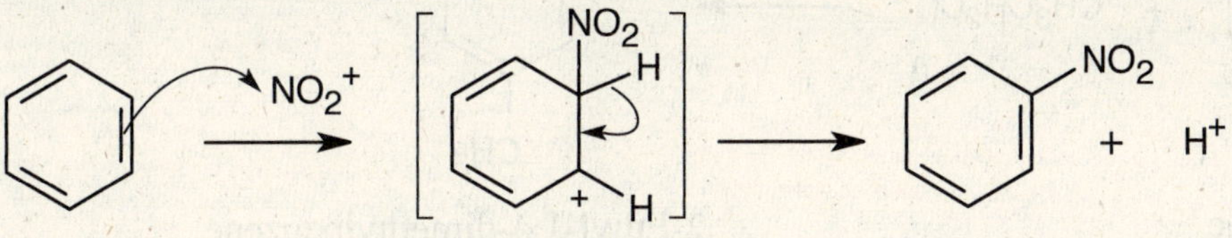

carbocation intermediate

**5.7**

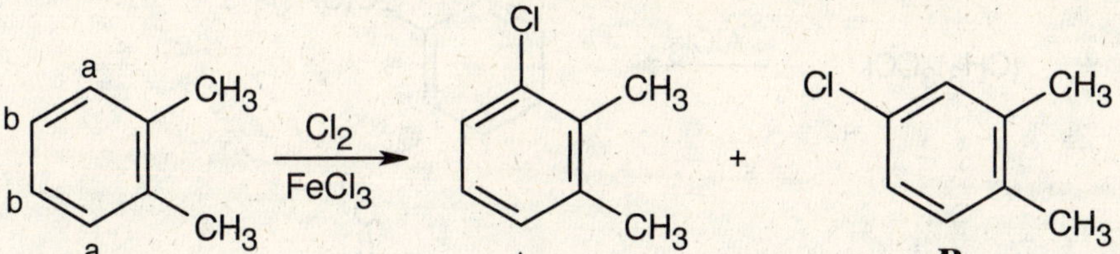

A $\quad$ B

*o*-Xylene

Chlorination at position "a" of *o*–xylene yields product **A**; chlorination at position "b" yields product **B**.

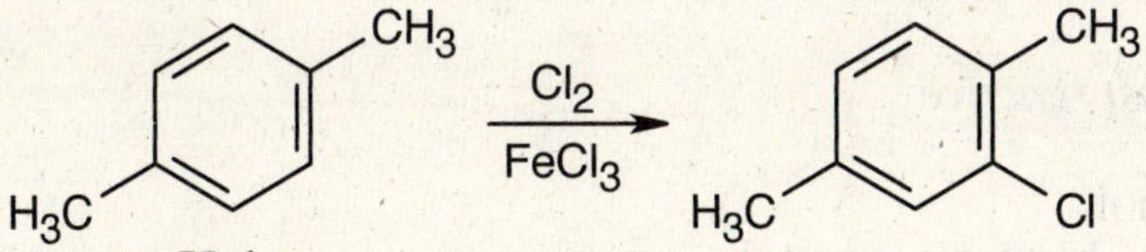

*p*-Xylene

Only one product results from chlorination of *p*–xylene because all possible sites for chlorination are equivalent.

**5.8**

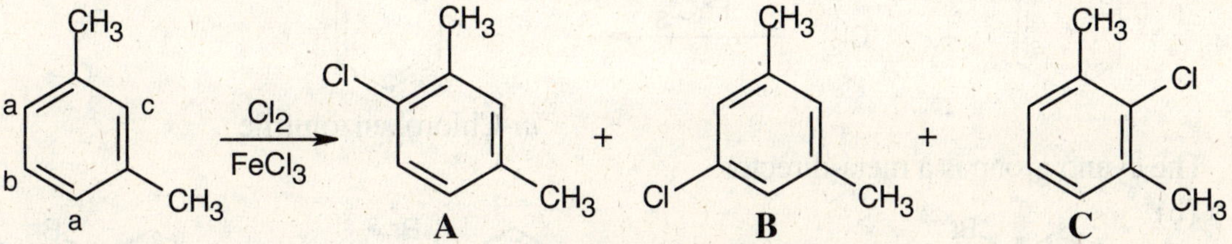

A $\quad$ B $\quad$ C

*m*-Xylene

Three products might form on chlorination of *m*-xylene. Product **C** is unlikely to form because substitution rarely occurs between two substituents with a meta relationship for steric reasons.

**5.9** Both reactions are Friedel–Crafts alkylations.

(a)

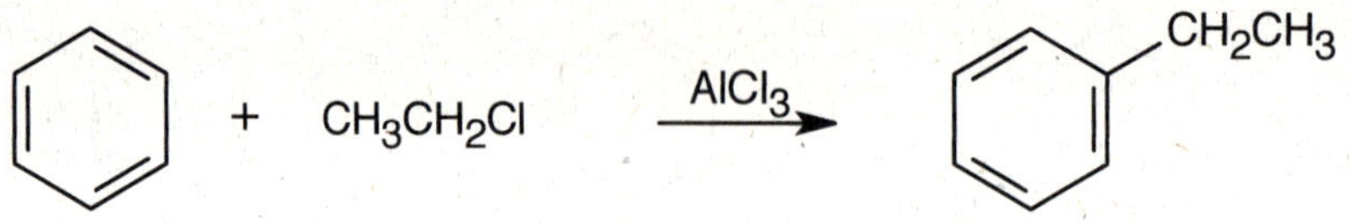


Ethylbenzene

(b)

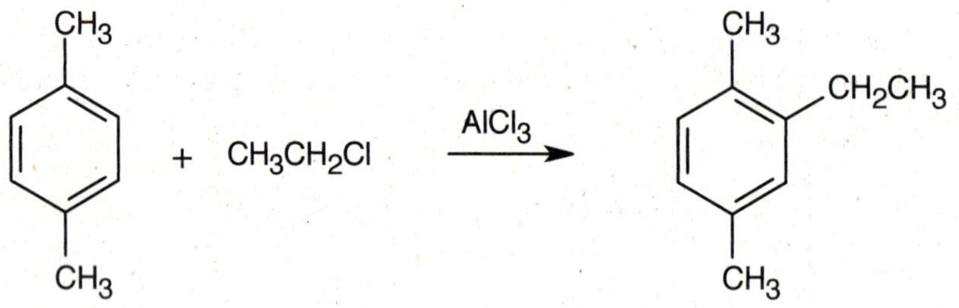


*p*-Xylene 2-Ethyl-1,4-dimethylbenzene

**5.10**

(a)

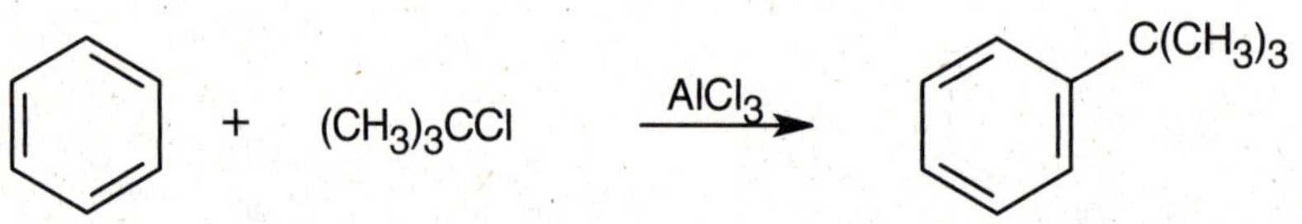


(b)

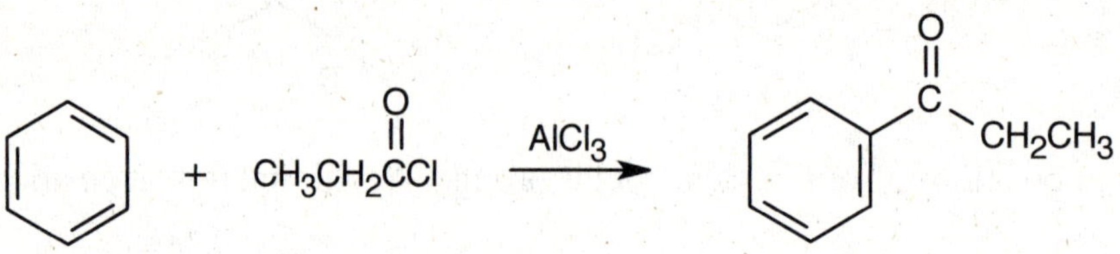


**5.11** Figure 5.6 lists groups in order of the strength of their activating or deactivating effects.

*Least reactive* ---> *Most reactive*

(a) Nitrobenzene, toluene, phenol
(b) Benzoic acid, chlorobenzene, benzene, phenol
(c) Benzaldehyde, bromobenzene, benzene, aniline

**5.12**

(a)

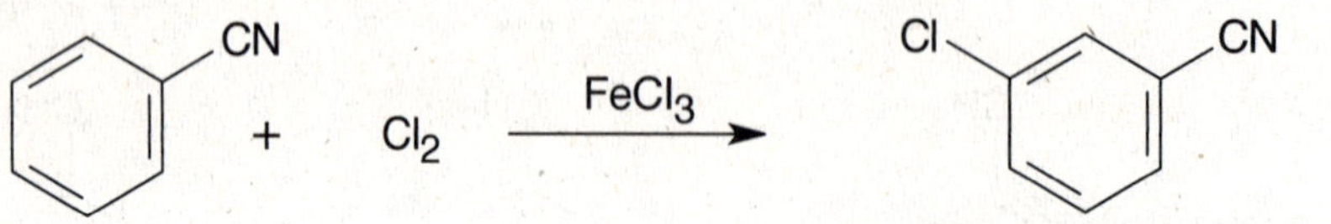


*m*-Chlorobenzonitrile

The cyano group is a meta-director.

(b)

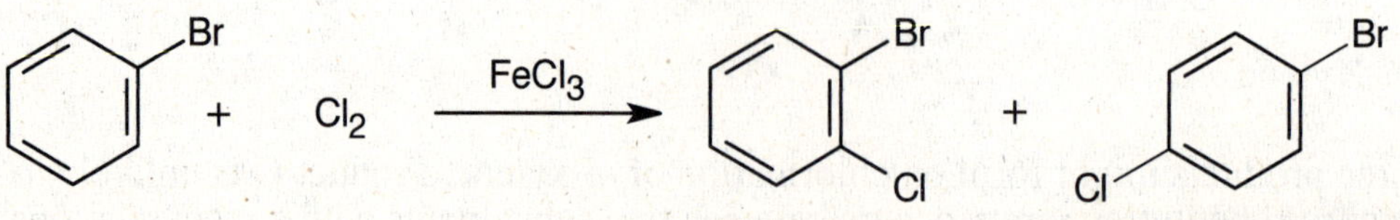


*o*-Bromochlorobenzene *p*-Bromochlorobenzene

The bromo group is an ortho-para director.

**5.13** The substituent already present determines the orientation of the second substituent.

(a)

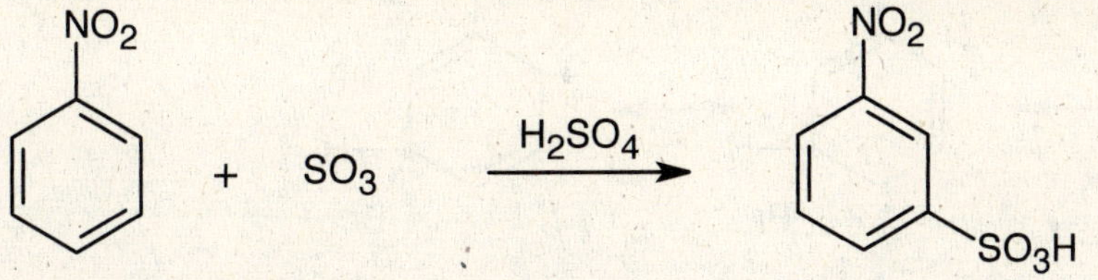


(b)

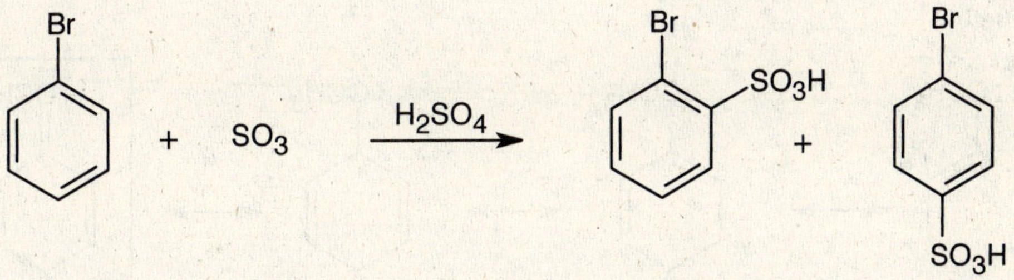


(c)

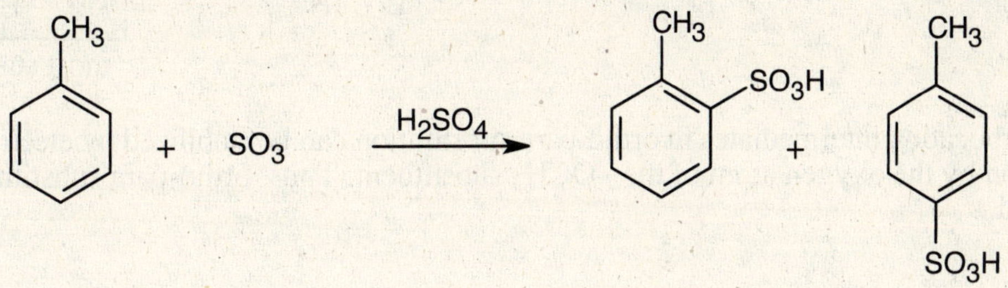


(d)

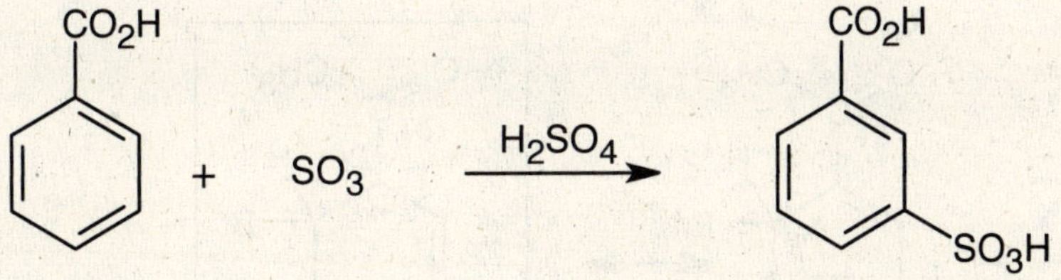


(e)

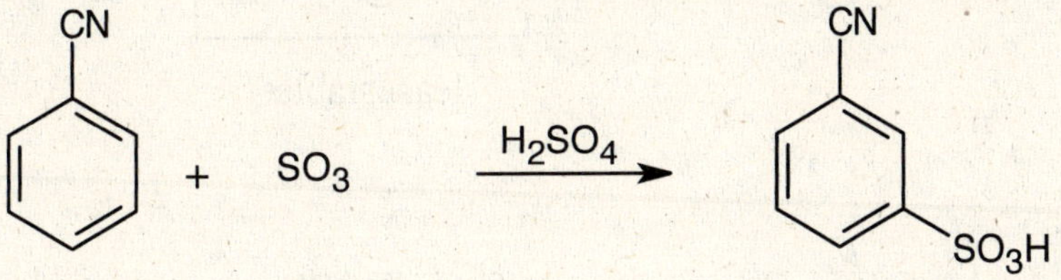


**5.14**

(a) Ortho attack:

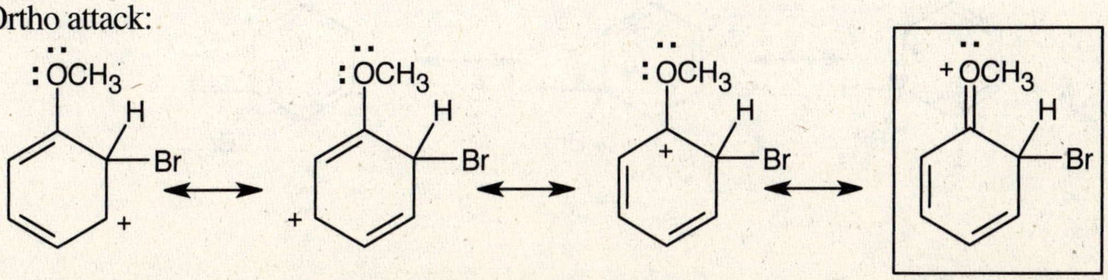


most stable

(b) Meta attack:

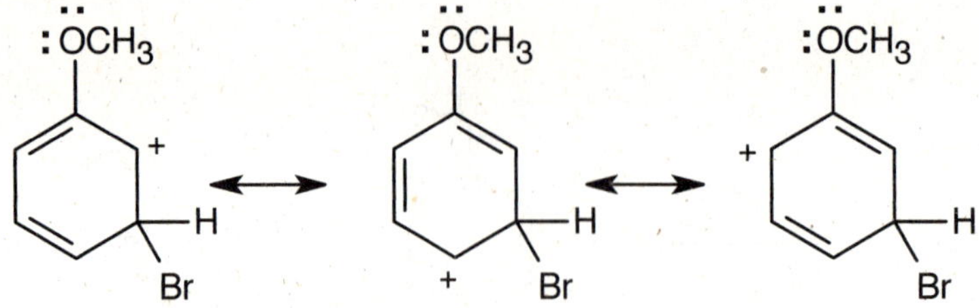

(c) Para attack:

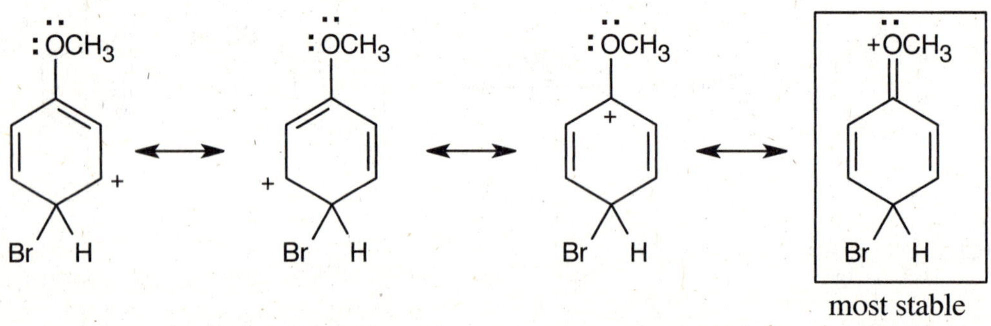

The carbocation intermediates in ortho-para substitution can be stabilized by electron donation by the oxygen atom of the $-OCH_3$ substituent. Thus, ortho-para substitution is favored.

**5.15**

(a) Ortho attack:

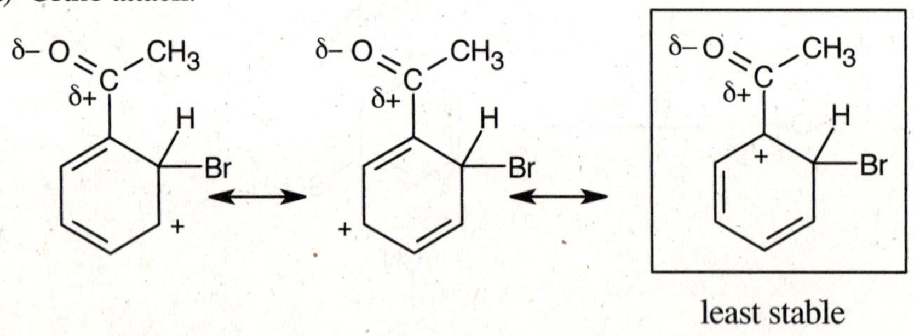

(b) Meta attack:

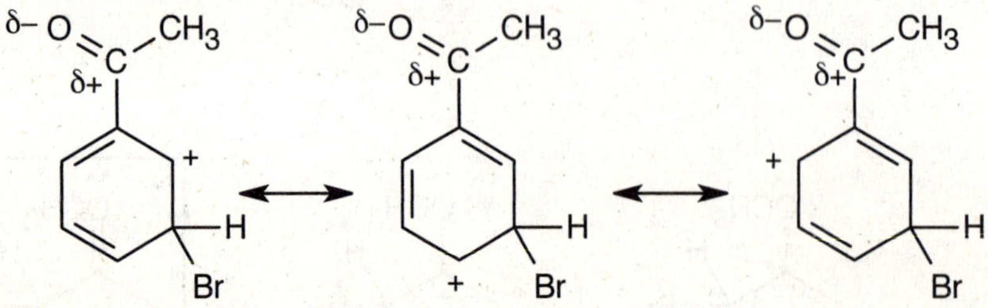

(c) Para attack:

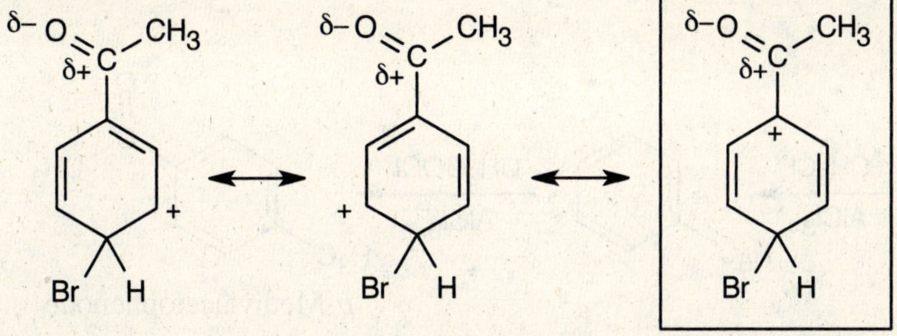

least stable

The indicated carbocation intermediates of ortho and para attack are least stable because they place a positive charge next to a positively polarized carbon atom. Thus, meta substitution is favored.

**5.16** Oxidation takes place at the benzylic position.

(a)

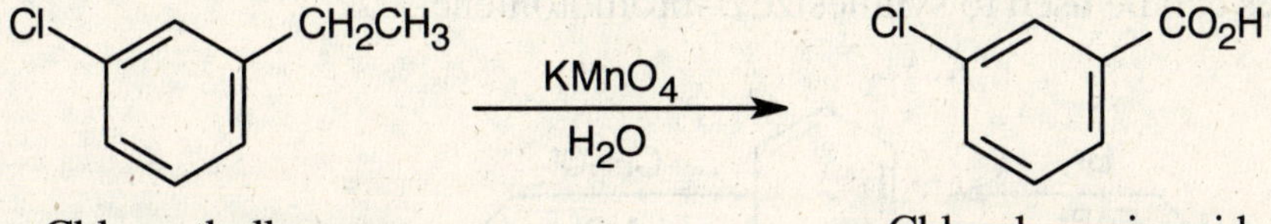

*m*-Chloroethylbenzene → *m*-Chlorobenzoic acid

(b)

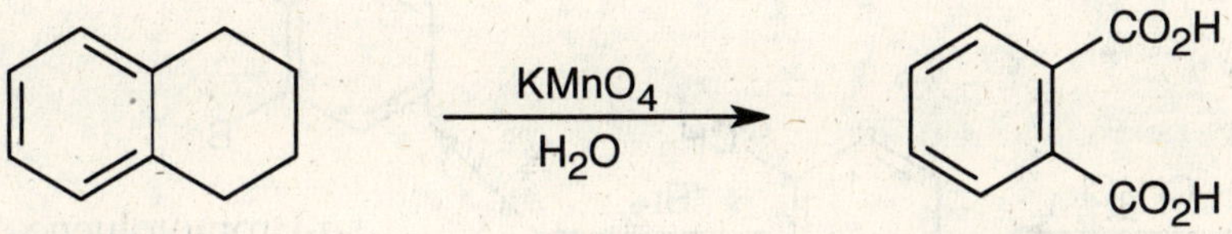

Tetralin → *o*-Benzenedicarboxylic acid (Phthalic acid)

**5.17**

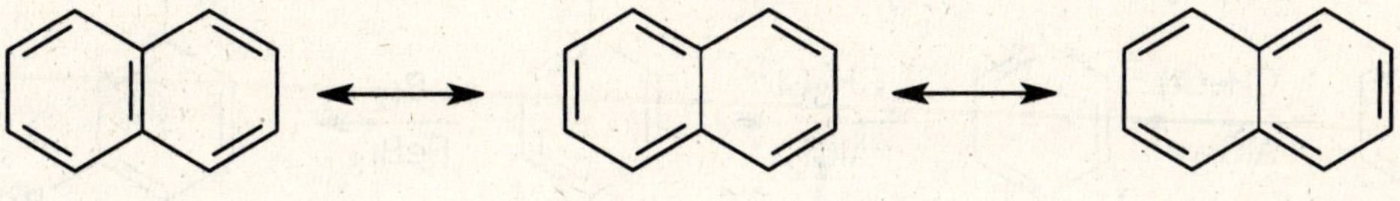

**5.18** In order to synthesize the desired product, you must perform the reactions in the correct order. Assume that you can separate ortho and para isomers.

(a)

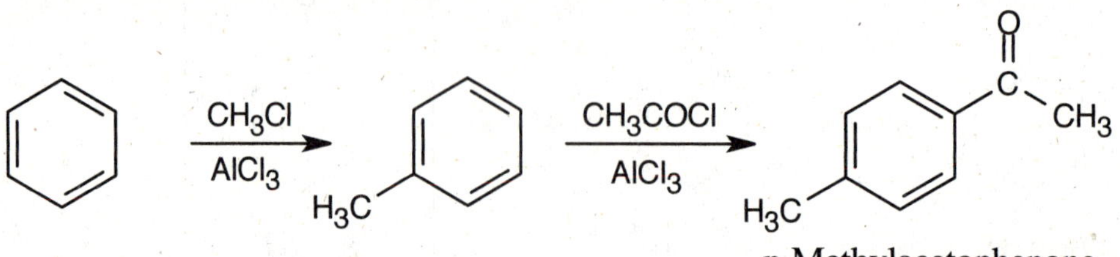


*p*-Methylacetophenone

(b)

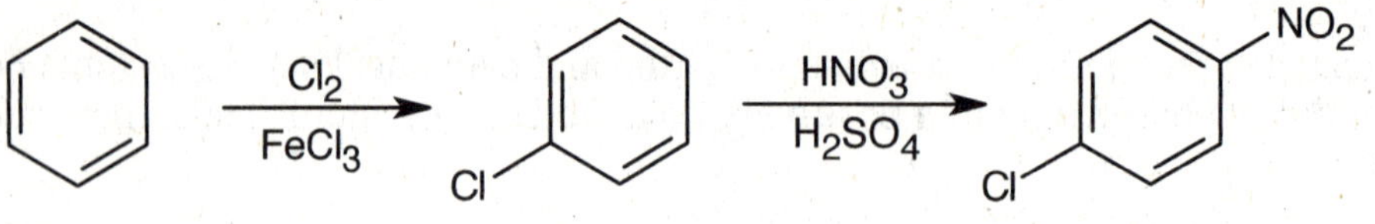


*p*-Chloronitrobenzene

**5.19** (a) Two routes can be used to synthesize *o*–bromotoluene.

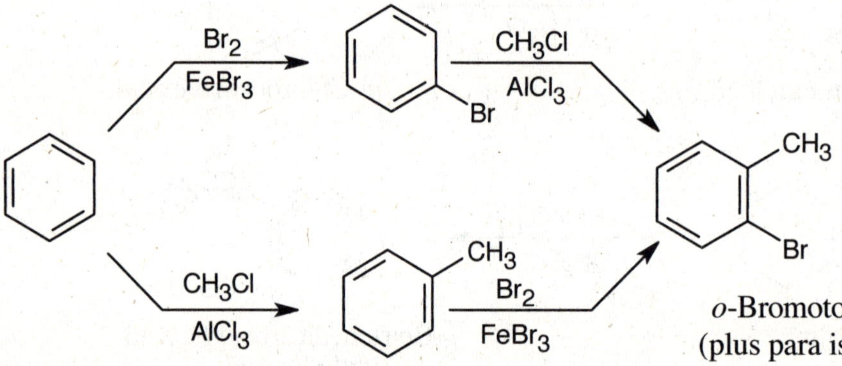


*o*-Bromotoluene (plus para isomer)

(b)

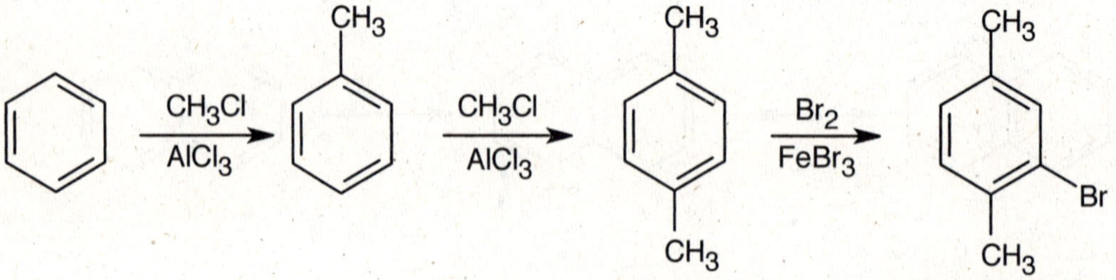


(plus ortho isomer)

2-Bromo-1,4-dimethylbenzene

**5.20**

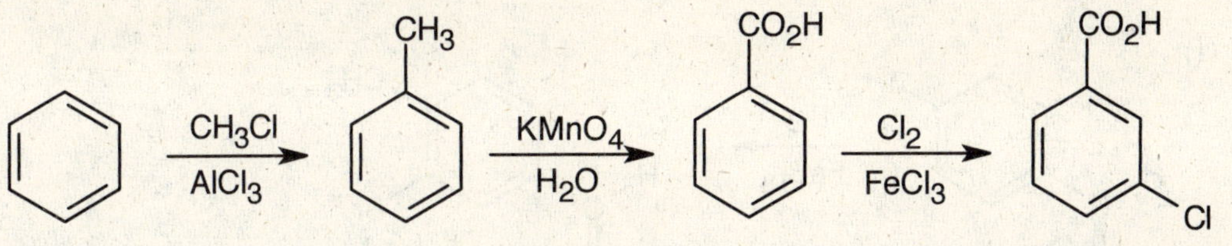

*m*-Chlorobenzoic acid

If the ring is chlorinated first, the functional groups in the product will have an *o,p* orientation.

## Visualizing Chemistry

**5.21** Three resonance forms for the carbocation of the formula $C_{13}H_9$ are shown below, and more can be drawn. These forms show that the positive charge of the carbocation can be stabilized in the same way as an allylic or benzylic carbocation is stabilized – by overlap with the neighboring π electrons of the ring system.

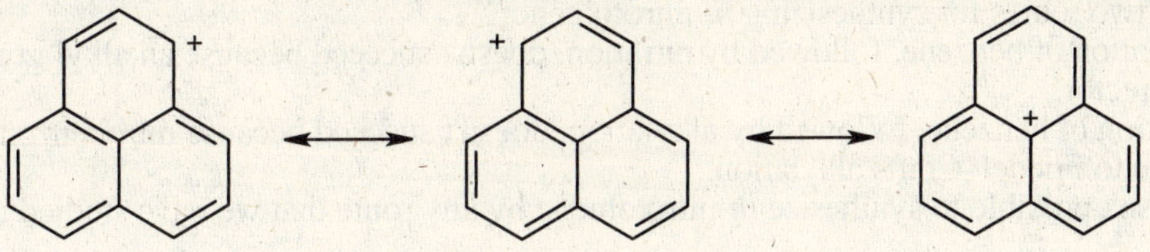

**5.22**

(a)

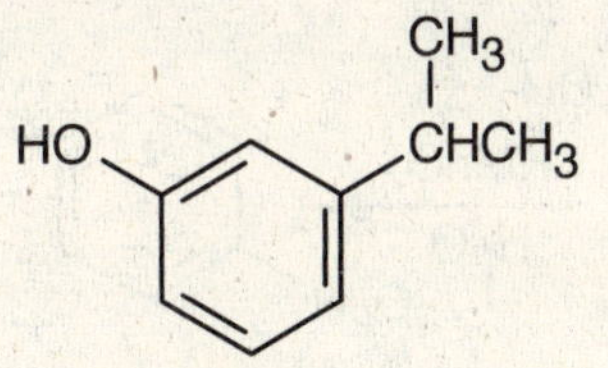

*m*-Isopropylphenol

(b)

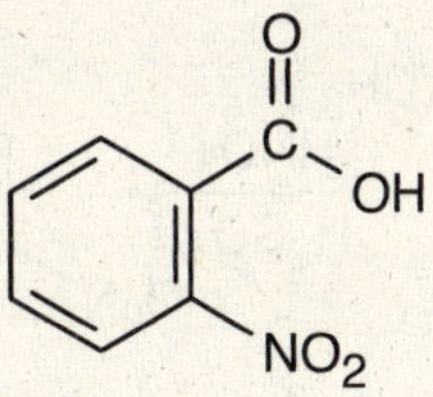

*o*-Nitrobenzoic acid

**5.23** (a) The methoxyl group is an *o,p*-director.

(i)

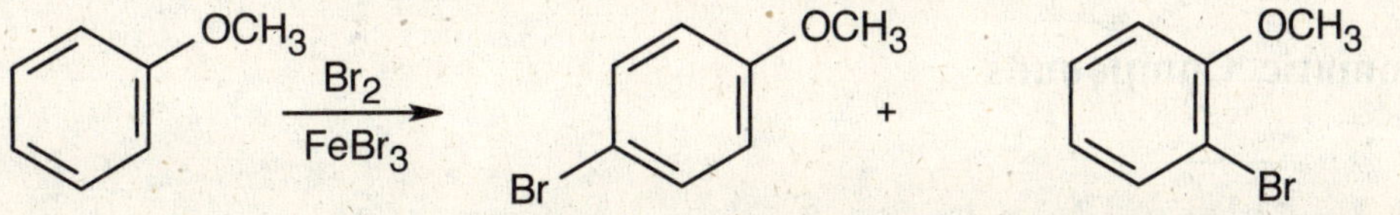

*p*-Bromomethoxybenzene *o*-Bromomethoxybenzene

(ii)

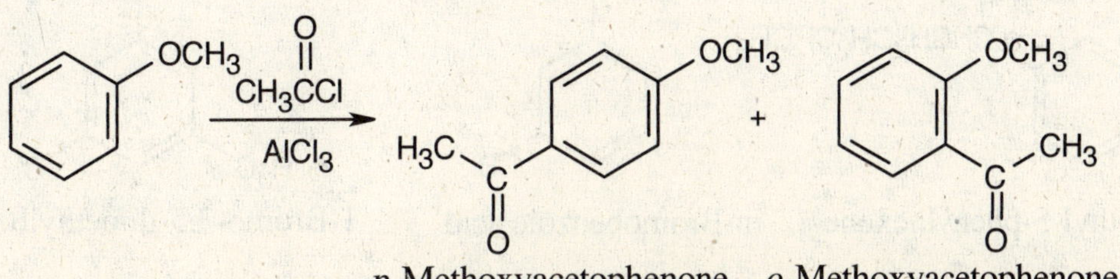

*p*-Methoxyacetophenone *o*-Methoxyacetophenone

(b) Both functional groups direct substituents to the same position.

(i)

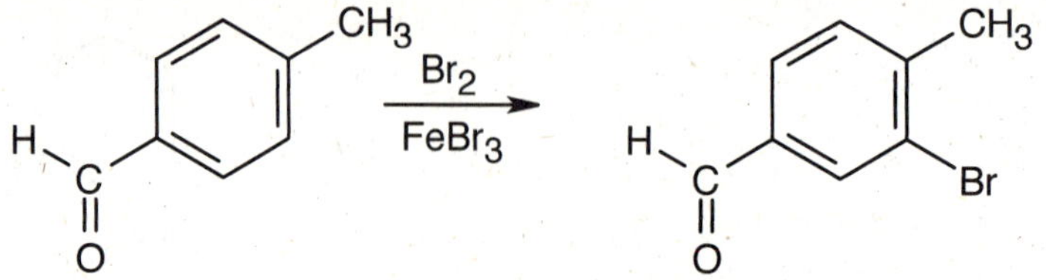

3-Bromo-4-methylbenzaldehyde

(ii)

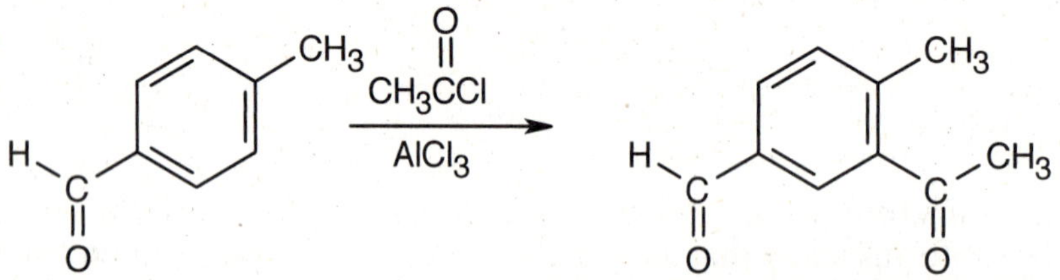

3-Acetyl-4-methylbenzaldehyde

**5.24** Imagine two routes for synthesizing *m*-nitrotoluene:
(1) Alkylation of benzene, followed by nitration, doesn't succeed because an alkyl group is an *o,p*-director.
(2) Nitration of benzene, followed by alkylation, doesn't succeed because nitrobenzene is unreactive to Friedel–Crafts alkylation.
Thus, it isn't possible to synthesize *m*-nitrotoluene by any route that we have studied in this chapter.

**5.25**

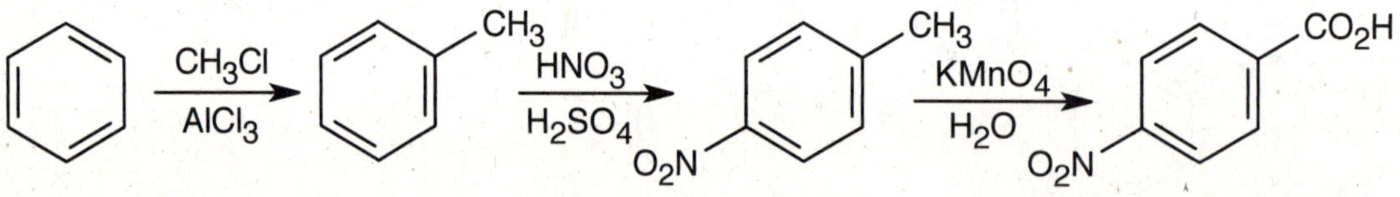

*p*-Nitrobenzoic acid

Groups must be introduced in the correct order in order to arrive at the desired product.

## Additional Problems

### Naming Aromatic Compounds

**5.26**

(a)

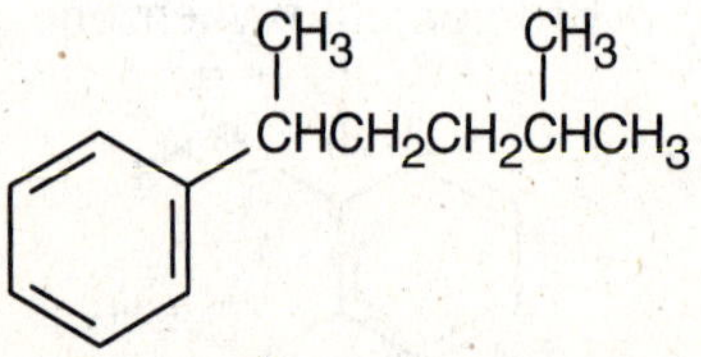

2-Methyl-5-phenylhexane

(b)

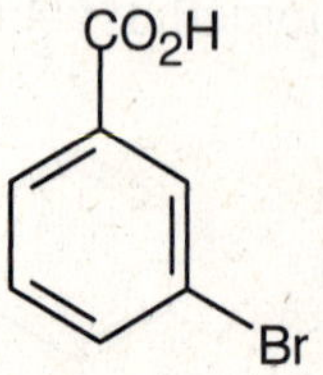

m-Bromobenzoic acid

(c)

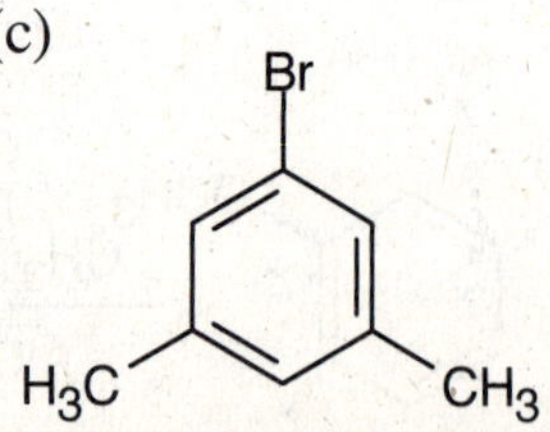

1-Bromo-3,5-dimethylbenzene

(d)

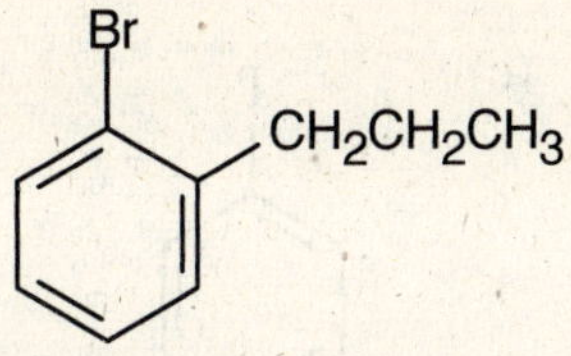

o-Bromopropylbenzene

(e)

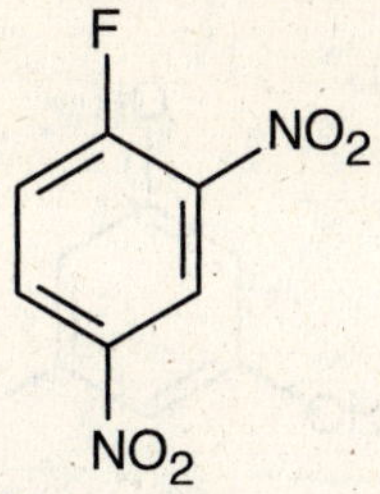

1-Fluoro-2,4-dinitrobenzene

(f)

p-Chloroaniline

**5.27**

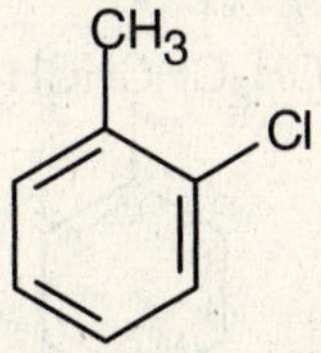

*o*-Chlorotoluene

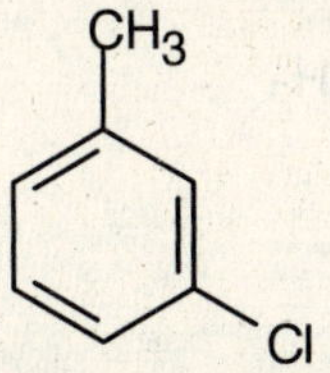

*m*-Chlorotoluene

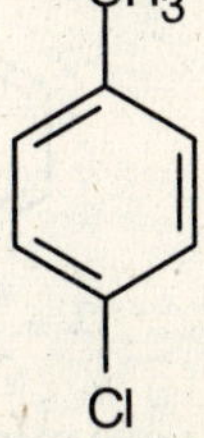

*p*-Chlorotoluene

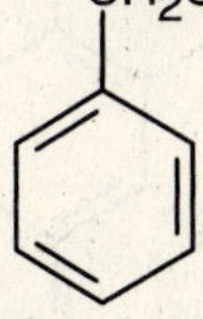

Benzyl chloride *or* (Chloromethyl)-benzene

**5.28**

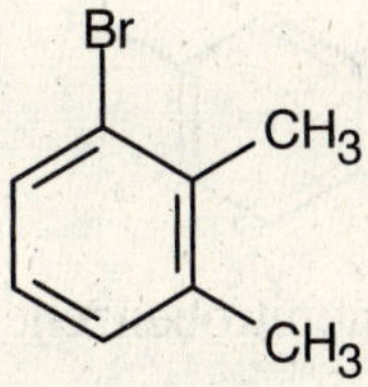

1-Bromo-2,3-dimethyl-benzene

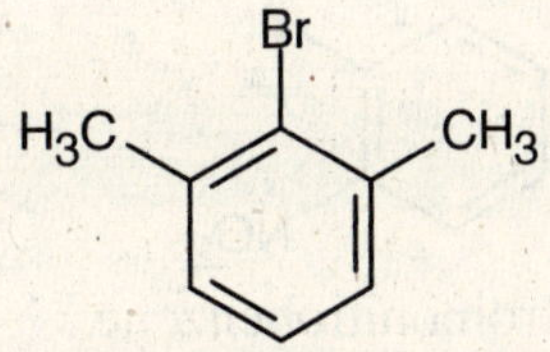

2-Bromo-1,3-dimethyl-benzene

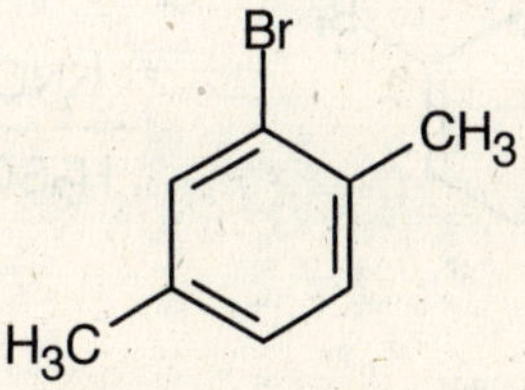

2-Bromo-1,4-dimethyl-benzene

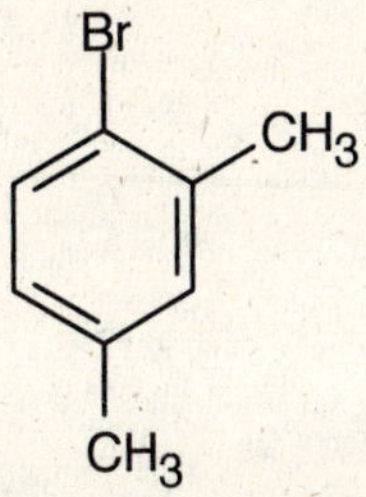

1-Bromo-2,4-dimethyl-benzene

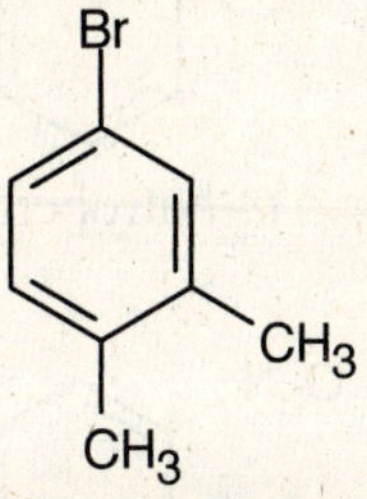

4-Bromo-1,2-dimethyl-benzene

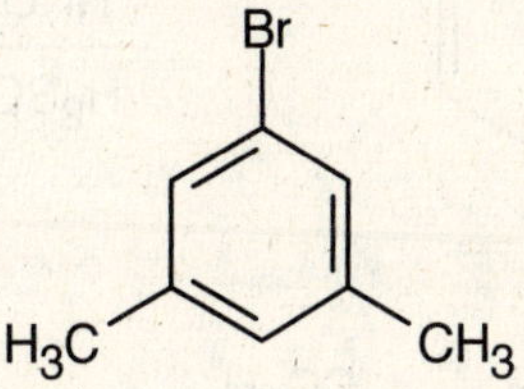

1-Bromo-3,5-dimethyl-benzene

**5.29**

(a)

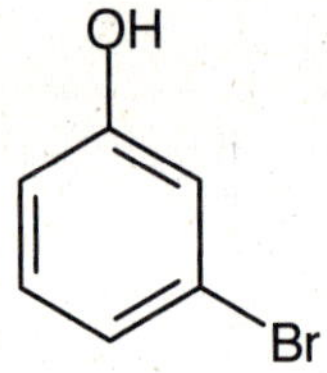


*m*-Bromophenol

(b)

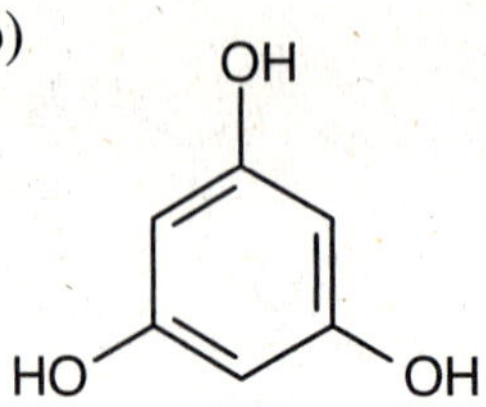


Benzene-1,3,5-triol

(c)

*p*-Iodonitrobenzene

(d)

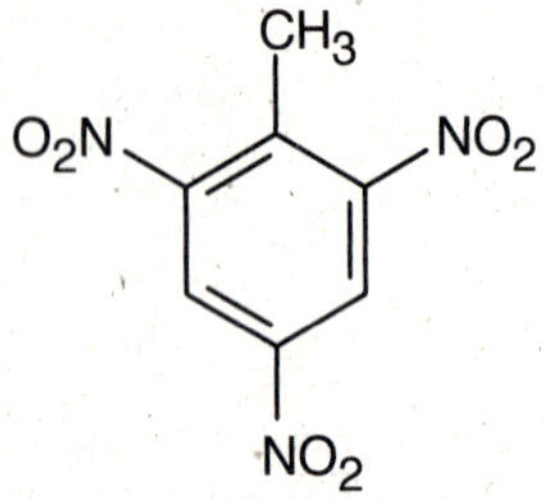


2,4,6-Trinitrotoluene

(e)

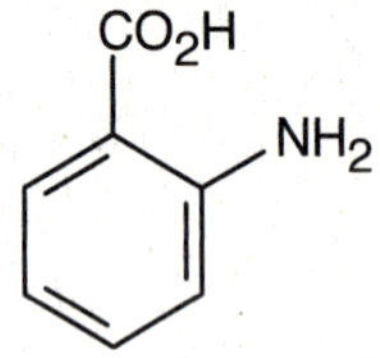


*o*-Aminobenzoic acid

(f)

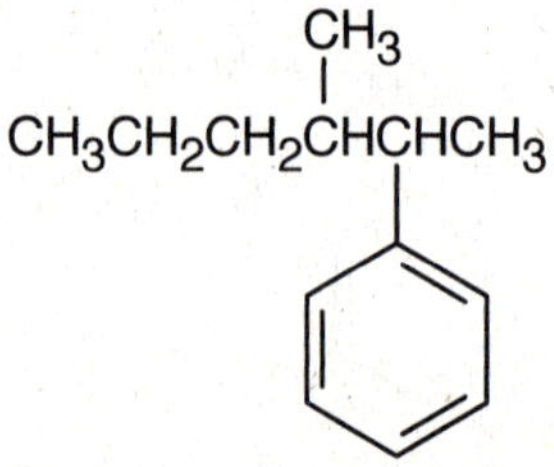


3-Methyl-2-phenylhexane

## Reactions and Substituent Effects

**5.30**

(a)

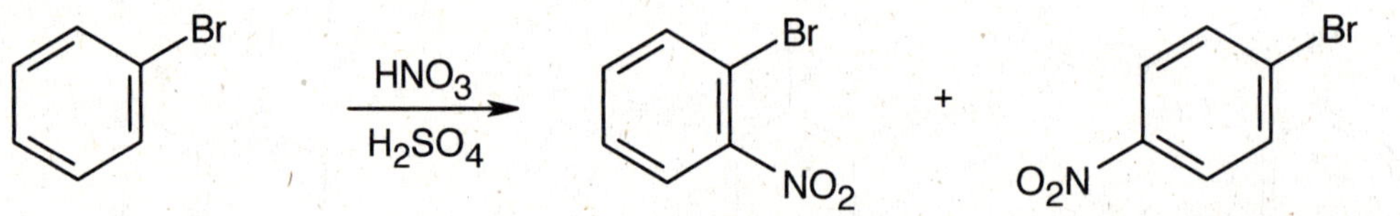


*o*-Bromonitrobenzene *p*-Bromonitrobenzene

(b)

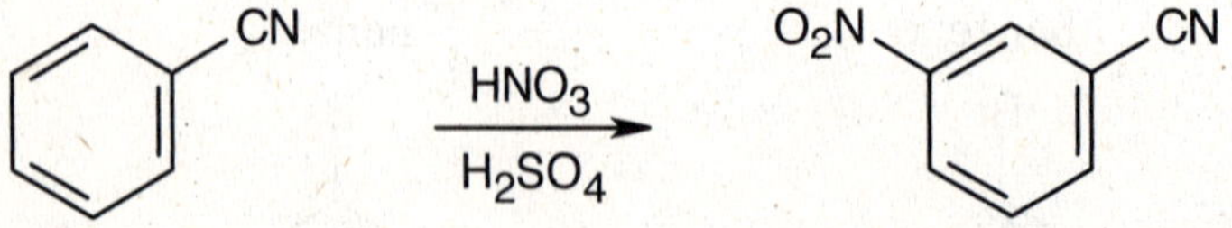


*m*-Nitrobenzonitrile

(c)

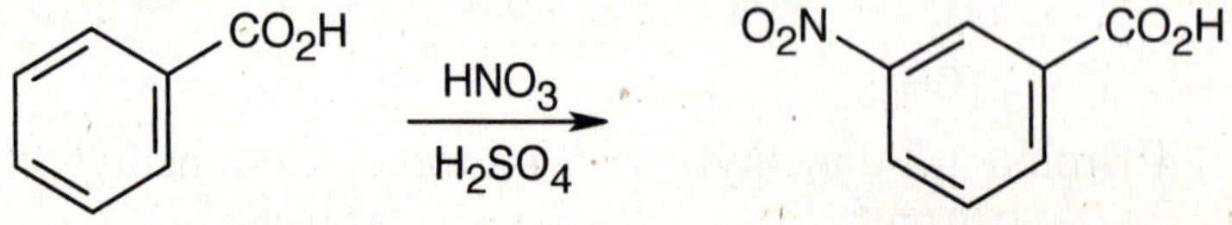


*m*-Nitrobenzoic acid

(d)

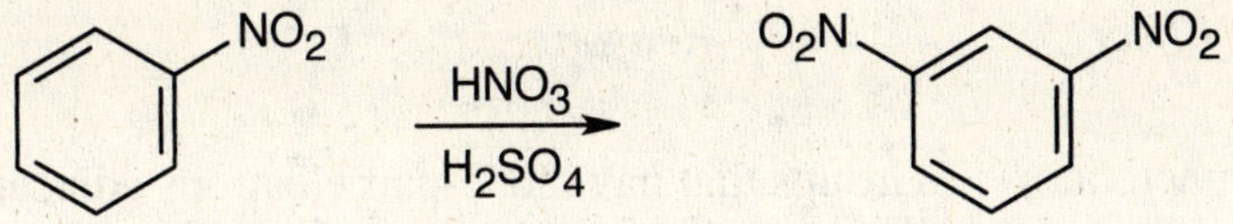

*m*-Dinitrobenzene

(e)

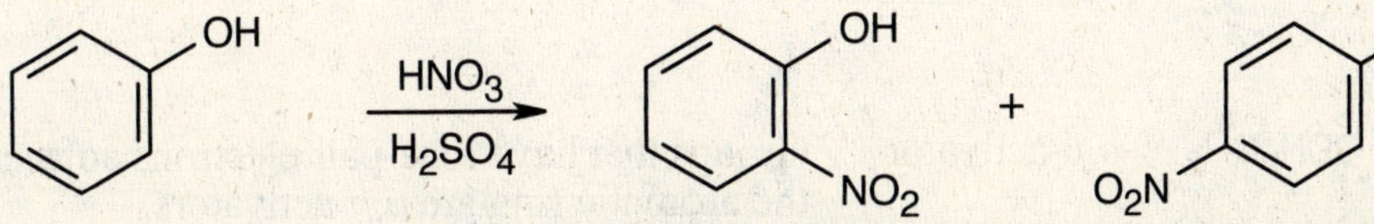

*o*-Nitrophenol *p*-Nitrophenol

(f)

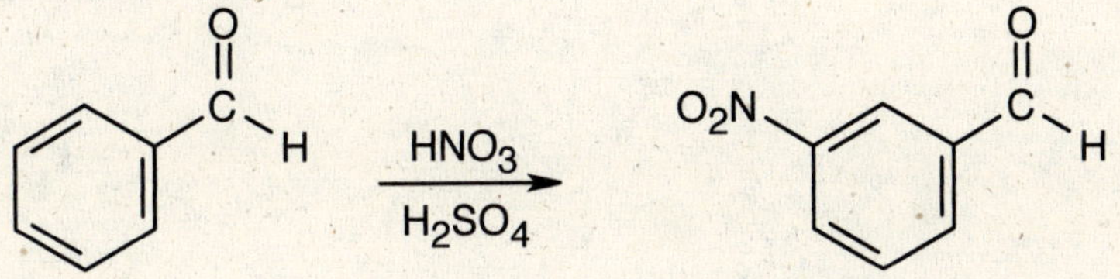

*m*-Nitrobenzaldehyde

**5.31** Only phenol (e) reacts faster than benzene.

**5.32**

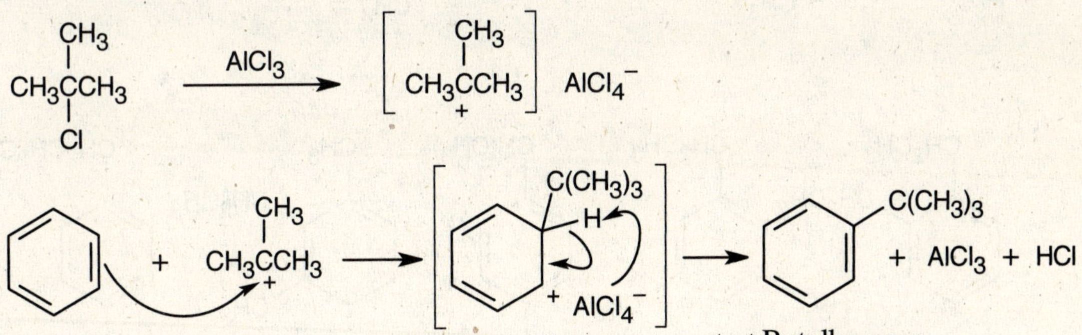

*tert*-Butylbenzene

**5.33**

| | *Group* | *Effect* | *Reason* |
|---|---|---|---|
| (a) | $-\ddot{N}(CH_3)_2$ | *o,p*-activator | Groups that have lone-pair electrons adjacent to the aromatic ring are *o,p*-activators. |
| (b) | 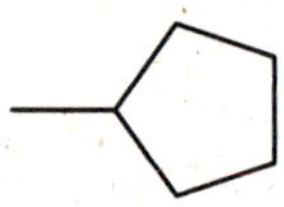 | *o,p*-activator | Alkyl groups are *o,p*-activators. |
| (c) | $-\ddot{\underset{\cdot\cdot}{O}}CH_2CH_3$ | *o,p*-activator | Groups that have lone-pair electrons adjacent to the aromatic ring are *o,p*-activators. |
| (d) | 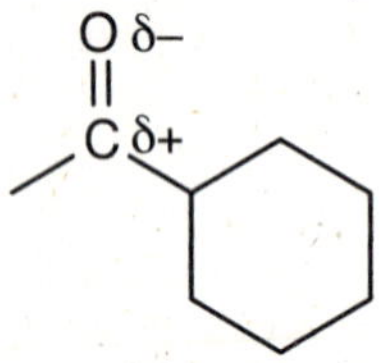 | *m*-deactivator | Groups that have a positively polarized atom adjacent to the aromatic ring are *m*-deactivators. |

**5.34**

(a)

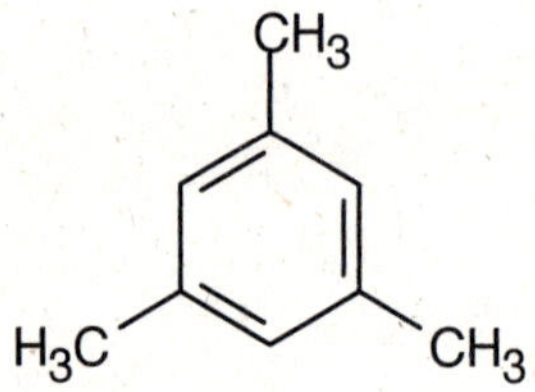


(b)

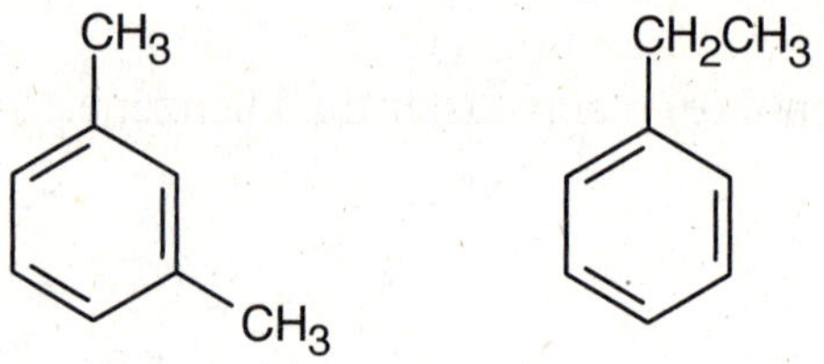


(c)

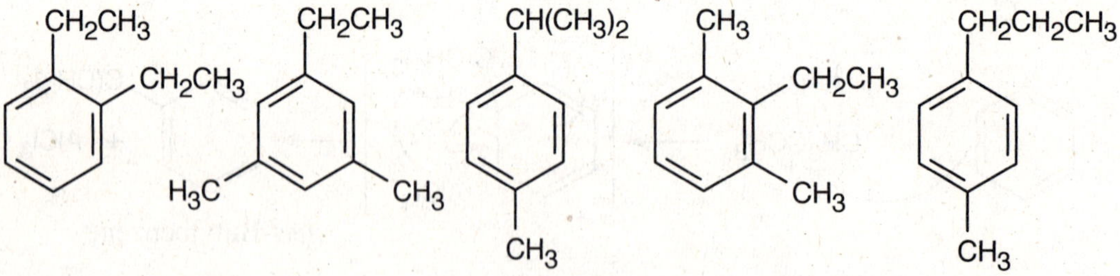

**5.35**

(a)

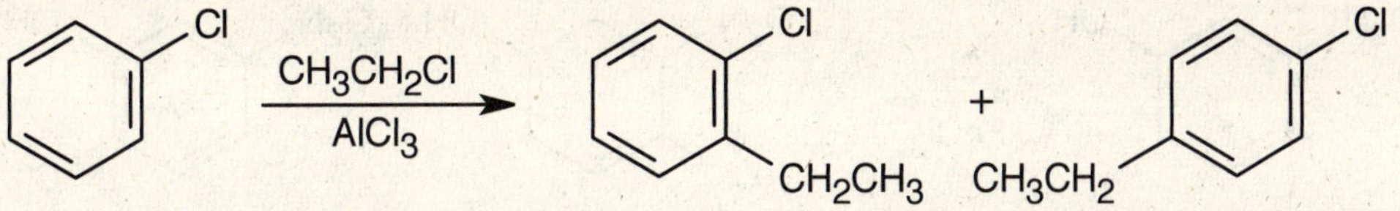

(b)

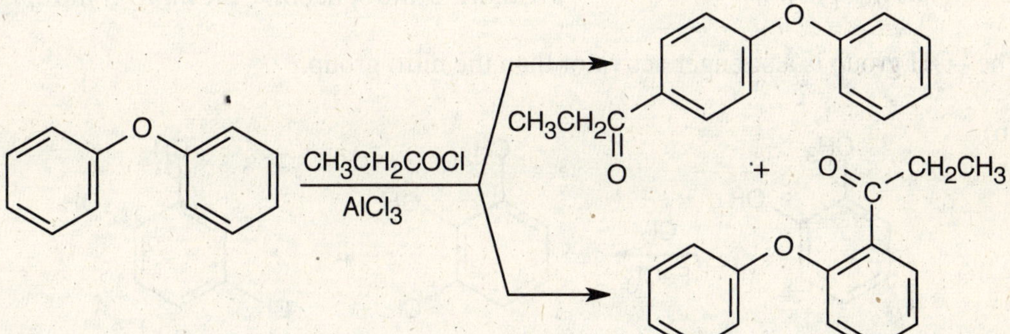

(c)

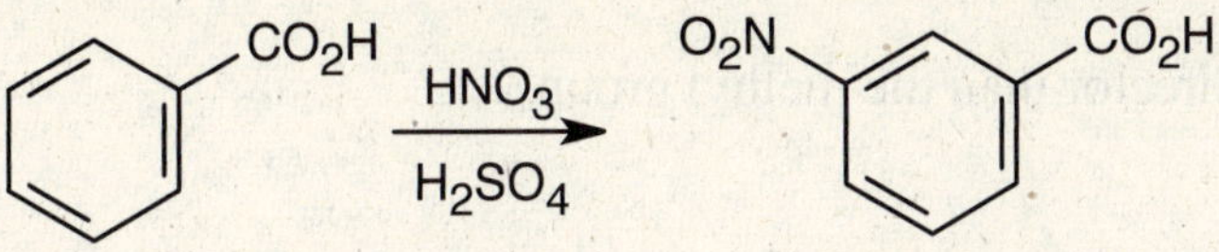

(d)

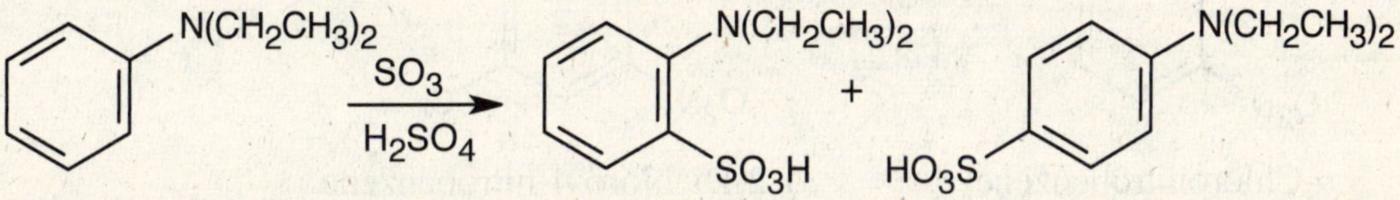

**5.36**

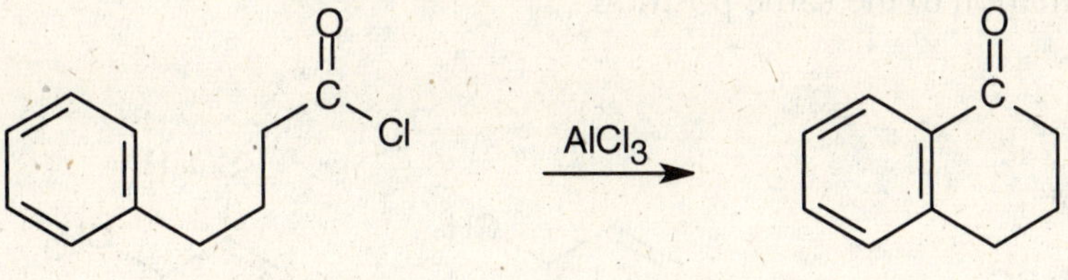

**5.37**

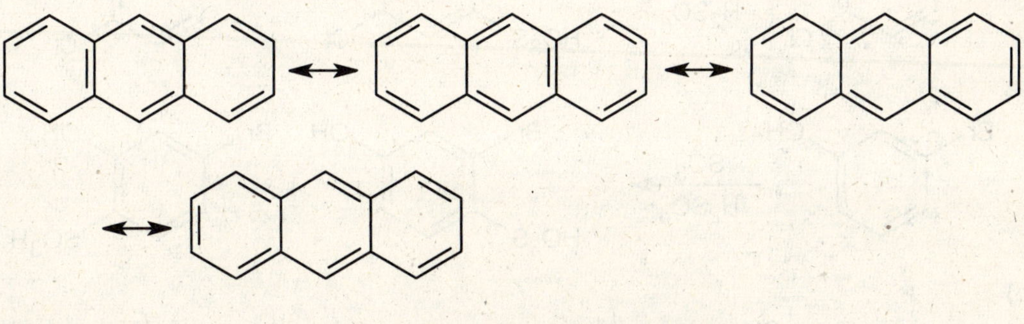

**5.38** *Most reactive* ————————————> *Least reactive*

Anisole > Toluene > *p*-Bromotoluene > Bromobenzene

Nitrobenzene does not undergo Friedel–Crafts acylation.

**5.39**

(a)

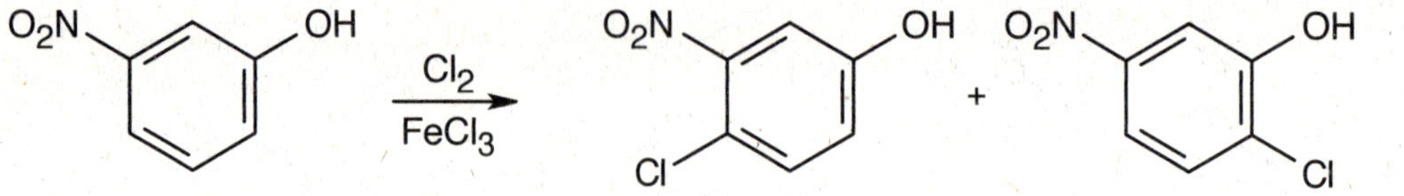


*m*-Nitrophenol 4-Chloro-3-nitrophenol 2-Chloro-5-nitrophenol

The –OH group is a stronger activator than the nitro group.

(b)

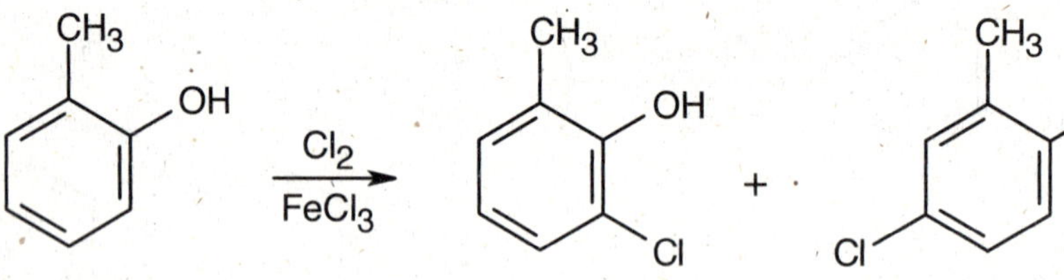


*o*-Methylphenol 2-Chloro-6-methylphenol 4-Chloro-2-methylphenol

The –OH group is a stronger director than the methyl group.

(c)

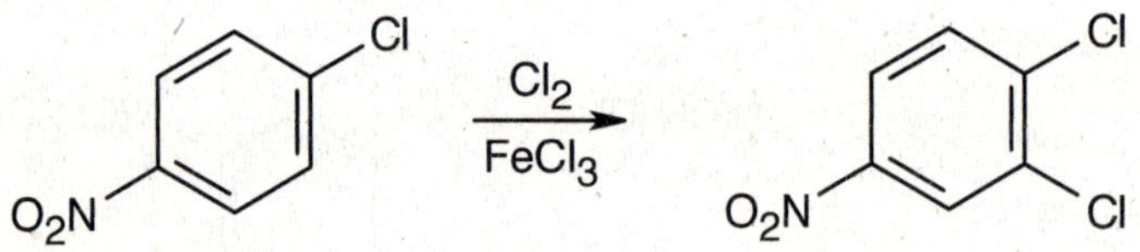


*p*-Chloronitrobenzene 1,2-Dichloro-4-nitrobenzene

Both groups direct chlorination to the same position.

**5.40**

(a)

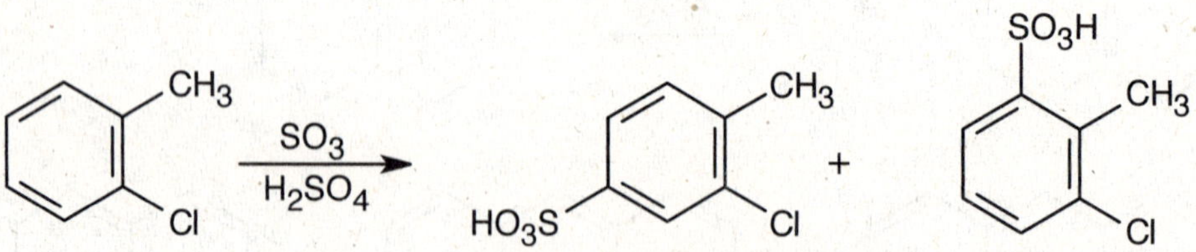


(b)

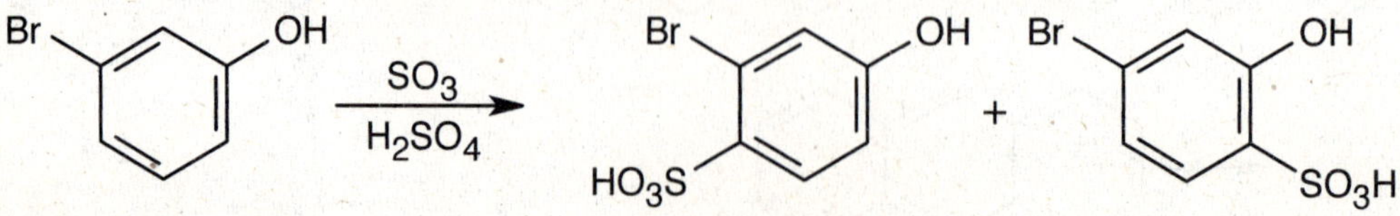


(c)

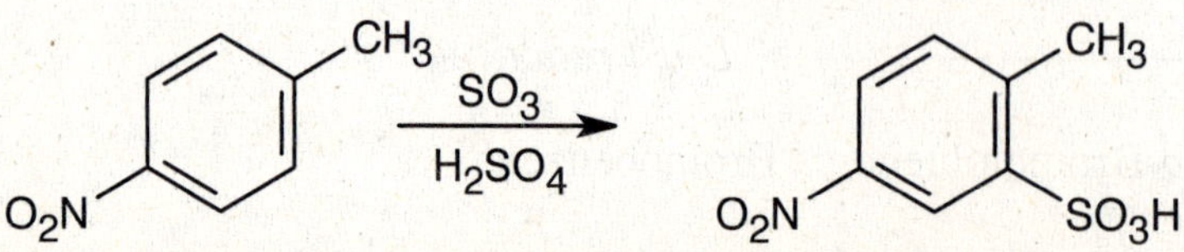

**5.41** *Most reactive ----> Least reactive*

(a) Benzene > chlorobenzene > *o*-dichlorobenzene
(b) Phenol > nitrobenzene > *p*-bromonitrobenzene
(c) *o*-Dimethylbenzene > fluorobenzene > benzaldehyde

Two deactivating substituents make a ring less reactive than one deactivating substituent.

**5.42** Oxidation takes place at the benzylic position.

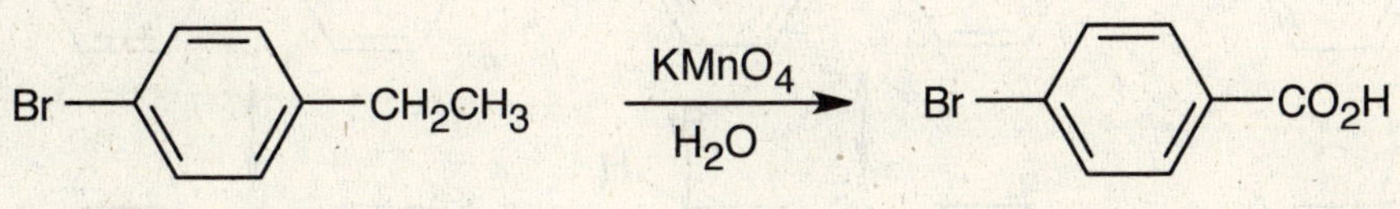


*p*-Bromoethylbenzene *p*-Bromobenzoic acid

## Mechanisms

**5.43**

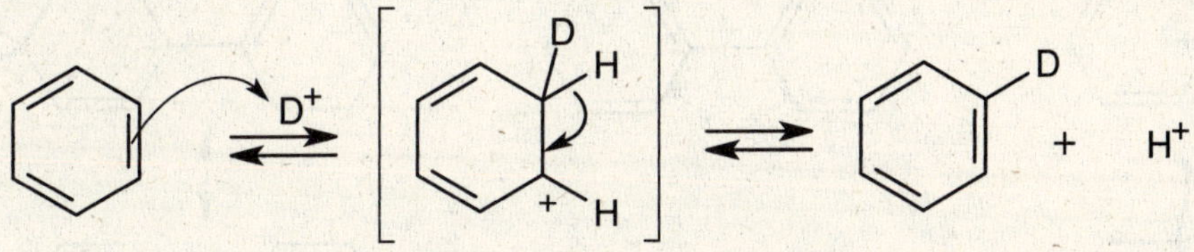


Benzene can be protonated by strong acids. The resulting intermediate can lose either deuterium or hydrogen. If $H^+$ is lost, deuterated benzene is produced. Attack on $D^+$ can occur at all positions of the ring and leads to eventual replacement of all hydrogens by deuterium. Only the first step is shown.

**5.44**

$CH_3Cl + AlCl_3 \longrightarrow [CH_3^+ AlCl_4^-]$ Formation of carbocation

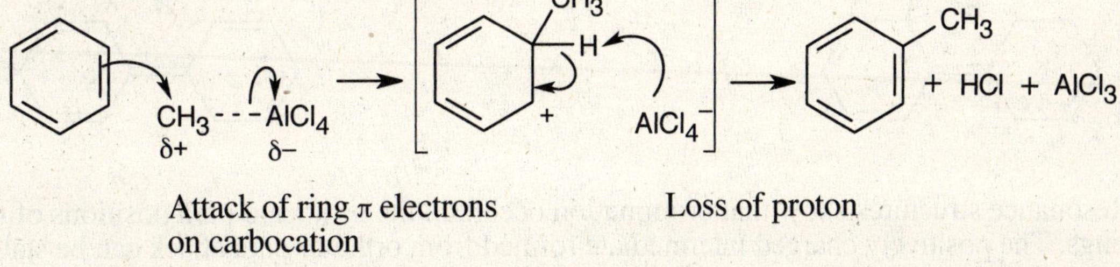


Attack of ring π electrons on carbocation

Loss of proton

**5.45**

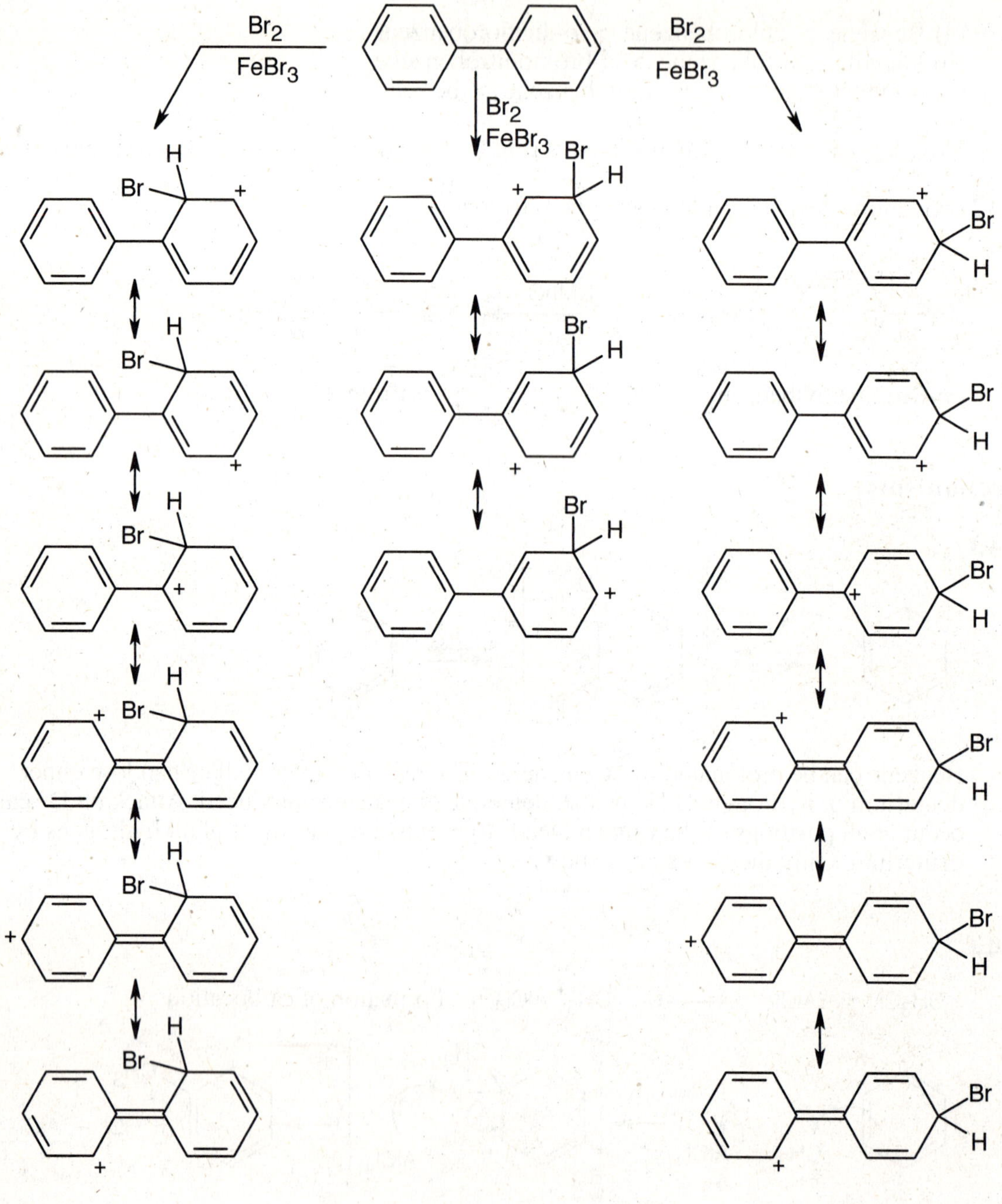


Resonance structures show that bromination occurs in the ortho and para positions of the rings. The positively charged intermediate formed from ortho or para attack can be stabilized by resonance contributions from the second ring of biphenyl, but this stabilization is not possible for meta attack. Note that one of the intermediates from ortho/para attack is a benzylic carbocation.

**5.46** Attack occurs on the unsubstituted ring because bromine is a deactivating substituent. Attack occurs at the ortho and para positions of the ring because the positively charged intermediate can be stabilized by resonance contributions from bromine and from the second ring (see Problem 5.45).

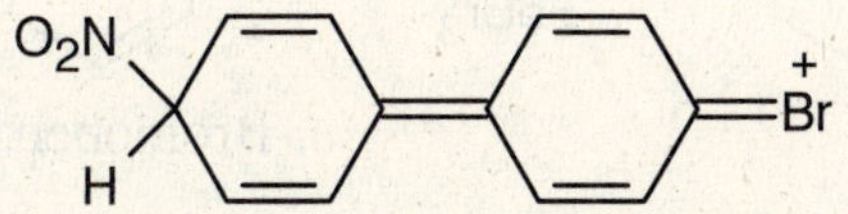


## Synthesis

**5.47** Resonance forms for the intermediate from attack at C1:

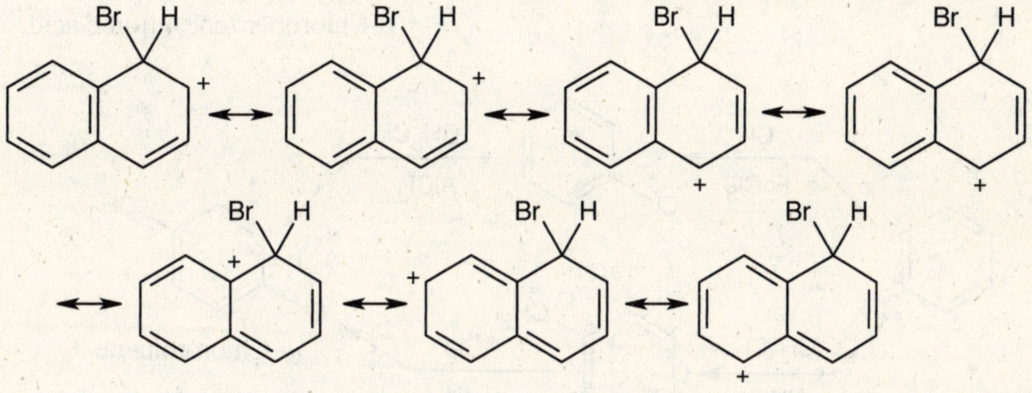


Resonance forms for the intermediate from attack at C2:

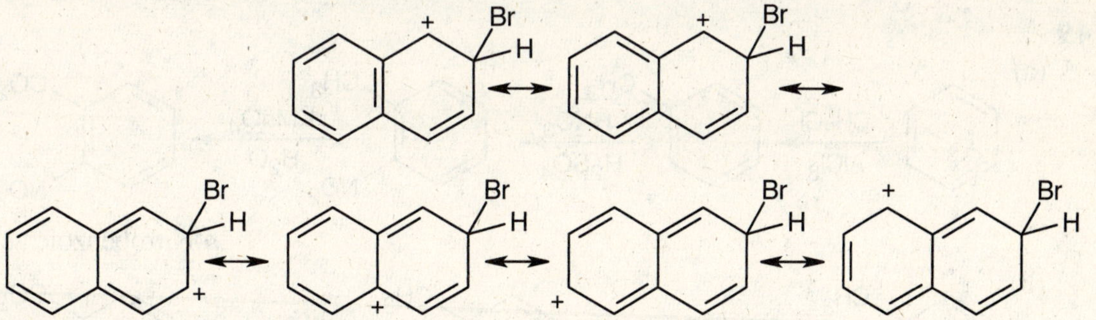


There are seven resonance forms for attack at C1 and six for attack at C2. For C1 attack, the second ring is fully aromatic in four of the resonance forms. In the other three forms, the positive charge has been delocalized into the second ring, destroying the ring's aromaticity. For C2 attack, the second ring is fully aromatic in only the first two forms. Since stabilization is lost when aromaticity is disturbed, the intermediate from C2 attack is less stable than the intermediate from C1 attack, and C1 attack is favored.

**5.48**

(a)

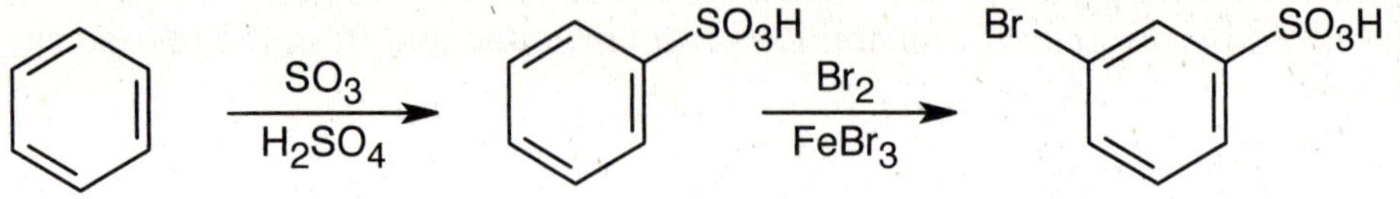

*m*-Bromobenzenesulfonic acid

(b)

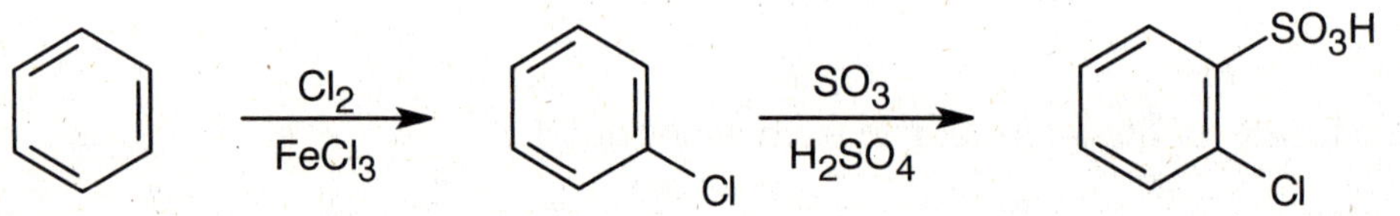

*o*-Chlorobenzenesulfonic acid

(c)

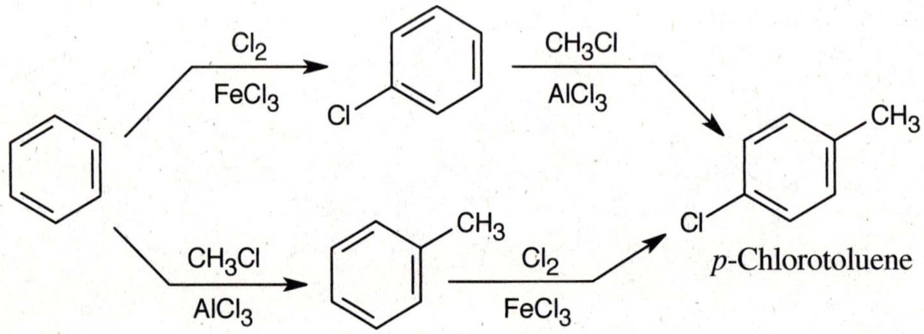

*p*-Chlorotoluene

Either route is acceptable.

**5.49**

(a)

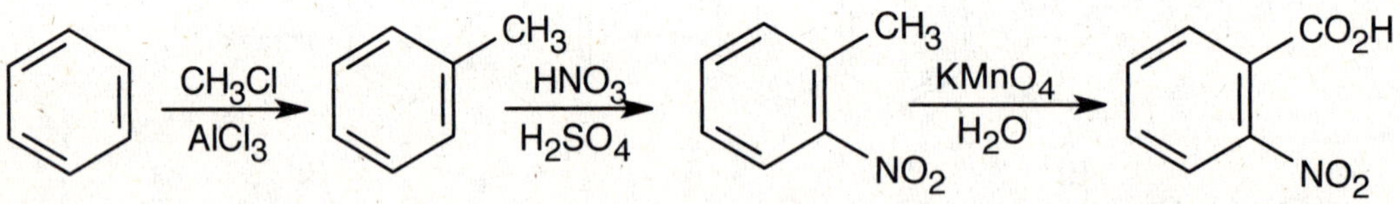

*o*-Nitrobenzoic acid

(b)

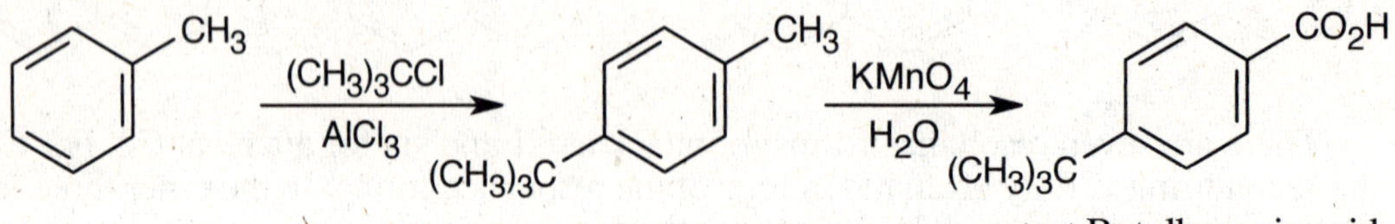

*p*-*tert*-Butylbenzoic acid

## General Problems

**5.50**

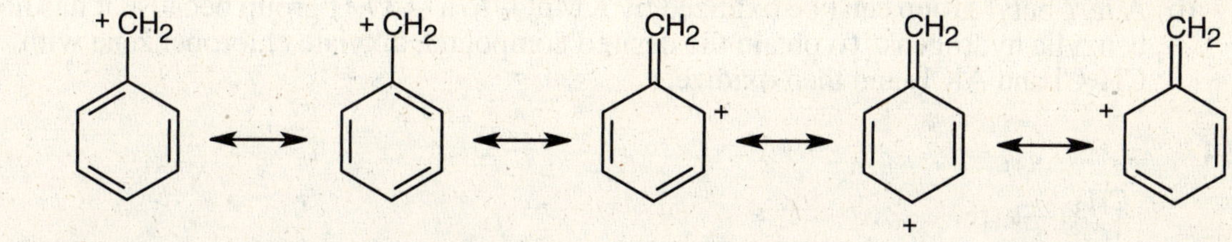

A benzylic carbocation is stabilized because its positive charge can be delocalized over the π system of the aromatic ring.

**5.51**

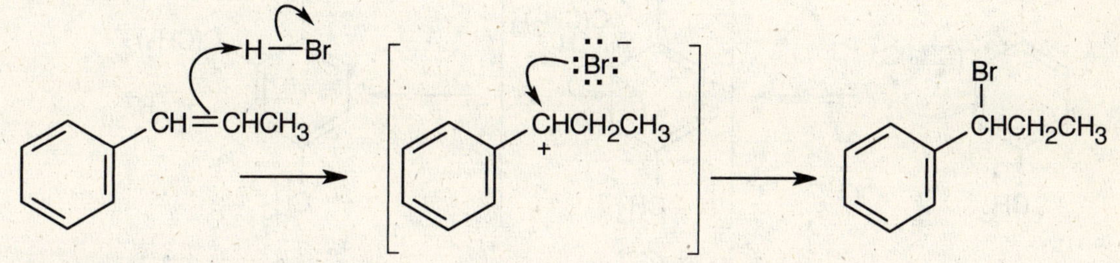

Protonation of the double bond at carbon 2 of 1-phenylpropene leads to an intermediate that can be stabilized by resonance involving the aromatic ring.

**5.52**

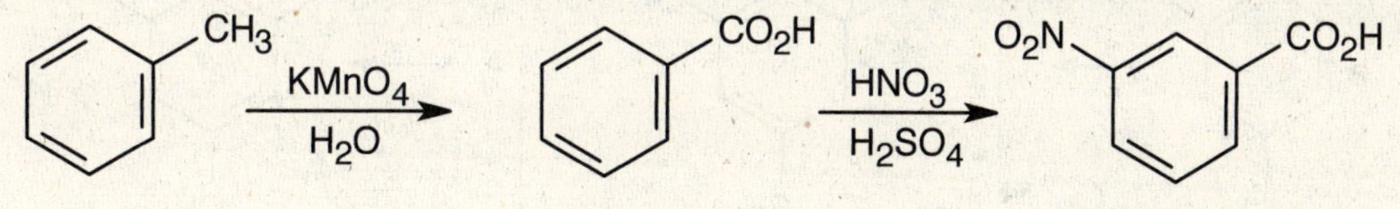

*m*-Nitrobenzoic acid

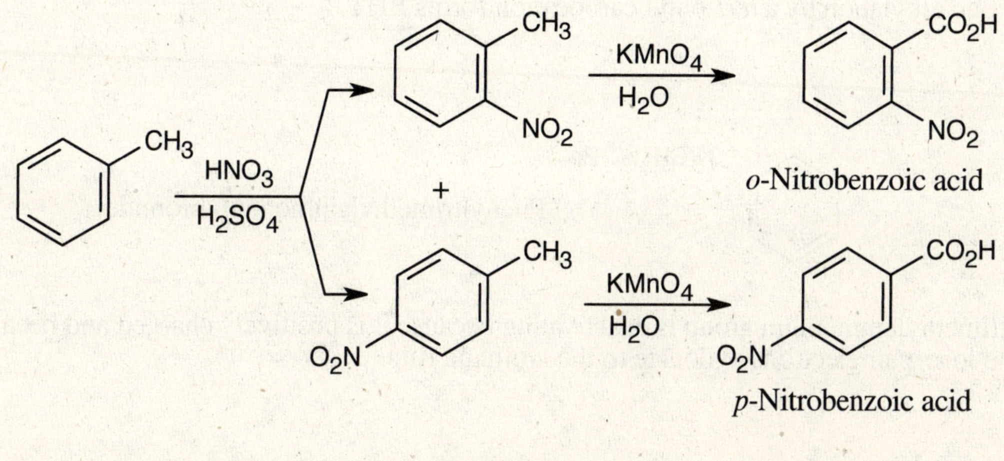

*o*-Nitrobenzoic acid

*p*-Nitrobenzoic acid

**5.53** (a) Chlorination of toluene occurs at the ortho and para positions. To synthesize the given product, first oxidize toluene to benzoic acid and then chlorinate.

(b) A *tert*-butyl group can't be oxidized by $KMnO_4$ to a $-CO_2H$ group because it has no benzylic hydrogens. To obtain the desired compound, alkylate chlorobenzene with $CH_3Cl$ and $AlCl_3$ and then oxidize.

**5.54**

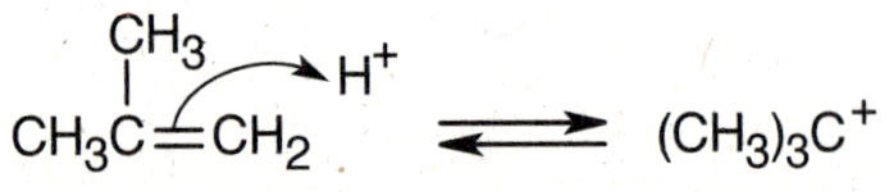


Acid protonates 2-methylpropene, forming a *tert*-butyl carbocation.

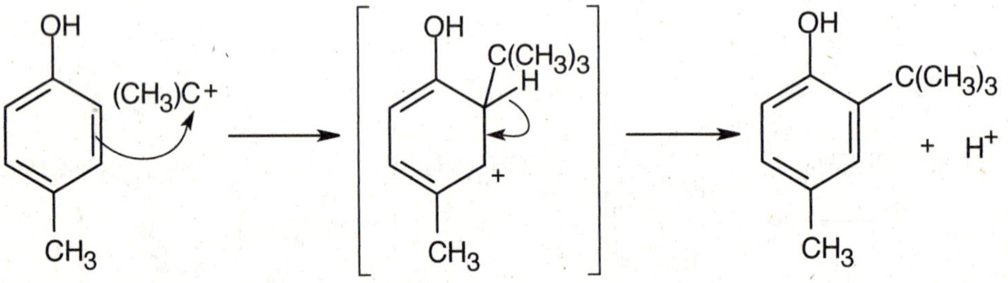


The *tert*-butyl carbocation acts as an electrophile and alkylates *p*-cresol. Alkylation occurs ortho to the –OH group for both steric and electronic reasons.

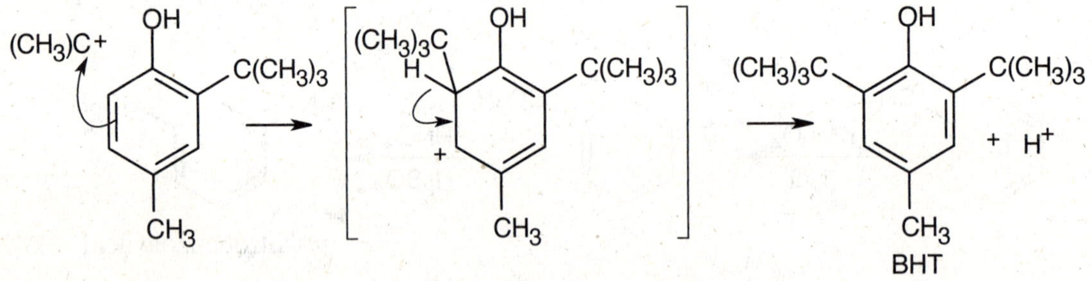


A second alkylation by a *tert*-butyl carbocation forms BHT.

**5.55**

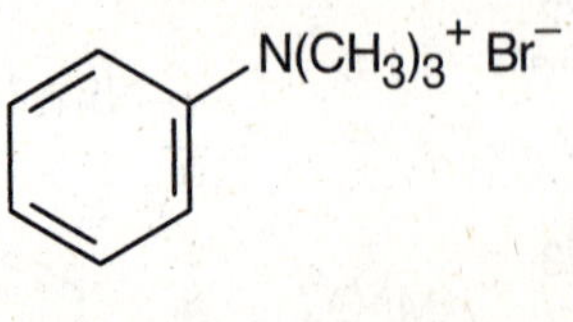


Phenyltrimethylammonium bromide

The trimethylammonium group is deactivating because it is positively charged and because it has no lone-pair electrons to donate to the aromatic ring.

**5.56**

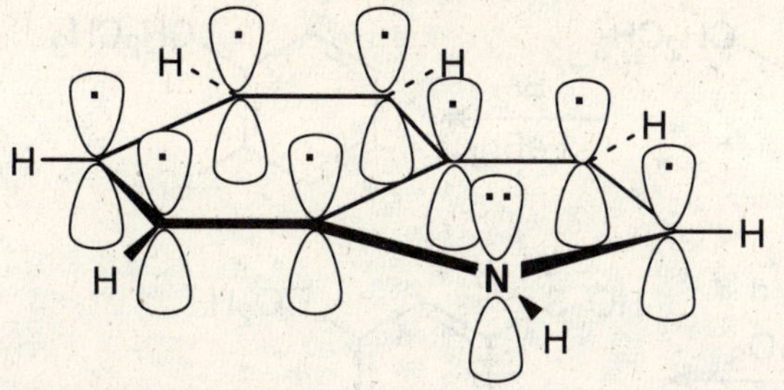


**Indole**

Indole, like naphthalene, has ten π electrons in two rings and is aromatic. Two π electrons come from the nitrogen atom, whereas in naphthalene all come from the ring carbons.

**5.57**

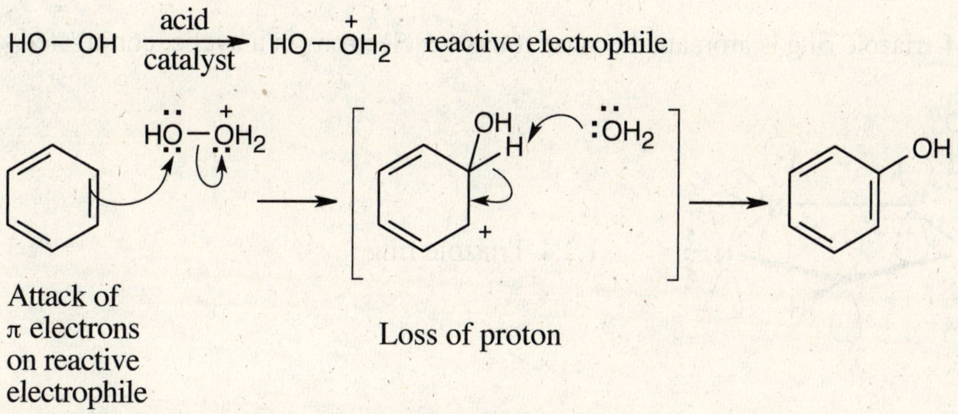


The reactive electrophile (protonated $H_2O_2$) is equivalent to $^+OH$.

**5.58** Problem 5.51 shows the mechanism of the addition of HBr to 1-phenylpropene and shows how the aromatic ring stabilizes the carbocation intermediate. Similar resonance forms can be drawn for the intermediates of reaction of the substituted styrenes with HBr. For the methoxyl-substituted styrene, an additional form can be drawn in which positive charge can be stabilized by the methoxyl group. For the nitro-substituted styrene, no additional form is possible. In addition, one of the resonance forms of the nitro-substituted styrene intermediate is not important because it places two positive charges next to each other.

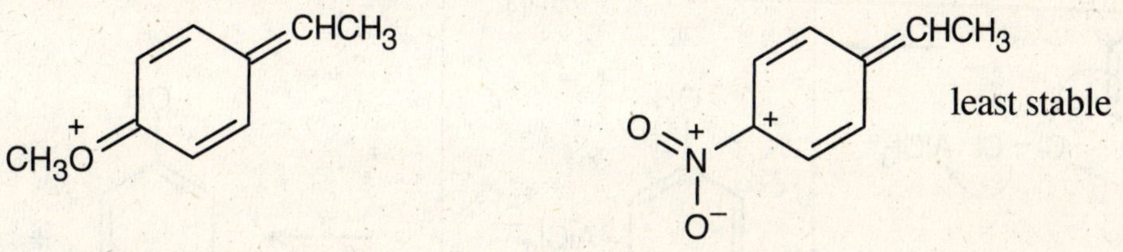


Thus, the intermediate resulting from addition of HBr to the methoxyl-substituted styrene is more stable, and reaction of *p*-methoxystyrene is faster.

**5.59**

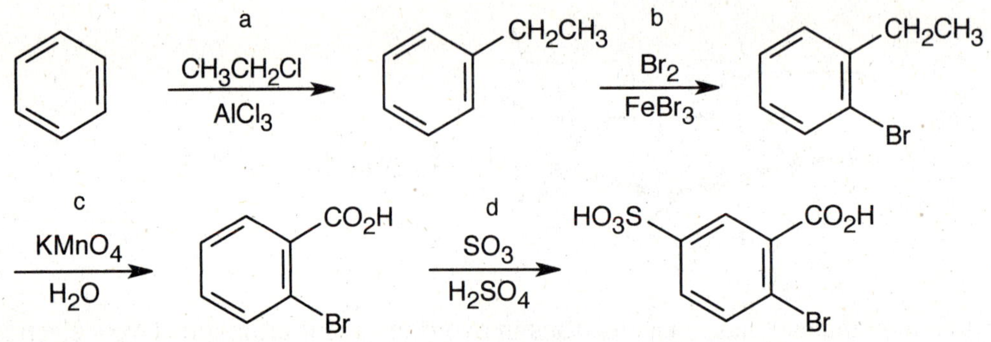


## In the Medicine Cabinet

**5.60** The 1,2,4-triazole ring is aromatic because it has 6 π electrons in a cyclic, conjugated system.

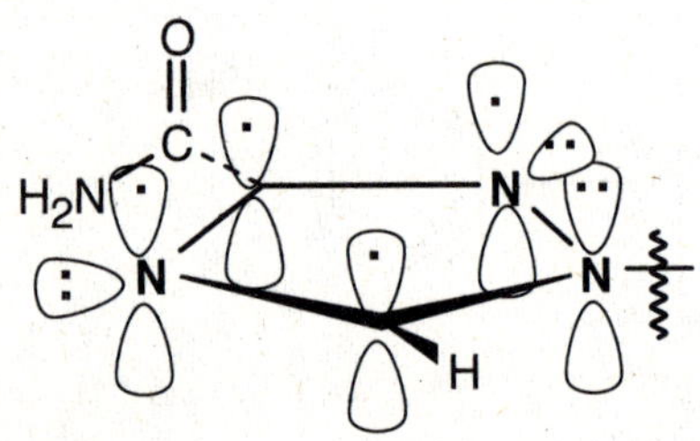


1,2,4-Triazole ring

## In the Field

**5.61** Monochlorination gives a mixture of ortho-and para-chlorophenols. Further chlorination yields 2,4-dichlorophenol.

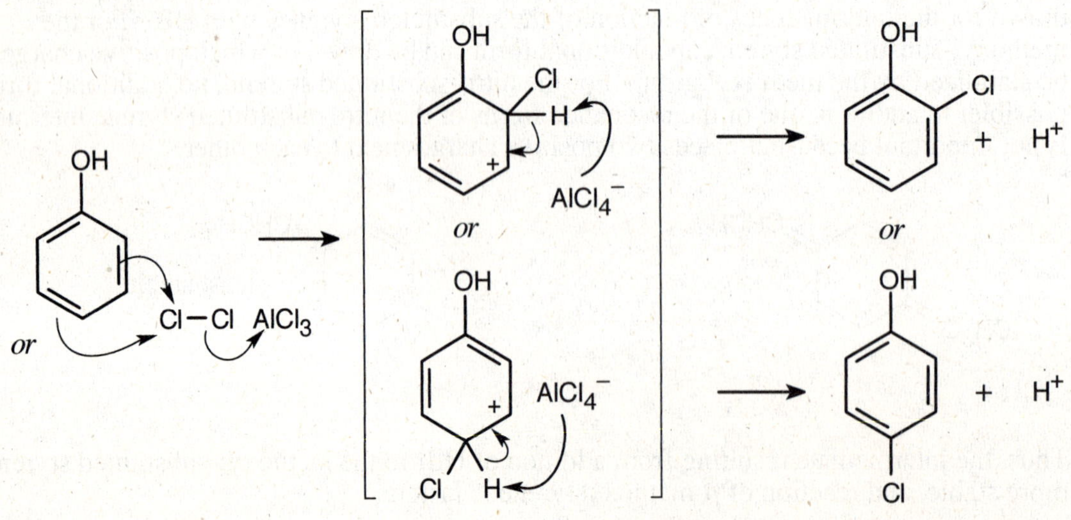


Resonance structures of the intermediates are shown below.

Ortho attack:

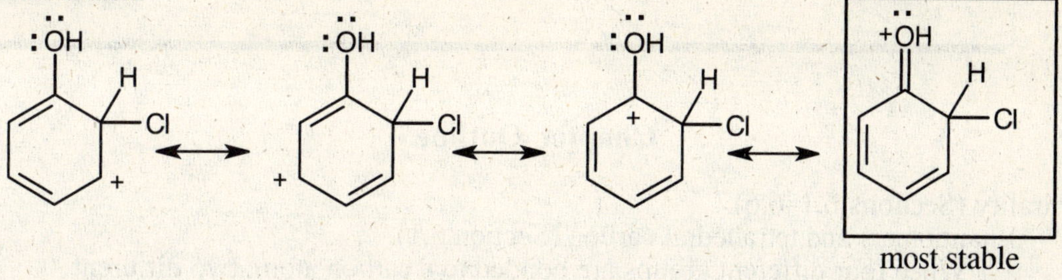

Para attack:

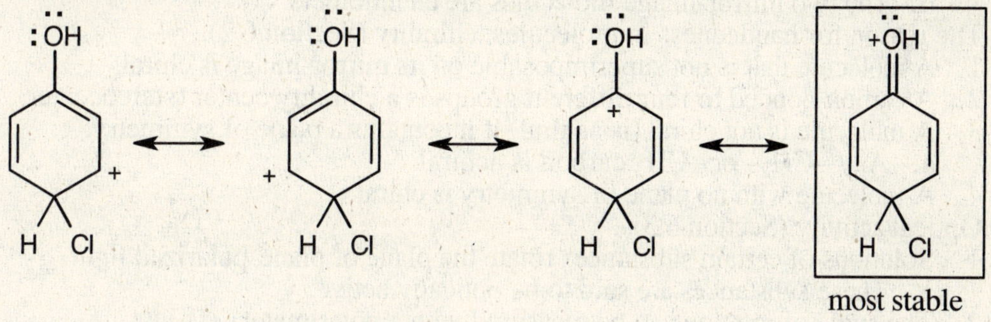

The second chlorine is directed to the correct location because the orienting effect of –OH outweighs the effect of –Cl.

**5.62**

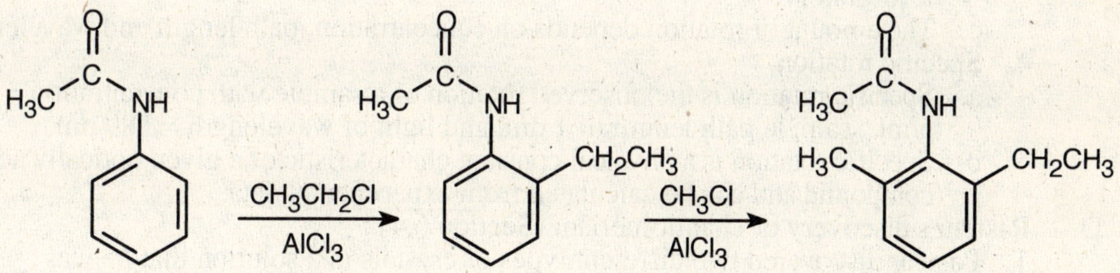

# Chapter 6 – Stereochemistry at Tetrahedral Centers

## Chapter Outline

I. Chirality (Sections 6.1–6.6).
  A. Enantiomers and tetrahedral carbon (Section 6.1).
    1. When four different groups are bonded to a carbon atom, two different arrangements are possible.
      a. These arrangements are mirror images.
      b. The two mirror-image molecules are enantiomers.
  B. The reason for handedness in molecules: chirality (Section 6.2).
    1. A molecule that is not superimposable on its mirror-image is chiral.
    2. A carbon bonded to four different groups is a chirality center (stereocenter).
    3. A molecule is not chiral (is achiral) if it contains a plane of symmetry.
      a. Any $–CH_2–$ or $–CH_3$ carbon is achiral.
    4. A molecule with no plane of symmetry is chiral.
  C. Optical activity (Section 6.3).
    1. Solutions of certain substances rotate the plane of plane-polarized light.
      a. These substances are said to be optically active.
    2. The angle of rotation can be measured with a polarimeter.
    3. The direction of rotation can also be measured.
      a. A compound whose solution rotates plane-polarized light to the right is dextrorotatory.
      b. A compound whose solution rotates plane-polarized light to the left is levorotatory.
      c. The amount of rotation depends on concentration, path length and wavelength.
    4. Specific rotation.
      a. Specific rotation is the observed rotation of a sample with concentration = 1 g/mL, sample path length of 1 dm, and light of wavelength = 589 nm.
      b. Specific rotation is a physical constant characteristic of a given optically active compound and can be calculated from experimental data.
  D. Pasteur's discovery of enantiomerism (Section 6.4).
    1. Pasteur discovered two different types of crystals in a solution that he was evaporating.
    2. The crystals were mirror images.
    3. Solutions of each of the two types of crystals were optically active, and their specific rotations were equal in magnitude but opposite in sign.
    4. Pasteur postulated that some molecules are handed and thus discovered the phenomenon of enantiomerism.
  E. Specification of configurations of stereoisomers (Section 6.5).
    1. Rules for assigning configurations at a stereocenter:
      a. Assign priorities to each group bonded to the carbon by using Cahn–Ingold–Prelog rules (Section 3.4).
      b. Orient the molecule so that the group of lowest priority is pointing to the rear.
      c. Draw a curved arrow from group 1 to group 2 to group 3 (highest to lowest priority).
      d. If the arrow is clockwise, the stereocenter has *R* configuration, and if the arrow is counterclockwise, the stereocenter has *S* configuration.
    2. The sign of optical rotation is not related to *R,S* designation.

II. Compounds with more than one stereocenter (Sections 6.6–6.8).
  A. Enantiomers and diastereomers (Section 6.6).
    1. A molecule with $n$ chirality centers can have a maximum of $2^n$ stereoisomers.
    2. A molecule with two chirality centers can have four possible stereoisomers.
      a. The stereoisomers group into two pairs of enantiomers.
      b. A stereoisomer from one pair is the diastereomer of a stereoisomer from the other pair.
    3. Diastereomers are stereoisomers that are not mirror images.
      a. Enantiomers have different configurations at all chirality centers, whereas diastereomers have different configurations at some, but not all, chirality centers.
  B. Meso compounds (Section 6.7).
    1. A meso compound occurs when a compound with two chirality centers possesses a plane of symmetry.
    2. A meso compound is achiral despite having two chirality centers.
    3. The physical properties of meso compounds, diastereomers and racemic mixtures differ from each other and from the properties of enantiomers.
  C. Racemic mixtures and the resolution of enantiomers (Section 6.8).
    1. A racemic mixture is a 50:50 mixture of two enantiomers.
      a. Racemic mixtures show zero optical rotation.
    2. Some racemic mixtures can be resolved (separated) into their component enantiomers.
      a. If a racemic mixture of a carboxylic acid reacts with a chiral amine, the product ammonium salts are diastereomers.
      b. The diastereomeric salts differ in chemical and physical properties and can be separated.
      c. The original enantiomers can be recovered by acidification.

III. A review of isomerism (Section 6.9).
  A. Constitutional isomers differ in connections between atoms.
    1. Skeletal isomers have different carbon skeletons.
    2. Functional isomers contain different functional groups.
    3. Positional isomers have functional groups in different positions.
  B. Stereoisomers have the same connections between atoms, but different geometry.
    1. Enantiomers have a mirror-image relationship.
    2. Diastereomers are non-mirror-image stereoisomers.
      a. Configurational diastereomers.
      b. Cis–trans isomers differ in the arrangement of substituents in a double bond or ring.

IV. Chirality in Nature (Section 6.10).
  A. Different stereoisomers have different biological properties.
  B. Often, only one enantiomer of a drug is effective.
  C. A chiral environment makes enantiomers behave as if they were diastereomers.

## Solutions to Problems

**6.1** Chiral: screw, shoe
Not chiral: soda can, screwdriver

**6.2** Draw each compound, and identify all carbons that are *not* chirality centers. These carbons include:

$CH_3–$, $–CH_2–$, $–CX_2–$, $–C=C–$, $–C≡C–$, $–C=O$, all benzene ring carbons

Cross out these carbons. If all carbons are crossed out, the compound is achiral. If any carbons remain, they should be bonded to four different groups, and the compound is chiral. (Carbon chirality centers are starred.)

(a)

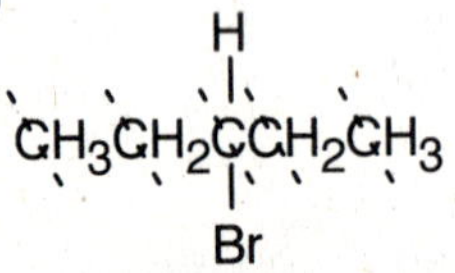

achiral

(b)

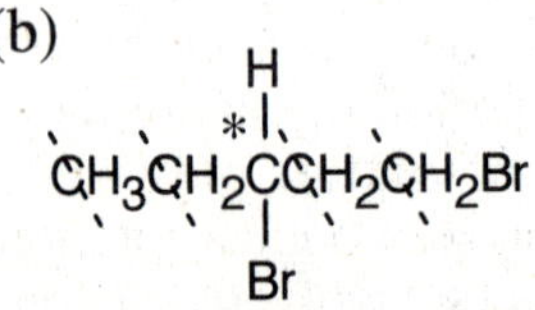

chiral

Carbon 3 is bonded to four different groups: $–CH_2CH_3$, $–CH_2CH_2Br$, $–Br$, $–H$

(c)

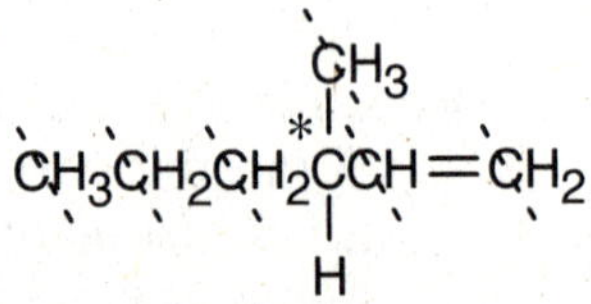

chiral

(d)

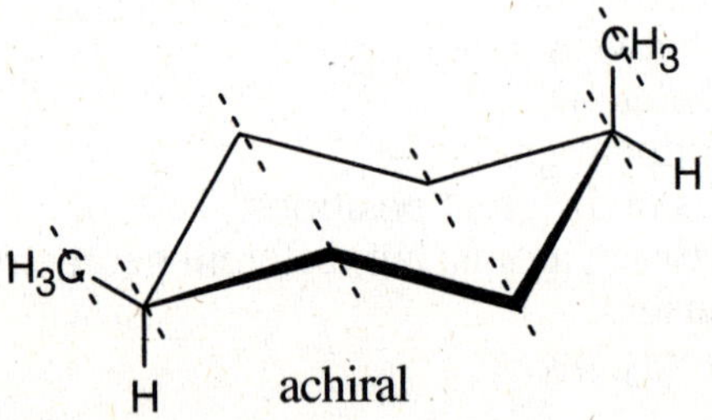

achiral

All carbons are bonded to at least two identical groups.

**6.3** Refer to Problem 6.2 for a list of carbons that can't be chirality centers.

(a)

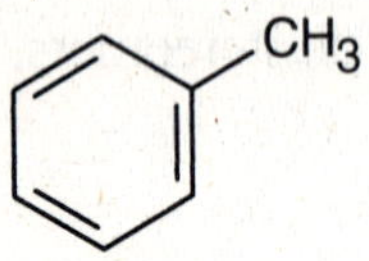

Toluene
achiral

(b)

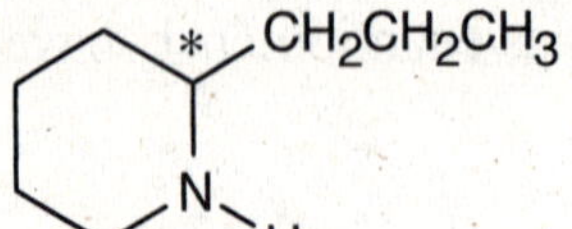

Coniine
chiral

(c)

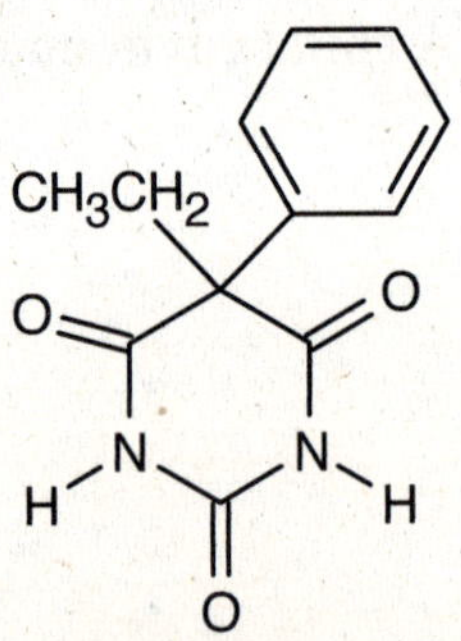

Phenobarbital
achiral

**6.4**

Alanine

**6.5**

(a)

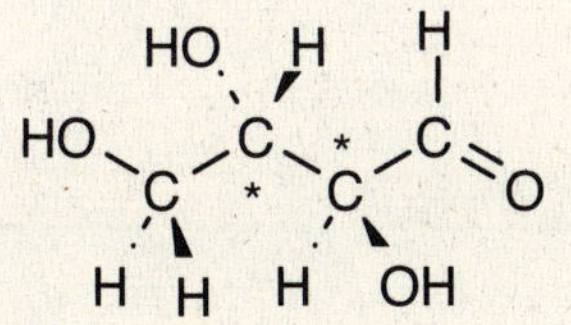

Threose

(b)

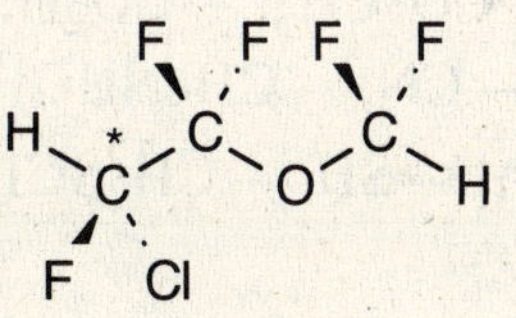

Enflurane

**6.6** Cocaine is levorotatory. (Levorotatory compounds have a minus sign in front of the degree of rotation, whereas dextrorotatory compounds have a plus sign.)

**6.7**

Use the formula $[\alpha]_D = \dfrac{\alpha}{l \times C}$, where

$[\alpha]_D$ = specific rotation

$\alpha$ = observed rotation

$l$ = path length of cell (in dm)

$C$ = concentration (in g/mL)

In this problem: $\alpha = +1.21°$

$l$ = 5.00 cm = 0.500 dm

$C$ = 1.50 g/10.0 mL = 0.150 g/mL

$$[\alpha]_D = \frac{+1.21°}{0.500 \text{ dm} \times 0.150 \text{ g/mL}} = +16.1$$

**6.8** Use the sequence rules in Section 3.4.

(a) By Rule 1, –H is of lowest priority, and –OH is of highest priority. By Rule 2, $-CH_2CH_2OH$ is of higher priority than $-CH_2CH_3$.

*Highest* ⟶ *Lowest*

$-OH$, $-CH_2CH_2OH$, $-CH_2CH_3$, $-H$

(b) By Rule 3, $-CO_2H$ is considered as $-C(O-)(O-)-OH$. Because 3 oxygens are attached to a $-CO_2H$ carbon and only one oxygen is attached to $-CH_2OH$, $-CO_2H$ is of higher priority than $-CH_2OH$. $-CO_2CH_3$ is of higher priority than $-CO_2H$ by Rule 2, and –OH is of highest priority by Rule 1.

*Highest* ⟶ *Lowest*

(b) $-OH$, $-CO_2CH_3$. $-CO_2H$, $-CH_2OH$
(c) $-NH_2$, $-CN$, $-CH_2NHCH_3$, $-CH_2NH_2$
(d) $-SSCH_3$, $-SH$, $-CH_2SCH_3$, $-CH_3$

**6.9** The following scheme may be used to assign *R, S* configurations to chirality centers:

*Step 1.* For each chirality center, rank substituents by the priority rules. Give the number 4 to the lowest priority substituent. For (a):

| *Substituent* | *Priority* |
|---|---|
| –Br | 1 |
| $-CO_2H$ | 2 |
| $-CH_3$ | 3 |
| –H | 4 |

*Step 2.* Imagine yourself looking at the molecule, with the group of lowest priority pointing to the back. Your viewpoint would be behind the plane of the paper, looking out (the arrow indicates the viewpoint). From that viewpoint, you would see –Br on the right, $-CO_2H$ on the left, and the methyl group at the bottom. Draw the molecule as you see it, and note the direction of the arrow that travels from group 1 to group 2 to group 3. The direction of rotation is counterclockwise, and the configuration is *S*.

(a)

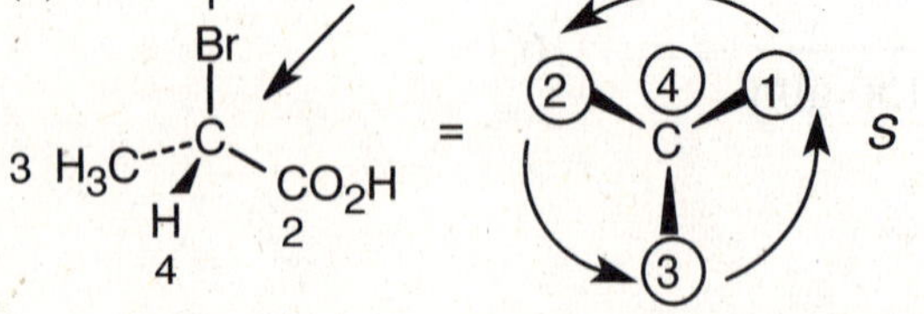


(b)

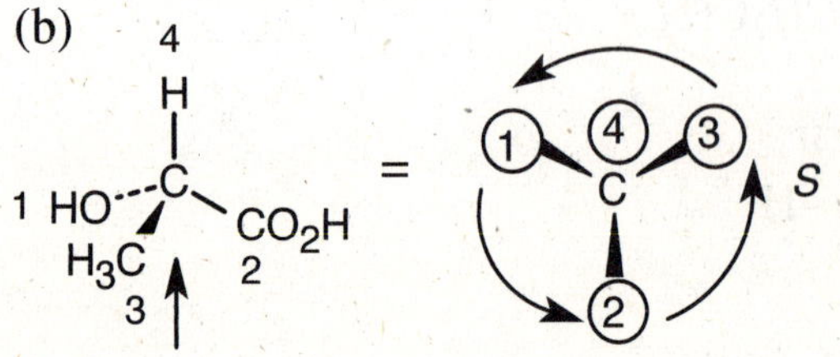


(c)

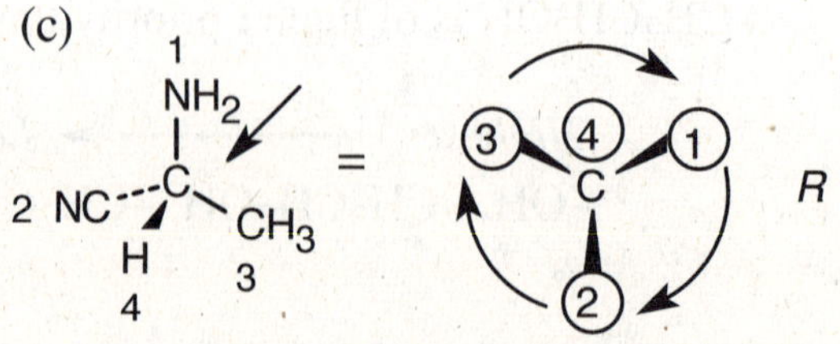

**6.10** As in the previous problem, assign priorities to the substituents, giving the number 4 to the lowest priority substituent. Orient the lowest priority group toward the back, and arrange the three other groups, as we have done previously, as spokes on a steering wheel, with a counterclockwise rotation (because we need to draw an *S* enantiomer). Then, tilt the drawing until it is a tetrahedral representation.

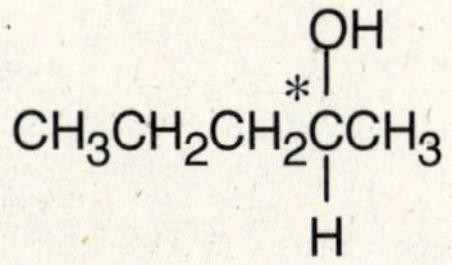
(*S*)-Pentan-2-ol

| *Substituent* | *Priority* |
|---|---|
| $-OH$ | 1 |
| $-CH_2CH_2CH_3$ | 2 |
| $-CH_3$ | 3 |
| $-H$ | 4 |

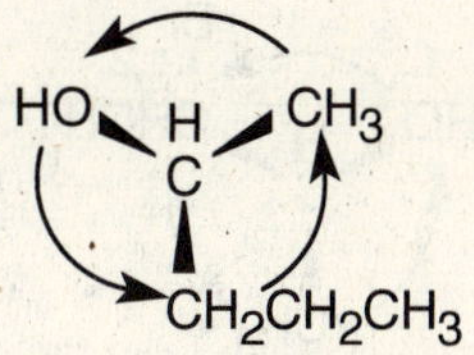

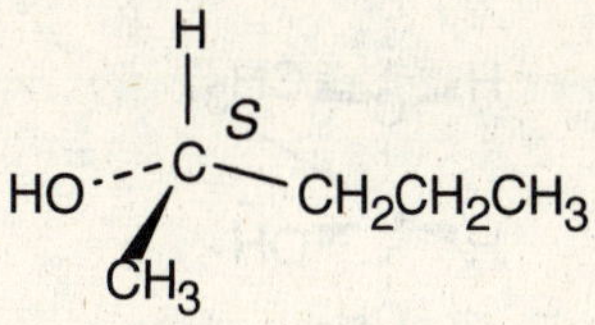

**6.11** Fortunately, the drawing of methionine is shown in the correct orientation.

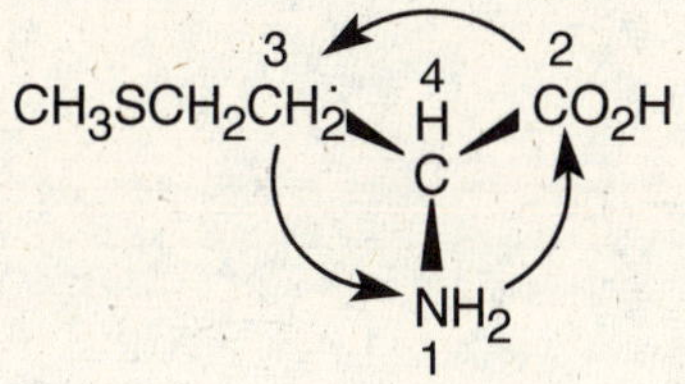
(*S*)-Methionine

**6.12** *R, S* assignments for more complicated molecules can be made by using the same method used in Problem 6.9. It is especially helpful to use molecular models when a compound has more than one chirality center.

*Step 1*. Assign priorities to groups at the *top* chirality center.

| *Substituent* | *Priority* |
|---|---|
| –Br | 1 |
| $–CH(OH)CH_3$ | 2 |
| $–CH_3$ | 3 |
| –H | 4 |

*Step 2*. Orient the model so that the lowest priority group of the first chirality center points to the rear. If you aren't using a model, orient yourself so that you are 180° from the lowest priority group. You would be looking out of the page, upward to the left.

*Step 3.* Note the direction of rotation of the arrow that travels from to group 1 to group 2 to group 3. The rotation is clockwise, and the configuration is *R*.

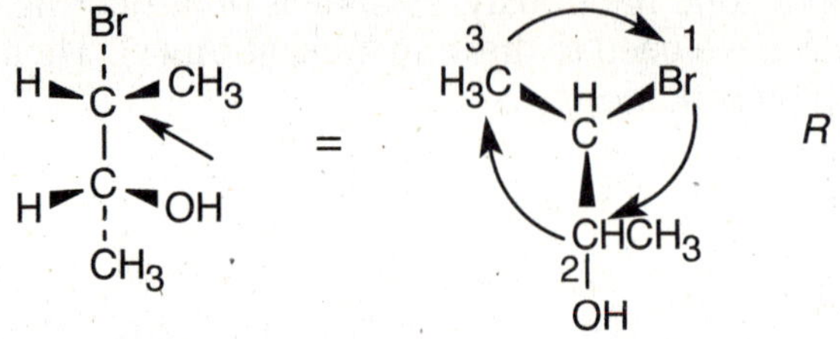

*Step 4.* Repeat steps 1–3 for the next chirality center.

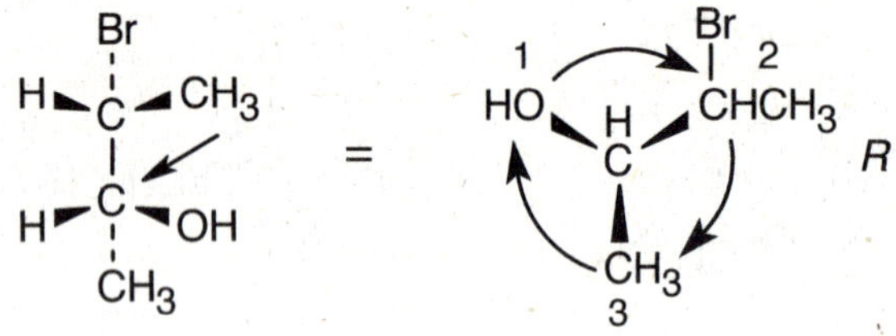

(a) *R,R* (b) *S,R* (c) *R,S*

**6.13** Molecules (b) and (c) are enantiomers (mirror images). Molecule (a) is the diastereomer of (b) and (c).

**6.14** For (a):

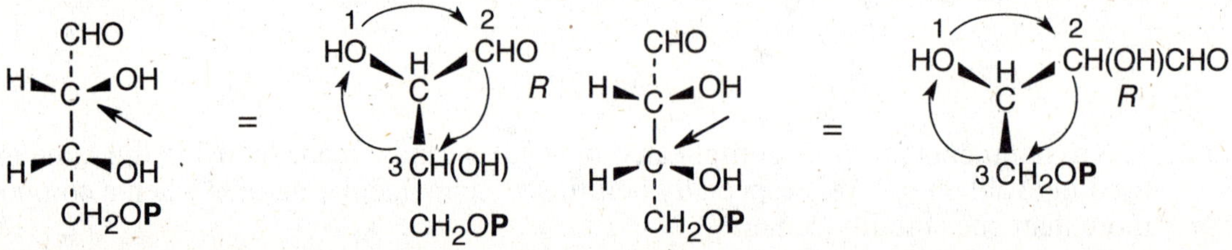

(a) *R,R* (b) *S,R* (c) *R,S* (d) *S,S*

a, d are enantiomers and are diastereomeric with b, c.
b, c are enantiomers and are diastereomeric with a, d.
Structure (a) is D-erythrose, structure (d) is its enantiomer, and structures (b) and (c) are its diastereomers.
Note: **P** = $-PO_3^{2-}$

**6.15**

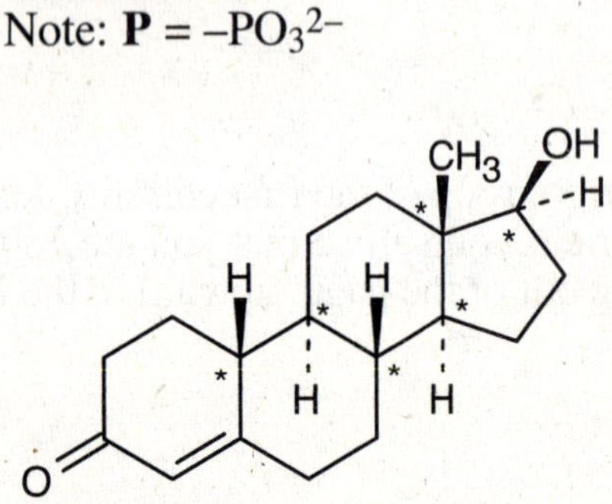

Nandrolone

Nandrolone has six chirality centers (starred) and can have, in principle, $2^6 = 64$ stereoisomers.

**6.16**

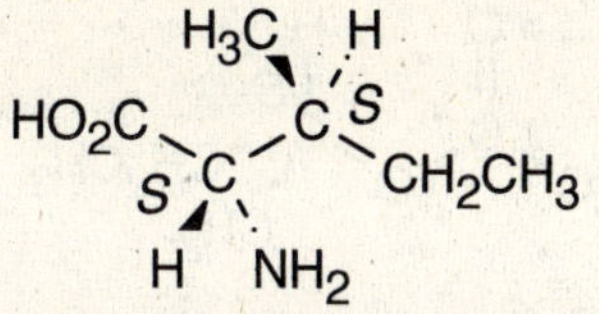

Isoleucine

**6.17** To decide if a structure represents a meso compound, try to locate a plane of symmetry that divides the molecule into two halves that are mirror images. Molecular models may be helpful.

(a)

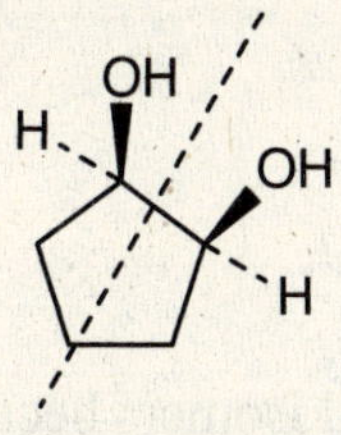

plane of symmetry

meso

(c)

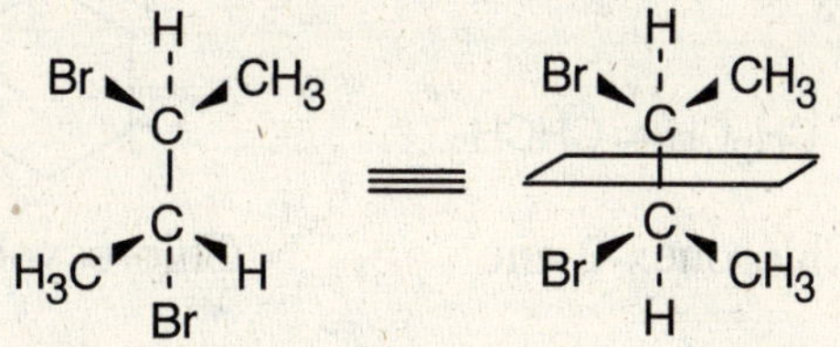

plane of symmetry

meso

Structure (b) does not represent a meso compound.

**6.18** To have a meso form, a molecule must have a plane of symmetry. 2,3-Dibromobutane can exist as a pair of enantiomers *or* as a meso compound, depending on the configurations at carbons 2 and 3.

(a)

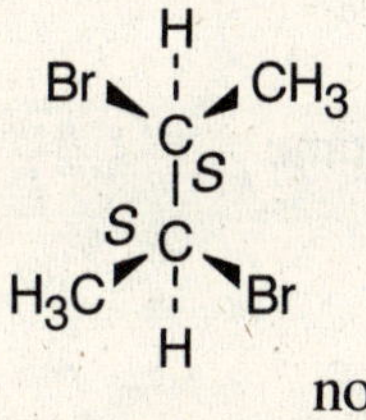

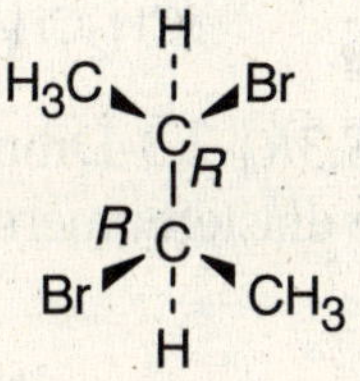

not meso

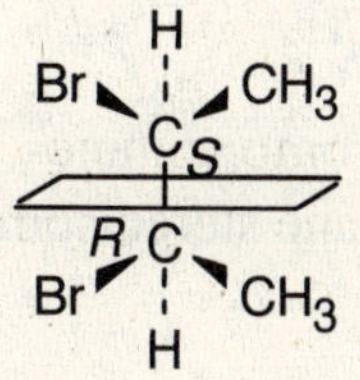

plane of symmetry

meso

(b) 2,3-Dibromopentane has no symmetry plane and thus can't exist in a meso form.

(c) 2,4-Dibromopentane can exist in a meso form.

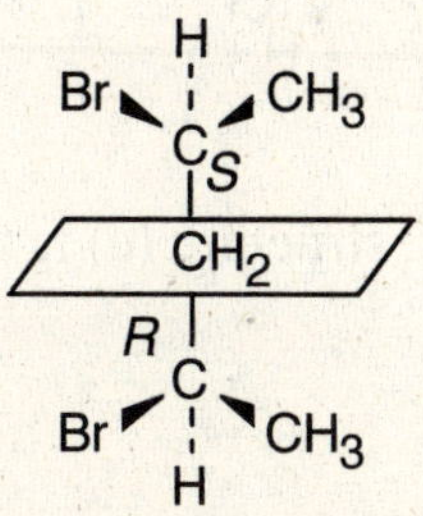

plane of symmetry

2,4-Dibromopentane can also exist as a pair of enantiomers [(2*R*,4*R* and 2*S*,4*S*)] that are not meso compounds.

**6.19**

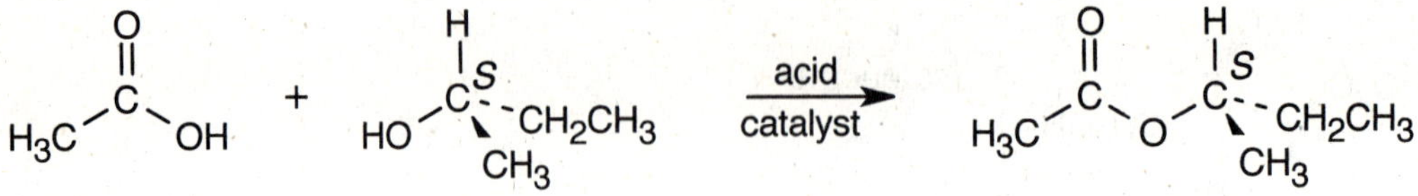

The product is the pure *S* ester because no bonds at the chiral carbon are broken or rearranged during the reaction..

**6.20**

(a)

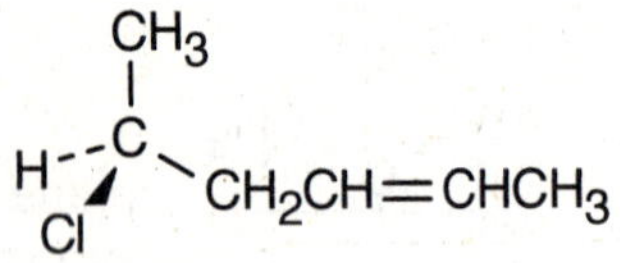

(*S*)-5-Chlorohex-2-ene

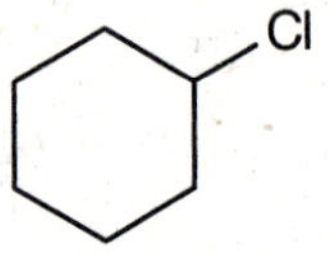

Chlorocyclohexane

These two compounds are constitutional isomers. They are skeletal isomers because they have different carbon skeletons.

(b)

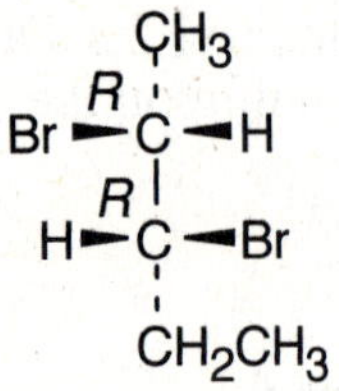

(2*R*,3*R*)-2,3-Dibromopentane

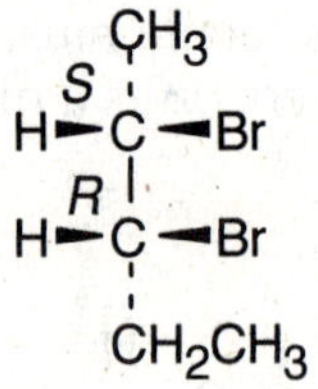

(2*S*,3*R*)-2,3-Dibromopentane

The two dibromopentane stereoisomers are diastereomers.

## Visualizing Chemistry

**6.21**

(a)

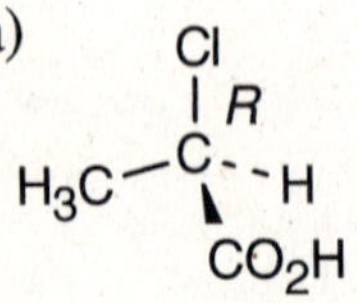

(b)

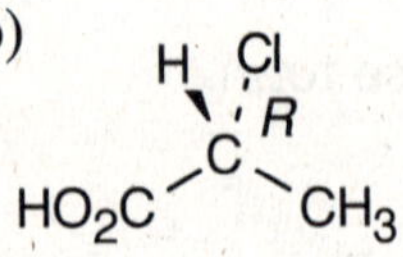

(c)

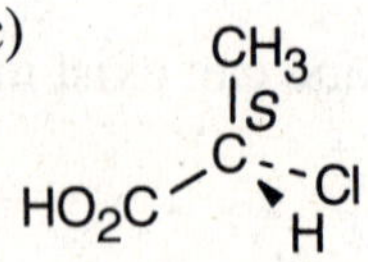

(d)

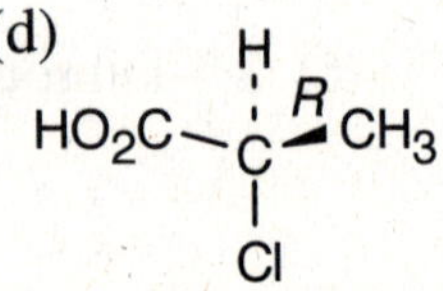

Structures (a), (b), and (d) are identical (*R* enantiomer); structure (c) is the *S* enantiomer.

**6.22**

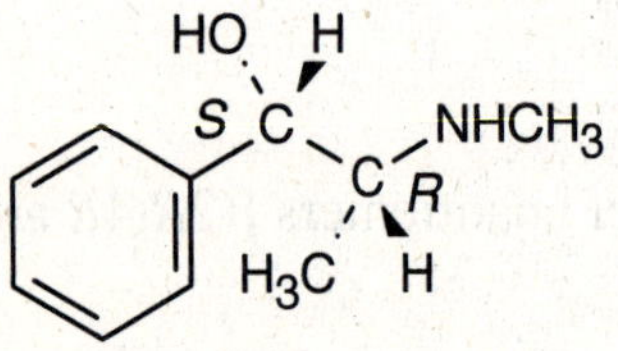

Pseudoephedrine

**6.23**

(a)

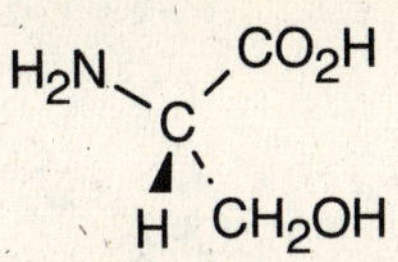

(*S*)-Serine

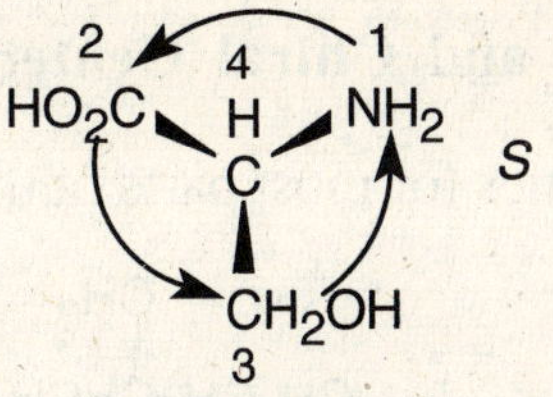

(b)

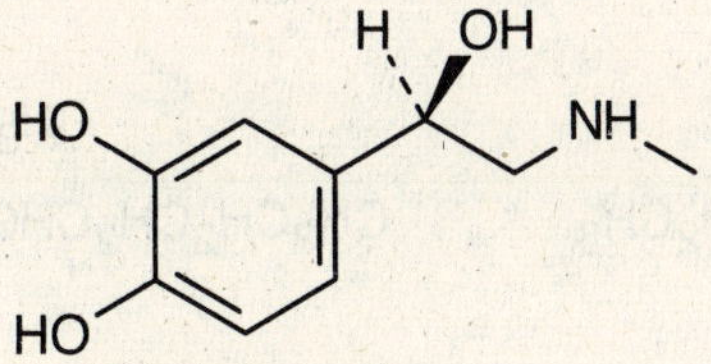

(*R*)-Adrenaline

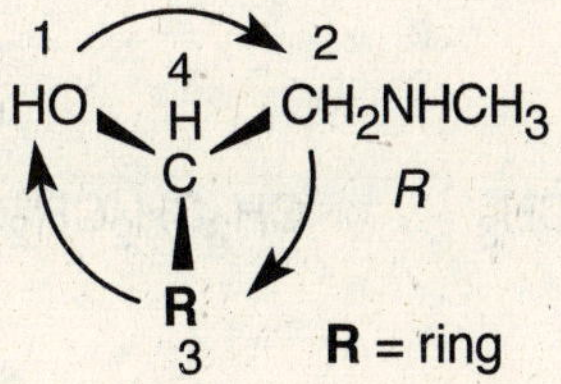

**6.24**

(a)

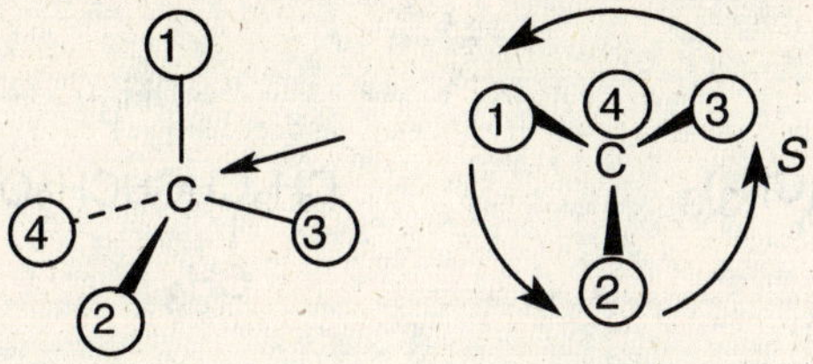

(b)

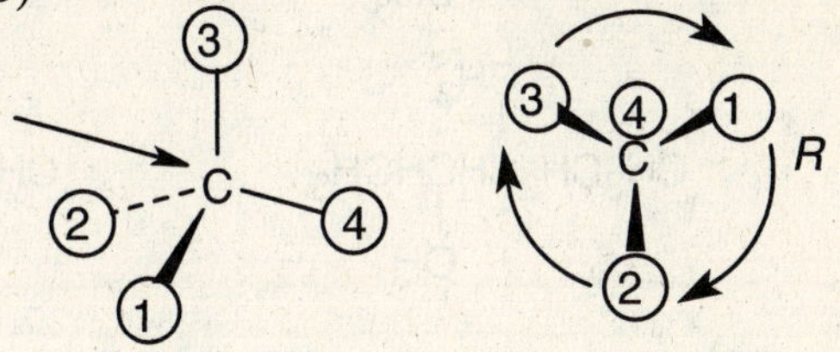

(c)

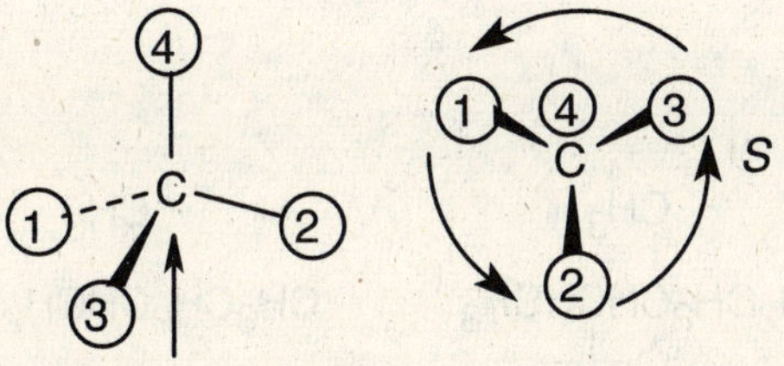

**6.25**

(a)

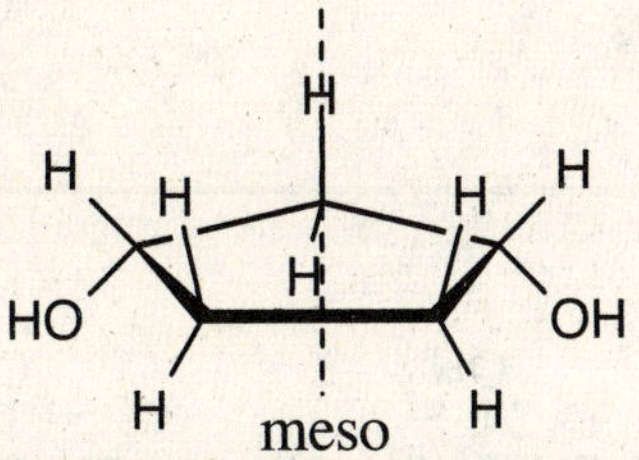

(b)

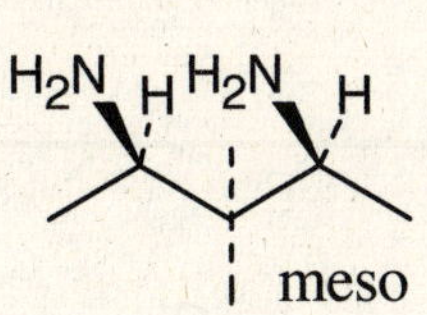

(c)

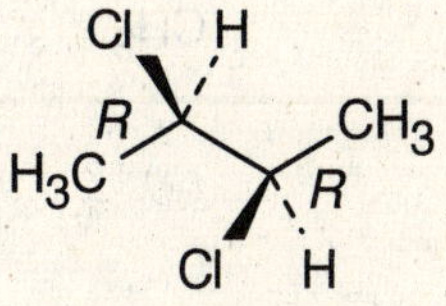

not meso

## Additional Problems

### Identifying Chiral Molecules and Chiral Centers

**6.26** There are several possibilities for most parts of this problem.

(a)

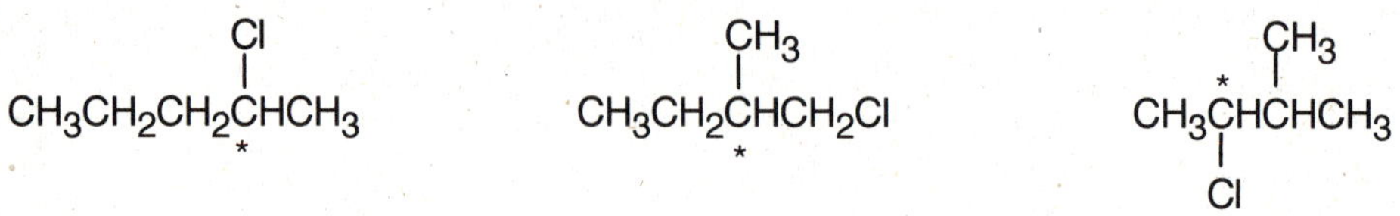

(b)

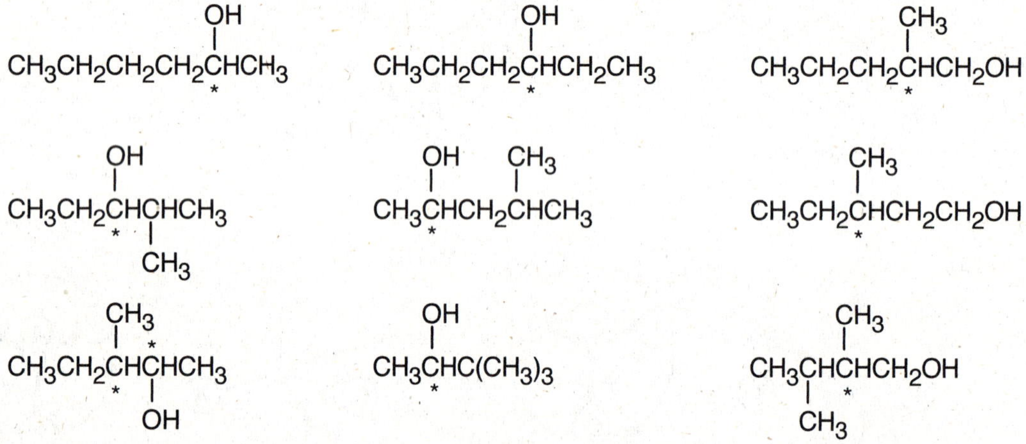

(c)

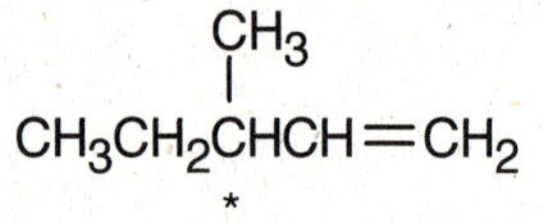

(d)

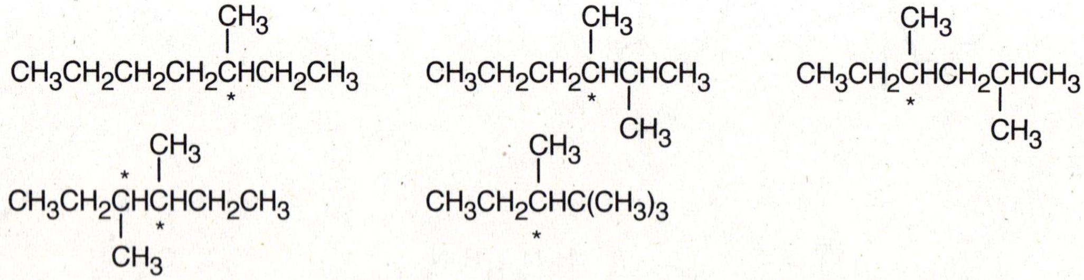

**6.27**

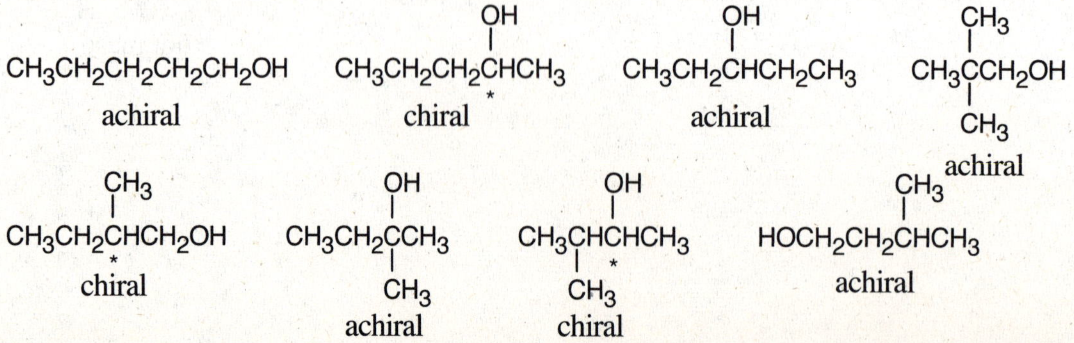

**6.28**

(a)

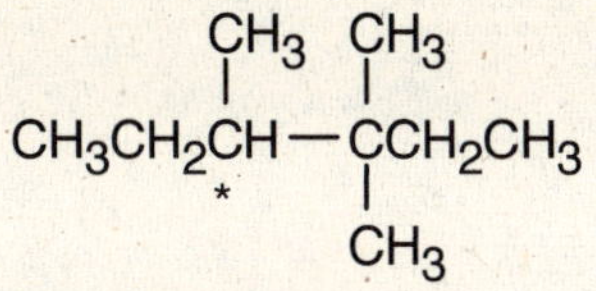

chiral

(b)

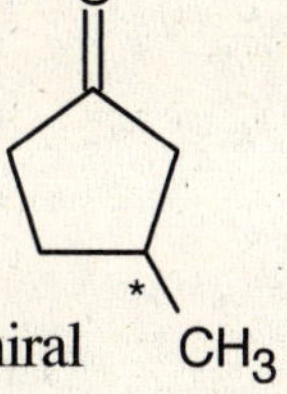

chiral

(c)

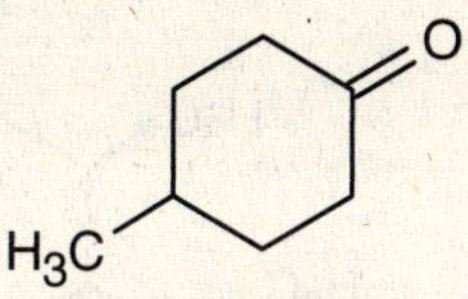

achiral

(d)

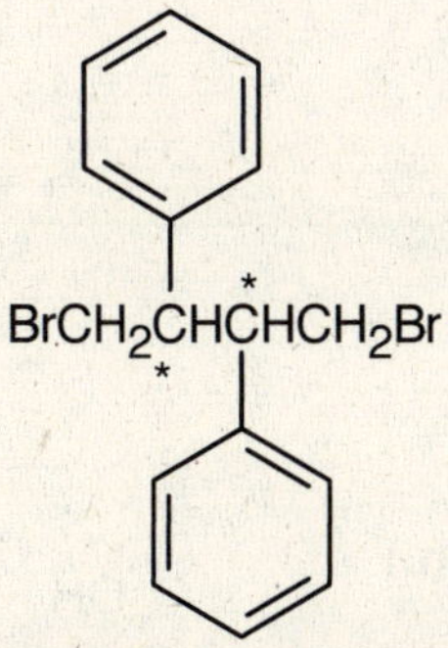

(e)

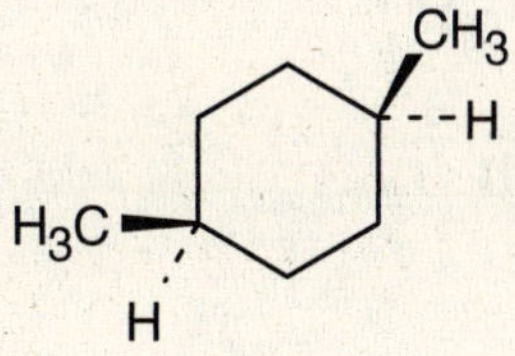

achiral

(f)

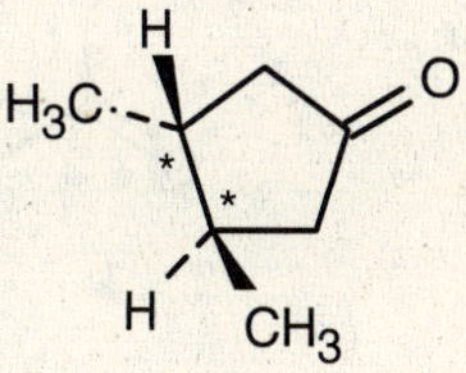

chiral

Compound (d) exists either as a pair of chiral enantiomers or as an achiral meso compound.

**6.29** Chiral: (d) golf club, (e) monkey wrench (because of the threads).
Achiral: (a) basketball, (b) fork, (c) wine glass, (f) snowflake.

**6.30**

(a)

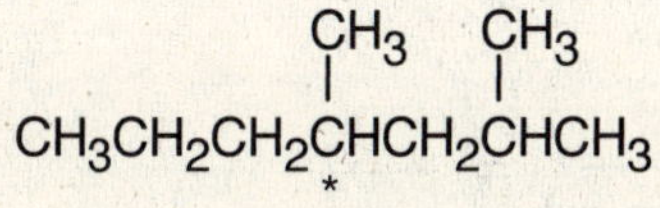

2,4-Dimethylheptane
chiral

(b)

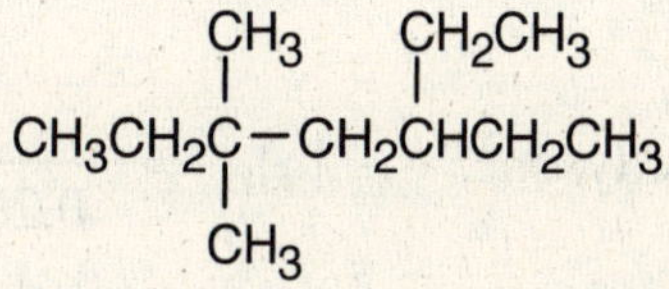

5-Ethyl-3,3-dimethylheptane
achiral

(c)

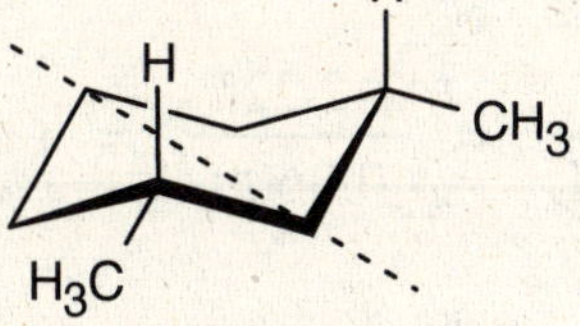

*cis*-1,3-Dimethylcyclohexane
achiral

(d)

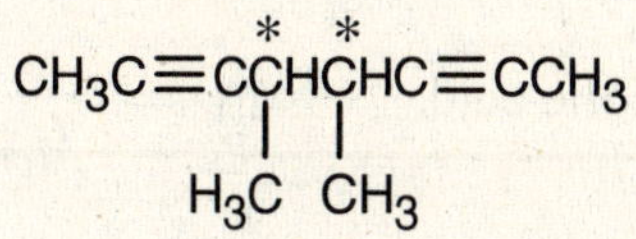

4,5-Dimethyl-2,6-octadiyne
chiral

The chirality of (d) depends on the configurations at both of the chirality centers. The (*R,R*) and (*S,S*) isomers are chiral enantiomers; the (*R,S*) isomer is an achiral meso compound.

**6.31**

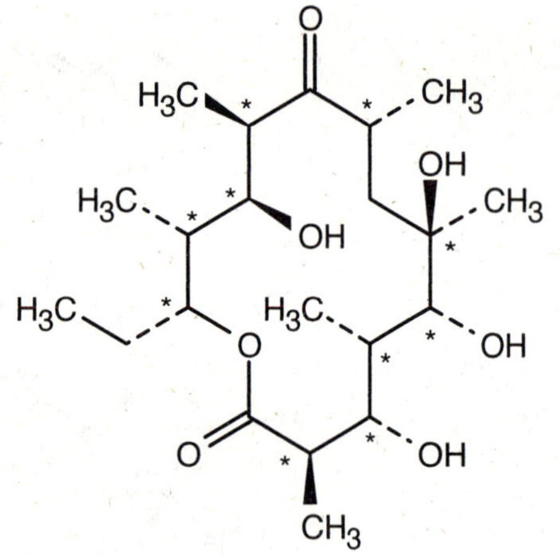

Erythronolide B

**6.32**

(a)

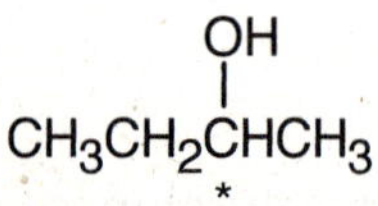

(b)

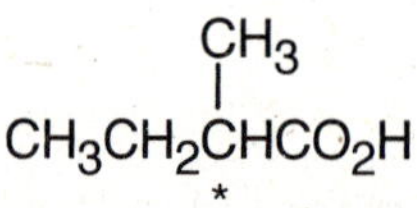

(c)

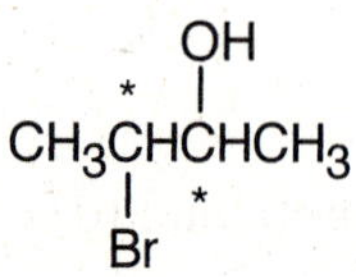

For (b) and (c), other compounds are also acceptable answers.

## Optical Rotation

**6.33**

$$\text{For ecdysone:} \quad [\alpha]_D = \frac{+0.087°}{0.200 \text{ dm} \times 0.00700 \text{ g/mL}} = +62$$

**6.34** (*R*)-Serine has a specific rotation of +6.83 because enantiomers differ only in the sign of their specific rotations.

**6.35**

Refer to Problem 6.7 for the formula for calculating $[\alpha]_D$. In this problem, C = 3.00 g ÷ 5.00 mL = 0.600 g/mL, and $l$ = 1.00 cm = 0.100 dm.

$$\text{For cholic acid:} \quad [\alpha]_D = \frac{+2.22°}{0.100 \text{ dm} \times 0.600 \text{ g/mL}} = \frac{+2.22°}{0.0600} = +37.0$$

## Assigning *R,S* Configuration to Chirality Centers

**6.36**

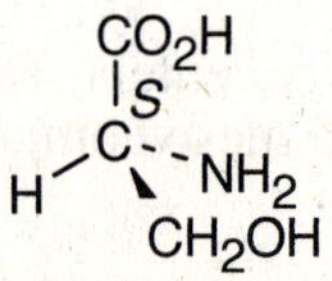

(*S*)-Serine

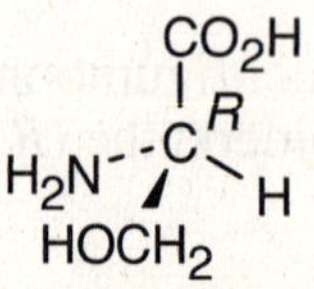

(*R*)-Serine

**6.37** The stereoisomer pictured is (*S*)-lactic acid.

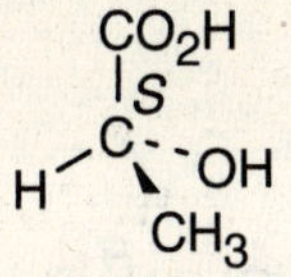

(*S*)-Lactic acid

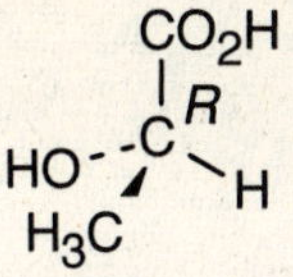

(*R*)-Lactic acid

**6.38**

*Highest priority* ⟶ *Lowest priority*

(a) $-OCH_3$, $-OH$, $-CH_3$, $-H$

(b) $-Br$, $-Cl$, $-CH_2Br$, $-CH_3$

(c) $-C(CH_3)_3$, $-CH{=}CH_2$, $-CH(CH_3)_2$, $-CH_2CH_3$

(d) $-OCH_3$, $-CO_2CH_3$, $-COCH_3$, $-CH_2OCH_3$

**6.39**

(a)

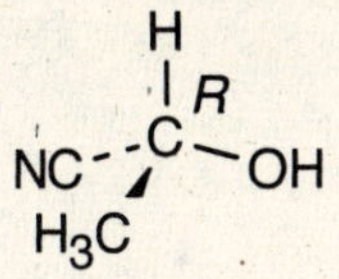

(b)

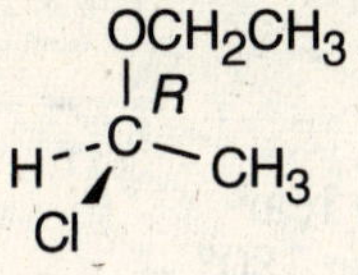

(c)

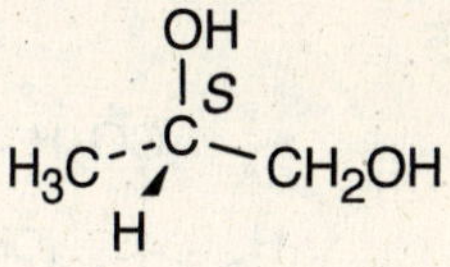

**6.40**

*Highest priority* ⟶ *Lowest priority*

(a)

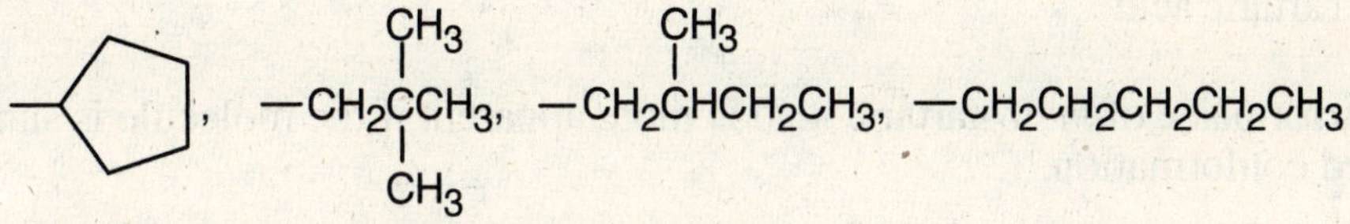

(b) $-SO_3H$, $-SH$, $-OCH_2CH_2OH$, $-NH_2$

**6.41**

(a)

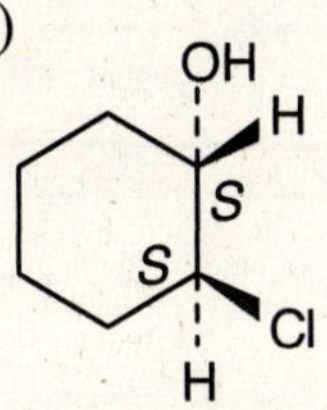

(b)

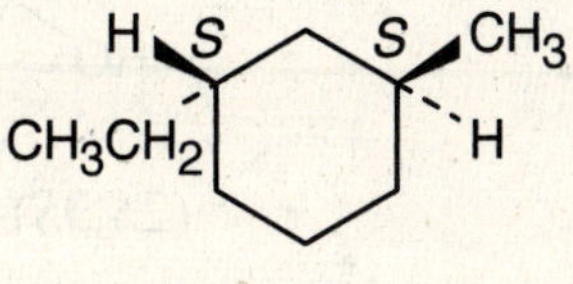

(c)

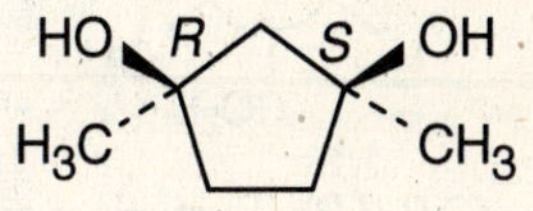

**6.42**

(a)

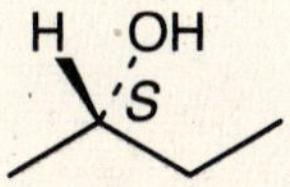

(b)

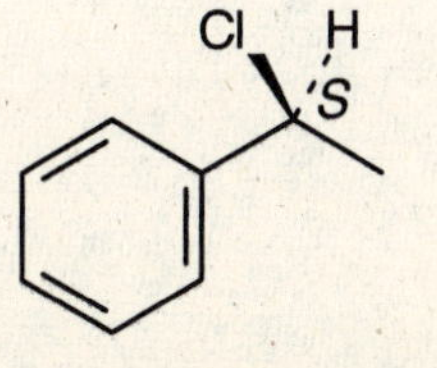

(c)

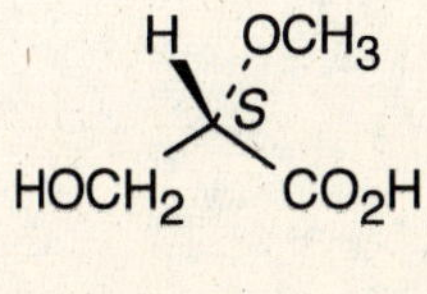

## Stereochemical Relationships

### 6.43–6.44

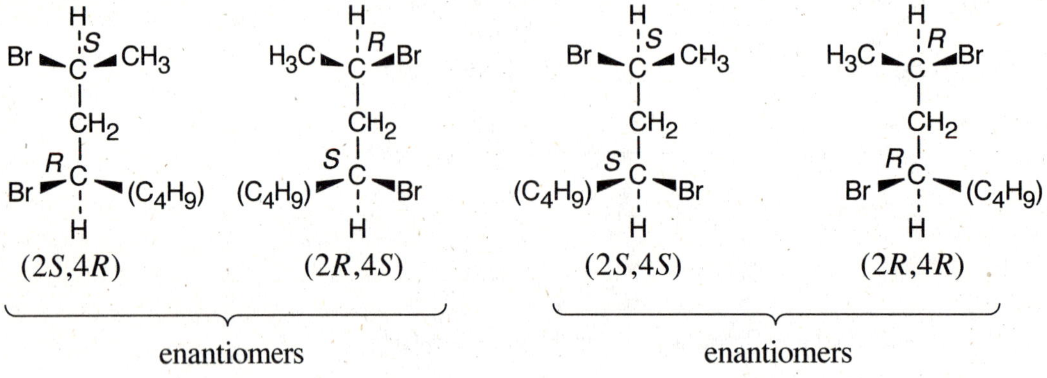


The (2*R*, 4*S*) stereoisomer is the enantiomer of the (2*S*, 4*R*) stereoisomer. The (2*S*, 4*S*) and (2*R*, 4*R*) stereoisomers are diastereomers of the (2*S*,4*R*) stereoisomer.

### 6.45

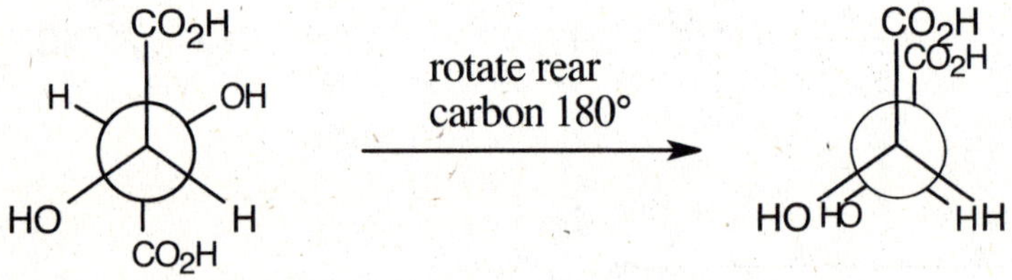


*meso*-Tartaric acid

The mirror plane of *meso*-tartaric acid is more apparent if the molecule is shown in its eclipsed conformation.

### 6.46

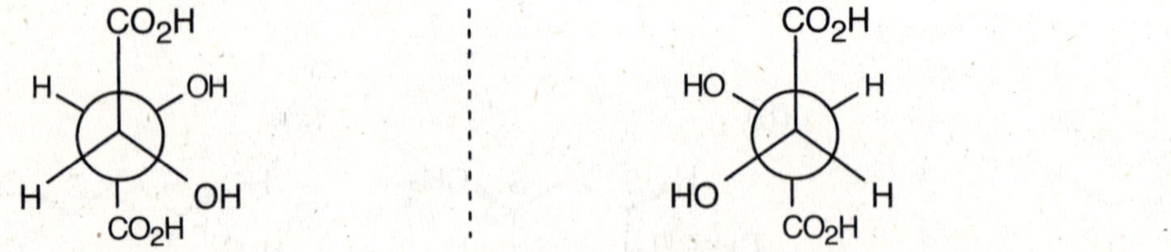


(2*R*,3*R*)-Tartaric acid (2*S*,3*S*)-Tartaric acid

The enantiomeric tartaric acids are mirror images of each other. Unlike the meso isomer, neither of the above enantiomers contains a mirror plane.

**6.47**

(a)

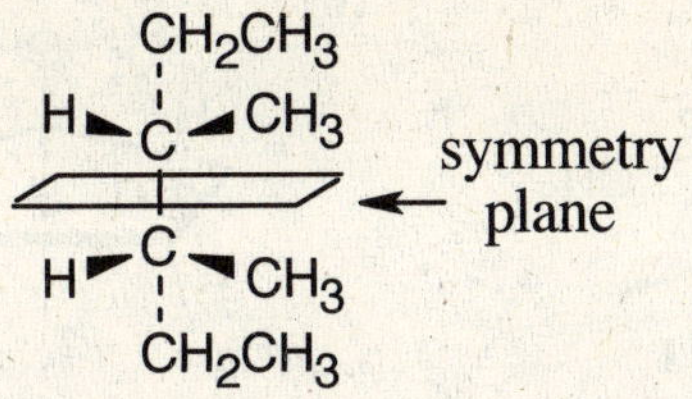

(b)

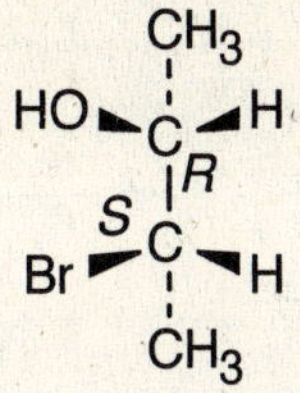

**6.48**

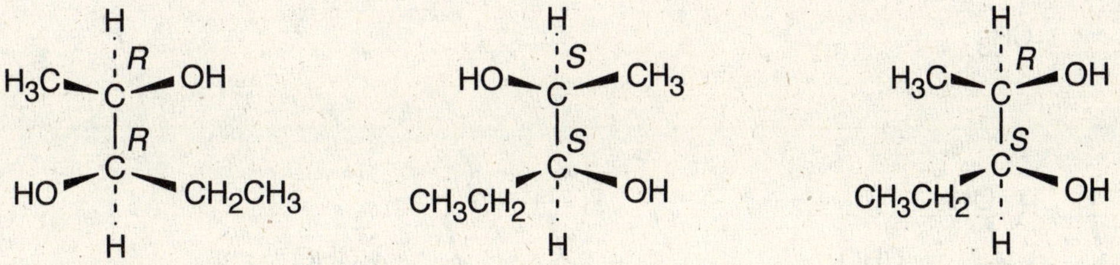

(2*R*,3*R*) (2*S*,3*S*) (2*R*,3*S*)

The specific rotations of the (2*R*,3*R*) and (2*S*,3*S*) enantiomers are equal in magnitude and opposite in sign. The specific rotations of the (2*R*,3*S*) and (2*R*,3*R*) diastereomers are not related.

**6.49**

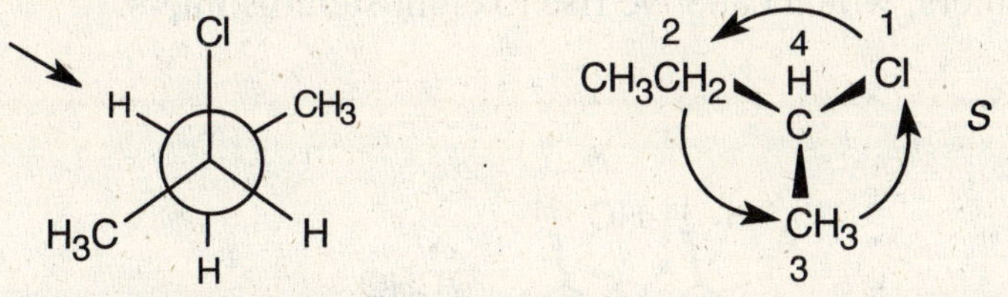

(*S*)-2-Chlorobutane

**6.50**

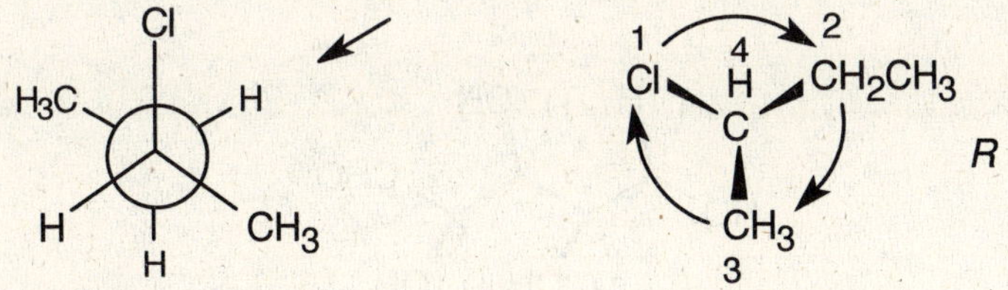

(*R*)-2-Chlorobutane

## General Problems

**6.51**

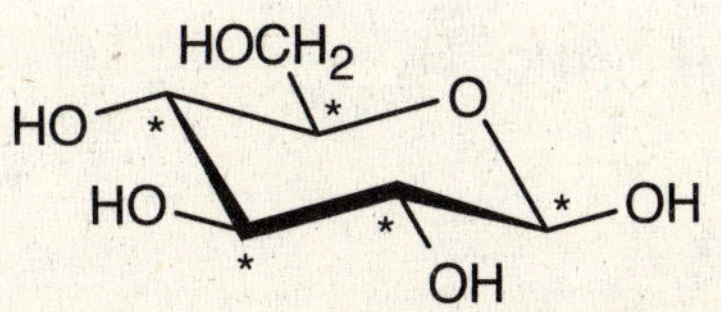

β-Glucose

The number of stereoisomers of a chiral compound is given by $2^n$, where $n$ equals the number of chirality centers present. Here, $n = 5$ and $2^n = 32$. Glucose thus has 32 possible stereoisomers.

**6.52**

(a)

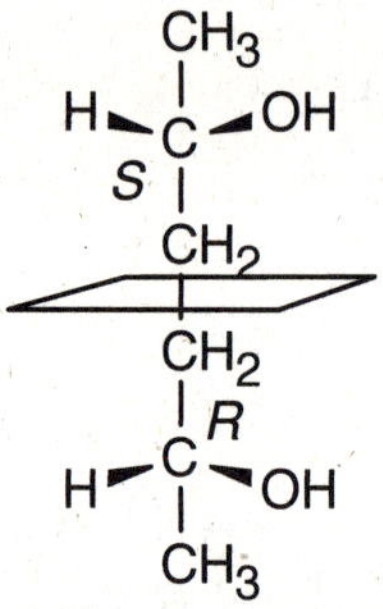

(b)

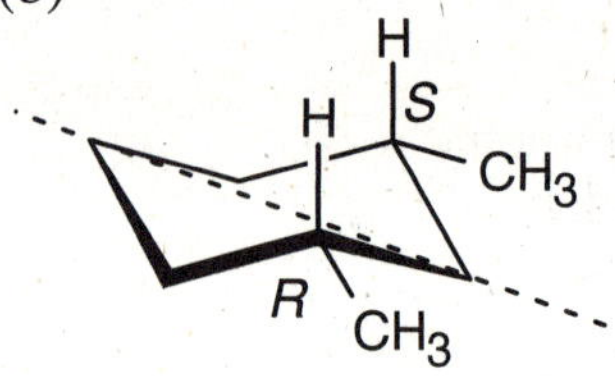

(c)

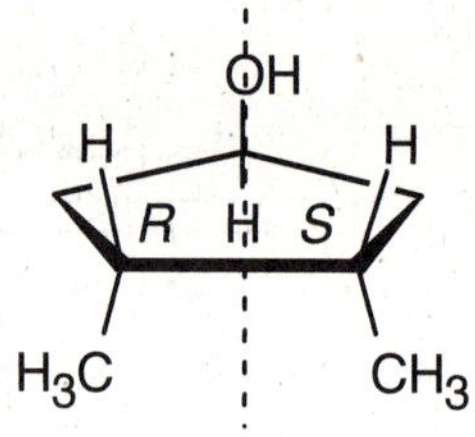

**6.53–6.54**

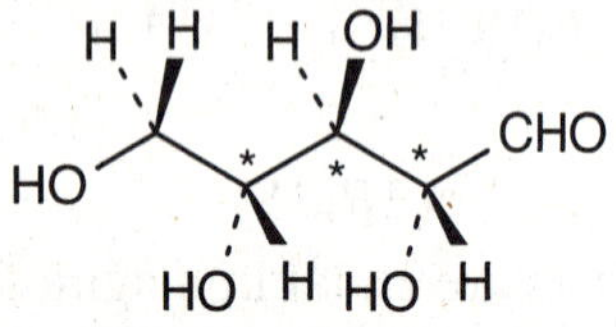

Ribose

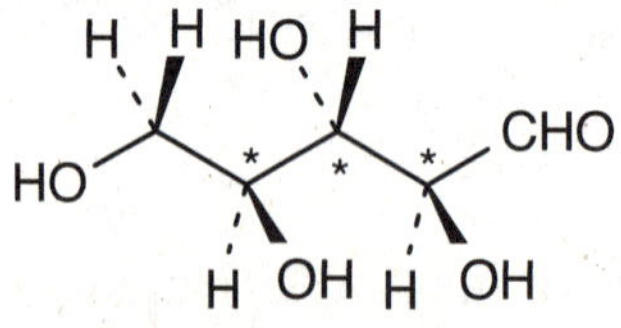

Enantiomer of ribose

Ribose has three chirality centers, which can give rise to eight stereoisomers.

**6.55** Ribose has six diastereomers.

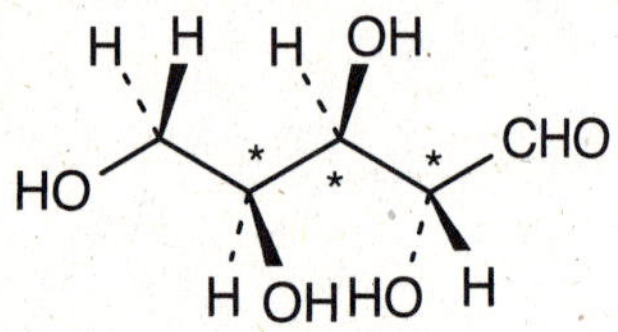

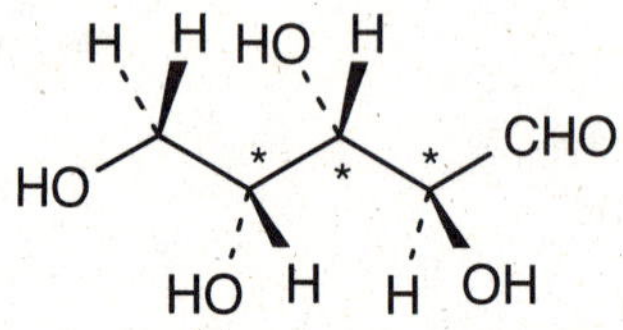

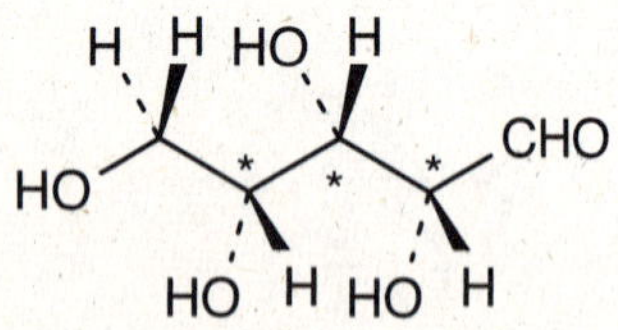

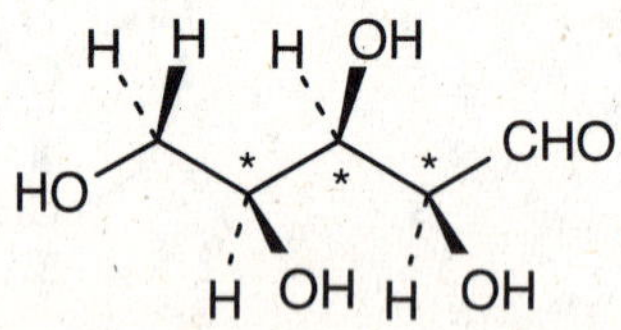

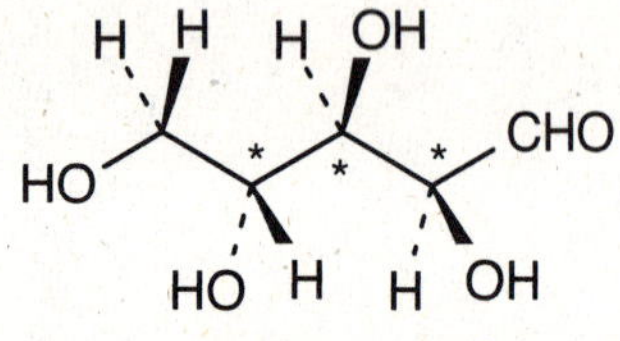

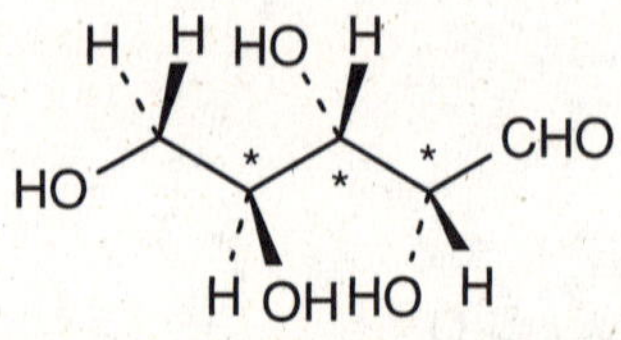

**6.56**

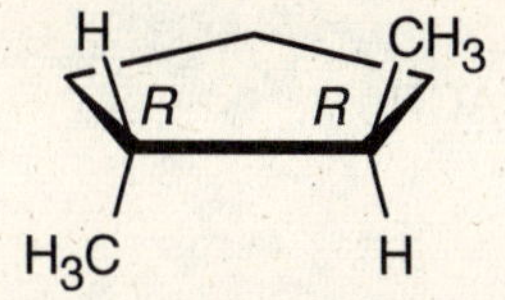


**A**

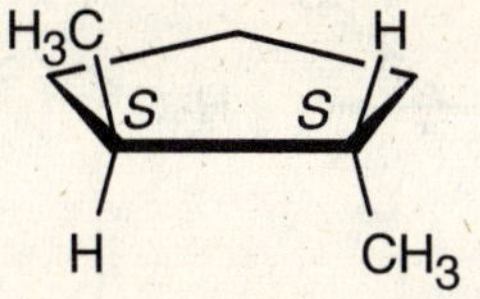


**B**

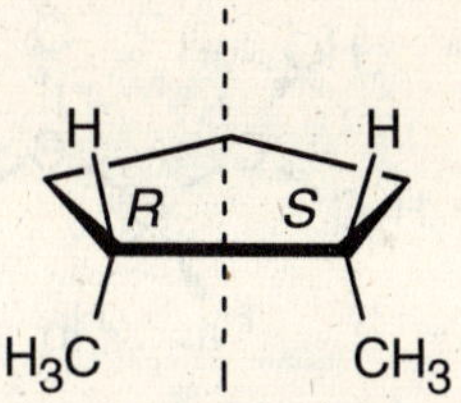


**C** meso

**A** and **B** are enantiomers and are chiral. Compound **C** is their diastereomer and is a meso compound.

**6.57**

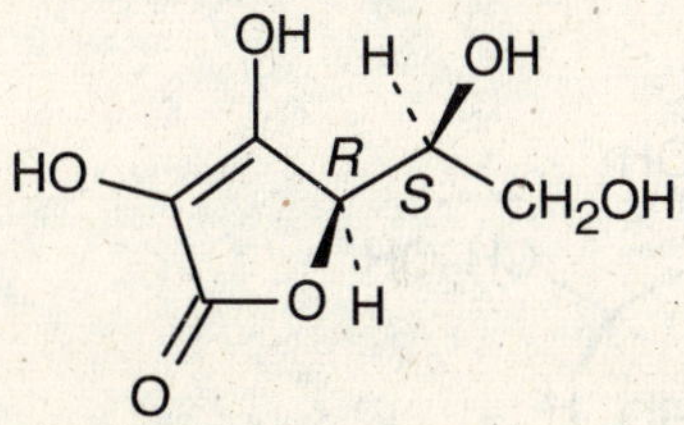


Ascorbic acid

**6.58**

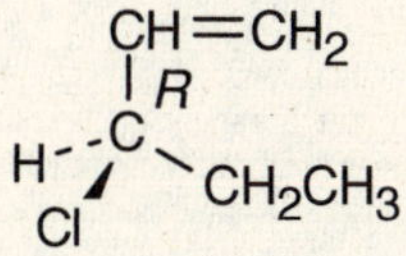


(*R*)-3-Chloropent-1-ene

**6.59**

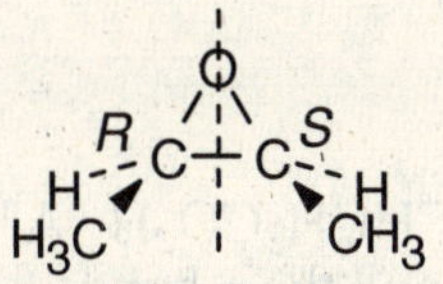


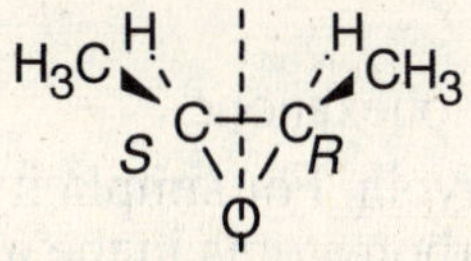


Peroxycarboxylic acids can attack either the "top" side or the "bottom" side of a double bond. The epoxide resulting from top side attack, pictured above, has two chirality centers, but because it has a plane of symmetry it is a meso compound. The epoxide resulting from bottom side attack is identical to epoxide resulting from top-side attack.

**6.60**

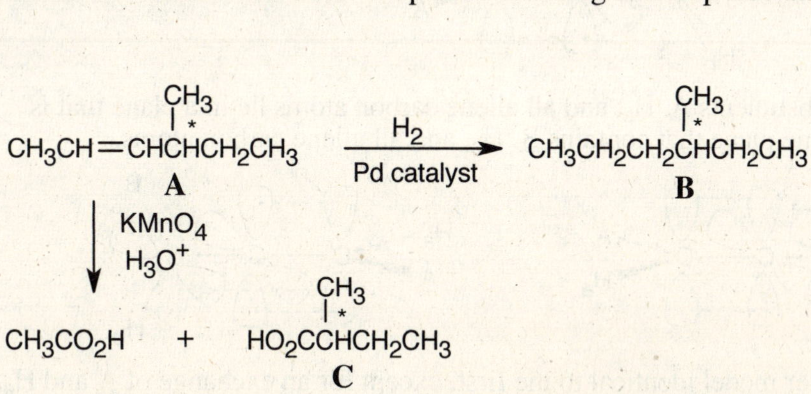

**6.61**

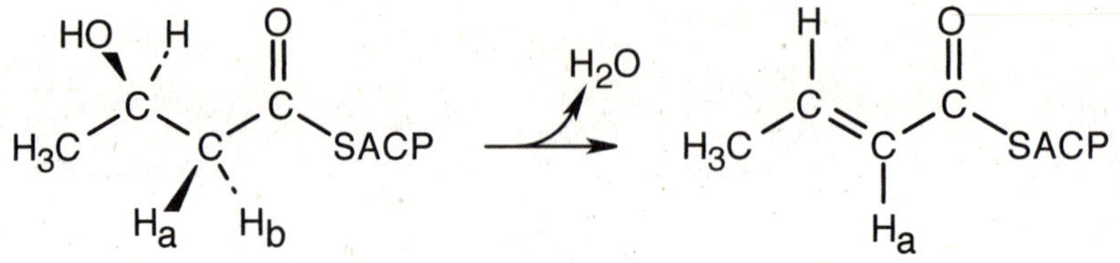


(R)-3-Hydroxybutyryl ACP *trans*-Crotonyl ACP

Notice the position of the –OH group relative to the two hydrogens. Hydrogen "b" is removed.

**6.62** Ribitol is an optically inactive meso compound. Catalytic hydrogenation converts the aldehyde functional group into a hydroxyl group and makes the two halves of ribitol mirror images of each other.

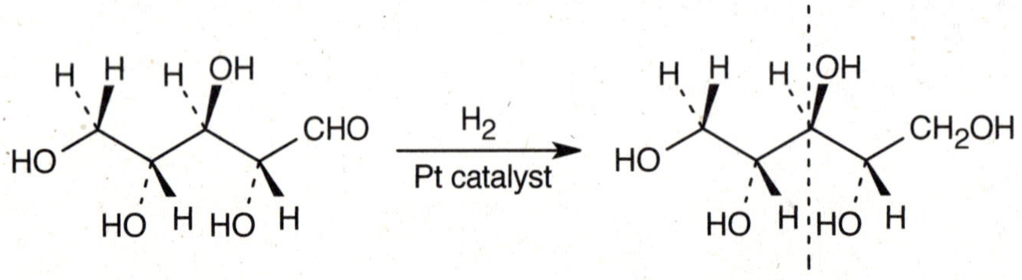


Ribose Ribitol

**6.63**

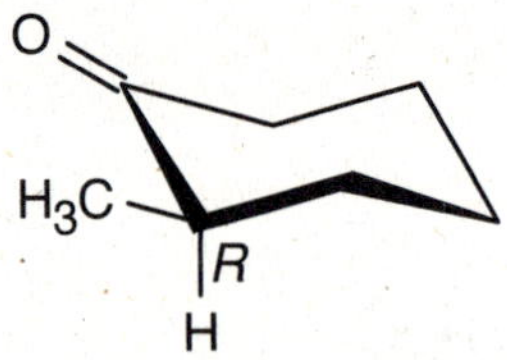


(*R*)-2-Methylcyclohexanone

**6.64** Make a model of mycomycin. For simplicity, call $–CH{=}CHCH{=}CHCH_2CO_2H$ "A" and $–C{\equiv}CC{\equiv}CH$ "B". The carbon atoms in the allene group are linear and the $\pi$ bonds formed are perpendicular to each other. Attach substituents to the $sp^2$ carbons.

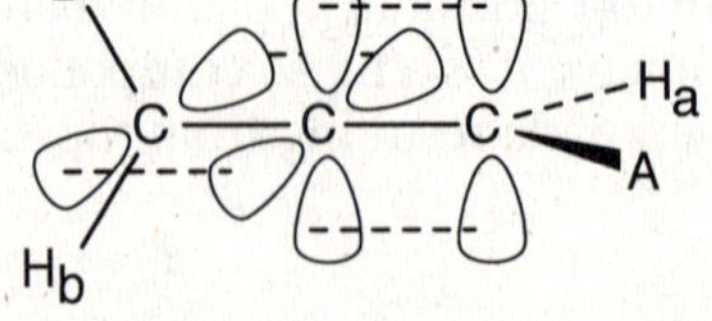


Notice that the substituents A, $H_a$, and all allene carbon atoms lie in a plane that is perpendicular to the plane that contains B, $H_b$, and all allene carbon atoms.

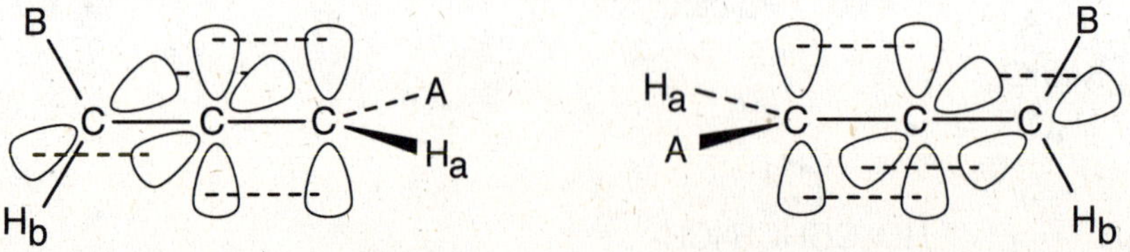


Now, make another model identical to the first, except for an exchange of A and $H_a$. This new allene is not superimposable on the original allene; the two allenes are mirror images. The two allenes are enantiomers and are chiral because they possess no plane of symmetry.

**In the Medicine Cabinet**

**6.65** (a)

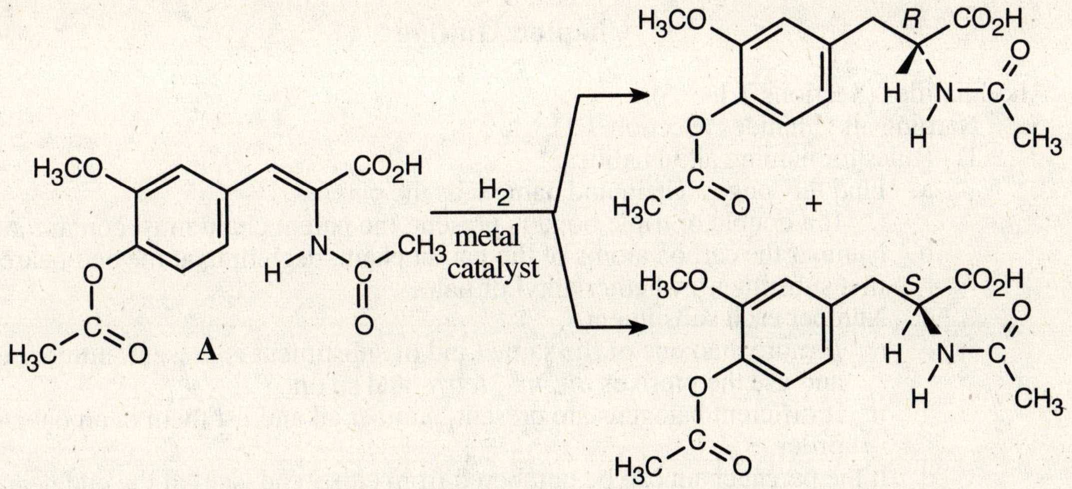

(b) Other than the sign of specific rotation, the enantiomers have identical physical properties.

(c) In a chiral environment, the enantiomers have different chemical properties and are not equally effective as drugs.

(d)

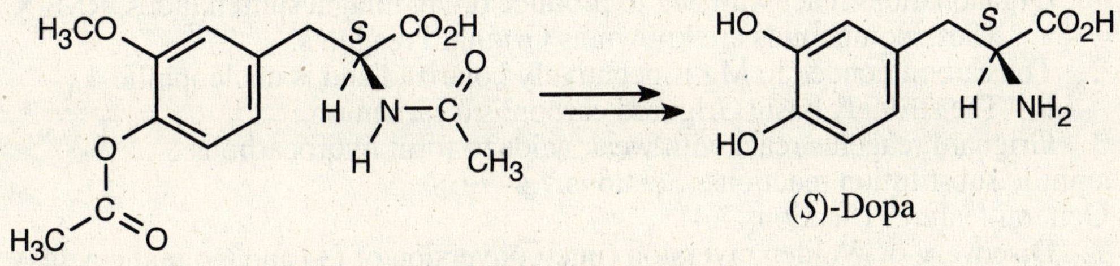

A chiral catalyst allows hydrogenation to take place on only one side of the double bond to produce a single enantiomer.

**In the Field**

**6.66** (a)

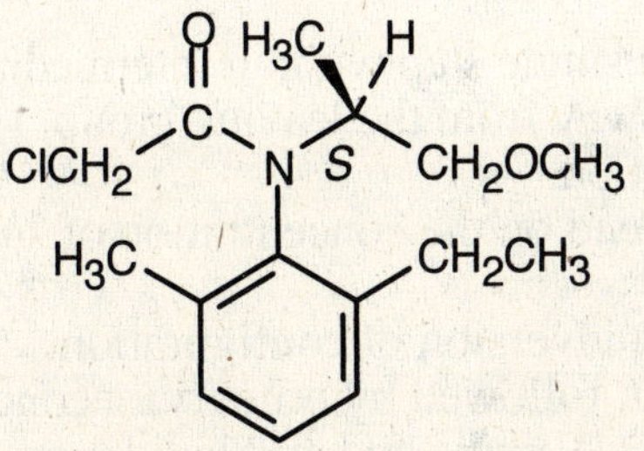

Metolachlor

(b) The enantiomers differ in activity because they interact differently with a chiral enzyme.

# Chapter 7 – Organohalides: Nucleophilic Substitutions and Eliminations

## Chapter Outline

I. Alkyl halides (Sections 7.1–7.3).
   A. Naming alkyl halides (Section 7.1).
      1. Rules for naming alkyl halides:
         a. Find the longest chain and name it as the parent.
            i. If a double or triple bond is present, the parent chain must contain it.
         b. Number the carbon atoms of the parent chain, beginning at the end nearer the first substituent, whether alkyl or halo.
         c. Number each substituent.
            i. If more than one of the same kind of substituent is present, number each, and use the prefixes *di-*, *tri-*, *tetra-* and so on.
            ii. If different halogens are present, number all and list them in alphabetical order.
         d. If the parent chain can be numbered from either end, start at the end nearer the substituent that has alphabetical priority.
      2. Some alkyl halides are named by first citing the name of the alkyl group and then citing the halogen (non-IUPAC).
   B. Preparation of alkyl halides (Sections 7.2).
      1. Radical chlorination of alkanes.
      2. Addition of $X_2$ (X = Cl, Br) or HX (X = Cl, Br, I) to alkenes.
      3. Reaction of alcohols with HX, $SOCl_2$ or $PBr_3$.
   C. Grignard reagents from alkyl halides (Section 7.3).
      1. Organohalides react with Mg to produce organomagnesium halides, RMgX.
         a. These compounds are known as Grignard reagents.
      2. The carbon bonded to Mg is negatively polarized and is nucleophilic.
         a. The strongly basic Grignard carbon is a carbanion.
      3. Grignard reagents react with weak acids to form hydrocarbons.

II. Nucleophilic substitution reactions (Sections 7.4–7.6).
   A. General features (Sections 7.4).
      1. Discovery of Walden inversion (interconversion of (+) and (–) malic acids).
         a. This discovery meant that one or more reactions must have occurred with inversion of configuration at the chirality center.
      2. Characteristics of nucleophilic substitution reactions.
         a. Nucleophilic substitutions occur by two pathways – $S_N2$ and $S_N1$.
         b. In all nucleophilic substitutions, a nucleophile reacts with a substrate and substitutes for a leaving group.
         c. Many types of products can be prepared by nucleophilic substitution reactions.
   B. $S_N2$ reactions (Section 7.5).
      1. A $S_N2$ reaction takes place in a single step without intermediates.
      2. The nucleophile attacks 180° away from the leaving group.
      3. Characteristics of the $S_N2$ reaction.
         a. Rates of $S_N2$ reactions depend on the concentration of substrate and of nucleophile.
         b. $S_N2$ reactions proceed with inversion of configuration.
         c. Methyl halides react fastest, followed by primary, secondary, and tertiary halides, which react slowest. Vinylic and aryl halides are unreactive.
         d. Substrates with better leaving groups (stable anions) react faster.

C. $S_N1$ reactions (Section 7.6).
   1. An $S_N1$ reaction usually occurs only with tertiary substrates under neutral or acidic conditions in hydroxylic solvents.
   2 An $S_N1$ reaction takes place in two steps.
      a. The leaving group dissociates to form a carbocation intermediate.
      b. The intermediate reacts with the nucleophile to form product.
   3. Characteristics of $S_N1$ reactions.
      a. The rate of an $S_N1$ reaction depends only on the concentration of substrate.
      b. The rate of reaction depends on the stability of the carbocation intermediate.
      c. The product of an $S_N1$ reaction is racemic.
      d. The best leaving groups give stable anions.

III. Elimination reactions (Sections 7.7–7.8).
   A. Characteristics of elimination reactions (Section 7.7).
      1. An alkene can form when an alkyl halide reacts with a base.
      2. The major product usually has the most highly substituted double bond (Zaitsev's Rule).
   B. E2 reactions.
      1. E2 reactions take place when a primary or secondary alkyl halide is treated with a strong base.
      2. The reaction takes place in one step.
   C. E1 and E1cB reactions (Section 7.8).
      1. An E1 reaction takes place in two steps.
      2. The intermediate carbocation loses $H^+$ to give an alkene.
      3. The best E1 substrates are also the best $S_N1$ substrates.
   D. E1cB reactions.
      1. An E1cB reaction takes place through a carbanion intermediate.
      2. This route takes place when a substrate has a poor leaving group.

IV. A summary of reactivity: $S_N1$, $S_N2$, E1, E2 (Section 7.9).
   A. A primary substrate reacts by either an $S_N2$ route (with a good nucleophile), an E2 route (with a strong base), or an E1cB route.
   B. A secondary substrate reacts by both $S_N2$ and E2 routes to give a mixture of products.
      1. An E1cB reaction takes place if the leaving group is two carbons away from a carbonyl group.
   C. A tertiary substrate reacts by an E2 pathway (with a strong base) or by a mixture of $S_N1$ and E1 pathways (with neutral or acidic nucleophiles).

V. Substitution and elimination reactions in living organisms (Section 7.10).
   A. Biological methylations are substitution reactions.
   B. Eliminations (especially E1cB) occur frequently in metabolic pathways.

## Solutions to Problems

**7.1**

(a)
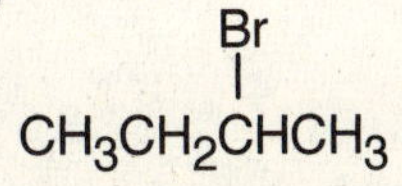
2-Bromobutane

(b)
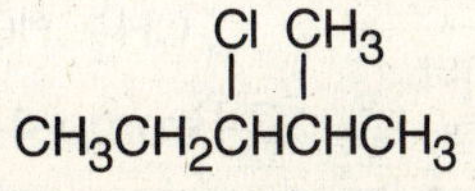
3-Chloro-2-methylpentane

(c)
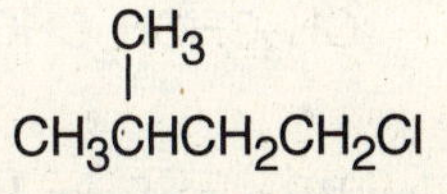
1-Chloro-3-methylbutane

(d)
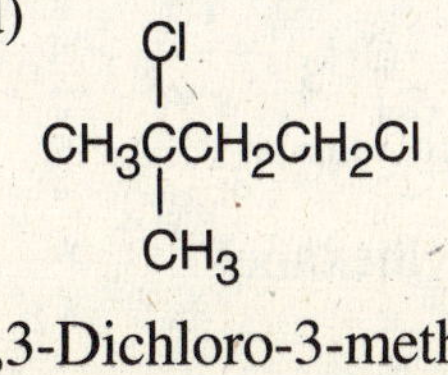
1,3-Dichloro-3-methylbutane

(e)
$BrCH_2CH_2CH_2CH_2Cl$
1-Bromo-4-chlorobutane

(f)
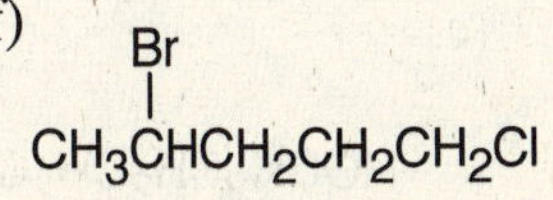
4-Bromo-1-chloropentane

**7.2**

(a)

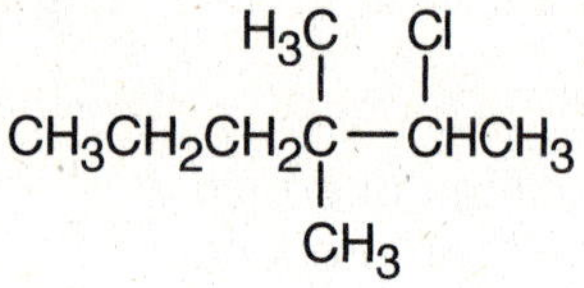

2-Chloro-3,3-dimethylhexane

(b)

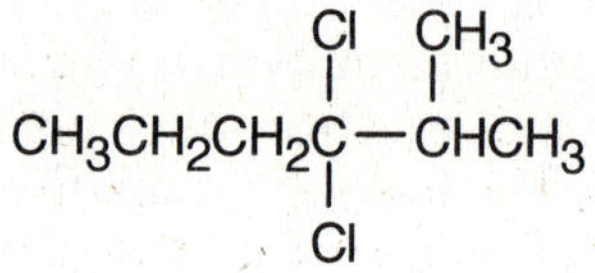

3,3-Dichloro-2-methylhexane

(c)

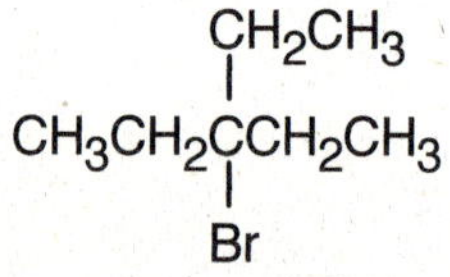

3-Bromo-3-ethylpentane

(d)

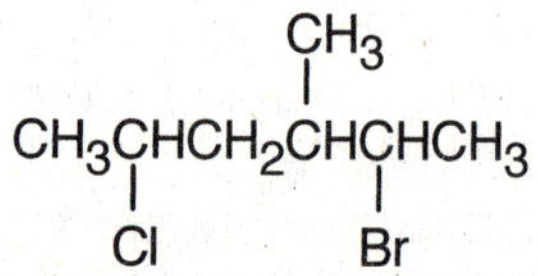

2-Bromo-5-chloro-3-methylhexane

**7.3**

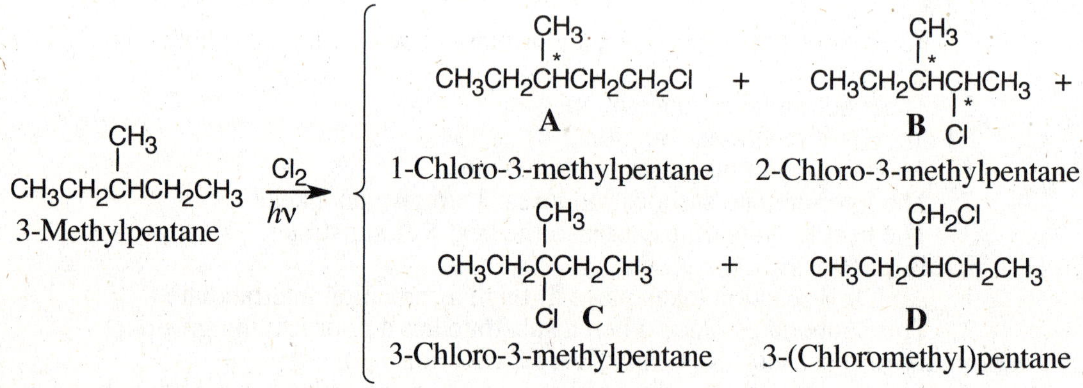

3-Methylpentane; A: 1-Chloro-3-methylpentane; B: 2-Chloro-3-methylpentane; C: 3-Chloro-3-methylpentane; D: 3-(Chloromethyl)pentane

Products **A** and **B**, in which a carbon is bonded to four different groups, are chiral.

**7.4** Tertiary alkyl halides can be prepared from the appropriate alcohols by reaction with HX.

(a)

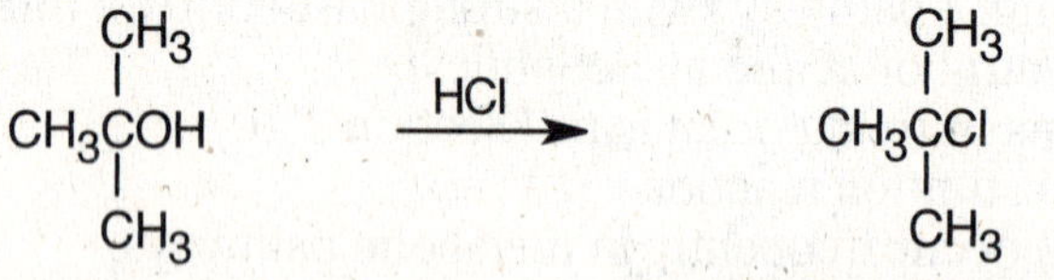

2-Chloro-2-methylpropane

(b)

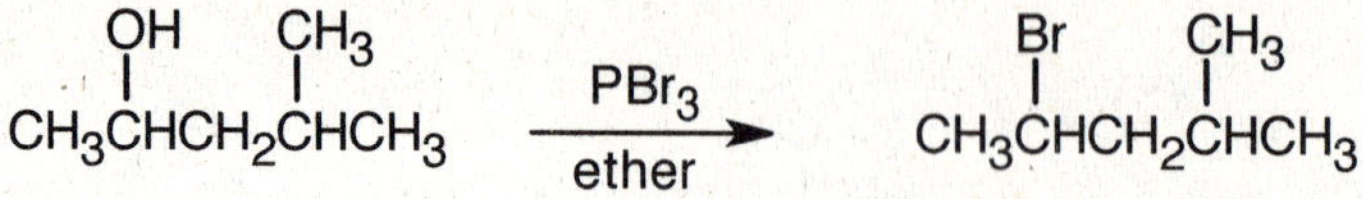

2-Bromo-4-methylpentane

Use $PBr_3$ to form bromoalkanes from primary and secondary alcohols.

(c)

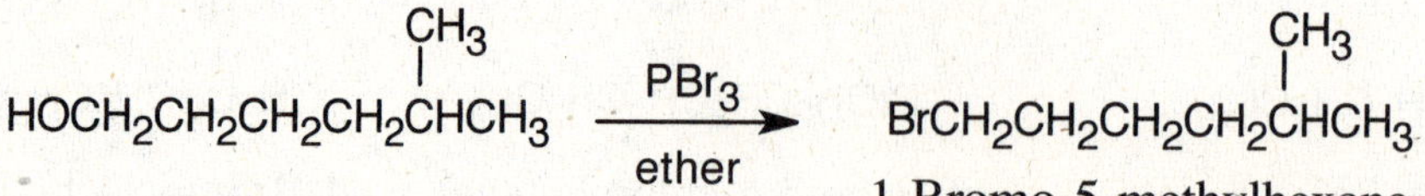

1-Bromo-5-methylhexane

(d)

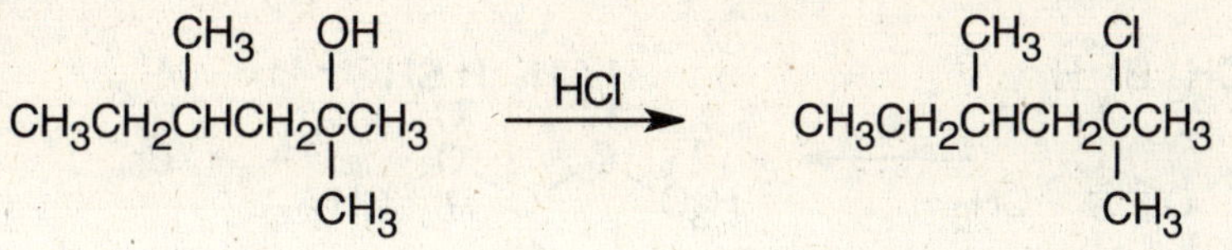

2-Chloro-2,4-dimethylhexane

**7.5**

(a)

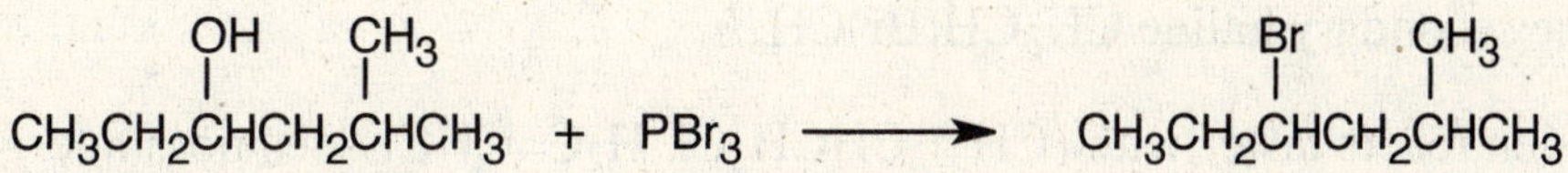

(b)

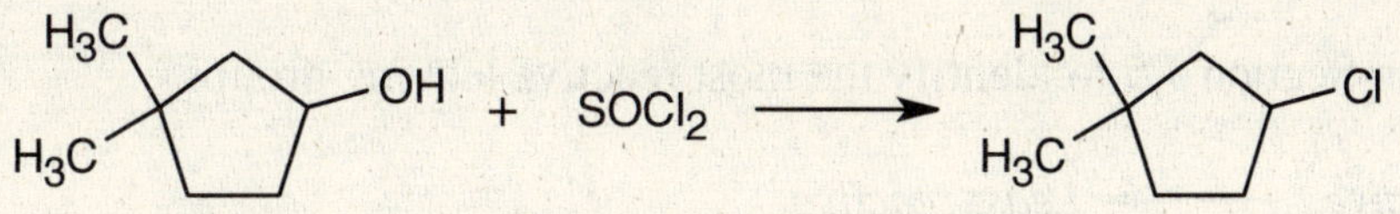

Thionyl chloride is used to form a chloride from a primary or secondary alcohol.

**7.6**

(a)

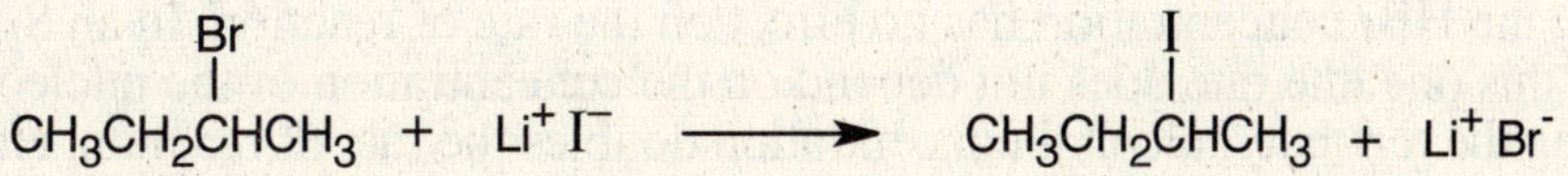

(b)

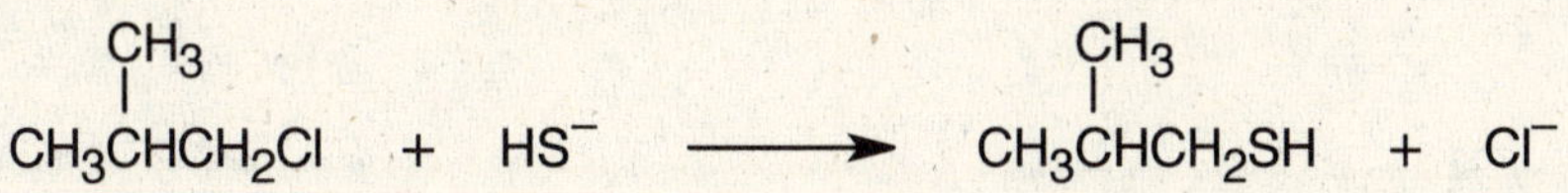

(c)

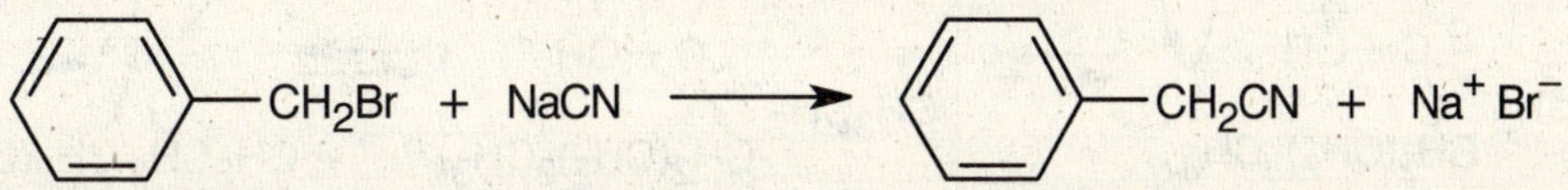

**7.7**

(a)

$$CH_3CH_2CH_2CH_2Br \; + \; Na^+ \,^-OH \longrightarrow CH_3CH_2CH_2CH_2OH \; + \; Na^+ \, Br^-$$

(b)

$$(CH_3)_2CHCH_2CH_2Br \; + \; Na^+ \, N_3^- \longrightarrow (CH_3)_2CHCH_2CH_2N_3 \; + \; Na^+ \, Br^-$$

**7.8** (a) If $[CH_3I]$ is tripled, the reaction rate is tripled.

(b) If both $[CH_3I]$ and $[CH_3CO_2Na]$ are doubled, the reaction rate is quadrupled.

**7.9** As described in Worked Example 7.4, identify the leaving group and the chirality center. Draw the product carbon skeleton, inverting the configuration at the chirality center, and replace the leaving group with the nucleophilic reactant.

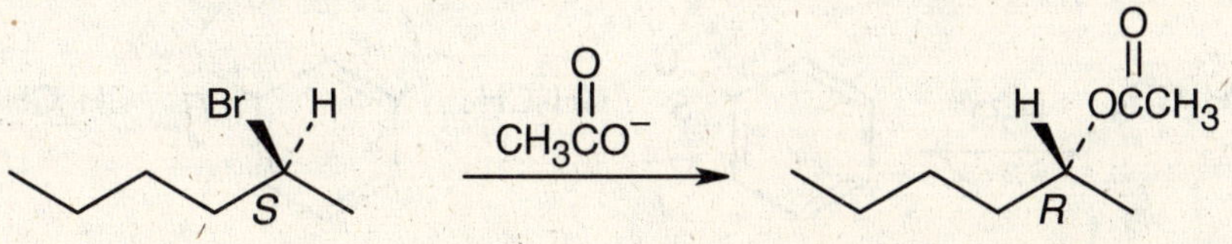

**7.10**

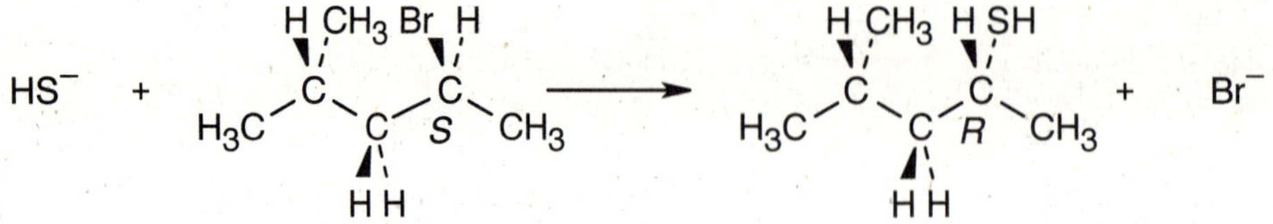

**7.11** (a) Reaction of cyanide ion proceeds faster with the primary halide $CH_3CH_2CH_2Br$ than with the secondary halide $CH_3CH(Br)CH_3$.

(b) Iodide ion reacts faster with $(CH_3)_2CHCH_2Cl$. $H_2C{=}CHCl$ does not undergo $S_N2$ reactions.

**7.12** Use the chart in Section 7.5 to identify the most reactive leaving groups.

*Most reactive* ⟶ *Least reactive*

$CH_3{-}I > CH_3{-}Br >> CH_3{-}F$

**7.13** (a) Tripling the HBr concentration has no effect on the rate of reaction. In an $S_N1$ reaction such as this one, the rate does not depend on the concentration of the nucleophile.
(b) Doubling the *tert*-butyl alcohol concentration doubles the rate of reaction. Halving the HBr concentration has no effect on the rate. Thus, the overall rate is doubled.

**7.14**

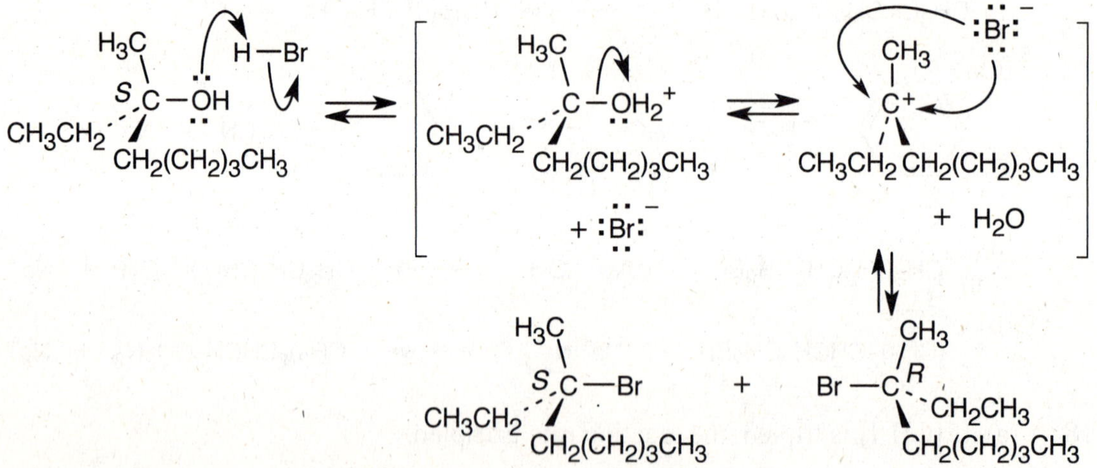

Attack by $Br^-$ can occur on either side of the planar, achiral carbocation intermediate. The resulting product is a racemic mixture.

**7.15** The *S* substrate reacts with water to form a mixture of *R* and *S* alcohols. The ratio of enantiomers is close to 50:50.

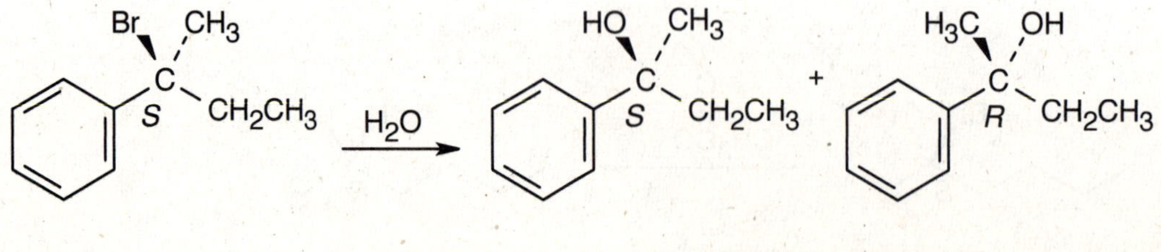

**7.16** The major product has the more substituted double bond (Zaitsev's Rule).

(a)

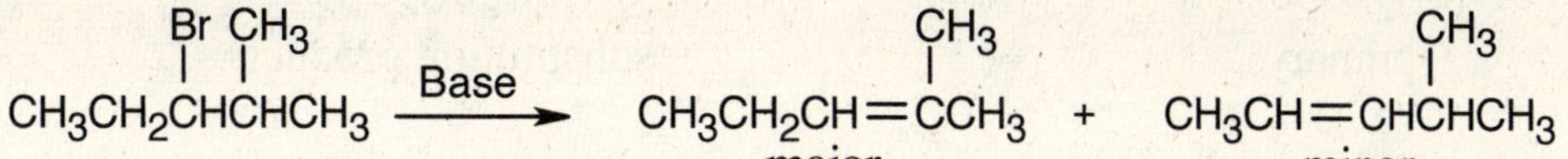

major (trisubstituted double bond) — minor (disubstituted double bond)

(b)

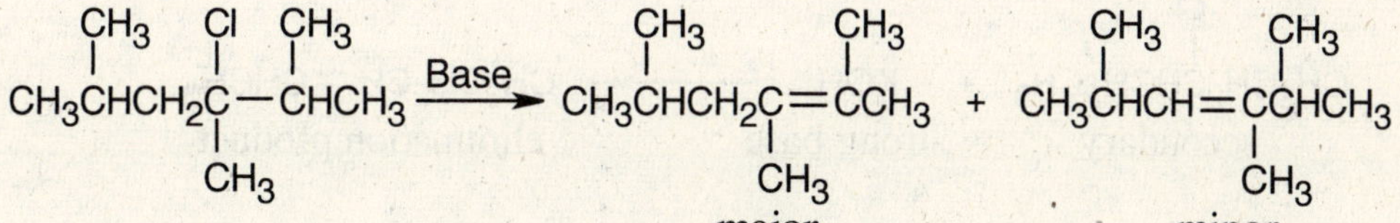

major (tetrasubstituted double bond) — minor (trisubstituted double bond)

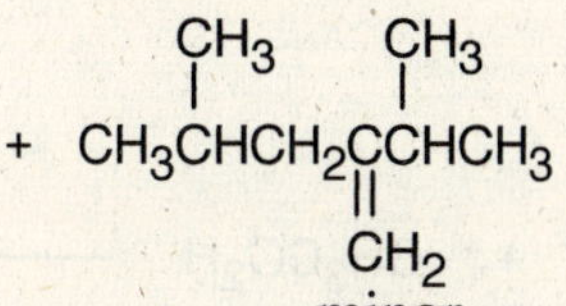

minor (disubstituted double bond)

(c)

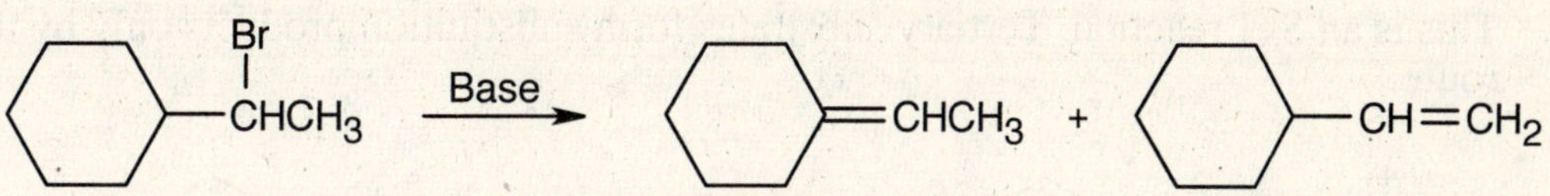

major (trisubstituted double bond) — minor (monosubstituted double bond)

**7.17**

(a)

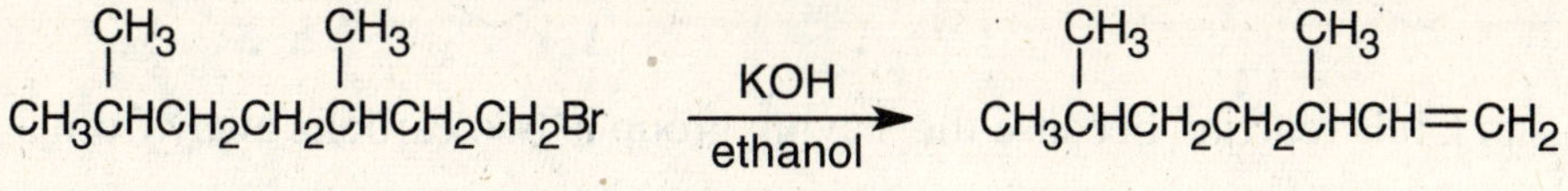

Only this alkyl halide gives the desired alkene as the major product.

(b)

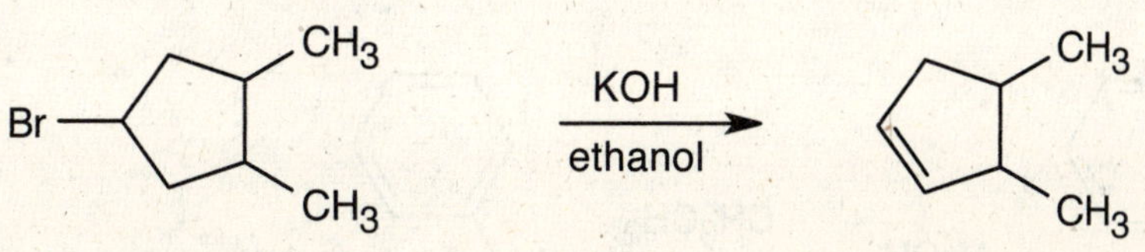

**7.18** The rate of E1 reaction would triple if the concentration of the alkyl halide were tripled because the reaction is unimolecular.

**7.19**

(a)

$CH_3CH_2CH_2CH_2Br$ + $NaN_3$ ⟶ $CH_3CH_2CH_2CH_2N_3$

primary — substitution product

The reaction occurs by a $S_N2$ mechanism because the substrate is primary, the nucleophile is nonbasic, and the product is a substitution product.

(b)

$CH_3CH_2CH(Cl)CH_2CH_3$ + KOH ⟶ $CH_3CH_2CH{=}CHCH_3$

secondary — strong base — elimination product

This is an E2 reaction since a secondary halide reacts with a strong base to yield an elimination product.

(c)

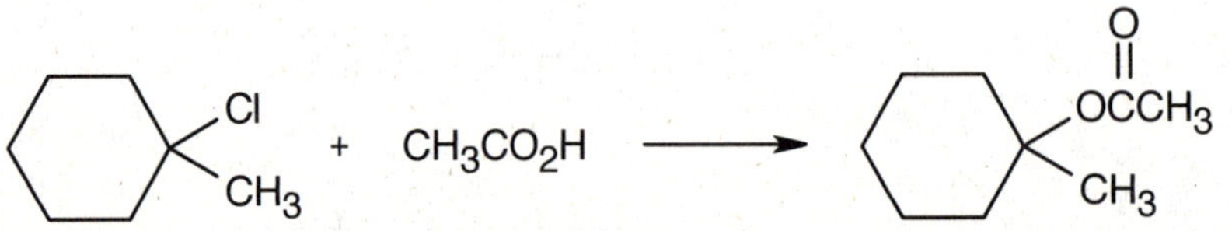

tertiary — substitution product

This is an $S_N1$ reaction. Tertiary substrates form substitution products only by the $S_N1$ route.

(d)

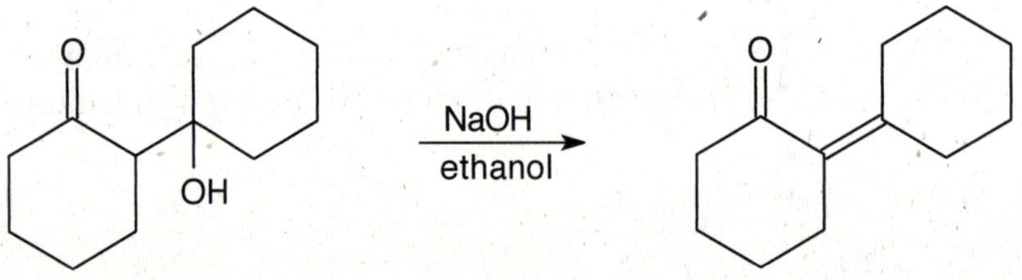

This is an E1cB reaction because the leaving group is two carbons away from a carbonyl group.

## Visualizing Chemistry

**7.20**

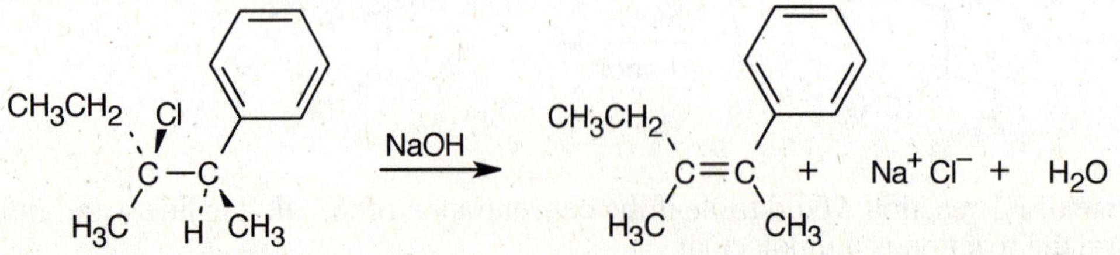

According to Zaitsev's rule, the product with the most substituted double bond is most likely to form. Notice also that the double bond that results from elimination is conjugated with the aromatic ring.

**7.21**

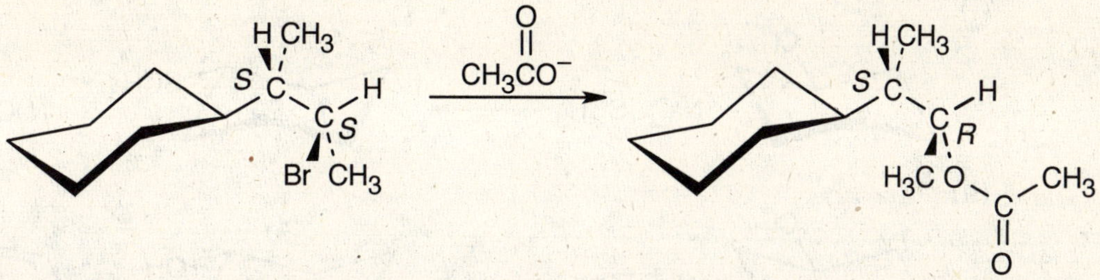

Reaction of the secondary bromide with the weakly basic acetate nucleophile occurs by an $S_N2$ route, with inversion of configuration, to produce the *R* acetate.

**7.22**

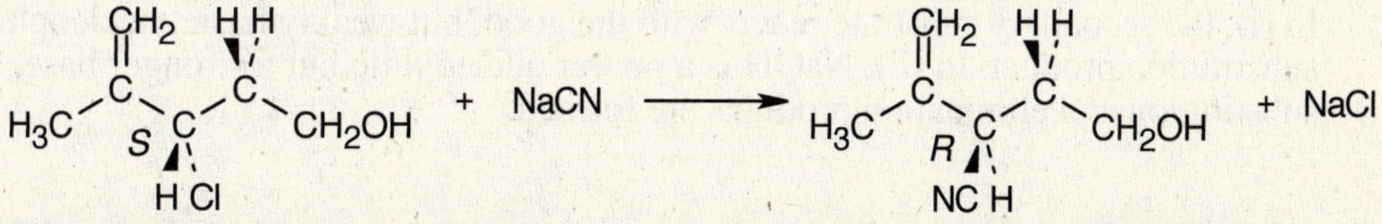

The *S* substrate has a secondary allylic chloride group and a primary hydroxyl group. $S_N2$ reaction occurs at the secondary carbon to give the *R* cyano product because hydroxide is a poor leaving group.

**7.23**

(a)

(i) $CH_3CH_2Cl + Na^+ \ ^-SCH_3 \longrightarrow CH_3CH_2SCH_3 + NaCl$

(ii) $CH_3CH_2Cl + Na^+ \ ^-OH \longrightarrow CH_3CH_2OH + NaCl$

Both reactions yield $S_N2$ substitution products because the substrate is primary and both nucleophiles are good.

(b)

(i)

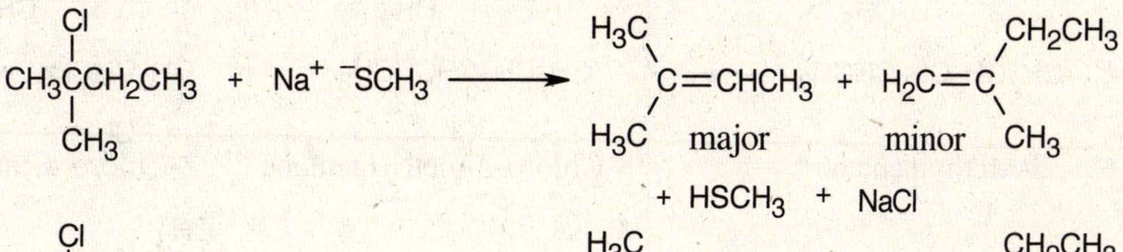

(ii)

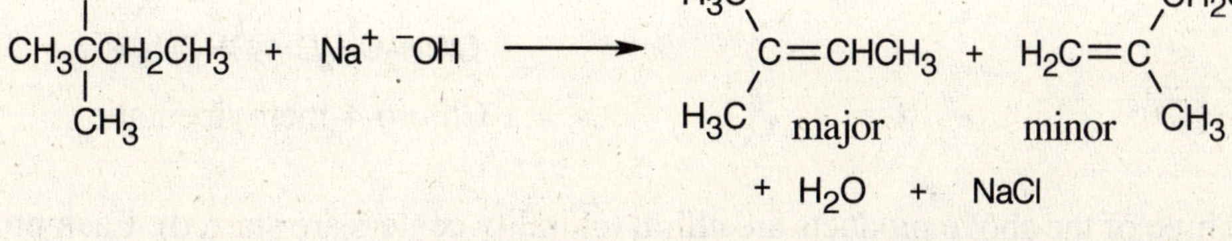

The substrate is tertiary, and the nucleophiles are basic. Two elimination products are expected; the major product has the more substituted double bond, in accordance with Zaitsev's rule.

(c)

(i)

(ii)

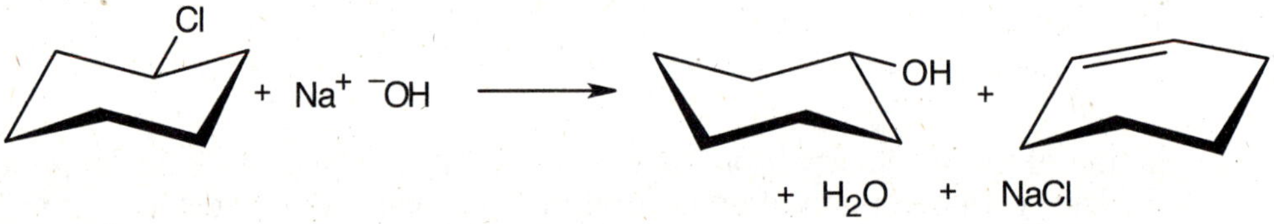


In (i), the secondary substrate reacts with the good, but weakly basic, nucleophile to yield substitution product. In (ii), NaOH is a poorer nucleophile but a stronger base, and both substitution and elimination products are formed.

**7.24** The name of the compound is (*R*)-2-bromopentane. Reaction of (*S*)-pentan-2-ol with $PBr_3$ to form (*R*)-2-bromopentane occurs with a change in stereochemistry because the configuration at the chirality center is inverted from *S* to *R*.

## Additional Problems

### Naming Alkyl Halides

**7.25**

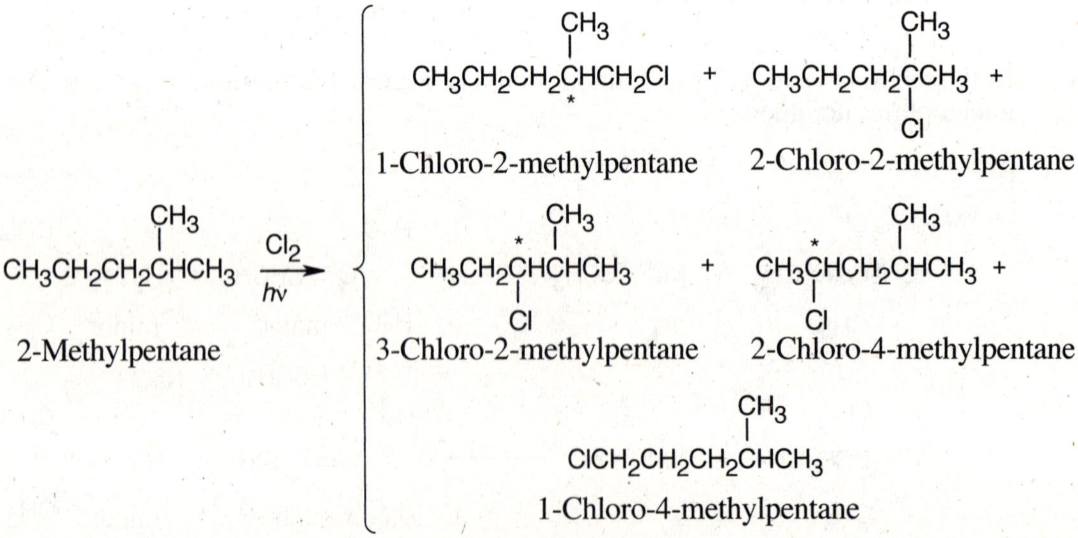


Three of the above products are chiral (chirality centers are starred). Each product occurs as a racemic mixture of enantiomers.

**7.26**

(a)

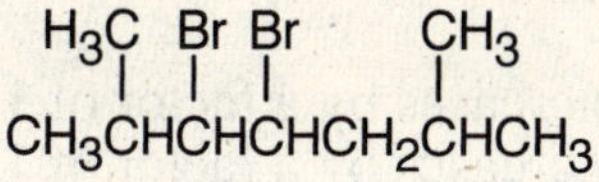

3,4-Dibromo-2,6-dimethylheptane

(b)

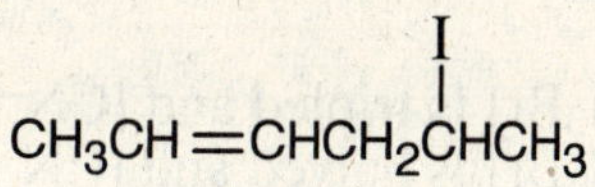

5-Iodohex-2-ene

(c)

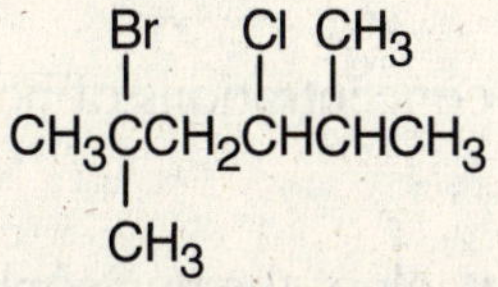

2-Bromo-4-chloro-2,5-dimethylhexane

(d)

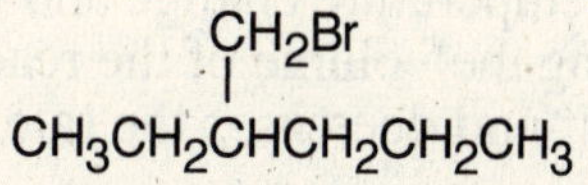

3-(Bromomethyl)hexane

(e) $ClCH_2CH_2CH_2C{\equiv}CCH_2Br$

1-Bromo-6-chlorohex-2-yne

**7.27**

(a)

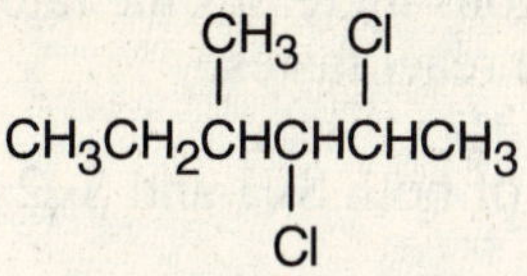

2,3-Dichloro-4-methylhexane

(b)

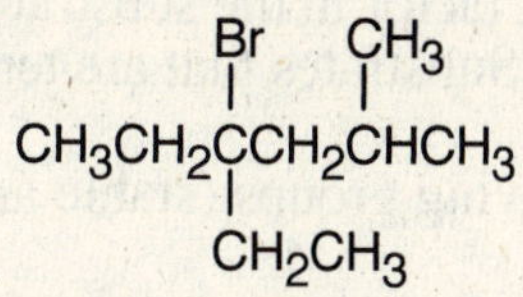

4-Bromo-4-ethyl-2-methylhexane

(c)

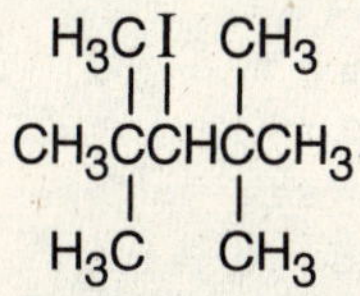

3-Iodo-2,2,4,4-tetramethylpentane

## Characteristics of $S_N1$ and $S_N2$ Reactions

**7.28**

| | *Reacts faster* | *Reacts slower* | *Reason* |
|---|---|---|---|
| (a) | 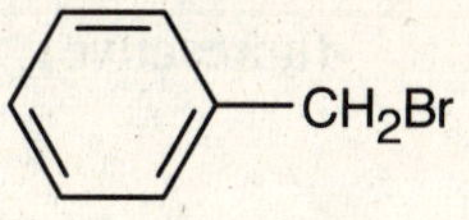 | 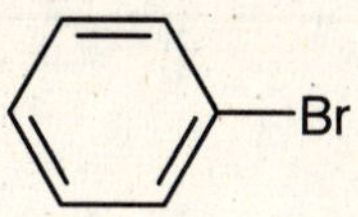 | Primary, benzylic halides react rapidly in $S_N2$ reactions. Aryl halides are unreactive in $S_N2$ reactions. |
| (b) | $CH_3Cl$ | $(CH_3)_3Cl$ | Primary halides react faster than tertiary halides in $S_N2$ reactions. |
| (c) | $H_2C{=}CHCH_2Br$ | $CH_3CH{=}CHBr$ | Primary, allylic halides react rapidly in $S_N2$ reactions. Vinyl halides are unreactive in $S_N2$ reactions. |

**7.29** Remember that the rate of an $S_N2$ reaction depends on the concentrations of both the substrate and the nucleophile.

(a) If $[CH_3Br]$ is tripled and $[CN^-]$ is halved, the rate increases by a factor of 1.5.
(b) If $[CH_3Br]$ is halved and $[CN^-]$ is tripled, the rate increases by a factor of 1.5.
(c) If $[CH_3Br]$ is tripled and $[CN^-]$ is doubled, the rate increases by a factor of 6.
(d) If the reaction temperature is raised, the rate increases; the size of the increase depends on the temperature change and on other factors.
(e) Doubling the volume of the reacting solution halves the concentrations of both $CH_3Br$ and $CN^-$ and decreases the rate by a factor of 4.

**7.30** (a) *$S_N2$:* In an $S_N2$ reaction, the rate-limiting step involves attack of the nucleophile on the substrate. Consequently, any factor that makes approach of the nucleophile more difficult slows down the rate of reaction. Especially important is the degree of crowding at the reacting carbon atom. Tertiary carbon atoms are too crowded to allow $S_N2$ substitution to occur. Even steric hindrance one carbon atom away from the reacting site causes a drastic slowdown in the rate of reaction.

*$S_N1$:* The rate-limiting step in an $S_N1$ reaction involves formation of a carbocation. Any structural factor in the substrate that stabilizes carbocations increases the rate of reaction. Substrates that are tertiary, allylic, or benzylic react fastest.

(b) Good leaving groups (stable anions) increase the rates of both $S_N1$ and $S_N2$ reactions.

**7.31** Use the Table in Section 7.5 if you need help.

| | *Better leaving group* | *Poorer leaving group* |
|---|---|---|
| (a) | $Br^-$ | $F^-$ |
| (b) | $Cl^-$ | $NH_2^-$ |
| (c) | $I^-$ | $OH^-$ |

**7.32**

*Most reactive* ⟶ *Least reactive*

$S_N1$ reactivity:

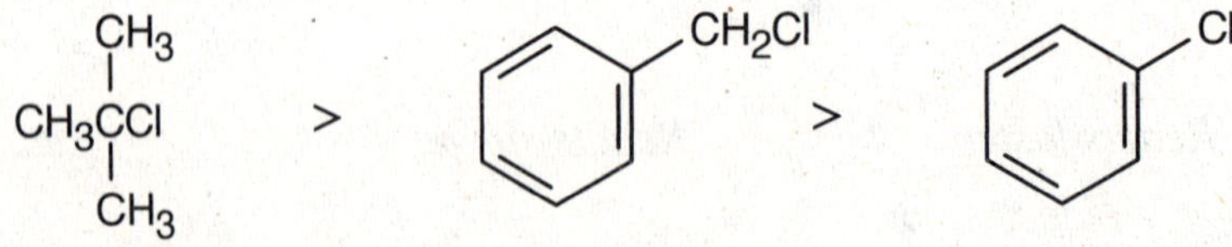

(most stable carbocation) (unreactive)

$S_N2$ reactivity:

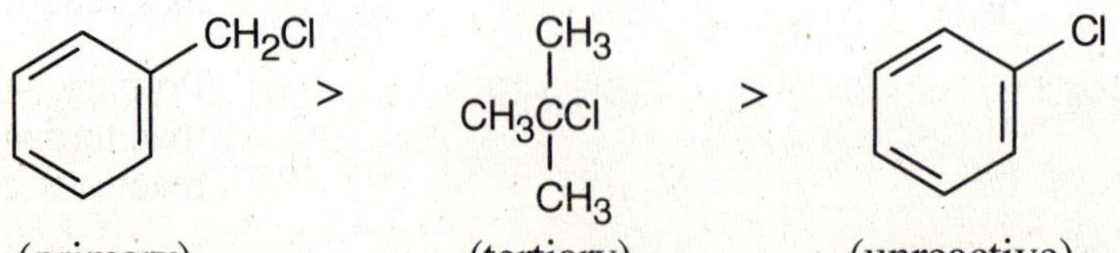

(primary) (tertiary) (unreactive)

**7.33** $S_N2$ reactivity:

*Most reactive* ⟶ *Least reactive*

(a)

$CH_3CH_2CH_2Cl$ (primary carbon atom) > $CH_3CH_2CH(Cl)CH_3$ (secondary carbon atom) > $(CH_3)_3CCl$ (tertiary carbon atom)

(b)

$CH_3Br$ > $(CH_3)_2CHCH_2Br$ (primary carbon atom) > $(CH_3)_2CHCH(Br)CH_3$ (secondary carbon atom)

**7.34** In an $S_N1$ reaction, the rate depends on the concentration of the substrate and is independent of the concentration of the nucleophile.

(a) If $[(CH_3)_3CBr]$ is doubled and $[CH_3OH]$ is halved, the rate increases by a factor of 2 (remember that $[CH_3OH]$ doesn't affect the reaction rate).
(b) If $[(CH_3)_3CBr]$ is halved and $[CH_3OH]$ is doubled, the rate is halved.
(c) If both $[(CH_3)_3CBr]$ and $[CH_3OH]$ are tripled, the rate is tripled.
(d) If the reaction temperature is lowered, the rate decreases.

## Synthesis

**7.35**

(a)

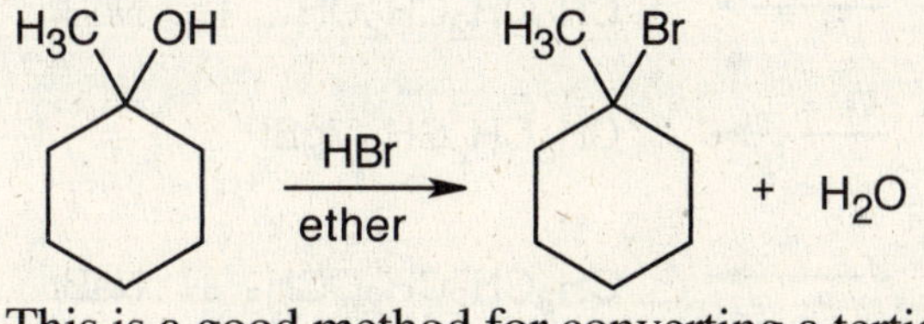

This is a good method for converting a tertiary alcohol into a bromide.

(b)

$$CH_3CH_2CH_2CH_2OH \xrightarrow{SOCl_2} CH_3CH_2CH_2CH_2Cl$$

This is a good method for converting a primary or secondary alcohol into a chloride.

(c)

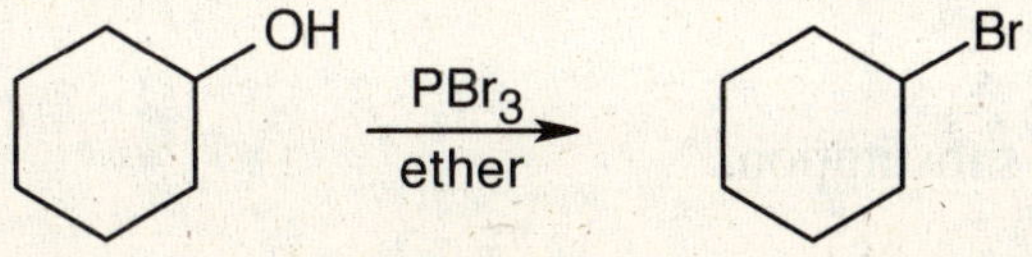

This is a good method for converting a primary or secondary alcohol into a bromide.

(d)

$$CH_3CH_2CH(Br)CH_3 \xrightarrow[\text{Ether}]{Mg} \underset{\textbf{A}}{CH_3CH_2CH(MgBr)CH_3} \xrightarrow{H_2O} \underset{\textbf{B}}{CH_3CH_2CH_2CH_3}$$

**7.36**

(a)

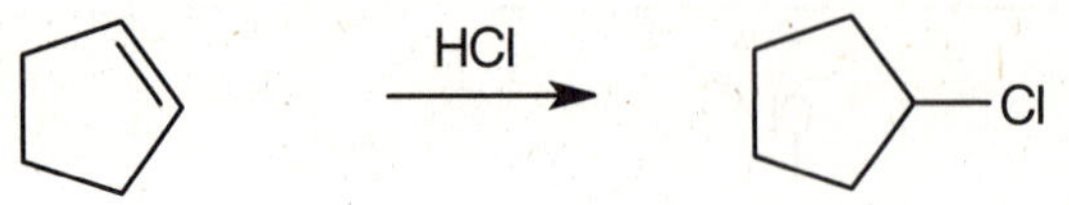

Chlorocyclopentane

(b)

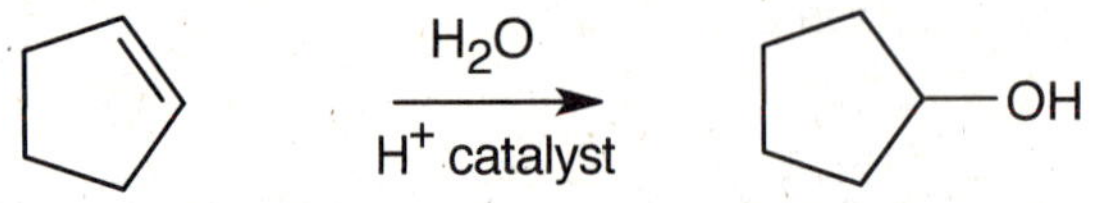

Cyclopentanol

(c)

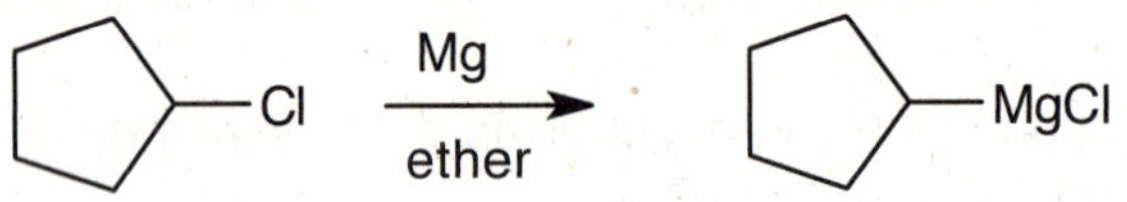

from (a) Cyclopentylmagnesium chloride

(d)

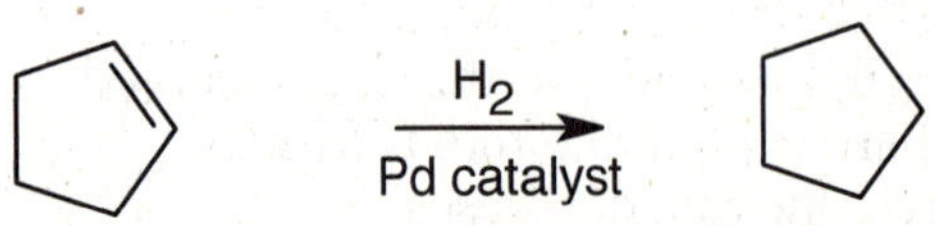

Cyclopentane

**7.37**

(a) $CH_3CH_2CH_2Br$ + NaI ⟶ $CH_3CH_2CH_2I$ + NaBr

(b) $CH_3CH_2CH_2Br$ + NaCN ⟶ $CH_3CH_2CH_2CN$ + NaBr

(c) $CH_3CH_2CH_2Br$ + NaOH ⟶ $CH_3CH_2CH_2OH$ + NaBr

(d) $CH_3CH_2CH_2Br$ + Mg ⟶ $CH_3CH_2CH_2MgBr$

(e) $CH_3CH_2CH_2Br$ + $NaOCH_3$ ⟶ $CH_3CH_2CH_2OCH_3$ + NaBr

**7.38** (a) The reaction of cyanide anion with a tertiary halide is more likely to yield the elimination product 3-methylpent-2-ene than substitution product.

(b) Use $PBr_3$ to convert a primary alcohol into a primary bromide.

(c) Reaction of a tertiary alcohol with HBr gives mainly substitution product along with a lesser amount of elimination product.

**7.39** All of these reactions proceed by $S_N2$ substitution.

(a) $CH_3CH_2OH$ + $PBr_3$ ⟶ $CH_3CH_2Br$

(b) $CH_3CH_2CH_2CH_2Br$ + $Na^+ \, ^-CN$ ⟶ $CH_3CH_2CH_2CH_2CN$ + $Na^+ \, Br^-$

(c) $CH_3Br$ + $Na^+ \, ^-OC(CH_3)_3$ ⟶ $CH_3OC(CH_3)_3$ + $Na^+ \, Br^-$

The reaction of $CH_3O^-$ with $(CH_3)_3CBr$ causes elimination, not substitution.

(d) $CH_3CH_2CH_2I + Na^{+}\,^{-}N{=}\overset{+}{N}{=}N^{-} \longrightarrow CH_3CH_2CH_2N{=}\overset{+}{N}{=}N^{-} + Na^{+}\,I^{-}$

(e) $CH_3CH_2I + Na^{+}\,^{-}SH \longrightarrow CH_3CH_2SH + Na^{+}\,I^{-}$

(f) $CH_3Br + Na^{+}\,^{-}O\overset{O}{\overset{\|}{C}}CH_3 \longrightarrow CH_3O\overset{O}{\overset{\|}{C}}CH_3 + Na^{+}\,Br^{-}$

## Substitutions and Eliminations

**7.40**

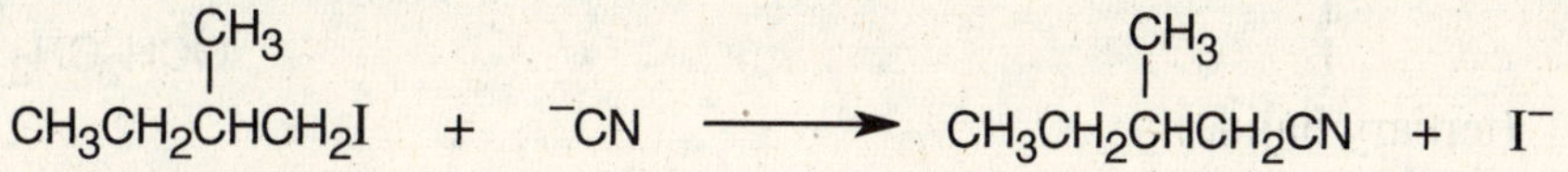

(primary halide)

This is a $S_N2$ reaction, in which reaction rate depends on the concentration of both alkyl halide and nucleophile.

(a) Halving the concentration of cyanide and doubling the concentration of alkyl halide does not change the reaction rate.

(b) Tripling the concentrations of both cyanide and alkyl halide causes a ninefold increase in reaction rate.

**7.41** 3-Bromo-3-methylhexane is one of many possibilities.

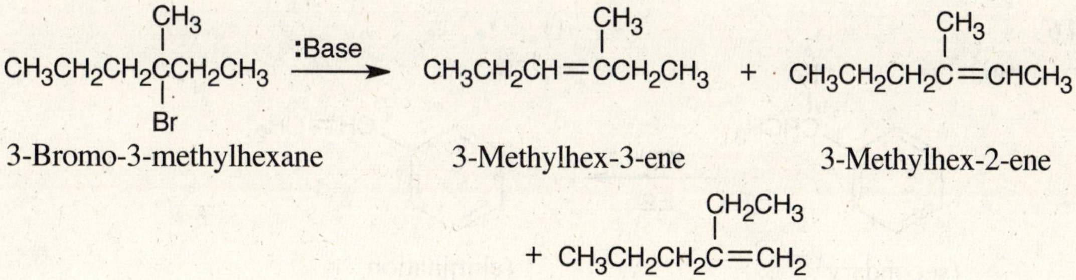

3-Bromo-3-methylhexane → 3-Methylhex-3-ene + 3-Methylhex-2-ene + 2-Ethylpent-1-ene

**7.42** Both *tert*-butyl chloride and *tert*-butyl bromide dissociate to form the same carbocation intermediate. In ethanol, this carbocation yields the same mixture of products in the same ratio, regardless of the starting material.

**7.43**

(a)

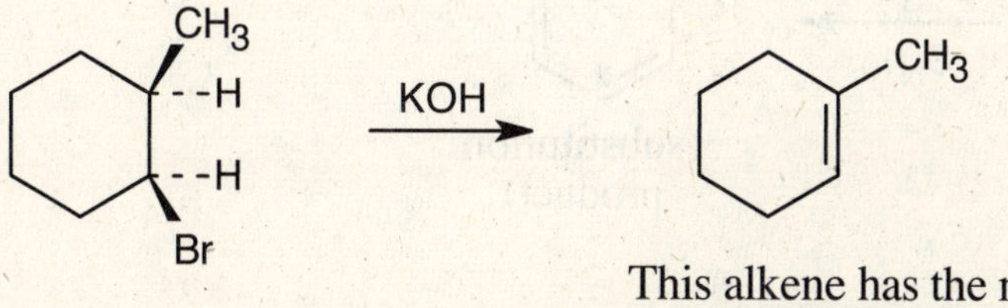

This alkene has the more substituted double bond.

(b)

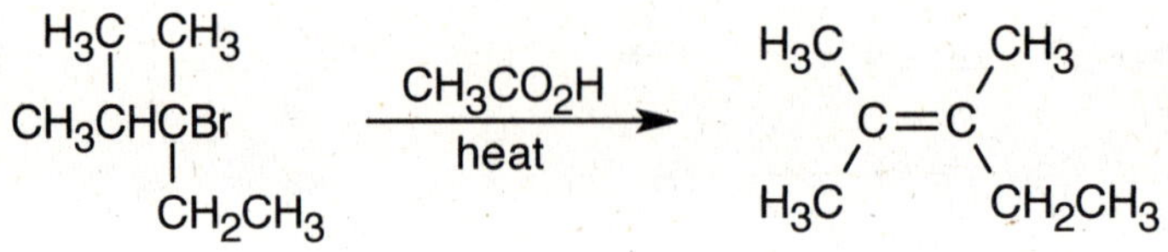

This alkene has the more substituted double bond.

**7.44**

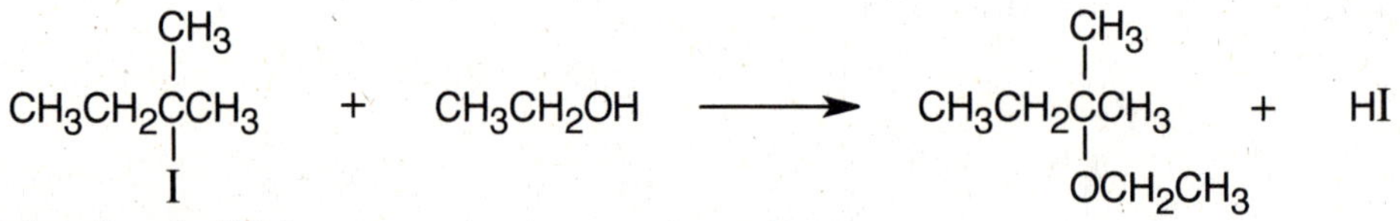

(tertiary halide)

This is an $S_N1$ reaction, whose reaction rate depends only on the concentration of 2-iodo-2-methylbutane. Tripling the concentration of alkyl halide triples the rate of reaction.

**7.45**

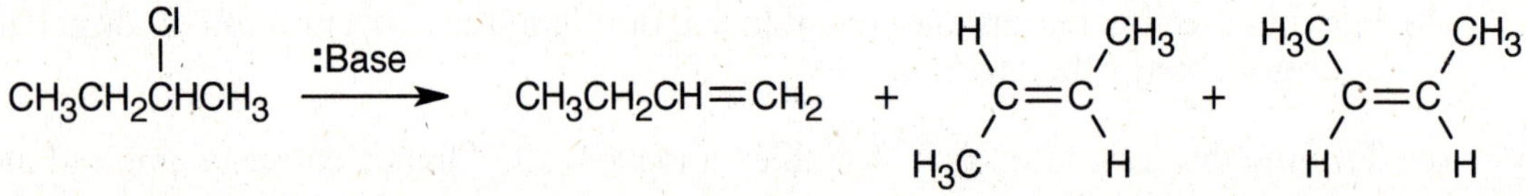

All other isomers of $C_4H_9Cl$ yield only one alkene product on treatment with strong base.

**7.46**

(a)

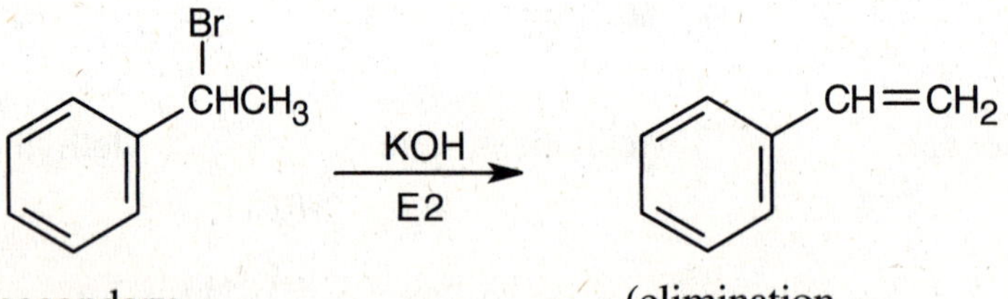

(secondary benzylic halide) (elimination product)

This is an E2 reaction.

(b)

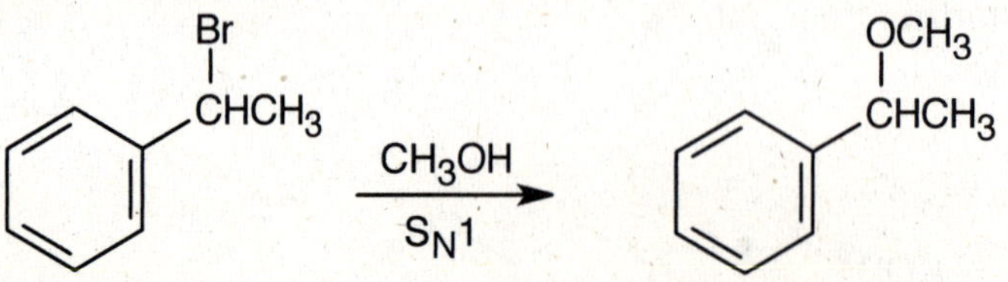

(secondary benzylic halide) (substitution product)

This is an $S_N1$ reaction.

## General Problems

**7.47** The intermediate benzyl radical is stabilized by resonance

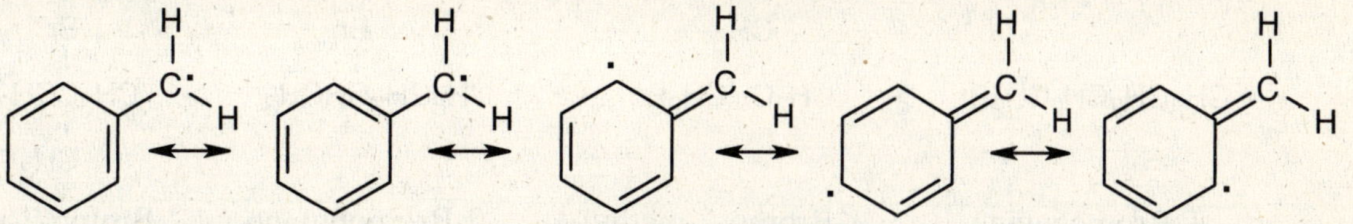

Five resonance forms contribute to the stability of the benzyl radical.

**7.48**

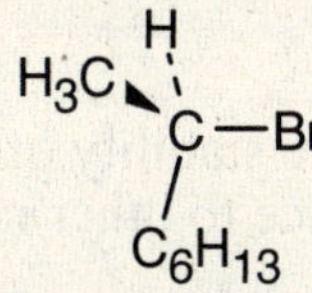

(*R*)-2-Bromooctane

(*R*)-2-Bromooctane is a secondary bromoalkane, which undergoes $S_N2$ substitution. Since $S_N2$ reactions proceed with inversion of configuration, the configuration at the carbon chirality center is inverted. (This does not necessarily mean that all *R* isomers become *S* isomers after an $S_N2$ reaction. The *R,S* designation refers to the priorities of groups,which may change when the nucleophile is varied.)

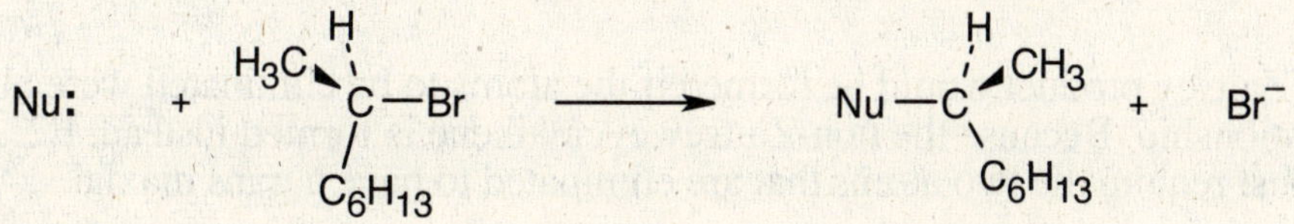

| | *Nucleophile* | *Product* |
|---|---|---|
| (a) | $^-CN$ | 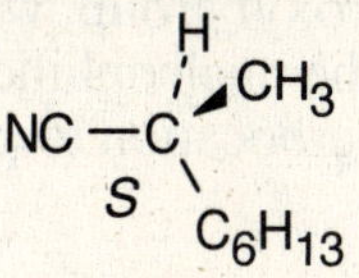 |
| (b) | $CH_3CO_2^-$ | 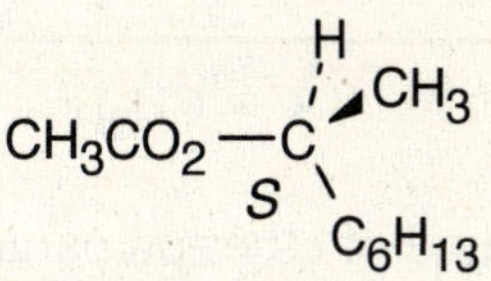 |
| (c) | $Br^-$ | 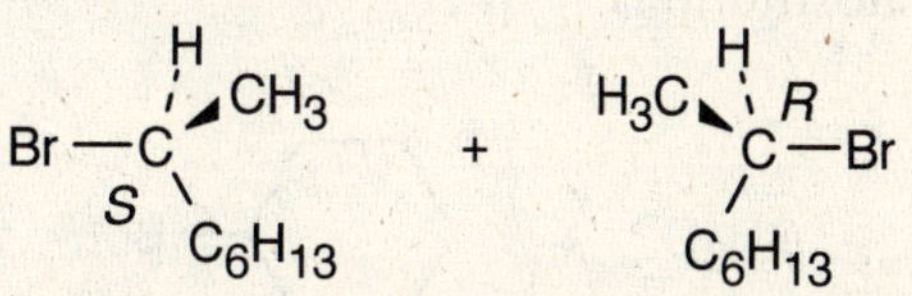 |

2-Bromooctane is 100% racemized after 50% of the original (*R*)-2-bromooctane has reacted with Br.

**7.49** $S_N2$ reactivity:

*Most reactive* → *Least reactive*

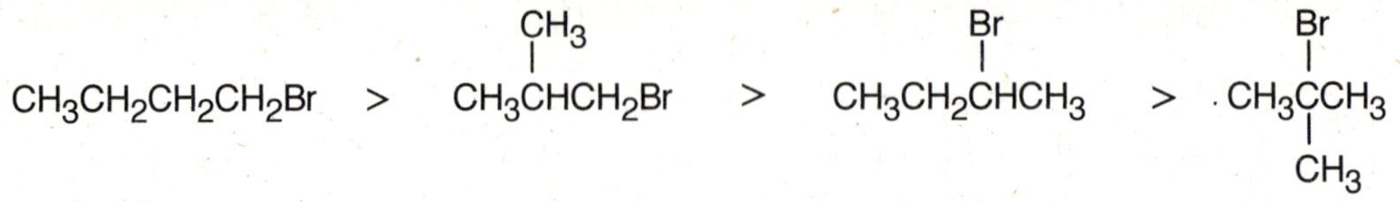


1-Bromobutane — 1-Bromo-2-methylpropane — 2-Bromobutane — 2-Bromo-2-methylpropane

**7.50**

Two resonance forms contribute to the relative stability of the allyl radical. Because it is more stable, this radical is formed in preference to other radicals.

**7.51**

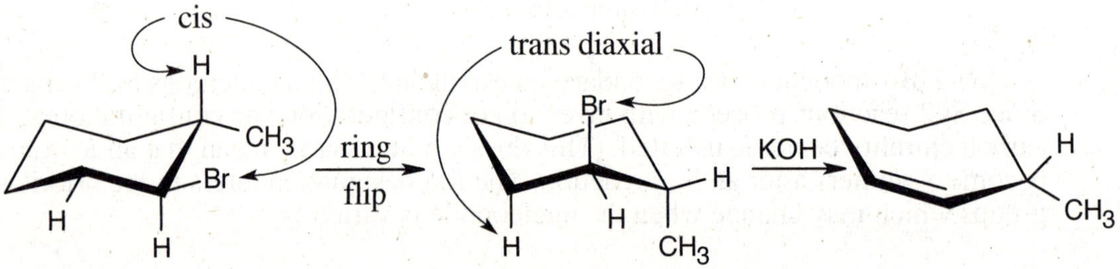


The expected Zaitsev product would be formed if the atoms to be eliminated were allowed have a cis relationship. Because the non-Zaitsev cycloalkene is formed instead, E2 elimination must require the two atoms that are eliminated to have a trans diaxial relationship.

**7.52** The chiral tertiary alcohol (*R*)-3-methylhexan-3-ol reacts with HBr by an $S_N1$ pathway. HBr protonates the hydroxyl group, which dissociates to yield a planar, achiral carbocation. Attack by the nucleophilic bromide anion can occur with equal probability from either side of the carbocation to produce (±)3-bromo-3-methylhexane.

**7.53**

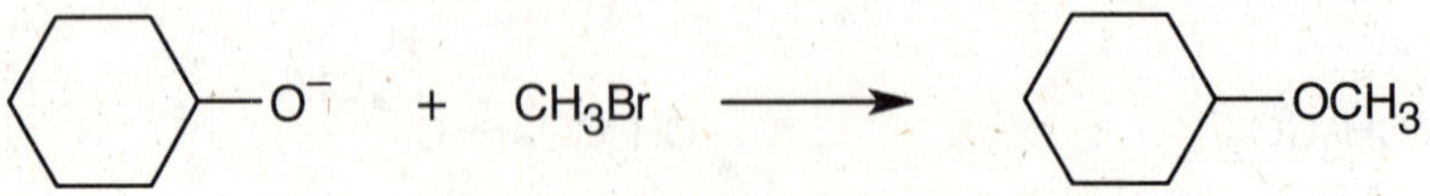


This is an excellent method of ether preparation since bromomethane is very reactive in $S_N2$ substitutions.

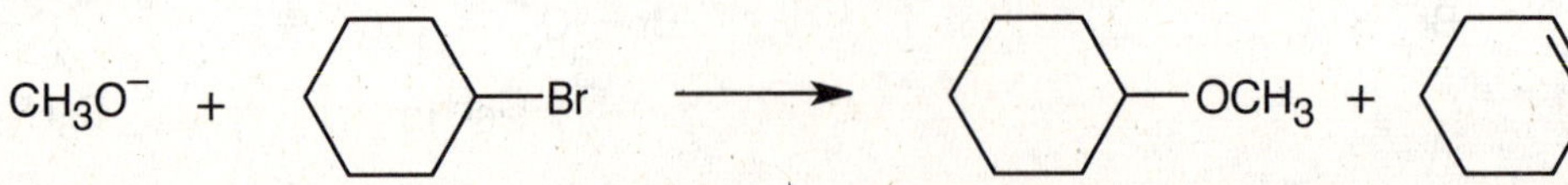


Reaction of a secondary haloalkane with a basic nucleophile yields both substitution and elimination products. This is obviously a less satisfactory method of ether preparation.

**7.54**

(1) $CH_3CH_2OH + PBr_3 \longrightarrow CH_3CH_2Br$

(2) $CH_3CH_2O^- + CH_3CH_2Br \longrightarrow CH_3CH_2OCH_2CH_3 + Br^-$

**7.55**

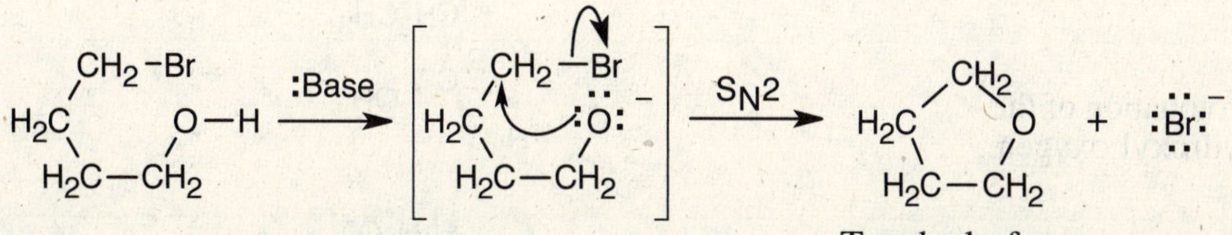

Tetrahydrofuran

The product is the cyclic ether tetrahydrofuran.

**7.56**

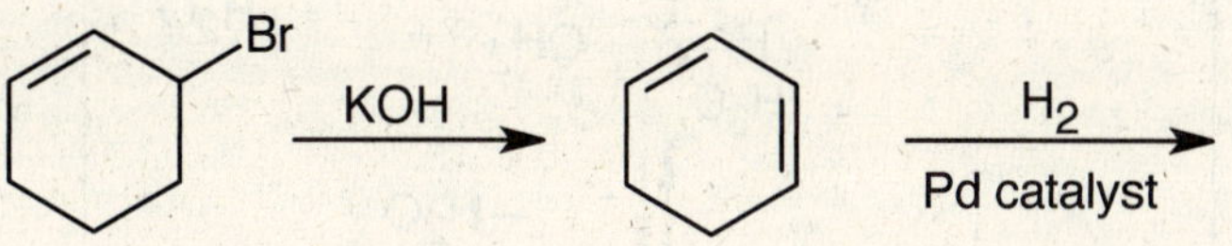

**7.57**

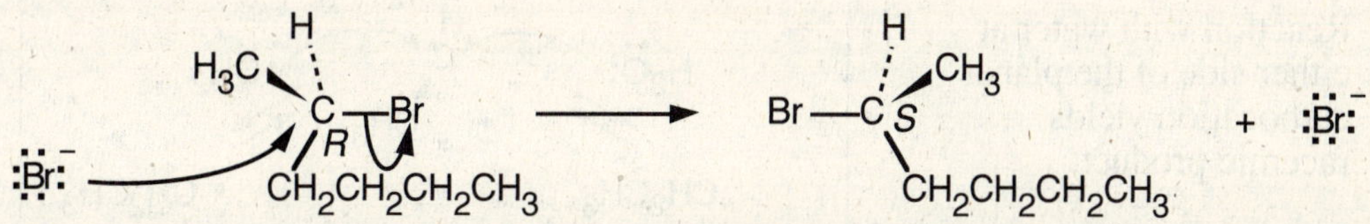

Reaction of one molecule of (*R*)-2-bromohexane with one bromide ion produces one molecule of (*S*)-2-bromohexane. Reaction of 50% of the *R* starting material gives a mixture of 50% *S* enantiomer plus 50% unreacted *R* starting material – a racemic mixture. Thus, after 50% of the *R* starting material has reacted, the product is 100% racemized.

**7.58** A Grignard reagent can't be prepared from a compound containing an acidic functional group because the Grignard reagent is immediately quenched by the proton source. For example, the $-CO_2H$, $-OH$, $-NH_2$, and $RC{\equiv}CH$ functional groups are too acidic to allow for preparation of a Grignard reagent.

**7.59** Since butan-2-ol is a secondary alcohol, substitution can occur by either a $S_N1$ or a $S_N2$ route, depending on reaction conditions. Two factors favor an $S_N1$ mechanism in this case. (1) The reaction is run under solvolysis (solvent as nucleophile) conditions. (2) Dilute acid converts a poor leaving group ($^-OH$) into a good leaving group ($H_2O$), which departs easily.

Protonation of the hydroxyl oxygen..

is followed by loss of water to form a planar carbocation.

Reaction with water at either side of the planar carbocation yields racemic product.

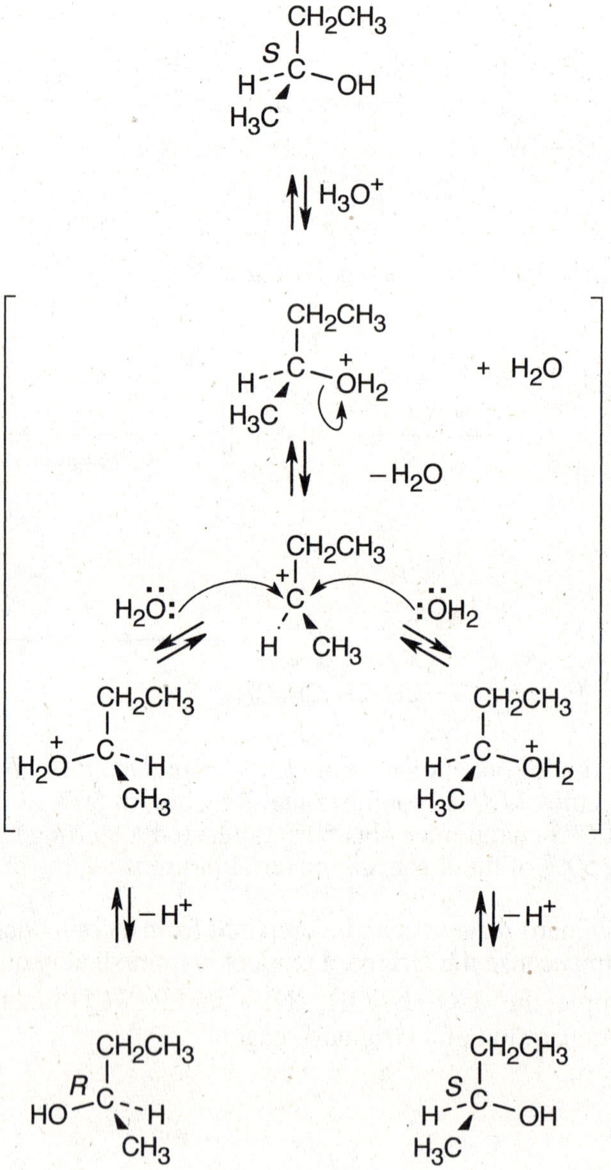

**7.60**

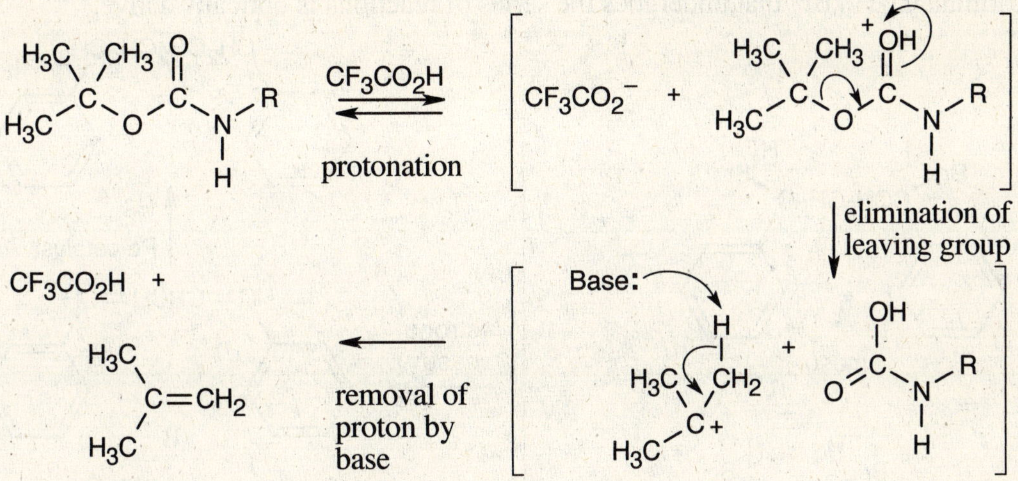


This reaction proceeds by an E1 mechanism.

**7.61** Base brings about an E1cB elimination reaction that has homocysteine as the leaving group.

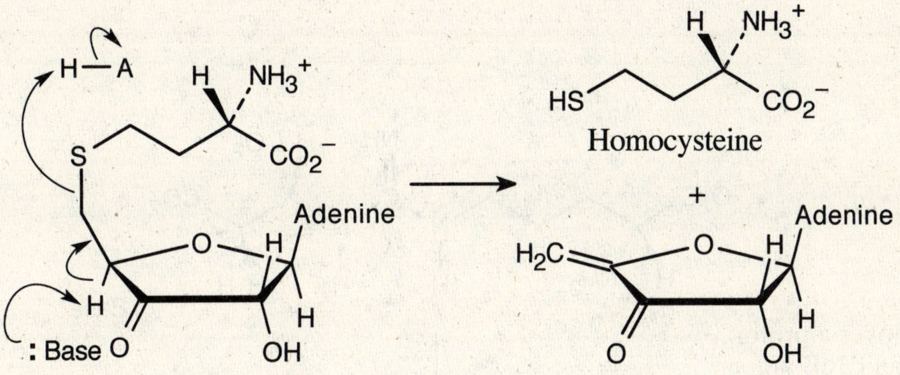


In reactions involving complicated substrates, concentrate on the reacting groups without letting the rest of the molecule distract you.

**7.62** Two optically inactive structures are possible for compound **A**. Any other structure of the formula $C_{16}H_{16}Br_2$ that undergoes the series of reactions is optically active.

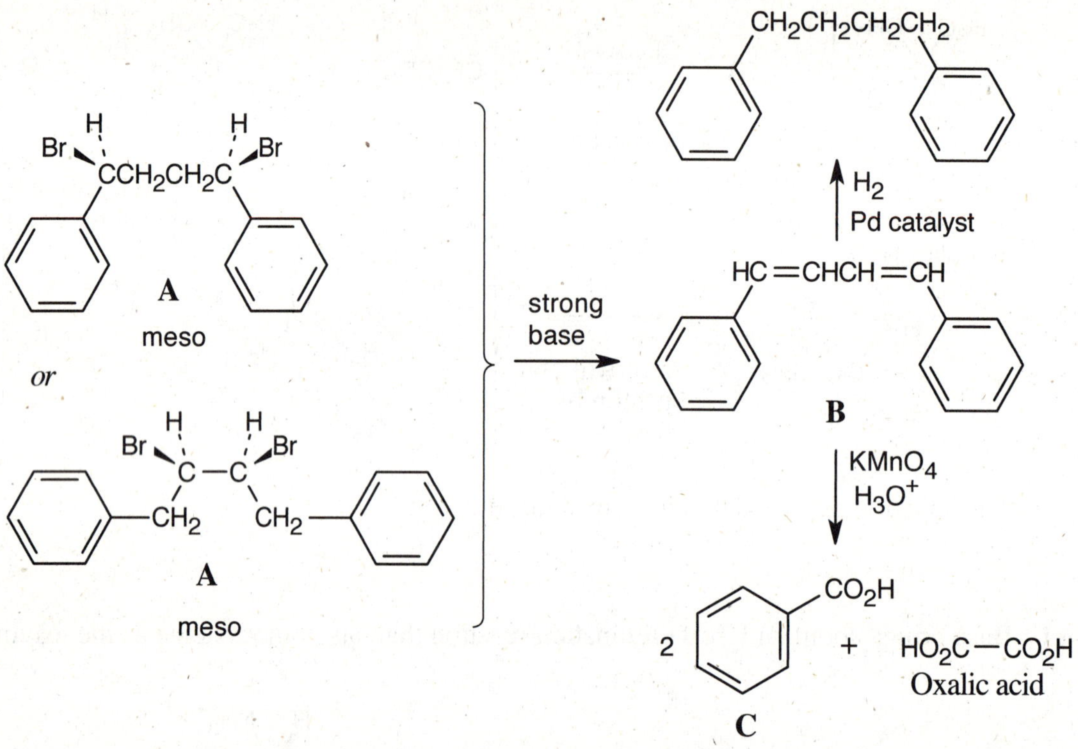


**7.63**

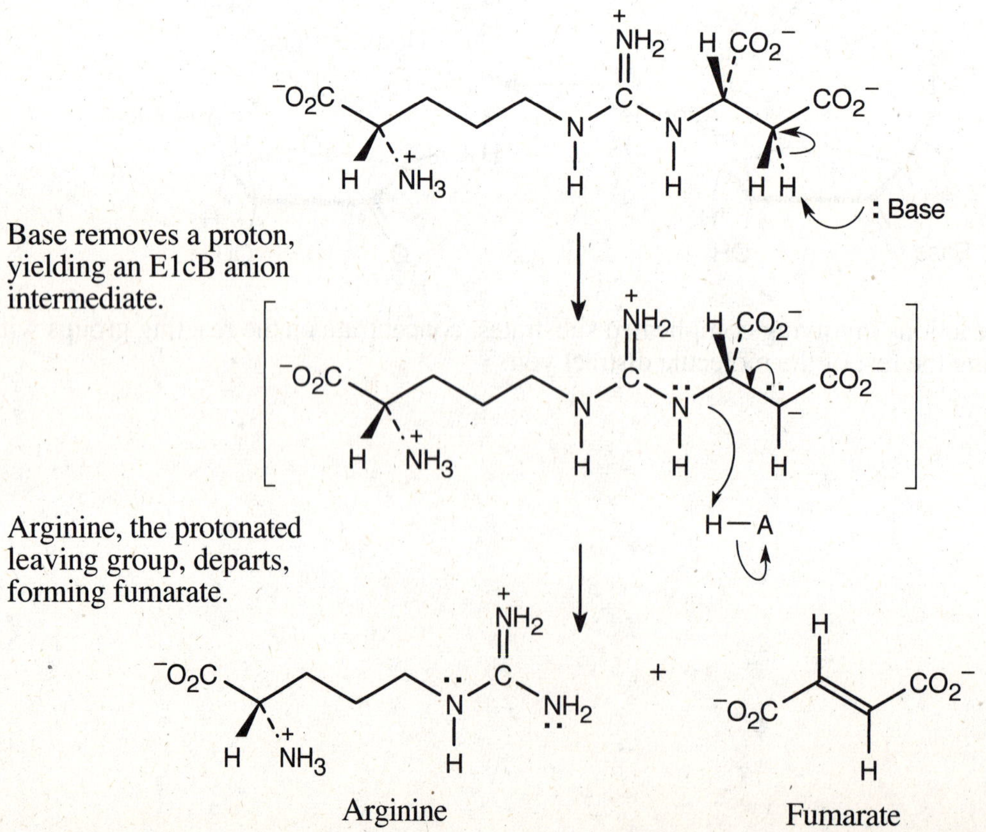

## In the Medicine Cabinet

**7.64** (a)

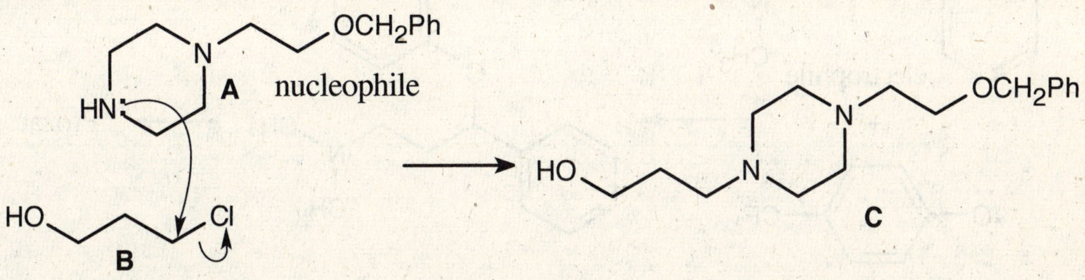


electrophile

(b)

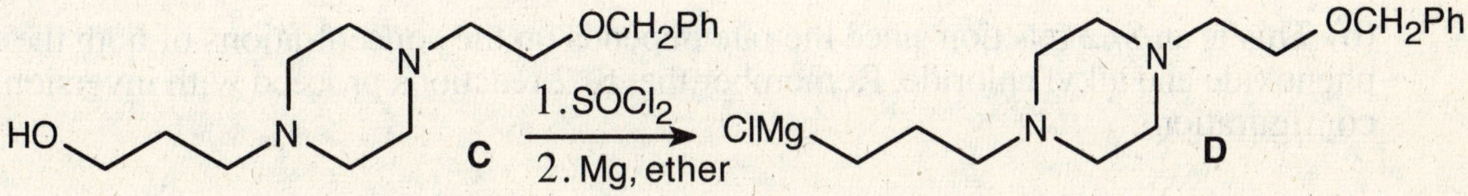


(c)

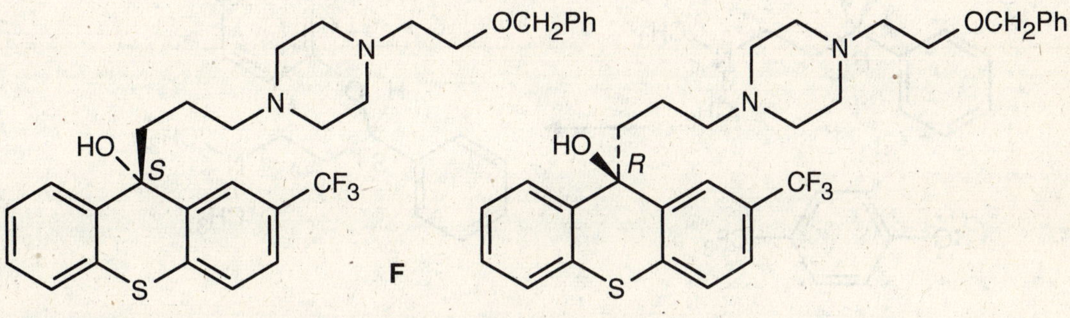


(d) The stereoisomers are *E*,*Z* double bond isomers and are diastereomers.

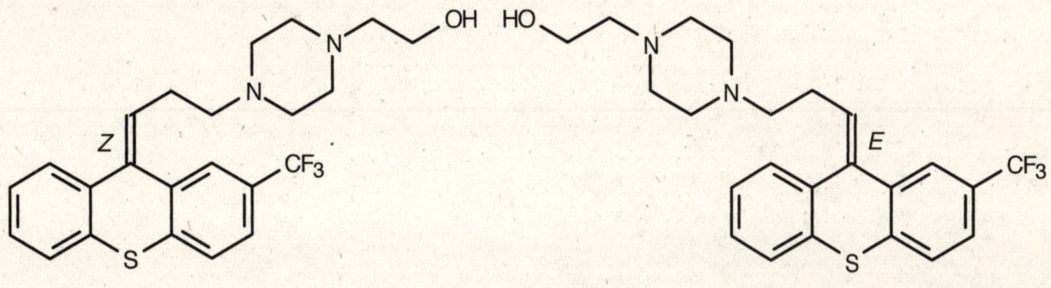

**7.65**

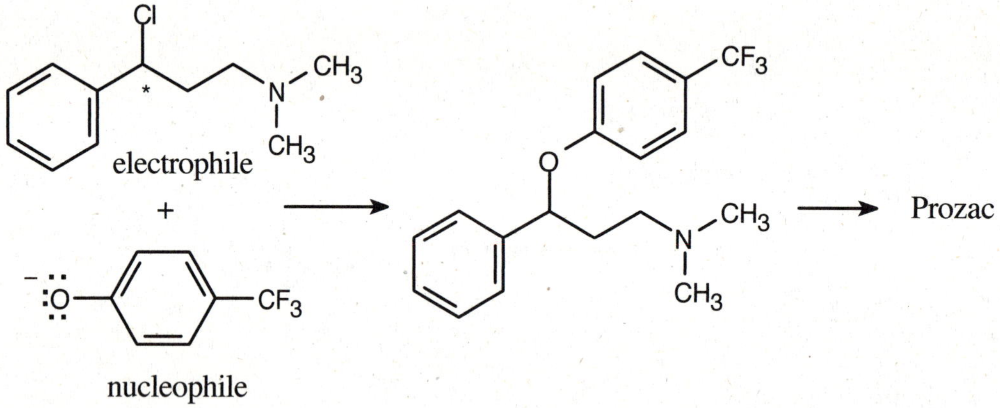


(a) The nucleophile is the phenoxide anion, and the electrophile is the alkyl chloride.

(b) This is an $S_N2$ reaction since the rate depends on the concentrations of both the phenoxide and alkyl chloride. Remember that $S_N2$ reactions proceed with inversion of configuration.

(c)

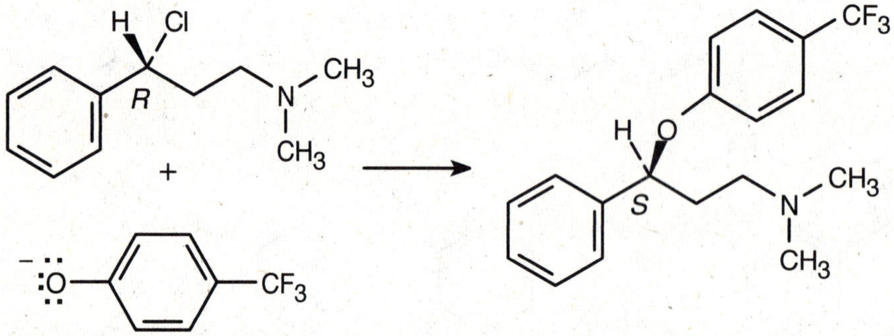

## In the Field

**7.66** (a) It is harder to remove electrons from a benzene ring bonded to a $-CF_3$ group. The electron-withdrawing group makes the benzene ring more electron-poor and less able to be oxidized.
(b) The first two compounds have a (trifluoromethyl)phenyl group bonded to a heterocyclic ring containing nitrogen and oxygen. One would expect them to function similarly - perhaps to inhibit pigment synthesis. The other three groups have the structural similarities that are circled and inhibit lipid synthesis.

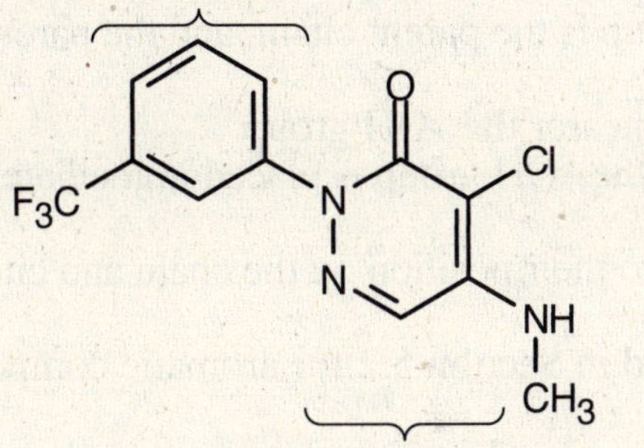


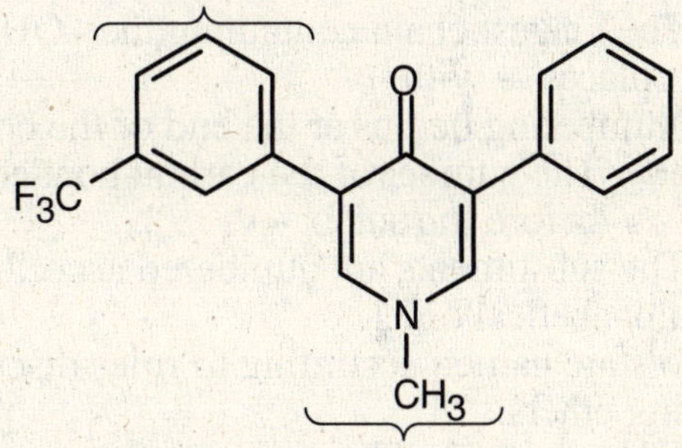


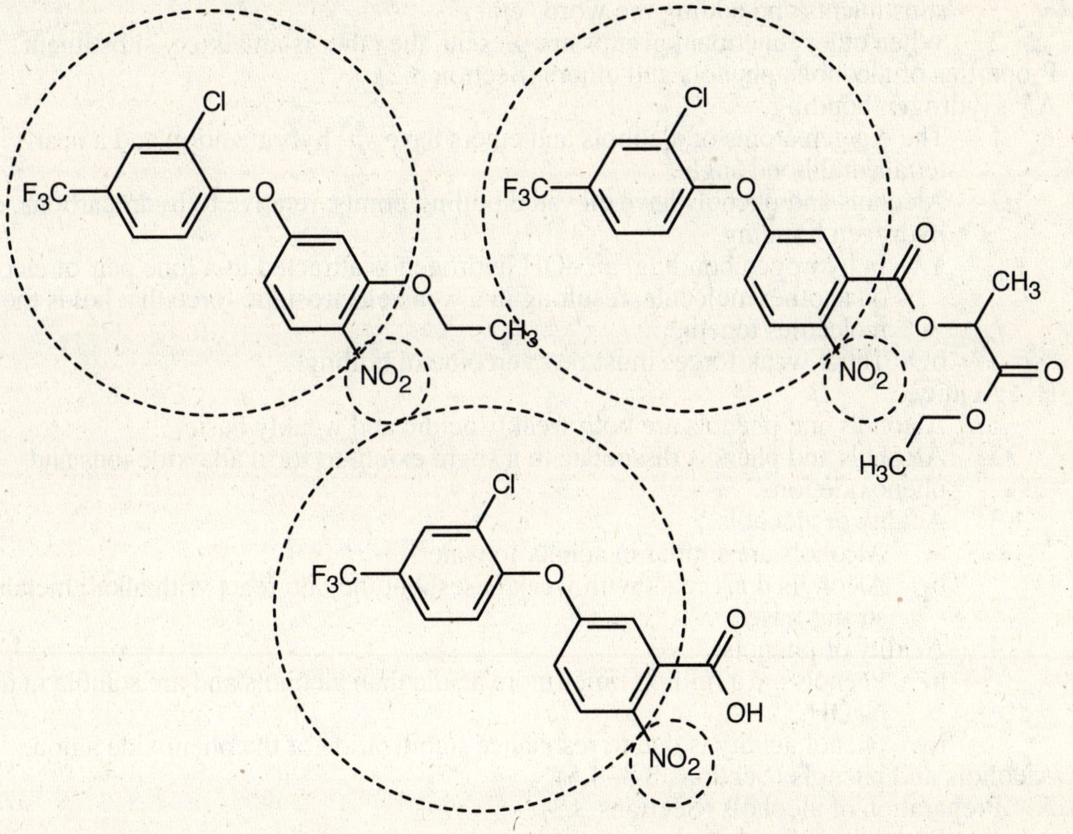

# Chapter 8 – Alcohols, Phenols, and Ethers, and Their Sulfur Analogs

## Chapter Outline

I. Naming alcohols, phenols, and ethers (Section 8.1).
   A. Alcohols are classified as primary, secondary or tertiary, depending on the number of carbon substituents bonded to the –OH carbon.
   B. Rules for naming alcohols.
      1. The longest chain containing the –OH group is the parent chain, and the parent name replaces *-e* with *-ol.*
      2. Numbering begins at the end of the chain nearer the –OH group.
         a. The number of the carbon bonded to the –OH group is placed immediately before the suffix *-ol.*
      3. The substituents are numbered according to their position on the chain and cited in alphabetical order.
   C. Phenols are named according to rules discussed in Section 5.2 for aromatic compounds.
   D. Naming ethers.
      1. Ethers with no other functional groups are named by citing the two organic substituents and adding the word "ether".
      2. When other functional groups are present, the ether is an alkoxy substituent.

II. Properties of alcohols, phenols and ethers (Section 8.2).
   A. Hydrogen bonding.
      1. The oxygen atoms of alcohols and ethers have $sp^3$ hybridization and a nearly tetrahedral bond angle.
      2. Alcohols and phenols have elevated boiling points, relative to hydrocarbons, due to hydrogen bonding.
         a. In hydrogen bonding, an –OH hydrogen is attracted to a lone pair of electrons on another molecule, resulting in a weak electrostatic force that holds the molecules together.
         b. These weak forces must be overcome in boiling.
   B. Acidity.
      1. Alcohols and phenols are both weakly acidic and weakly basic.
      2. Alcohols and phenols dissociate to a slight extent to form alkoxide ions and phenoxide ions.
      3. Acidity of alcohols.
         a. Alcohols are similar in acidity to water.
         b. Alcohols don't react with weak bases, but they do react with alkali metals and strong bases.
      4. Acidity of phenols.
         a. Phenols are a million times more acidic than alcohols and are soluble in dilute NaOH.
         b. Phenol acidity is due to resonance stabilization of the phenoxide anion.

III. Alcohols and phenols (Sections 8.3–8.5).
   A. Preparation of alcohols (Sections 8.3).
      1. Reduction of carbonyl compounds.
         a. Aldehydes are reduced to primary alcohols, and ketones are reduced to secondary alcohols.
         b. $LiAlH_4$ and $NaBH_4$ are the most commonly used reducing reagents.
         c. The reaction involves addition of $H^-$, followed by addition of $H_3O^+$.
         d. Reduction of carboxylic acids and esters are slower and require $LiAlH_4$.
      2. Grignard reactions of carbonyl compounds.
         a. Grignard reagents react with carbonyl compounds to give alcohols.

b. Formaldehyde reacts with Grignard reagents to give primary alcohols, aldehydes give secondary alcohols, and ketones give tertiary alcohols.
c. Esters give tertiary alcohols in which two of the substituents come from the Grignard reagent.

B. Reactions (Sections 8.4–8.5).
1. Alcohols (Section 8.4).
a. Acid-catalyzed dehydration to yield alkenes.
i. Dehydration follows Zaitsev's Rule.
ii. Dehydration proceeds by an E1 mechanism.
b. Oxidation to yield carbonyl compounds.
i. Primary alcohols can be oxidized to carboxylic acids with $CrO_3$.
ii. Secondary alcohols can be oxidized to ketones with $CrO_3$.
iii. Primary alcohols can be oxidized to aldehydes with a periodinane reagent.
c. Conversion into ethers – Williamson ether synthesis.
i. Reaction of an alkoxide with a primary alkyl halide produces an ether.
ii. The alkoxide ion is formed by treating an alcohol with Na metal or NaH.
iii. The reaction proceeds by an $S_N2$ mechanism.
2. Phenols (Section 8.5).
a. Williamson ether synthesis.
b. Electrophilic aromatic substitution.
c. Oxidation to yield quinones.
i. Oxidation of a phenol with $Na_2Cr_2O_7$ yields a quinone.
ii. Quinones can be reduced to hydroquinones using $NaBH_4$ or $SnCl_2$.
iii. Oxidation–reduction reactions of quinones are important in living systems.

IV. Ethers (Section 8.6–8.7).
A. Preparation by Williamson ether synthesis (Section 8.6).
B. Acidic cleavage.
1. Cleavage reactions proceed either by a $S_N1$ or a $S_N2$ mechanism.
2. With primary and secondary ethers, the ether oxygen stays with the more substituted alkyl group ($S_N2$).
3. With tertiary ethers, oxygen stays with the less substituted alkyl group ($S_N1$).
C. Epoxides (Section 8.7).
1. Epoxides are prepared by reaction of an alkene with a peroxyacid.
2. Ring-opening reactions of epoxides.
a. Dilute aqueous acid converts an epoxide to a glycol.
i. Reaction takes place by $S_N2$ attack of $H_2O$ on a protonated epoxide.
b. Epoxide rings can also be opened by reaction with nucleophiles.

V. Thiols and sulfides (Section 8.8).
A. Naming thiols and sulfides.
1. Thiols (sulfur analogs of alcohols) are named by the same system as alcohols, with the suffix *-thiol* replacing *-ol.*
a. The –SH group is a mercapto group.
2. Sulfides (sulfur analogs of ethers) are named by the same system as ethers, with *sulfide* replacing *ether*.
a. The –SR group is an alkylthio group.
B. Preparation.
1. Thiols may be prepared by $S_N2$ displacement with a sulfur nucleophile.
2. Treatment of a thiol with base yields a thiolate anion, which can react with an alkyl halide to form a sulfide.
C. Reactions.
1. Thiols can be oxidized by $Br_2$ or $I_2$ to yield disulfides, RSSR.
a. The reaction can be reversed by treatment with zinc and acid.
b. The thiol-disulfide interconversion is biologically important.

## Solutions to Problems

**8.1–8.2**

(a)

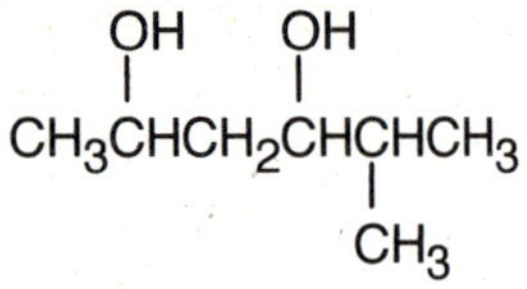

5-Methylhexane-2,4-diol
secondary alcohol

(b)

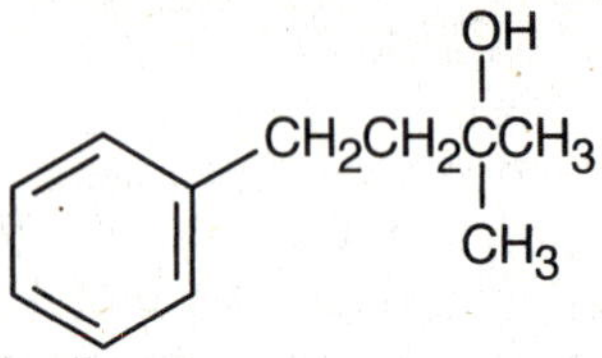

2-Methyl-4-phenylbutan-2-ol
tertiary alcohol

(c)

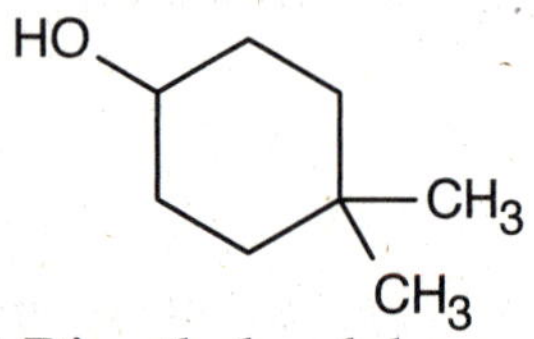

4,4-Dimethylcyclohexanol
secondary alcohol

(d)

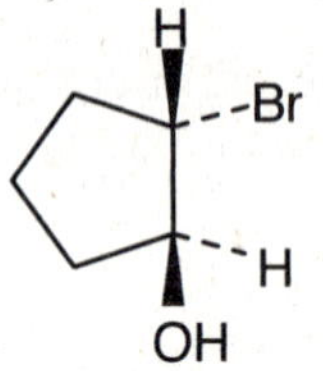

*trans*-2-Bromo-cyclopentanol
secondary alcohol

(e)

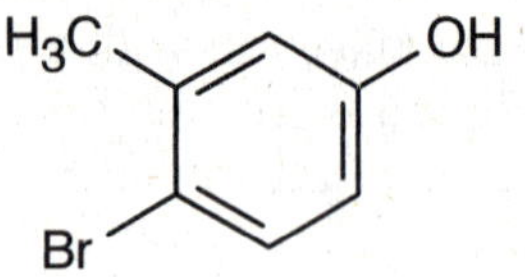

4-Bromo-3-methylphenol

(f)

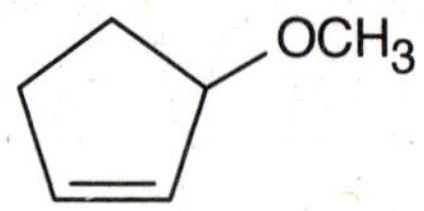

Cyclopent-3-enyl methyl ether

**8.3**

(a)

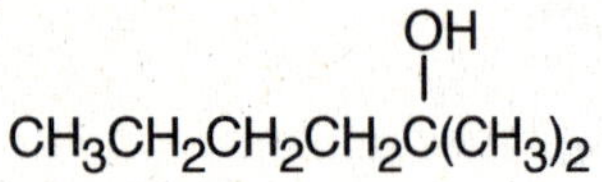

2-Methylhexan-2-ol

(b)

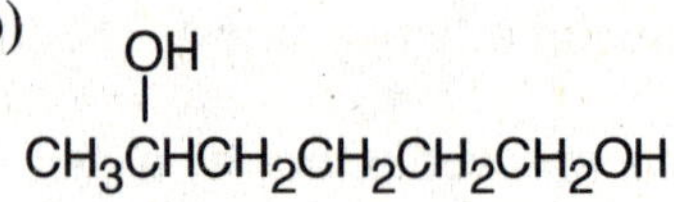

Hexane-1,5-diol

(c)

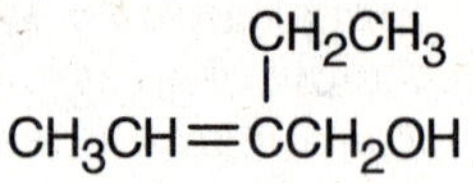

2-Ethylbut-2-en-1-ol

(d)

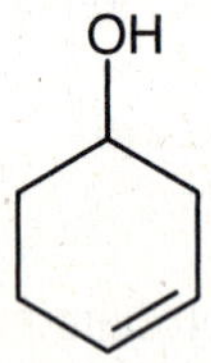

Cyclohex-3-en-1-ol

(e)

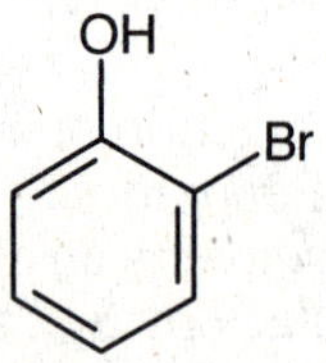

*o*-Bromophenol

(f)

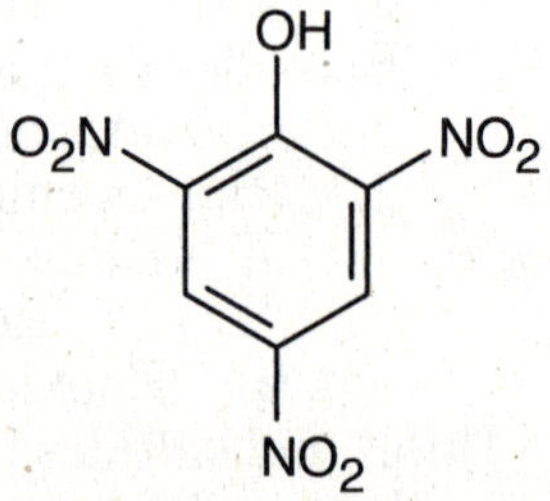

2,4,6-Trinitrophenol

**8.4**

(a)

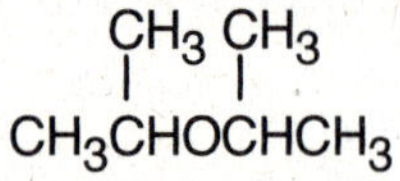

2-Isopropoxypropane
*or*
Diisopropyl ether

(b)

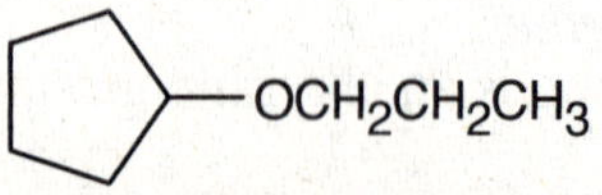

Propoxycyclopentane
*or*
Cyclopentyl propyl ether

(c)

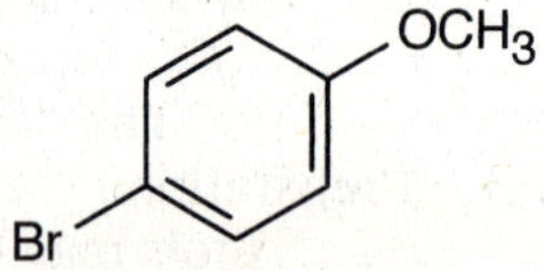

*p*-Bromoanisole
*or*
*p*-Bromomethoxybenzene
*or*
Methyl *p*-bromophenyl ether

(d)

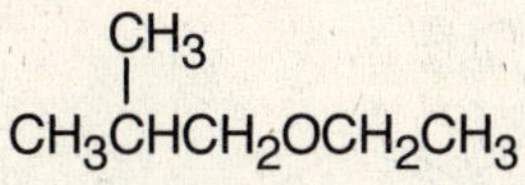

1-Ethoxy-2-methylpropane

*or*

Ethyl isobutyl ether

**8.5**

(a)

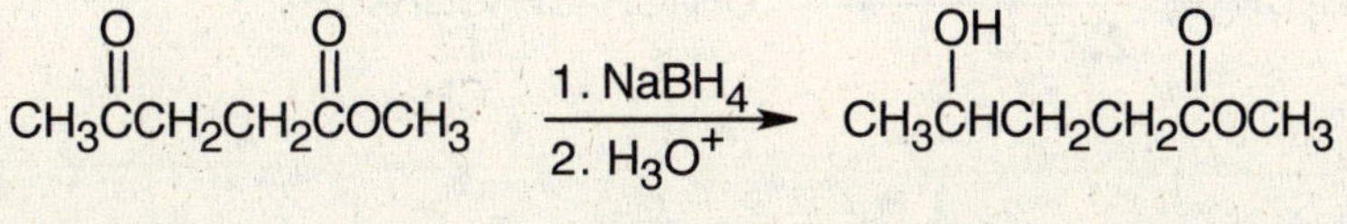

$NaBH_4$ reduces aldehydes and ketones without interfering with other functional groups.

(b)

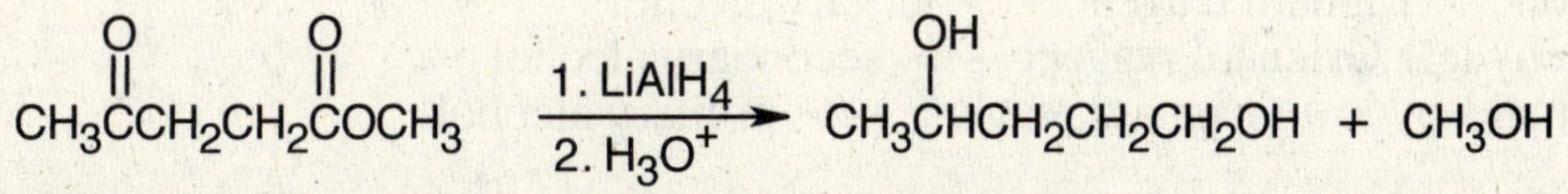

$LiAlH_4$, a stronger reducing agent, reduces both ketones and esters.

**8.6**

(a)

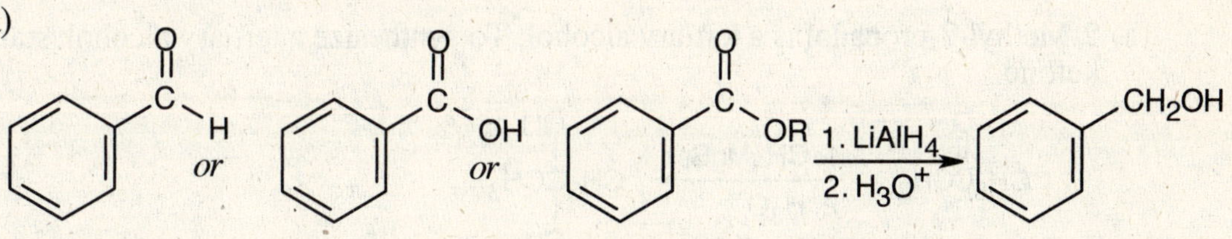

Benzyl alcohol is a product of reduction of an aldehyde, a carboxylic acid, or an ester.

(b)

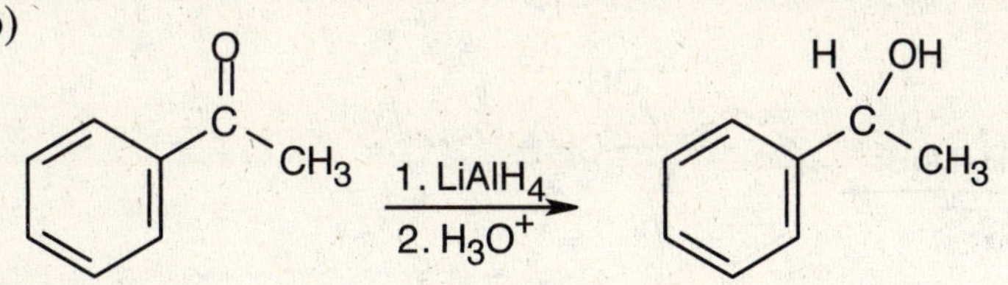

Reduction of a ketone yields this secondary alcohol.

(c)

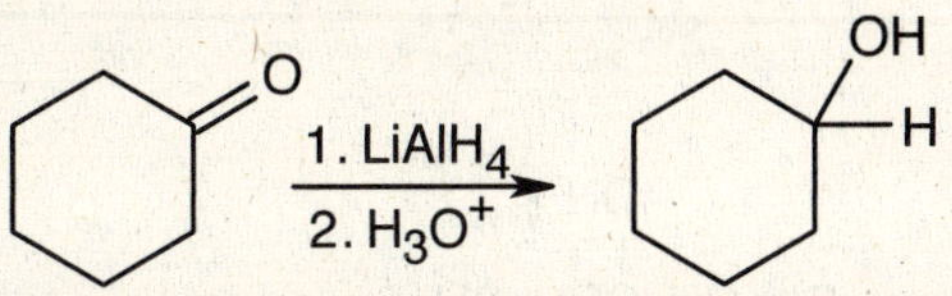

**8.7** All of the products have an –OH and a methyl group bonded to what was formerly a ketone carbon

(a)

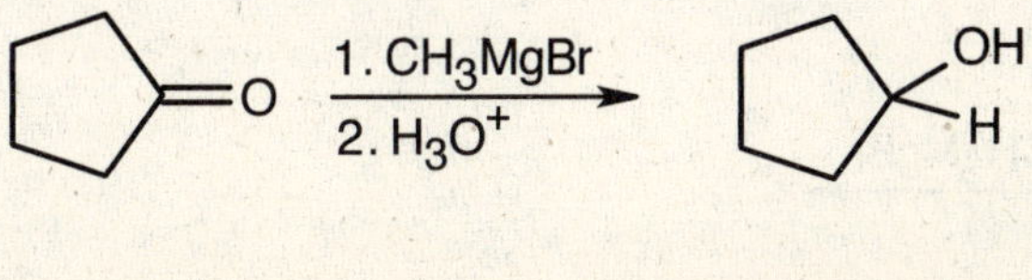

(b)

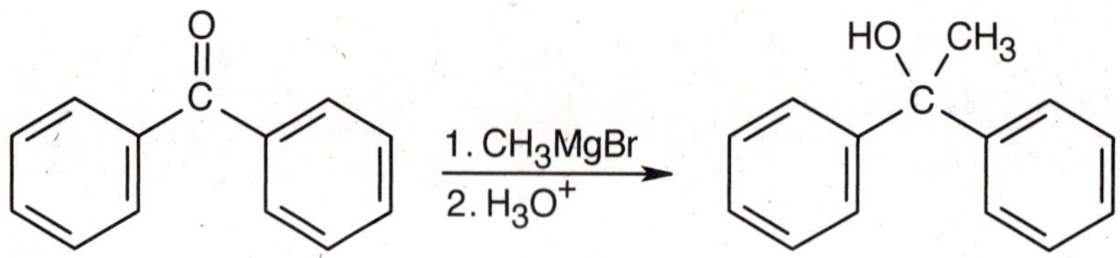


(c)

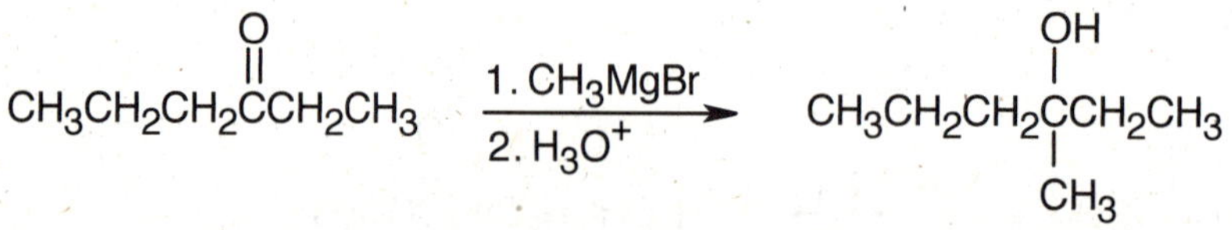


**8.8** When choosing the starting materials for a Grignard reaction, remember:

1. Ketone + Grignard reagent —> tertiary alcohol
   Aldehyde + Grignard reagent —> secondary alcohol
   Formaldehyde + Grignard reagent —> primary alcohol

2. More than one combination of carbonyl compound plus Grignard reagent may yield the same product.

(a) 2-Methyl-2-propanol is a tertiary alcohol. To synthesize a tertiary alcohol, start with a ketone.

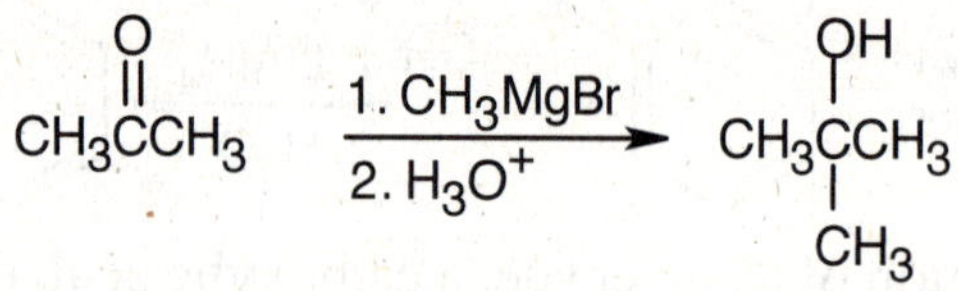


(b) Since 1-methylcyclohexanol is a tertiary alcohol, start with a ketone.

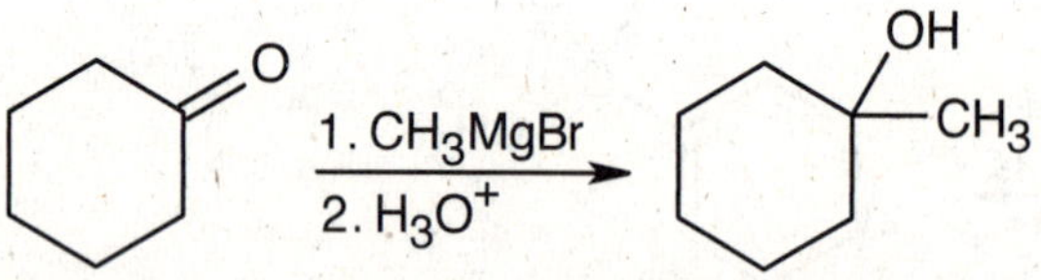


1-Methylcyclohexanol

(c) 3-Methyl-3-pentanol is a tertiary alcohol that can be synthesized by three different routes.

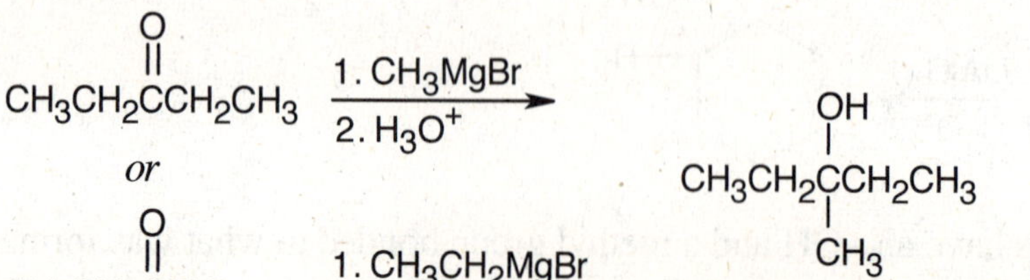


3-Methylpentan-3-ol

**8.9**

(a)

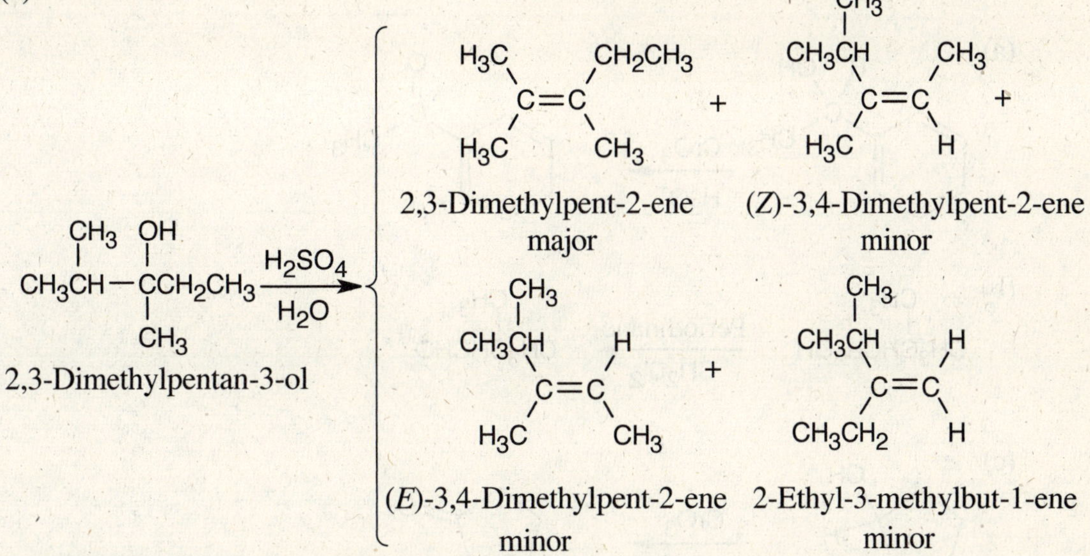

2,3-Dimethylpentan-3-ol

2,3-Dimethylpent-2-ene, major; (*Z*)-3,4-Dimethylpent-2-ene, minor; (*E*)-3,4-Dimethylpent-2-ene, minor; 2-Ethyl-3-methylbut-1-ene, minor

The alkene with the most substituted double bond is the major product.

(b)

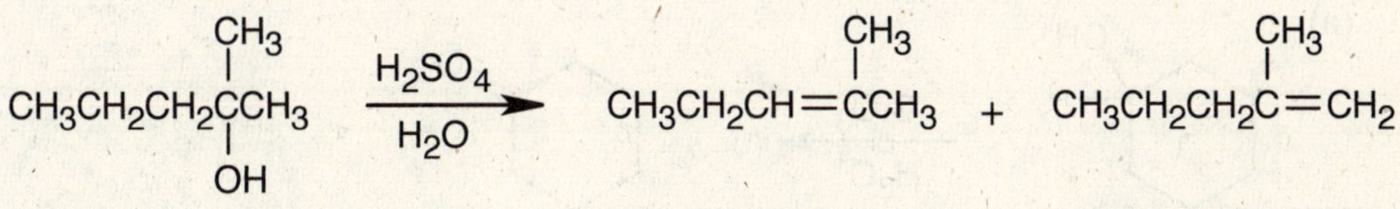

2-Methylpentan-2-ol → 2-Methylpent-2-ene (major) + 2-Methylpent-1-ene (minor)

**8.10**

(a)

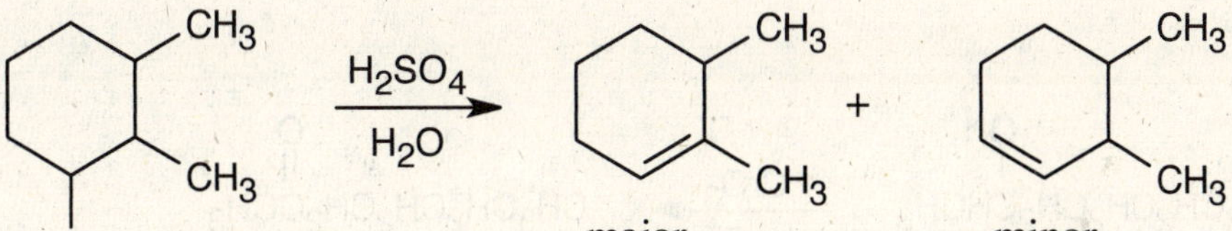

major minor

(b)

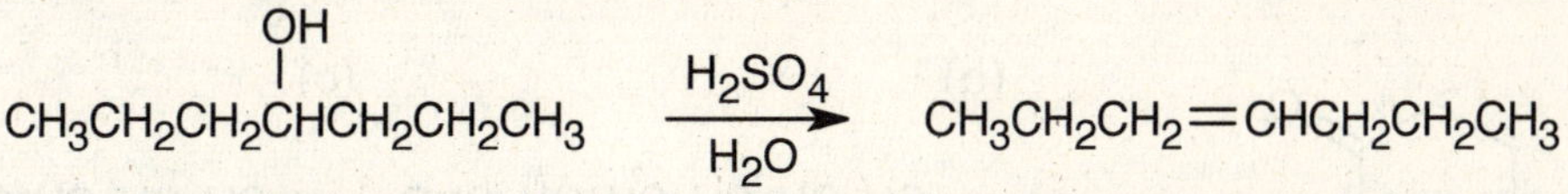

Whenever possible, start with an alcohol that gives only one product (the desired product).

**8.11** Aldehydes are synthesized by oxidation of primary alcohols, and ketones are synthesized by oxidation of secondary alcohols.

(a)

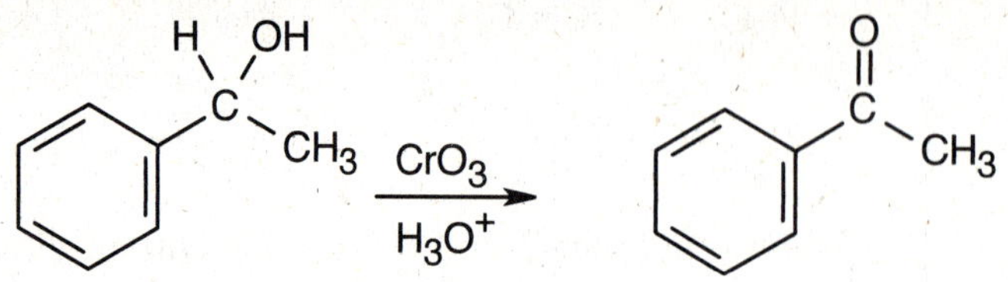

(b)

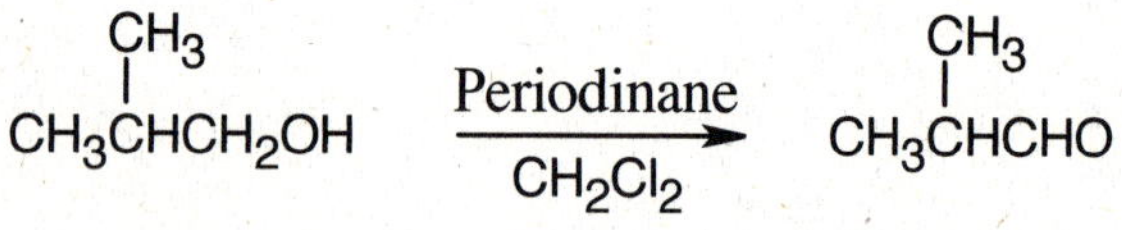

(c)

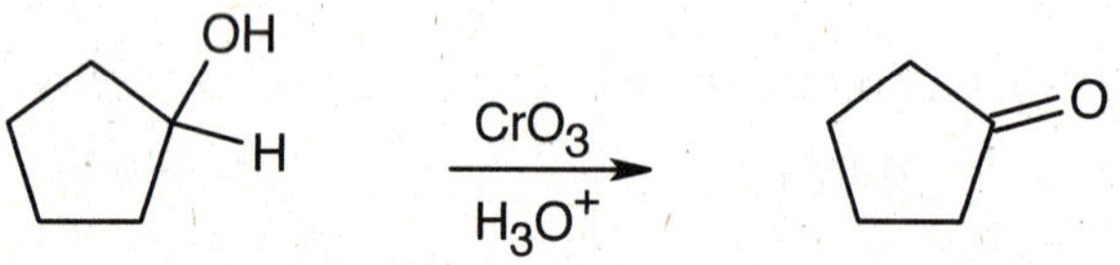

**8.12**

(a)

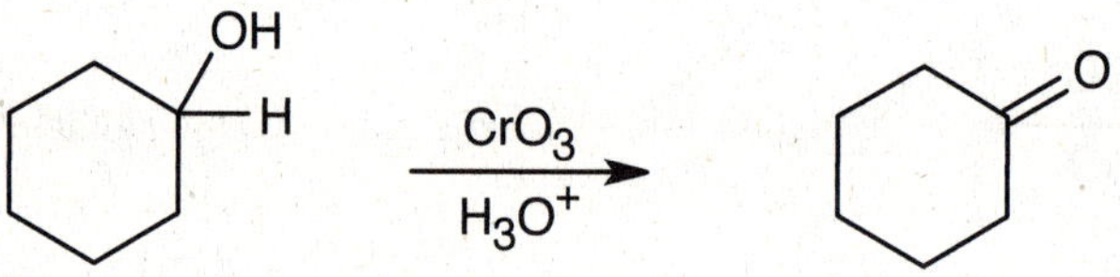

(b)

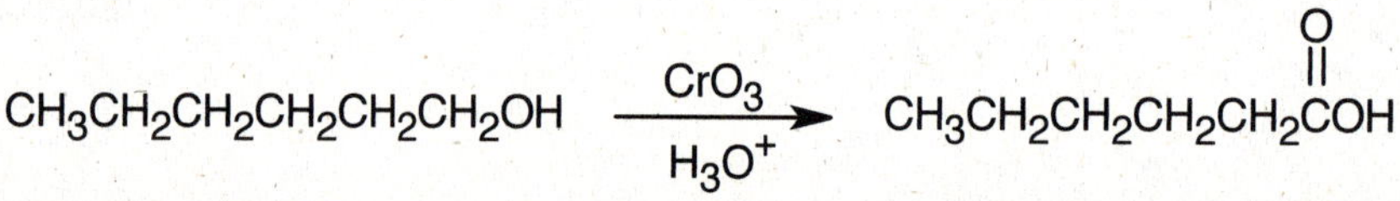

(c)

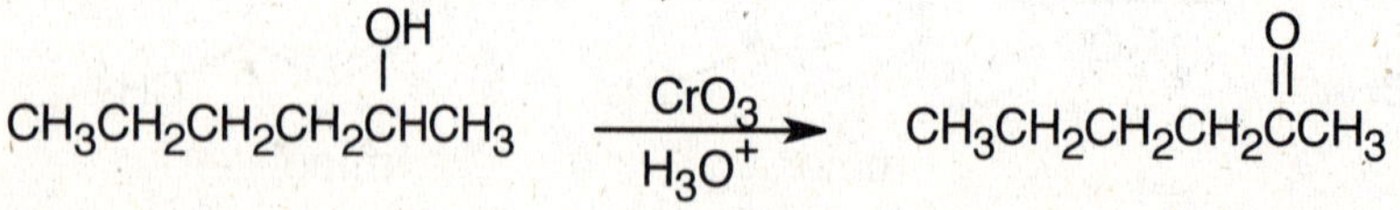

**8.13**

(a)

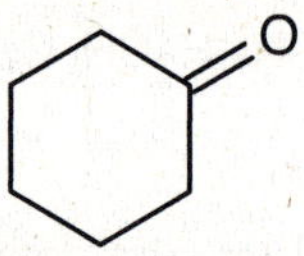

(b) $CH_3CH_2CH_2CH_2CH_2CHO$

(c)

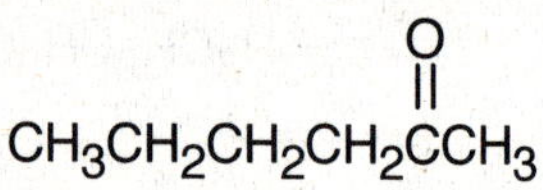

**8.14** Remember that the alkyl halide in the Williamson ether synthesis should be primary or methyl, in order to avoid competing elimination reactions. The alkoxide anions shown are formed by treating the corresponding alcohols with Na or NaH.

(a)

$$\left.\begin{array}{l} CH_3CH_2CH_2O^- + CH_3Br \\ \quad\quad \textit{or} \\ CH_3CH_2CH_2Br + CH_3O^- \end{array}\right\} \longrightarrow CH_3CH_2CH_2OCH_3 + Br^-$$

Methyl propyl ether

(b)

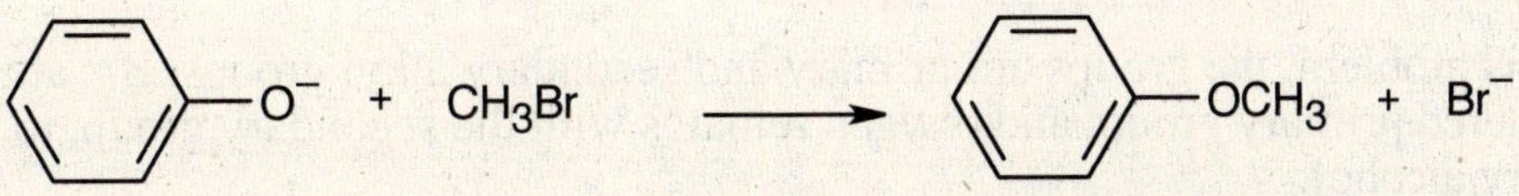


Methyl phenyl ether
(Anisole)

(c)

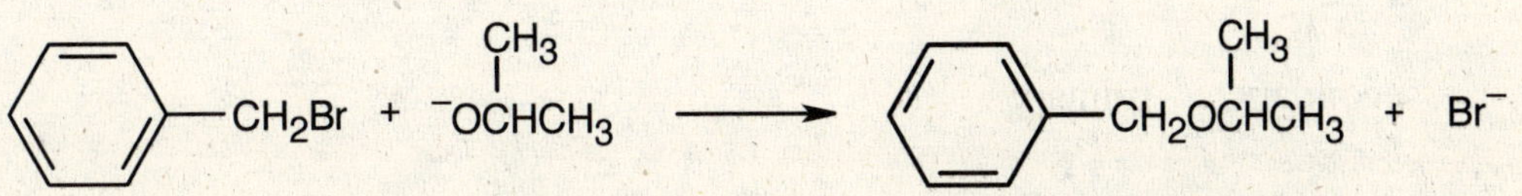


Benzyl isopropyl ether

**8.15**

*Least reactive* ⟶ *Most reactive*

$$(CH_3)_3CCl < CH_3CHBrCH_3 < CH_3CH_2Cl < CH_3CH_2Br$$

2-Chloro-2-methylpropane < 2-Bromopropane < Chloroethane < Bromoethane

The reactivity of alkyl halides in the Williamson ether synthesis is the same as their reactivity in any $S_N2$ reaction.

**8.16**

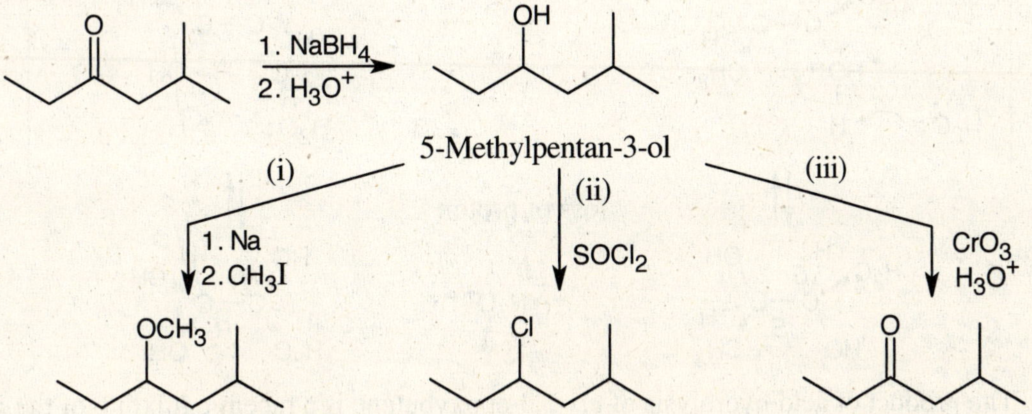

**8.17** (a) First, notice the substitution pattern of the ether. Bonded to the ether oxygen are a primary alkyl group and a tertiary alkyl group. When one group is tertiary, cleavage occurs by an $S_N1$ or E1 route to give either an alkene or a tertiary halide and a primary alcohol.

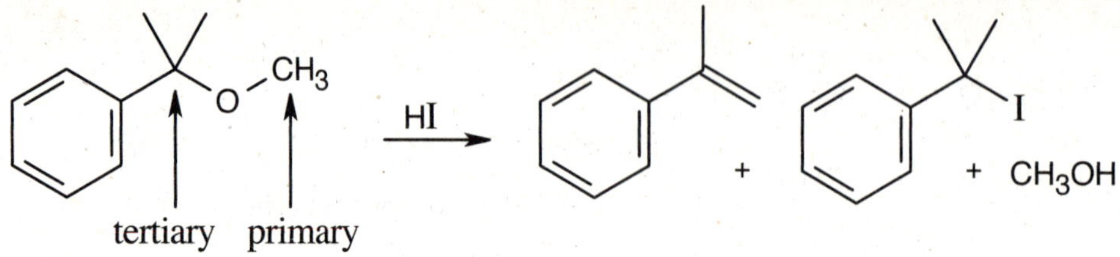


(b) In this problem, the groups are primary and secondary alkyl groups. $Br^-$ attacks at the less hindered primary group, and oxygen remains with the secondary group, to give a secondary alcohol.

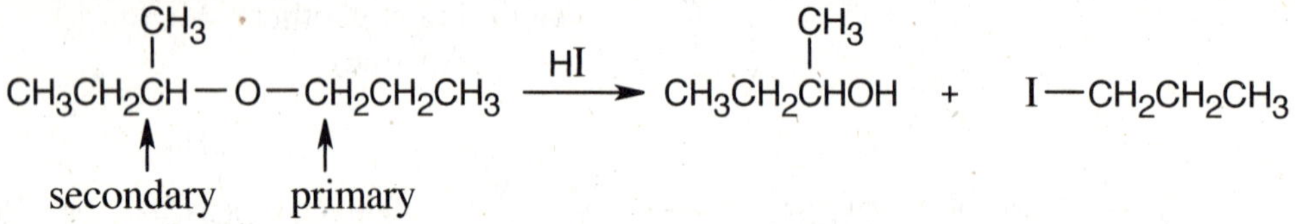


**8.18**

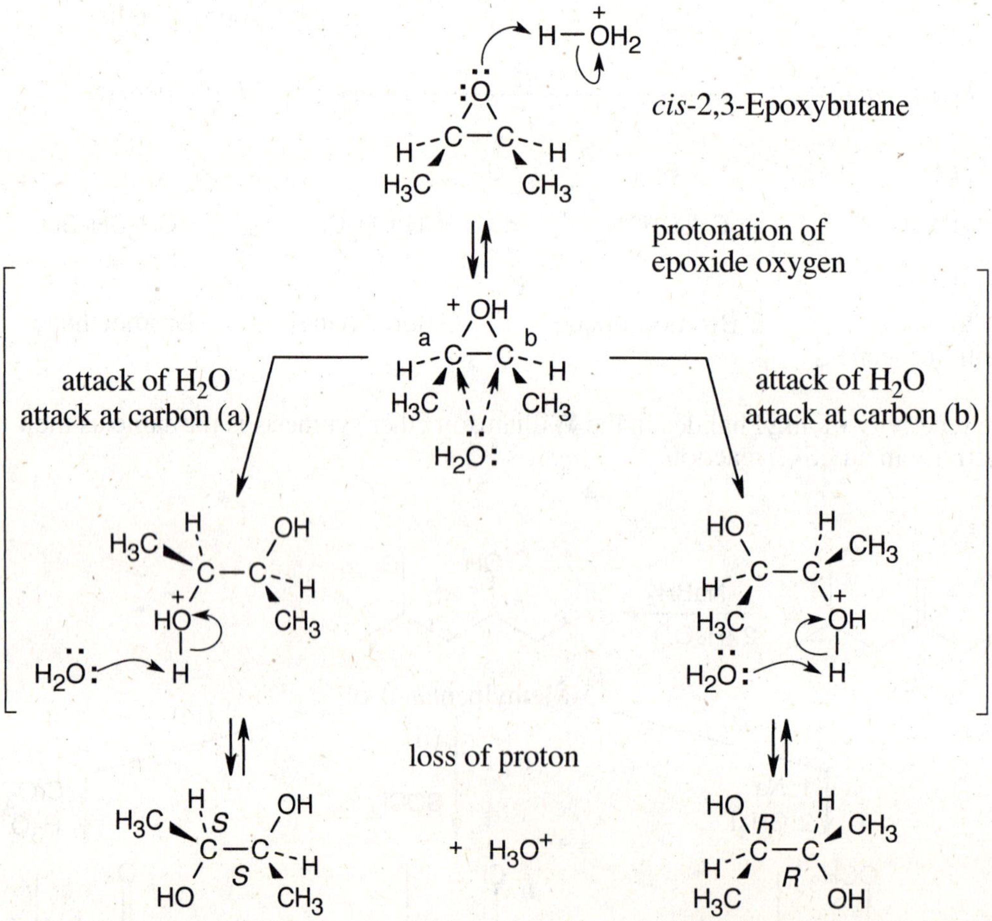


The product of acid hydrolysis of *cis*-2,3-epoxybutane is a racemic mixture of the *R*,*R* and *S*,*S* diol enantiomers.

**8.19**

(a)

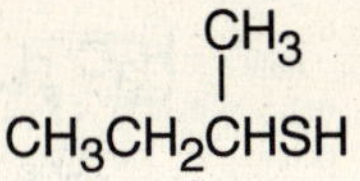

Butane-2-thiol

(b)

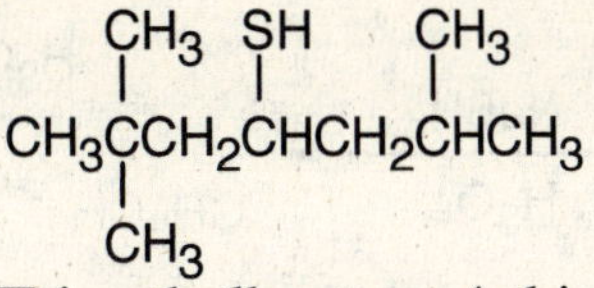

2,2,6-Trimethylheptane-4-thiol

(c)

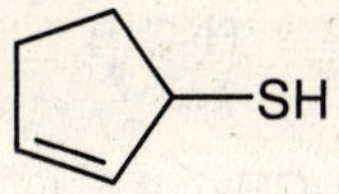

Cyclopent-2-ene-1-thiol

(d)

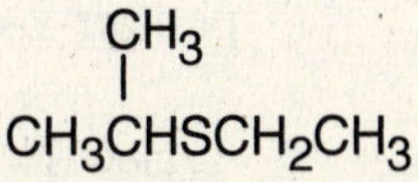

Ethyl isopropyl sulfide

(e)

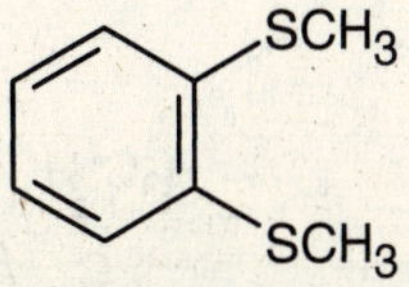

*o*-(Dimethylthio)benzene

**8.20**

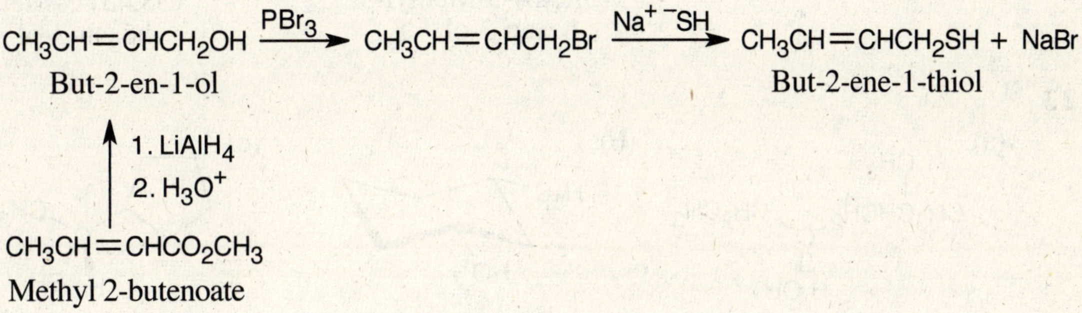

But-2-en-1-ol → But-2-ene-1-thiol; Methyl 2-butenoate

## Visualizing Chemistry

**8.21**

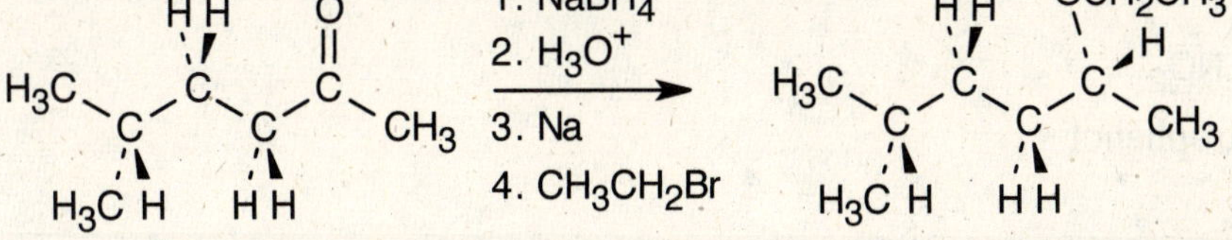

+

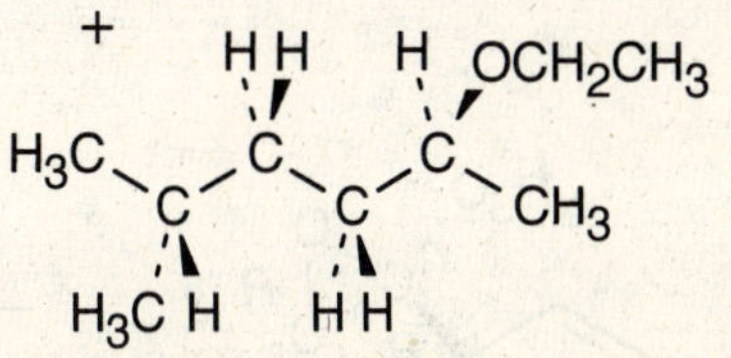

The product is a racemic mixture of enantiomers.

**8.22** The products of both reactions are diastereomers.

(a)

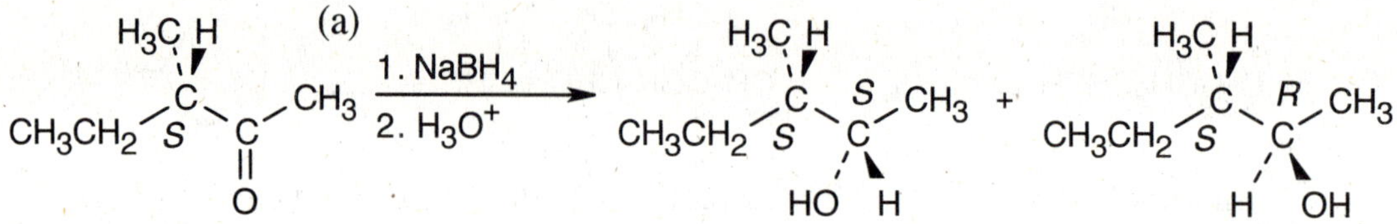


(2*S*,3*S*)-3-Methylpentan-2-ol (2*R*,3*S*)-3-Methylpentan-2-ol

(b)

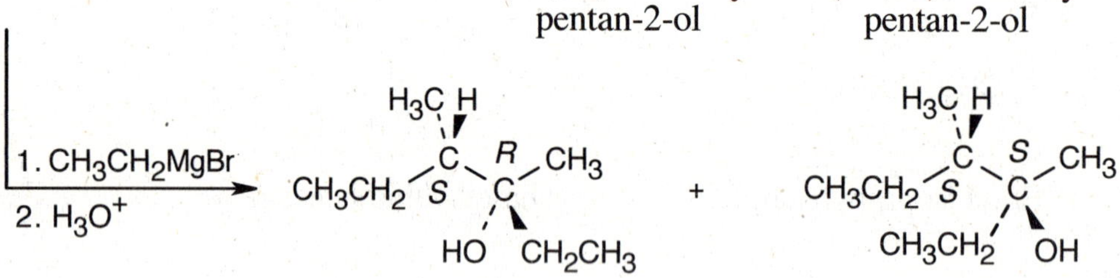


(3*R*,4*S*)-3-Methylhexan-3-ol (3*S*,4*S*)-3-Methylhexan-3-ol

**8.23**

(a)

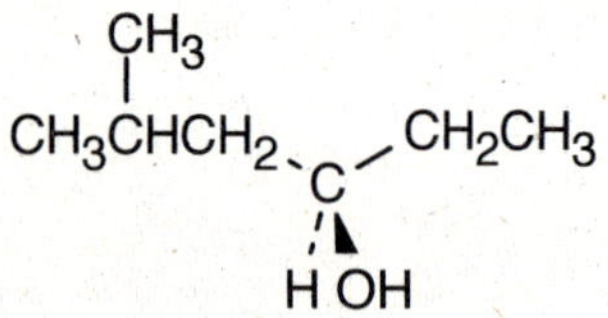

(*R*)-5-Methylhexan-3-ol

(b)

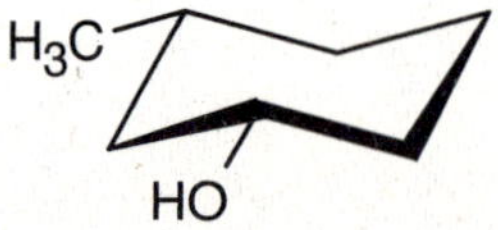

*cis*-3-Methylcyclohexanol

(c)

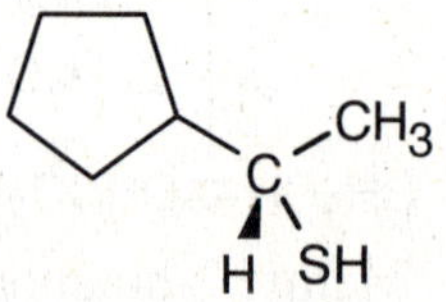

(*S*)-1-Cyclopentylethanethiol

(d)

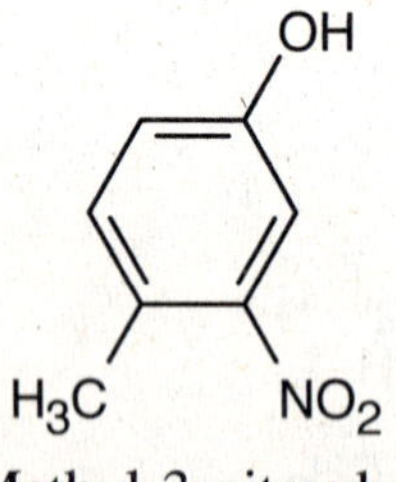

4-Methyl-3-nitrophenol

**8.24**

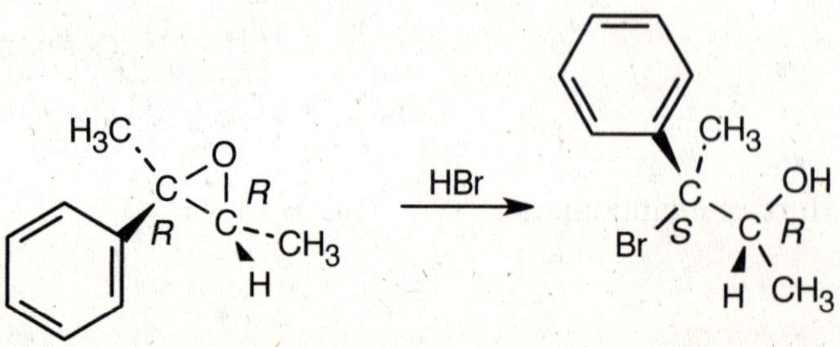


Acid-catalyzed opening of the epoxide ring takes place with the observed regiochemistry because the resulting carbocation-like transition state can be stabilized by the aromatic ring. Attack of bromine is an $S_N2$-like reaction that produces a single stereoisomer.

**8.25**

(a)

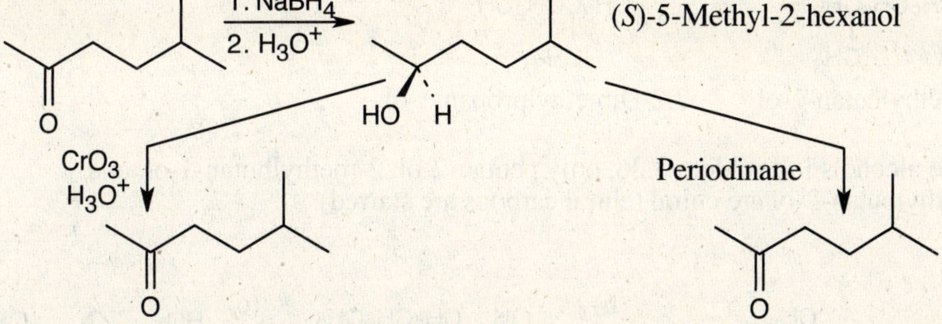


(b)

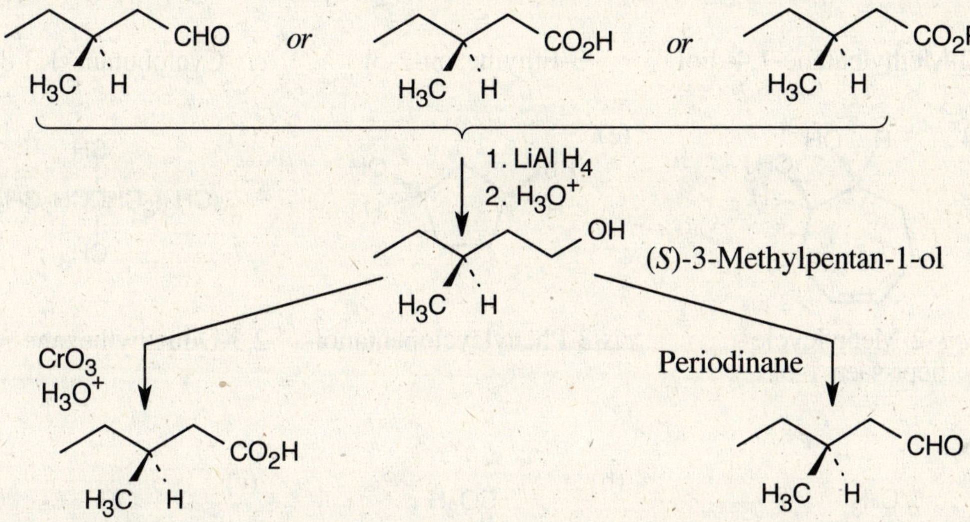


## Additional Problems

### Naming Alcohols and Ethers

**8.26**

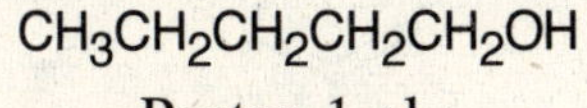


Pentan-1-ol

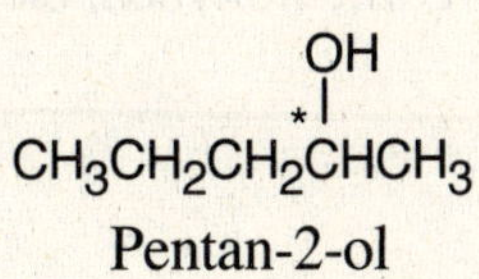


Pentan-2-ol

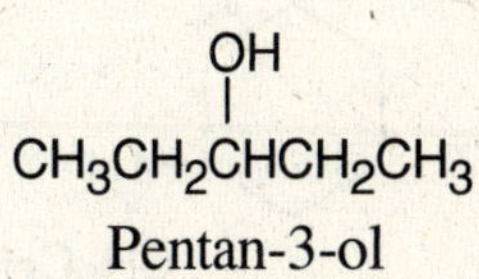


Pentan-3-ol

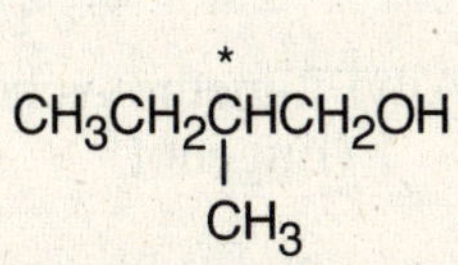


2-Methylbutan-1-ol

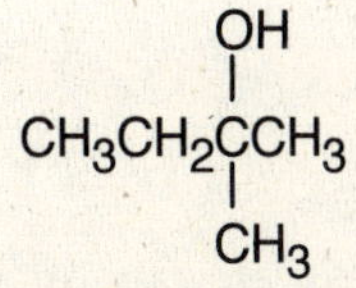


2-Methylbutan-2-ol

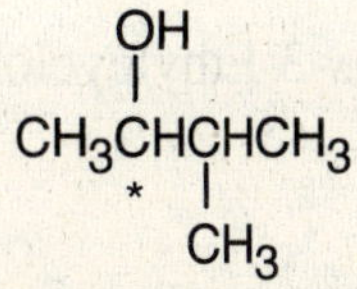


3-Methylbutan-2-ol

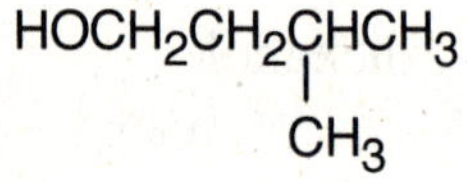

3-Methylbutan-1-ol

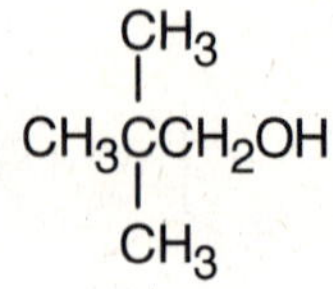

2,2-Dimethylpropan-1-ol

**8.27** Of the alcohols in Problem 8.26, only pentan-2-ol, 2-methylbutan-1-ol, and 3-methylbutan-2-ol are chiral (chiral carbons are starred).

**8.28**

(a)

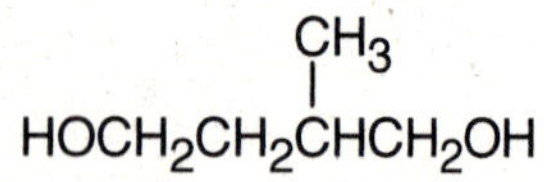

2-Methylbutane-1,4-diol

(b)

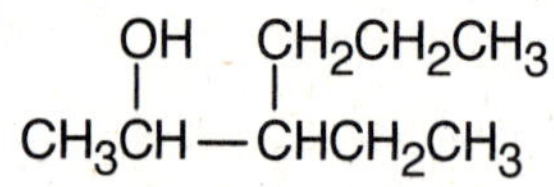

3-Ethylhexan-2-ol

(c)

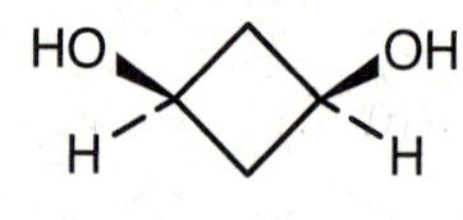

*cis*-Cyclobutane-1,3-diol

(d)

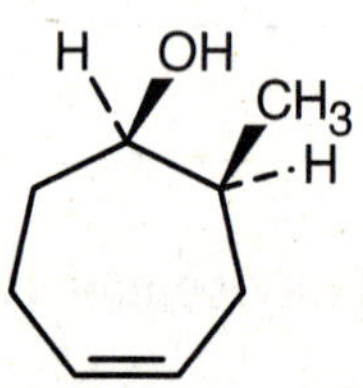

*cis*-2-Methylcyclo-hept-4-en-1-ol

(e)

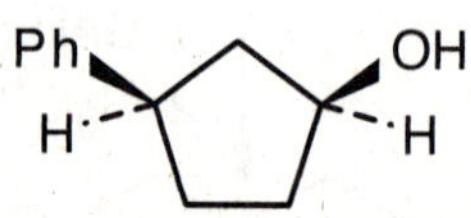

*cis*-3-Phenylcyclopentanol

(f)

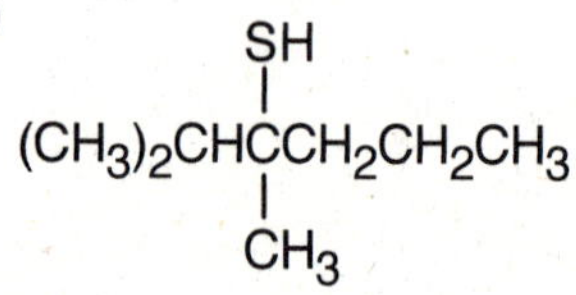

2,3-Dimethylhexane-3-thiol

**8.29**

(a)

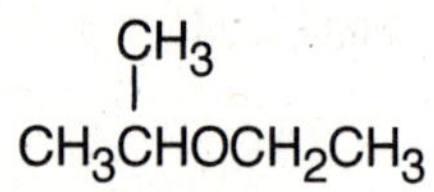

Ethyl isopropyl ether

(b)

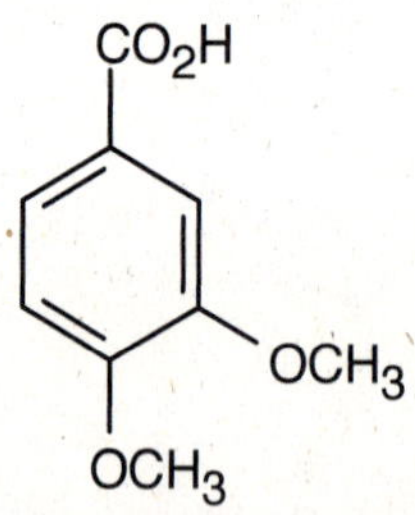

3,4-Dimethoxybenzoic acid

(c)

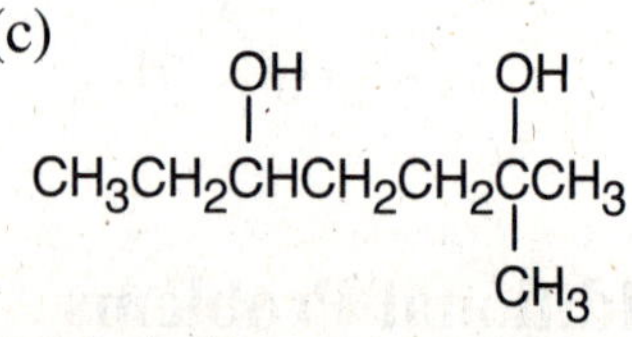

2-Methylheptane-2,5-diol

(d)

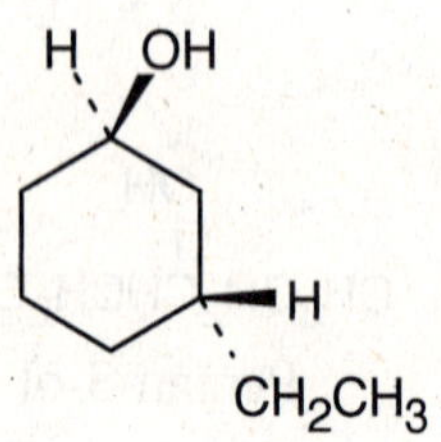

*trans*-3-Ethylcyclohexanol

(e)

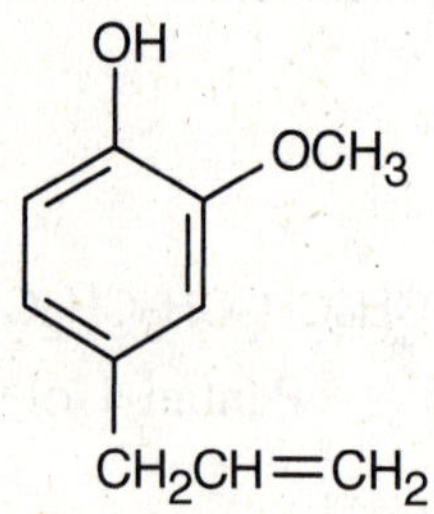

4-Allyl-2-methoxyphenol
(Eugenol)

**8.30**

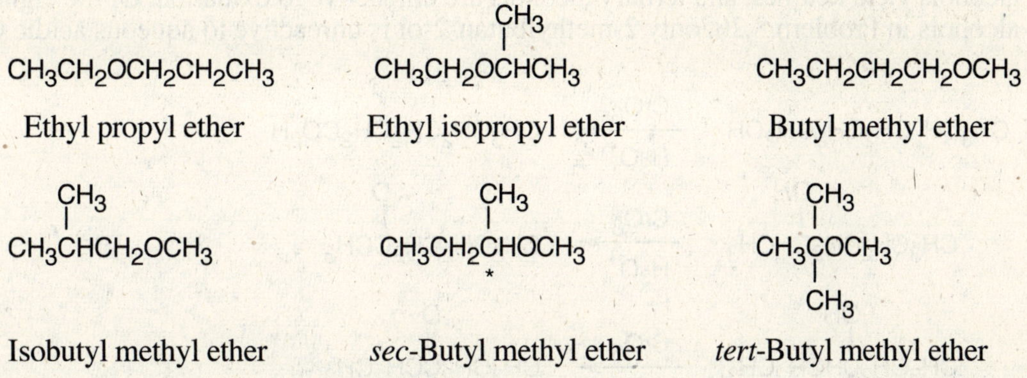


Only *sec*-butyl methyl ether is chiral.

## Reactions and Synthesis

**8.31**

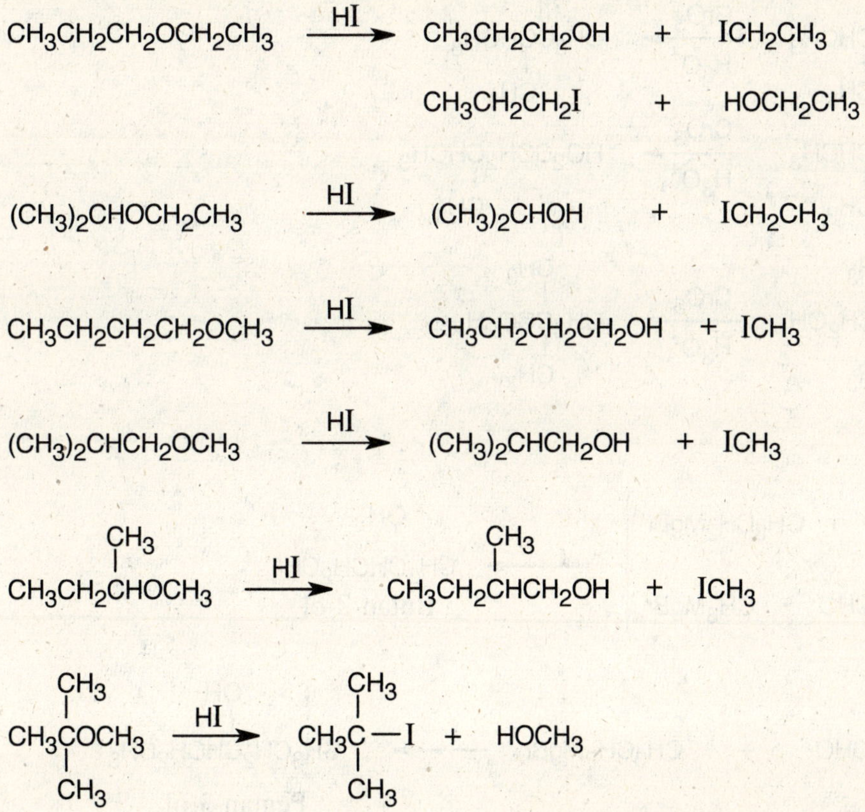

**8.32** Primary alcohols react with aqueous acidic $CrO_3$ to form carboxylic acids, secondary alcohols yield ketones, and tertiary alcohols are unreactive to oxidation. Of the eight alcohols in Problem 8.26, only 2-methylbutan-2-ol is unreactive to aqueous acidic $CrO_3$.

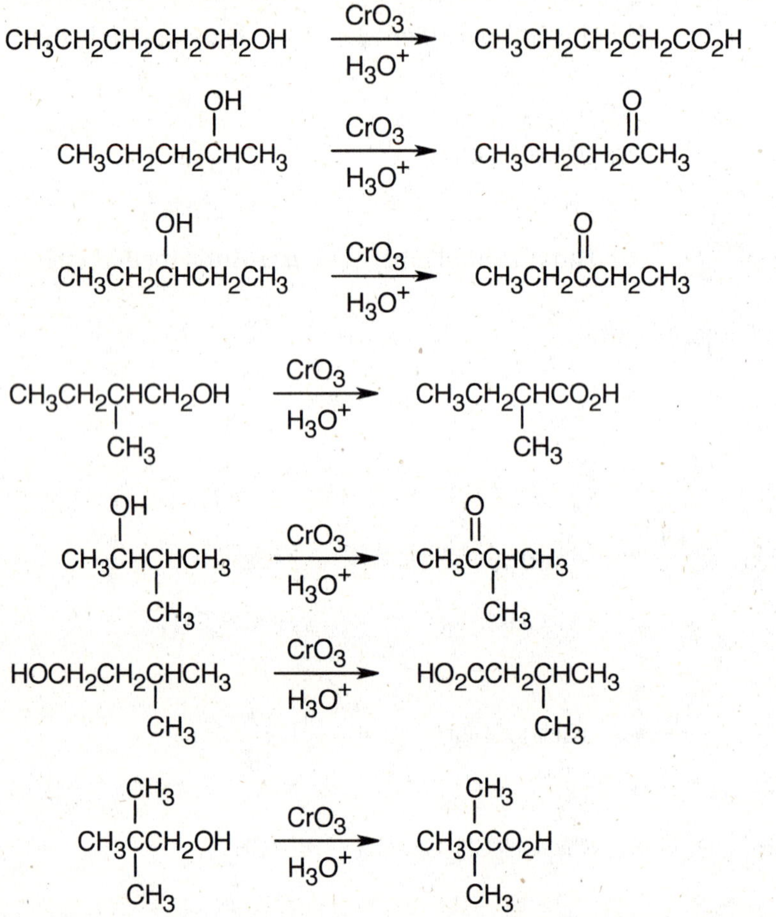

**8.33**

(a)

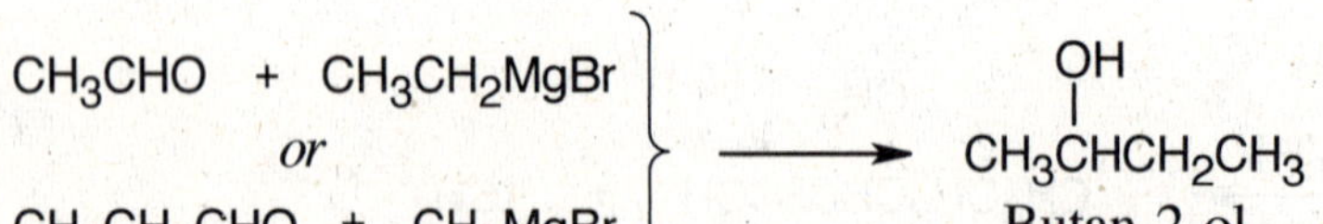

Butan-2-ol

(b)

Pentan-3-ol

(c)

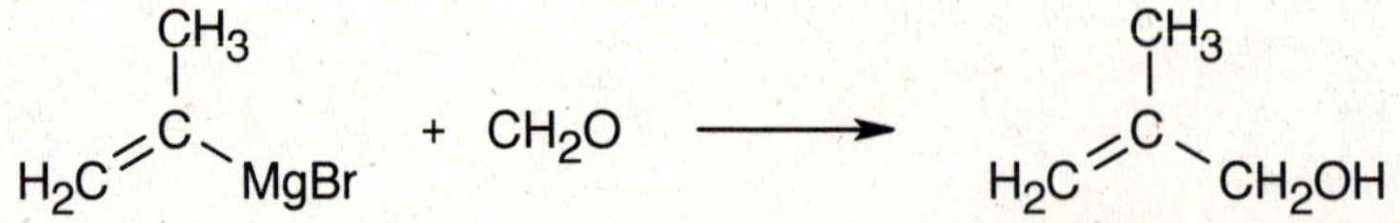

2-Methylprop-2-en-1-ol

(d)

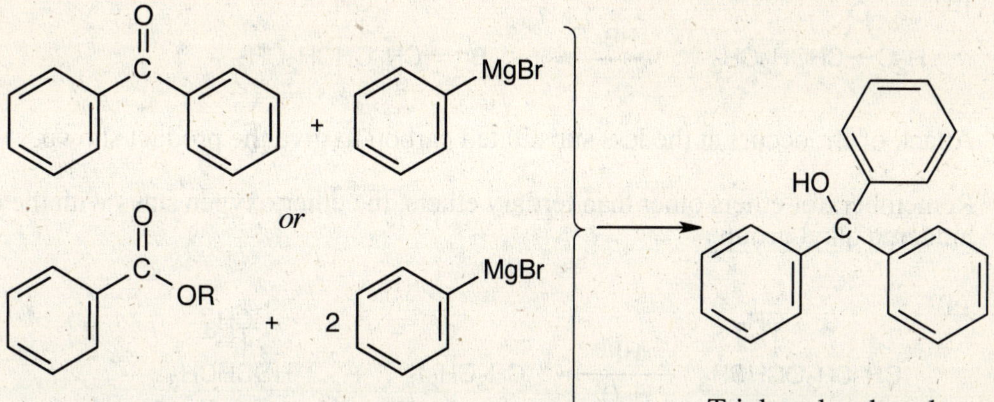


Triphenylmethanol

(e)

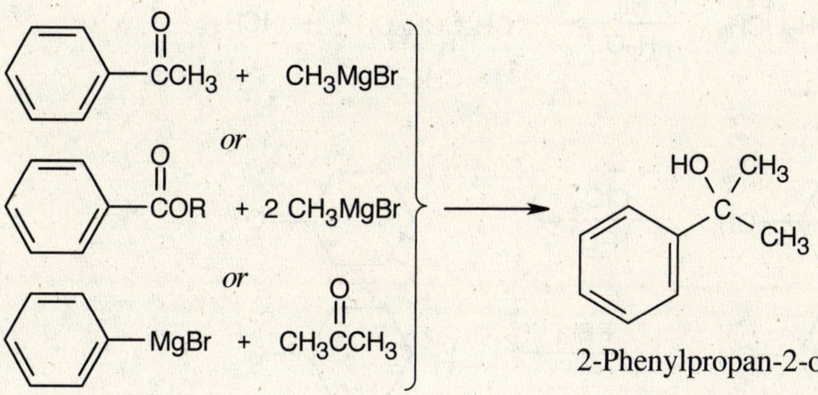


2-Phenylpropan-2-ol

(f)

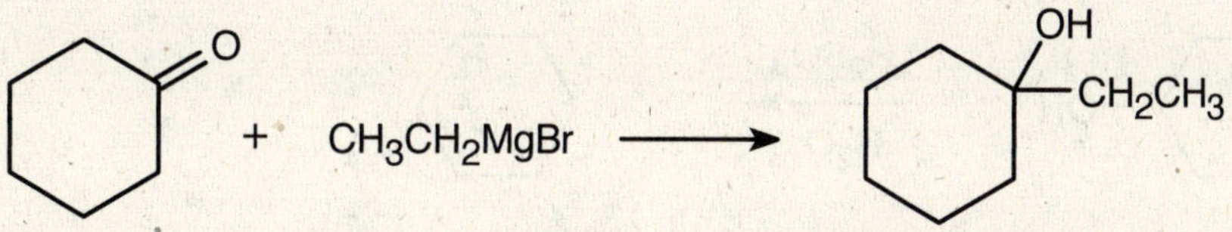


1-Ethylcyclohexanol

**8.34**

(a)

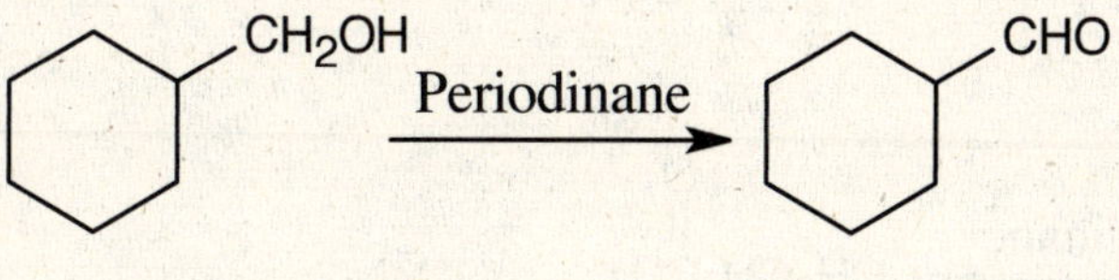


(b)

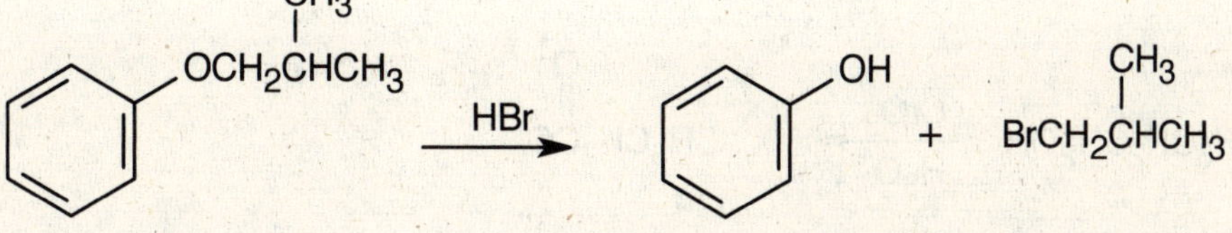


(c)

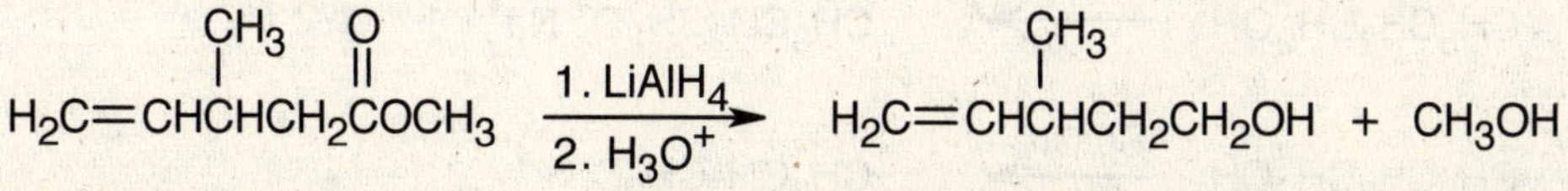

(d)

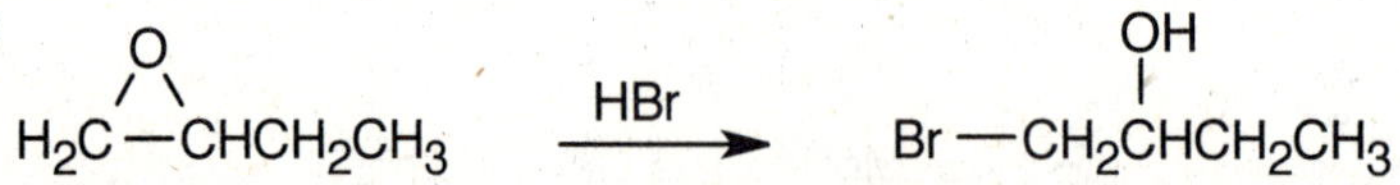


Attack of Br⁻ occurs at the less substituted carbon to give the product shown.

**8.35** Remember: for ethers other than tertiary ethers, the ether oxygen stays with the more hindered alkyl group.

(a)

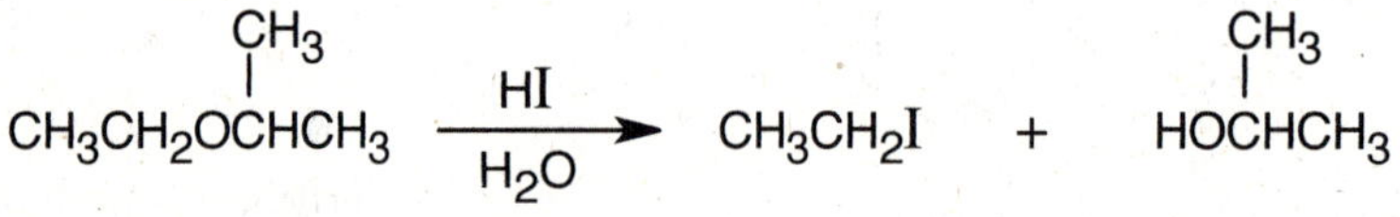


(b)

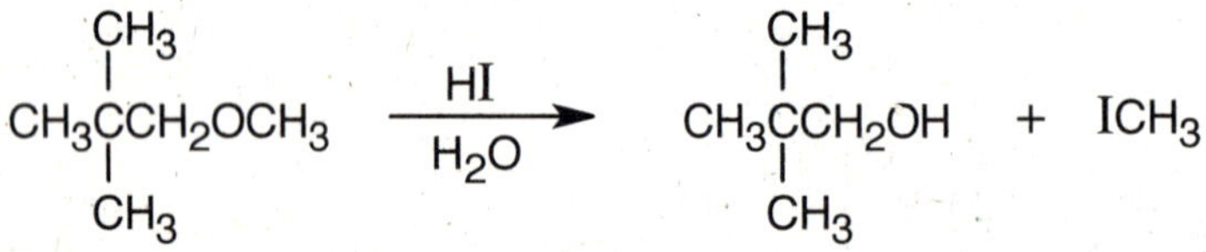


**8.36**

(a)

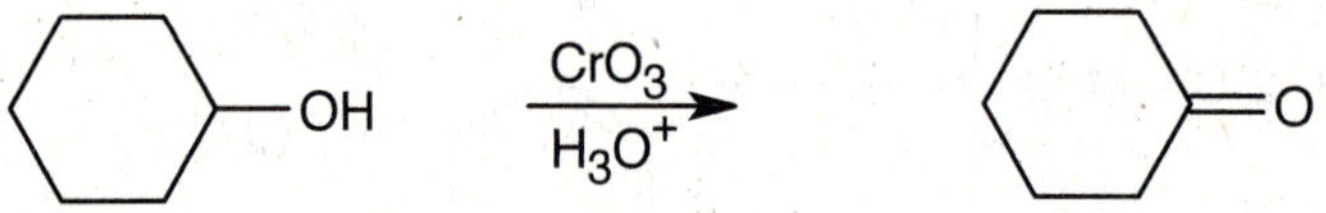


(b)

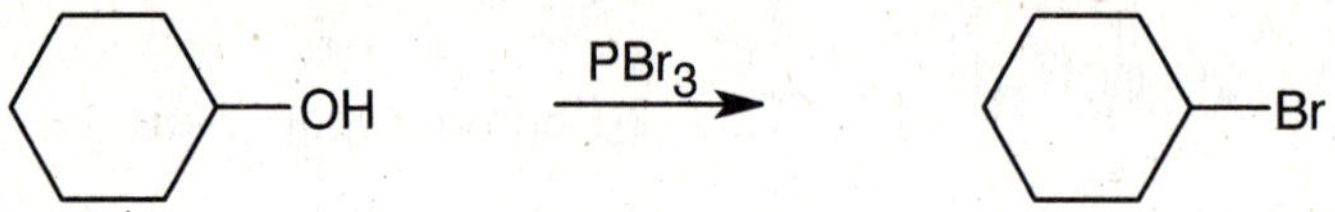


(c)

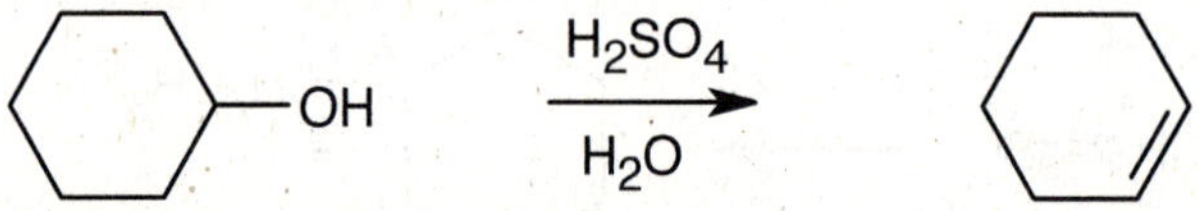


(d)

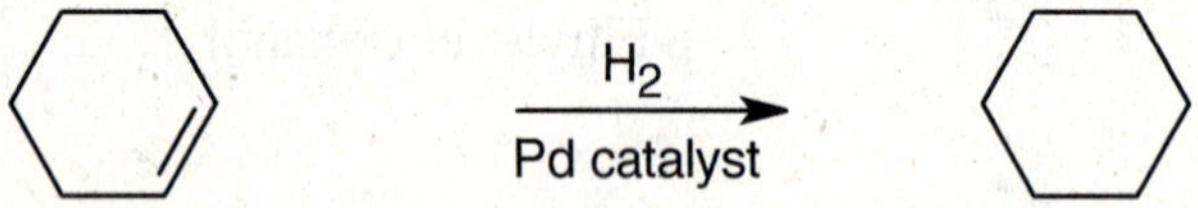


from (c)

**8.37**

(a)

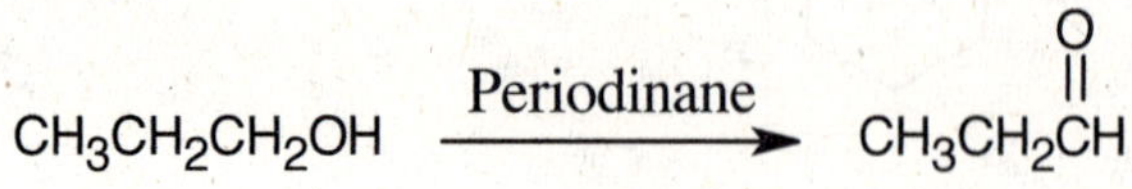


(b)

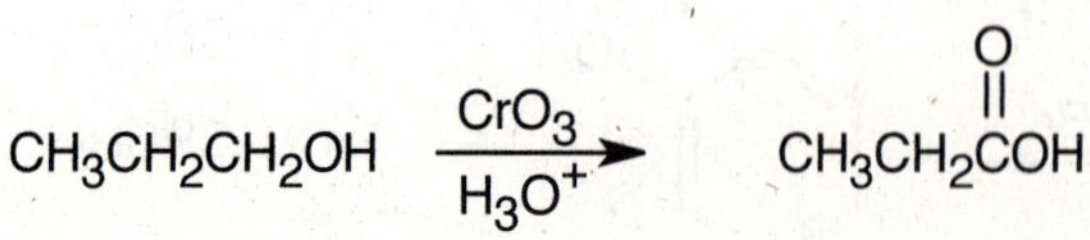


(c)

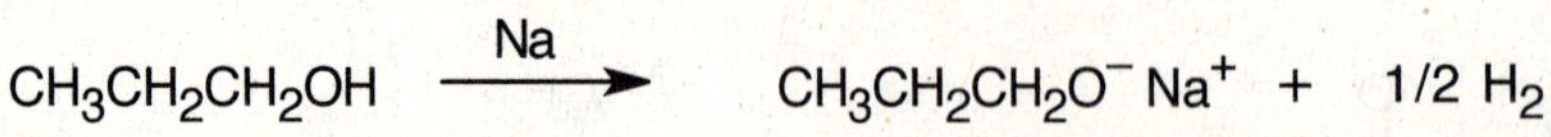


(d)

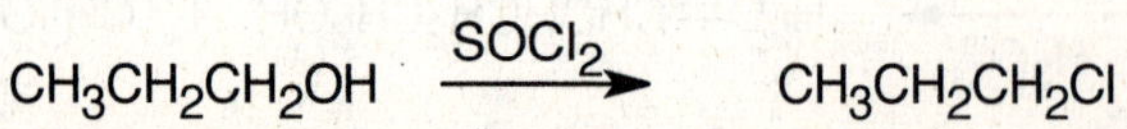

**8.38**

(a)

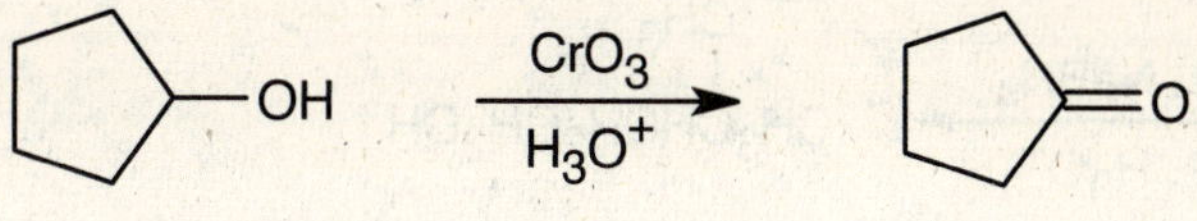

(b)

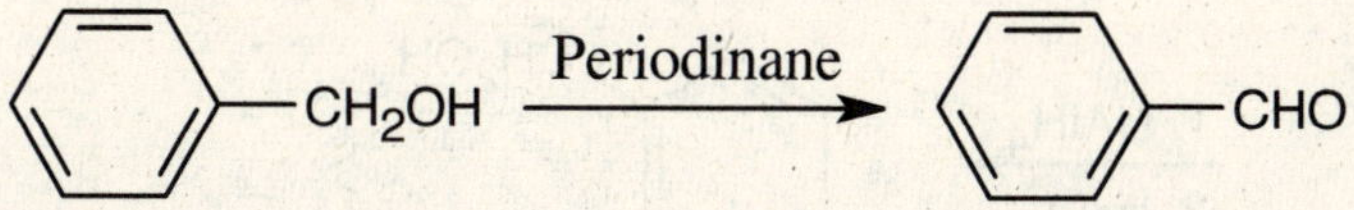

(c)

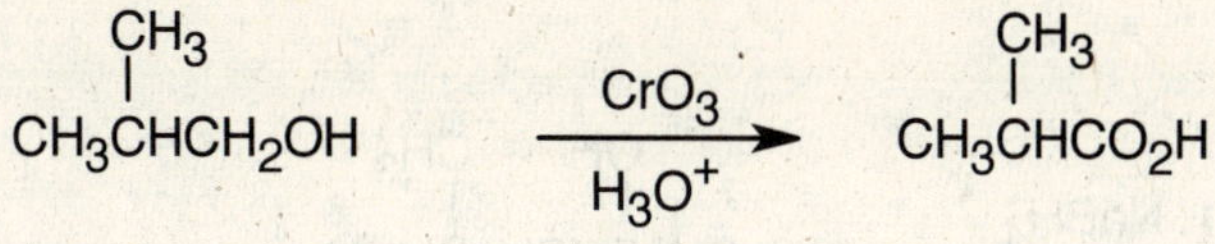

**8.39**

(a)

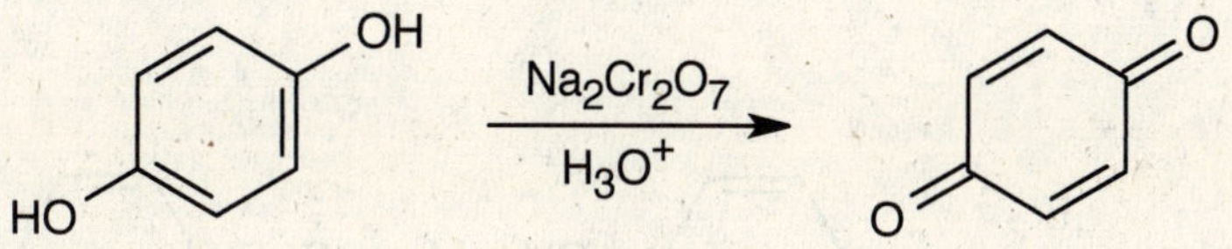

(b)

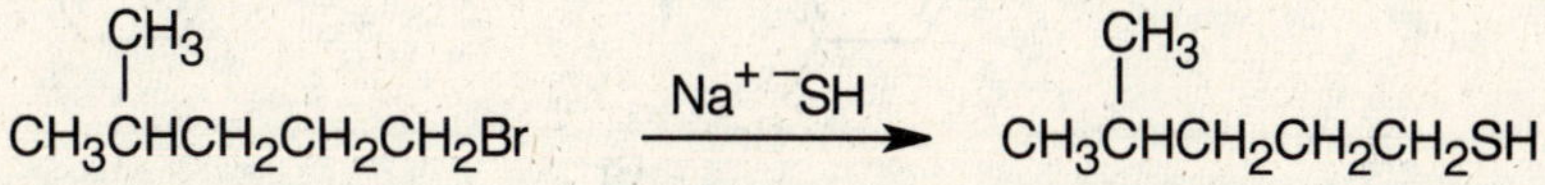

(c)

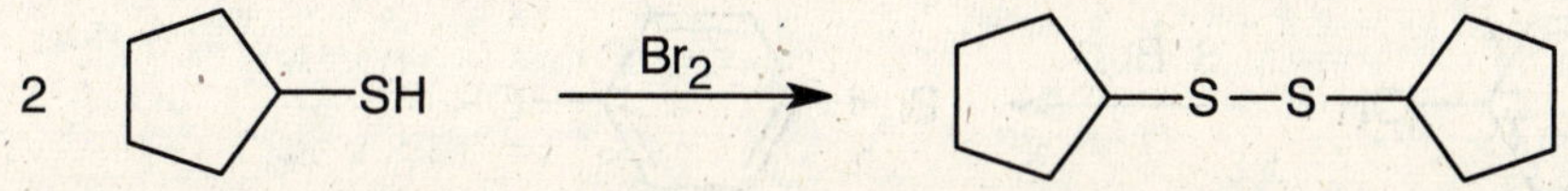

**8.40**

*Alcohol* | *Carbonyl precursor(s)*

(a)

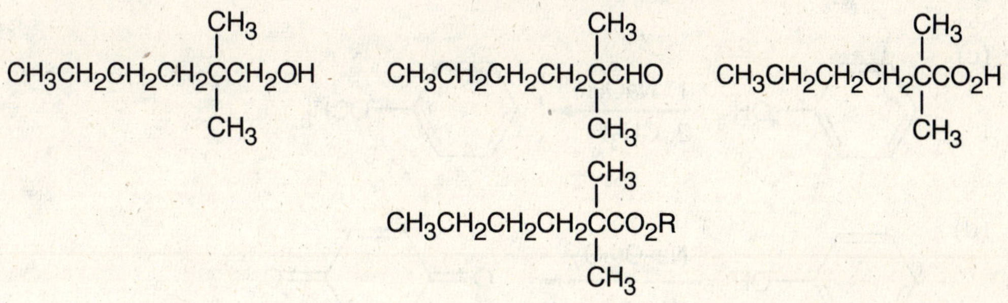

(b)

(c)

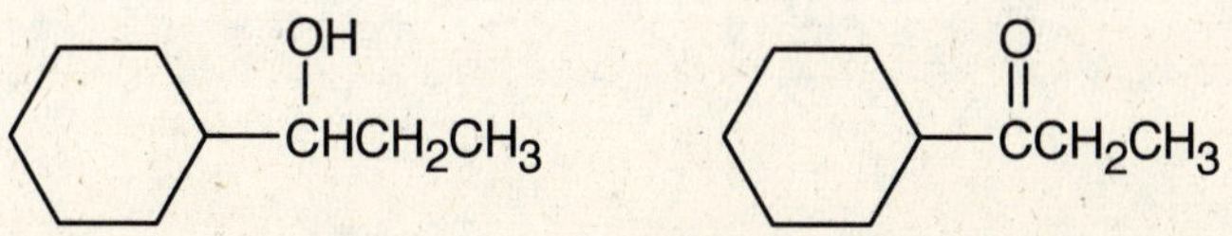

**8.41**

(a)

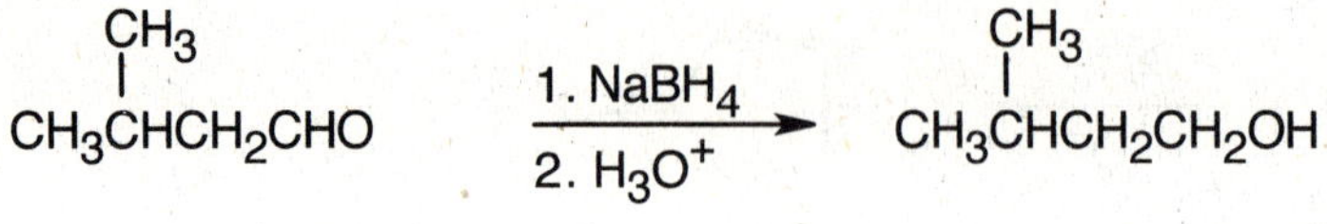


(b)

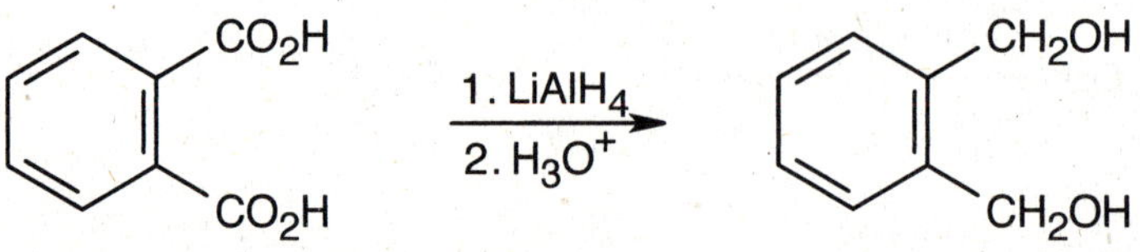


(c)

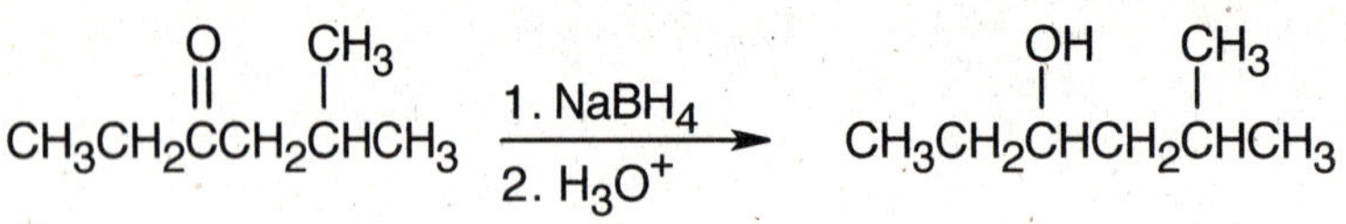


**8.42**

(a)

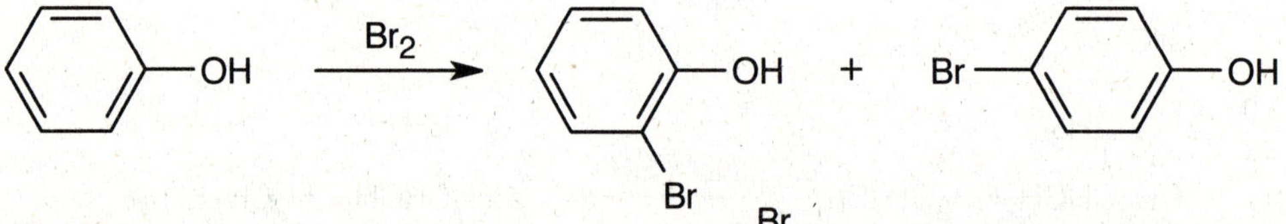


(b)

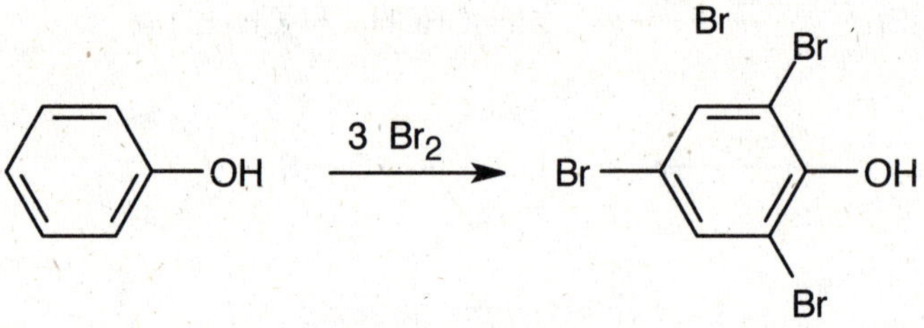


(c)

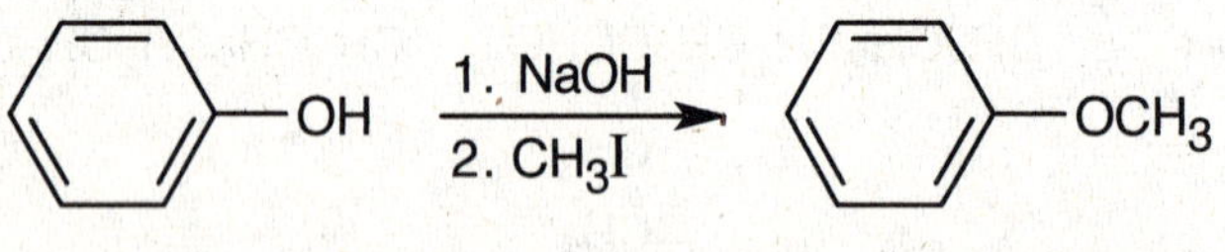


(d)

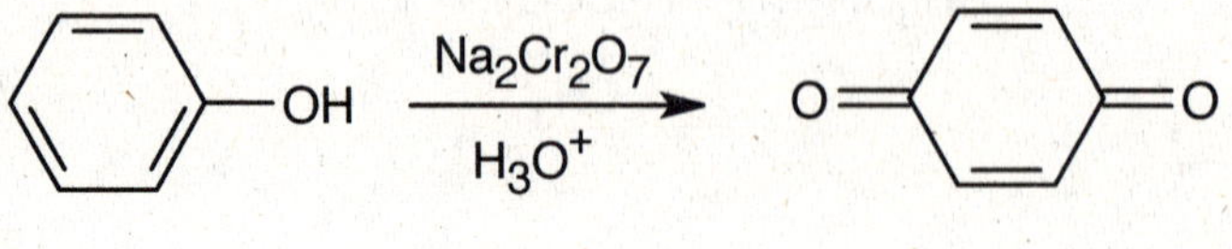

**8.43**

(a)

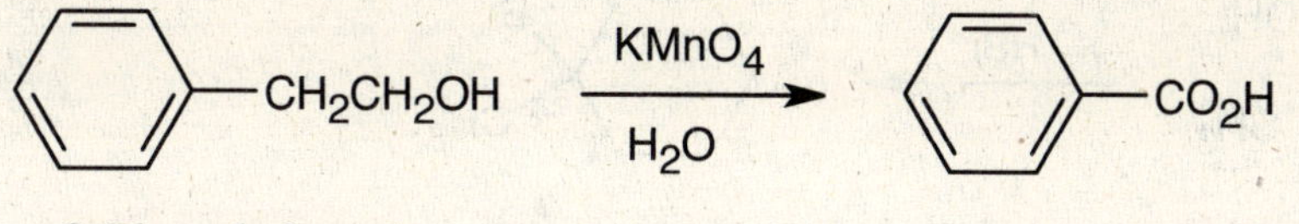

2-Phenylethanol

(b)

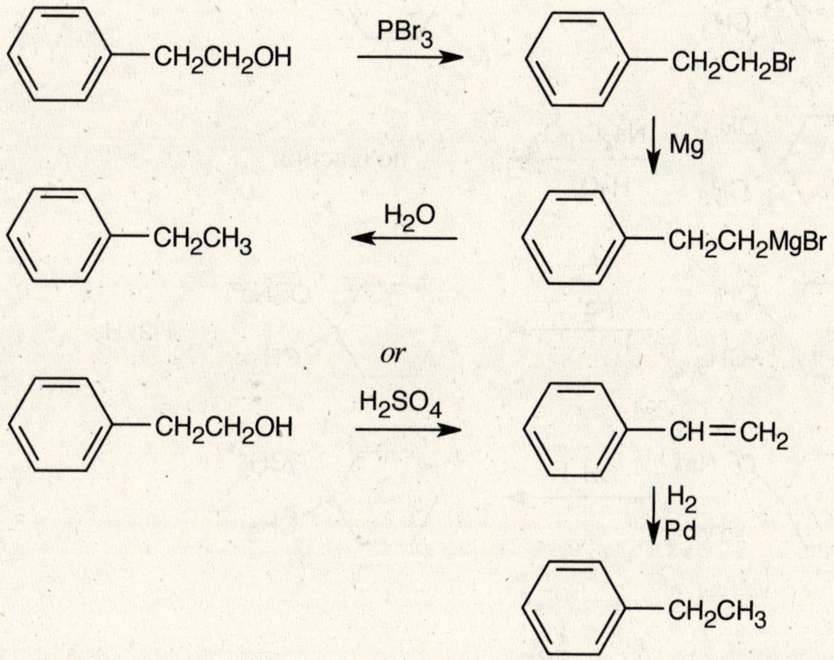

(c)

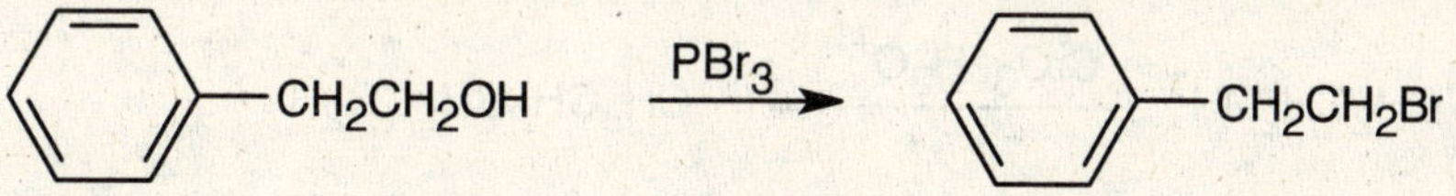

(d)

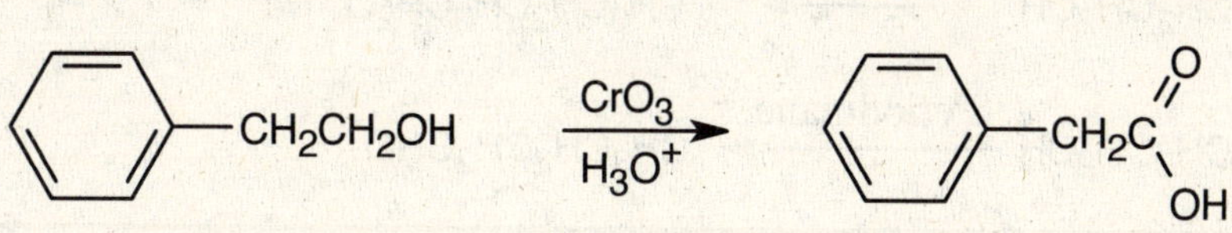

(e)

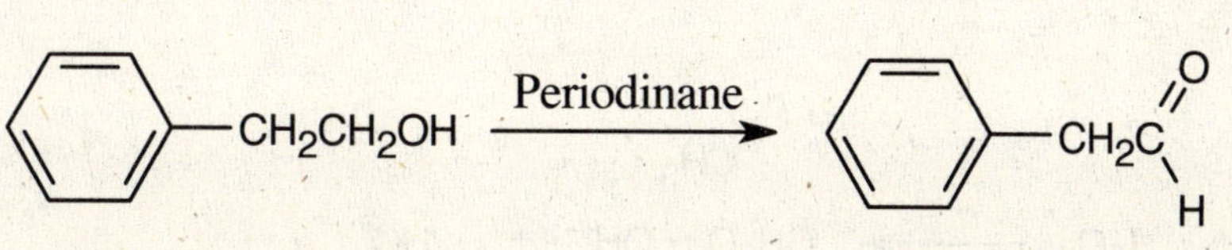

**8.44**

(a)

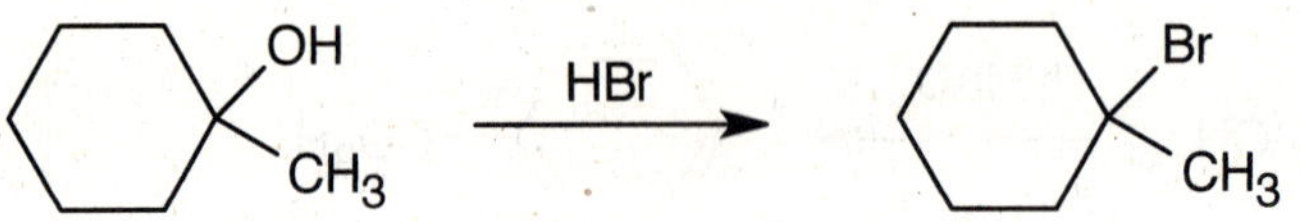

(b)

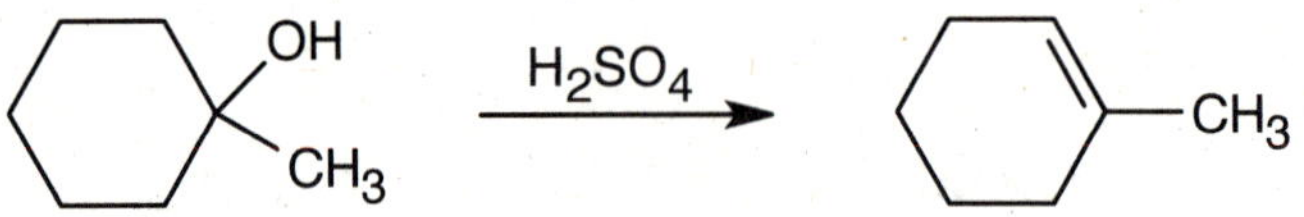

(c)

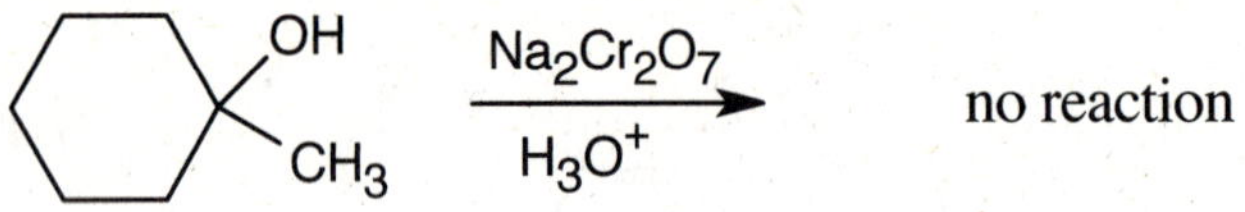

(d)

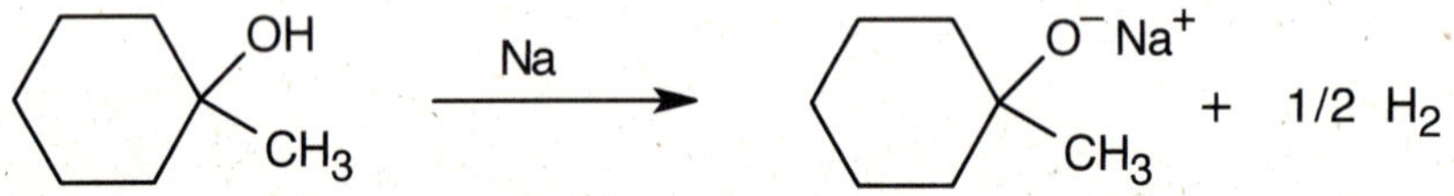

(e)

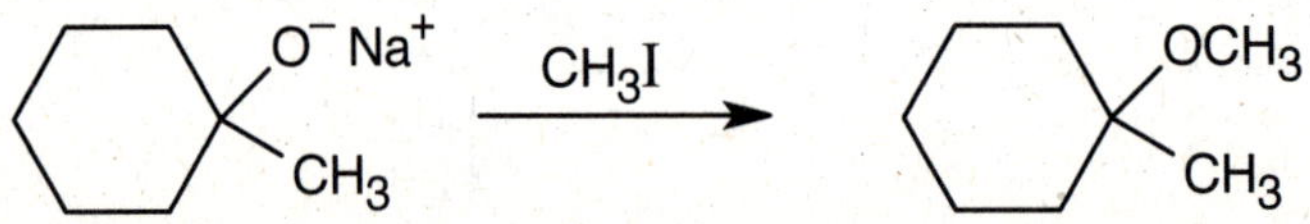

**8.45**

(a) $CH_3CH_2CH_2CH_2OH \xrightarrow{PBr_3} CH_3CH_2CH_2CH_2Br$

(b) $CH_3CH_2CH_2CH_2OH \xrightarrow{CrO_3,\ H_3O^+} CH_3CH_2CH_2CO_2H$

(c) $CH_3CH_2CH_2CH_2OH \xrightarrow{Na} CH_3CH_2CH_2CH_2O^-\ Na^+ + 1/2\ H_2$

(d) $CH_3CH_2CH_2CH_2OH \xrightarrow{\text{Periodinane}} CH_3CH_2CH_2CHO$

## Acidity

**8.46**

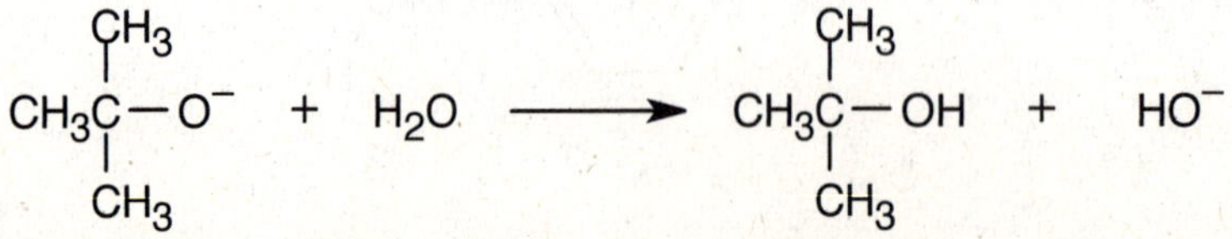

p$K_a$ = 15.7 — stronger acid; p$K_a$ = 18 — weaker acid

The reaction takes place as written because water is a stronger acid than *tert*–butyl alcohol.

**8.47**

*Least acidic* ⟶ *Most acidic*

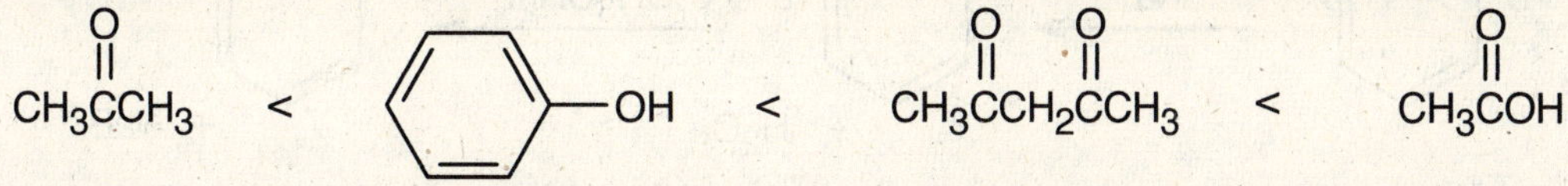

p$K_a$ = 19 p$K_a$ = 9.9 p$K_a$ = 9 p$K_a$ = 4.7

**8.48** To react completely with NaOH, an acid must have a p$K_a$ lower than the p$K_a$ of $H_2O$. Thus, all compounds in the previous problem except acetone react completely with NaOH.

**8.49** Only acetic acid will react with sodium bicarbonate because p$K_a$ of acetic acid is less than p$K_a$ of bicarbonate..

**8.50** Sodium bicarbonate reacts with acetic acid to produce carbonic acid, which breaks down to form $CO_2$. The resulting $CO_2$ bubbles indicate the presence of acetic acid. Phenol does not react with sodium bicarbonate.

## General Problems

**8.51**

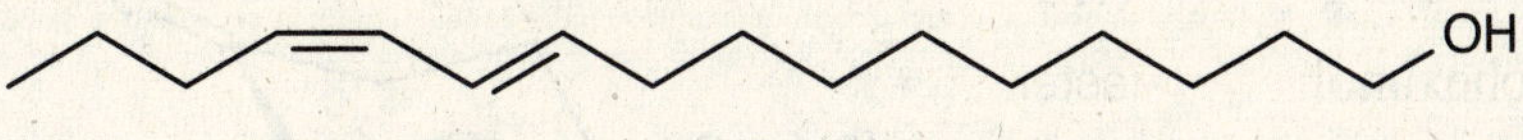

(10*E*,12*Z*)-Hexadeca-10,12-dien-1-ol

**8.52**

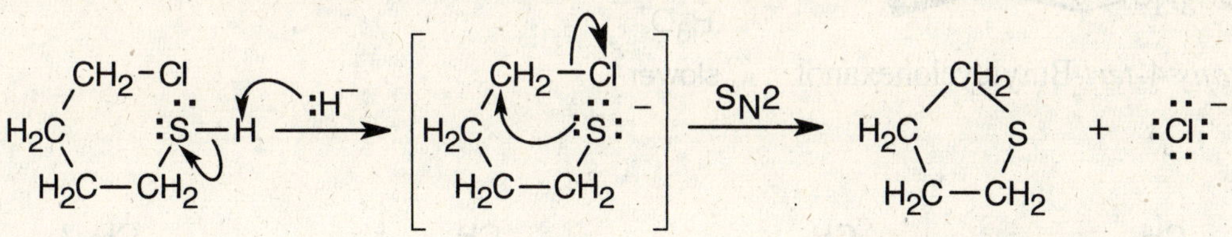

Tetrahydrothiophene

**8.53**

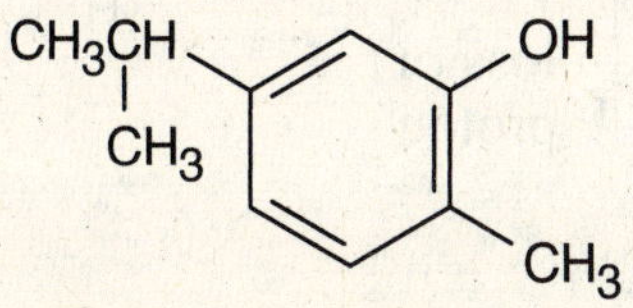

Carvacrol 5-Isopropyl-2-methylphenol

**8.54**

(a)

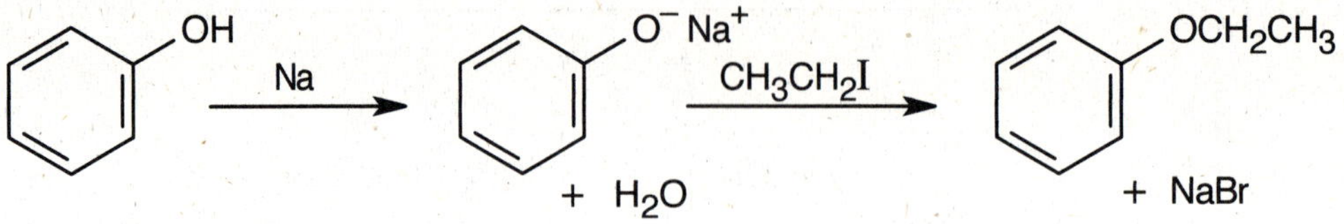


(b)

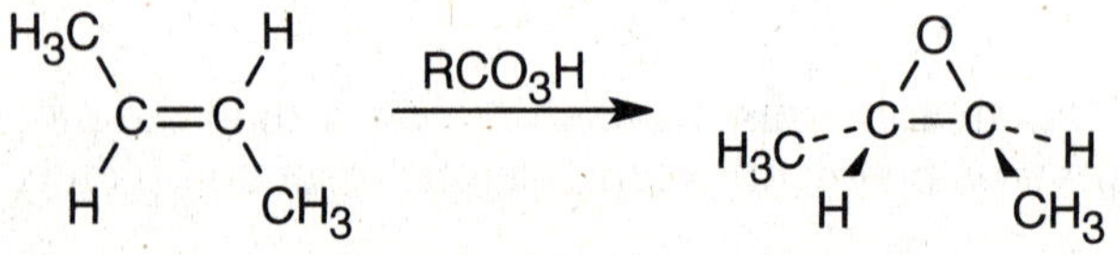


$RCO_3H$ = *m*-Chloroperbenzoic acid

**8.55** The hydroxyl group is axial in the cis isomer, which is expected to oxidize faster than the trans isomer. (Remember that the bulky *tert*-butyl group is always equatorial in the more stable conformation.)

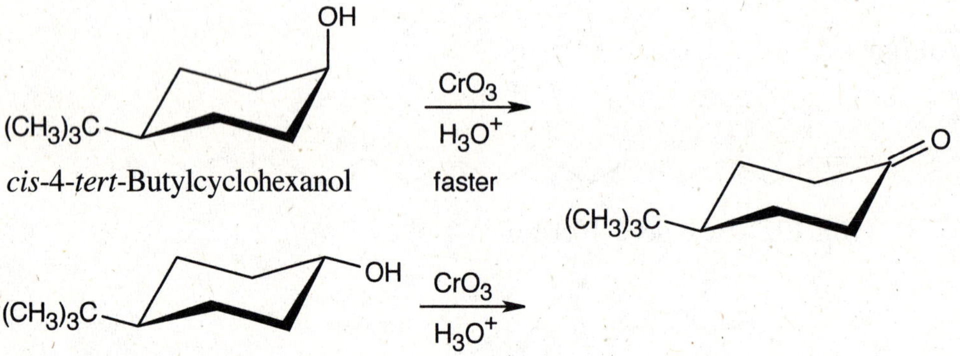


*cis*-4-*tert*-Butylcyclohexanol

*trans*-4-*tert*-Butylcyclohexanol

**8.56**

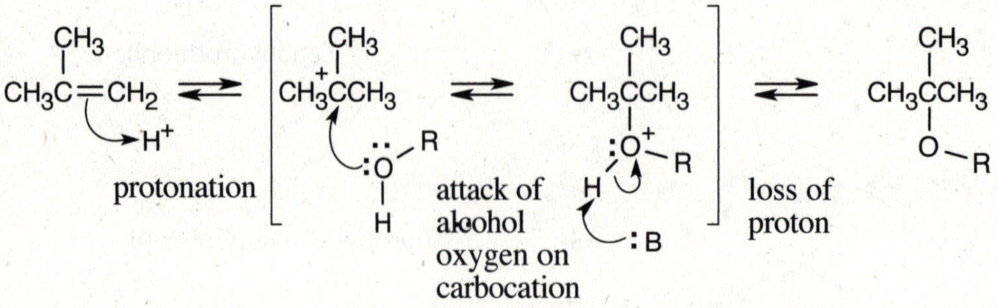

**8.57**

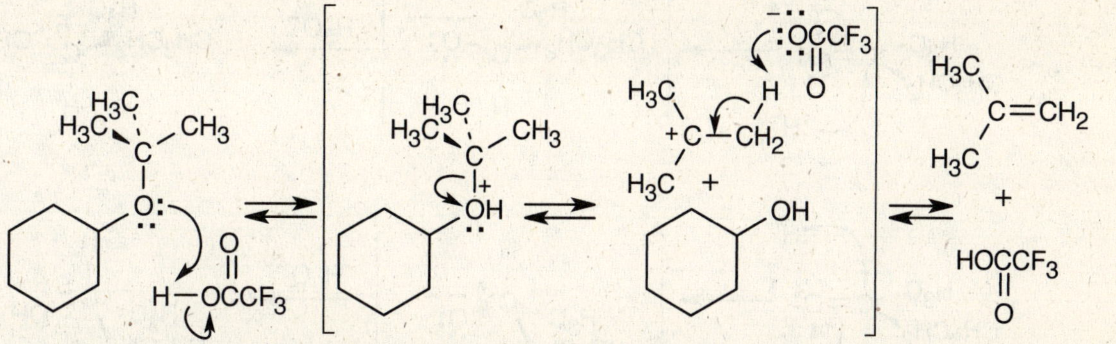

The first step of acid-catalyzed cleavage of ethers is protonation of the ether oxygen. The protonated intermediate forms cyclohexanol and a tertiary carbocation, which loses a proton to form 2-methylpropene. This is an example of an E1 elimination and is the reverse of the mechanism shown in the previous problem.

**8.58**

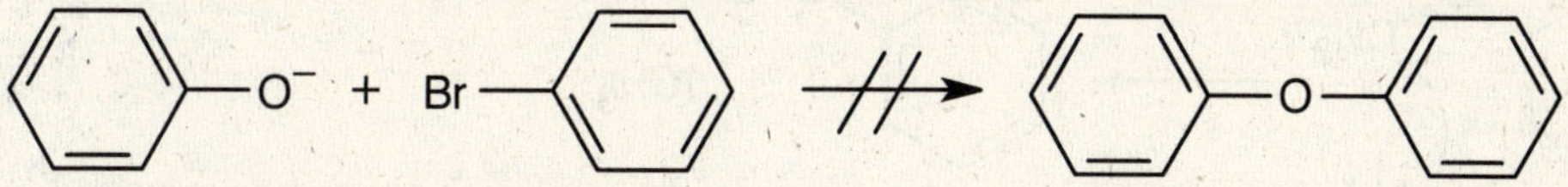

The Williamson ether synthesis is an $S_N2$ reaction between an alkoxide or phenoxide anion and an alkyl halide. It can't be used to synthesize diphenyl ether because an aryl halide ($C_6H_5Br$) doesn't undergo $S_N2$ reactions.

**8.59**

$$2\ CH_3Br\ +\ 2\ Na^+\ {}^-SH \longrightarrow 2\ CH_3SH\ +\ 2\ NaBr$$

$$2\ CH_3SH\ +\ Br_2 \longrightarrow CH_3SSCH_3\ +\ 2\ HBr$$

Dimethyl disulfide

**8.60**

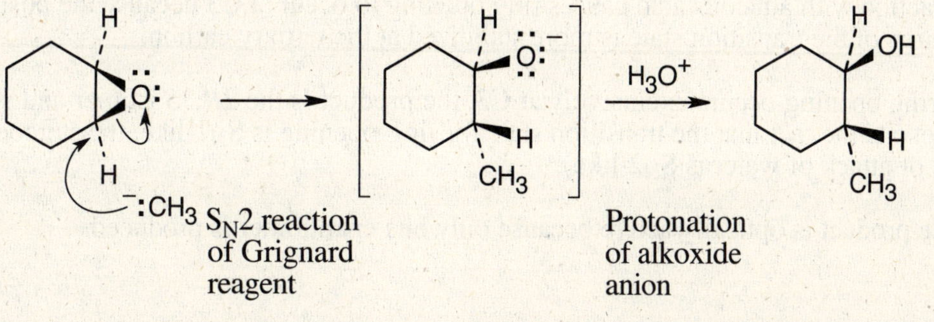


The methyl group and the hydroxyl group have a trans relationship.

**8.61**

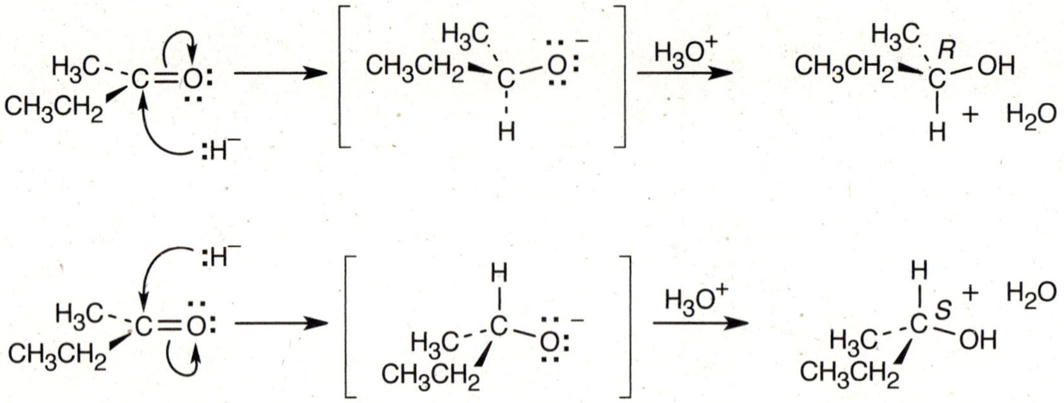

Attack can occur from either side of the planar carbonyl group to yield a racemic mixture of enantiomers.

**8.62**

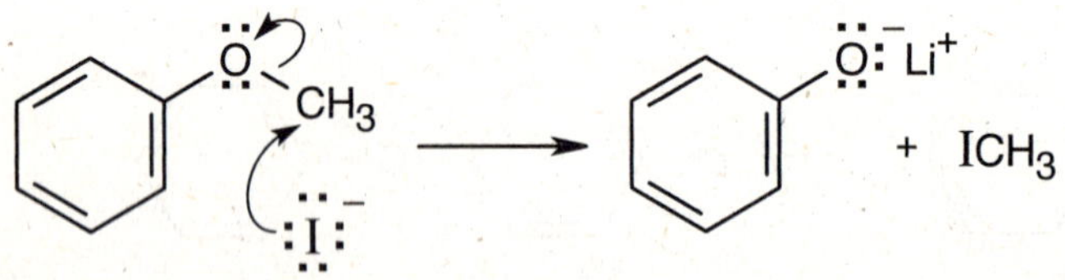

This reaction is an $S_N2$ displacement and can't occur at an aryl carbon.

**8.63**

(a) (b)

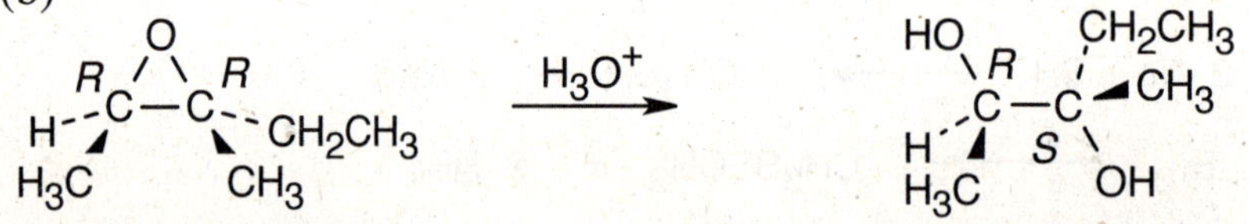

(2*R*,3*R*)-2,3-Epoxy-3-methylpentane (2*R*,3*S*)-3-Methylpentane-2,3-diol

Reaction with aqueous acid causes ring opening to occur at C3 because the positive charge of the transition state is more stabilized at the tertiary carbon.

(c) If ring opening occurs exclusively at C3, the product is the 2*R*,3*S* isomer and is chiral. (Despite the fact that the transition state for ring-opening is $S_N1$-like, the stereochemistry of attack of water is $S_N2$-like).

(d) The product is optically active because only one enantiomer is produced.

**8.64**

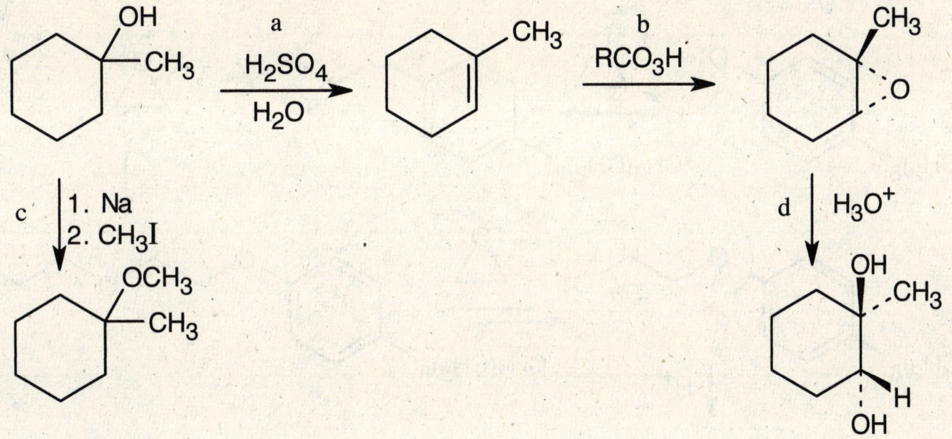


**8.65**

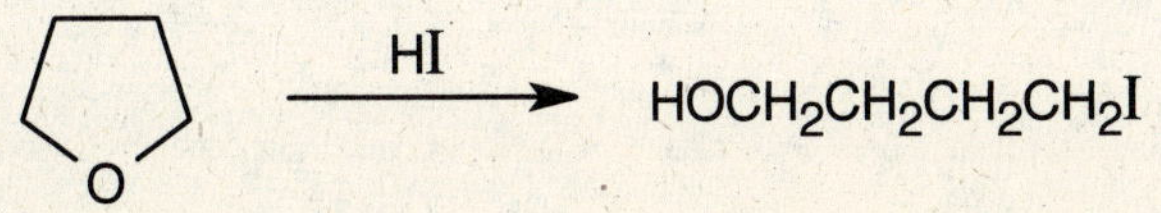


## In the Medicine Cabinet

**8.66**

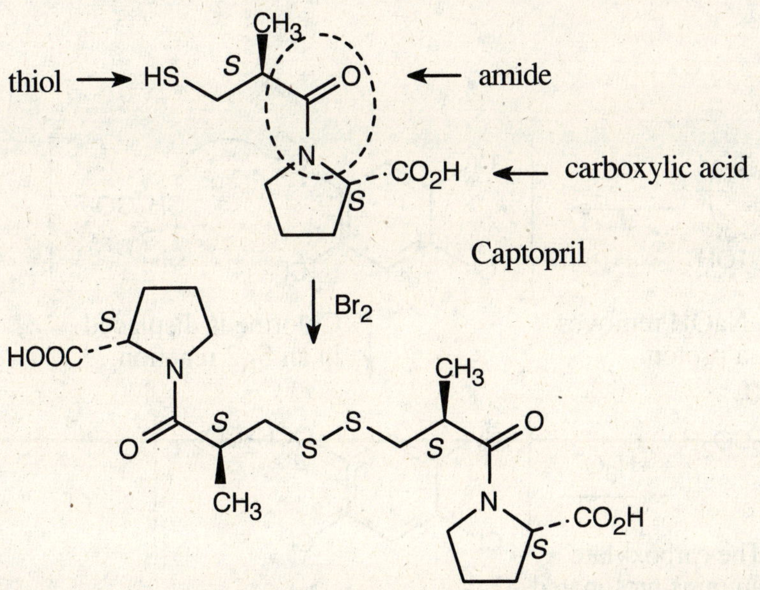

**8.67**

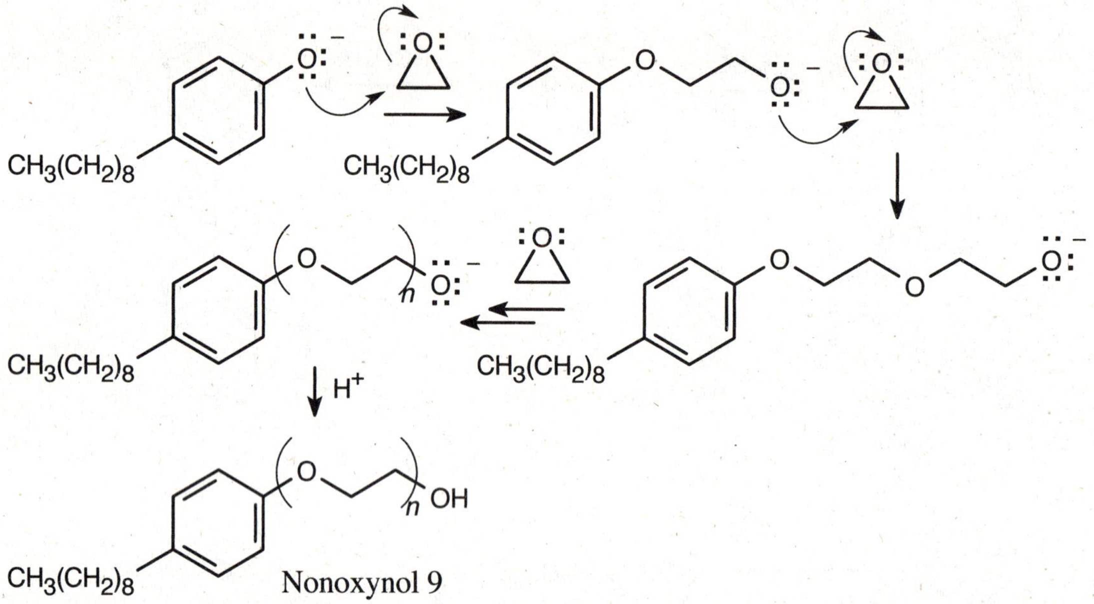


$S_N2$ attack of the phenoxide ion on the reactive ethylene oxide ring causes ring opening that yields a new alkoxide ion, which reacts with a second epoxide to form a second alkoxide ion. This process is repeated many times, until acid is added to quench the reaction.

## In the Field

**8.68**

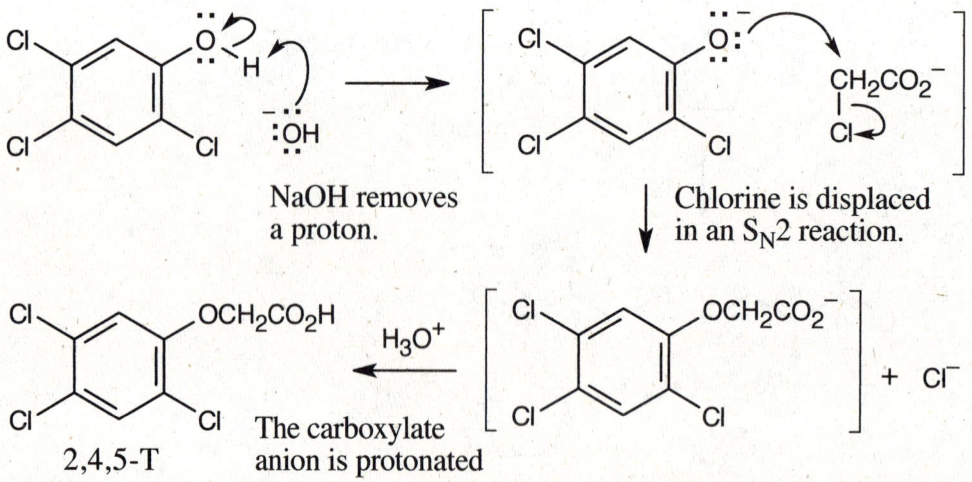


Treatment of 2,4,5-trichlorophenol with NaOH produces 2,4,5-trichlorophenoxide anion, which displaces chlorine from $ClCH_2CO_2^-$ $^+Na$ in an $S_N2$ reaction to yield 2,4,5-T.

# Chapter 9 – Aldehydes and Ketones: Nucleophilic Addition Reactions

## Chapter Outline

I. Introduction to carbonyl chemistry (Sections 9.1–9.2).
   A. The nature of carbonyl compounds (Section 9.1).
      1. All carbonyl compounds contain an acyl group, R–C=O, bonded to another group.
      2. There are two kinds of carbonyl compounds.
         a. The group bonded to the acyl group can't act as a leaving group (aldehydes and ketones).
         b. The group bonded to the acyl group can act as a leaving group.
   B. Structure and properties of the carbonyl group.
      1. The carbonyl carbon is $sp^2$ hybridized.
      2. The carbon–oxygen bond is polarized.
         a. The positively polarized carbonyl carbon can react with nucleophiles.
         b. The negatively polarized carbonyl oxygen can react with electrophiles.
      3. Most carbonyl-group reactions are a result of this polarization.
   C. Naming aldehydes and ketones (Section 9.2).
      1. Naming aldehydes.
         a. Aldehydes are named by replacing the *-e* of the corresponding alkane with *-al*.
         b. The parent chain must contain the –CHO group.
         c. The aldehyde carbon is always carbon 1.
         d. When the –CHO group is attached to a ring, the suffix *-carbaldehyde* is used.
      2. Naming ketones.
         a. Ketones are named by replacing the *-e* of the corresponding alkane with *-one*.
         b. Numbering starts at the end of the carbon chain nearer to the carbonyl carbon.
         c. The word *acyl* is used when a RCO– group is a substituent.

II. Preparation of aldehydes and ketones (Section 9.3).
   A. Aldehydes.
      1. Oxidation of primary alcohols using a periodinane.
   B. Ketones.
      1. Oxidation of secondary alcohols with $CrO_3$ or $NaCrO_7$.
      2. Hydration of terminal alkynes.
      3. Friedel–Crafts acylation of arenes.

III. Reactions of ketones and aldehydes (Sections 9.4–9.10).
   A. Oxidation of aldehydes (Section 9.4).
      1. Aldehydes can be easily oxidized to carboxylic acids.
   B. Nucleophilic addition reactions (Sections 9.5–9.10).
      1. General features (Section 9.5).
         a. A nucleophile attacks the carbonyl carbon, with simultaneous bond formation and bond breaking.
         b. Under basic conditions, negatively charged intermediates are favored.
         c. Under acidic conditions, positively charged intermediates are favored.
      2. Alcohol formation (Section 9.6).
         a. Reduction.
            i. Aldehydes and ketones are reduced by $NaBH_4$.
            ii. $H:^-$ adds to the carbonyl group, and the resulting negatively charged alkoxide intermediate is protonated.

b. Addition of Grignard reagents.
   i. Grignard reagents act as carbanions and undergo nucleophilic addition to aldehydes and ketones.
      (a). "R:−" adds to the carbonyl carbon to form a tetrahedral intermediate.
      (b). Water is added in a separate step to protonate the intermediate, yielding an alcohol.
   ii. Types of Grignard products.
      (a). Reaction of RMgX with formaldehyde yields primary alcohols.
      (b). Reaction of RMgX with aldehydes yields secondary alcohols.
      (c). Reaction of RMgX with ketones yields tertiary alcohols.
   iii. Limitations of the Grignard reaction.
      (a). Grignard reagents can't be prepared from reagents containing other reactive functional groups.
      (b). Grignard reagents can't be prepared from compounds having acidic hydrogens.

3. Addition of water – hydration (Section 9.7).
   a. Water can add reversibly to aldehydes and ketones to form a gem diol.
   b. Equilibrium usually favors the reactants.
   c. The stability of the product depends on the structure of the carbonyl compound.
   d. The process can be catalyzed by both acid and base and takes place in several steps.
4. Addition of alcohols – acetal formation (Section 9.8).
   a. Aldehydes and ketones react with alcohols to form acetals.
   b. The reaction is acid-catalyzed and is reversible.
   c. The initial product is a hemiacetal.
   d. Acetals can serve as protecting groups.
   e. Carbohydrates and other biomolecules contain acetals and hemiacetals.
5. Amines can add to aldehydes and ketones to form imines (Section 9.9).
6. Conjugate nucleophilic addition (1,4-addition) (Section 9.10).
   a. A nucleophile can add to the double bond of an α,β-unsaturated carbonyl compound.
   b. Addition occurs at the β carbon, and the initial product is a resonance-stabilized enolate anion.
   c. Conjugate addition occurs because the carbonyl group makes the β carbon electron-poor and reactive towards nucleophiles.

## Solutions to Problems

**9.1**

(a)

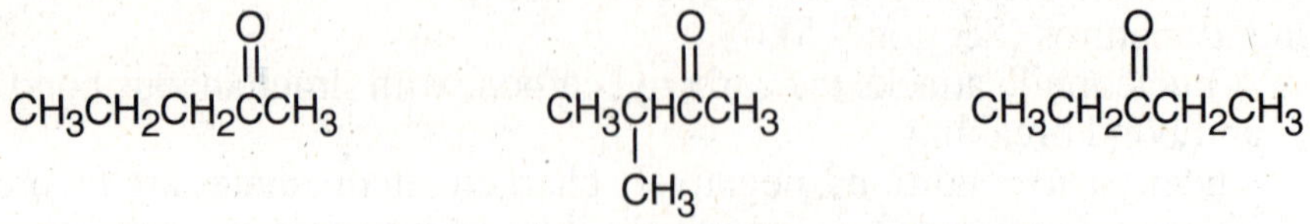

(b)

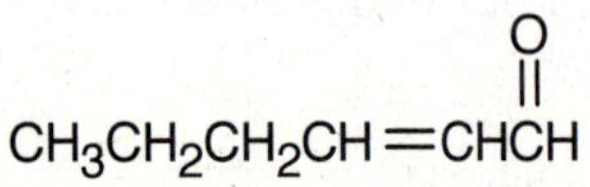

and many other possible answers.

(c)

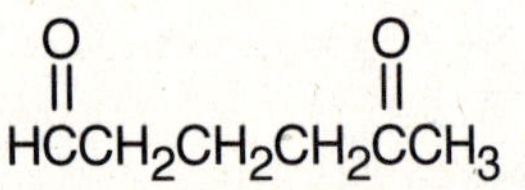

and other possible answers.

(d) 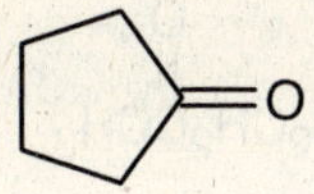 and many other possible answers.

**9.2**

(a) $CH_3CH_2\overset{O}{\overset{\|}{C}}CH(CH_3)_2$

2-Methylpentan-3-one

(b) 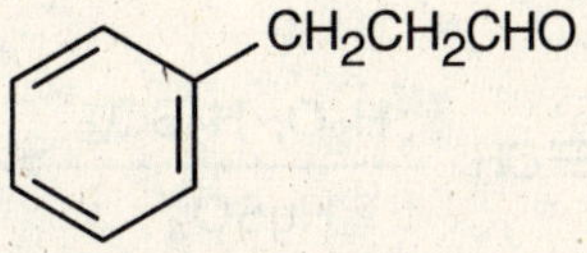

3-Phenylpropanal

(c) $CH_3\overset{O}{\overset{\|}{C}}CH_2CH_2CH_2\overset{O}{\overset{\|}{C}}CH_2CH_3$

Octane-2,6-dione

(d) 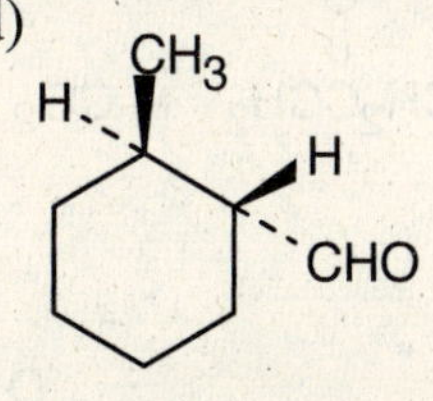

*trans*-2-Methylcyclohexane-carbaldehyde

(e) $CH_3CH{=}CHCH_2CH_2CHO$

Hex-4-enal

(f) 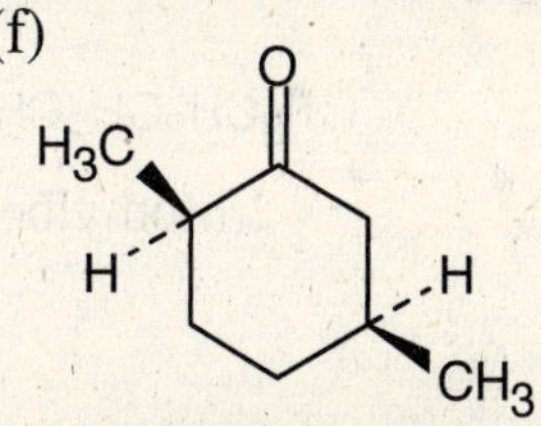

*cis*-2,5-Dimethyl-cyclohexanone

**9.3**

(a) $CH_3\overset{CH_3}{\overset{|}{C}}HCH_2CHO$

3-Methylbutanal

(b) $H_2C{=}\overset{CH_3}{\overset{|}{C}}CH_2CHO$

3-Methylbut-3-enal

(c) $CH_3\overset{Cl}{\overset{|}{C}}HCH_2\overset{O}{\overset{\|}{C}}CH_3$

4-Chloropentan-2-one

(d) 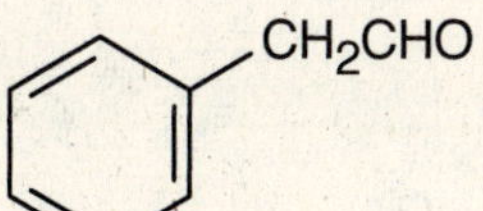

Phenylacetaldehyde

(e) 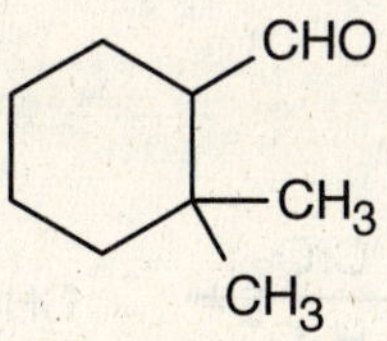

2,2-Dimethylcyclo-hexanecarbaldehyde

(f) 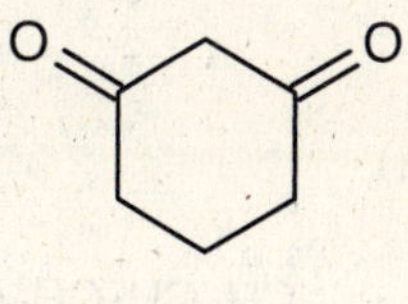

Cyclohexane-1,3-dione

**9.4**

(a)

$$CH_3CH_2CH_2CH_2CH_2OH \xrightarrow{\text{Periodinane}} CH_3CH_2CH_2CH_2\overset{O}{\overset{\|}{C}}H$$

Pentan-1-ol → Pentanal

(b) Pentan-1-ol is formed from pentanoic acid:

$$2\ CH_3CH_2CH_2CH_2\overset{O}{\overset{\|}{C}}OH \xrightarrow{1.\ 2\ LiAlH_4\quad 2.\ H_3O^+} \text{Pentan-1-ol}$$

(c)

$$CH_3CH_2CH_2CH_2CH{=}CHCH_2CH_2CH_2CH_3 \xrightarrow[H_3O^+]{KMnO_4} 2\ CH_3CH_2CH_2CH_2\overset{O}{\overset{\|}{C}}OH$$

Dec-5-ene

**9.5**

(a)

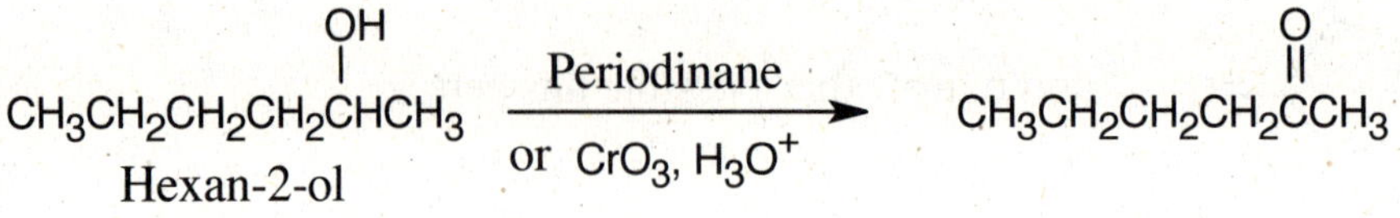


(b)

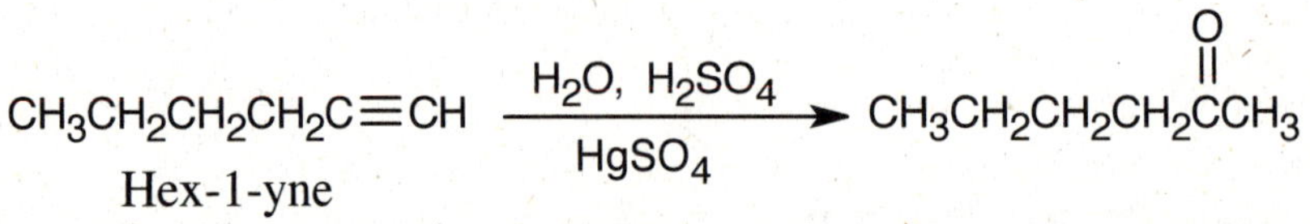


(c)

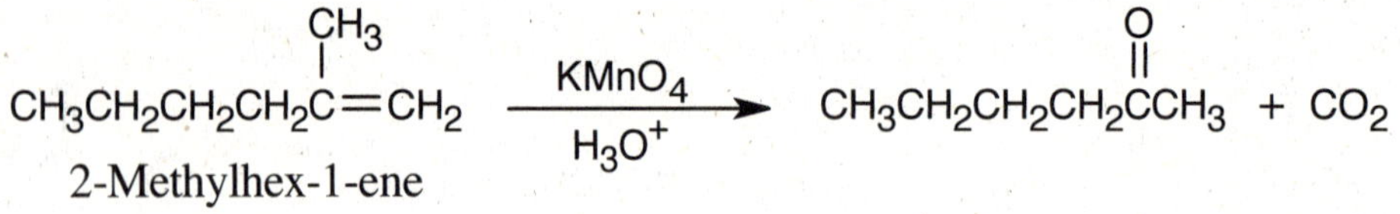


**9.6**

(a)

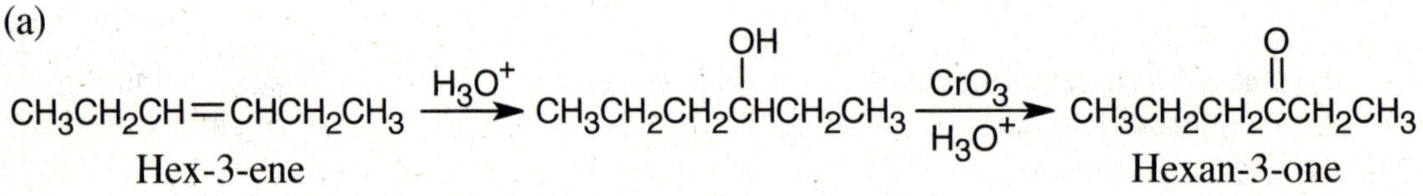


(b)

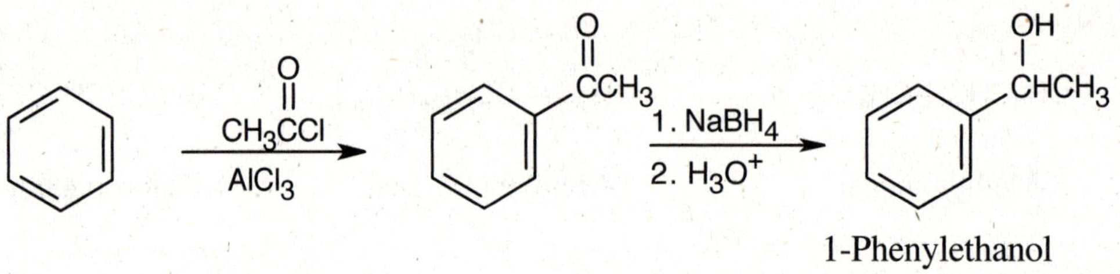


**9.7**

(a)

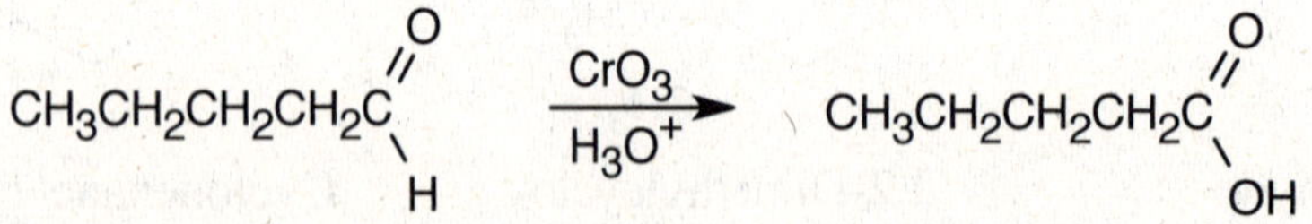


(b)

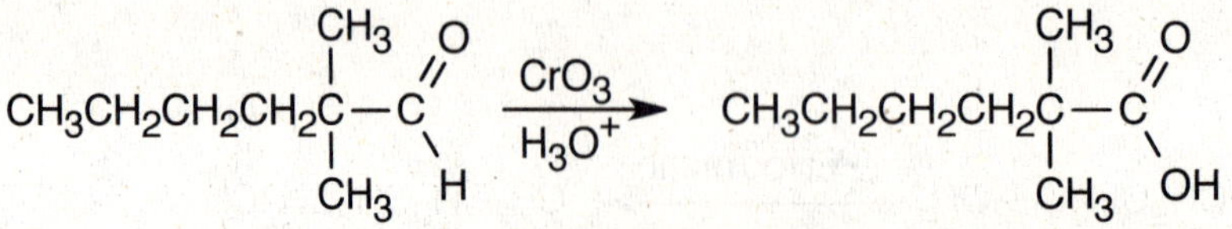


(c)

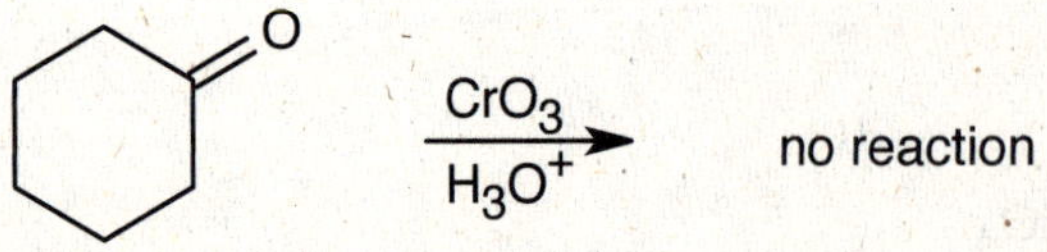

**9.8**

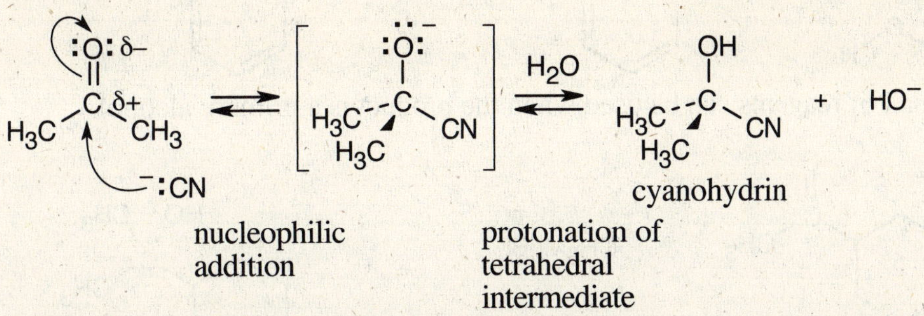


Cyanide anion adds to the positively polarized carbonyl carbon to form a tetrahedral intermediate. This intermediate is protonated to yield acetone cyanohydrin.

**9.9**

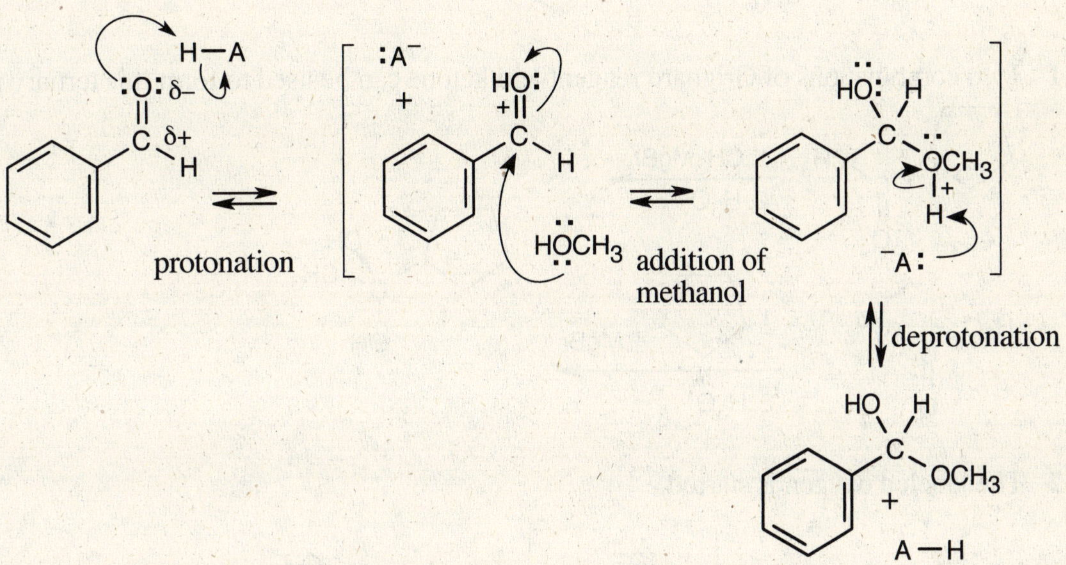


**9.10**

*Aldehyde/ Ketone* — *Grignard reagent* — *Product*

(a)

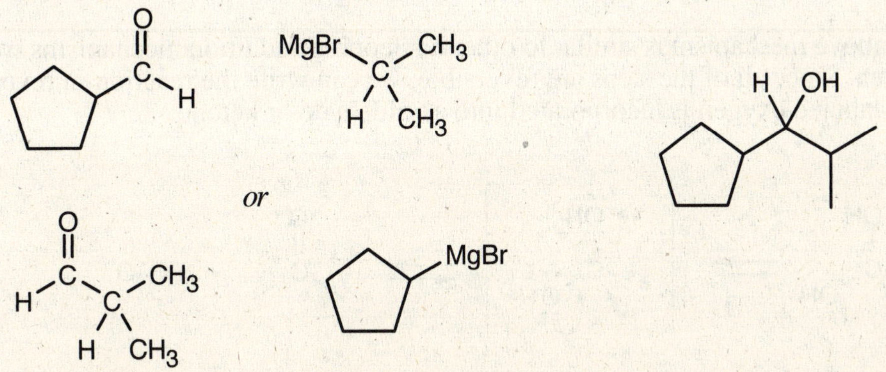

(b)

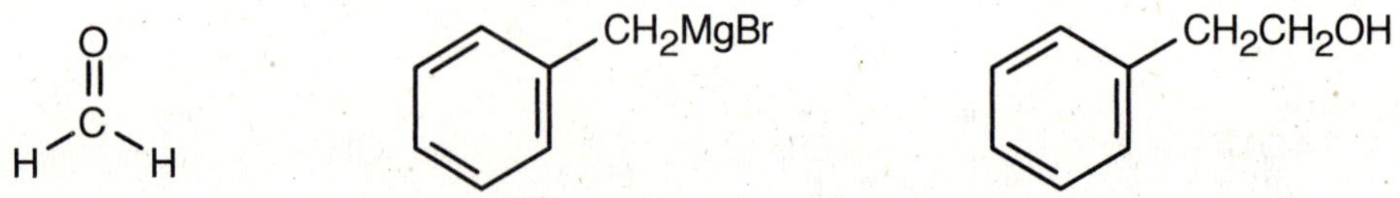

Only one pair of reagents can be used when the product is a primary alcohol.

(c)

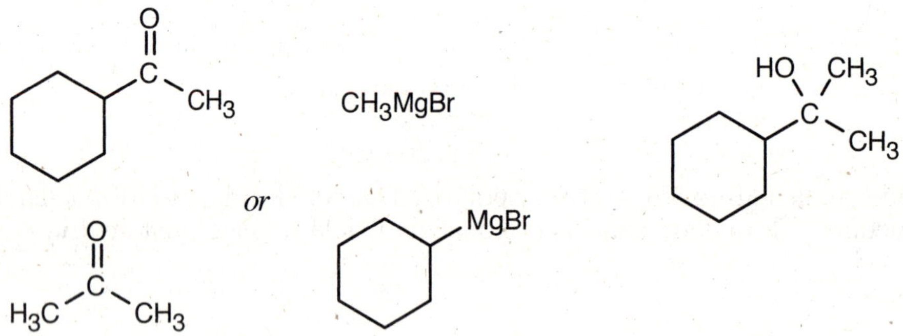

**9.11** Two combinations of Grignard reagent and ketone can be used to form this tertiary alcohol.

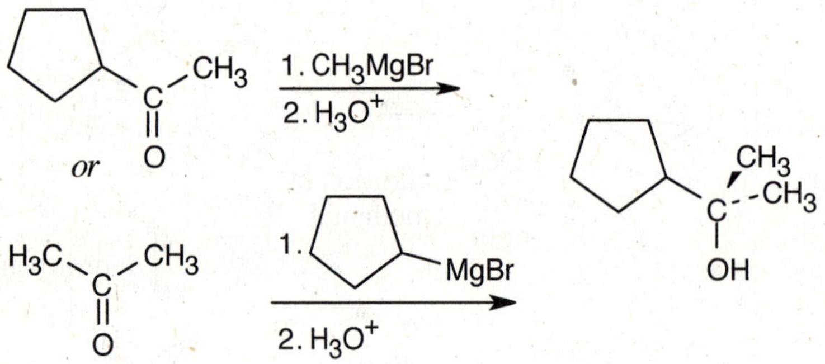

**9.12** The labeled oxygen is starred.

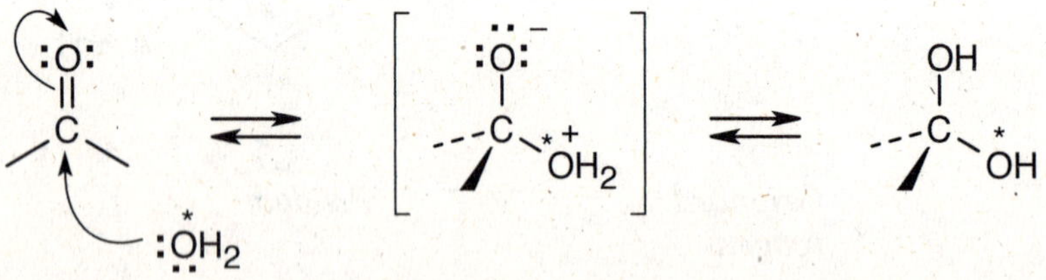

The above mechanism is similar to other nucleophilic addition mechanisms we have written. Since all of the steps are reversible, we can write the reaction in reverse to show how labeled oxygen is incorporated into an aldehyde or ketone.

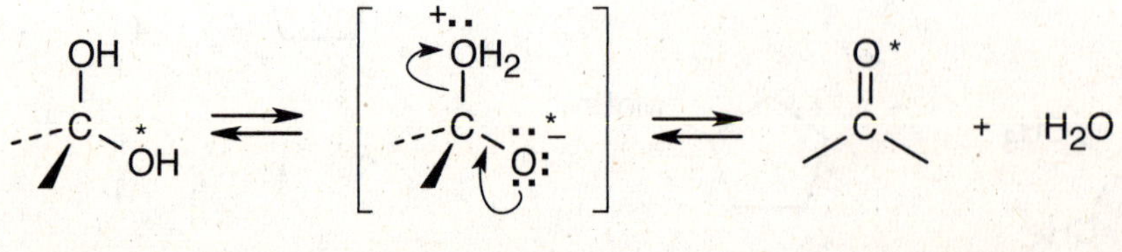

This exchange is very slow in water but proceeds more rapidly when either acid or base is present.

**9.13**

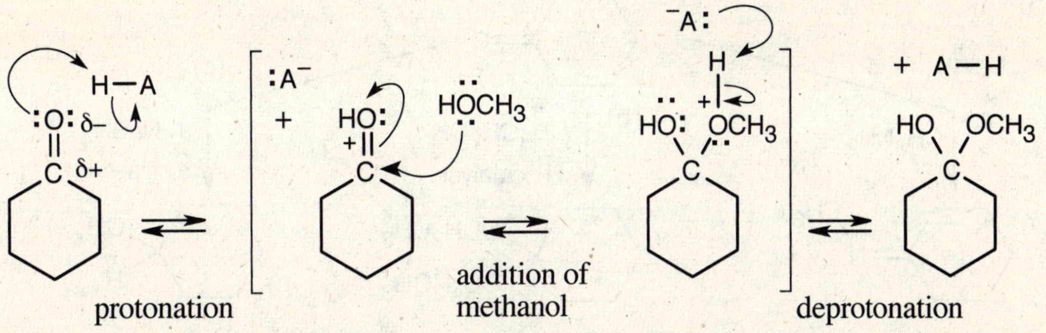


**9.14**

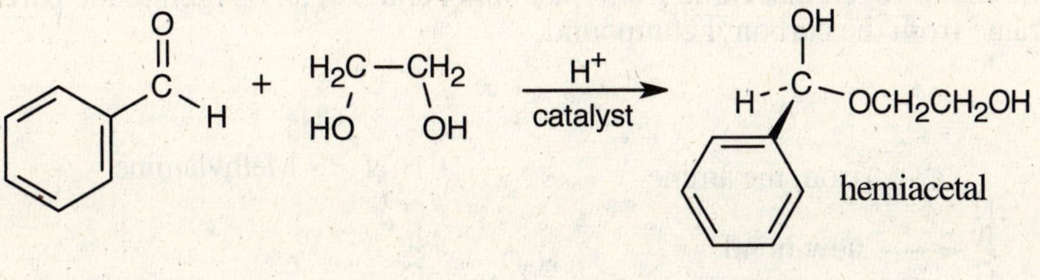


Nucleophilic addition of one hydroxyl group of ethylene glycol to the carbonyl carbon of benzaldehyde, followed by a proton shift, produces the hemiacetal.

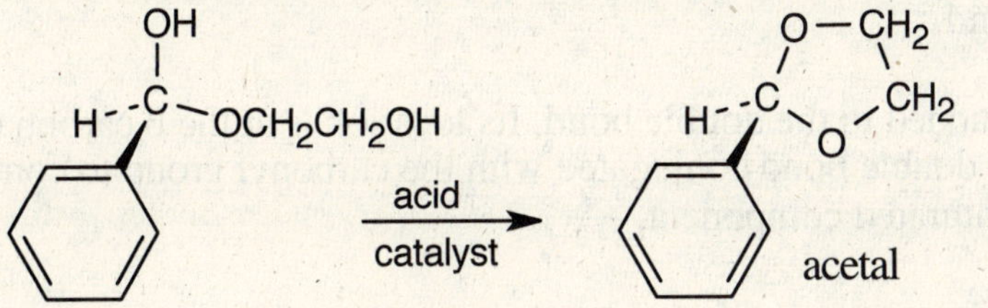


Protonation of one hydroxyl group, followed by displacement of water by the other hydroxyl group, gives the cyclic acetal.

**9.15**

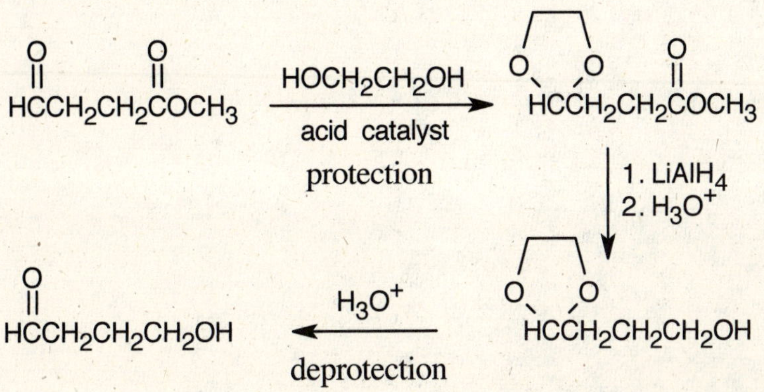

**9.16**

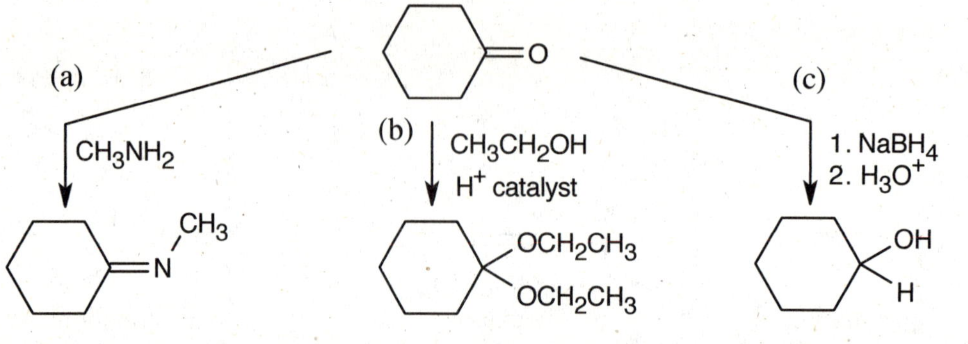


**9.17** Identify the bond formed between the amine and the carbonyl compound. Add two hydrogens to the nitrogen that came from the amine, and add an oxygen to the part of the bond that came from the carbonyl compound.

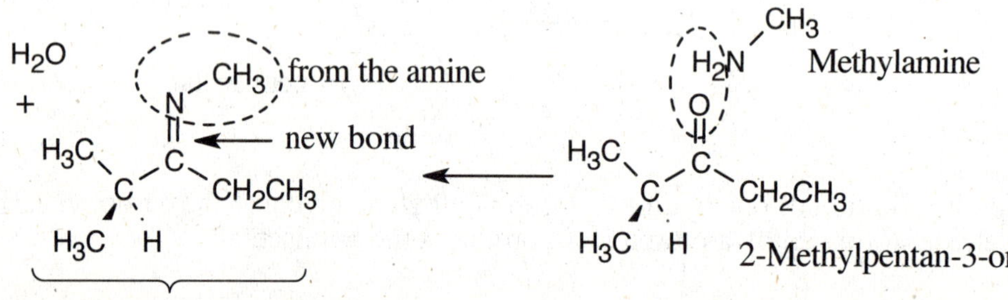


**9.18** Identify the group that has added to the double bond. Its location is at the ß carbon of the conjugated system. Draw a double bond conjugated with the carbonyl group to complete the structure of the α,β-unsaturated component.

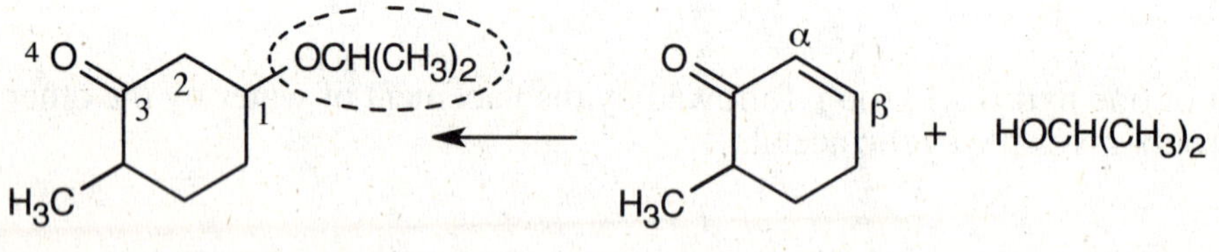

## Visualizing Chemistry

**9.19**

(a) 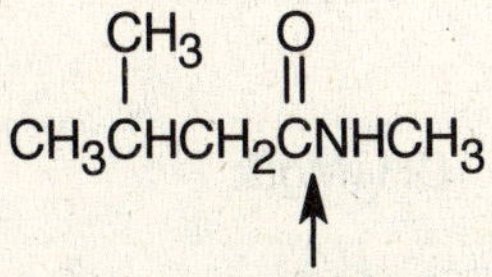


(b) 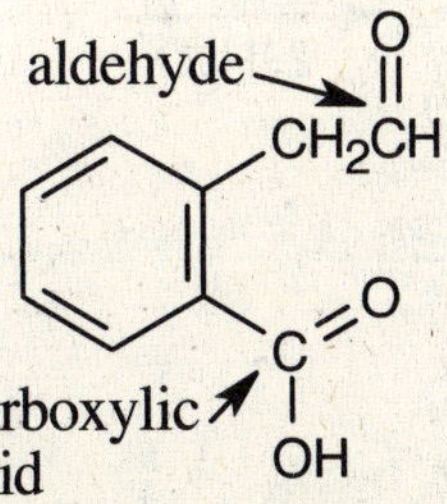


**9.20** According to the electrostatic potential maps, the carbonyl carbon of acetyl chloride is more electrophilic and the oxygen of acetone is more nucleophilic. This makes sense, because acetyl chloride has two electron-withdrawing groups that make its carbonyl carbon electron-poor and electrophilic. Because acetyl chloride has two electron-withdrawing groups, neither group is as nucleophilic as the carbonyl oxygen of acetone.

**9.21**

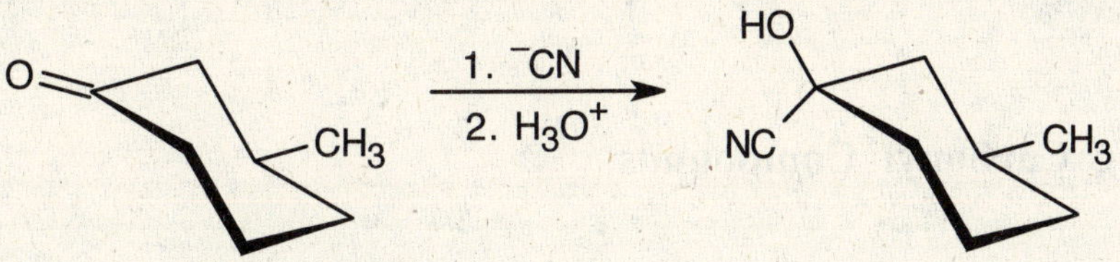


3-Methylcyclohexanone

**9.22** The amino alcohol intermediate results from the addition of an amine to a ketone.

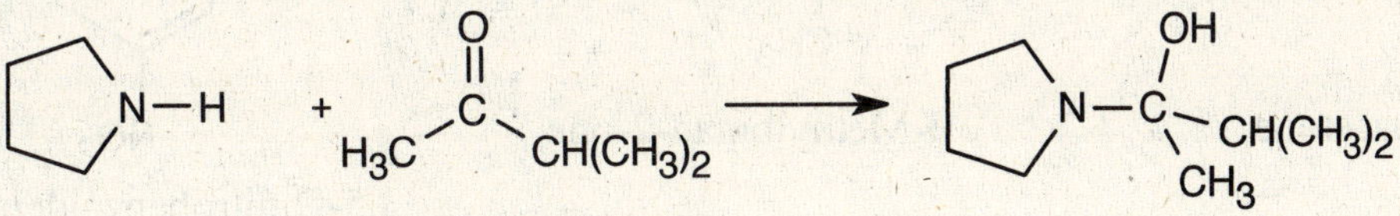

3-Methylbutan-2-one

**9.23**

(a) 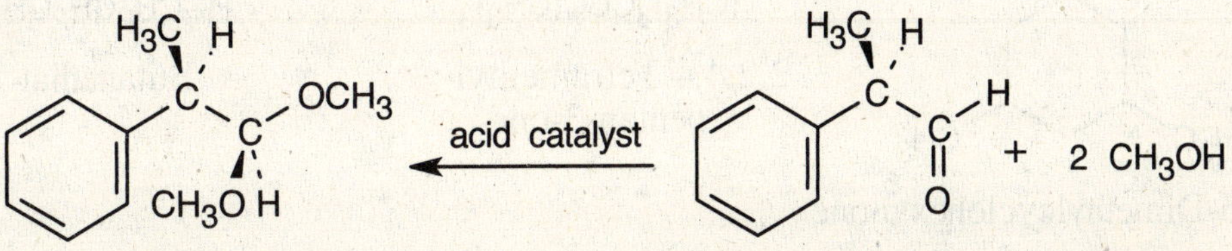


acetal (R)-2-Phenylpropanal Methanol

(b)

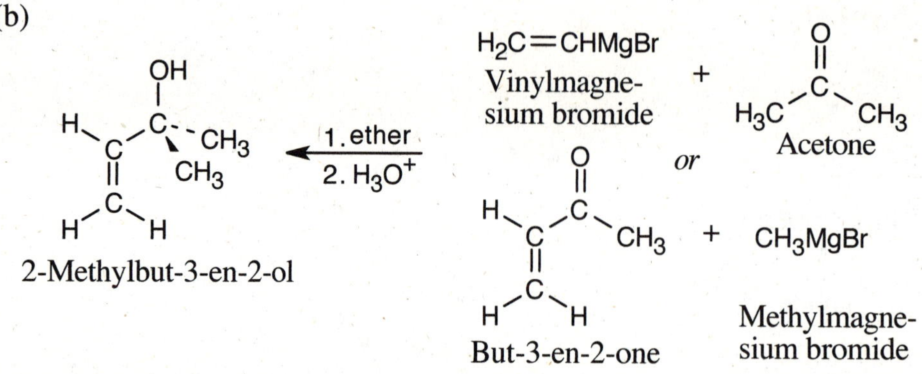


(c)

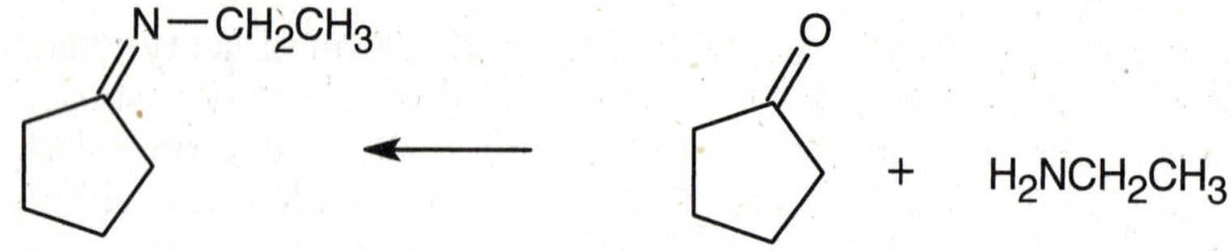


imine Cyclopentanone Ethylamine

## Additional Problems

### Identifying and Naming Carbonyl Compounds

**9.24**

(a)

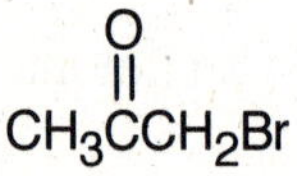

Bromoacetone

(b)

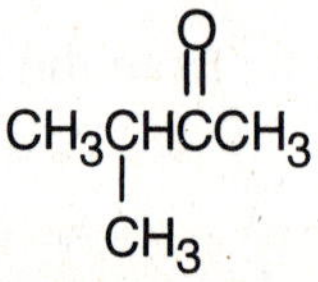

3-Methylbutan-2-one

(c)

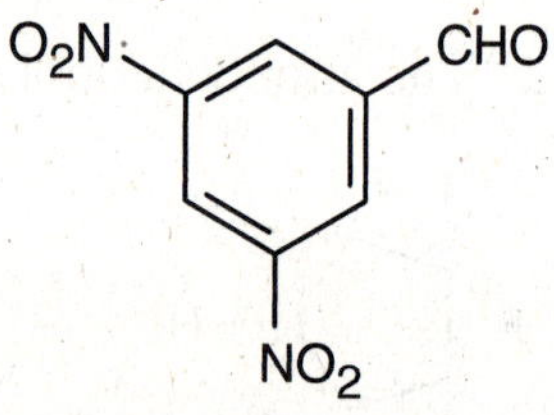

3,5-Dinitrobenzaldehyde

(d)

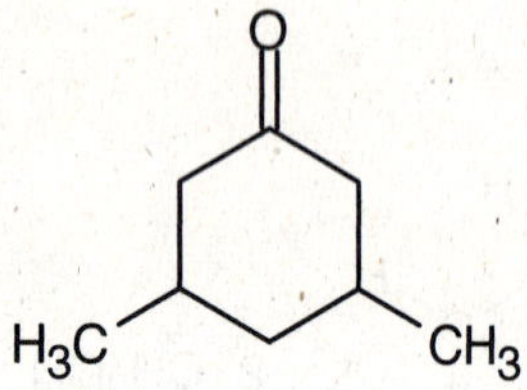

3,5-Dimethylcyclohexanone

(e)

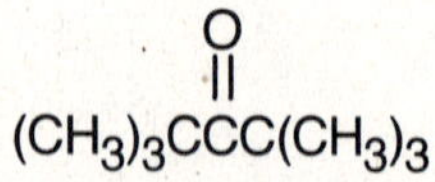

2,2,4,4-Tetramethylpentan-3-one

(f)

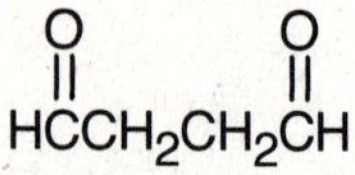

Butanedial

(g)

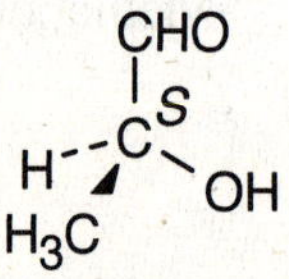

(*S*)-2-Hydroxypropanal

(h)

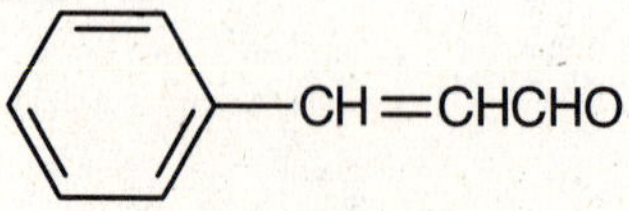

3-Phenylprop-2-enal

**9.25** Other structures are possible.

(a)

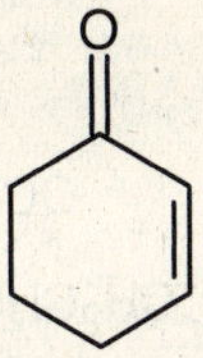

(b)

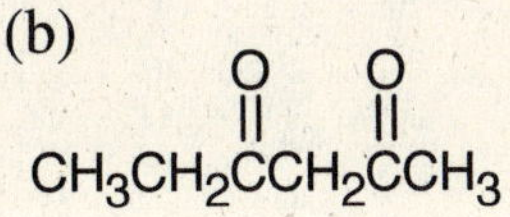

(c)

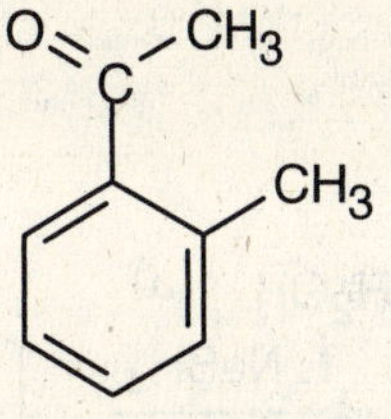

(d)

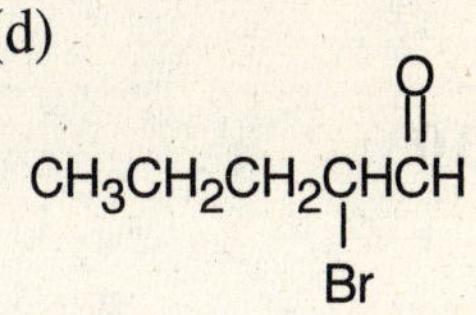

**9.26**

(a)

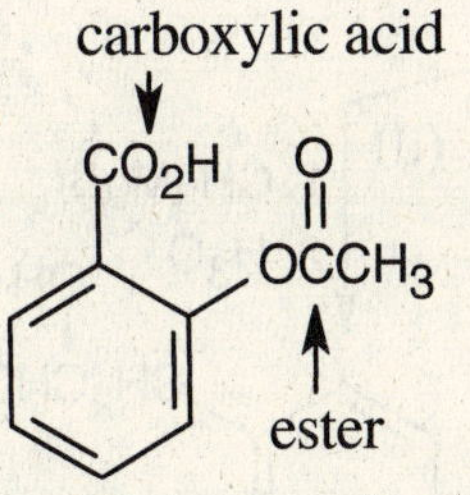


Aspirin

(b)

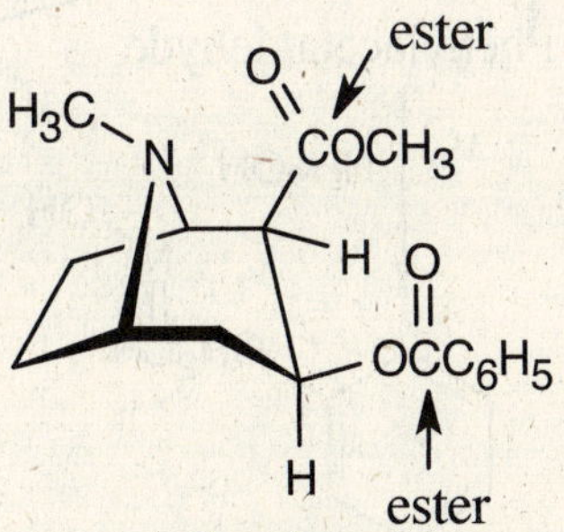


Cocaine

(c)

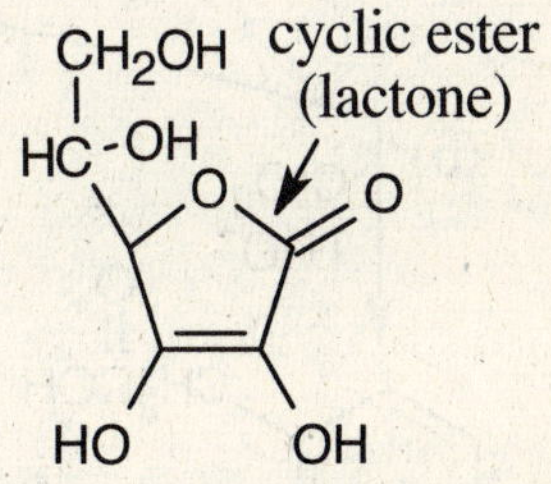


Ascorbic acid

**9.27**

(a)

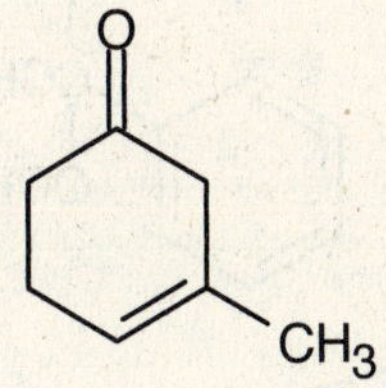

3-Methylcyclohex-3-enone

(b)

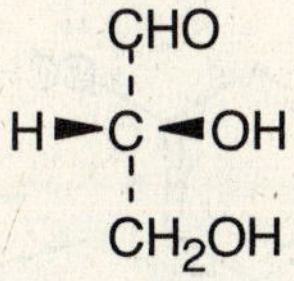

(*R*)-2,3-Dihydroxypropanal
(D-Glyceraldehyde)

(c)

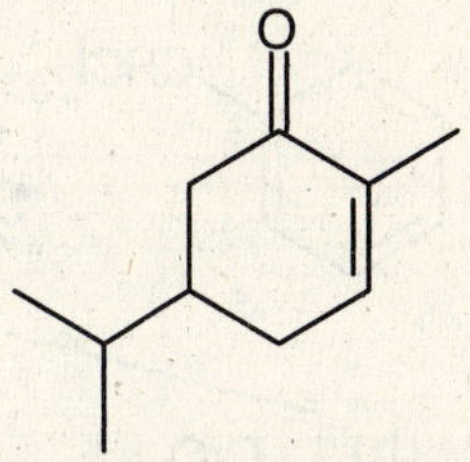

5-Isopropyl-2-methyl-
cyclohex-2-enone

(d)

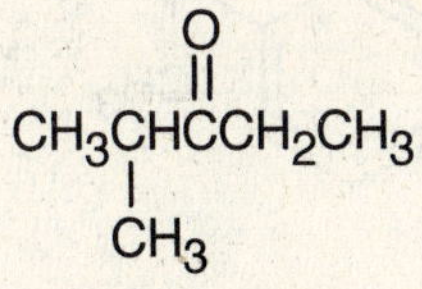

2-Methylpentan-3-one

(e)

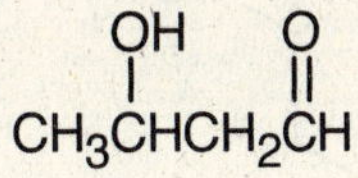

3-Hydroxybutanal

(f)

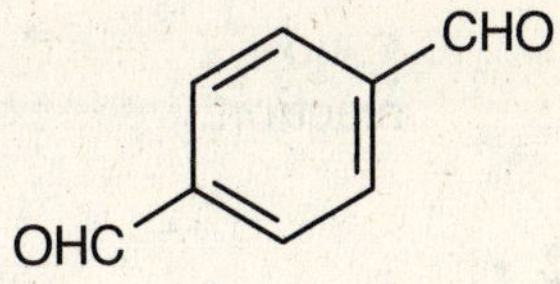

*p*-Benzenedicarbaldehyde

**9.28**

$CH_3CH_2CH_2CH_2CHO$

Pentanal

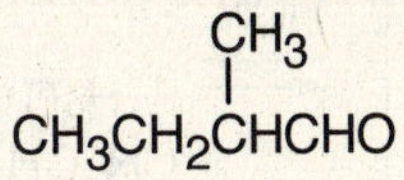

2-Methylbutanal

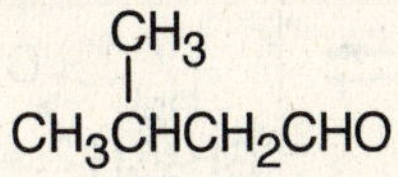

3-Methylbutanal

$(CH_3)_3CCHO$

2,2-Dimethylpropanal

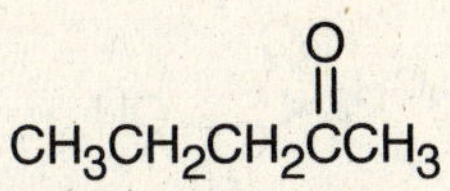

Pentan-2-one

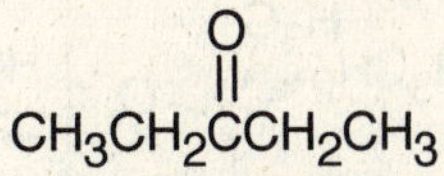

Pentan-3-one

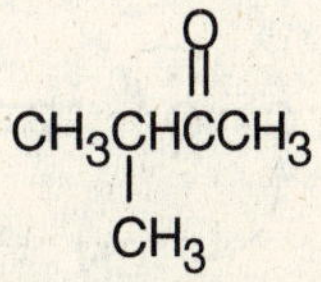

3-Methylbutan-2-one

**9.29** Of the compounds shown in Problem 9.28, only 2-methylbutanal is chiral.

## Reactions

**9.30**

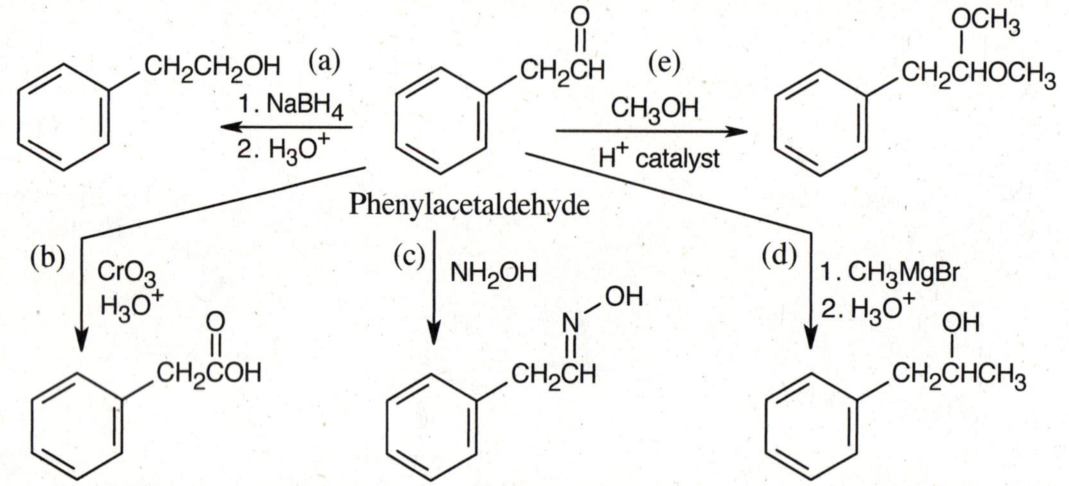

**9.31**

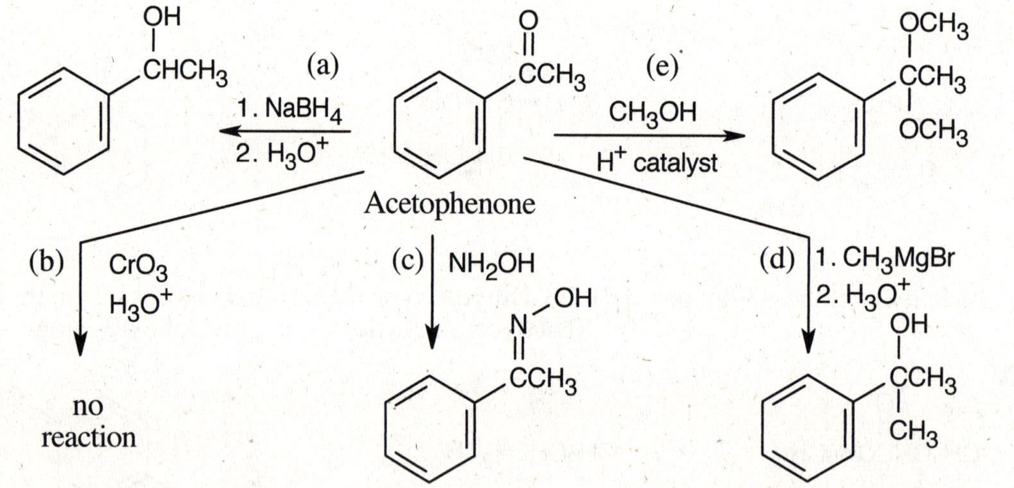

**9.32**

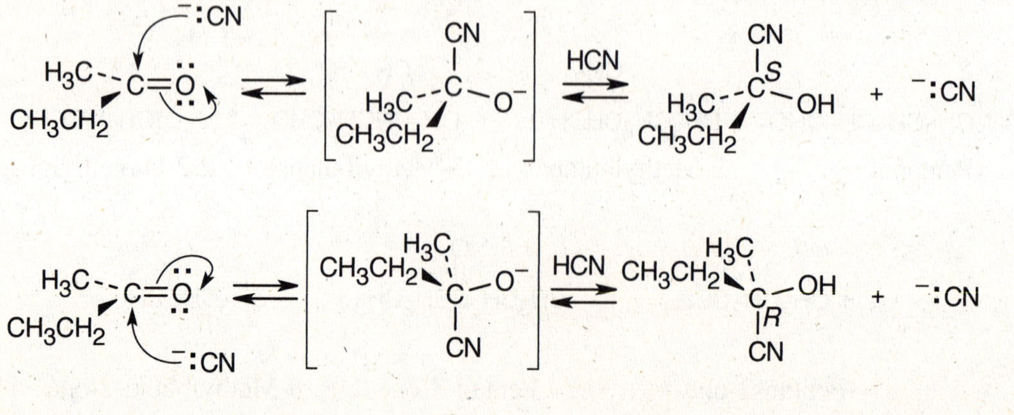

Attack occurs with equal probability from either side of the planar carbonyl group to yield a racemic product mixture that is optically inactive.

**9.33**

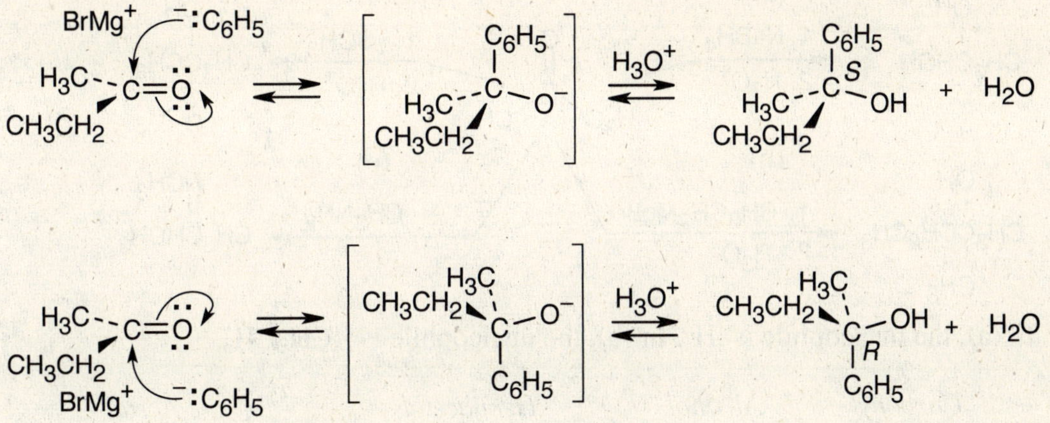

The product is a racemic mixture of *R* and *S* enantiomers.

**9.34**

(a)

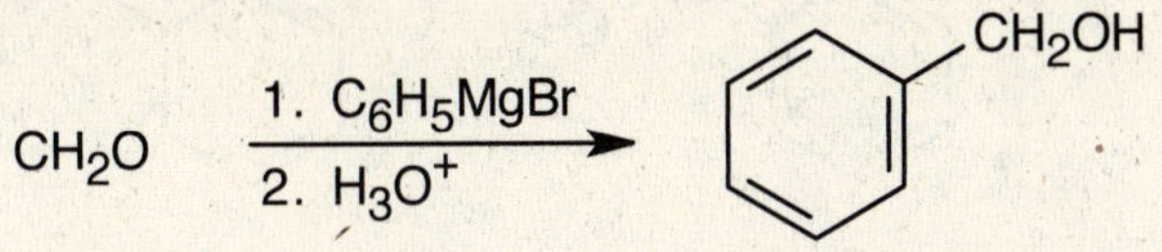

Benzyl alcohol

(b)

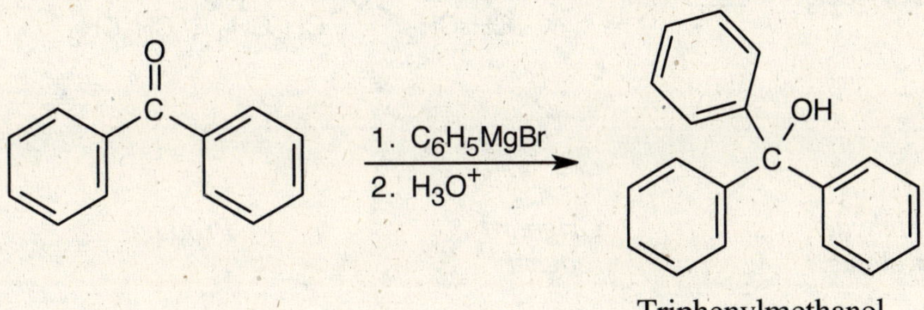

Triphenylmethanol

(c)

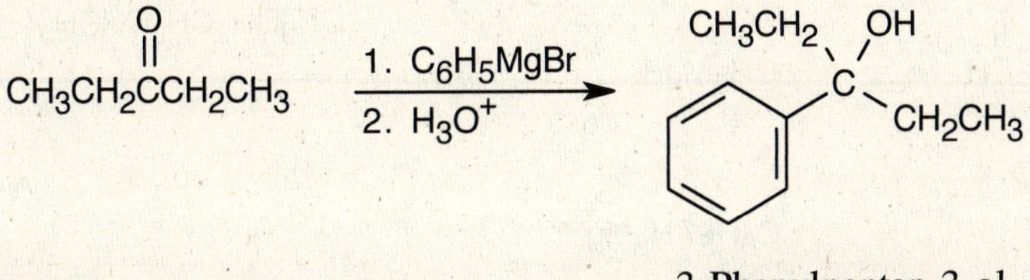

3-Phenylpentan-3-ol

**9.35**

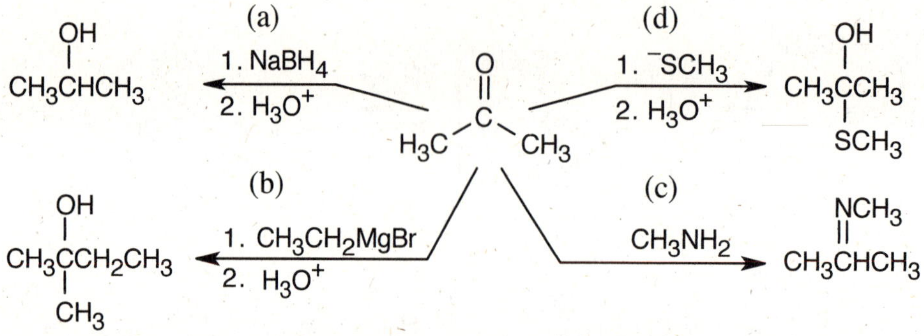

In (a), the nucleophile is $:H^-$. In (b), the nucleophile is $^-:CH_2CH_3$.

**9.36**

| | *Carbonyl compound* | *Alcohol* | *Hemiacetal* | *Acetal* |
|---|---|---|---|---|

(a)

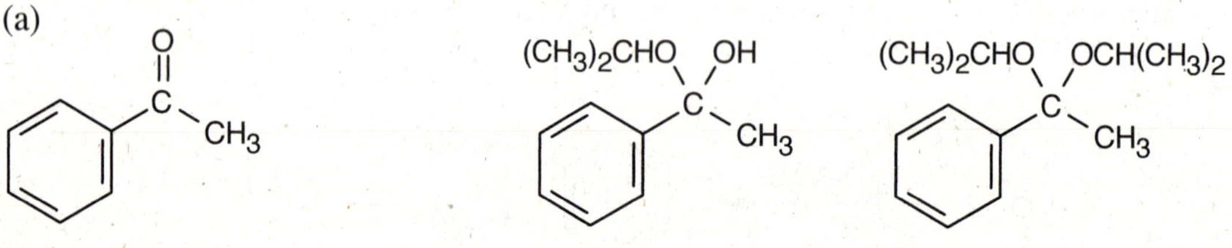

(b)

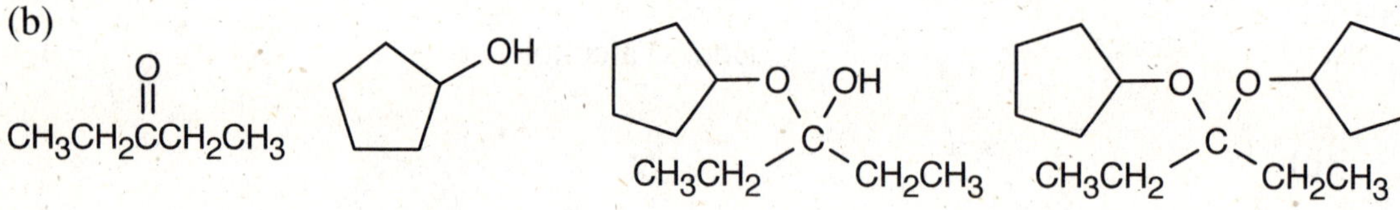

## Synthesis

**9.37**

| | *Aldehyde/ Ketone* | *Grignard reagent* | *Product* |
|---|---|---|---|

(a)

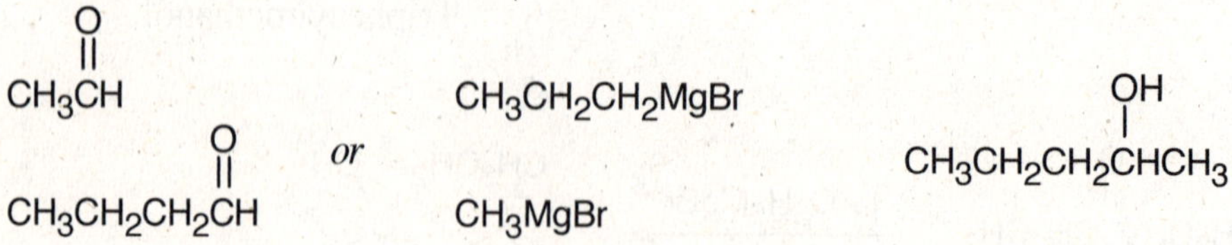

(b)

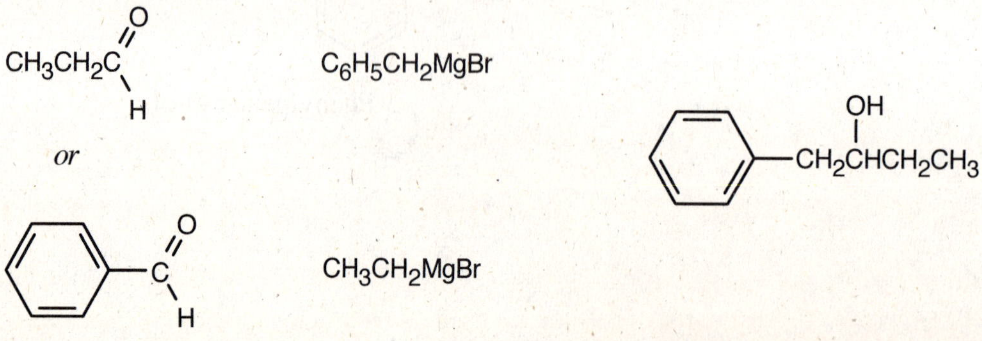

| *Aldehyde/ Ketone* | *Grignard reagent* | *Product* |
|---|---|---|
| (c) 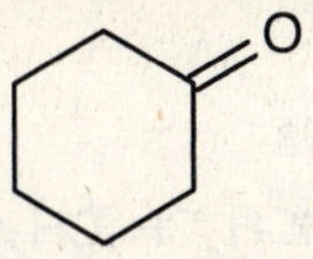 | $CH_3CH_2MgBr$ | 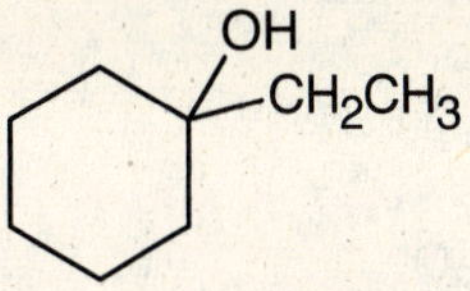 |
| (d) 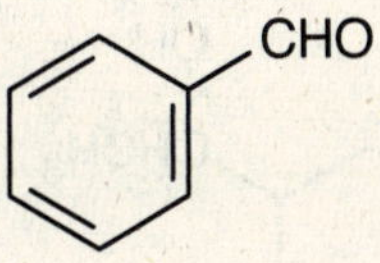 | $C_6H_5MgBr$ | 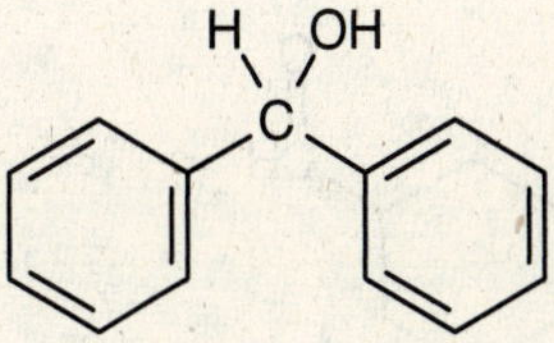 |

**9.38**

(a)

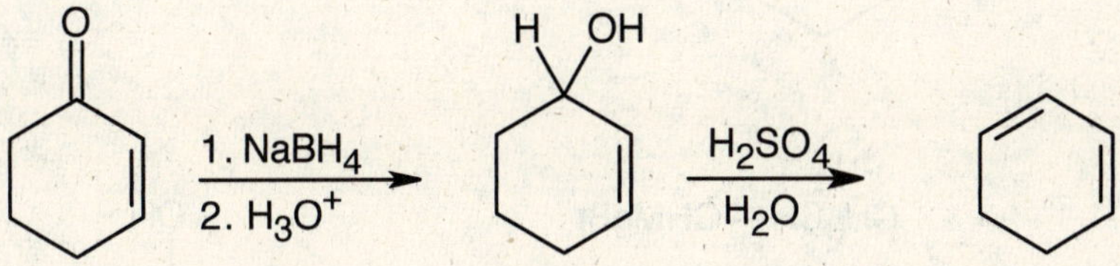


(b)

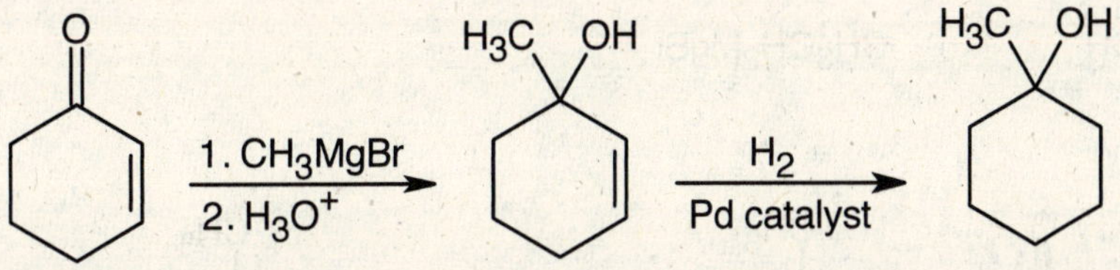


(c)

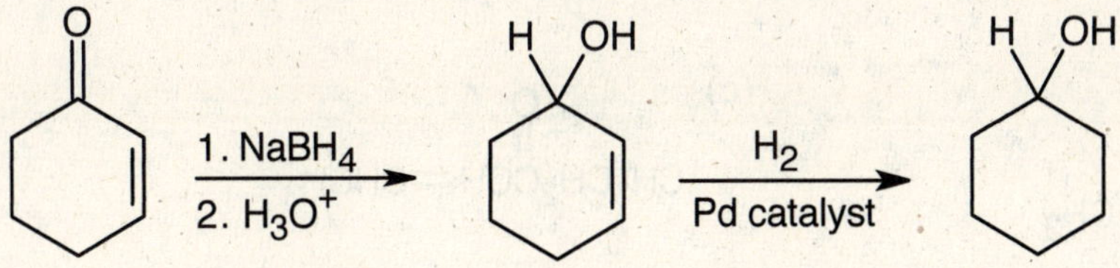


(d)

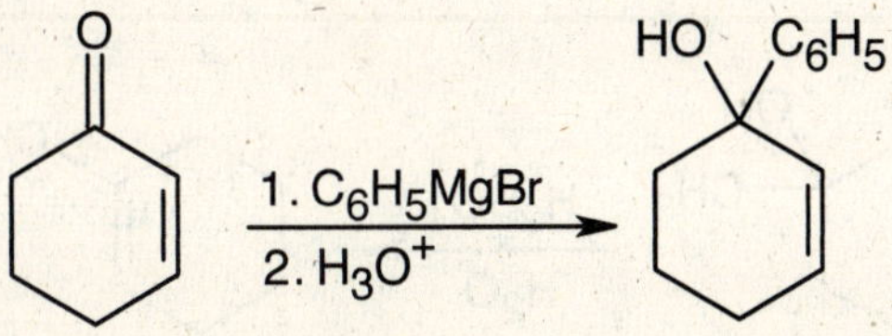

**9.39**

| *Aldehyde/ Ketone* | *Grignard reagent* | *Product* |
|---|---|---|

(a)

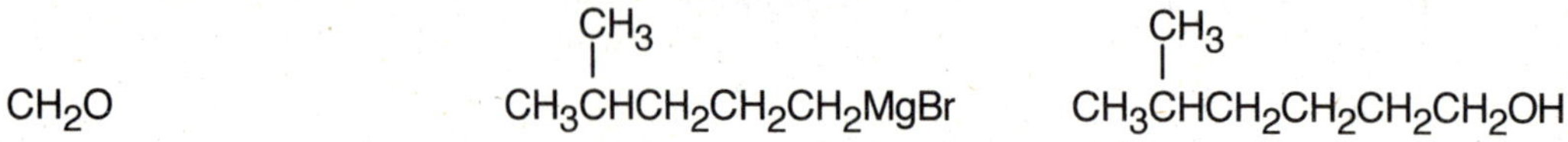

(b)

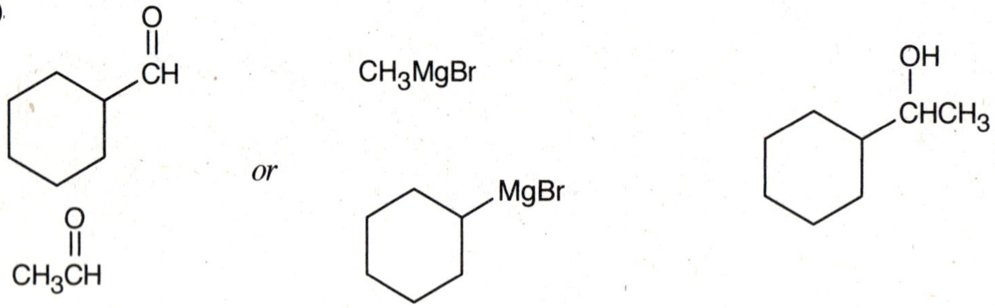

(c)

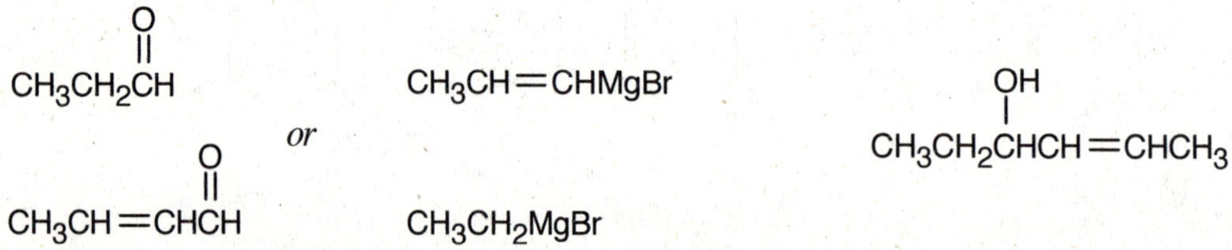

**9.40**

(a)

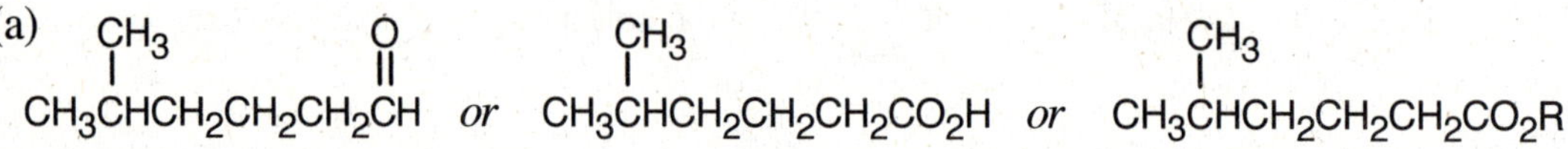

(b)

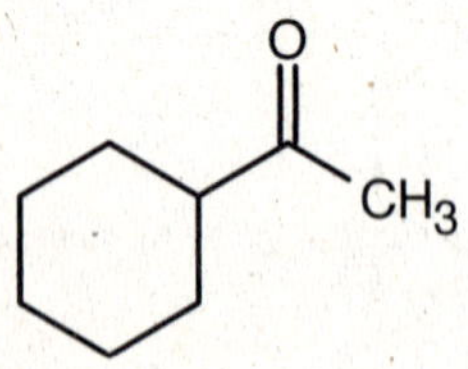

(c)

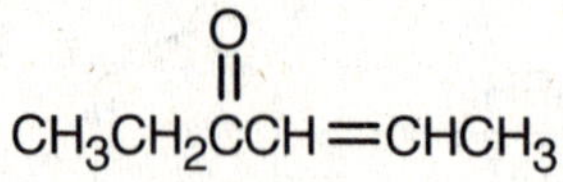

**9.41**

(a)

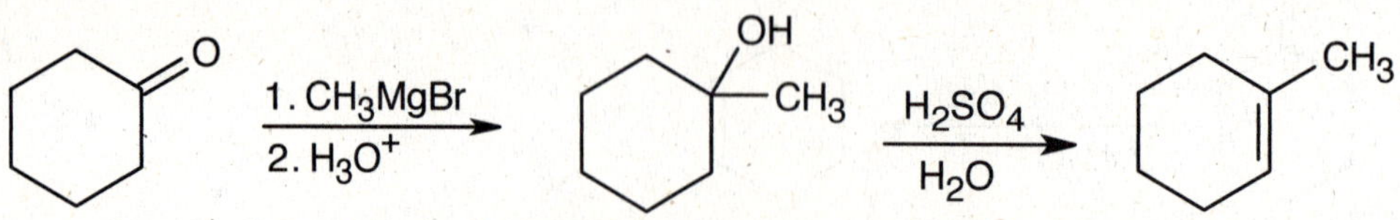

(b)

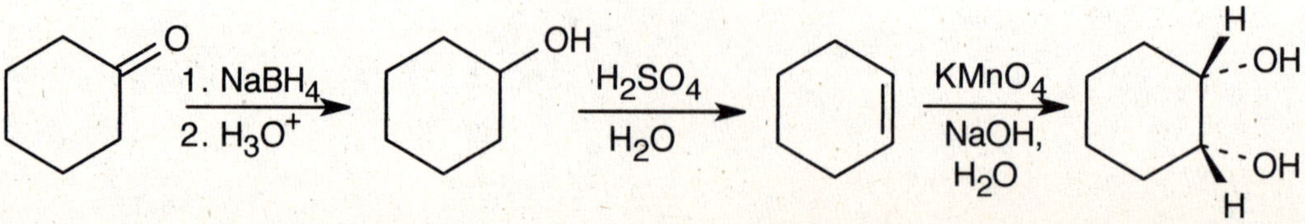

(c)

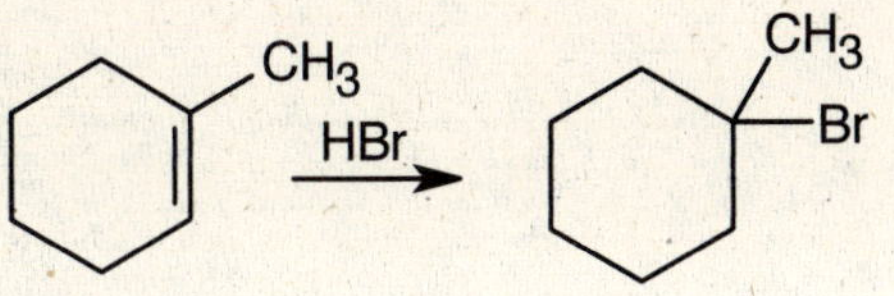

from (a)

(d)

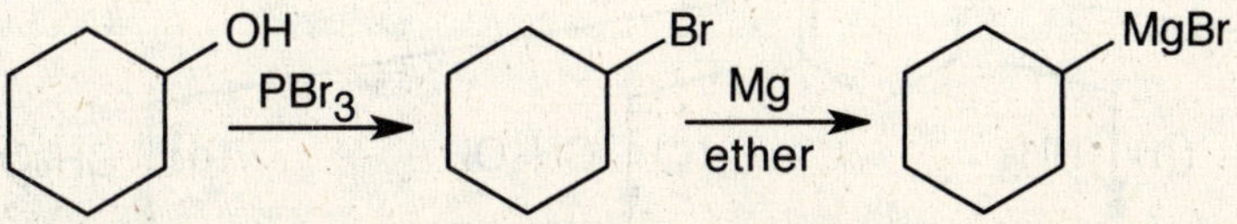

from (b)

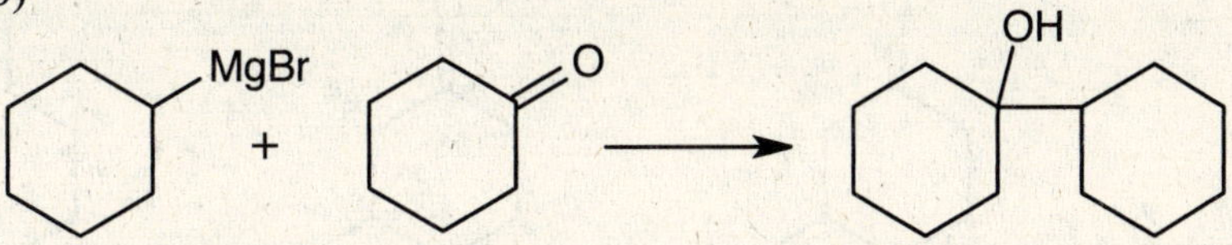

**9.42**

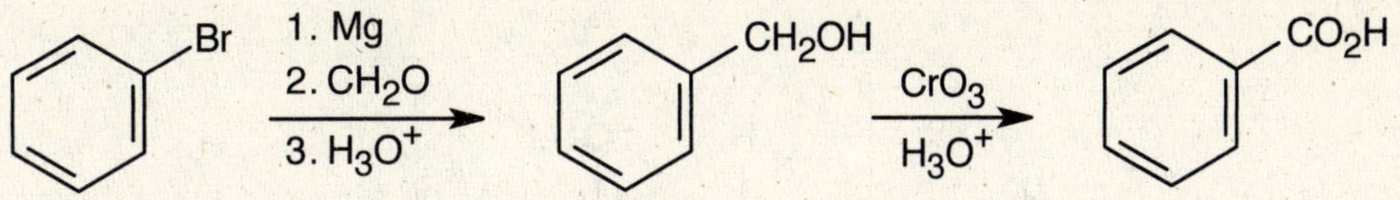

**9.43**

(a)

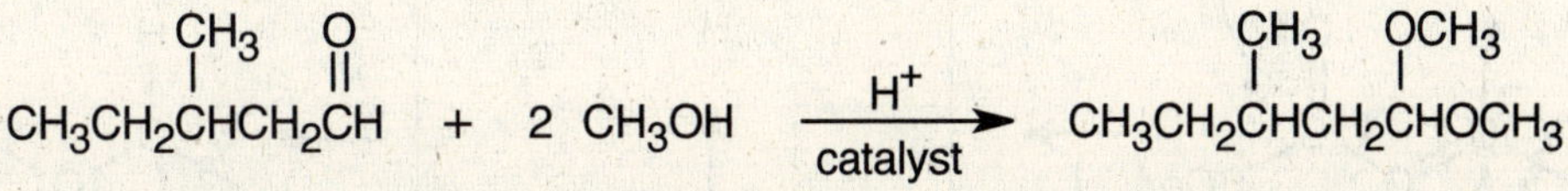

(b)

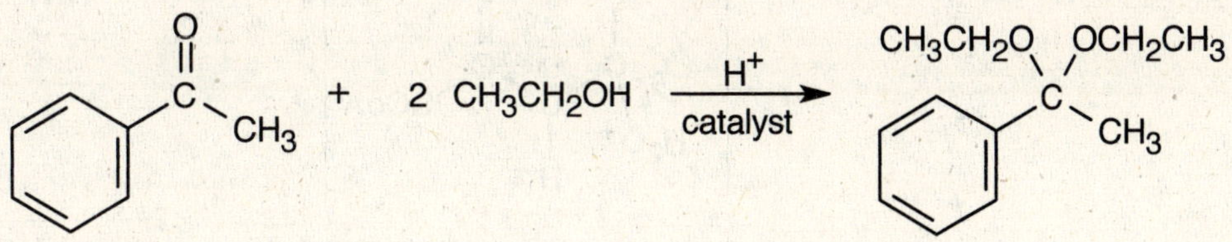

(c)

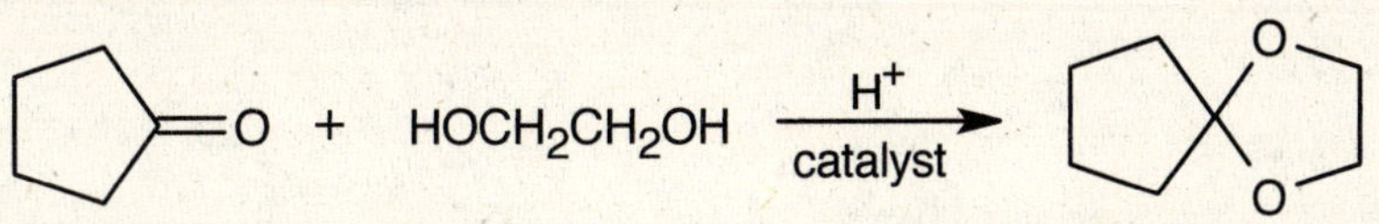

A cyclic acetal is formed from a ketone and a diol.

## General Problems

**9.44**

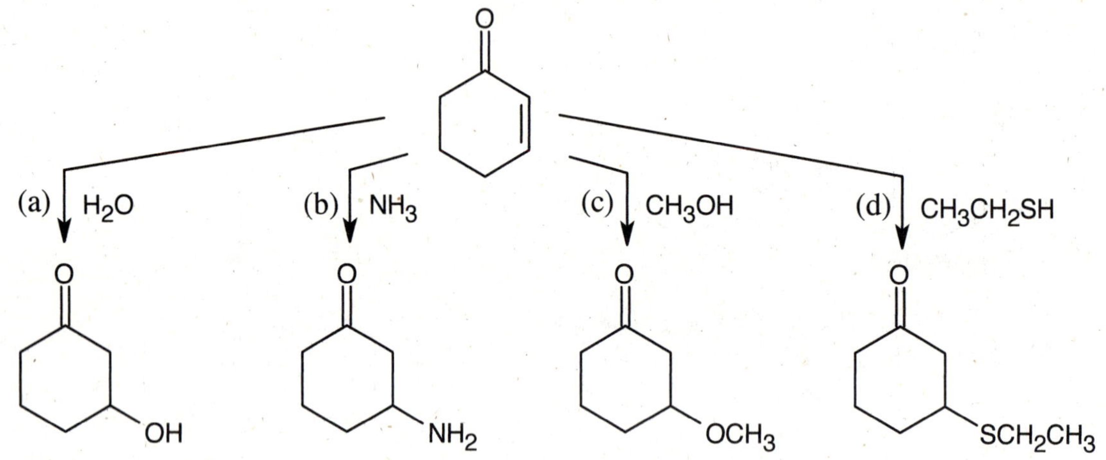


**9.45**

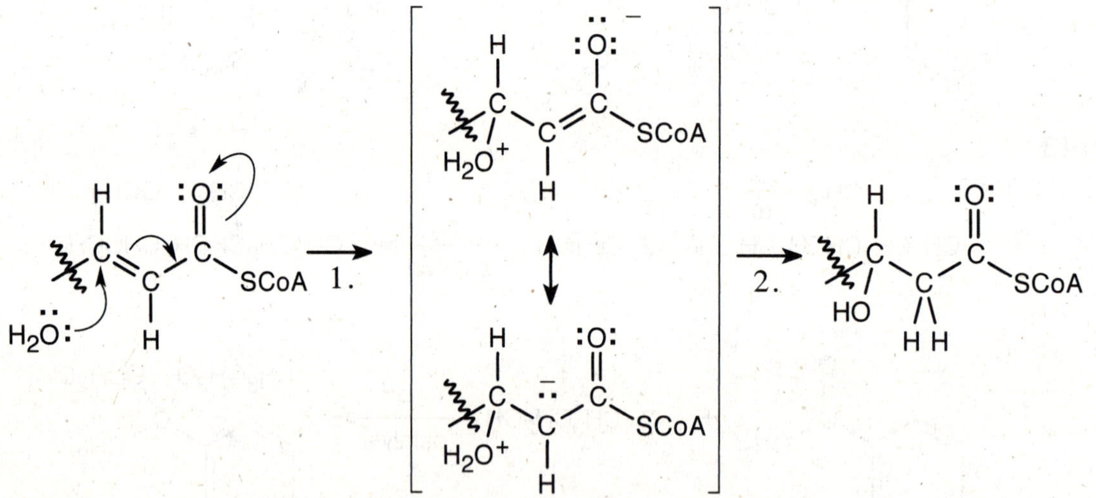


Conjugate addition of water (1) is followed by a proton shift (2) to give the β-hydroxyacyl CoA product.

**9.46**

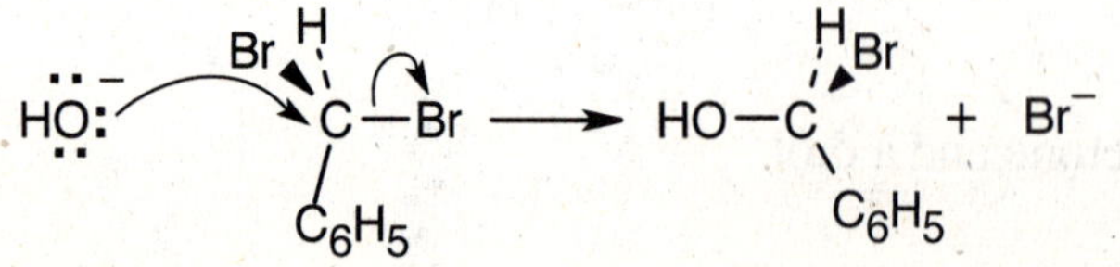

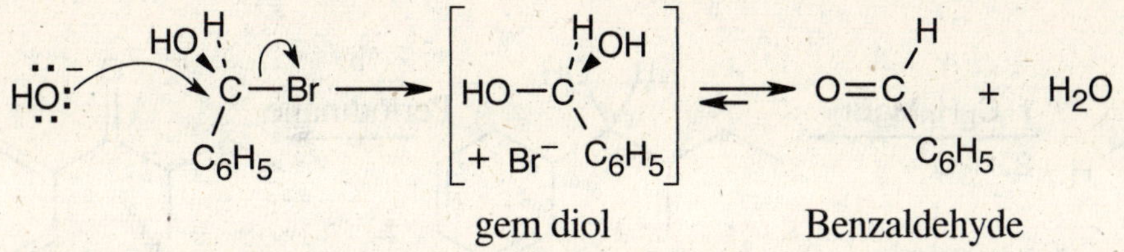

The initial product of two successive $S_N2$ reactions of $C_6H_5CHBr_2$ with hydroxide ion is a gem diol. Since the equilibrium between the gem diol and aldehyde favors the aldehyde, benzaldehyde is the observed product.

**9.47**

(a)

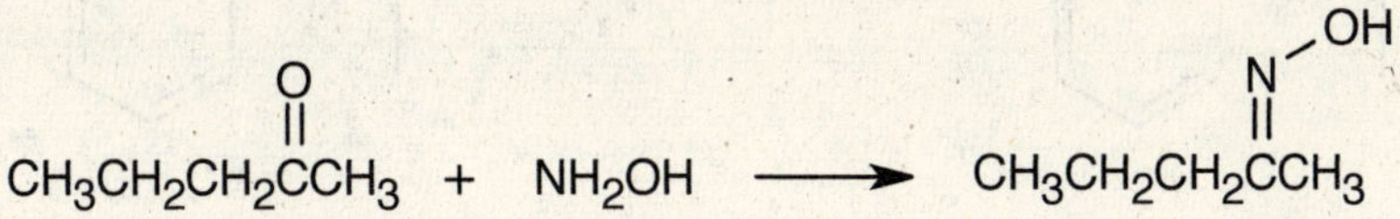

(b)

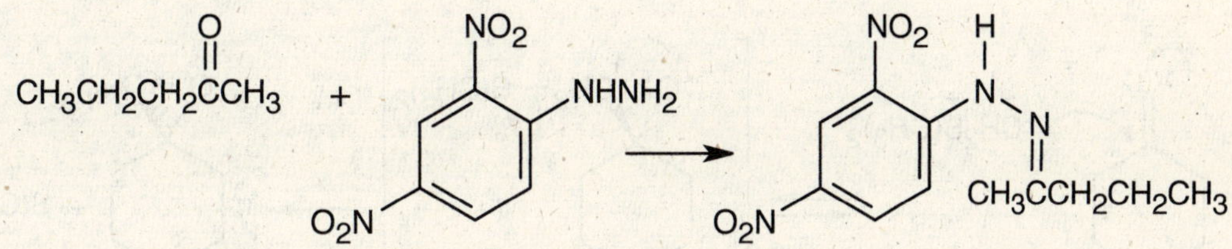

(c)

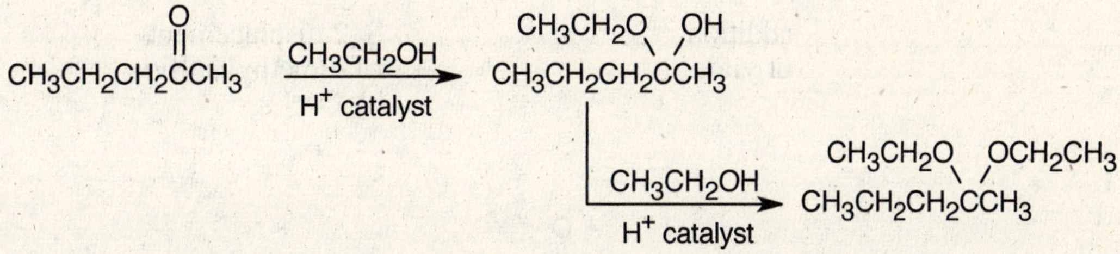

**9.48** Glucose is an aldehyde that exists as an equilibrium mixture of hemiacetal and free aldehyde. The open-chain aldehyde form of glucose is reduced in the same manner described in the text for other aldehydes to produce the polyalcohol sorbitol.

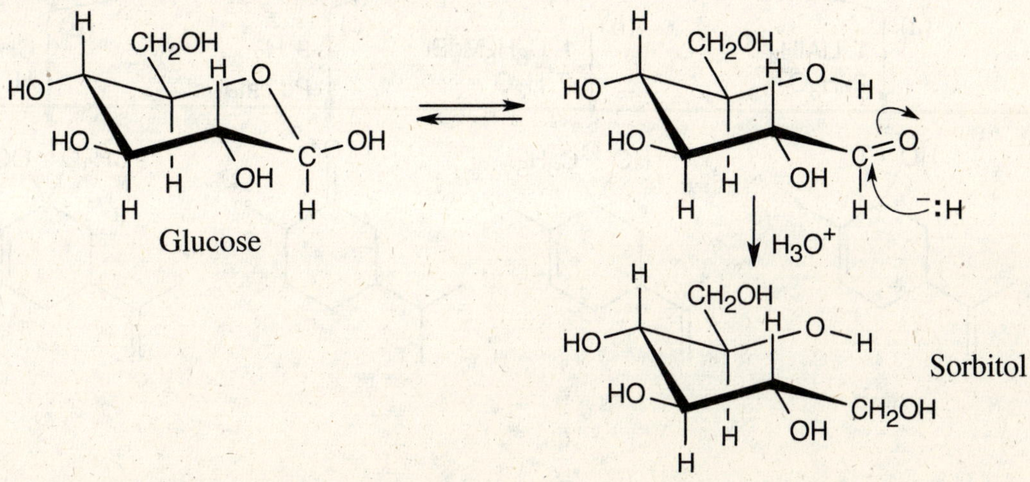

**9.49**

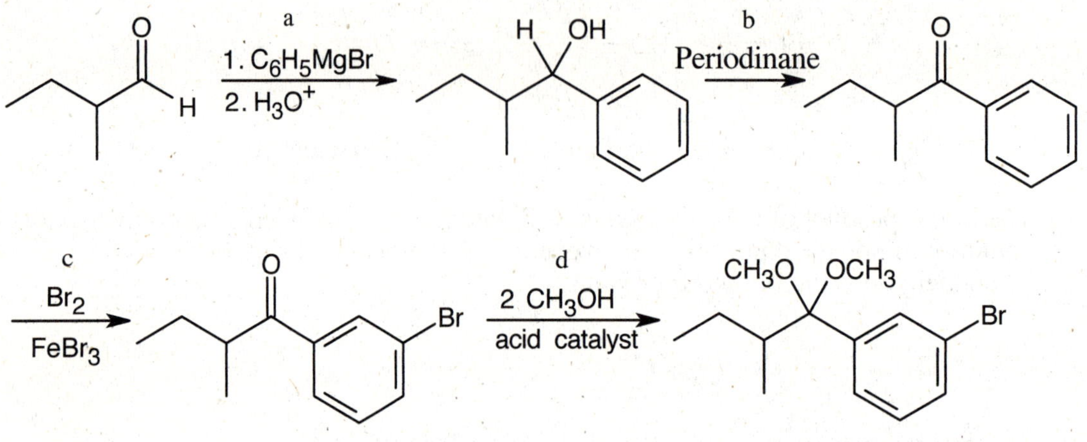
a
1. $C_6H_5MgBr$
2. $H_3O^+$
H OH
b
Periodinane
c
$Br_2$
$FeBr_3$
Br
d
2 $CH_3OH$
acid catalyst
$CH_3O$ $OCH_3$

**9.50**

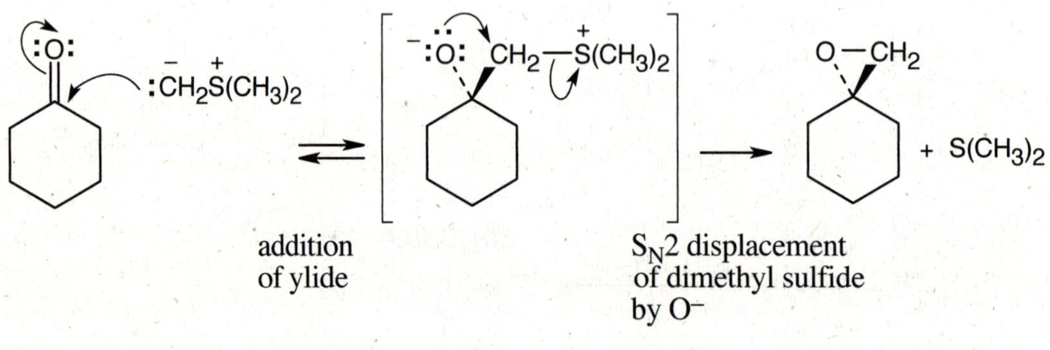
$:\overset{-}{C}H_2\overset{+}{S}(CH_3)_2$
$CH_2-\overset{+}{S}(CH_3)_2$
O—$CH_2$
+ $S(CH_3)_2$
addition
of ylide
$S_N2$ displacement
of dimethyl sulfide
by $O^-$

**9.51**

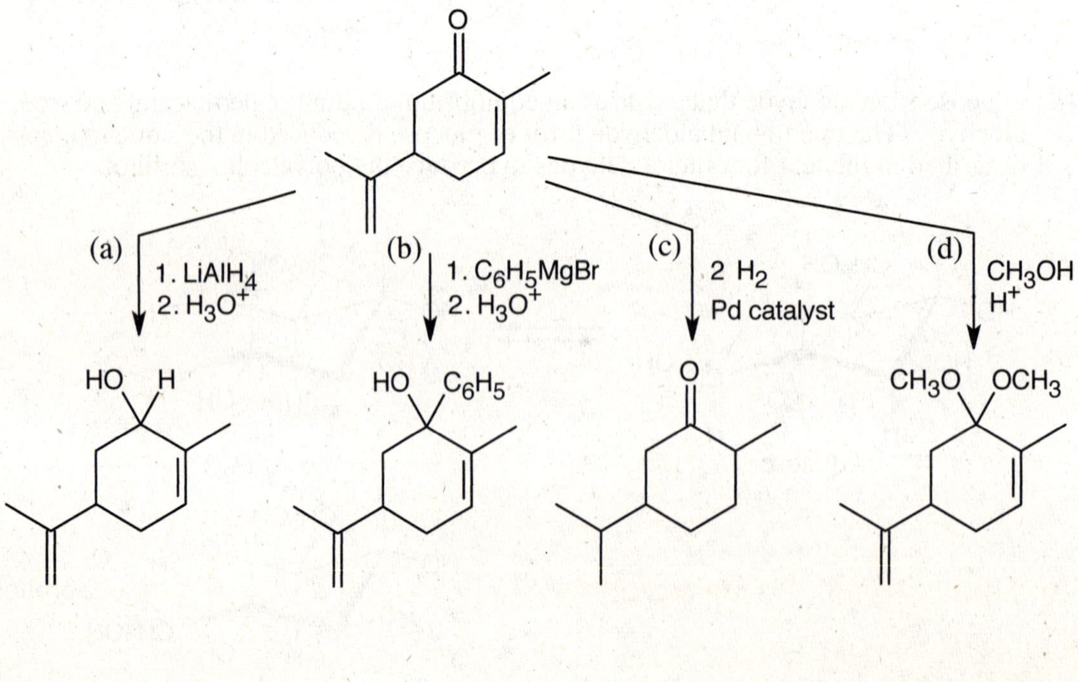
(a)
1. $LiAlH_4$
2. $H_3O^+$
(b)
1. $C_6H_5MgBr$
2. $H_3O^+$
(c)
2 $H_2$
Pd catalyst
(d)
$CH_3OH$
$H^+$
HO H
HO $C_6H_5$
$CH_3O$ $OCH_3$

**9.52**

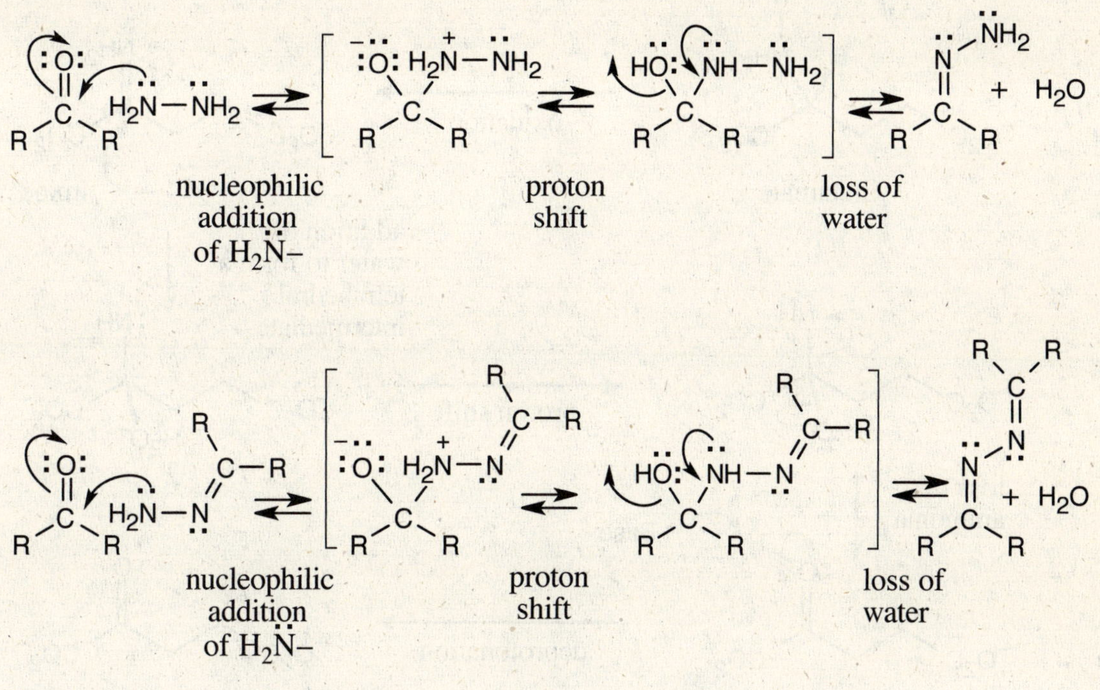


Each nitrogen of $H_2N–NH_2$ adds to a carbonyl group to give the azine.

**9.53**

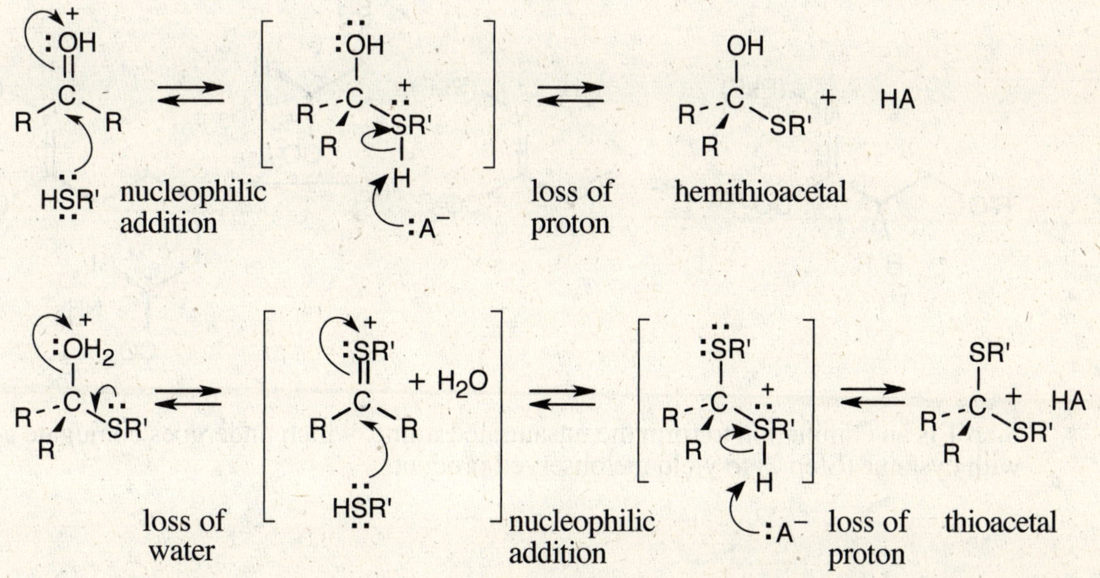


The reaction is similar to acetal formation, which occurs when a ketone or aldehyde reacts with an alcohol in the presence of an acid catalyst.

**9.54**

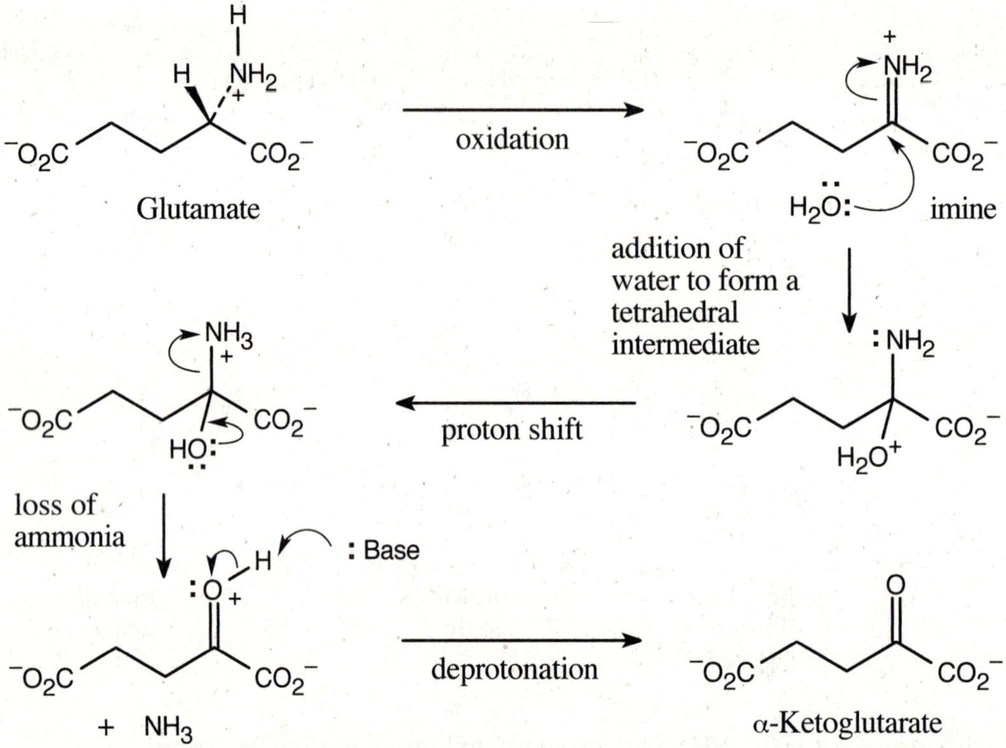


**9.55**

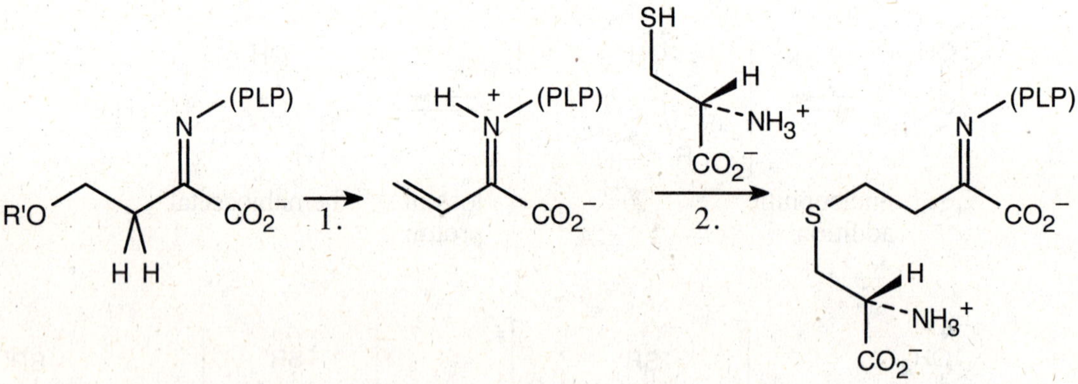


Step 1 is an elimination to form the unsaturated imine, which undergoes conjugate addition with cysteine (Step 2) to yield the observed product.

## In the Medicine Cabinet

**9.56**

Synthesis of 1-phenylpropan-1-ol:

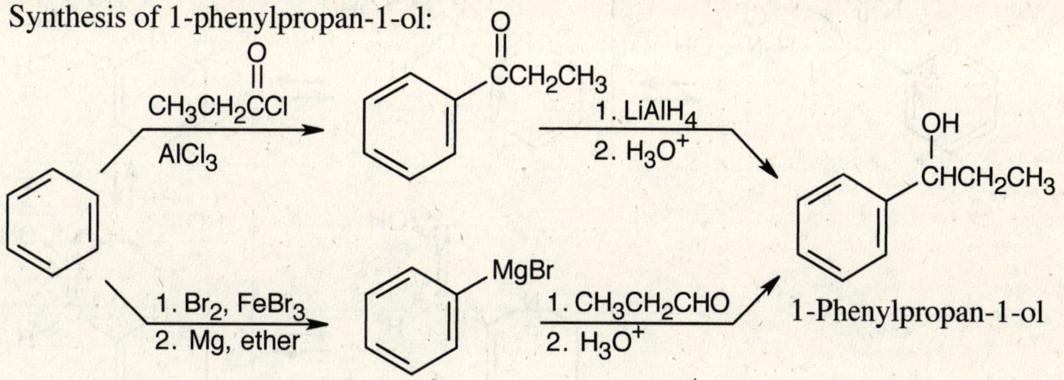


1-Phenylpropan-1-ol

Formation of Grignard reagent:

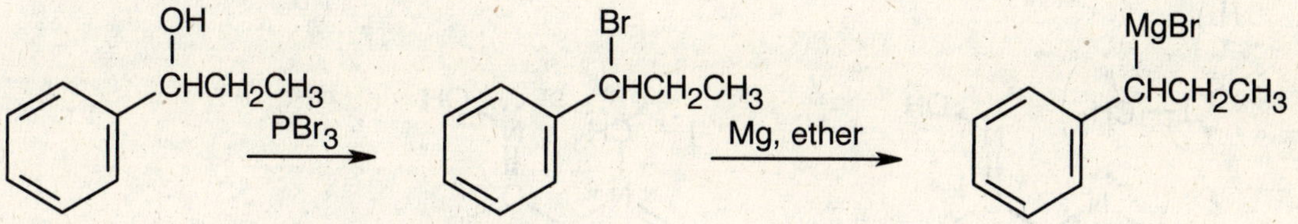


Addition of Grignard reagent to ketone:

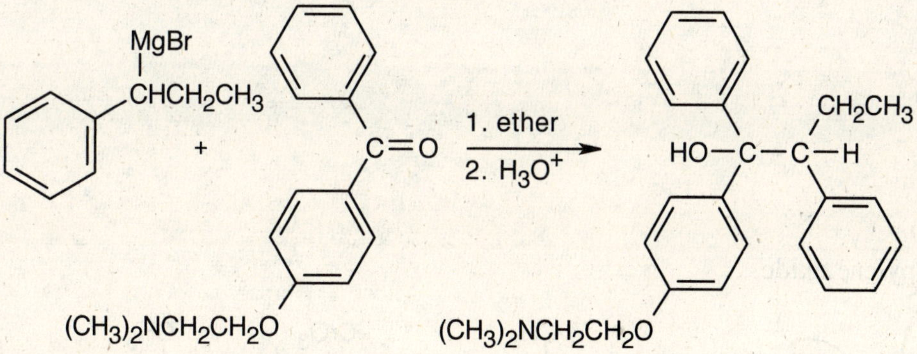


Dehydration to form Tamoxifen:

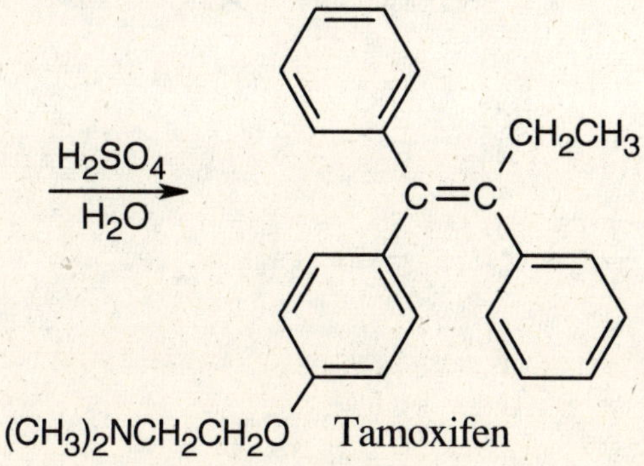


Tamoxifen

**9.57** (a)

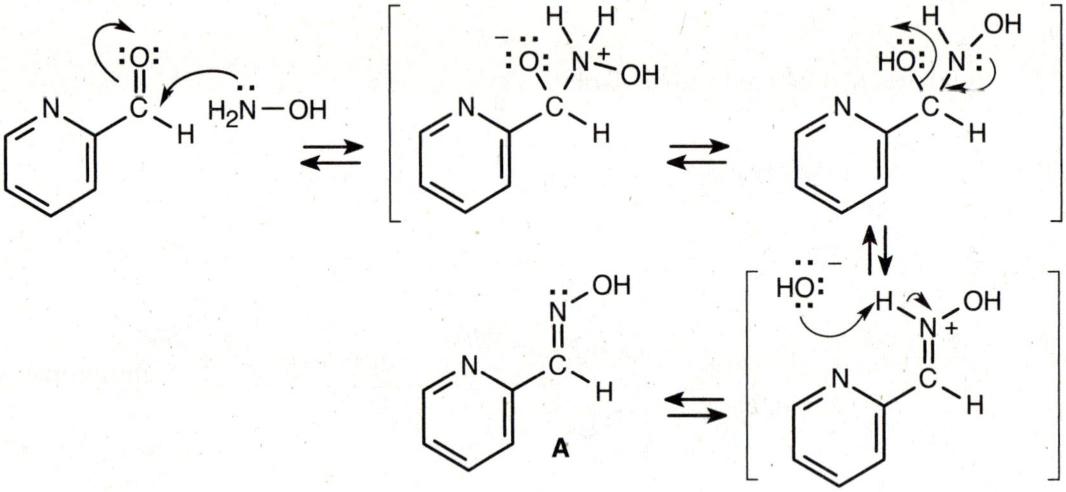

(b)

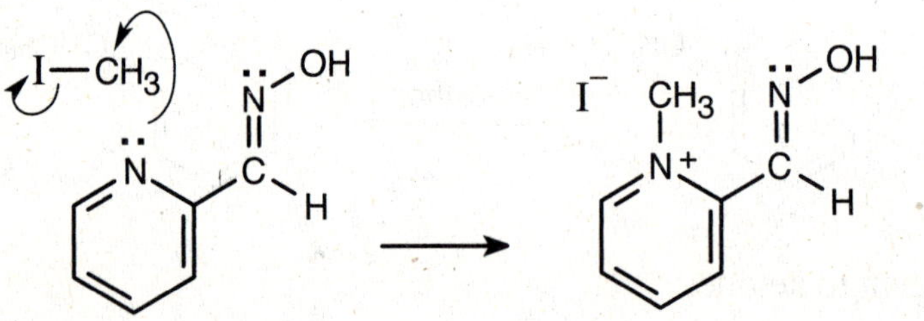

## In the Field

**9.58**

(a) – (c)

Propylene oxide

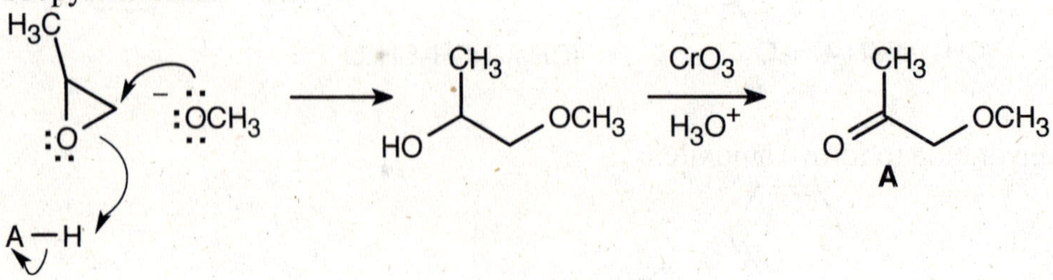

(d)

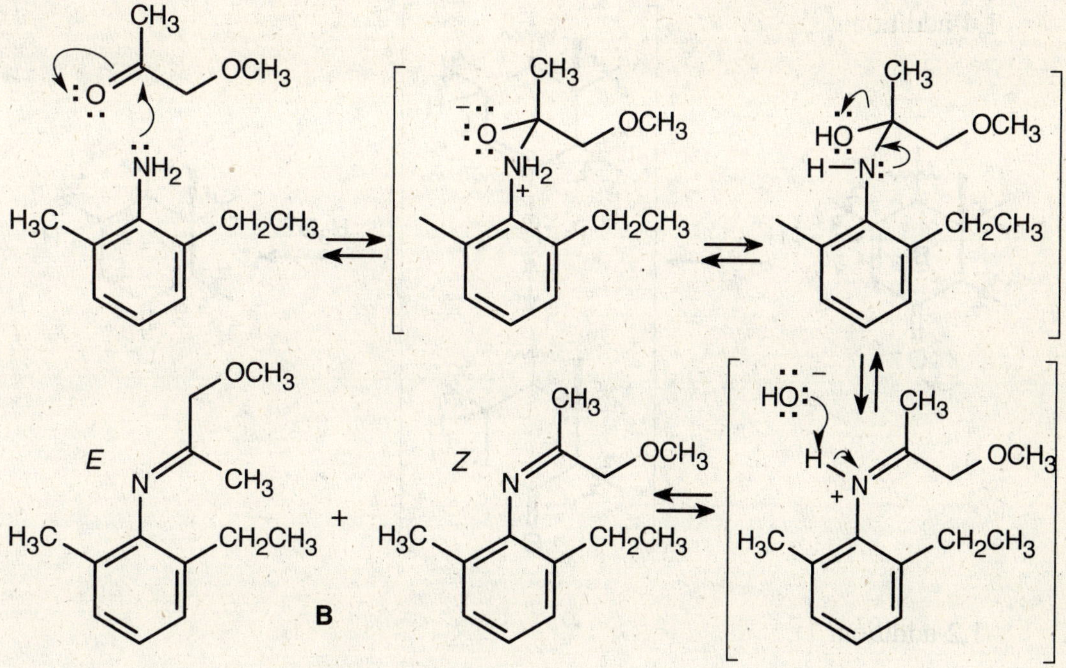


Imine **B** is a mixture of *E* and *Z* isomers.

**9.59**

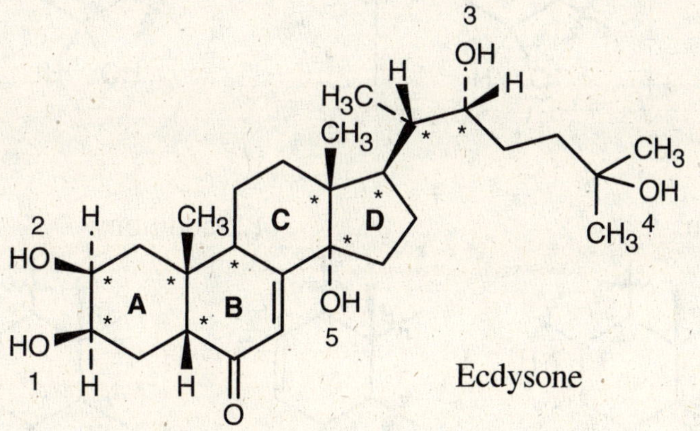


Ecdysone

(a) Hydroxyl groups 1,2,and 3 are secondary; groups 4 and 5 are tertiary.

(b) The ten chirality centers are starred.

(c)

1,4-addition:

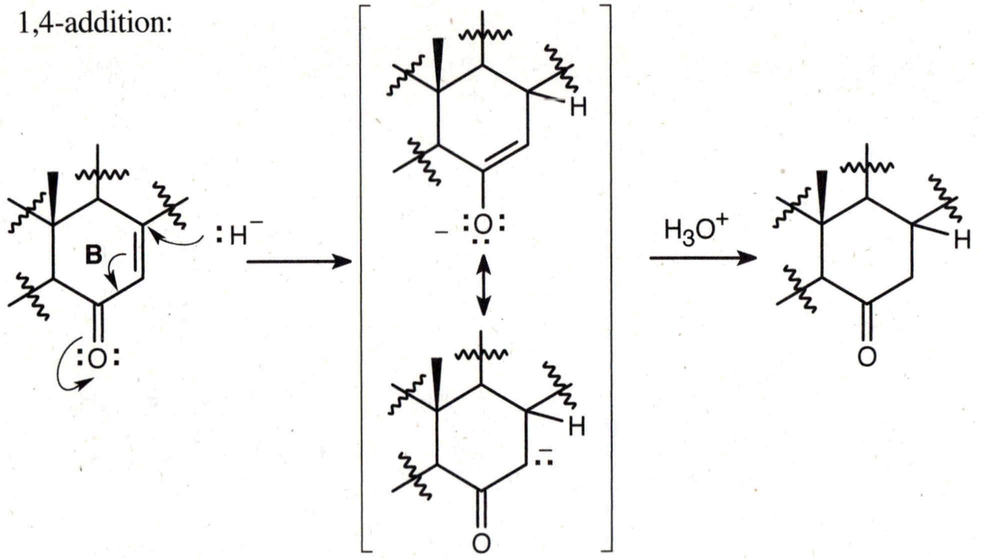

1,2-addition:

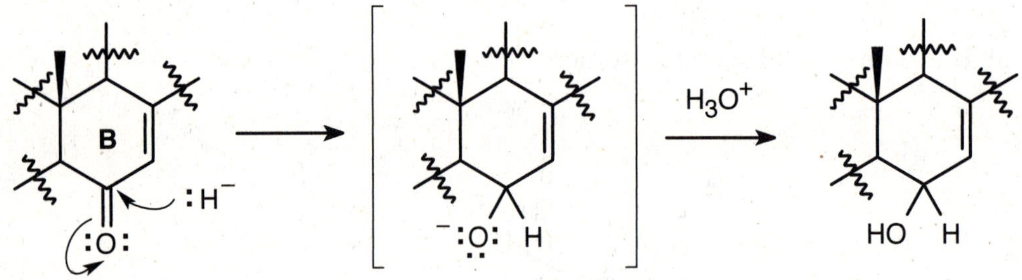

(d)

1,4-addition:

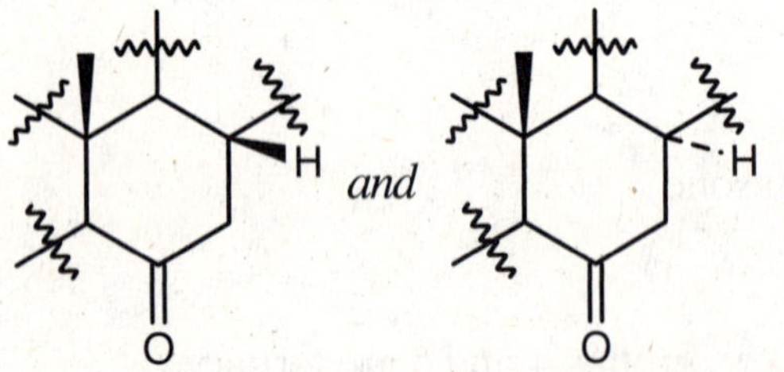

1,2-addition:

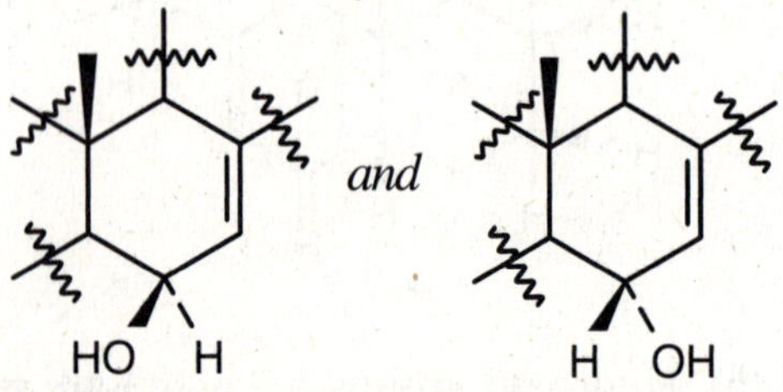

The products of 1,4-addition are diastereomers, as are the products of 1,2-addition. Often pairs like these, that have the same configuration at all chirality centers except one, are described as epimers. A 1,2-addition product is a functional group isomer of a 1,4-addition product.

# Chapter 10 – Carboxylic Acids and Derivatives: Nucleophilic Acyl Substitution Reactions

## Chapter Outline

I. Carboxylic acids and derivatives (Sections 10.1–10.2).
  A. Naming carboxylic acids (Section 10.1).
    1. Noncyclic carboxylic acids are named by replacing the *-e* of the corresponding alkane by *-oic acid.*
      a. The $–CO_2H$ carbon is C1.
    2. Compounds that have a carboxylic acid bonded to a ring are named by using the suffix *-carboxylic acid.*
    3. Many carboxylic acids have historical, nonsystematic names.
  B. Naming carboxylic acid derivatives.
    1. Acid halides.
      a. The acyl group is named first, followed by the halide.
      b. For acyclic compounds, the *-ic acid* of the carboxylic acid name is replaced by *–yl,* followed by the name of the halide.
      c. For cyclic compounds, the *-carboxylic acid* ending is replaced by *-carbonyl,* followed by the name of the halide.
    2. Acid anhydrides.
      a. Symmetrical anhydrides are named by replacing *acid* by *anhydride.*
      b. Unsymmetrical anhydrides are named by citing the two acids alphabetically, followed by *anhydride.*
    3. Amides.
      a. Amides with an unsubstituted $–NH_2$ group are named by replacing *-oic acid* by *-amide* or by replacing *-carboxylic acid* with *-carboxamide.*
      b. If nitrogen is substituted, the nitrogen substituents are named, and an *N-* is put before each.
    4. Esters.
      a. Esters are named by first identifying the alkyl group and then the carboxylic acid group, replacing *-oic acid* by *-oate.*
    5. Nitriles.
      a. Simple nitriles are named by adding *-nitrile* to the alkane name.
      b. More complex nitriles are named as derivatives of carboxylic acids by replacing *-oic acid* by *-onitrile* or by replacing *-carboxylic acid* by *-carbonitrile.*
  C. Occurrence and properties of carboxylic acids and derivatives (Section 10.2).
    1. Carboxylic acids are abundant in nature.
    2. Carboxylic acids are strongly associated because of hydrogen-bonding, and their boiling points are elevated.
    3. Esters and amides are both widespread in nature.
      a. Many polymers are polyesters or polyamides.
    4. Acid chlorides and anhydrides are too reactive to be found in nature but are widely used in the laboratory.

II. Carboxylic acids (Sections 10.3–10.4).
  A. Acidity (Section 10.3).
    1. Dissociation of carboxylic acids.
      a. Carboxylic acids react with bases to form salts that are water-soluble.
      b. Carboxylic acids dissociate slightly in dilute aqueous solution to give $H_3O^+$ and carboxylate anions.
        i. The $K_a$ values for most carboxylic acids are near $10^{-5}$, making them weaker than mineral acids but stronger than alcohols.

- c. The relative strength of carboxylic acids is due to resonance stabilization of the carboxylate anion.
  - i. Both carbon–oxygen bonds of carboxylate ions are the same length.
  - ii. The bond length is intermediate between single and double bonds.

B. Synthesis of carboxylic acids (Section 10.4).
1. Oxidation of alkylbenzenes.
2. Oxidation of primary alcohols and aldehydes.
3. Hydrolysis of nitriles.

III. Nucleophilic acyl substitution reactions (Sections 10.5–10.11).

A. Mechanism of nucleophilic acyl substitution reactions (Section 10.5).
1. A nucleophile adds to the polar carbonyl group.
2. The tetrahedral intermediate eliminates one of the two substituents originally bonded to it, resulting in a net substitution reaction.
3. Substitution, rather than addition, takes place because the acyl carbon is bonded to a group that can have a stable anion.
4. The more electron-poor the carbonyl carbon, the more reactive the derivative.
5. It is possible to convert more reactive derivatives into less reactive derivatives.
   - a. In order of decreasing reactivity: acid chlorides > acid anhydrides > thioesters > esters > amides.
6. The reactions in this chapter can be grouped into several categories, based on the nucleophile.
   - a. Hydrolysis reactions.
   - b. Alcoholysis.
   - c. Aminolysis.
   - d. Reduction.
   - e. Grignard reaction.

B. Reactions of carboxylic acids (Section 10.6).
1. Conversion to acid chlorides with $SOCl_2$.
2. Conversion to esters by acid-catalyzed reaction with an alcohol (Fischer esterification).
3. Conversion to amides with DCC.
4. Reduction to alcohols using $LiAlH_4$.

C. Acid halides (Section 10.7).
1. Preparation from carboxylic acids, using $SOCl_2$.
2. Reactions of acid halides.
   - a. Hydrolysis to give carboxylic acids.
   - b. Reaction with alcohols to give esters.
   - c. Reaction with amines to give amides.

D. Acid anhydrides (Section 10.8).
1. Preparation from acid halides by reaction of a carboxylate with an acid chloride.
2. Reactions (same as acid halides).
   - a. Only "half" of the anhydride molecule is used; the other half is the leaving group.

E. Esters (Section 10.9).
1. Preparation from acids and acid halides.
2. Reactions.
   - a. Hydrolysis to give carboxylic acids plus alcohols.
   - b. Reduction with $LiAlH_4$ to give primary alcohols.
   - c. Reaction with Grignard reagents to give tertiary alcohols.

F. Amides (Section 10.10).
1. Preparation from acid chlorides.
2. Reactions.
   - a. Hydrolysis to give carboxylic acids.
   - b. Reduction with $LiAlH_4$ to give amines.

G. Nitriles (Section 10.11).
 1. Preparation from alkyl halides.
 2. Reactions.
  a. Hydrolysis to give carboxylic acids.
  b. Reduction with $LiAlH_4$ to give amines.
  c. Reaction with Grignard reagents to give ketones.
H. Biological carboxylic acid derivatives (Section 10.12).
 1. Acyl phosphates and thioesters are important biological intermediates.
 2. Both types of derivatives are intermediate in reactivity.

IV. Polyamides and polyesters: step-growth polymers (Section 10.13).
 A. When a diamine and a dicarboxylic acid react, a polyamide (nylon) is formed.
 B. When a diacid and a diol react, a polyester is formed.
 C. These polymers are called step-growth polymers because each bond is formed independently of the others.
 D. Biodegradable polymers are polyesters that slowly hydrolyze.

## Solutions to Problems

**10.1**

(a)
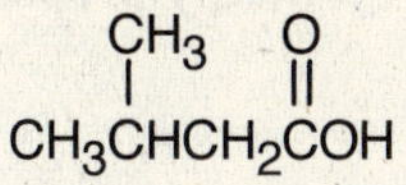
3-Methylbutanoic acid

(b)
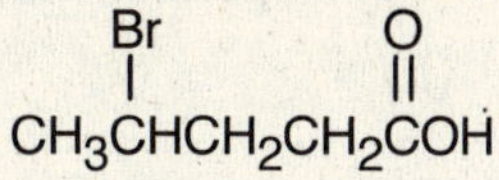
4-Bromopentanoic acid

(c)
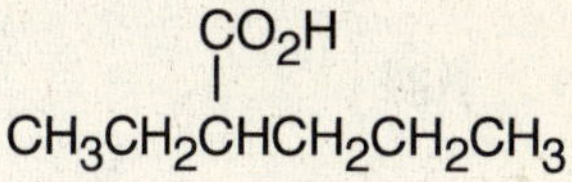
2-Ethylpentanoic acid

(d)
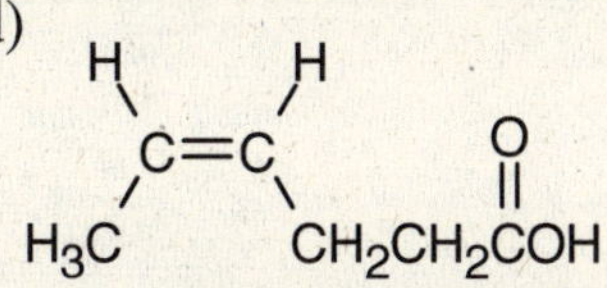
(*Z*)-4-Hexenoic acid

(e)
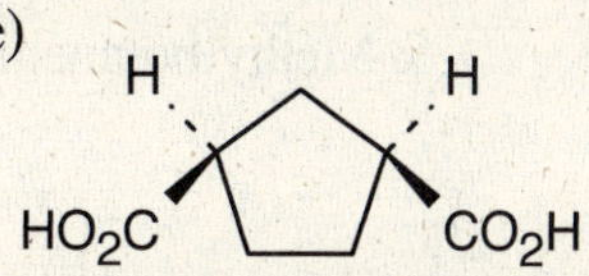
*cis*-1,3-Cyclopentane-dicarboxylic acid

**10.2**

(a)
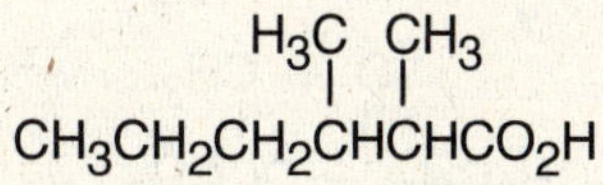
2,3-Dimethylhexanoic acid

(b)
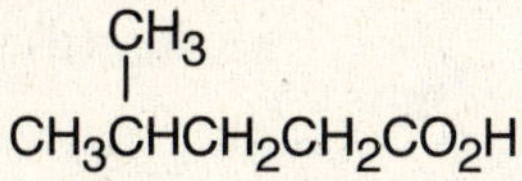
4-Methylpentanoic acid

(c)
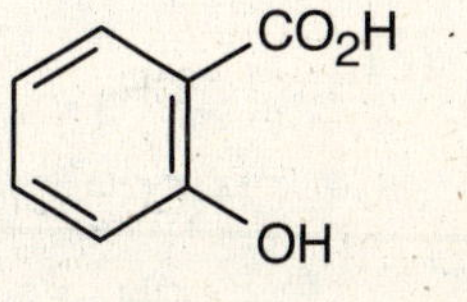
*o*-Hydroxybenzoic acid

(d)
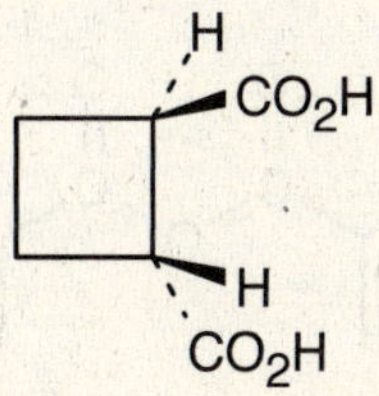
*trans*-Cyclobutane-1,2-dicarboxylic acid

**10.3**

(a)

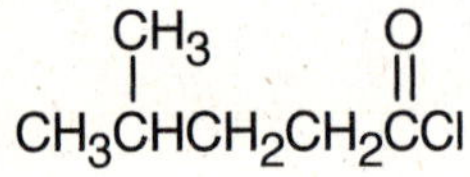

4-Methylpentanoyl chloride

(b)

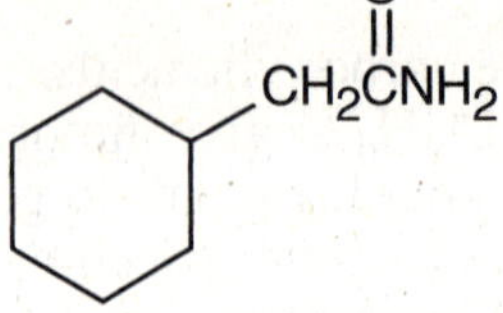

Cyclohexylacetamide

(c)

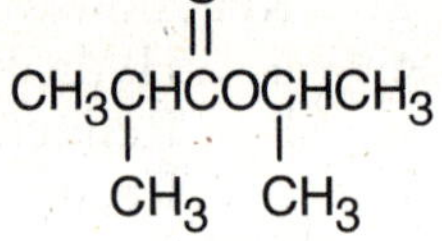

Isopropyl 2-methylpropanoate

(d)

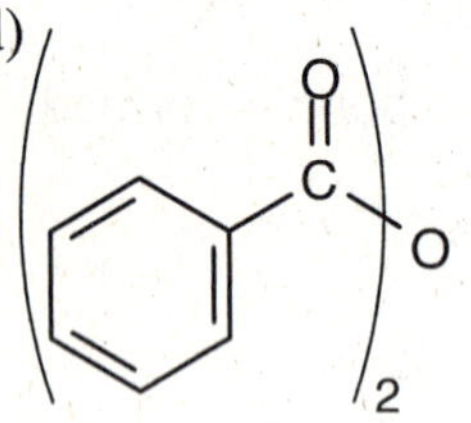

Benzoic anhydride

(e)

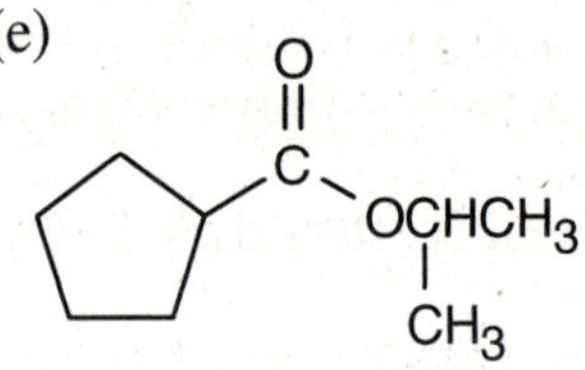

Isopropyl cyclopentanecarboxylate

(f)

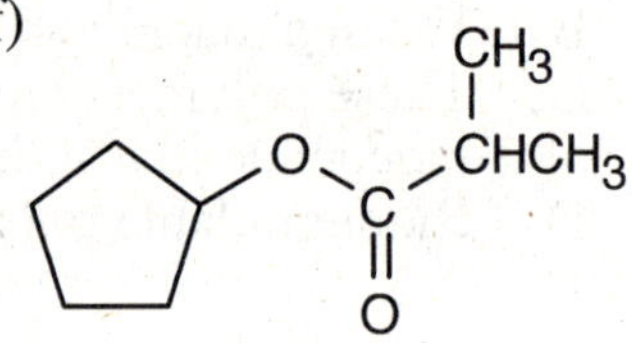

Cyclopentyl 2-methylpropanoate

(g)

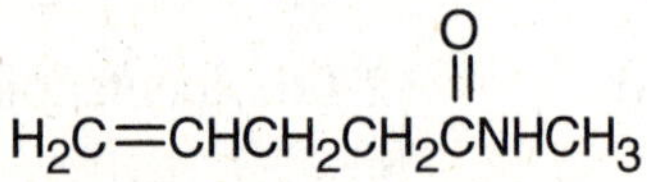

*N*-Methylpent-4-enamide

(h)

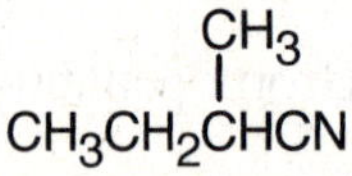

2-Methylbutanenitrile

**10.4**

(a)

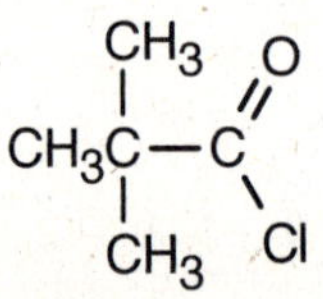

2,2-Dimethylpropanoyl chloride

(b)

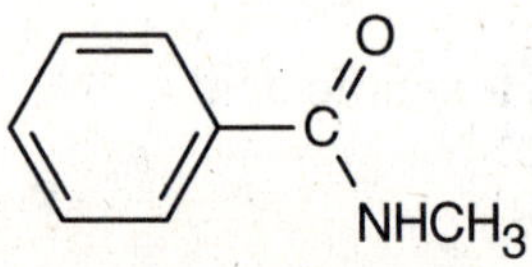

*N*-Methylbenzamide

(c)

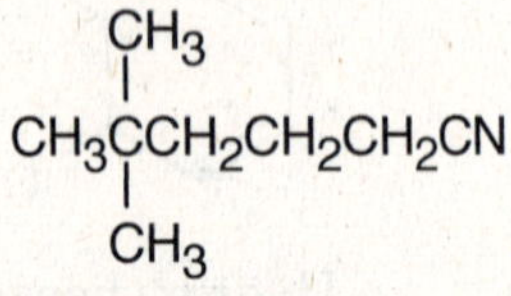

5,5-Dimethylhexanenitrile

(d)

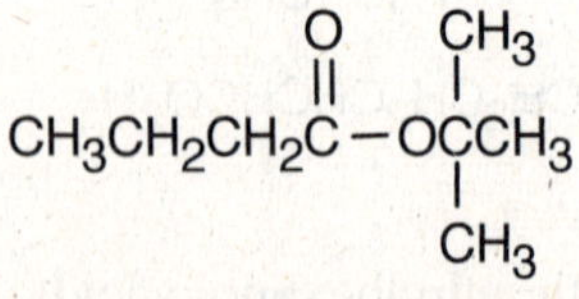

*tert*-Butyl butanoate

(e)

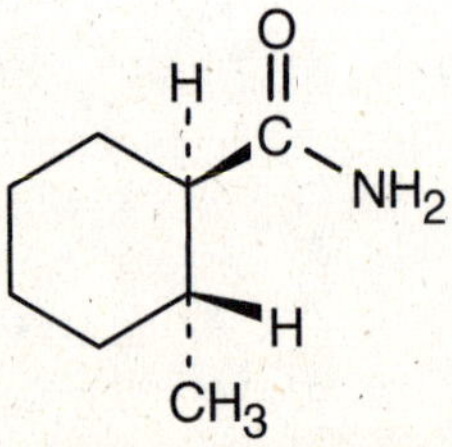

*trans*-2-Methylcyclohexanecarboxamide

(f)

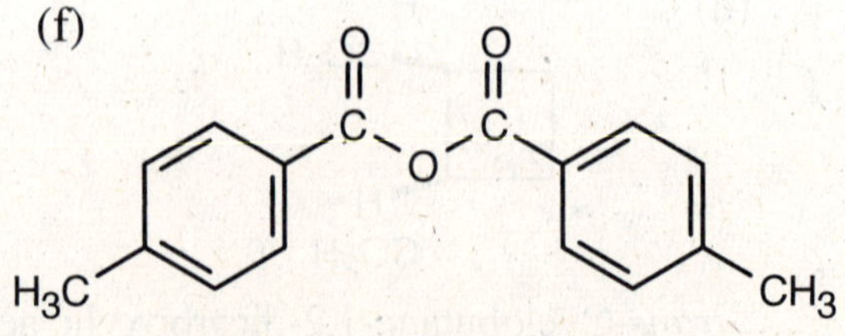

*p*-Methylbenzoic anhydride

(g) 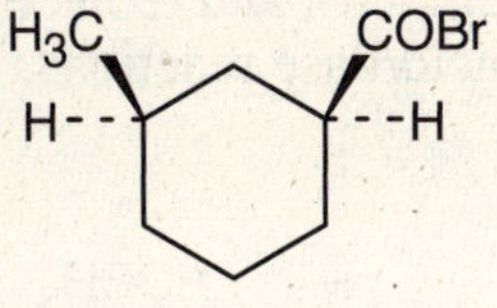

*cis*-3-Methylcyclohexanecarbonyl bromide

(h) 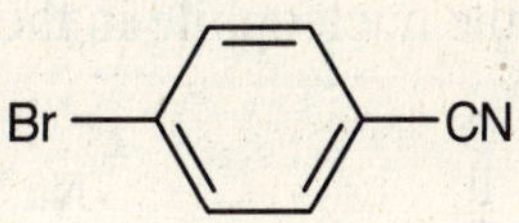

*p*-Bromobenzonitrile

**10.5**

(a)

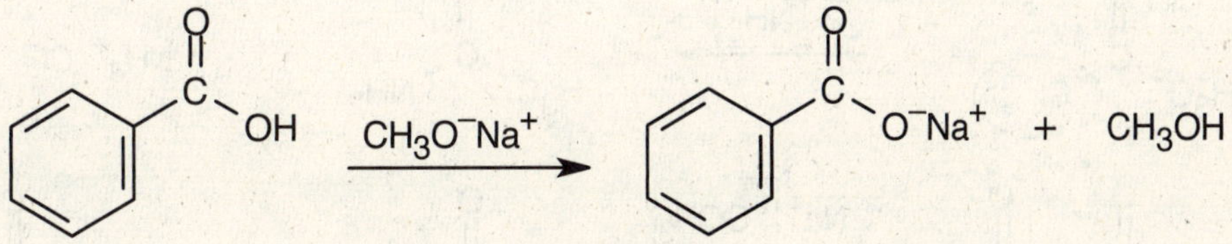

(b)

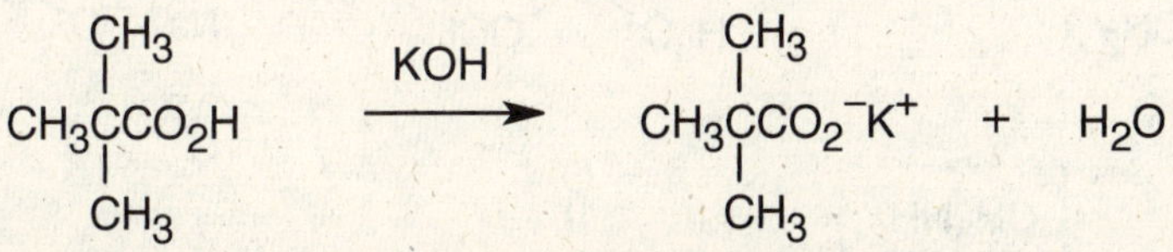

**10.6**

*Least acidic* ⟶ *Most acidic*

(a) Methanol < Phenol < *p*-Nitrophenol < Acetic acid < Sulfuric acid
(b) Ethanol < Benzoic acid < *p*-Cyanobenzoic acid

**10.7** You would expect lactic acid to be a stronger acid because the electron-withdrawing inductive effect of the hydroxyl group can stabilize the lactate anion.

**10.8**

$$CH_3I \xrightarrow{NaCN} CH_3CN \xrightarrow[2.\ H_3O^+]{1.\ ^-OH,\ H_2O} CH_3CO_2H$$

This reaction sequence can't be used to convert iodobenzene to benzoic acid because aryl halides don't undergo $S_N2$ substitution.

**10.9**

| | *More reactive* | *Less reactive* | *Reason* |
|---|---|---|---|
| (a) | $CH_3COCl$<br>Acid chloride | $CH_3CO_2CH_3$<br>Ester | Acid chlorides are more reactive (polar) than esters. |
| (b) | $CH_3CH_2CO_2CH_3$<br>Ester | $(CH_3)_2CHCONH_2$<br>Amide | Esters are more reactive than amides in nucleophilic acyl substitution reactions. |
| (c) | $CH_3CO_2COCH_3$<br>Acid anhydride | $CH_3CO_2CH_3$<br>Ester | Acid anhydrides are more reactive than esters in nucleophilic acyl substitution reactions. |
| (d) | $CH_3CO_2CH_3$<br>Ester | $CH_3CHO$<br>Aldehyde | Aldehydes don't undergo nucleophilic acyl substitution reactions. |

**10.10** Identify the nucleophile (boxed) and the leaving group (circled), and replace the leaving group by the nucleophile in the product. Remember the acid/base reactions.

(a)

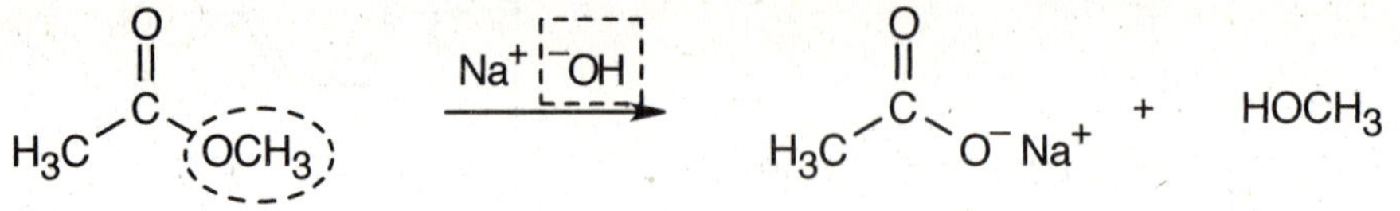

(b)

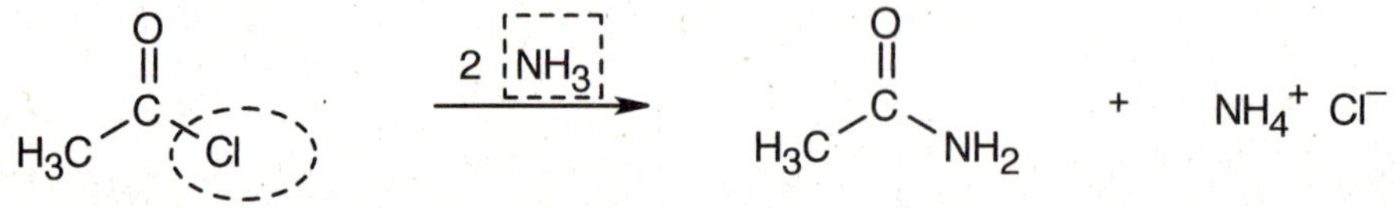

(c)

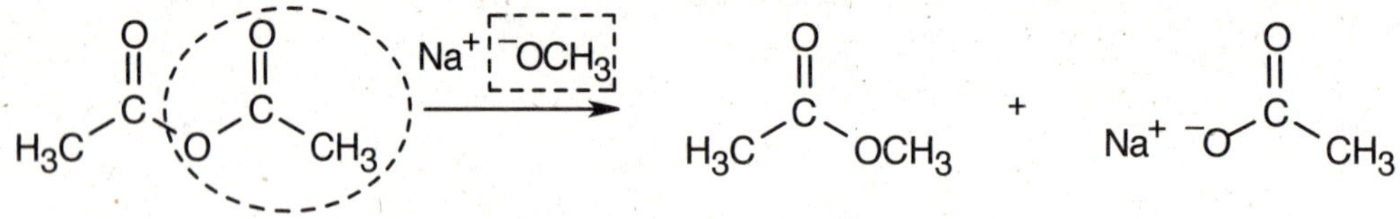

(d)

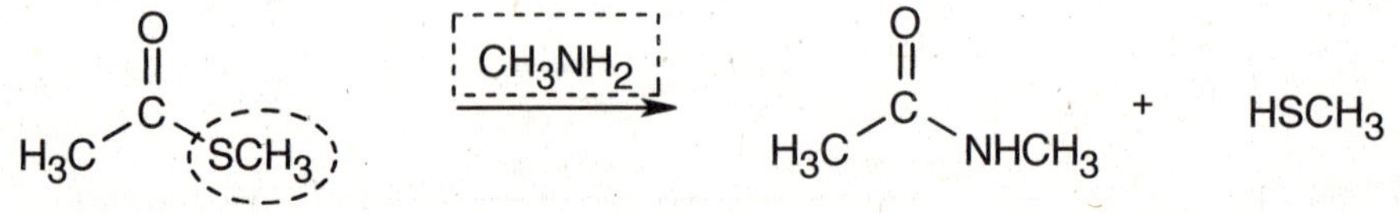

**10.11**

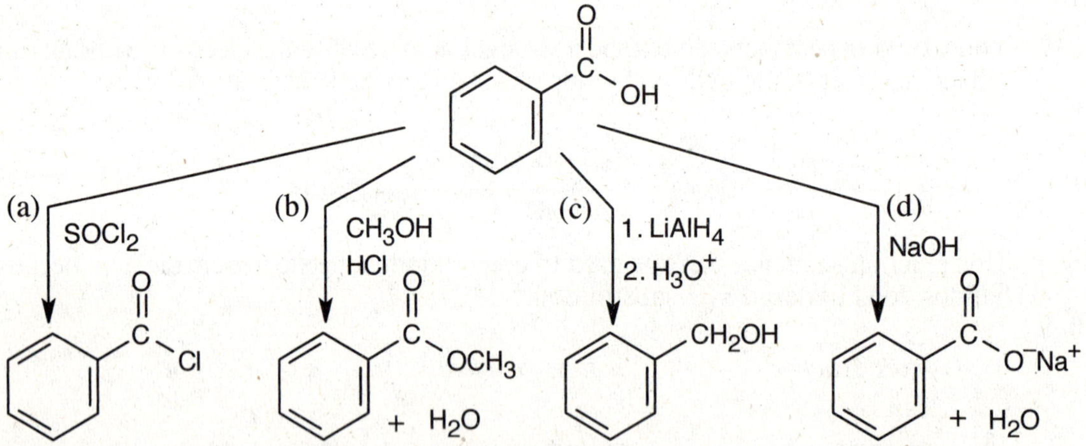

**10.12** Arrange the acid and the ester so that the parts of the water molecule to be lost are next to each other. Remove water, and draw the new bond.

(a)

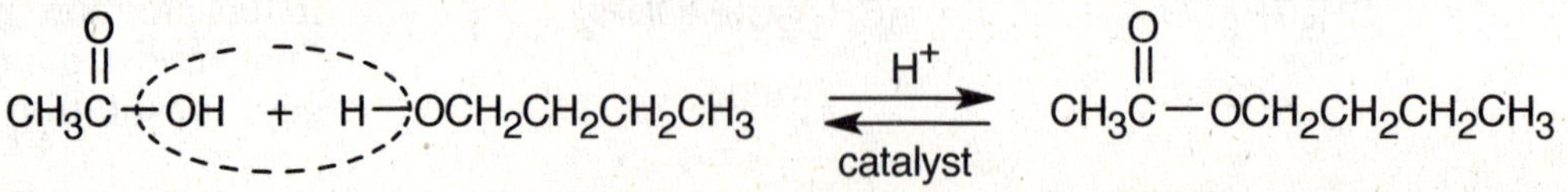

Acetic acid 1-Butanol Butyl acetate + $H_2O$

(b)

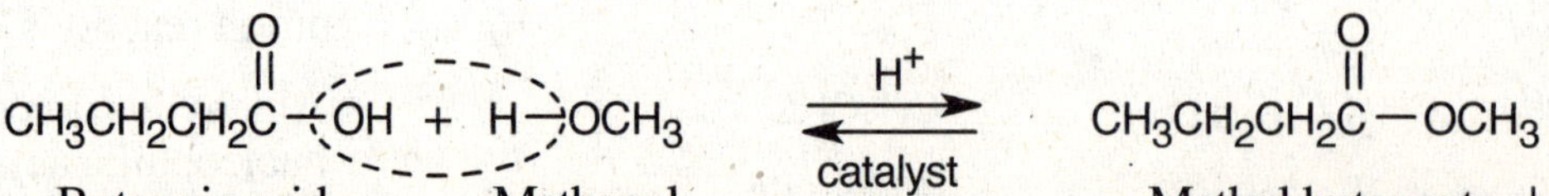

Butanoic acid Methanol Methyl butanoate + $H_2O$

(c)

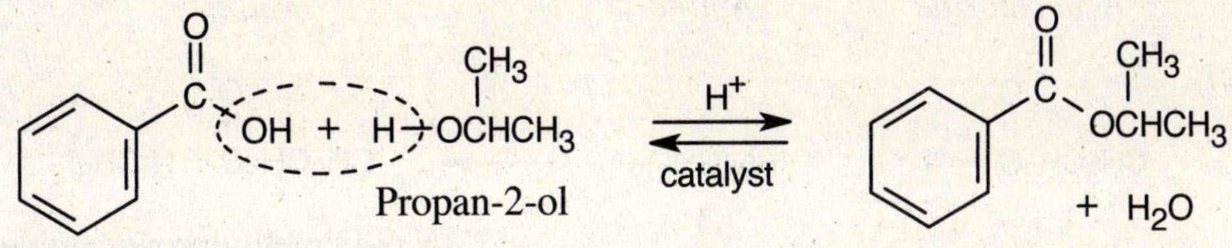


Benzoic acid Isopropyl benzoate

**10.13**

(a)

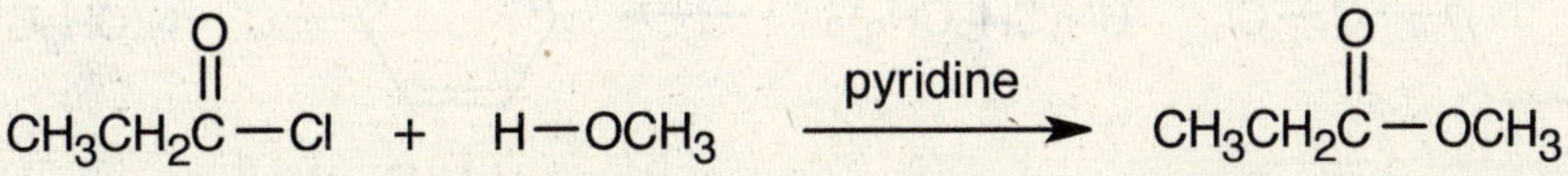


Methyl propanoate

(b)

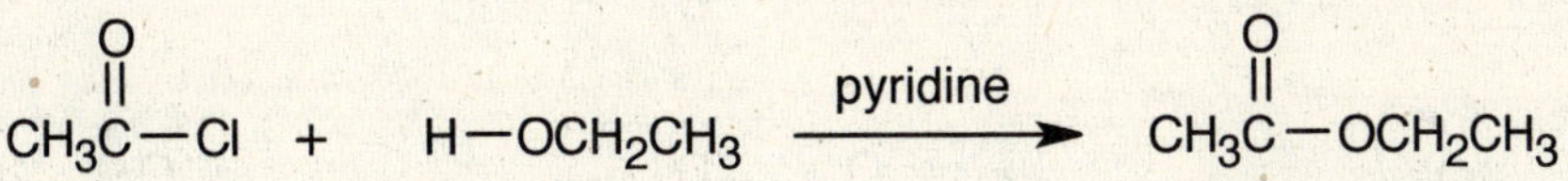


Ethyl acetate

(c)

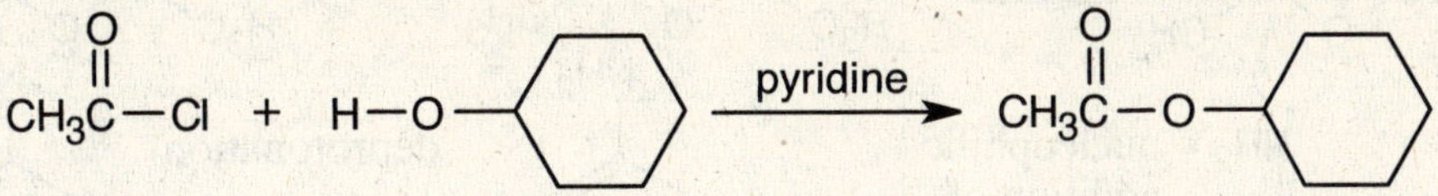


Cyclohexyl acetate

**10.14**

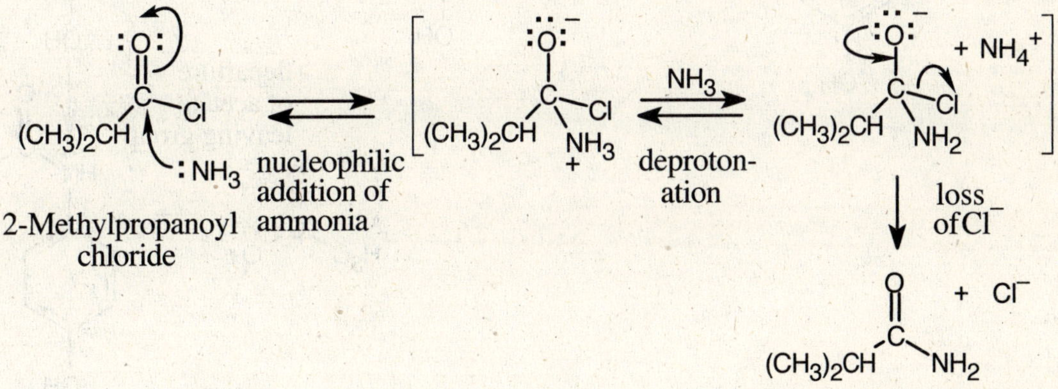


2-Methylpropanamide

**10.15**

*Acid chloride* *Amine* *Amide*

(a)

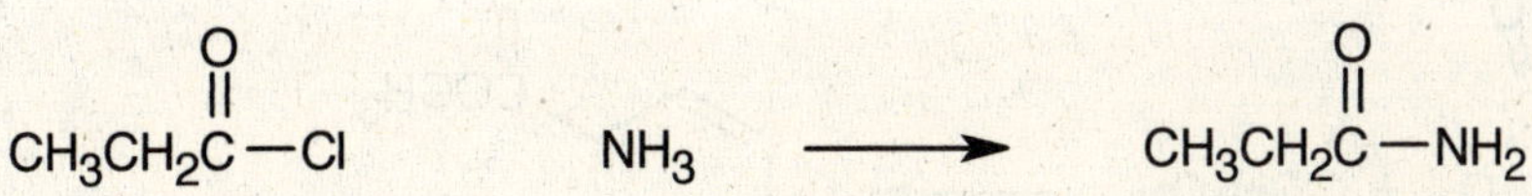

Propanamide

(b)

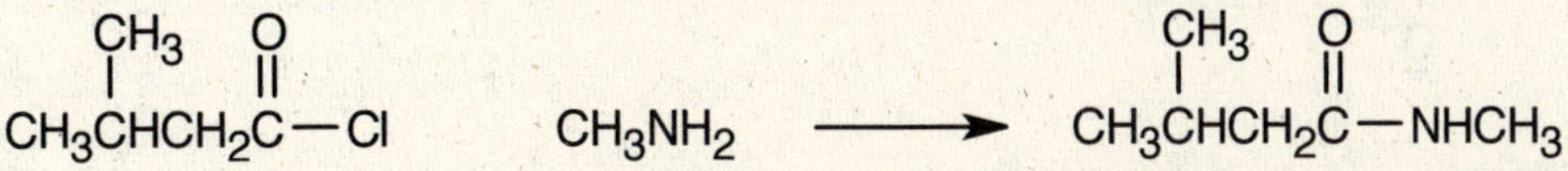

*N*-Methyl-3-methylbutanamide

*Acid chloride* *Amine* *Amide*

(c)

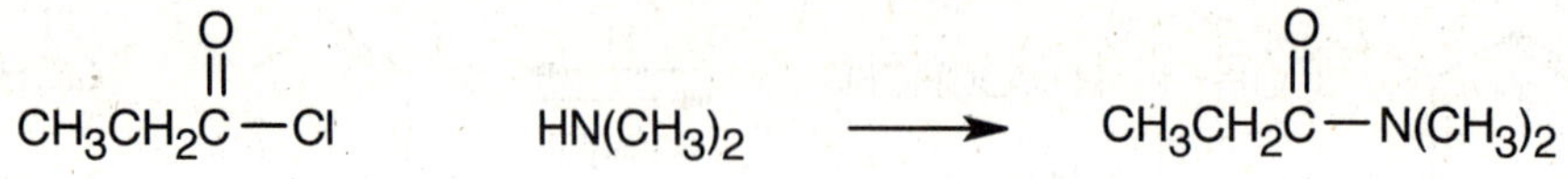

*N,N*-Dimethylpropanamide

(d)

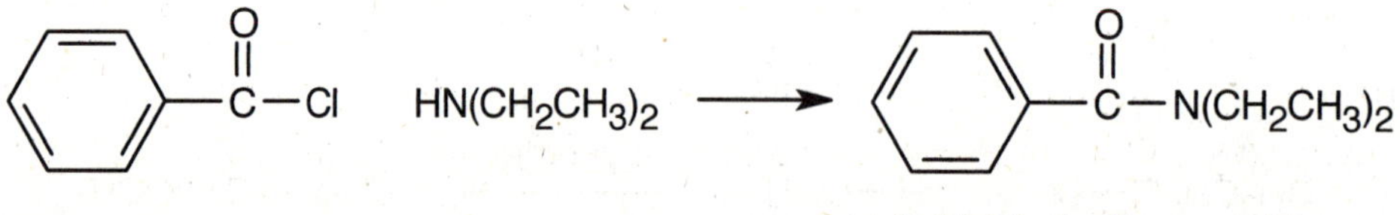

*N,N*-Diethylbenzamide

**10.16**

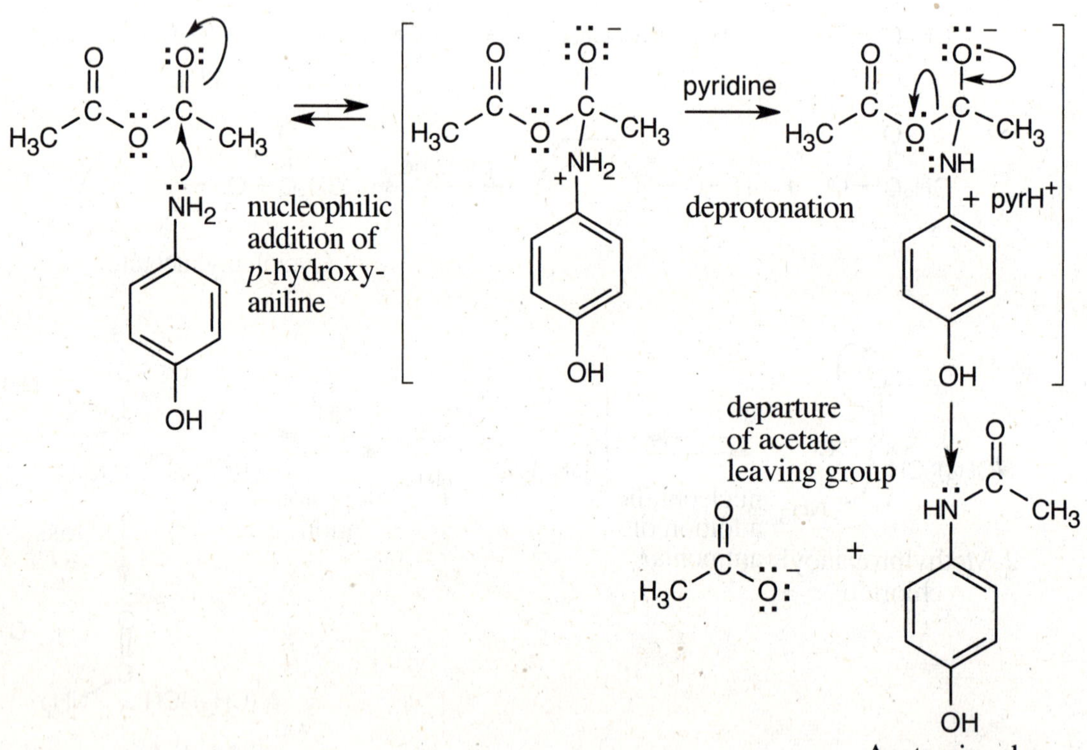


Acetaminophen

**10.17**

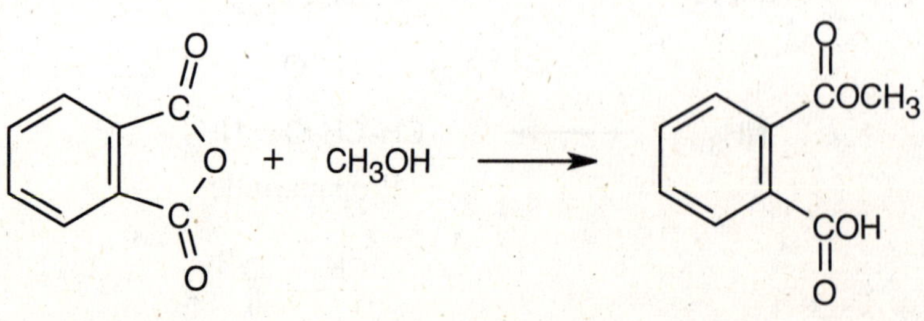

The second half of a cyclic anhydride becomes a carboxylic acid functional group.

**10.18** Locate the bond to be broken, break it, add –OH to the carbonyl carbon, and add –H to the alcohol oxygen.

(a)

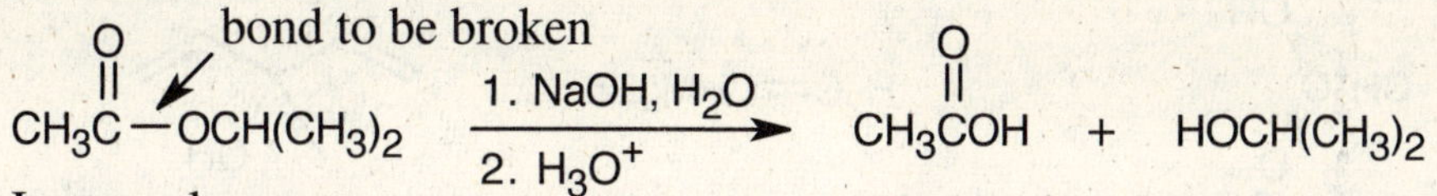


Isopropyl acetate — Acetic acid — Isopropyl alcohol

(b)

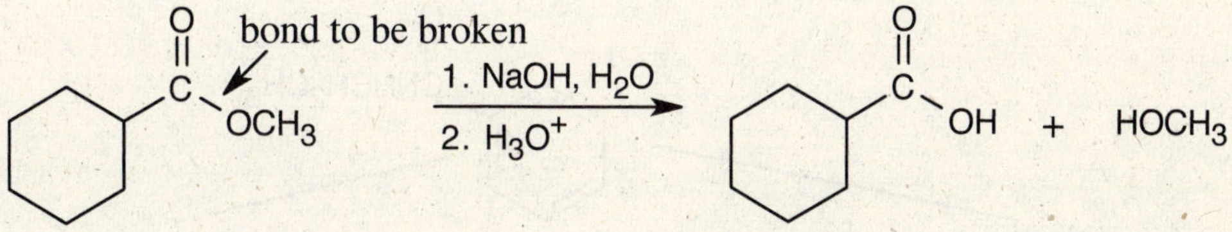


Methyl cyclohexanecarboxylate — Cyclohexane-carboxylic acid — Methanol

**10.19**

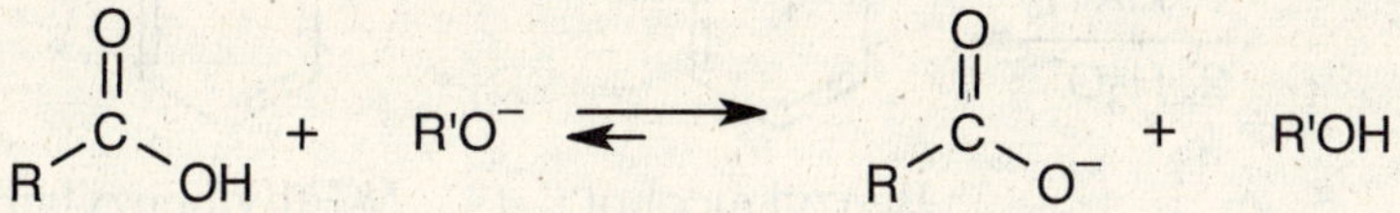


The principal reaction of a carboxylic acid and an alkoxide is an acid-base reaction, which yields an alcohol and a carboxylate anion. The negatively charged carboxylate group is not reactive toward nucleophiles, and thus the reverse of saponification does not occur.

**10.20** Reduction of an ester with $LiAlH_4$ produces two alcohols.

*Ester* — *Alcohols*

(a)

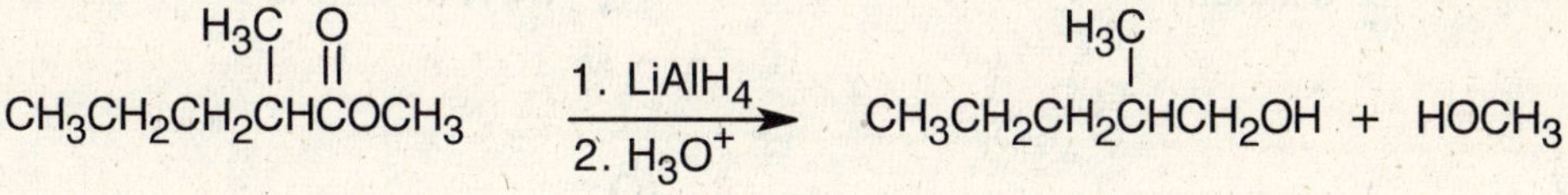


(b)

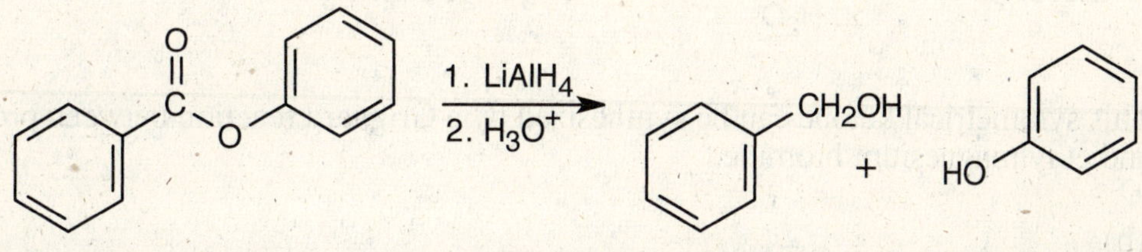


**10.21** Reaction of an ester with two equivalents of Grignard reagent yields a tertiary alcohol with two identical groups that come from the Grignard reagent.

*Ester* + *Grignard Reagent* ⟶ *Tertiary alcohol*

(a)

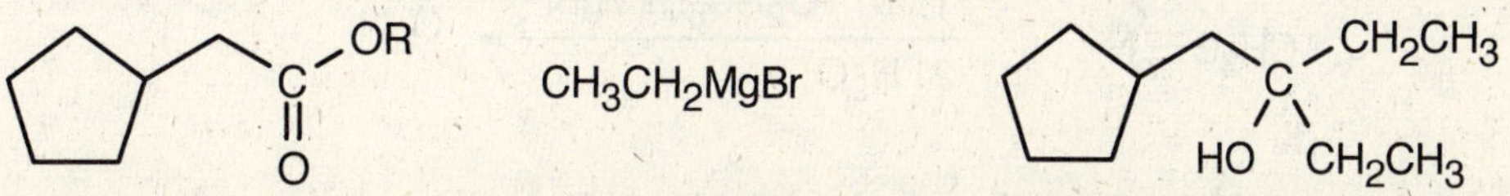

*Ester* + *Grignard Reagent* ⟶ *Tertiary alcohol*

(b)

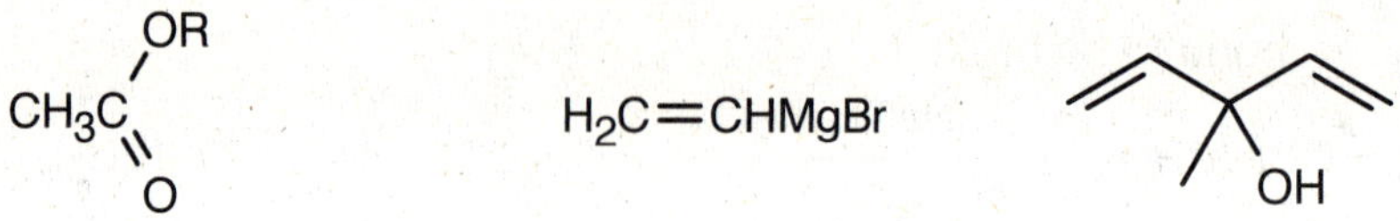


**10.22**

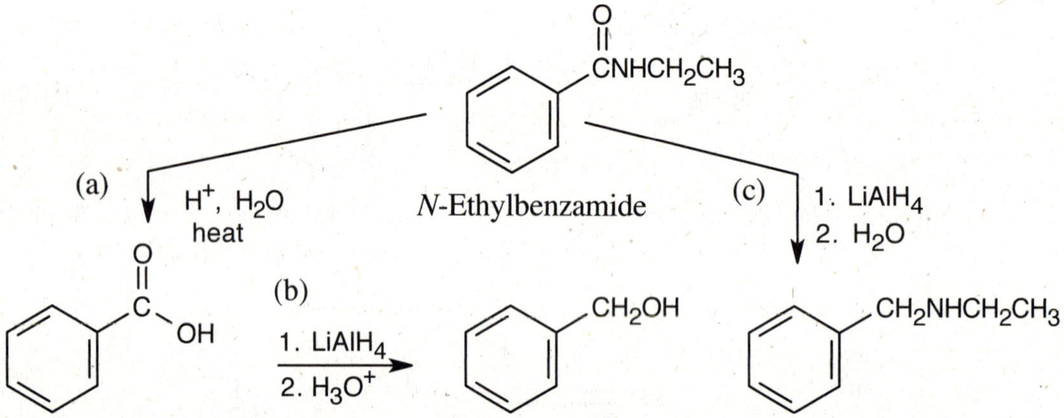


Benzoic acid Benzyl alcohol *N*-Ethylbenzylamine

**10.23**

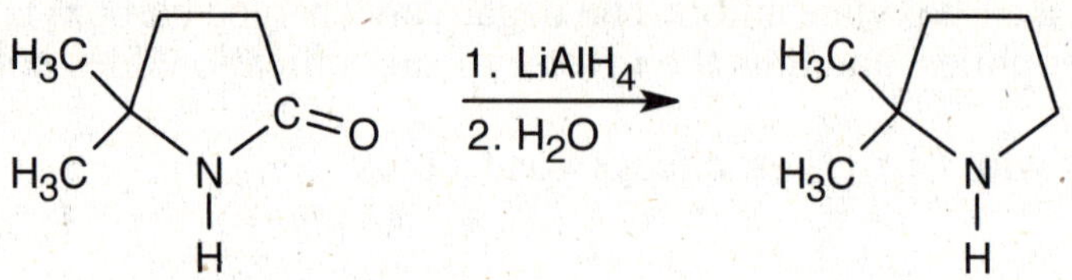


a lactam a cyclic amine

**10.24**

(a)

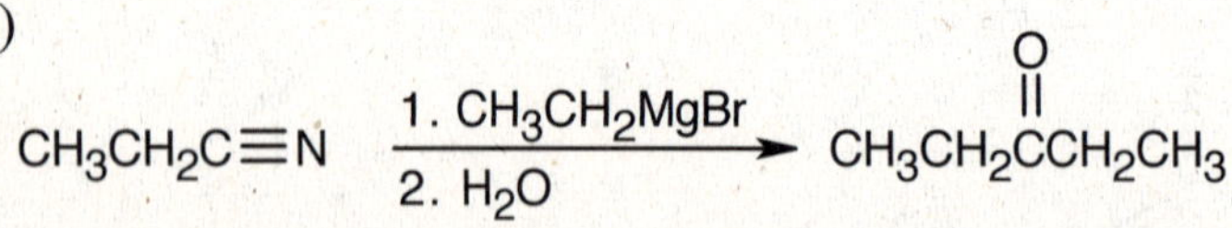


This symmetrical ketone can be synthesized by a Grignard reaction between propanenitrile and ethylmagnesium bromide.

(b)

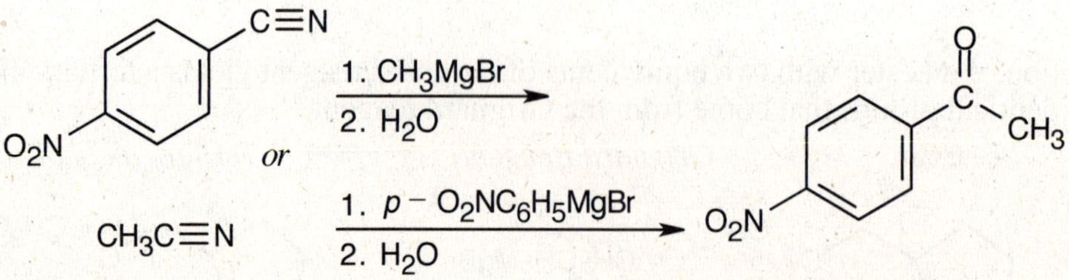


*p*-Nitroacetophenone can be synthesized by either of two Grignard routes.

**10.25**

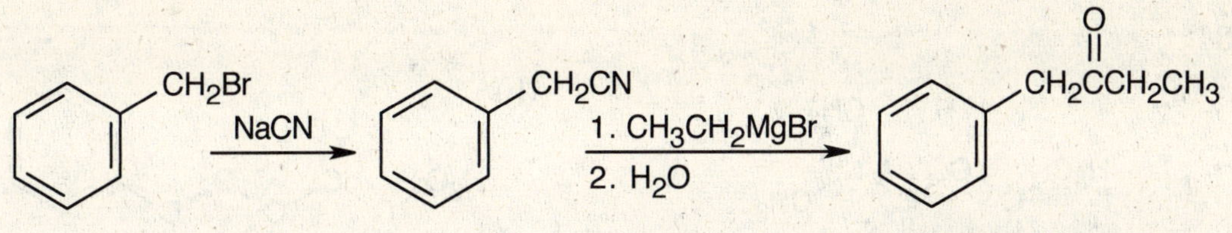


1-Phenyl-2-butanone

Once you realize that the product results from a Grignard reaction with a nitrile, this synthesis is easy.

**10.26** Even though the entire molecule of coenzyme A is biologically important, we are concerned in this problem only with the –SH group. The remainder of the structure is represented here as "R".

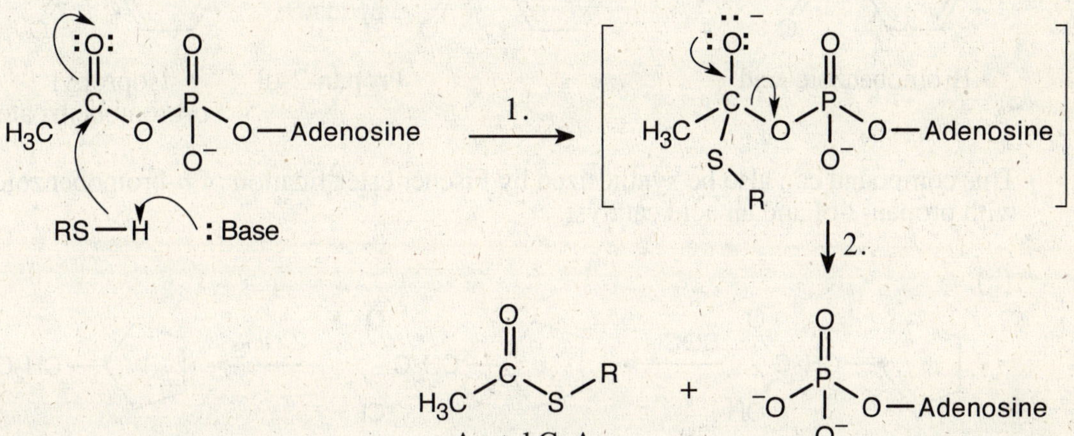


Acetyl CoA

**Step 1:** Nucleophilic addition of the –SR group of CoA (after deprotonation) to acetyl adenylate to form a tetrahedral intermediate.
**Step 2:** Loss of adenosine monophosphate.

**10.27**

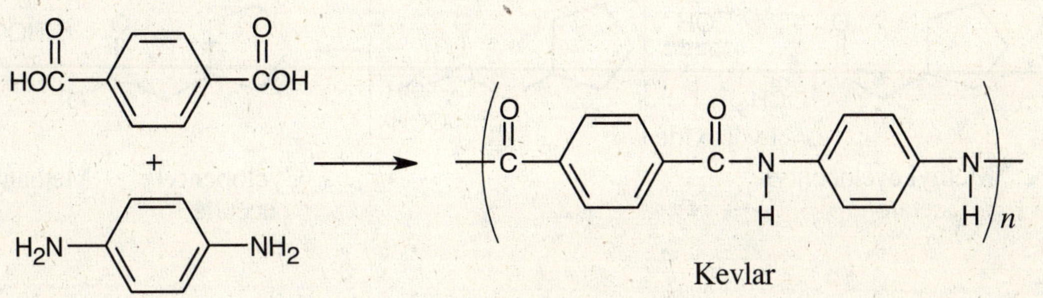


Kevlar

## Visualizing Chemistry

**10.28**

(a)

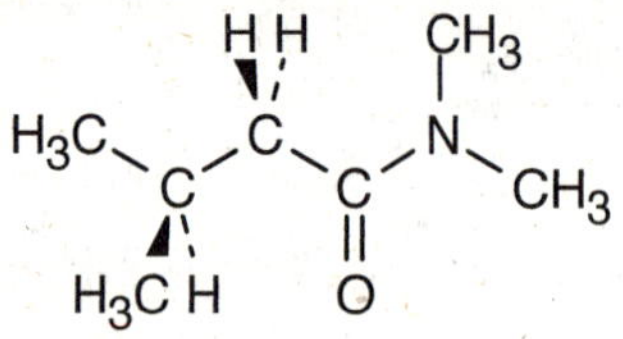

*N,N*-Dimethyl-3-methylbutanamide

(b)

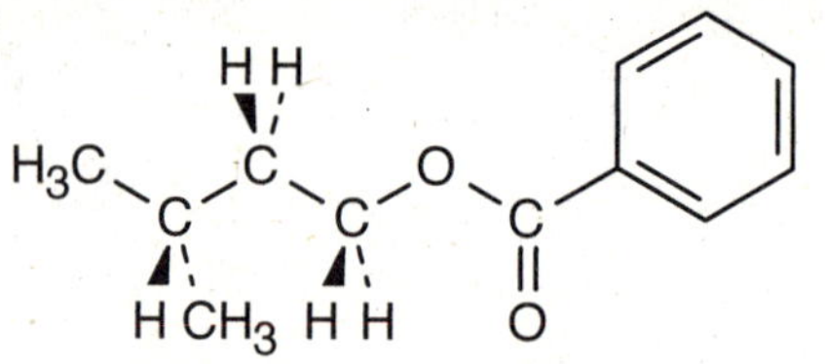

3-Methylbutyl benzoate

**10.29**

(a)

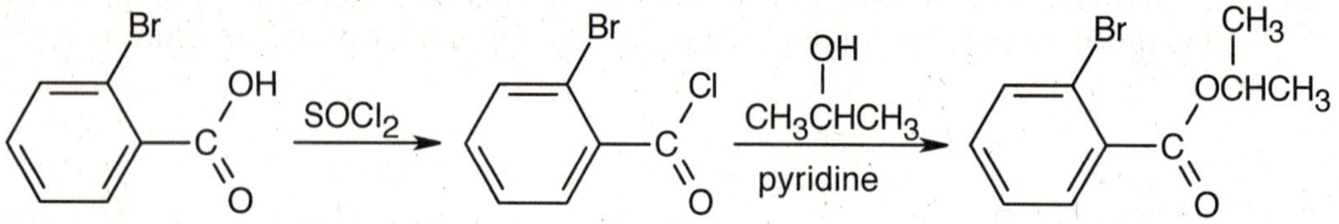

*o*-Bromobenzoic acid    Propan-2-ol    Isopropyl *o*-bromobenzoate

This compound can also be synthesized by Fischer esterification of *o*-bromobenzoic acid with propan-2-ol and an acid catalyst.

(b)

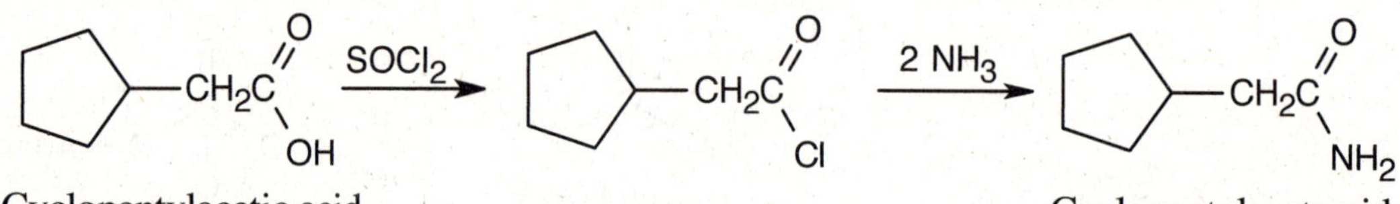

Cyclopentylacetic acid    Cyclopentylacetamide

**10.30** The structure represents the tetrahedral intermediate in the reaction of methyl cyclopentanecarboxylate with hydroxide, a nucleophile. The products are cyclopentanecarboxylate anion and methanol.

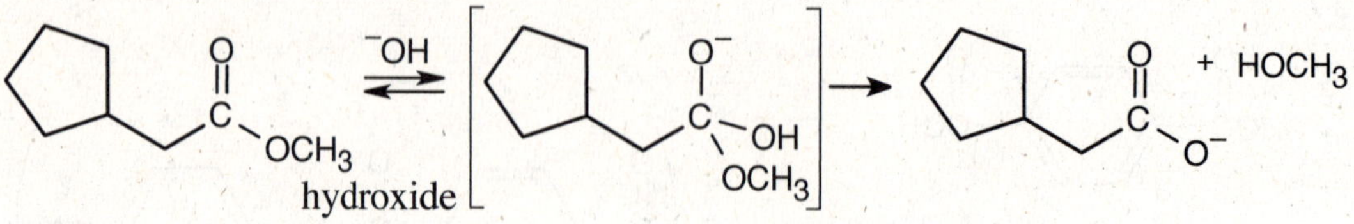

Methyl cyclopentylacetate    Cyclopentylacetate    Methanol

**10.31** According to the electrostatic potential maps, the carbonyl carbon of acetyl azide is more electron-poor (less red) and thus more reactive in nucleophilic acyl substitution reactions. Resonance donation of nitrogen lone-pair electrons to the carbonyl group is greater in an amide than in an acyl azide.

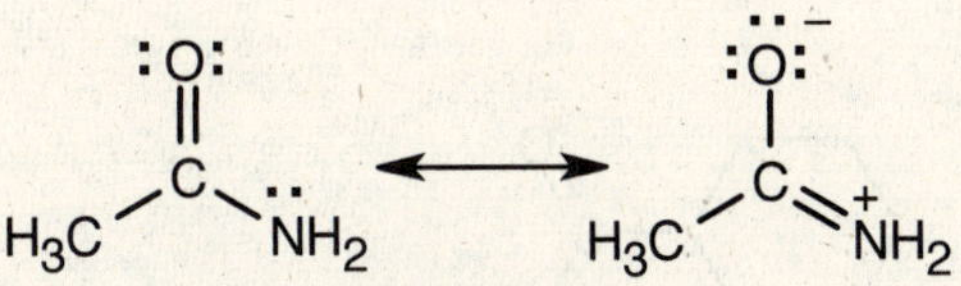

Acetamide

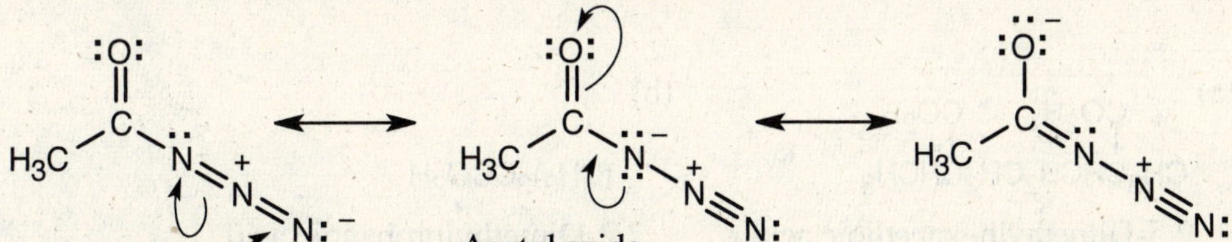

Acetyl azide

**10.32**

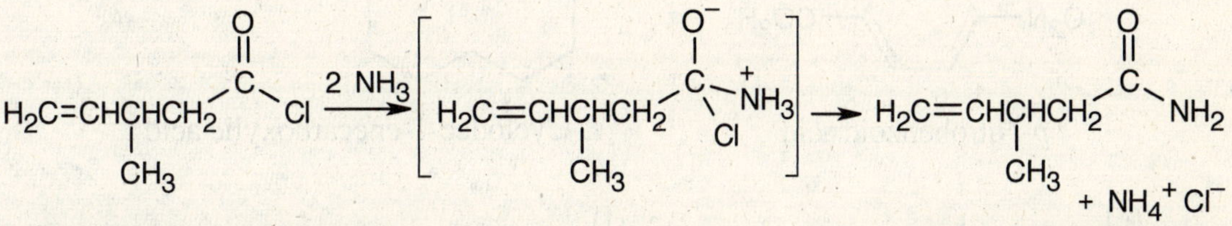

3-Methylpent-4-enoyl chloride  3-Methylpent-4-enamide

The starting material is 3-methylpent-4-enoyl chloride. Ammonia adds to give the observed tetrahedral intermediate, which eliminates $Cl^-$ to yield the above amide.

## Additional Problems

### Naming Carboxylic Acids and Derivatives

**10.33**

(a)

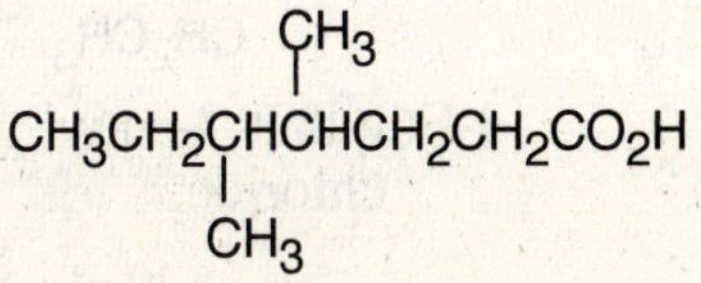

4,5-Dimethylheptanoic acid

(b)

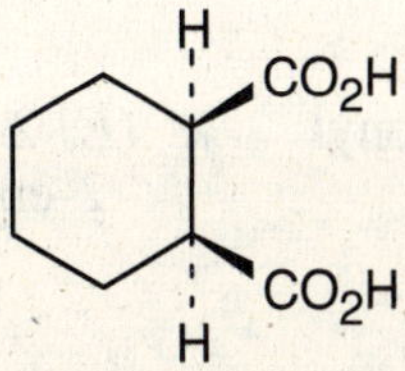

*cis*-Cyclohexane-1,2-dicarboxylic acid

(c) $HO_2CCH_2CH_2CH_2CH_2CH_2CO_2H$

Heptanedioic acid

(d) $(C_6H_5)_3CCO_2H$

Triphenylacetic acid

(e)

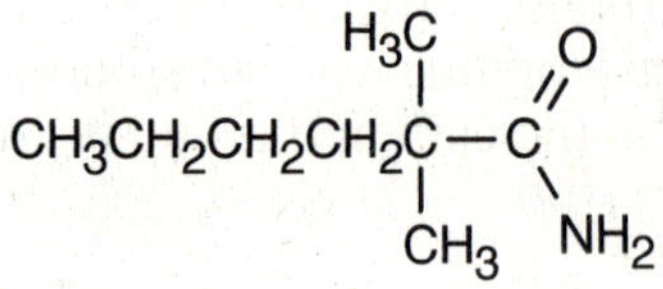

2,2-Dimethylhexanamide

(f)

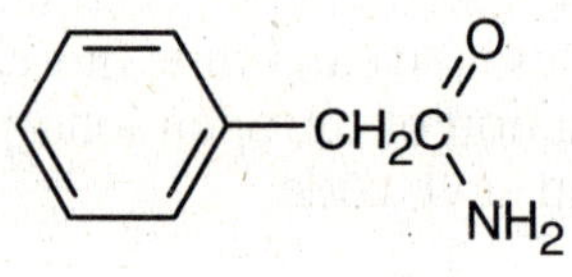

Phenylacetamide

(g)

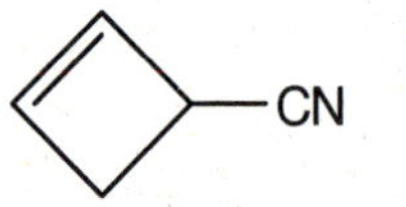

Cyclobut-2-enecarbonitrile

(h)

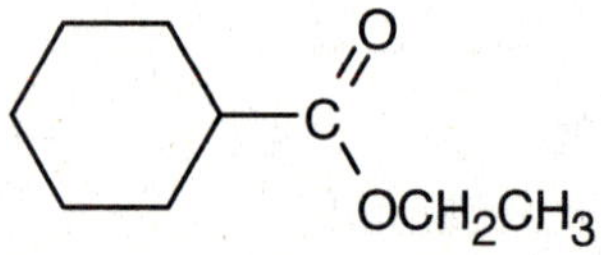

Ethyl cyclohexanecarboxylate

**10.34**

(a)

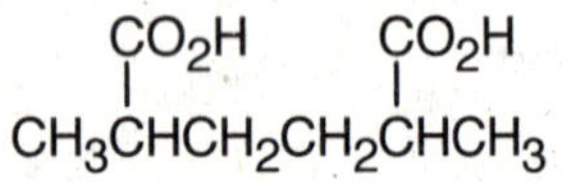

2,5-Dimethylhexanedioic acid

(b)

$(CH_3)_3CCO_2H$

2,2-Dimethylpropanoic acid

(c)

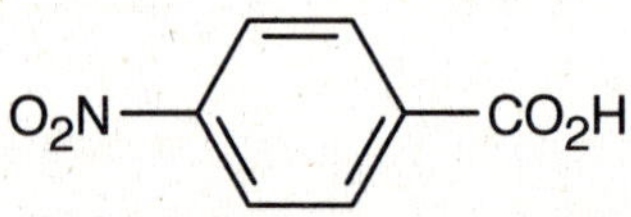

*p*-Nitrobenzoic acid

(d)

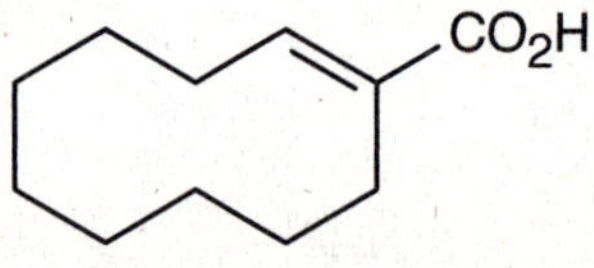

*E*-Cyclodec-1-enecarboxylic acid

(e)

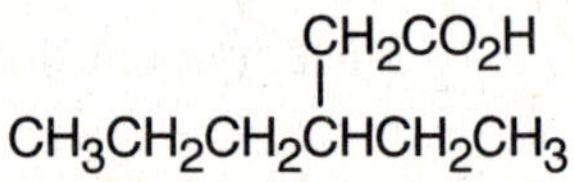

3-Ethylhexanoic acid

(f)

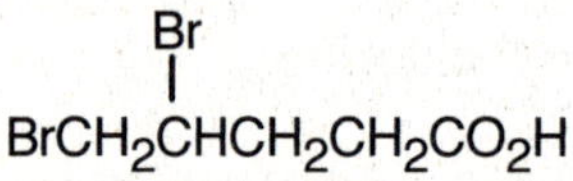

4,5-Dibromopentanoic acid

**10.35** Many other compounds having these formulas can be drawn.

(a)

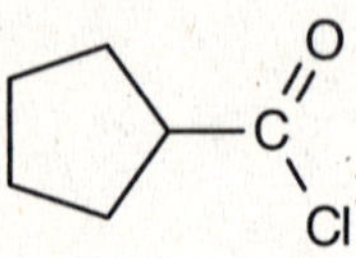

Cyclopentanecarbonyl chloride

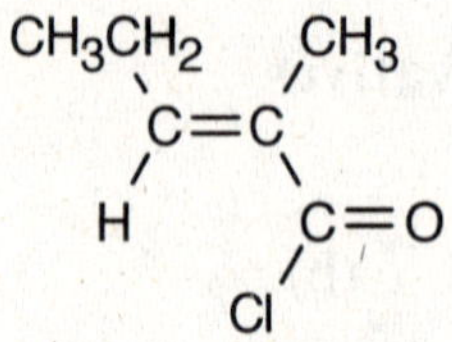

*(E)*-2-Methylpent-2-enoyl chloride

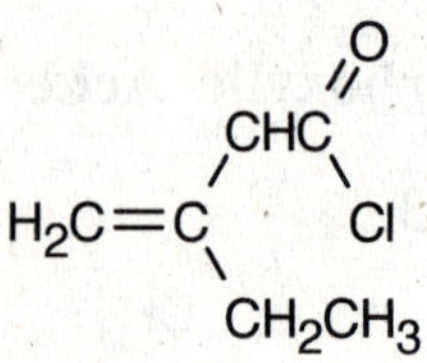

3-Ethylbut-3-enoyl chloride

(b)

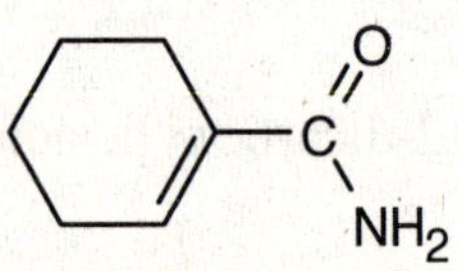

Cyclohex-1-enecarboxamide

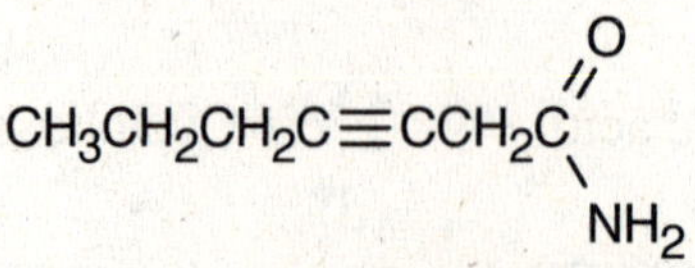

Hept-3-ynamide

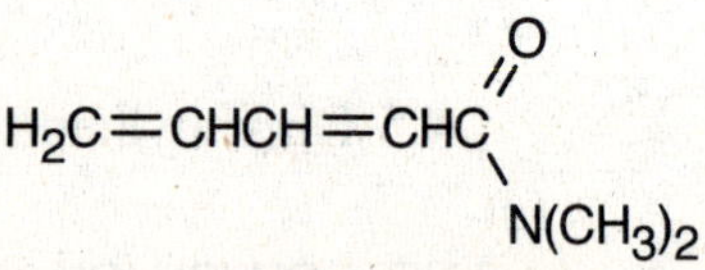

*N,N*-Dimethylpenta-2,4-dienamide

(c)

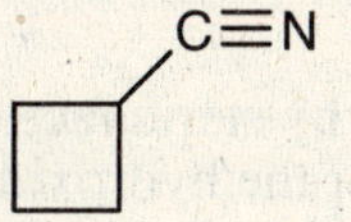

Cyclobutanecarbonitrile

$CH_3CH{=}CHCH_2C{\equiv}N$

Pent-3-enenitrile

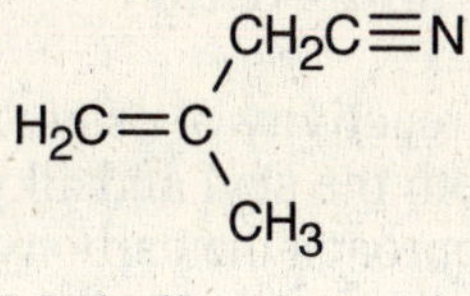

3-Methylbut-3-enenitrile

(d)

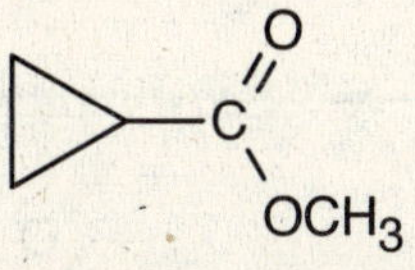

Methyl cyclopropane-carboxylate

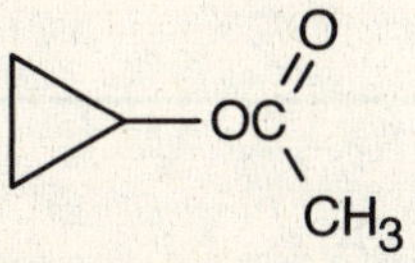

Cyclopropyl acetate

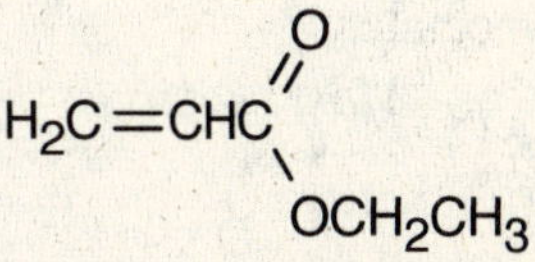

Ethyl propenoate

**10.36**

(a)

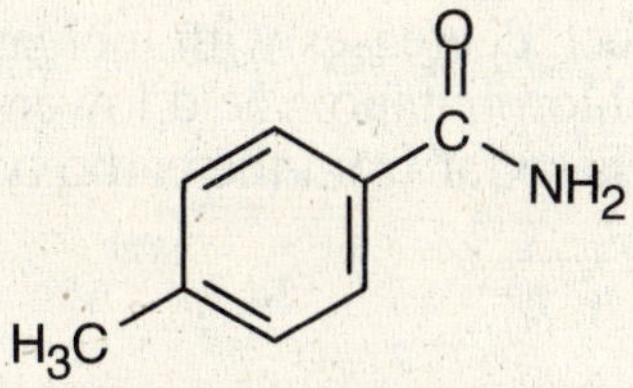

*p*-Methylbenzamide

(b)

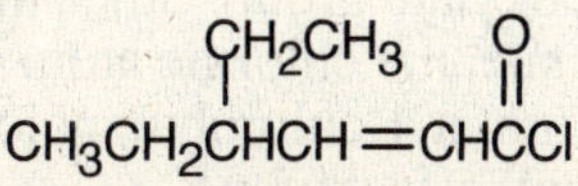

4-Ethyl-2-hexenoyl chloride

(c)

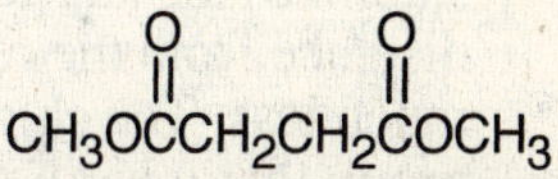

Dimethyl butanedioate
*or*
Dimethyl succinate

(d)

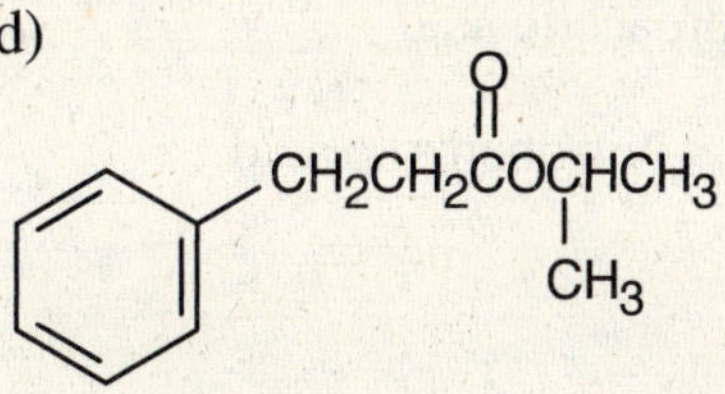

Isopropyl 3-phenylpropanoate

(e)

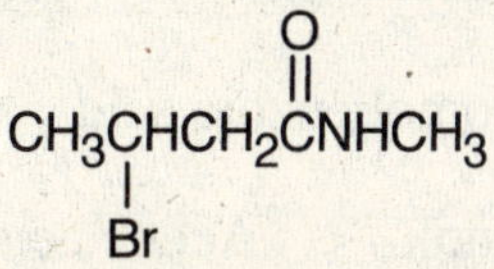

3-Bromo-*N*-methylbutanamide

(f)

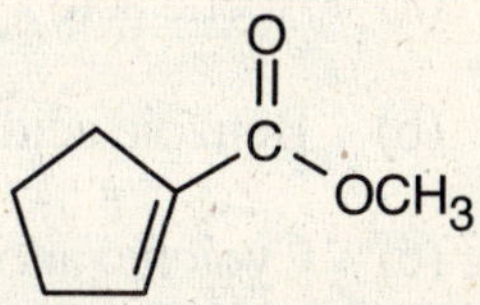

Methyl cyclopent-1-ene-carboxylate

**10.37**

$CH_3CH_2CH_2CH_2CH_2CO_2H$

Hexanoic acid

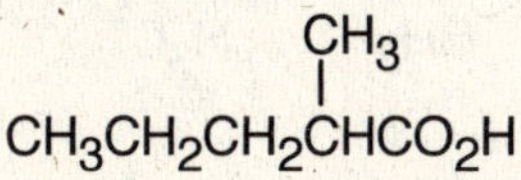

2-Methylpentanoic acid

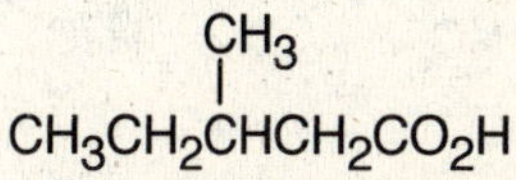

3-Methylpentanoic acid

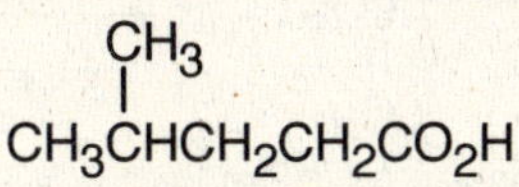

4-Methylpentanoic acid

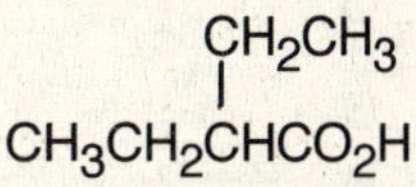

2-Ethylbutanoic acid

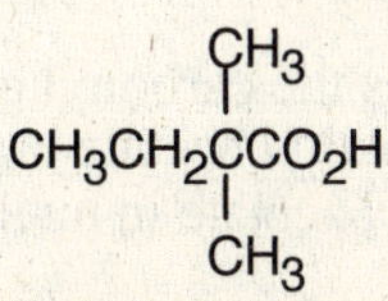

2,2-Dimethylbutanoic acid

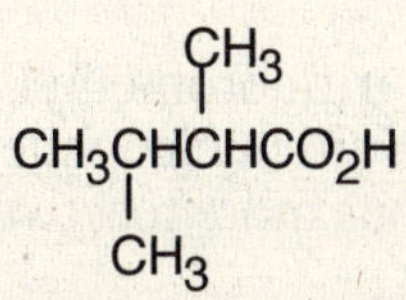

2,3-Dimethylbutanoic acid

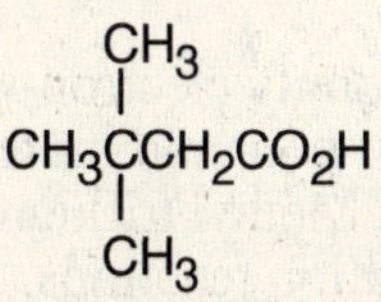

3,3-Dimethylbutanoic acid

2-Methylpentanoic acid, 3-methylpentanoic acid and 2,3-dimethylbutanoic acid are chiral.

## Reactivity and Acidity

**10.38** The reactivity of esters in saponification reactions is influenced by steric factors. Branching in both the acyl and alkyl portions of an ester makes it harder for the hydroxide nucleophile to approach the carbonyl carbon. This effect is less important in the alkyl portion of the ester than in the acyl portion (because alkyl branching is one atom farther away from the site of attack), but it is still measurable. The reactivity order for saponification of alkyl acetates:

*Most reactive* ⟶ *Least reactive*

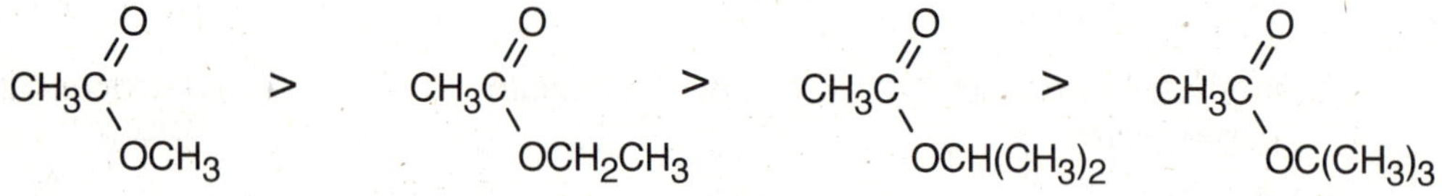

**10.39** The p$K_a$ of 2-chlorobutanoic acid is lowered by the electronegative chlorine atom, which stabilizes the adjacent carboxylate group. Since this effect decreases with increasing distance from the reaction site, the chlorine atom in 3-chlorobutanoic acid lowers the p$K_a$ much less. The chlorine atom in 4-chlorobutanoic acid is so far from the carboxylic acid group that it has a very small effect on p$K_a$.

**10.40**

*Least acidic* ⟶ *Most acidic*

(a) Acetic acid < Chloroacetic acid < Trifluoroacetic acid

(b) Benzoic acid < *p*-Bromobenzoic acid < *p*-Nitrobenzoic acid

(c) Cyclohexanol < Phenol < Acetic acid

**10.41**

*Most reactive* ⟶ *Least reactive*

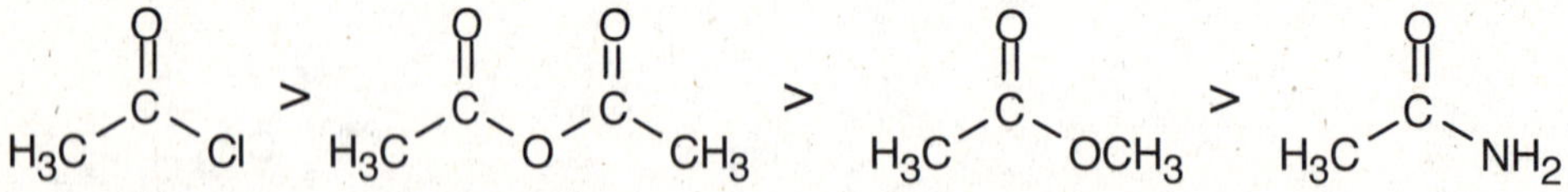

**10.42**

The strongly electron-withdrawing trifluoromethyl group makes the carbonyl carbon more electron-poor and more reactive toward nucleophiles than the methyl acetate carbonyl carbon. Methyl trifluoroacetate is thus more reactive than methyl acetate in nucleophilic acyl substitution reactions.

**10.43** The lower the p$K_a$, the stronger the acid. Thus, tartaric acid (p$K_a$ = 2.98) is a stronger acid than citric acid (p$K_a$ = 3.14).

## Reactions

### 10.44

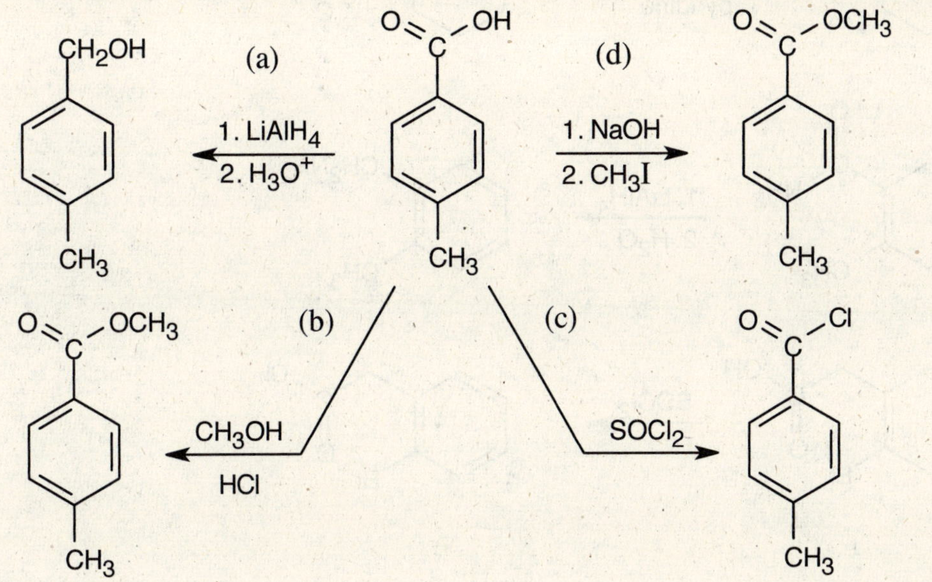


### 10.45

(a)

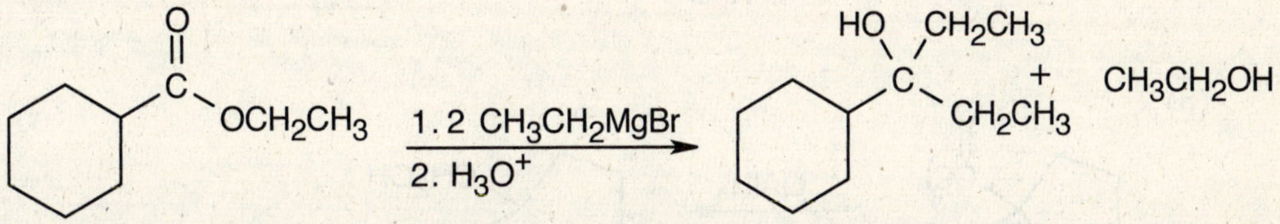


(b)

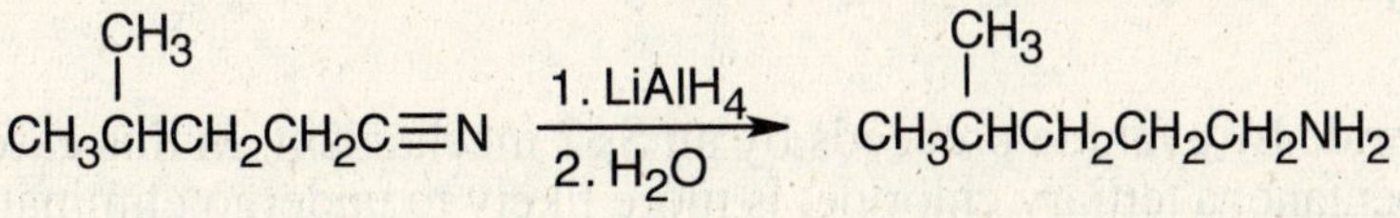


(c)

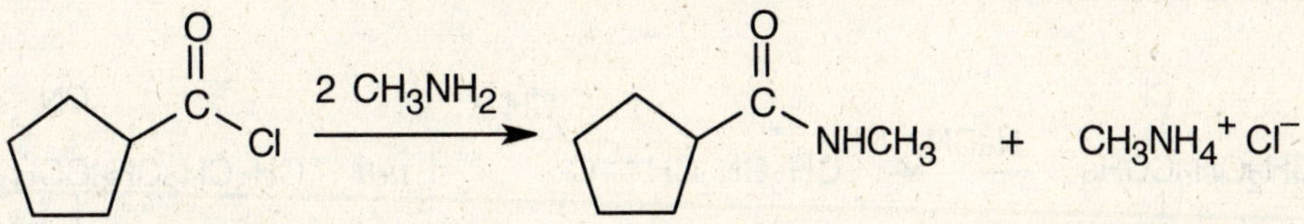


(d)

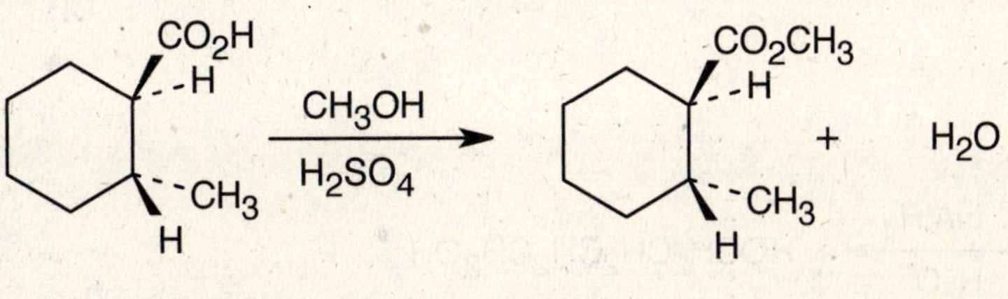


(e)

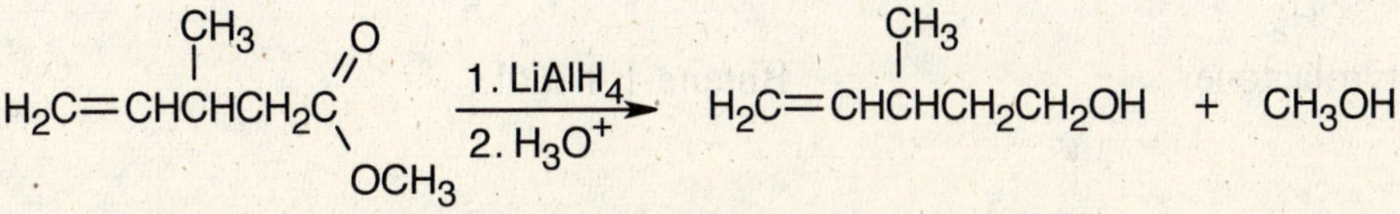

(f)

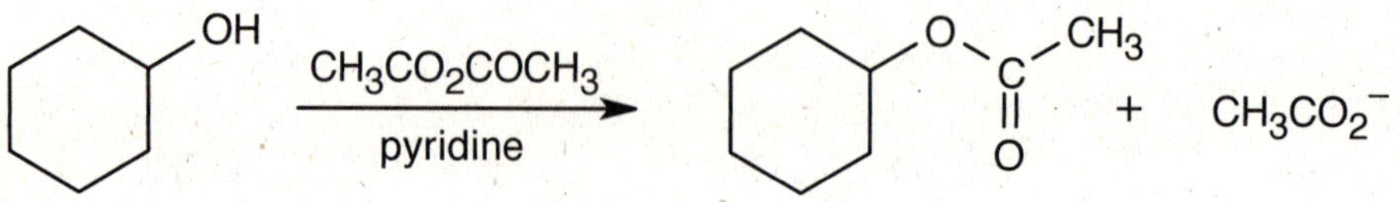

(g)

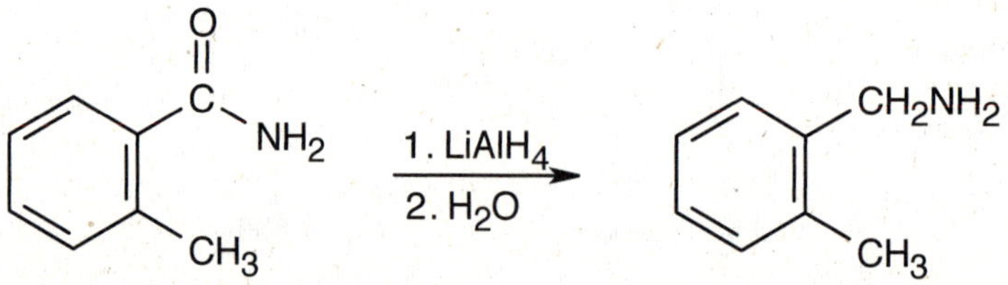

(h)

**10.46**

(a)

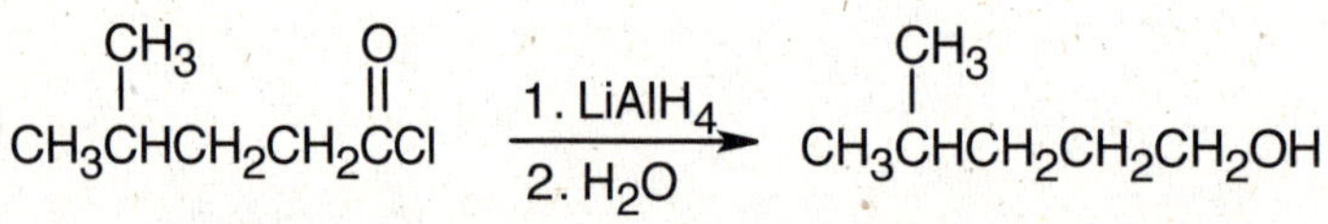

(b)

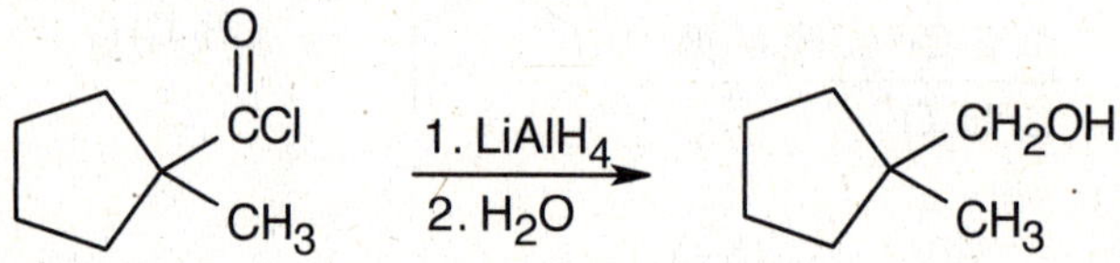

**10.47** Substitution of chloride by cyanide proceeds by an $S_N2$ mechanism. In this case, however, 2-chloro-2-methylpentane, a tertiary chloride, is more likely to undergo elimination to yield 2-methyl-2-pentene.

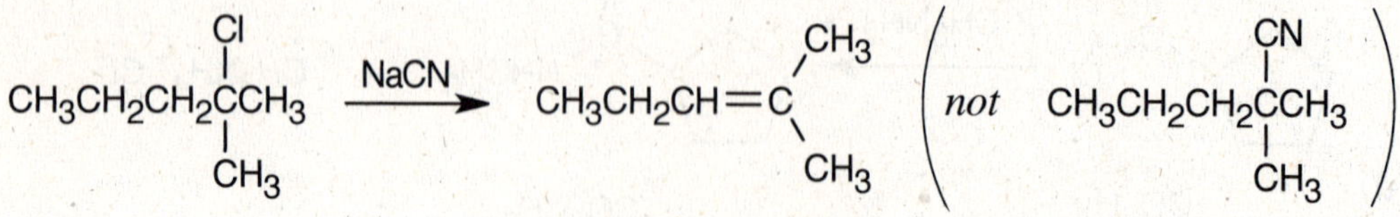

**10.48**

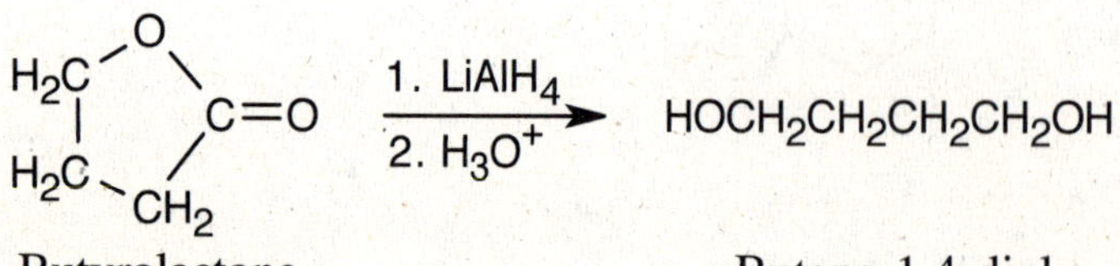

Butyrolactone Butane-1,4-diol

**10.49**

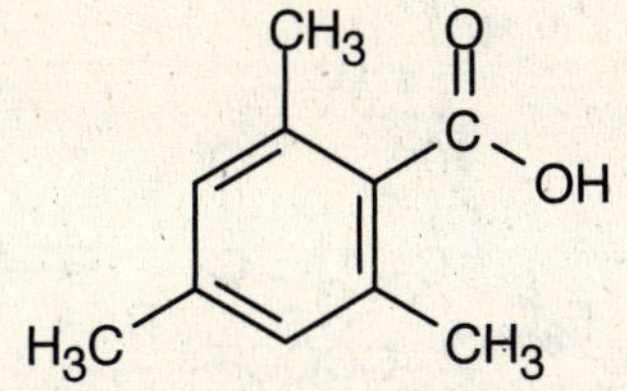

2,4,6-Trimethylbenzoic acid has two methyl groups ortho to the carboxylic acid functional group. These bulky methyl groups block the approach of the alcohol and prevent esterification from occurring under Fischer esterification conditions.

**10.50**

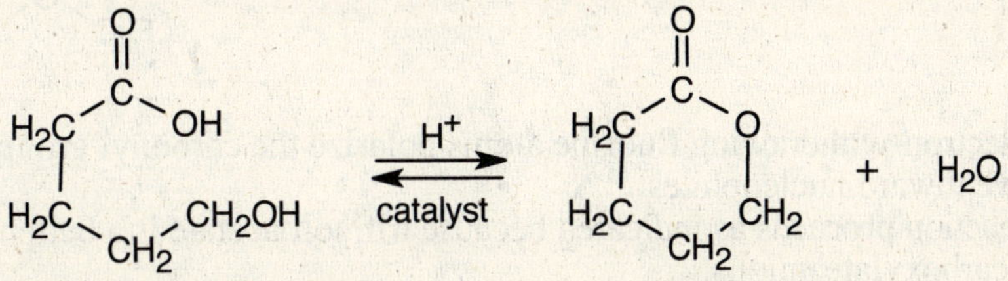

5-Hydroxypentanoic acid A *lactone*

## Mechanisms

**10.51**

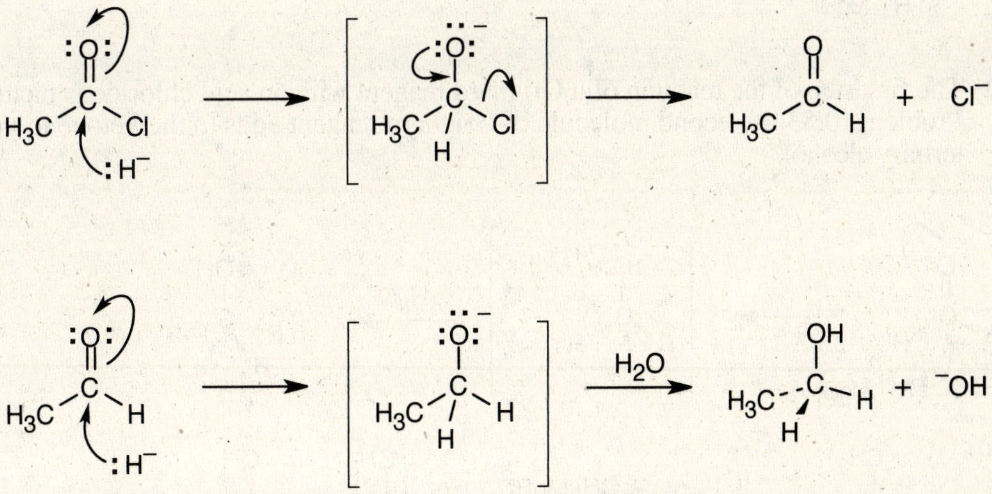

**10.52**

(a)

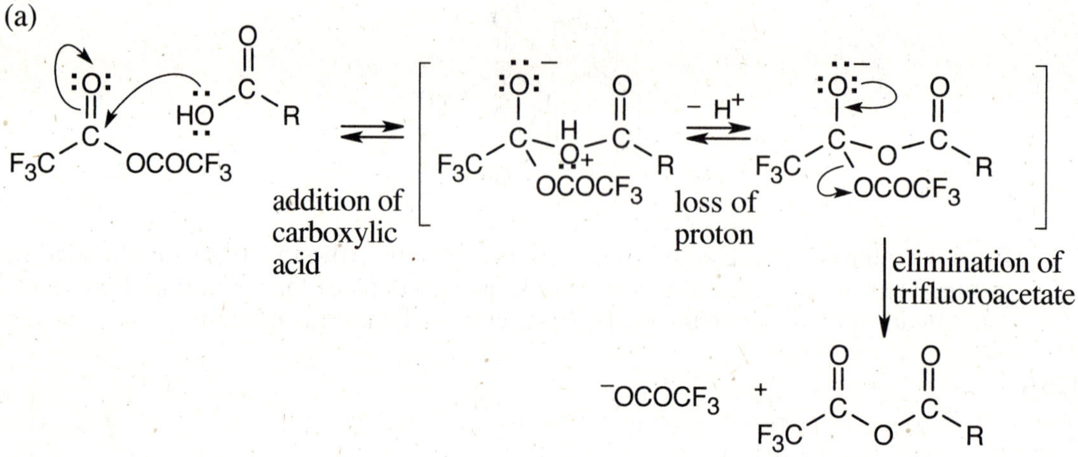


(b) The electron-withdrawing fluorine atoms polarize the carbonyl group, making it more reactive toward nucleophiles.

(c) The reaction proceeds as indicated because trifluoroacetate is a better leaving group than other carboxylate anions.

**10.53**

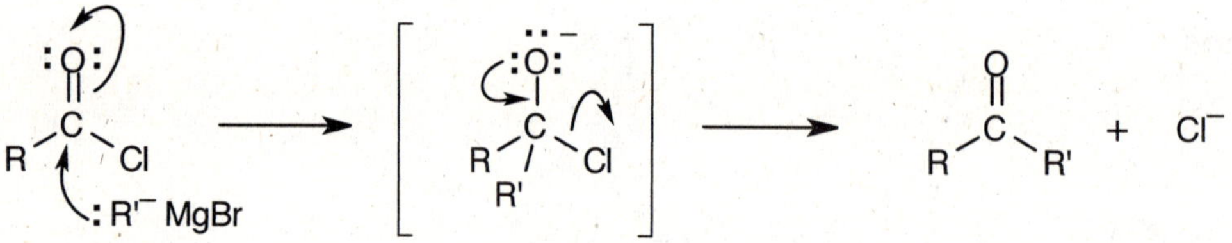

**10.54**(a) The first step of the reaction of a Grignard reagent with an acid chloride is pictured in Problem 10.53. A second molecule of Grignard reagent adds to the ketone to yield a tertiary alcohol.

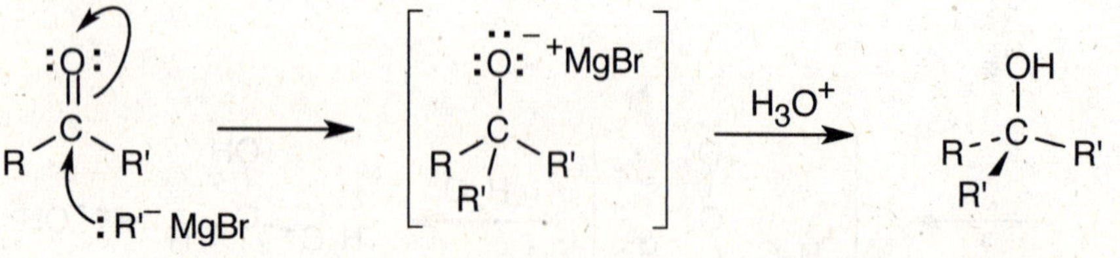

(b)

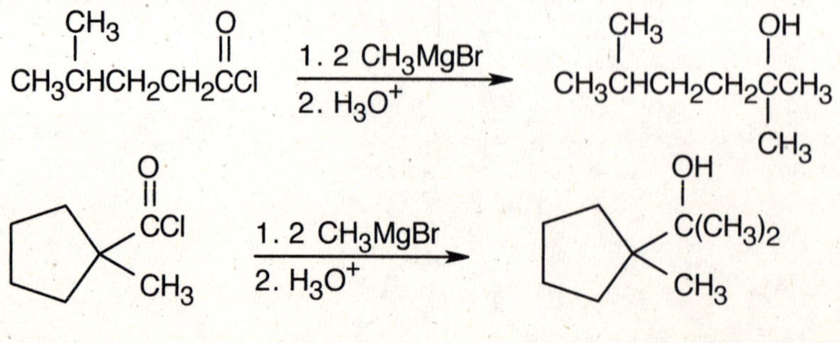

## Synthesis

### 10.55

(a)

$$CH_3CH_2CH_2CO_2H \xrightarrow[2.\ H_3O^+]{1.\ LiAlH_4} CH_3CH_2CH_2CH_2OH$$

(b)

$$\underset{\text{from (a)}}{CH_3CH_2CH_2CH_2OH} \xrightarrow{\text{Periodinane}} CH_3CH_2CH_2CHO$$

(c)

$$\underset{\text{from (a)}}{CH_3CH_2CH_2CH_2OH} \xrightarrow{PBr_3} CH_3CH_2CH_2CH_2Br$$

(d)

$$\underset{\text{from (c)}}{CH_3CH_2CH_2CH_2Br} \xrightarrow{NaCN} CH_3CH_2CH_2CH_2CN$$

(e)

$$\underset{\text{from (c)}}{CH_3CH_2CH_2CH_2Br} \xrightarrow[\text{ethanol}]{KOH} CH_3CH_2CH{=}CH_2$$

(f)

$$CH_3CH_2CH_2CO_2H \xrightarrow{SOCl_2} CH_3CH_2CH_2COCl \xrightarrow{2\ NH_3} CH_3CH_2CH_2CONH_2$$

$$CH_3CH_2CH_2CONH_2 \xrightarrow[2.\ H_2O]{1.\ LiAlH_4} CH_3CH_2CH_2CH_2NH_2$$

### 10.56

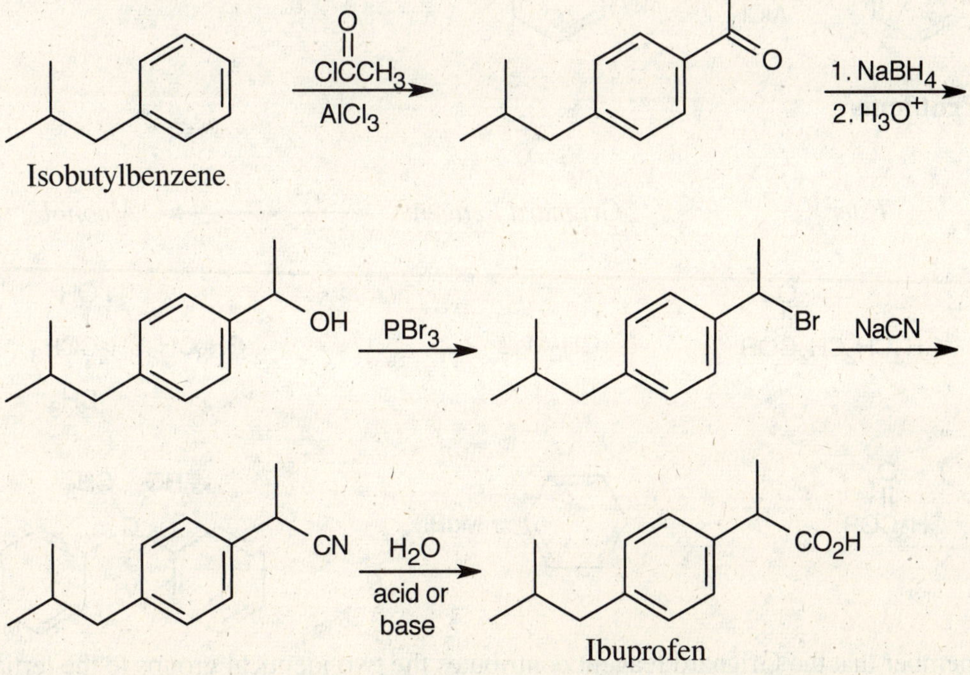

**10.57**

(a)

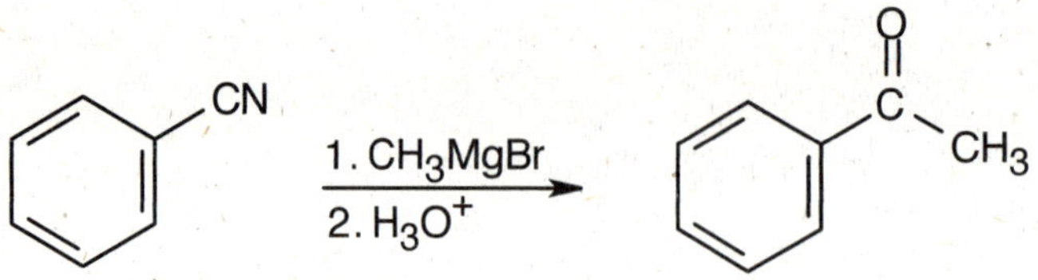

(b)

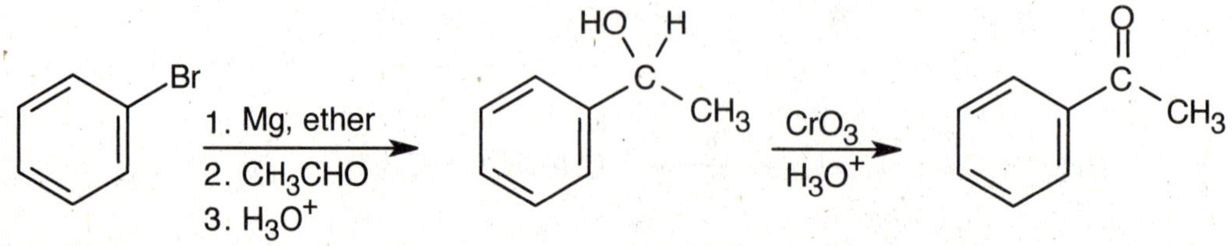

(c)

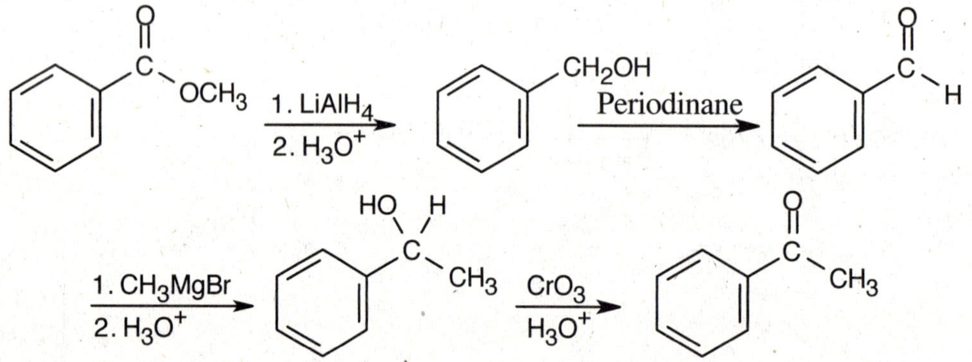

(d)

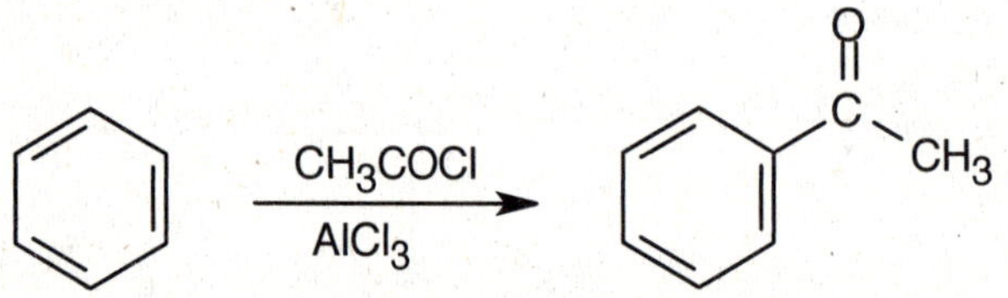

## General Problems

**10.58**

| | *Ester* | + | *Grignard Reagent* | → | *Alcohol* |
|---|---|---|---|---|---|
| (a) | $CH_3CH_2CH_2\overset{O}{\overset{\|\|}{C}}OR$ | | $CH_3MgBr$ | | $CH_3CH_2CH_2C(OH)(CH_3)CH_3$ |
| (b) | $CH_3\overset{O}{\overset{\|\|}{C}}OR$ | | 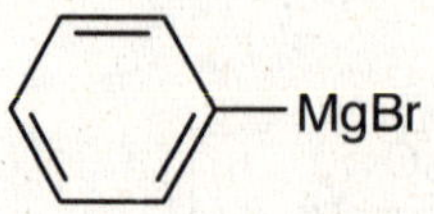 | | 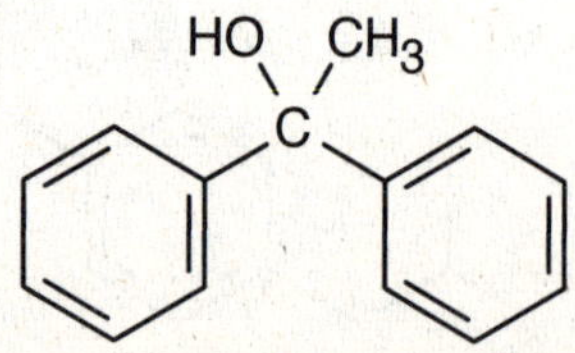 |

Remember that the Grignard reagent contributes the two identical groups to the tertiary alcohol.

**10.59** Dimethyl carbonate is a diester. Use your knowledge of the Grignard reaction to work your way through this problem.

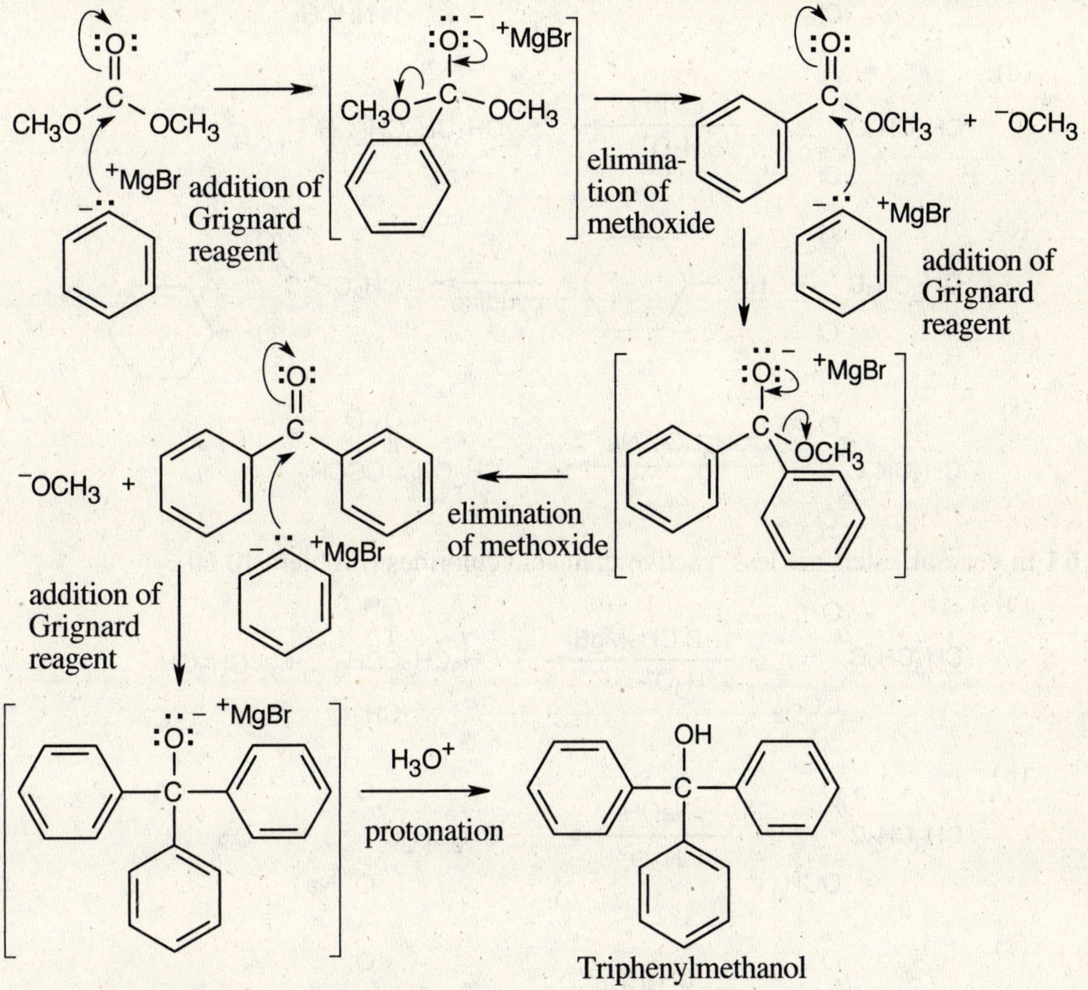


The overall reaction consists of three additions of phenylmagnesium bromide, two eliminations of methoxide and one protonation.

**10.60**

(a)

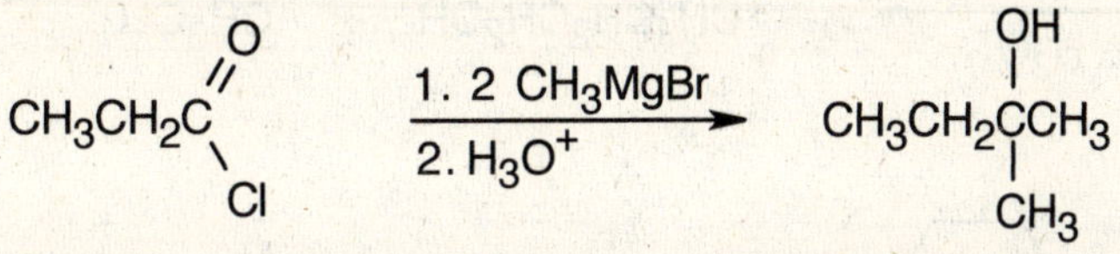


(b)

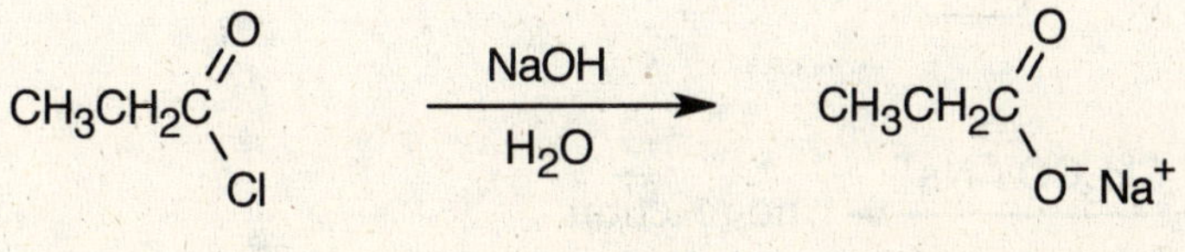

(c)

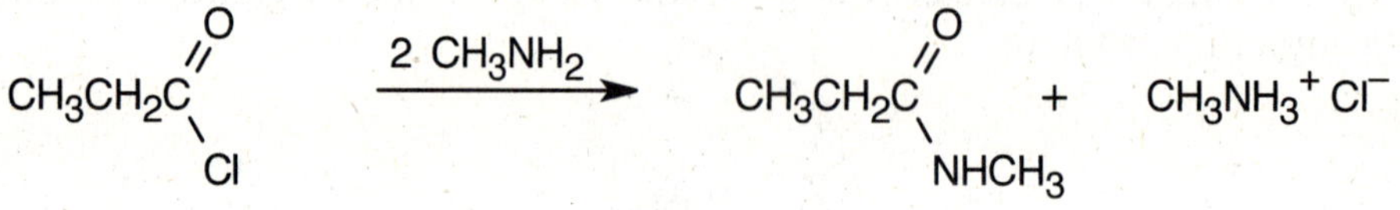

(d)

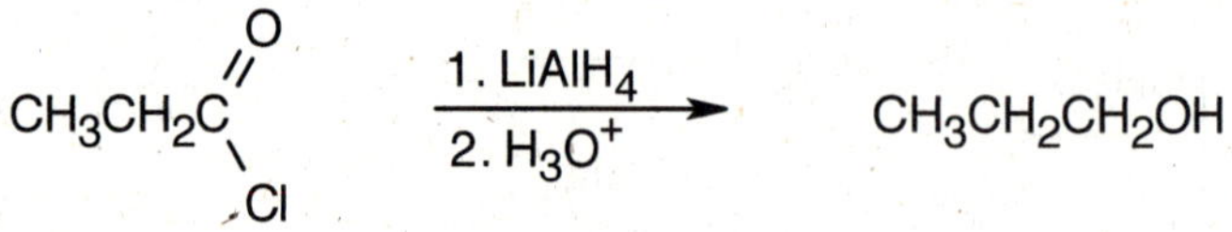

(e)

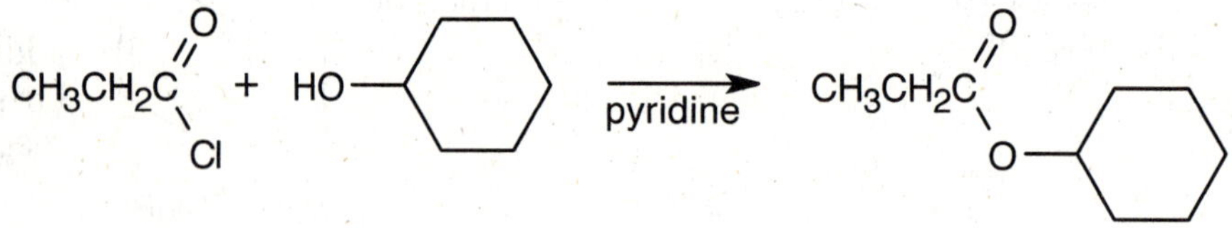

(f)

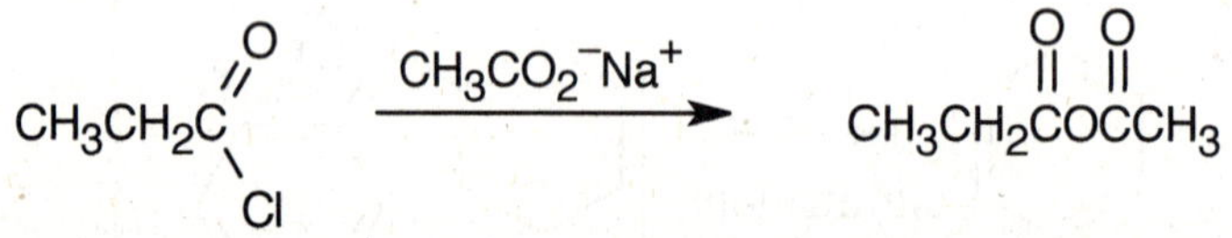

**10.61** In general, esters are less reactive than acid chlorides (Problem 10.60).

(a)

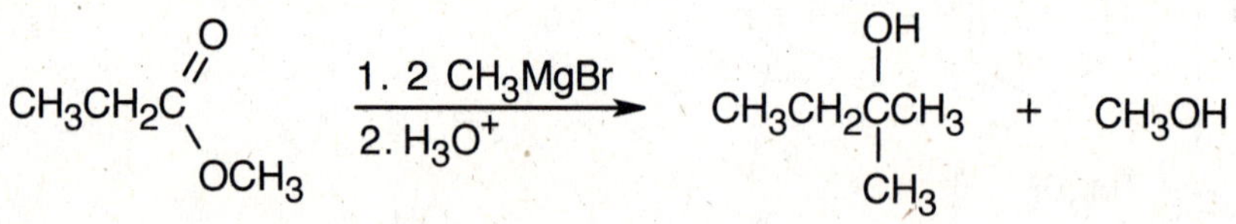

(b)

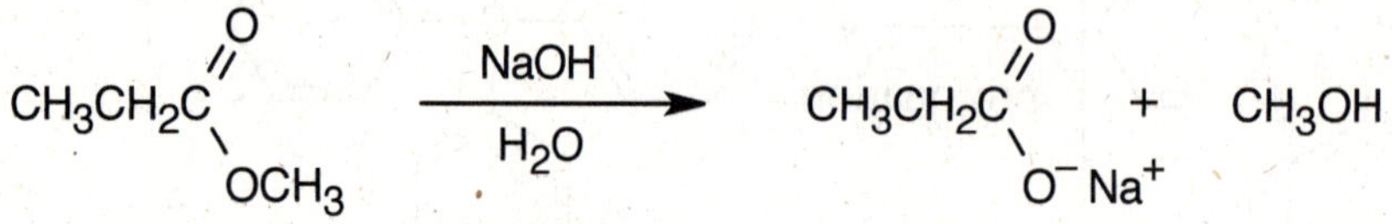

(c)

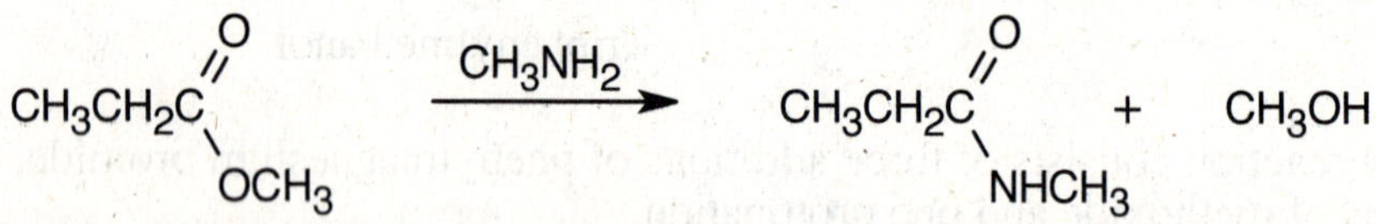

(d)

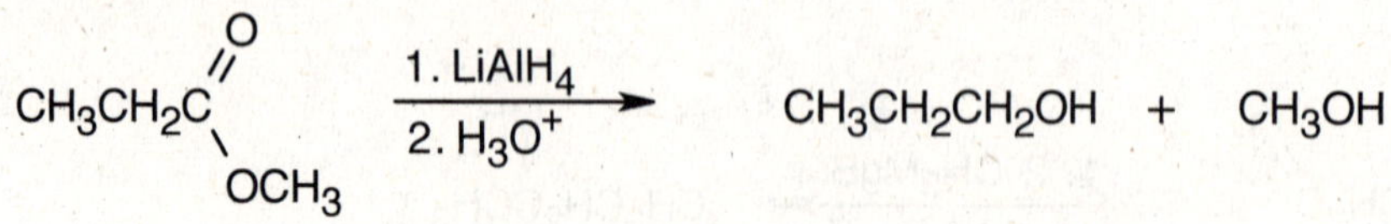

(e)

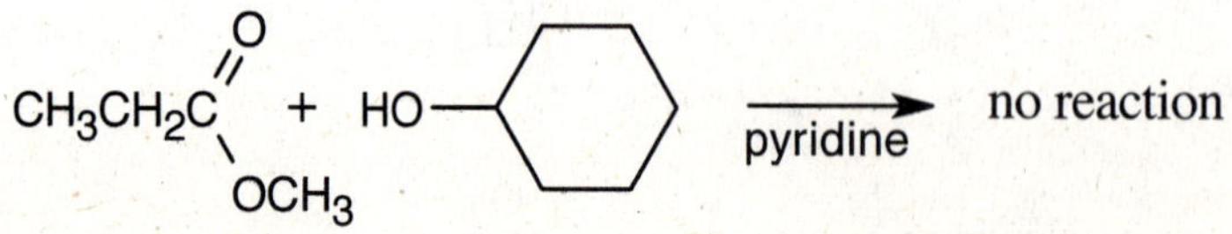

(f)

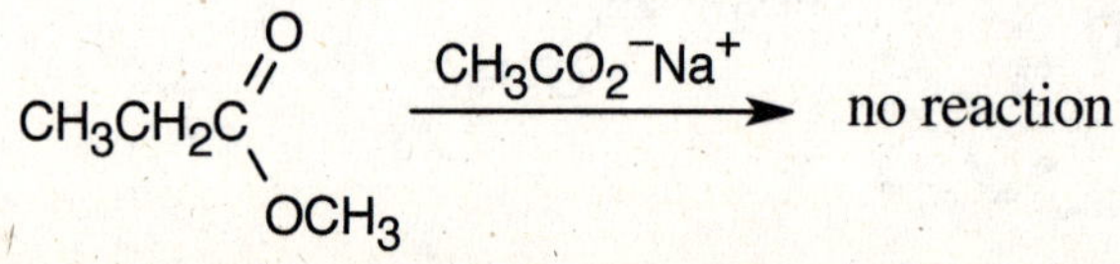

**10.62**

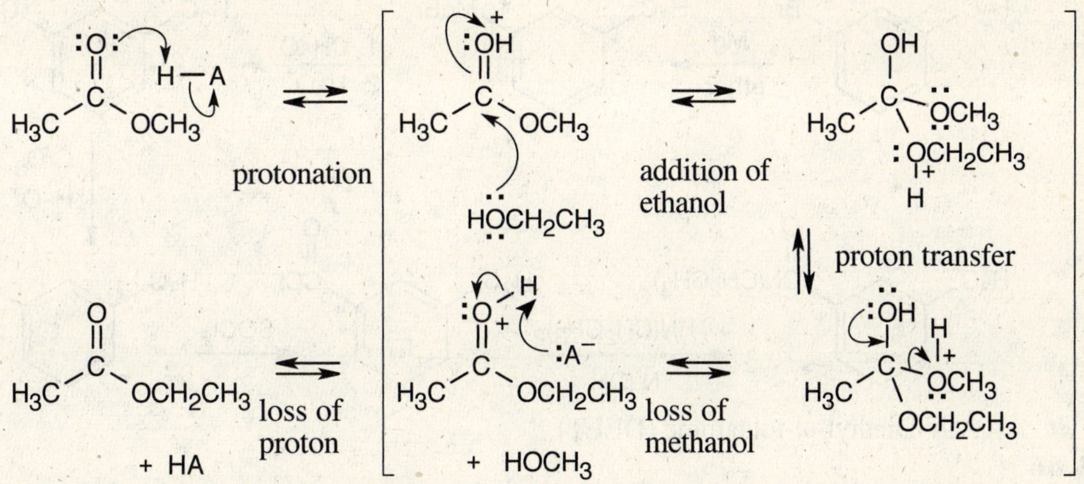


The above mechanism is very similar to the mechanism of Fischer esterification, illustrated in Section 10.9. Conversion of the methyl ester to the ethyl ester occurs because the large excess of the solvent, ethanol, shifts the equilibrium to favor formation of the ethyl ester.

**10.63**

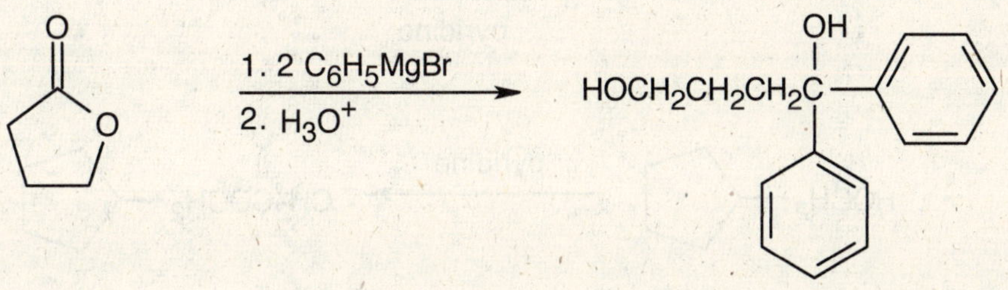


**10.64**

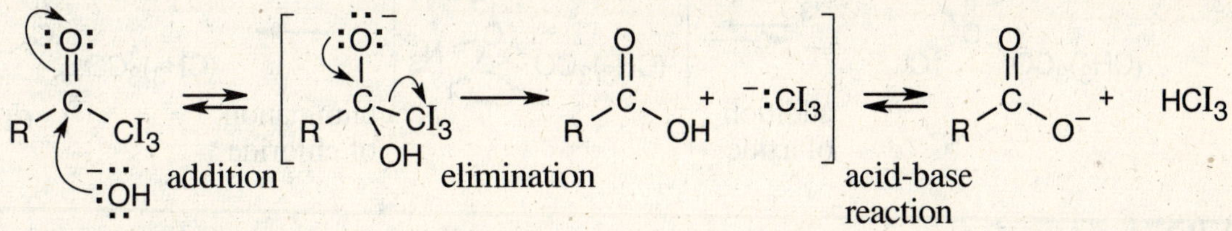


$I_3C:^-$ can act as a leaving group because the electron-withdrawing iodine atoms stabilize the carbanion.

**10.65**

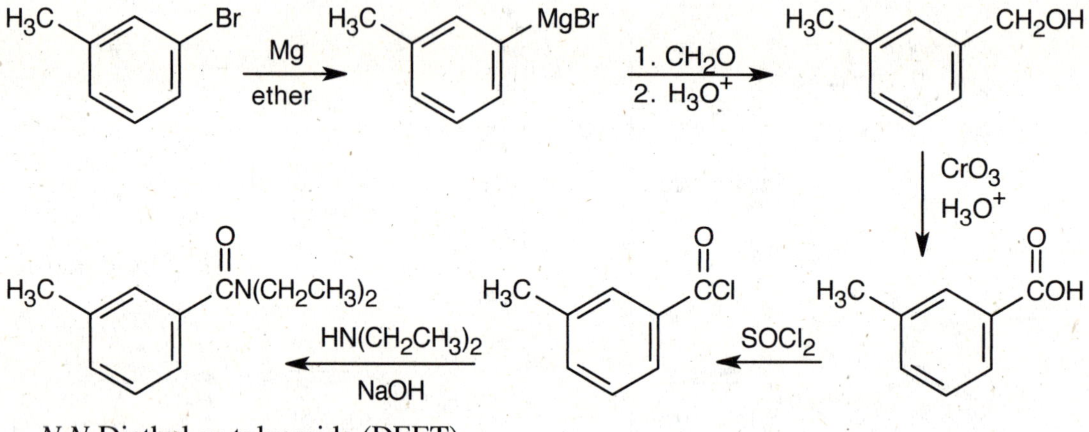

*N,N*-Diethyl-*m*-toluamide (DEET)

**10.66**

(a)

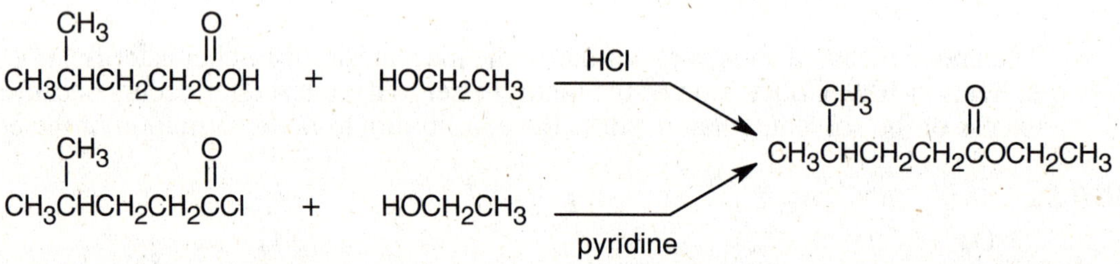

(b)

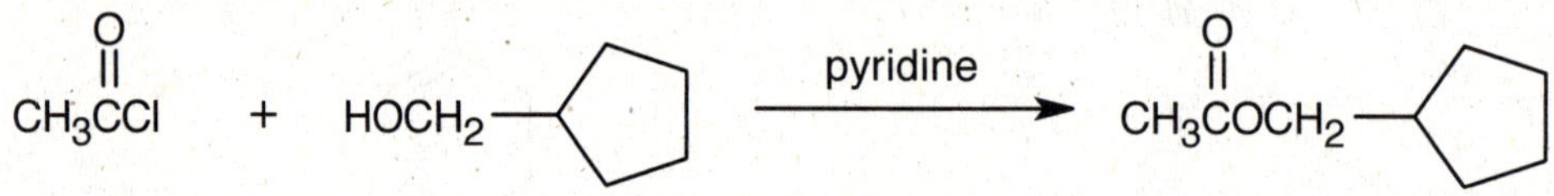

**10.67**

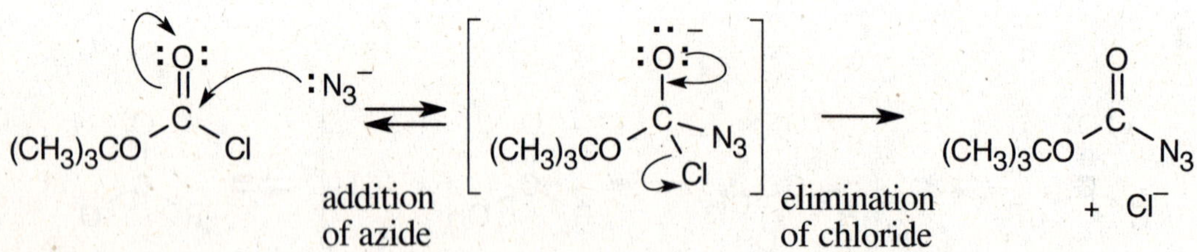

**10.68**

(a)

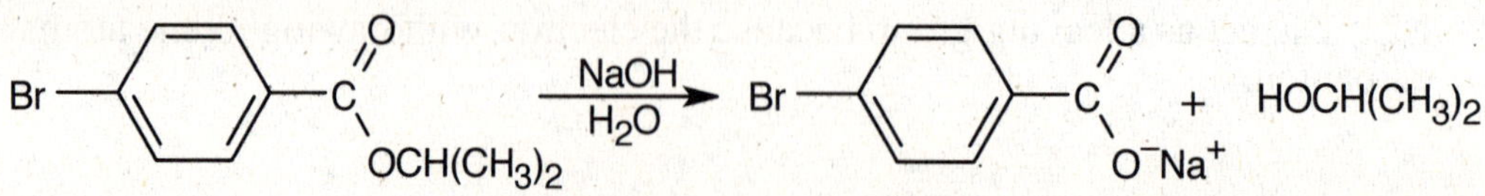

(b)

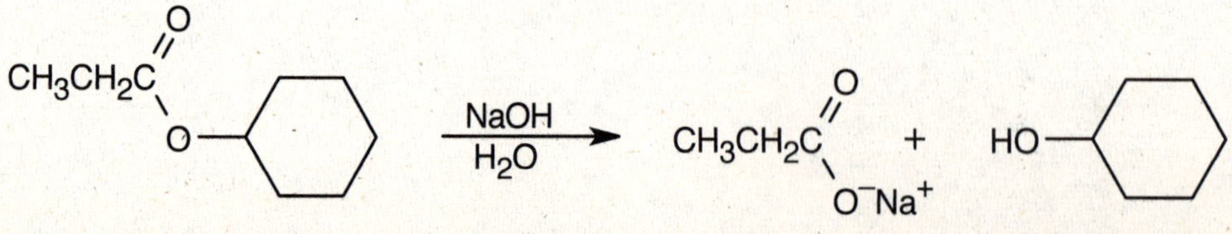

**10.69**

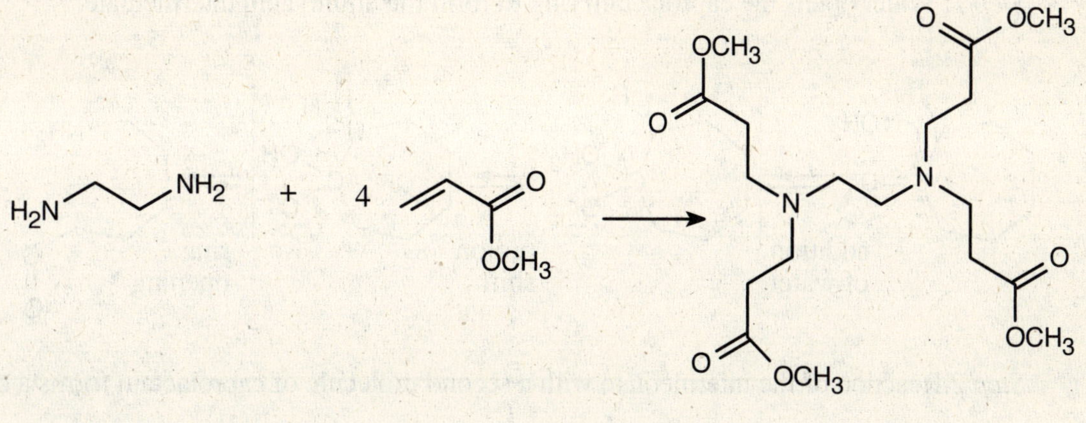


(a) Conjugate addition of the diamine to 4 equivalents of methyl acrylate forms the tetraester.

(b)

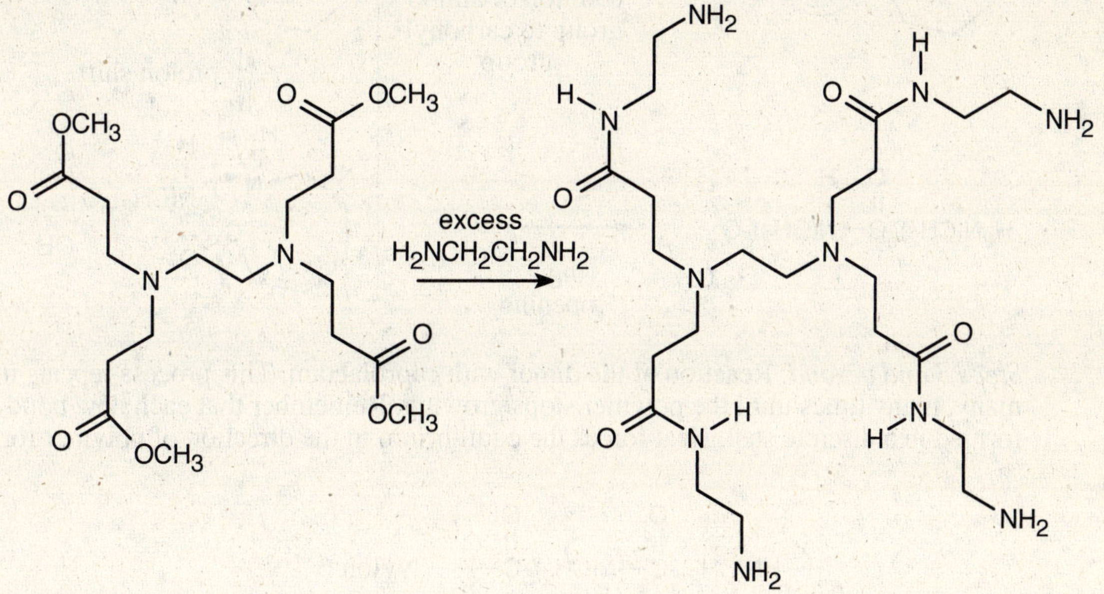


The product results from nucleophilic acyl substitution reactions at the four ester groups to form a tetraamide.

**10.70**

*Step 1*: Water opens the caprolactam ring to form the amino acid intermediate.

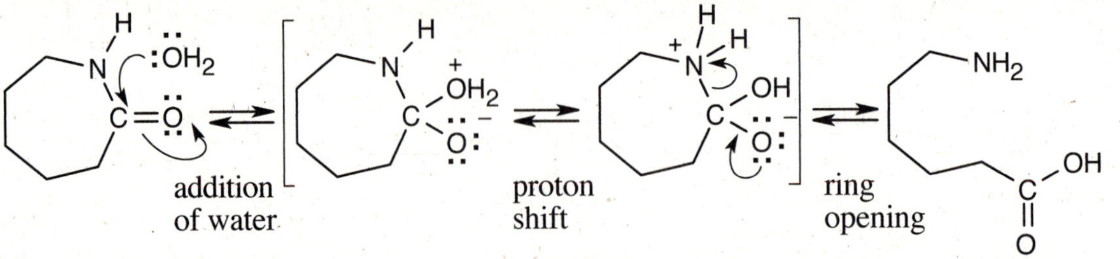


*Step 2*:Reaction of the intermediate with a second molecule of caprolactam forms a dimer.

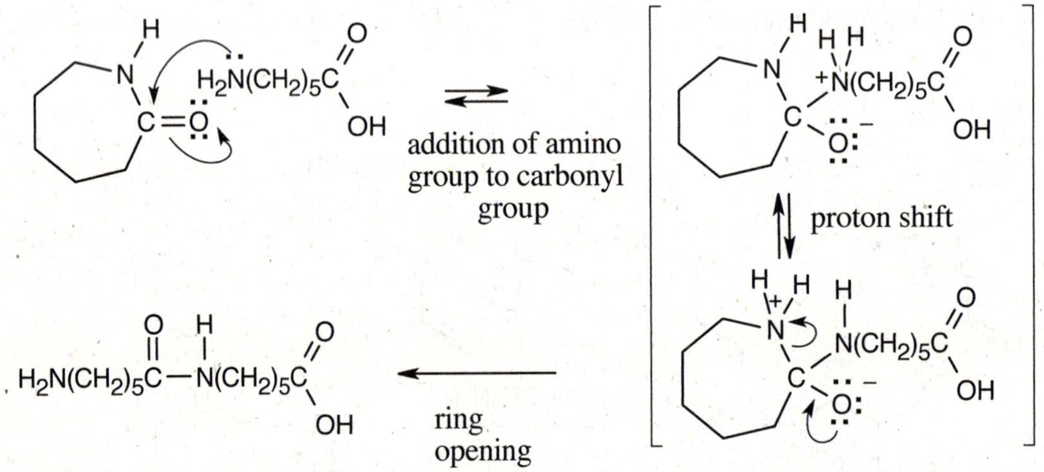


*Steps 3 and beyond*: Reaction of the dimer with caprolactam. This process repeats itself many, many times until the polymer stops growing. Remember that each new bond is formed in a discrete step. Heat forces the equilibrium in the direction of polymer formation.

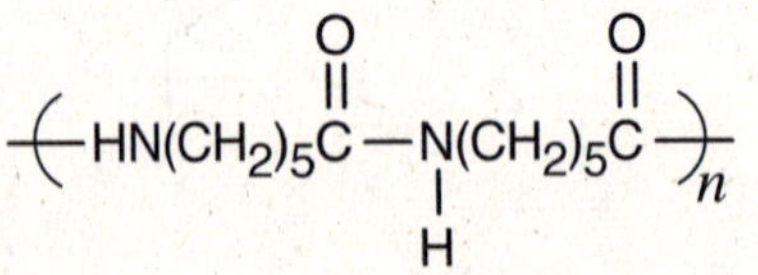 Nylon 6

**10.71**

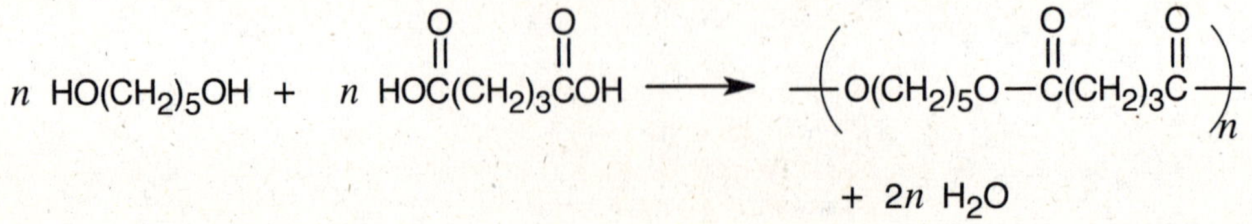

## In the Medicine Cabinet

**10.72**

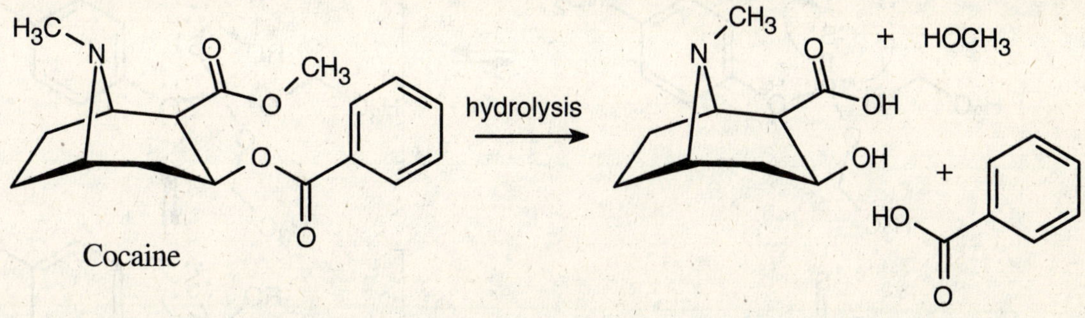


Cocaine

**10.73**

Nucleophilic acyl substitution:

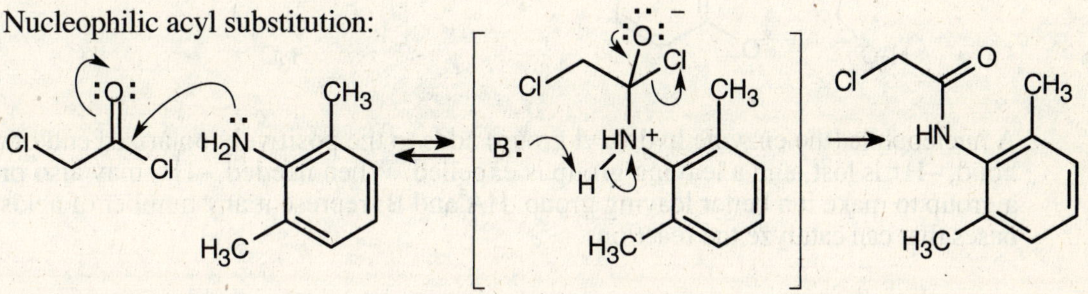


$S_N2$ Displacement

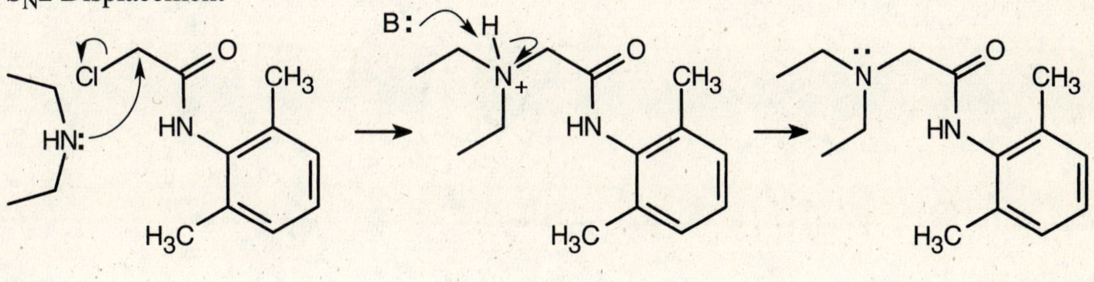


## In the Field

**10.74**

In the following reaction, –Ar represents naphthalene.

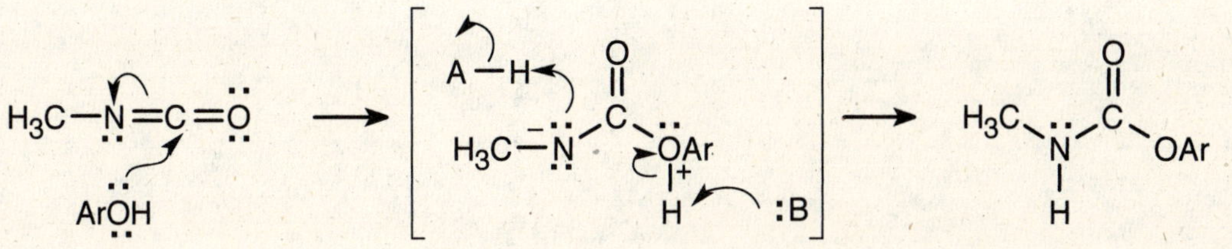

**10.75**

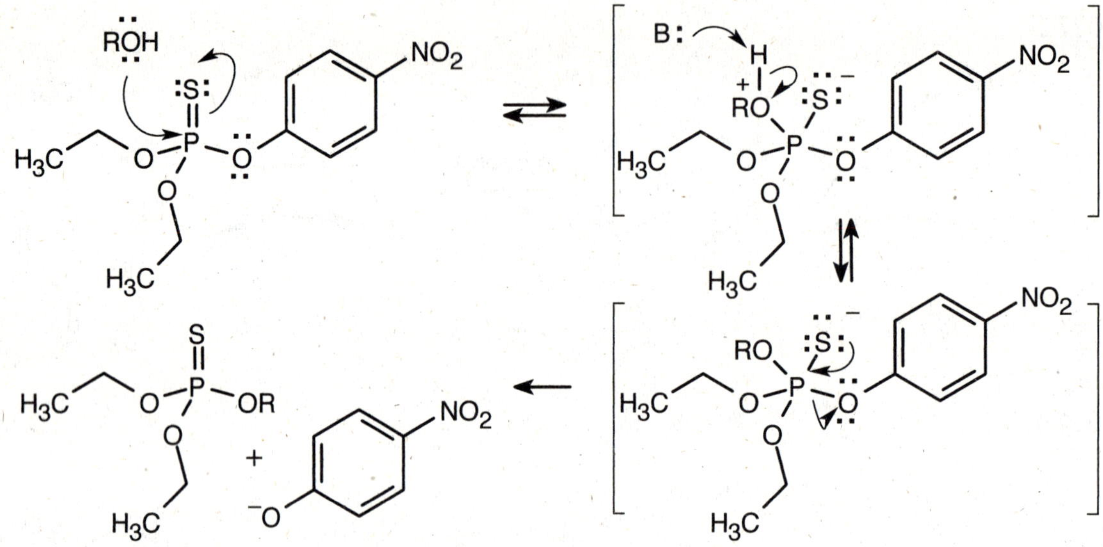


A nucleophile (the enzyme hydroxyl group) adds to the positively polarized end of a double bond, $-H^+$ is lost, and a leaving group is expelled. When needed, $-H^+$ may also protonate a group to make it a better leaving group. HA and B: represent any number of acids or bases that can catalyze the reactions.

# Chapter 11 – Carbonyl Alpha-Substitution Reactions and Condensation Reactions

## Chapter Outline

I. Keto-enol tautomerism (Section 11.1).
   A. Nature of tautomerism.
      1. Carbonyl compounds with hydrogens bonded to their α carbons equilibrate with their corresponding enols.
      2. This rapid equilibration is called tautomerism, and the individual isomers are tautomers.
      3. Unlike resonance forms, tautomers are different isomers.
      4. Despite the fact that very little of the enol isomer is present at room temperature, enols are very important because they are reactive.
   B. Mechanism of tautomerism.
      1. In acid catalysis, the carbonyl oxygen is protonated to form an intermediate that can lose a hydrogen from its α carbon to yield a neutral enol.
      2. In base-catalyzed enol formation, an acid–base reaction occurs between a base and an α hydrogen.
         a. The resultant enolate anion is protonated to yield an enol.
         b. Protonation can occur either on carbon (keto tautomer) or on oxygen (enol tautomer).
      3. Only hydrogens on the α positions of carbonyl compounds are acidic.

II. Enols and enolates (Sections 11.2–11.6).
   A. Enols (Sections 11.2–11.3).
      1. Reactivity of enols (Section 11.2).
         a. The electron-rich double bonds of enols cause them to behave as nucleophiles.
         b. The electron-donating enol –OH groups make enols more reactive than alkenes.
         c. When an enol reacts with an electrophile, the initial adduct loses a proton from oxygen to give a substituted carbonyl compound.
      2. Reactions of enols: alpha halogenation of aldehydes and ketones (Section 11.3).
         a. Aldehydes and ketones can be halogenated at their α positions by reaction of $X_2$ in acidic solution.
         b. α-Bromoketones are useful in syntheses because they can be dehydrobrominated by base treatment to form α,β-unsaturated ketones.
   B. Enolates (Sections 11.4–11.6).
      1. Enolate anion formation (Section 11.4).
         a. Hydrogens α to a carbonyl group are weakly acidic.
            i. This stability is due to overlap of a vacant *p* orbital of the enolate with the carbonyl group *p* orbitals, allowing the carbonyl group to stabilize the negative charge by resonance.
            ii. The two resonance forms aren't equivalent; the form with the negative charge on oxygen is the major contributor to the resonance hybrid.
         b. Strong bases such as LDA and sodium amide are needed for enolate anion formation.
         c. When a hydrogen is flanked by two carbonyl groups, it is much more acidic.
      2. Reactivity of enolate anions (Section 11.5).
         a. Enolates are more useful than enols for two reasons:
            i. Unlike enols, stable solutions of enolates are easily prepared.
            ii. Enolates are more reactive than enols because they are more nucleophilic.

- b. Enolates can react either at carbon or at oxygen.
  - i. Reaction at carbon yields an α-substituted carbonyl compound.
  - ii. Reaction at oxygen yields an enol derivative.
- 3. Alkylation reactions of enolates (Section 11.6).
  - a. General features.
    - i. Alkylations are useful because they form a new carbon–carbon bond.
    - ii. Alkylations have the same limitations as $S_N2$ reactions; the alkyl groups must be methyl, primary, allylic or benzylic.
  - b. The malonic ester synthesis.
    - i. The malonic ester synthesis is used for preparing a carboxylic acid from a halide while lengthening the carbon chain by two atoms.
    - ii. Diethyl malonate is useful because its enolate is easily prepared by reaction with sodium ethoxide.
    - iii. Since diethyl malonate has two acidic hydrogens, two alkylations can take place.
    - iv. Heating in aqueous HCl causes hydrolysis and decarboxylation of the alkylated malonate to yield a substituted monocarboxylic acid.
      - (a). Decarboxylations are common only to β-keto acids and malonic acids.

III. Carbonyl condensation reactions (Sections 11.7–11.11).

- A. Mechanism of carbonyl condensation reactions (Section 11.7).
  - 1. Carbonyl condensation reactions take place between two carbonyl components.
    - a. One component (the nucleophilic donor) is converted to its enolate and undergoes an α-substitution reaction.
    - b. The other component (the electrophilic acceptor) undergoes nucleophilic addition.
  - 2. Many kinds of carbonyl compounds undergo carbonyl condensation reactions.
- B. The aldol reaction (Sections 11.8–11.9).
  - 1. Characteristics of the aldol reaction (Section 11.8).
    - a. The aldol condensation is a base-catalyzed dimerization of two aldehydes or ketones.
    - b. The reaction can occur between two components that have α hydrogens.
    - c. For simple aldehydes, the equilibrium favors products, but for other aldehydes and ketones, the equilibrium favors reactants.
  - 2. Dehydration of aldol products (Section 11.9).
    - a. Aldol products are easily dehydrated to yield conjugated enones.
    - b. Dehydration can be catalyzed by both acid and base.
    - c. Reaction conditions for dehydration are only slightly more severe than for condensation.
    - d. Often, dehydration products are isolated directly from condensation reactions.
- C. The Claisen condensation (Section 11.10).
  - 1. Treatment of an ester with 1 equivalent of base yields a β-keto ester.
  - 2. The reaction is reversible and has a mechanism similar to that of the aldol reaction.
  - 3. A major difference from the aldol condensation is the expulsion of an alkoxide ion from the tetrahedral intermediate of the initial Claisen adduct to yield an acyl substitution product.
- D. Biological carbonyl reactions (Section 11.11).
  - 1. Many biomolecules are synthesized by carbonyl condensation reactions.
  - 2. Glycolysis (breakdown of glucose) involves several carbonyl-group reactions.
  - 3. Acetyl CoA is the major building block for synthesis of biomolecules.
    - a. Acetyl CoA can act as an electrophilic acceptor by being attacked at its carbonyl group.
    - b. Acetyl CoA can act as a nucleophilic donor by loss of its acidic α hydrogen.

## Solutions to Problems

**11.1–11.2** Acidic hydrogens in the keto form of each of these compounds are bold.

| | *Keto Form* | *Enol Form* | *Number of Acidic Hydrogens* |
|---|---|---|---|
| (a) | 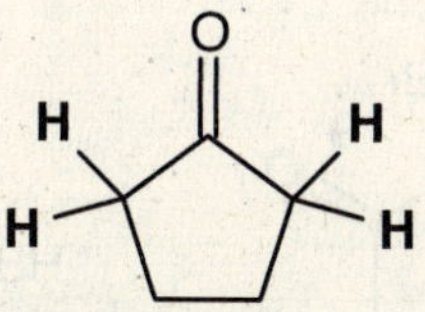 | 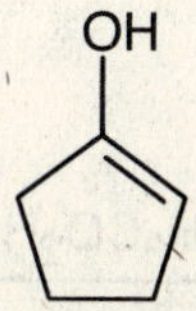 | 4 |
| (b) | 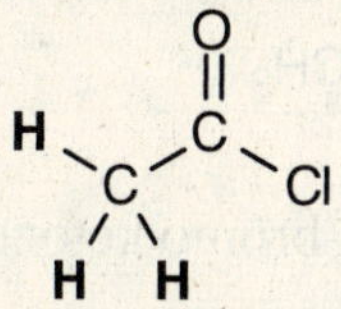 | 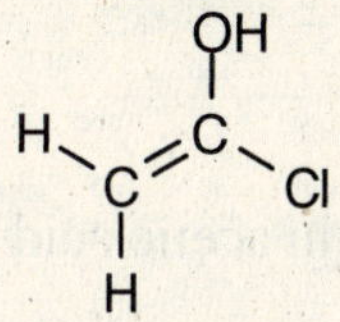 | 3 |
| (c) | 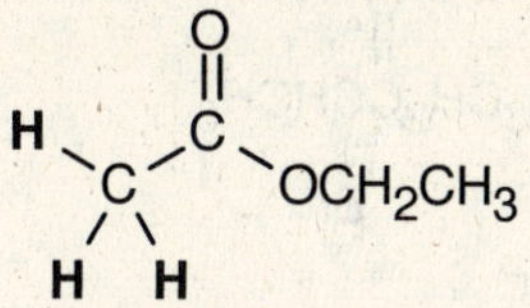 | 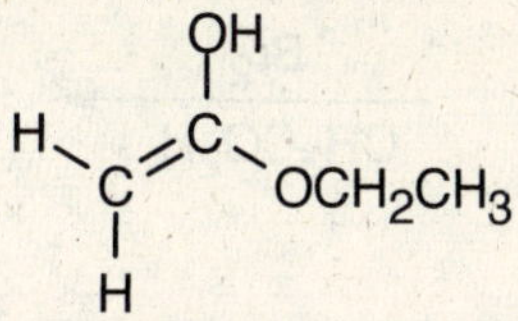 | 3 |
| (d) | 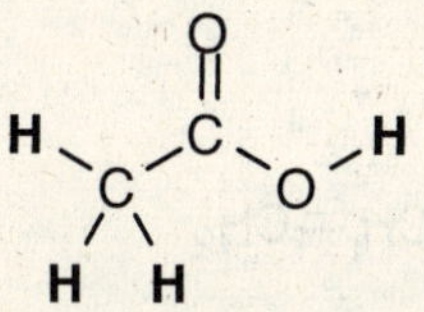 | 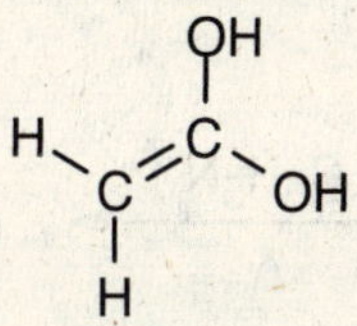 | 4 |
| (e) | 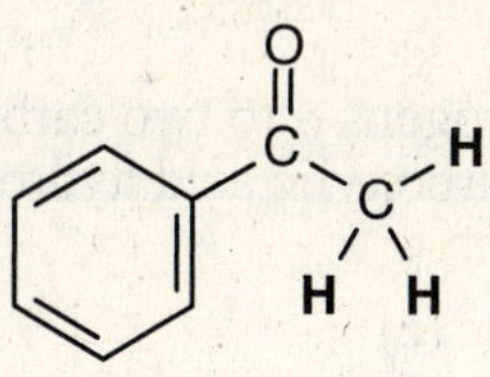 | 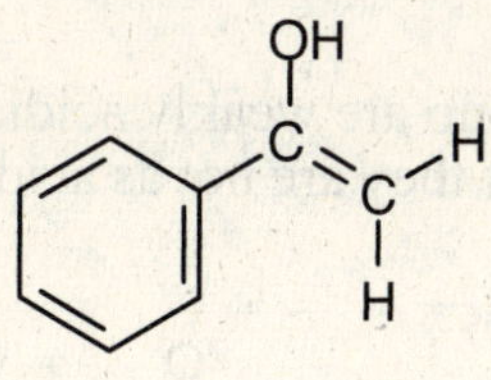 | 3 |

**11.3**

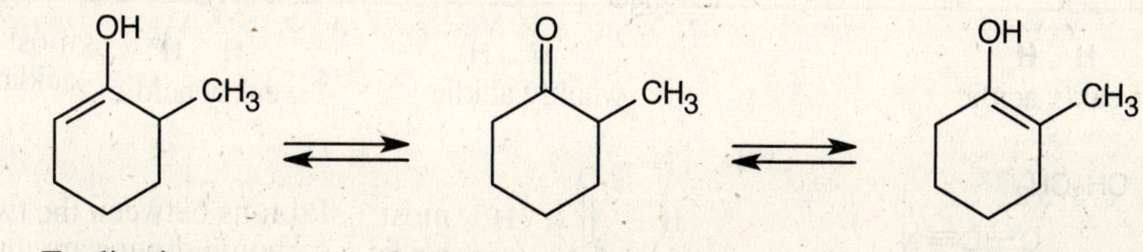

Enolization toward carbon 6 — 2-Methylcyclohexanone — Enolization toward carbon 2

Enolization can occur in either direction from the carbonyl group. Two different enols are formed because carbon 2 has a methyl substituent and carbon 6 does not.

**11.4** As suggested in Worked Example 11.2, locate the acidic protons and replace one of them with a halogen in order to show the product of the α-substitution reaction.

(a)

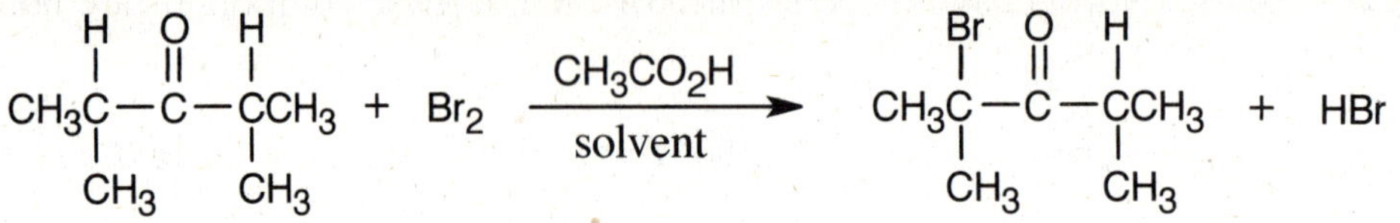


(b)

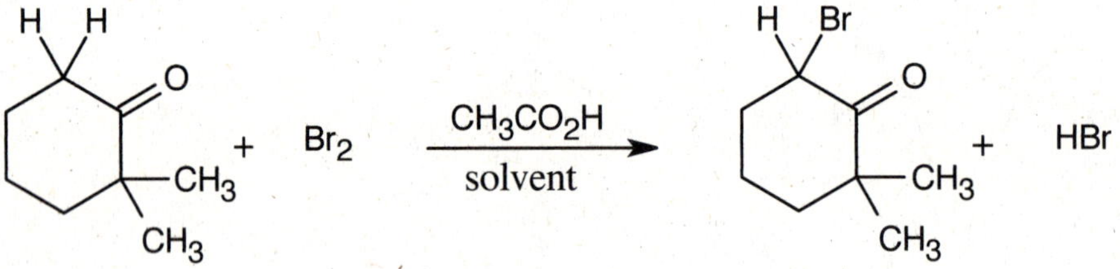


**11.5** *Step 1.* Treat pentan-3-one with $Br_2$ in acetic acid to form the α-bromo ketone.

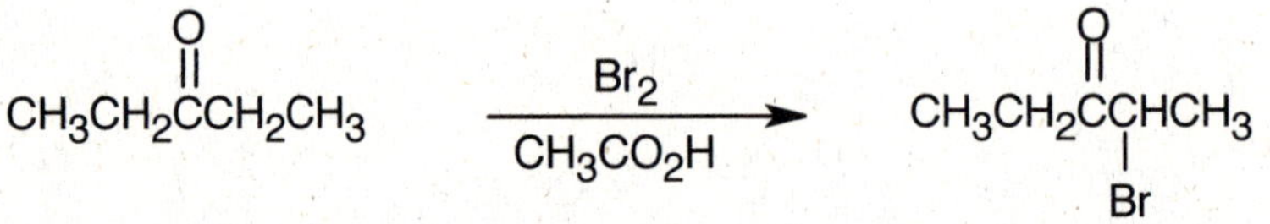

*Step 2.* Heat the α-bromo ketone in pyridine to form pent-1-en-3-one. This is an E2 dehydrohalogenation

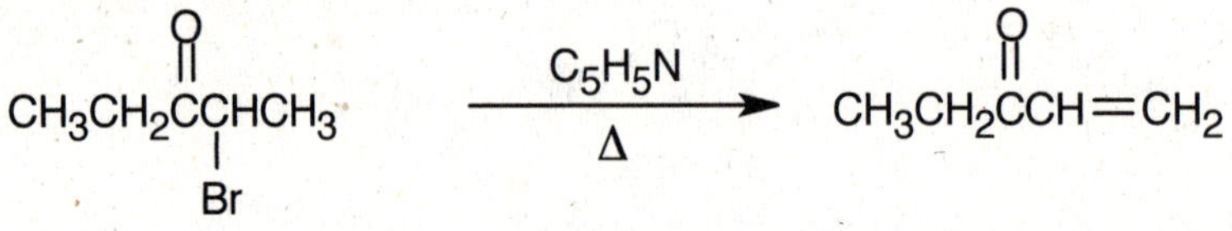

**11.6** Hydrogens α to one carbonyl group are weakly acidic. Hydrogens α to two carbonyl groups are much more acidic, but they are not as acidic as carboxylic acid hydrogens.

(a)

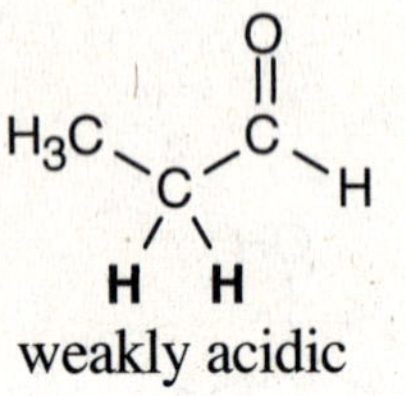


(b)

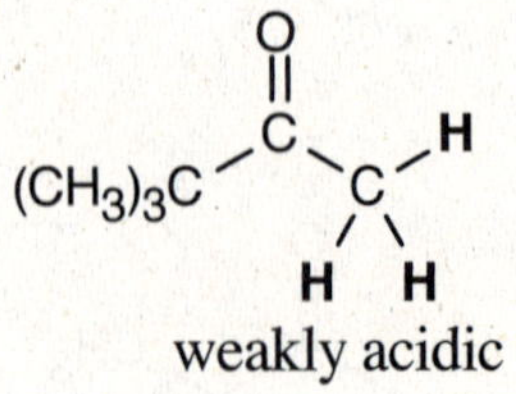


(c)

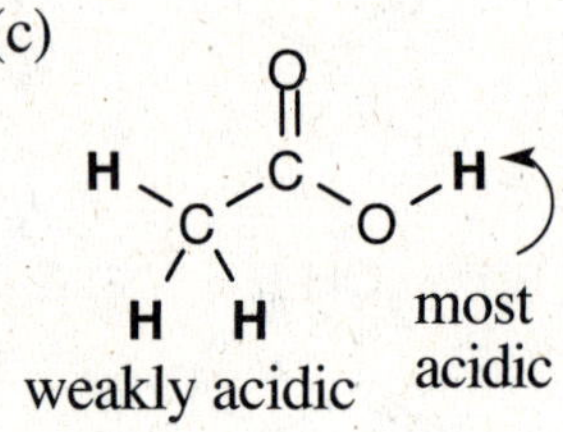


(d)

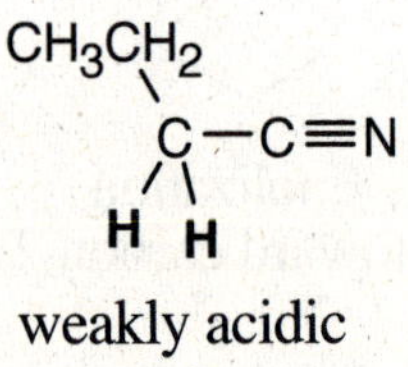


(e)

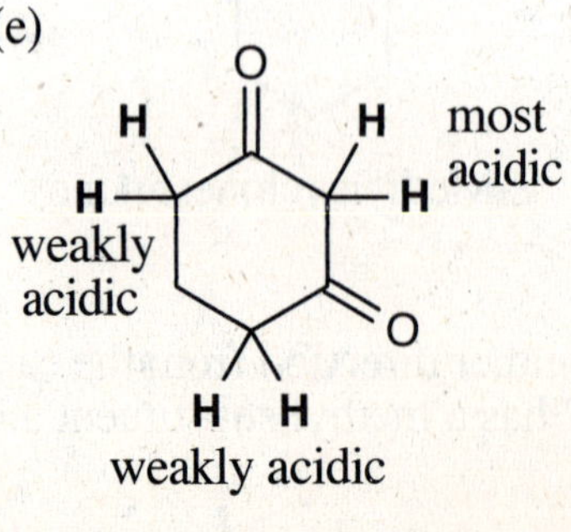


Protons between the two carbonyl groups are much more acidic than protons next to only one carbonyl group.

**11.7**

(a)

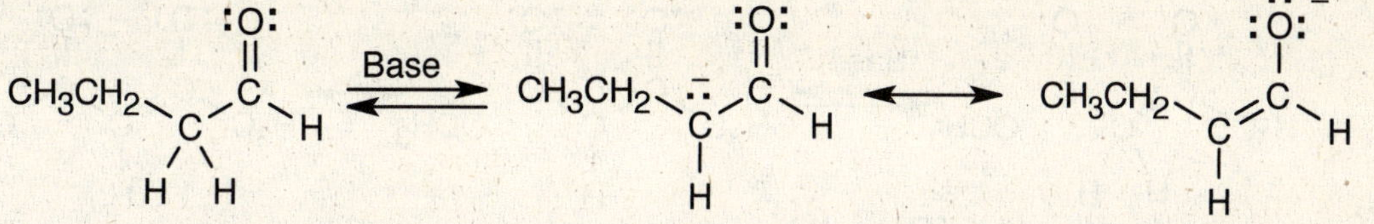


(b)

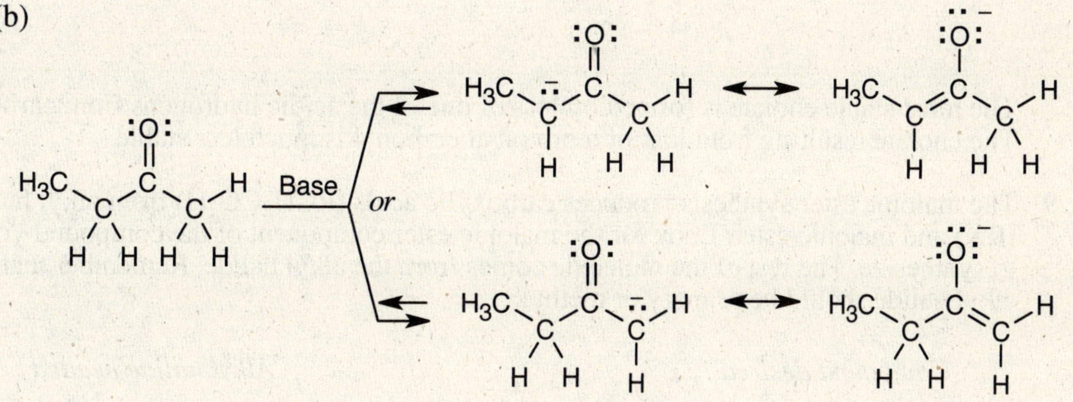


(c)

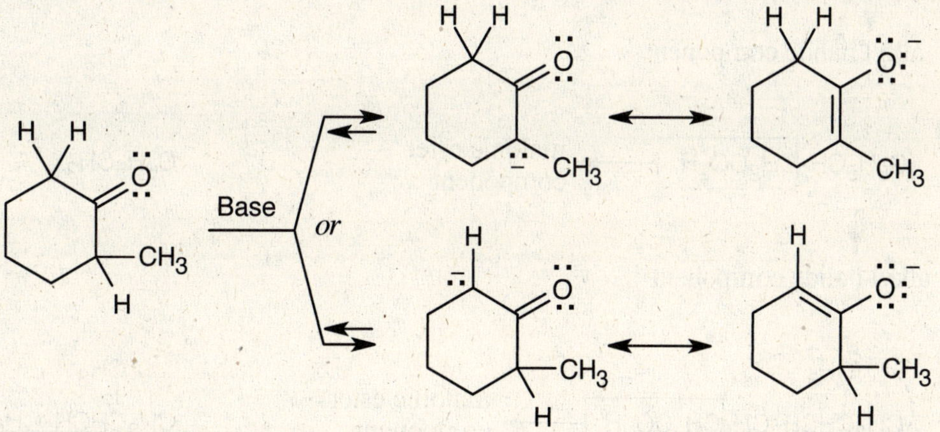


In parts (b) and (c), two different enolate ions can form. *E/Z* double bond isomers of enolate anions in (a) and (b) are also formed.

**11.8**

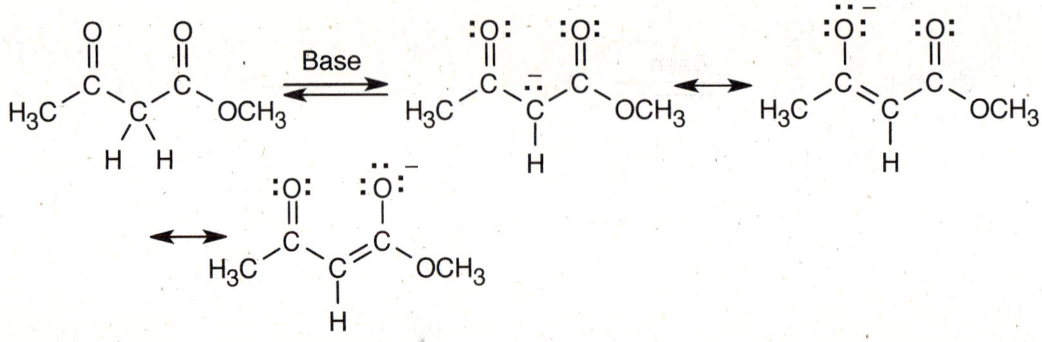


The most stable enolate is formed by loss of one of the acidic hydrogens from carbon 2. The enolate resulting from loss of a proton at carbon 4 is much less stable.

**11.9** The malonic ester synthesis produces carboxylic acids ($RCH_2CO_2H$) from alkyl halides (RX) and malonic ester. Look for the malonic ester component of the compound you want to synthesize. The rest of the molecule comes from the alkyl halide. Remember that the alkyl halide should be primary or methyl.

*Compound desired* — *Alkyl halide needed*

(a)

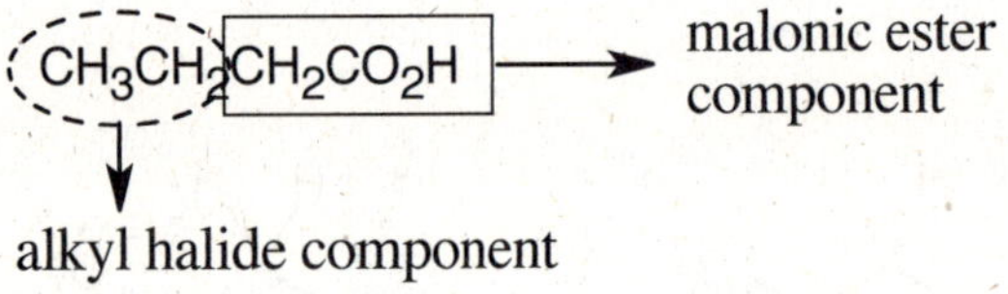


Alkyl halide needed: $CH_3CH_2X$

(b)

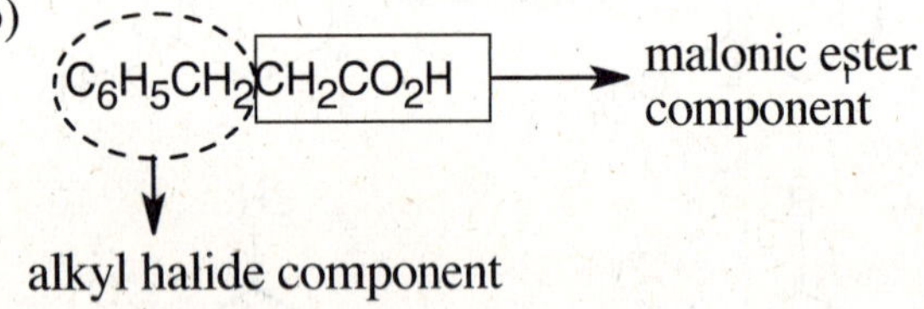


Alkyl halide needed: $C_6H_5CH_2X$

(c)

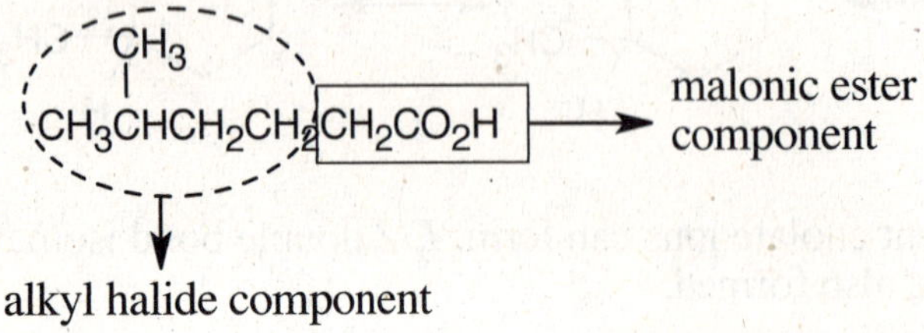


Alkyl halide needed: $CH_3CH(CH_3)CH_2CH_2X$

**11.10** As in Problem 11.9, locate the malonic ester component of the target molecule. The rest of the molecule comes from the alkyl halide or halides.

Notice in this problem, and in other problems that involve the malonic ester synthesis, the space-saving use of "Et" when an ethyl group is part of a molecule. Also notice that sodiomalonic ester is prepared by the following reaction:

$$CH_2(CO_2Et)_2 \;+\; Na^+\;{}^-OEt \;\longrightarrow\; Na^+\;{}^-\!:CH(CO_2Et)_2 \;+\; EtOH$$

(a)

from halide $(CH_3)_2CHCH_2$ — $CH_2CO_2H$ from malonic ester

$$CH_2(CO_2Et)_2 \xrightarrow[2.\ (CH_3)_2CHCH_2Br]{1.\ Na^+\ ^-OEt} (CH_3)_2CHCH_2{-}CH(CO_2Et)_2 + NaBr$$

$$\downarrow H_3O^+,\ \text{heat}$$

$$(CH_3)_2CHCH_2{-}CH_2CO_2H + CO_2 + 2\ EtOH$$

4-Methylpentanoic acid

(b)

$CH_3CH_2CH_2$ — $CHCO_2H$ (with $CH_3$ on the CH) from malonic ester

from halides $CH_3CH_2CH_2$, $CH_3$

$$CH_2(CO_2Et)_2 \xrightarrow[2.\ CH_3CH_2CH_2Br]{1.\ Na^+\ ^-OEt} CH_3CH_2CH_2{-}CH(CO_2Et)_2 + NaBr$$

$$\downarrow 1.\ Na^+\ ^-OEt;\ 2.\ CH_3Br$$

$$CH_3CH_2CH_2{-}CHCO_2H\ (\text{CH}_3) \xleftarrow[\text{heat}]{H_3O^+} CH_3CH_2CH_2{-}C(CH_3)(CO_2Et)_2 + NaBr$$

2-Methylpentanoic acid

$+\ CO_2 + 2\ EtOH$

**11.11** This carboxylic acid is composed of the same alkyl groups that we used in the previous problem. The target carboxylic acid is 2,4-dimethylpentanoic acid, and the alkyl groups are methyl and 2-methylpropyl (isobutyl). The order of alkylation can be reversed.

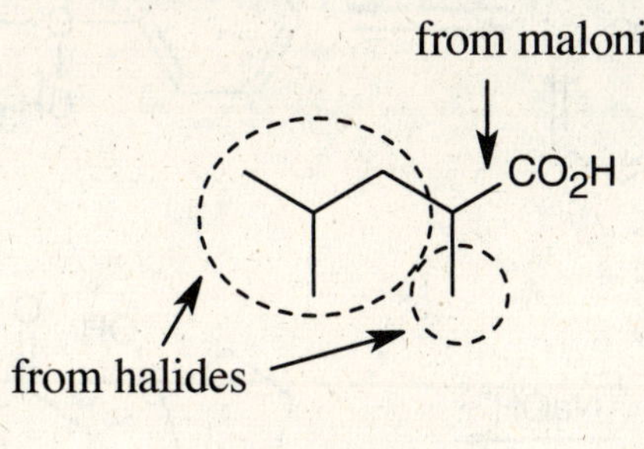


$$CH_2(CO_2Et)_2 \xrightarrow[2.\ CH_3CH(CH_3)CH_2Br]{1.\ Na^+\ ^-OEt} CH_3CH(CH_3)CH_2{-}CH(CO_2Et)_2 + NaBr$$

$$\downarrow 1.\ Na^+\ ^-OEt;\ 2.\ CH_3Br$$

$$CH_3CH(CH_3)CH_2{-}CH(CH_3)CO_2H \xleftarrow[\text{heat}]{H_3O^+} CH_3CH(CH_3)CH_2{-}C(CH_3)(CO_2Et)_2 + NaBr$$

2,4-Dimethylpentanoic acid

$+\ CO_2 + 2\ EtOH$

**11.12** Compounds that can undergo the aldol reaction must have acidic α-hydrogen atoms. 2,2-Dimethylyclohexanone (a) is the only compound in this problem capable of undergoing the aldol reaction.

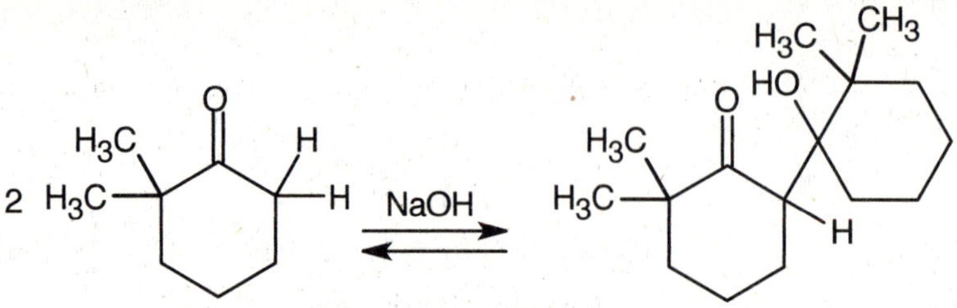

No α-hydrogens:

(b)

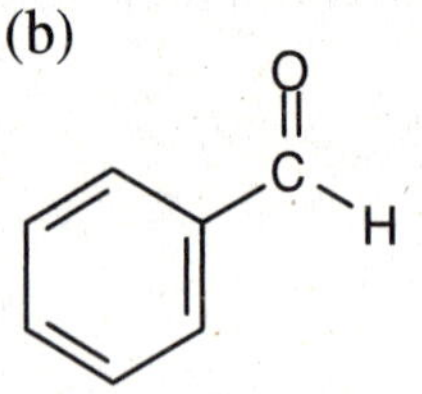

Benzaldehyde

(c)

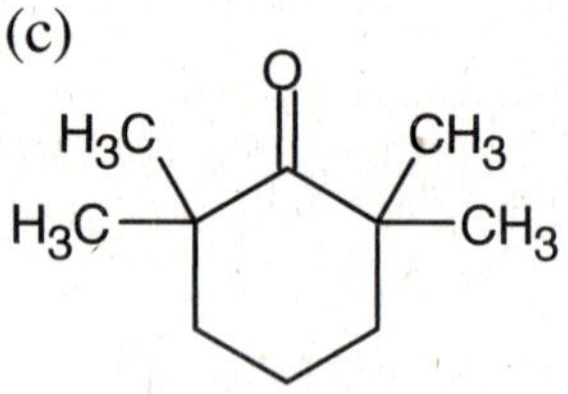

2,2,6,6-Tetramethyl-cyclohexanone

(d)

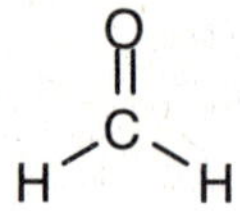

Formaldehyde

**11.13** It is easier to draw the product of an aldol reaction if the two components are drawn so that the new bond connects them.

(a)

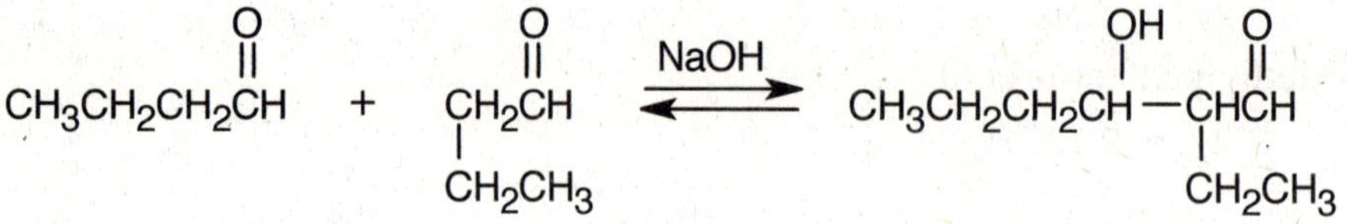

(b)

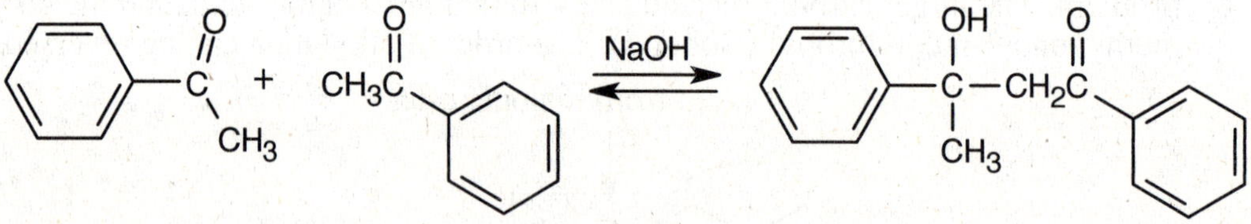

(c)

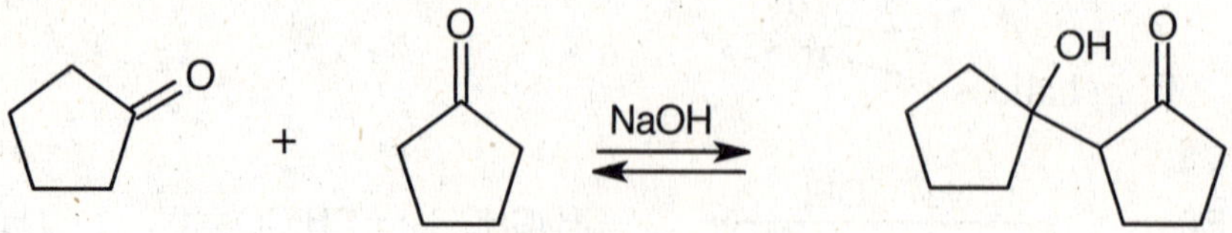

**11.14**

(a)

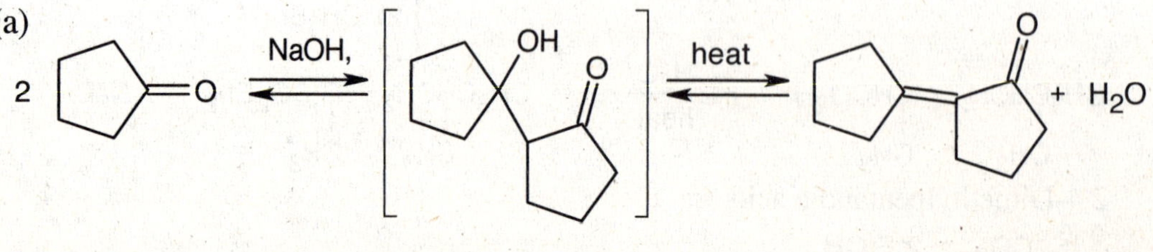

(b)

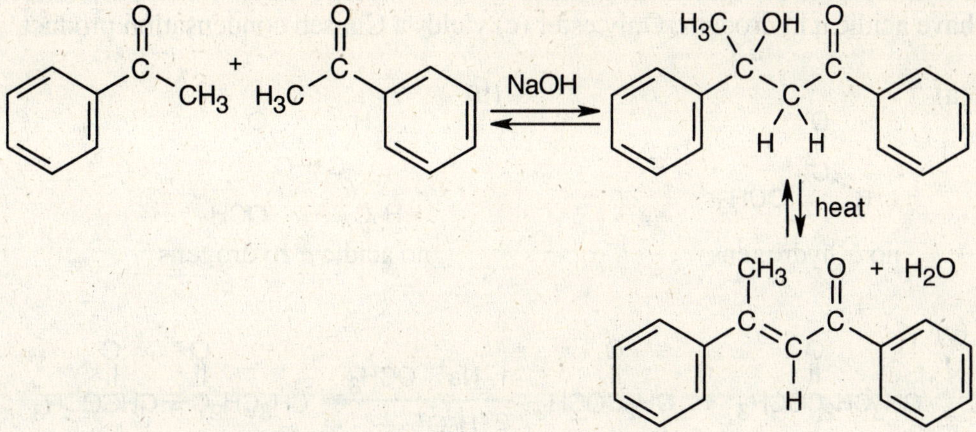

(c)

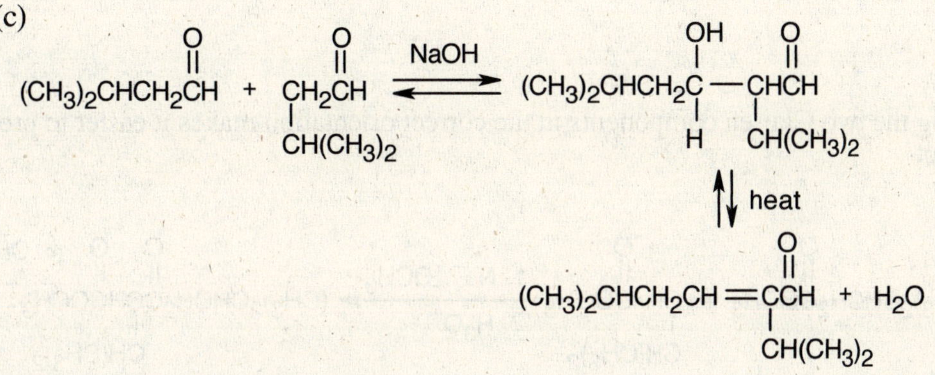

**11.15** Two different enolate anions can be formed from butan-2-one because of the presence of two different kinds of alpha hydrogens. Each enolate can react with a second molecule of butan-2-one to yield two different enones.

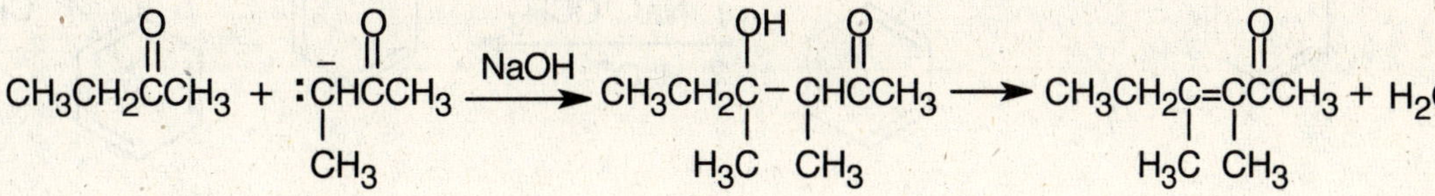

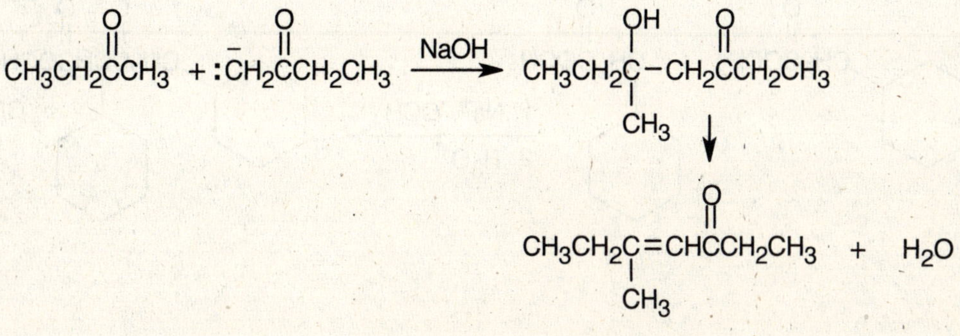

**11.16** As in the aldol condensation, compounds that undergo Claisen condensation reactions must have acidic α hydrogens. Only ester (c) yields a Claisen condensation product.

(a)

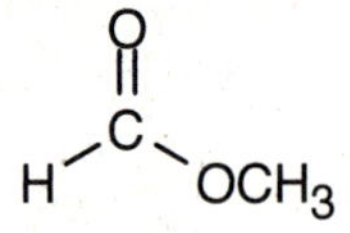

no α hydrogens

(b)

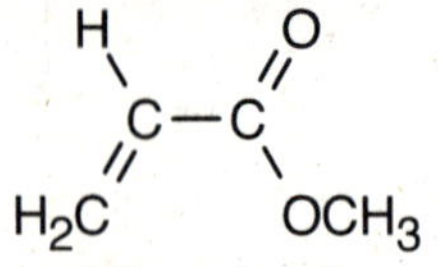

no acidic α hydrogens

(c)

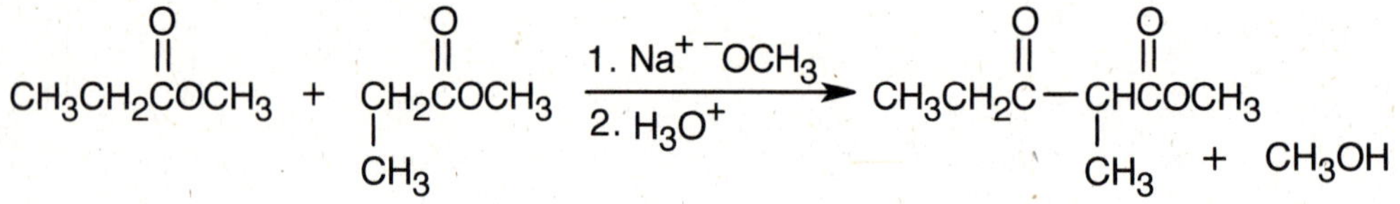

**11.17** Writing the two Claisen components in the correct orientation makes it easier to predict the product.

(a)

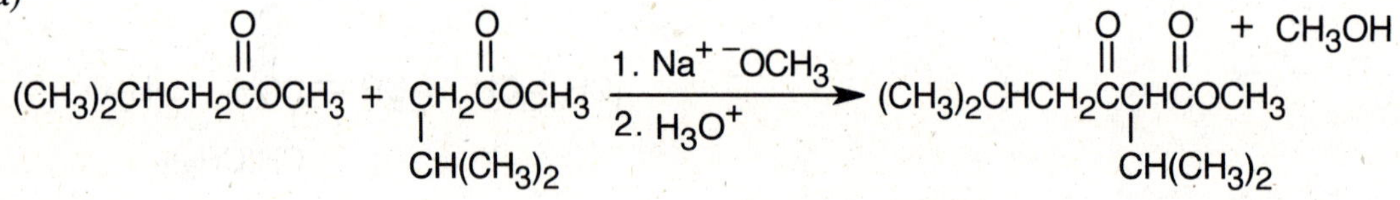

(b)

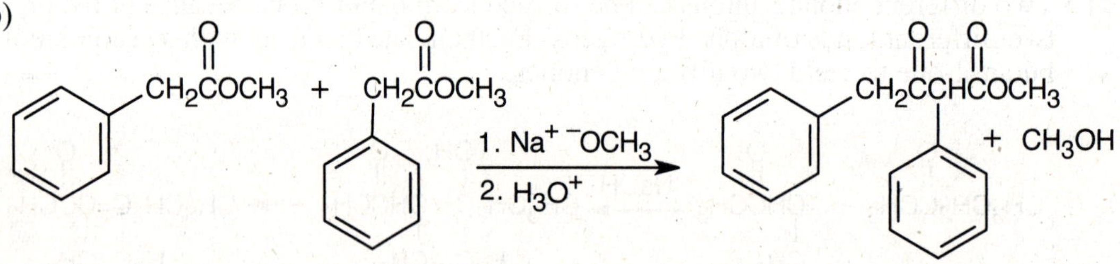

(c)

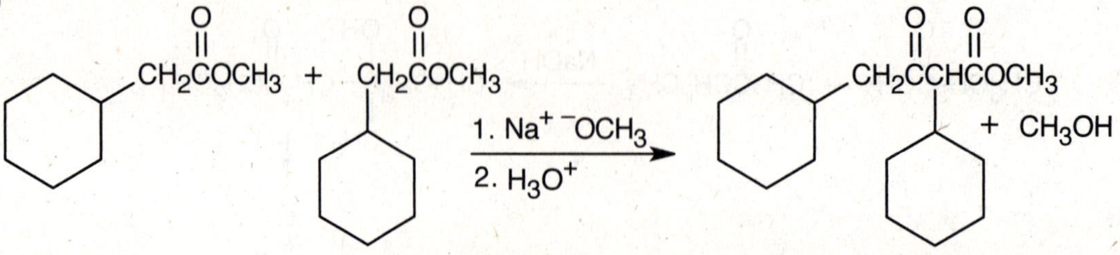

## Visualizing Chemistry

**11.18** The enolate of methyl phenylacetate adds to a second molecule of methyl phenylacetate to form the intermediate that is pictured. Elimination of methoxide (circled) and acidification give the product shown.

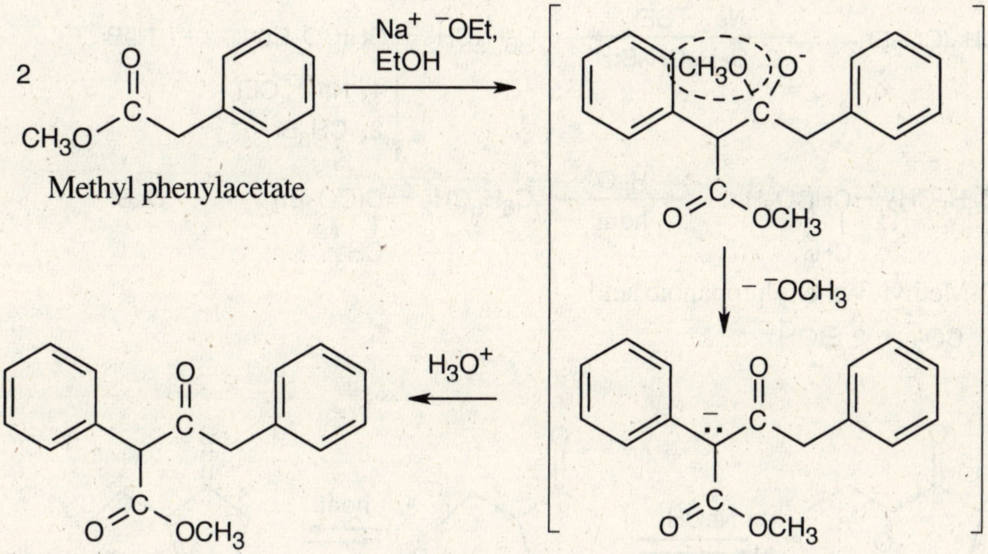


**11.19**

(a)

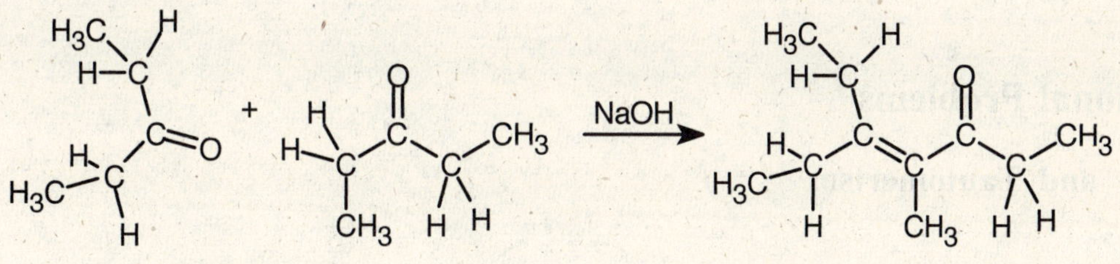


Pentan-3-one

(b)

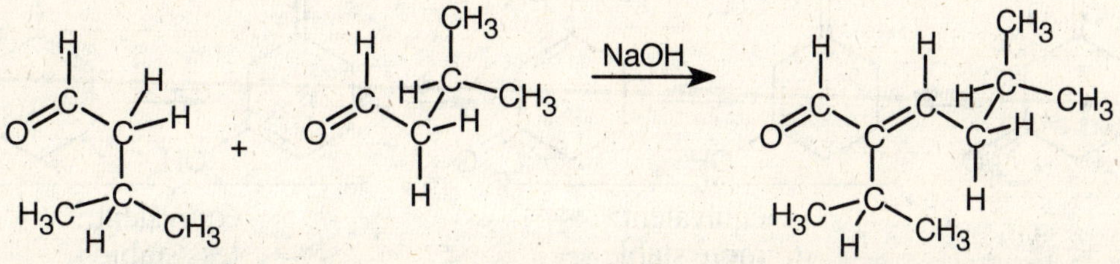


3-Methylbutanal

**11.20** This product is formed from the reaction of malonic ester with both benzyl bromide and bromomethane.

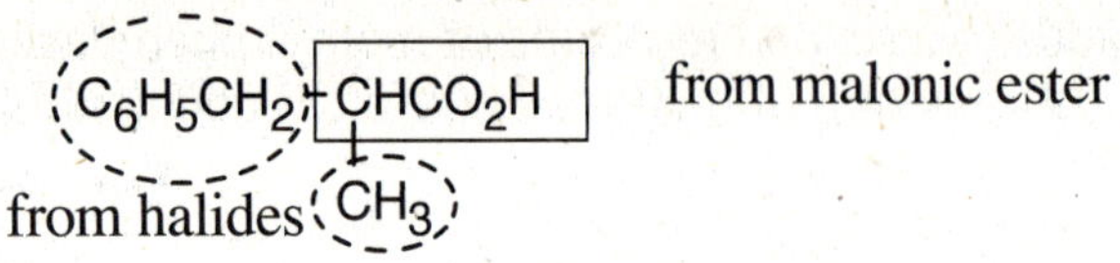


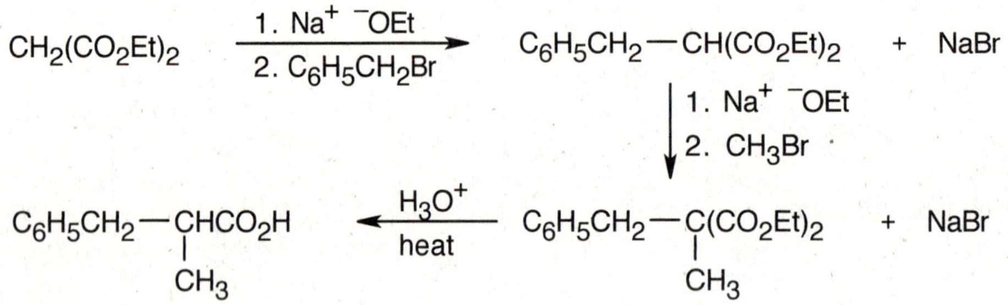


2-Methyl-3-phenylpropanoic acid

+ $CO_2$ + 2 EtOH

**11.21**

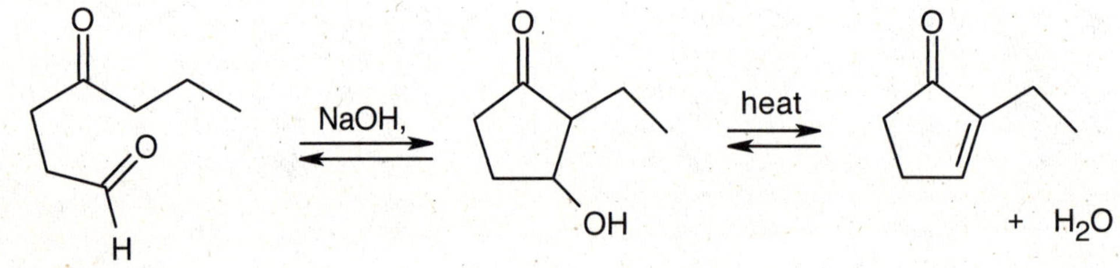


4-Oxoheptanal

## Additional Problems

### Acidity and Tautomerism

**11.22**

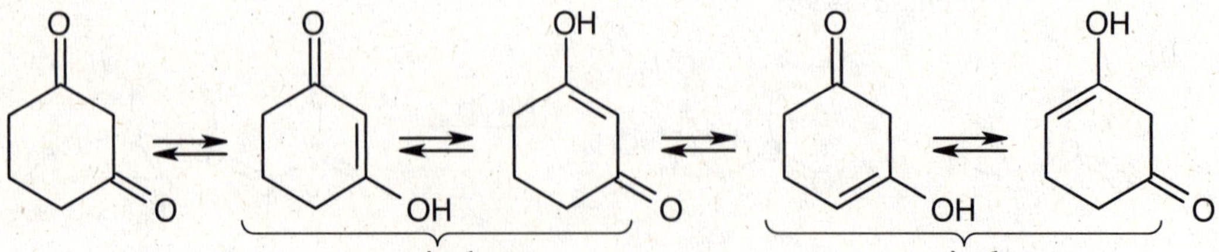


equivalent; more stable

equivalent; less stable

Four enol forms can be drawn.The first two monoenols are more stable because the enol double bond is conjugated with the carbonyl group.

**11.23**

*Least acidic* ⟶ *Most acidic*

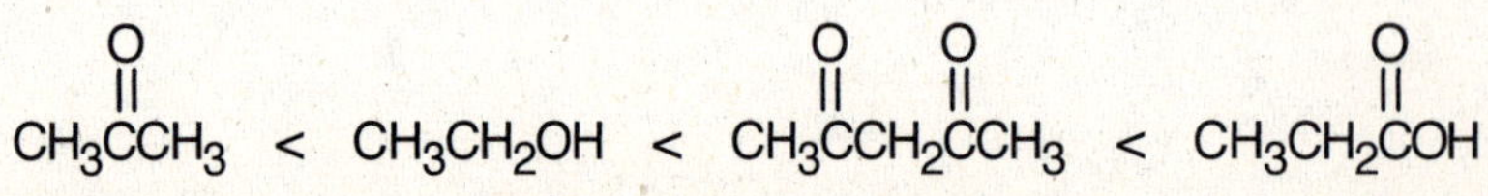

**11.24** Pentane-2,4-dione is enolized to a greater extent than acetone because its enol form is more stable than the enol form of acetone. This enol stability is due to two factors: (1) conjugation of the enol double bond with the second carbonyl group, and (2) hydrogen bonding between the enol hydrogen and the second carbonyl group.

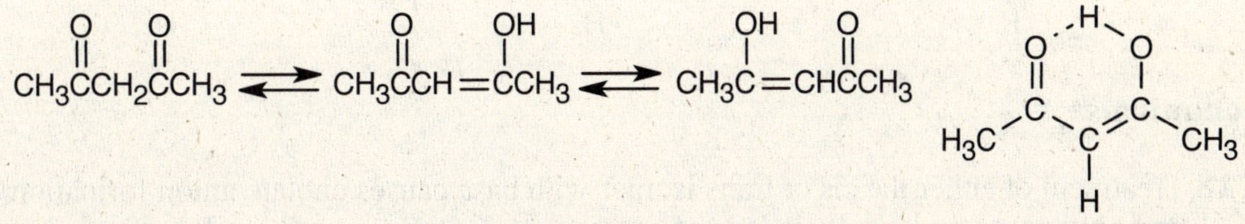

**11.25** Acidic hydrogens are bold. The most acidic hydrogens are the two between the carbonyl groups in (b) and the hydroxyl hydrogen in (c). The hydrogens in (c) that are bonded to the methyl group are acidic (draw resonance forms to prove it).

(a)

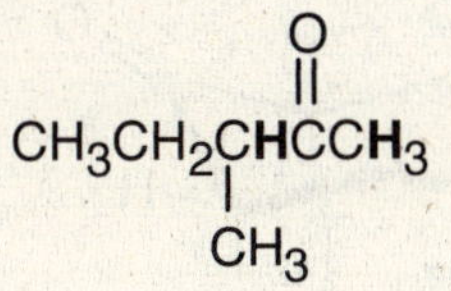

(b)

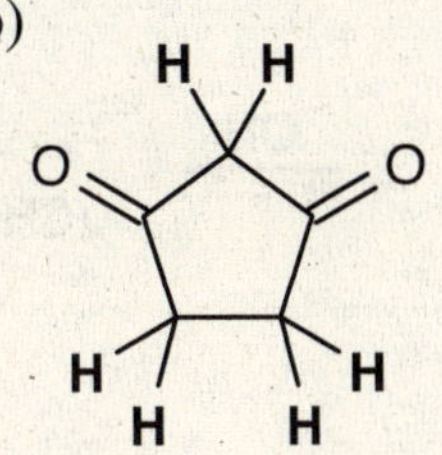

(c)

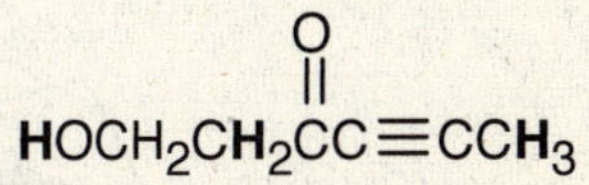

(d)

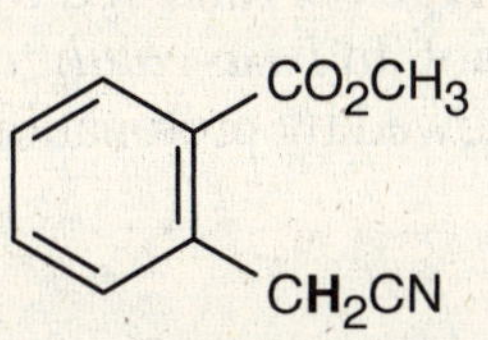

(e)

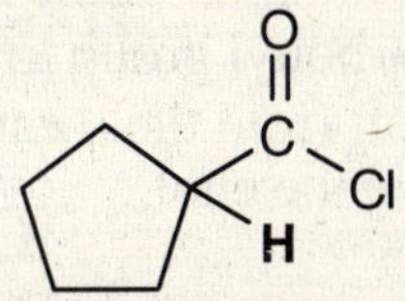

(f)

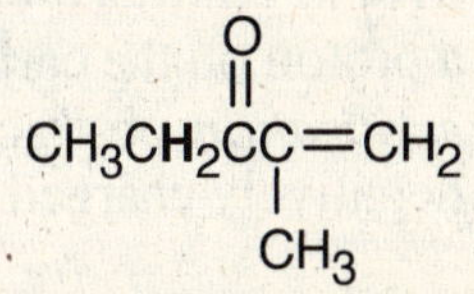

**11.26** An enolate anion is generally more reactive than an enol because an enolate is negatively charged and is more nucleophilic than a neutral enol.

**11.27**

(a)

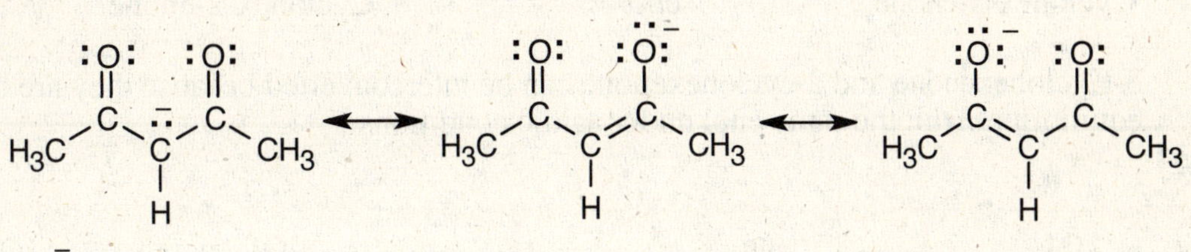

(b)

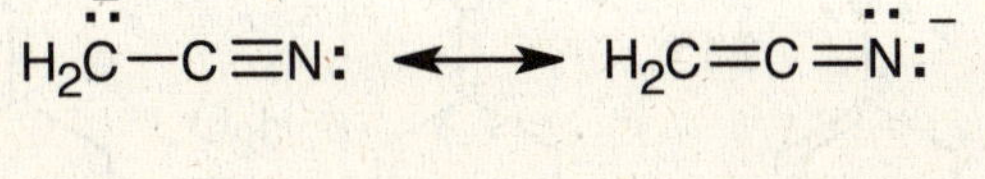

(c)

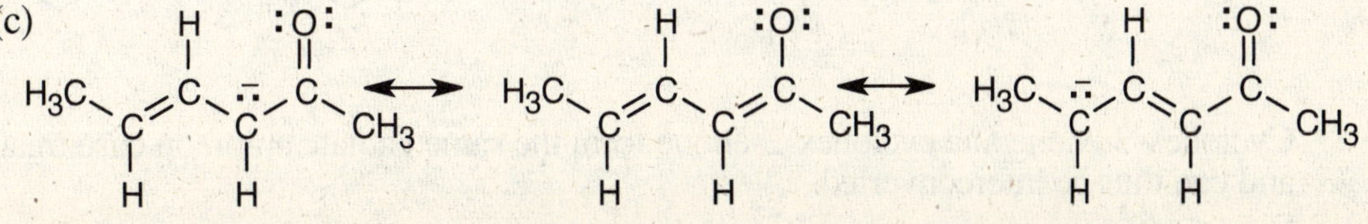

(d)

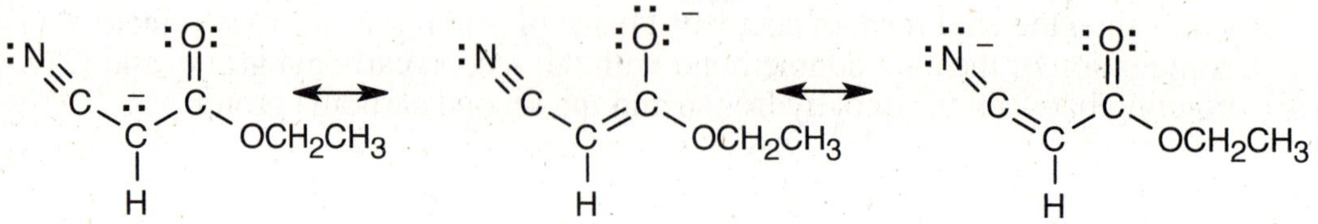

## Mechanisms

**11.28** Treatment of either the cis or trans isomer with base causes enolate anion formation α to the carbonyl group and results in loss of configuration at the α-position. Reprotonation at carbon 2 produces either of the diastereomeric 4-*tert*-butyl-2-methylcyclohexanones. In both diastereomers the *tert*-butyl group of carbon 4 occupies the equatorial position for steric reasons. The methyl group of the cis isomer is also equatorial; the methyl group of the trans isomer is axial. The trans isomer is less stable because of interactions of the axial methyl group with the ring hydrogens.

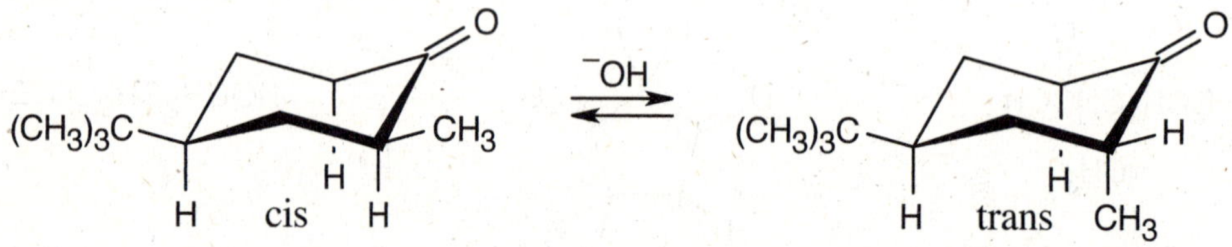

**11.29** In *acid-catalyzed enolization*, protonation of the carbonyl oxygen is followed by the loss of a proton on the carbon α to the carbonyl group to give the enol. In *base-catalyzed enolization*, abstraction of a proton from the α carbon is followed by protonation of the negatively charged oxygen to give the enol.

**11.30**

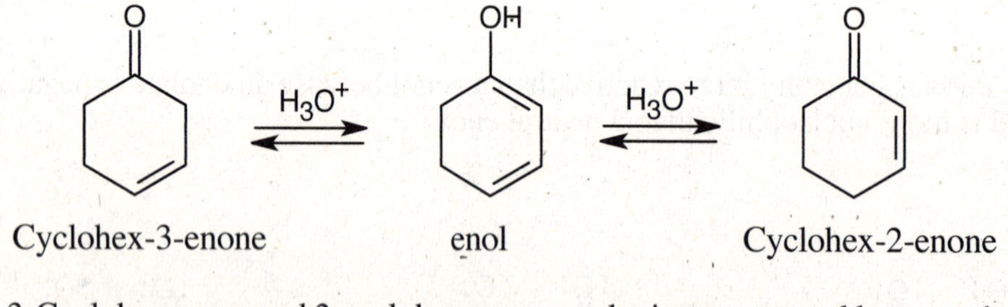

3-Cyclohexenone and 2-cyclohexenone can be interconverted because they are both in equilibrium with the same enol under acidic conditions.

**11.31**

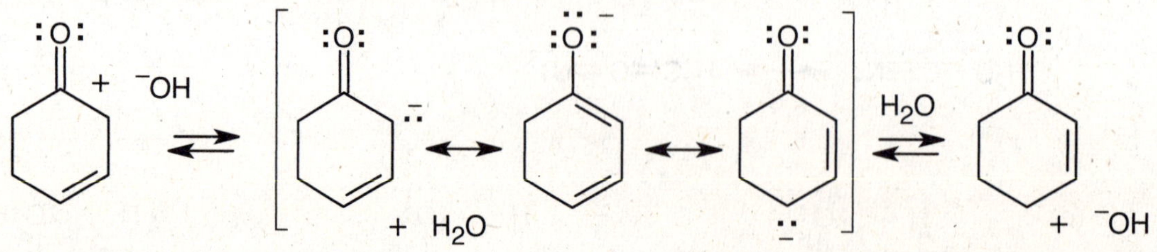

Cyclohex-3-enone and cyclohex-2-enone form the same enolate anion on base treatment and can thus be interconverted.

**11.32**

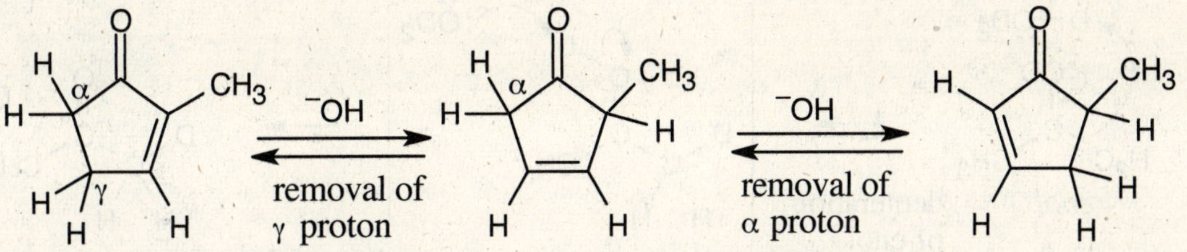


As in Problem 11.31, the isomers are in equilibrium through their common enolate.

**11.33**

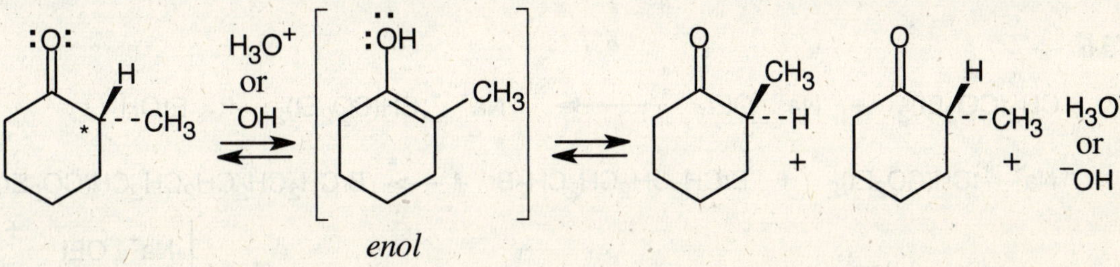


Carbon 2 loses its chirality in the step in which the enol double bond is formed. Protonation occurs with equal probability from either side of $sp^2$-hybridized carbon 2, resulting in racemic product.

**11.34**

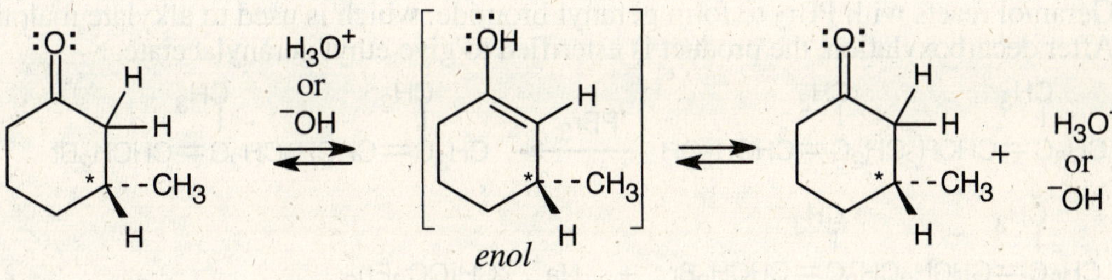


(*R*)-3-Methylcyclohexanone is not racemized by acid or base because its chirality center is not involved in the enolization reaction.

**11.35**

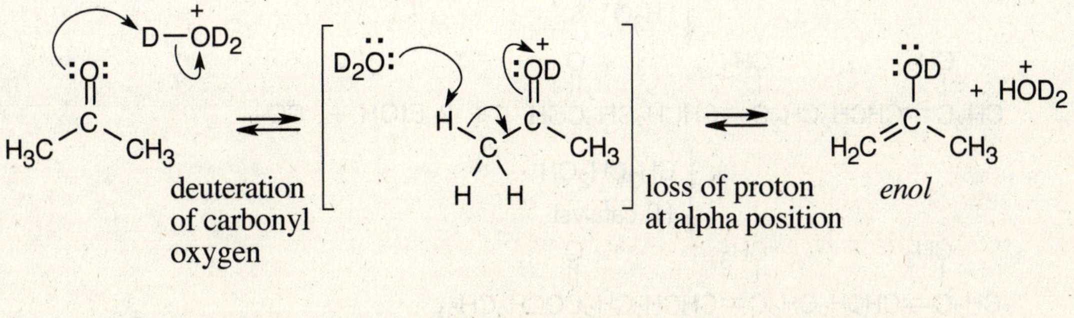

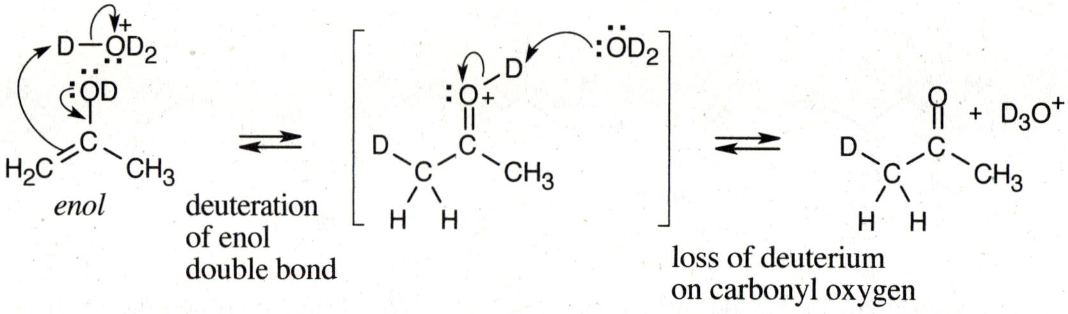


## Malonic Ester Synthesis

**11.36**

$$CH_2(CO_2Et)_2 + Na^+ \, ^-OEt \longrightarrow Na^+ \, ^-{:}CH(CO_2Et)_2 + EtOH$$

$$Na^+ \, ^-{:}CH(CO_2Et)_2 + BrCH_2CH_2CH_2CH_2Br \longrightarrow BrCH_2CH_2CH_2CH_2CH(CO_2Et)_2 + NaBr$$

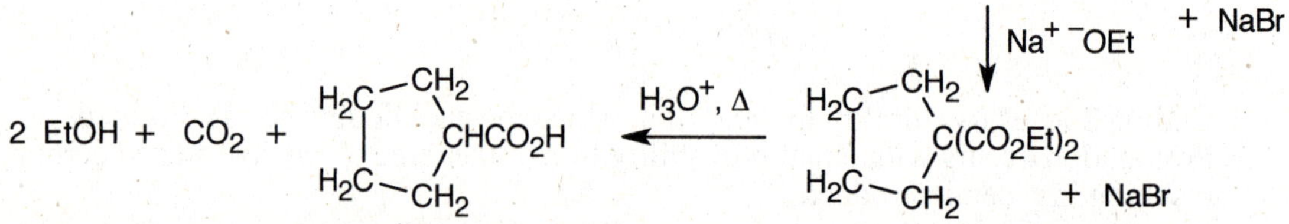

Cyclopentanecarboxylic acid

**11.37** Geraniol reacts with $PBr_3$ to form geranyl bromide, which is used to alkylate malonic ester. After decarboxylation, the product is esterified to give ethyl geranylacetate.

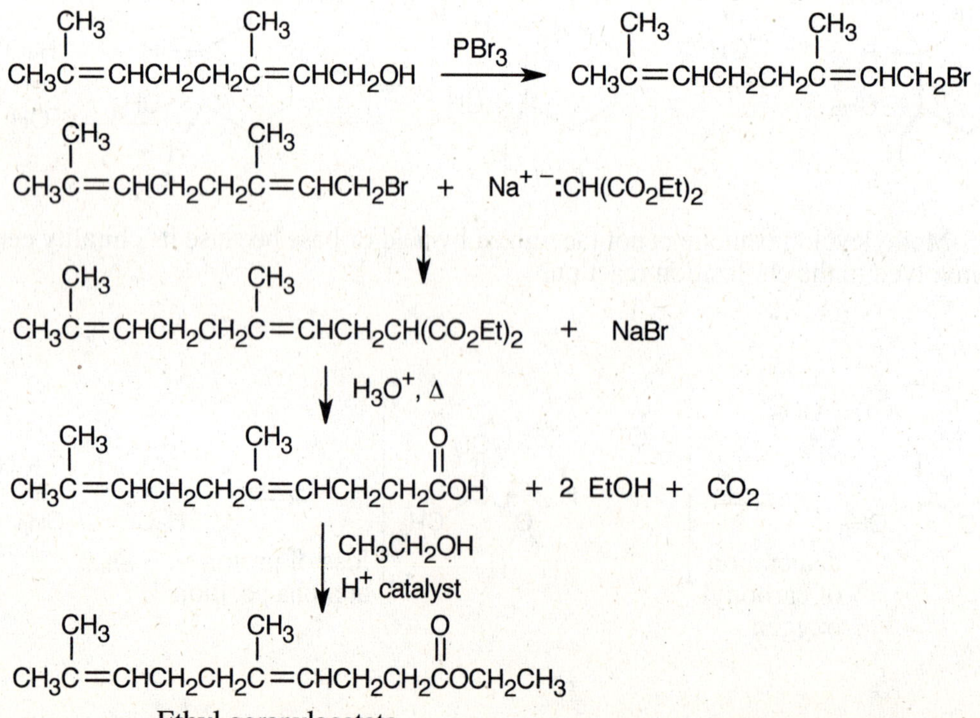

Ethyl geranylacetate

**11.38** First, convert malonic ester into its sodium salt.

$$CH_2(CO_2Et)_2 + Na^+ \,^-OEt \longrightarrow Na^+ \,^-{:}CH(CO_2Et)_2 + EtOH$$

(a)

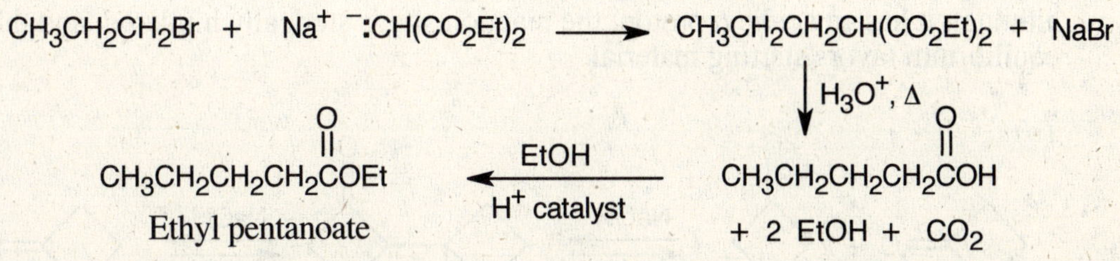


(b)

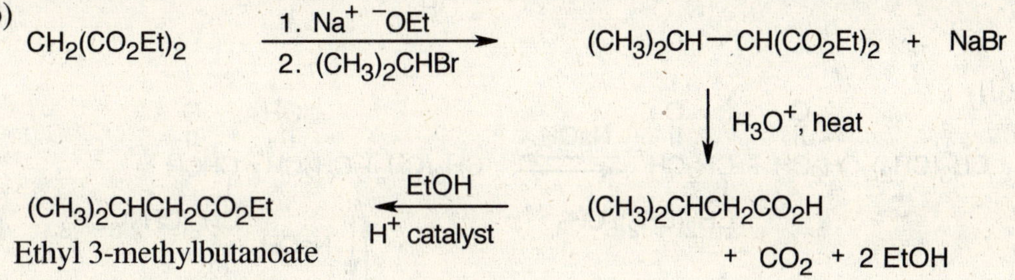


Some elimination product will also form.

(c)

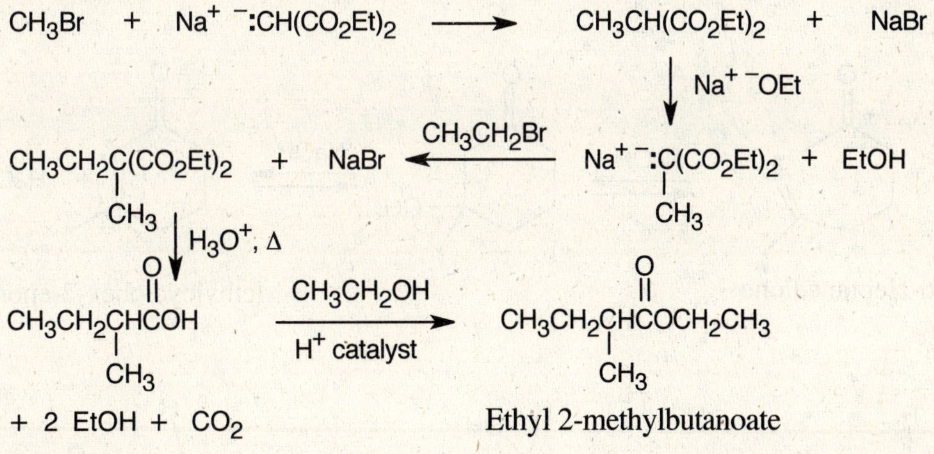


(d)

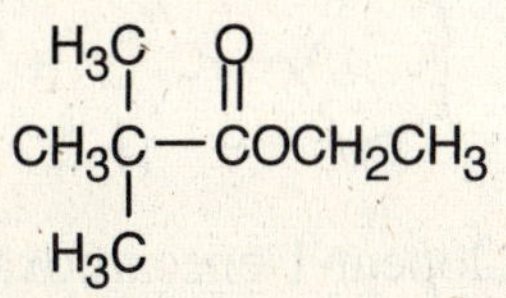

Ethyl 2,2-dimethylpropanoate

This ester can't be prepared by the malonic ester route because it is trisubstituted at the $\alpha$ carbon.

## Aldol Condensation

**11.39** Compounds (a), (c) and (f) do not undergo aldol condensation because they don't have acidic protons α to the carbonyl group. Compound (e) also doesn't form aldol products; although it has an acidic α proton, the reaction site is sterically hindered, and the equilibrium favors starting material.

(b)

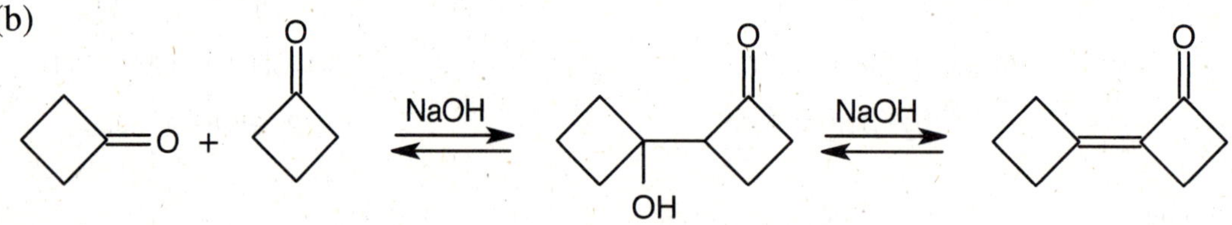


(d)

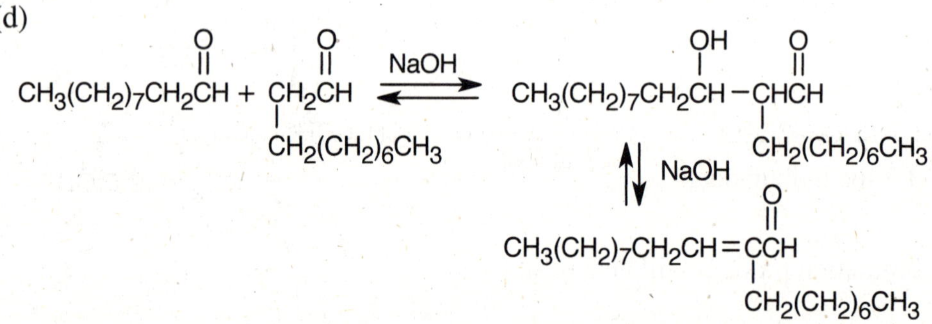


**11.40**

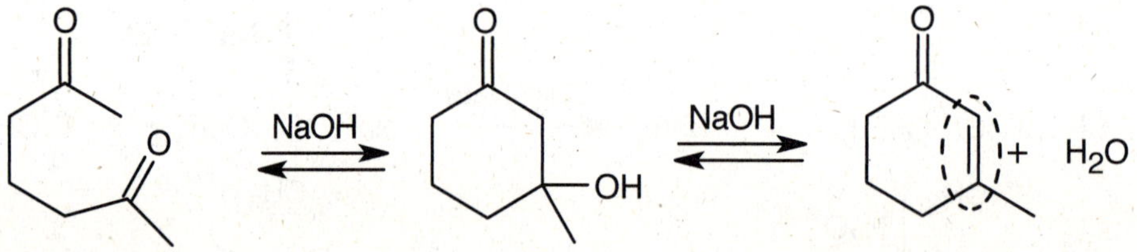


2,6-Heptanedione 3-Methylcyclohex-2-enone

**11.41**

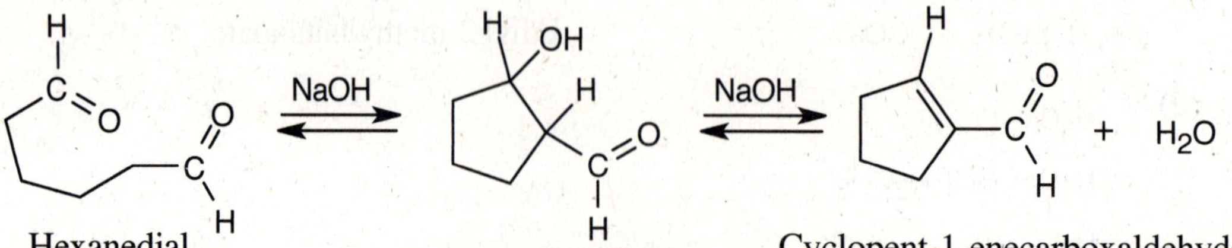


Hexanedial Cyclopent-1-enecarboxaldehyde

**11.42** Try to think backwards from the product to recognize the aldol components. In these products, the double bond is the new bond formed.

(a)

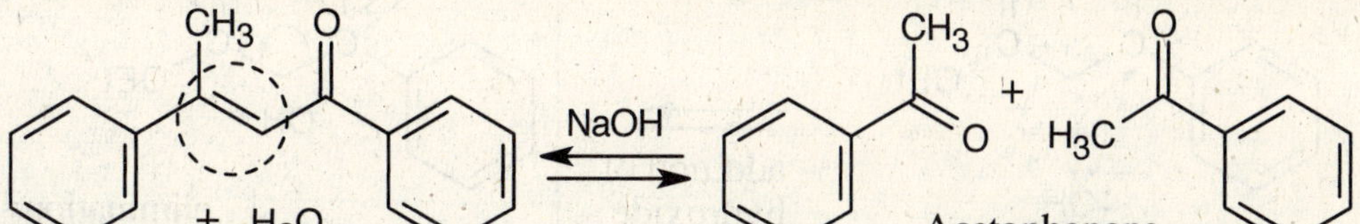

Acetophenone

(b)

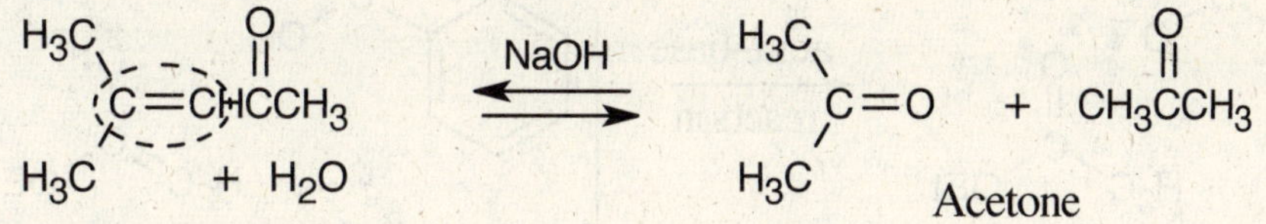

Acetone

(c)

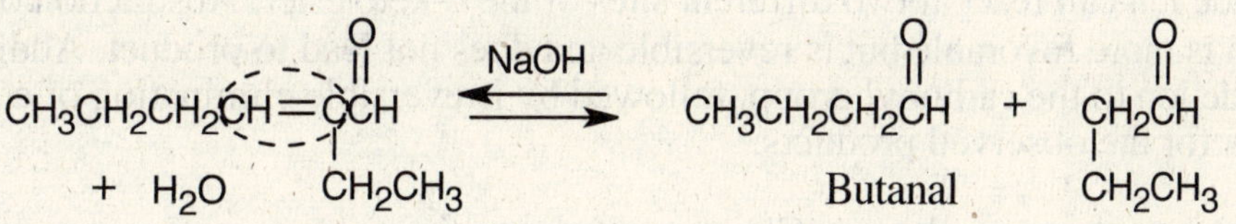

Butanal

(d)

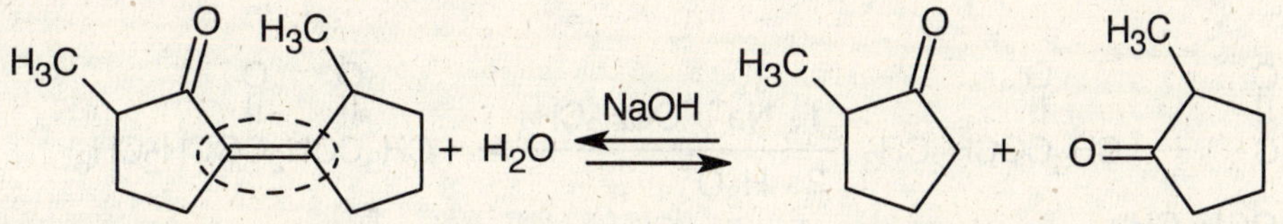

2-Methylyclopentanone

## Claisen Condensation

**11.43** As in the previous problem, a Claisen *self* condensation between two molecules of ethyl acetate yields one product.

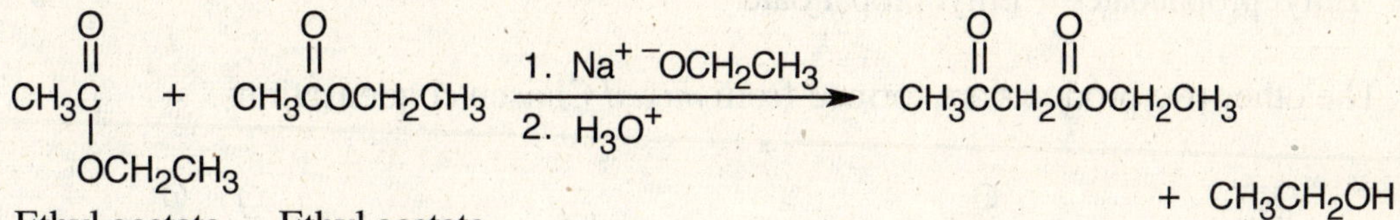

Ethyl acetate Ethyl acetate

The other product results from a *mixed* Claisen condensation.

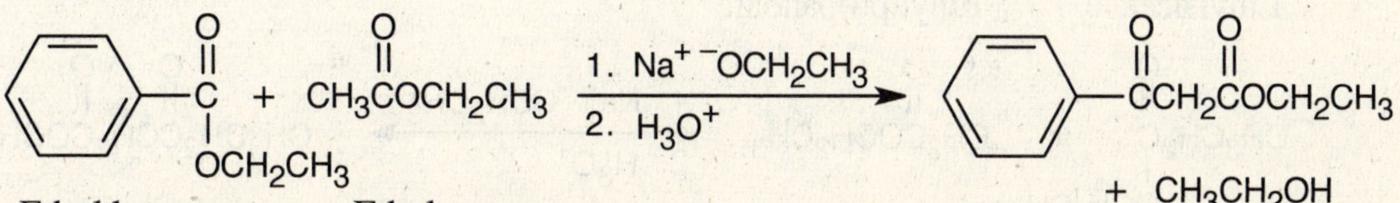

Ethyl benzoate Ethyl acetate

Ethyl benzoate does not undergo Claisen self-condensation because it has no α hydrogens.

**11.44**

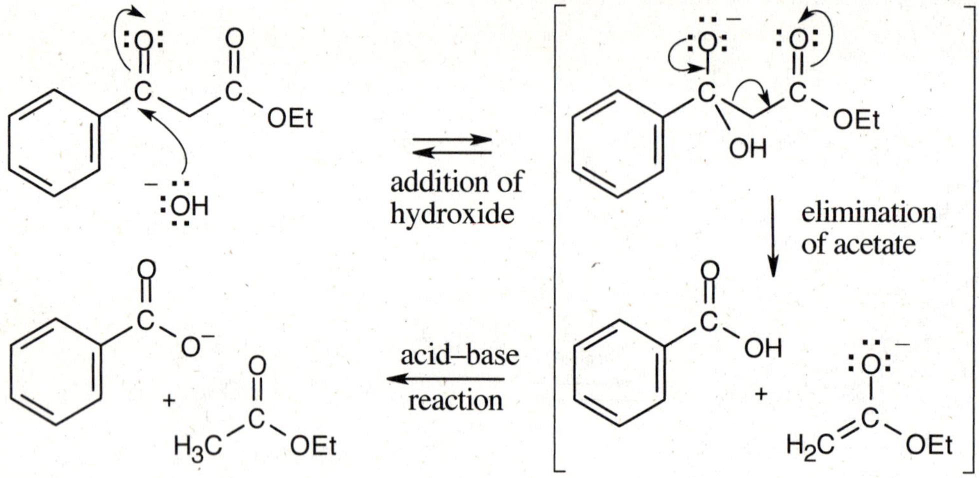


Hydroxide ion can react at two different sites of the ß-keto ester. Abstraction of the acidic α-proton is more favorable but is reversible and does not lead to product. Addition of hydroxide ion to the carbonyl group, followed by irreversible elimination of ethyl acetate, accounts for the observed products.

**11.45** Two of the products result from Claisen *self*-condensation.

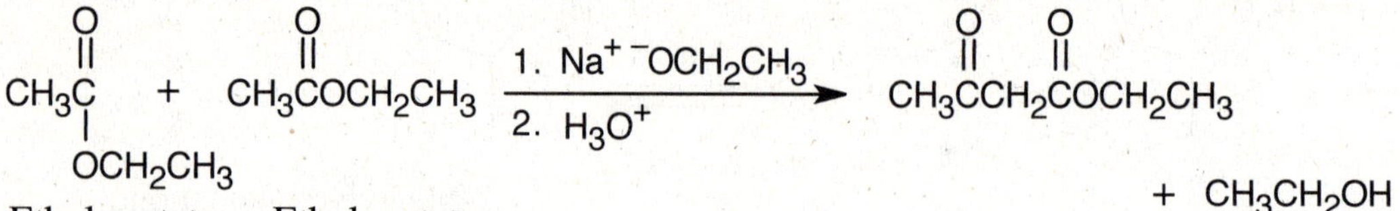

Ethyl acetate Ethyl acetate

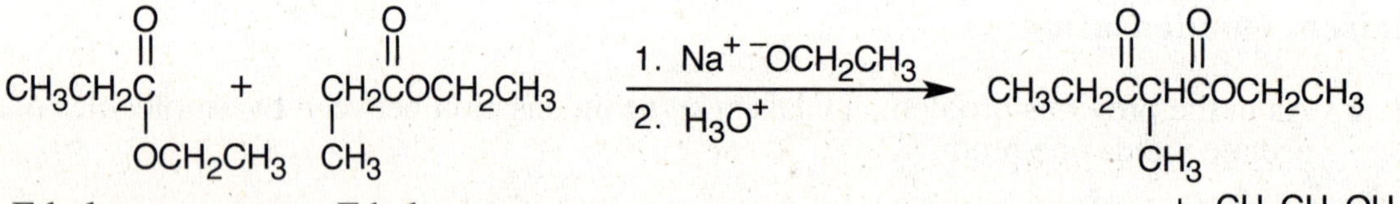

Ethyl propanoate Ethyl propanoate

The other two products are formed from *mixed* Claisen condensations.

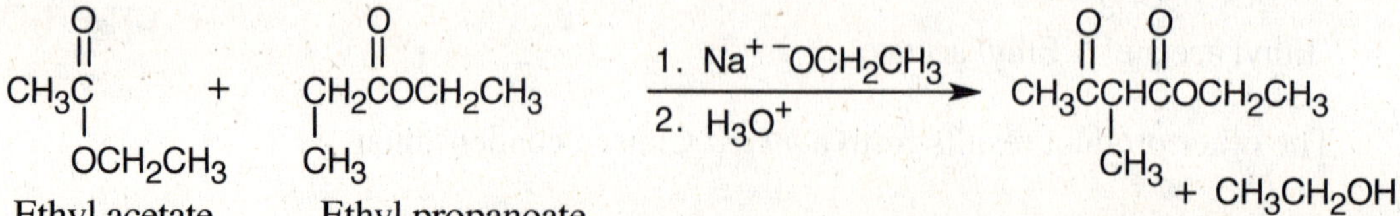

Ethyl acetate Ethyl propanoate

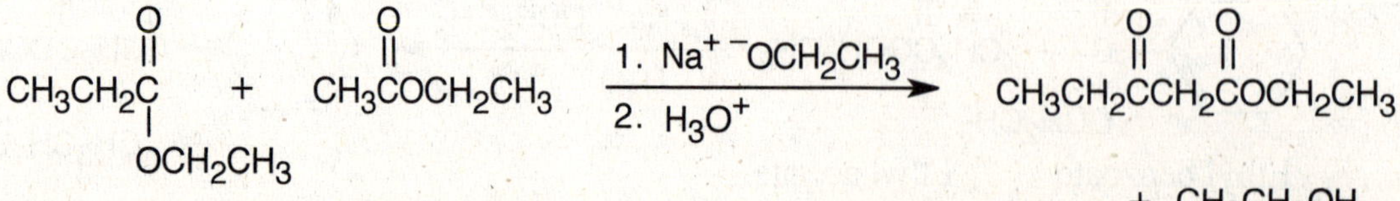

Ethyl propanoate Ethyl acetate

## General Problems

**11.46** As in the malonic acid synthesis, you should identify the components of the target compound. The "acetone component" (methyl ketone) comes from acetoacetic ester; the other component comes from a primary alkyl halide.

For all parts of this problem, the ethyl acetoacetate anion is prepared by the following reaction:

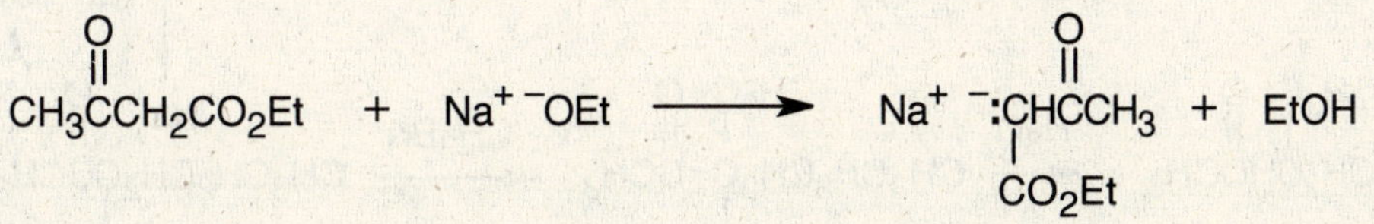

(a)

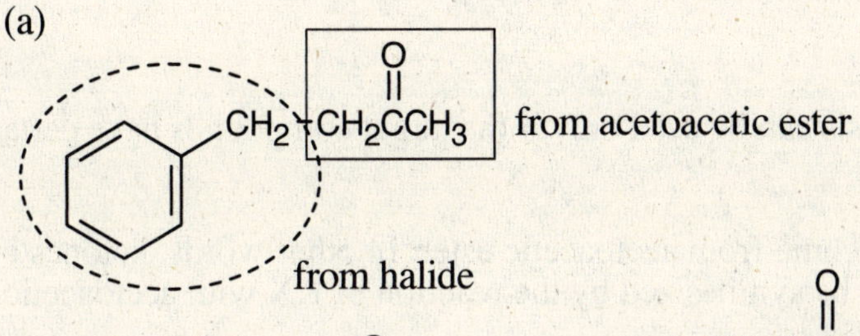


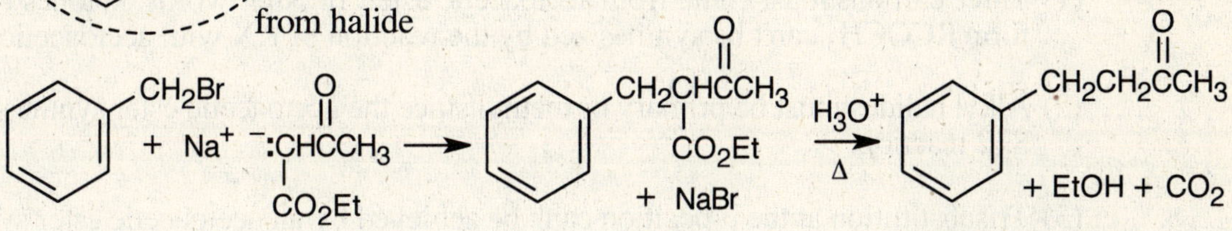

4-Phenylbutan-2-one

(b)

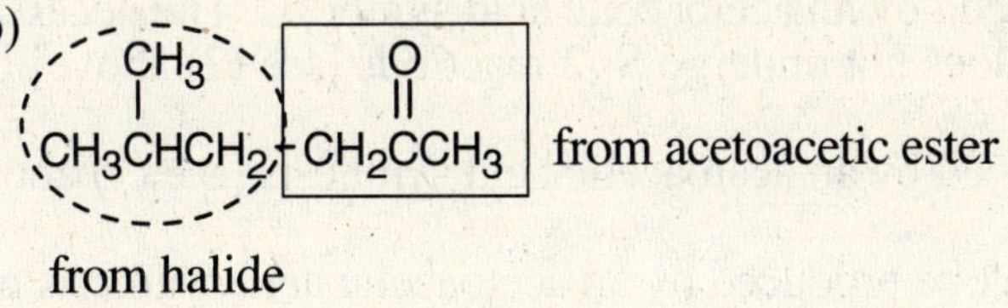


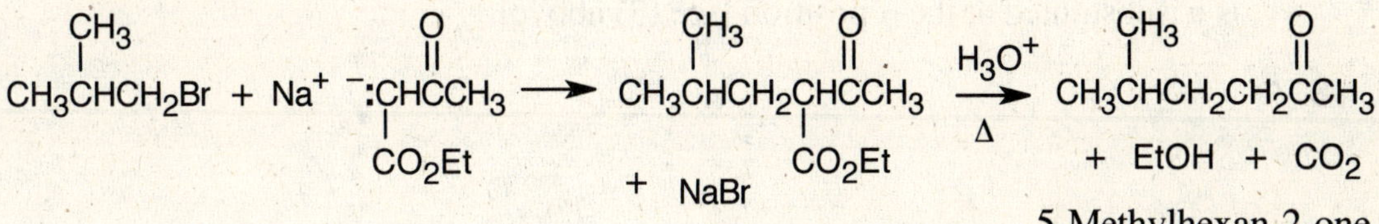

5-Methylhexan-2-one

(c)

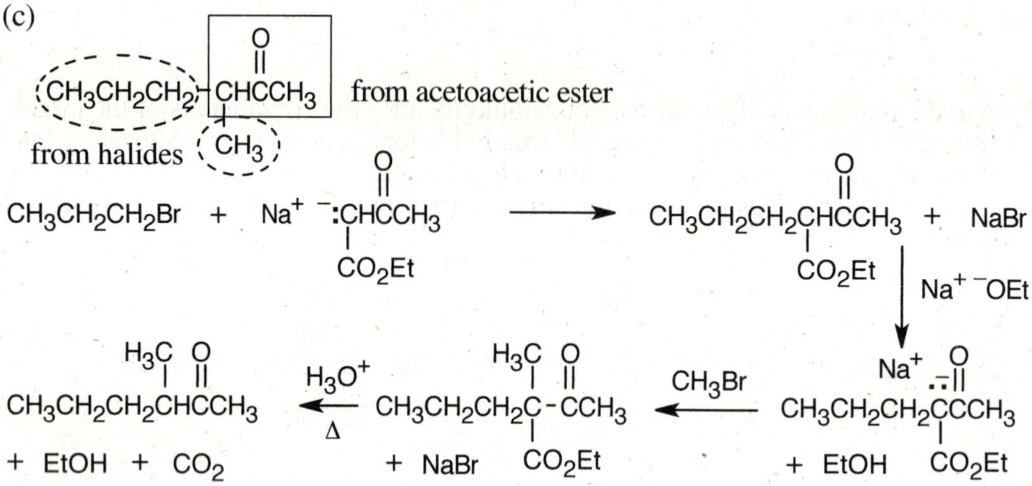


3-Methylhexan-2-one

**11.47** The acetoacetic ester synthesis can be used only if the target compounds have certain characteristics.

(1) Three carbons must come from acetoacetic ester. In other words, ketones of the form $RCOCH_3$ can't be synthesized by the reaction of RX with acetoacetic ester.

(2) Alkyl halides must be primary or methyl since the acetoacetic ester synthesis is an $S_N2$ reaction.

(3) Trisubstitution at the $\alpha$ position can't be achieved by an acetoacetic ester synthesis.

(a) Butan-2-one is produced by the reaction of sodium acetoacetate with $CH_3Br$.

(b) Phenylacetone can't be produced by an acetoacetic acid synthesis. The necessary halide component, bromobenzene, does not undergo $S_N2$ reactions [see (2) above].

(c) Acetophenone can't be produced by an acetoacetic acid synthesis [see (1) above].

(d) 3,3-Dimethylbutan-2-one can't be produced by an acetoacetic acid synthesis because it is trisubstituted at the $\alpha$ position [see (3) above].

**11.48**

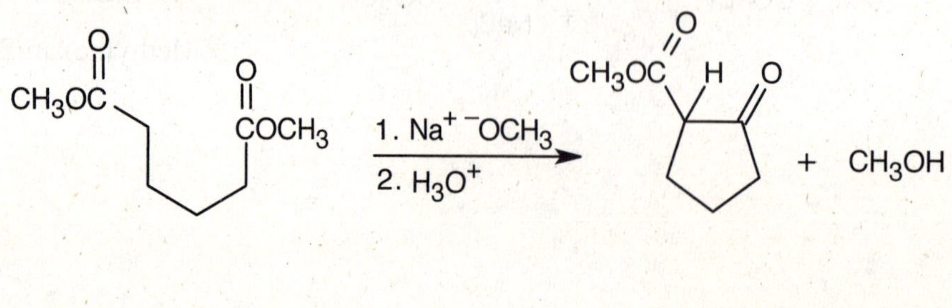

**11.49**

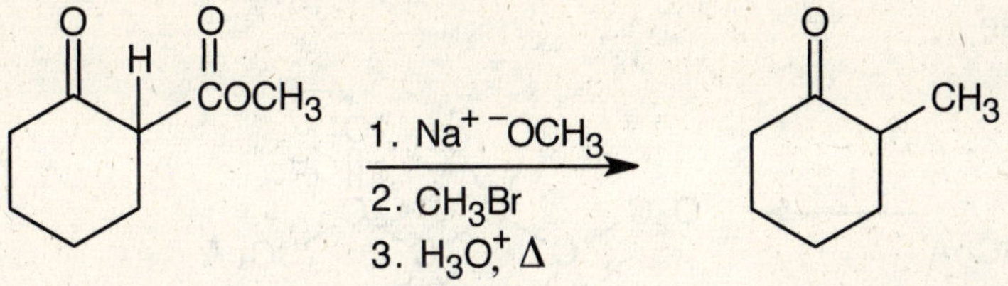

**11.50**

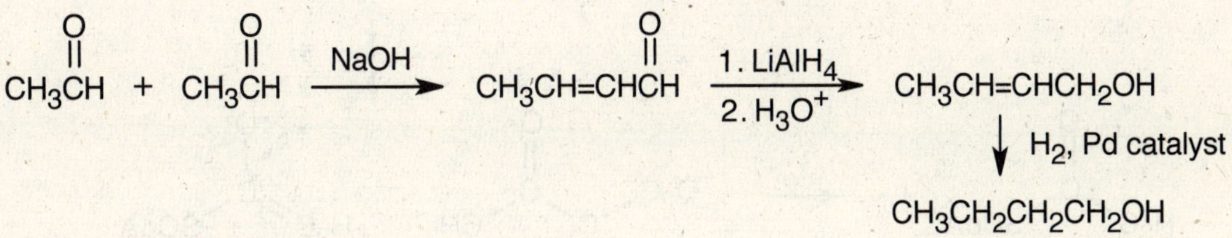

Butan-1-ol

**11.51** Aldol self-condensation:

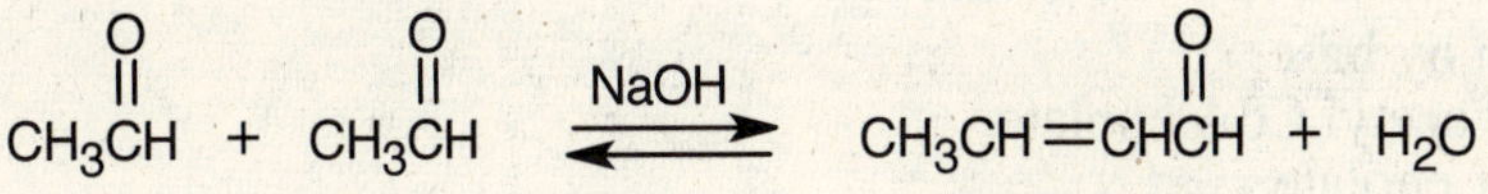

But-2-enal

Mixed aldol reaction:

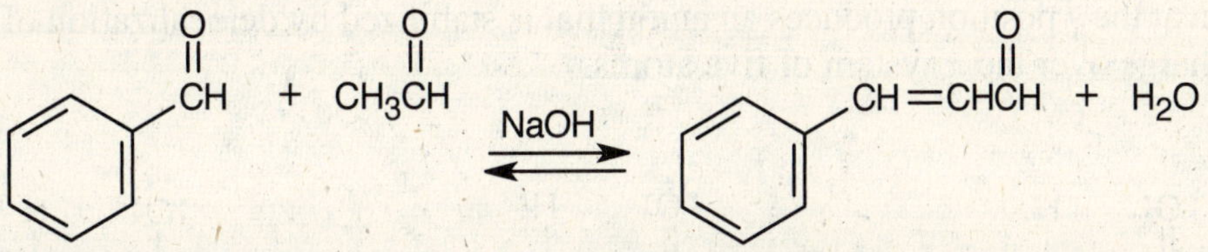

Cinnamaldehyde

**11.52** Malonic ester has two acidic protons and can be alkylated twice, to yield a compound of the structure $R_2C(CO_2C_2H_5)_2$. Decarboxylation then gives $R_2CHCO_2H$. The $\alpha$ proton is only weakly acidic and is no longer activated by two adjacent ester groups. Thus, trialkylation does not occur.

**11.53**

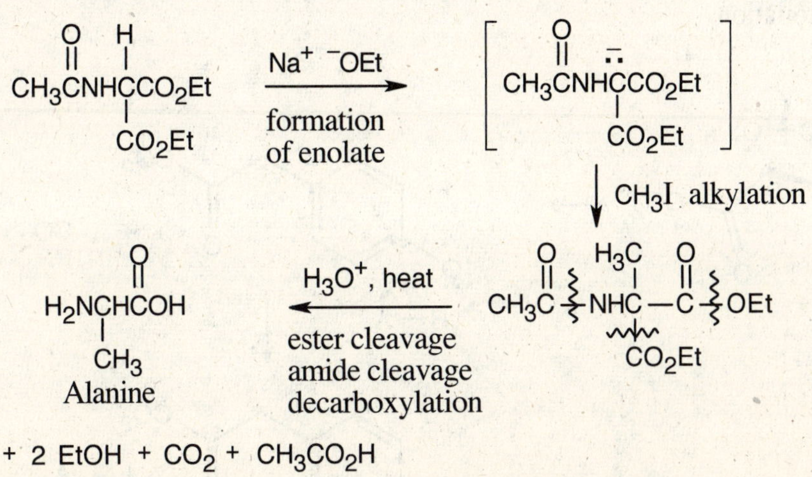

Acid cleaves both ester bonds, as well as the amide bond.

**11.54** This sequence is a reverse aldol reaction.

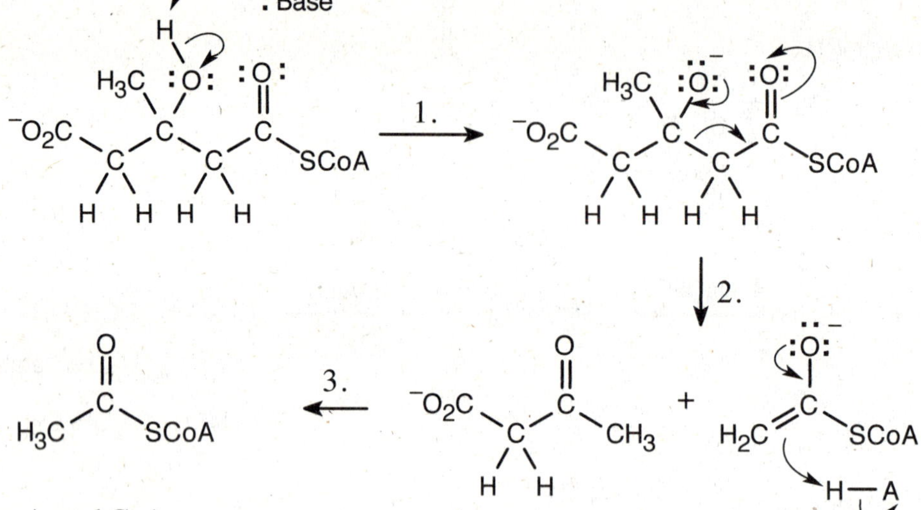


Acetyl CoA Acetoacetate

**Step 1:** Deprotonation by base.
**Step 2:** Elimination of acetyl CoA enolate.
**Step 3:** Protonation of enolate.

**11.55**

Enolization at the $\gamma$ position produces an anion that is stabilized by delocalization of the negative charge over the $\pi$ system of five atoms.

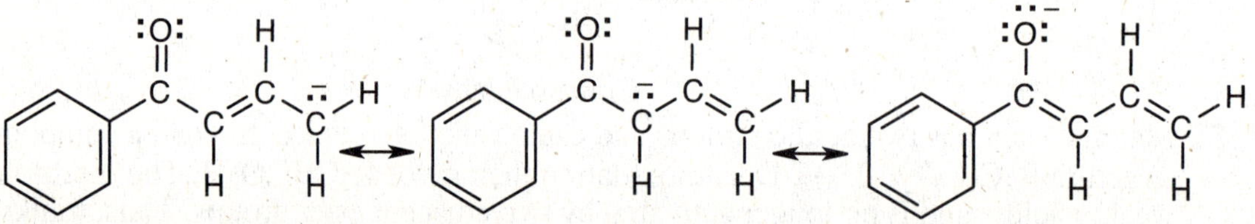

**11.56** Decarboxylation, which takes place because of the stability of the resulting anion, is followed by protonation.

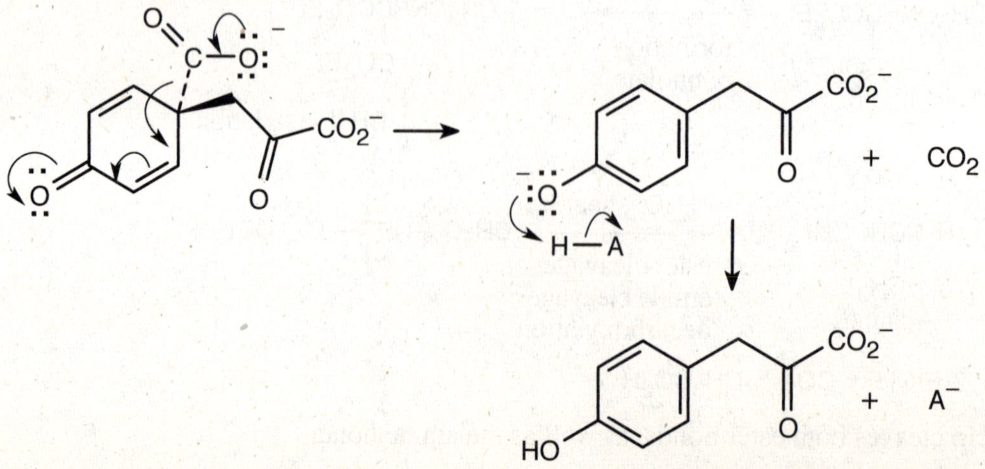

**11.57**

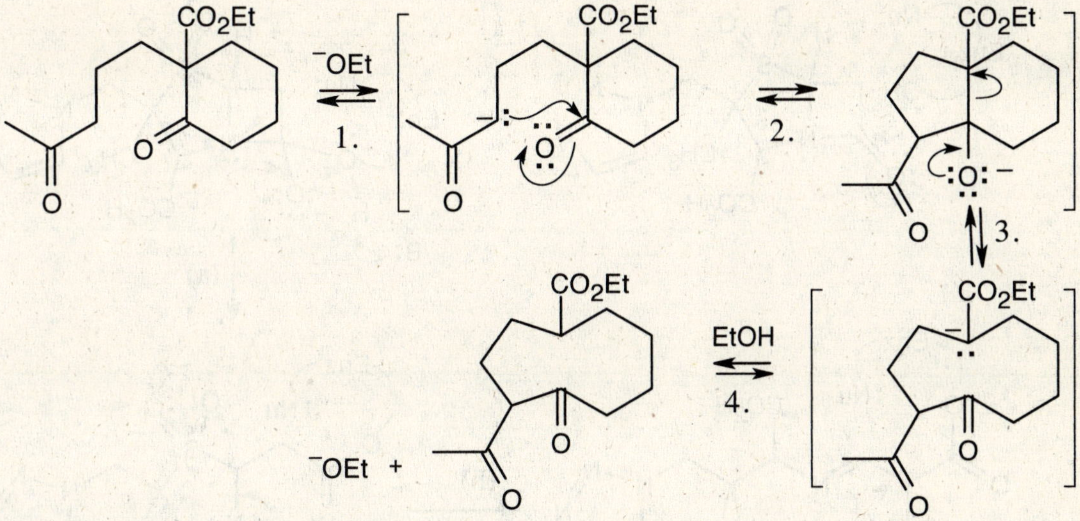


**Step 1:** Enolate formation.

**Step 2:** Intramolecular aldol condensation

**Step 3:** Retro-aldol

**Step 4:** Protonation

## In the Medicine Cabinet

**11.58**

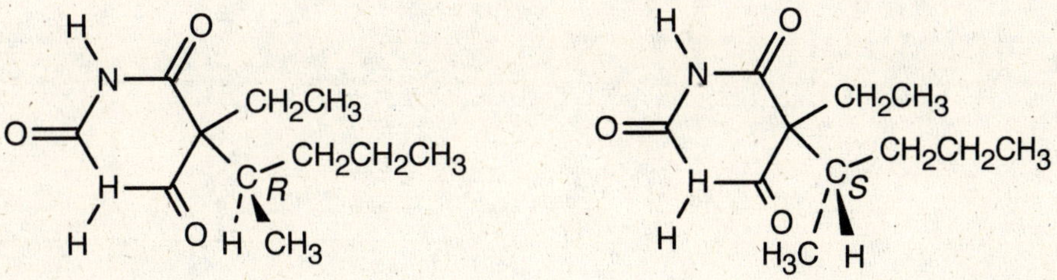


In the chiral cellular environment, the enantiomers are likely to have different biological activity.

**11.59**

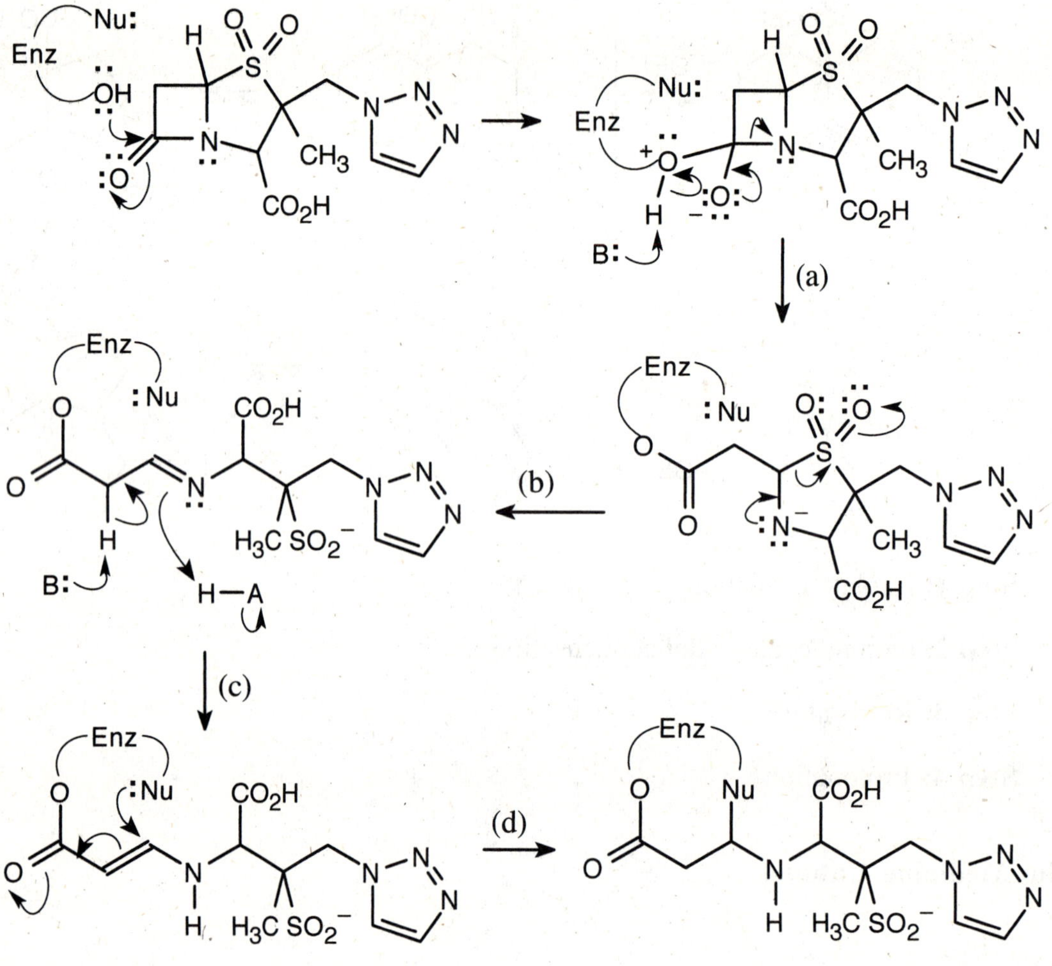

## In the Field

**11.60**

(a)

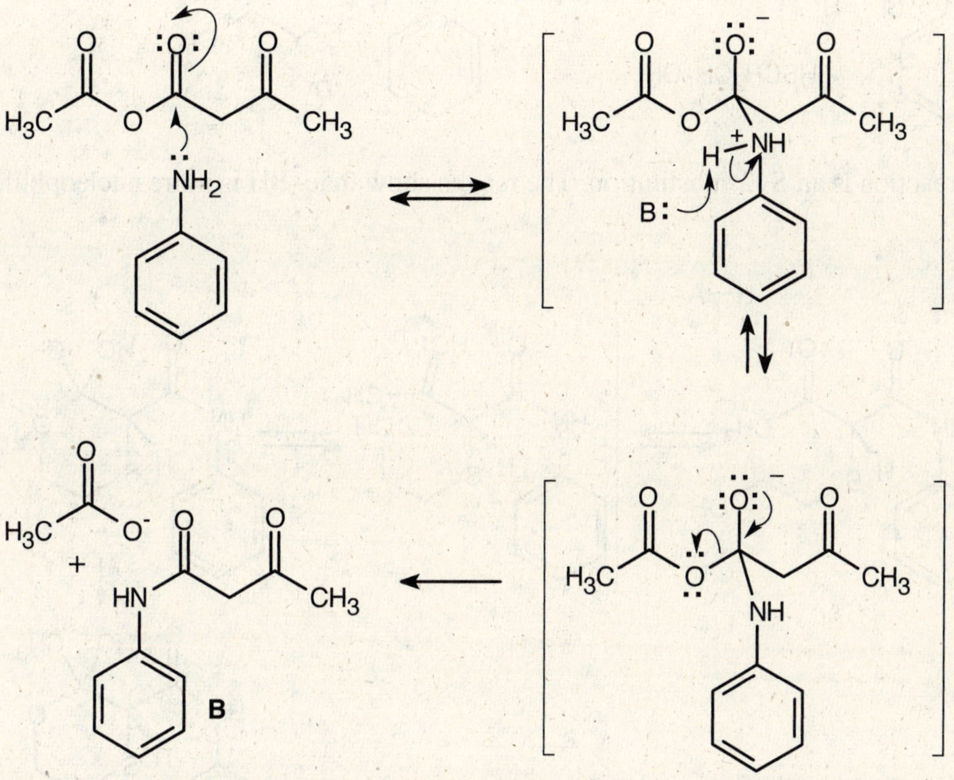


(b)

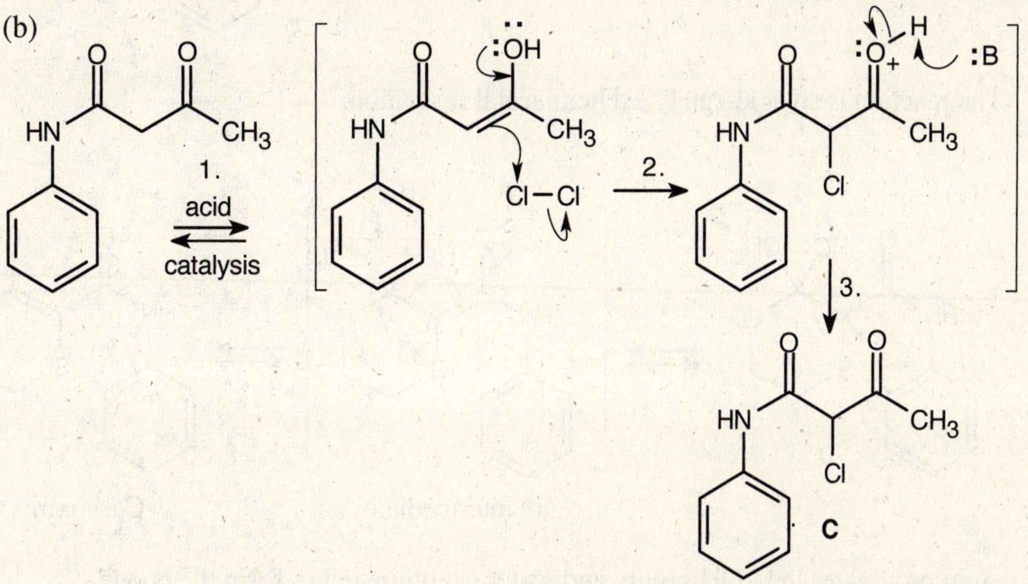


The steps: (1) acid-catalyzed enolization; (2) electrophilic addition of –Cl to the double bond; (3) deprotonation.

(c)

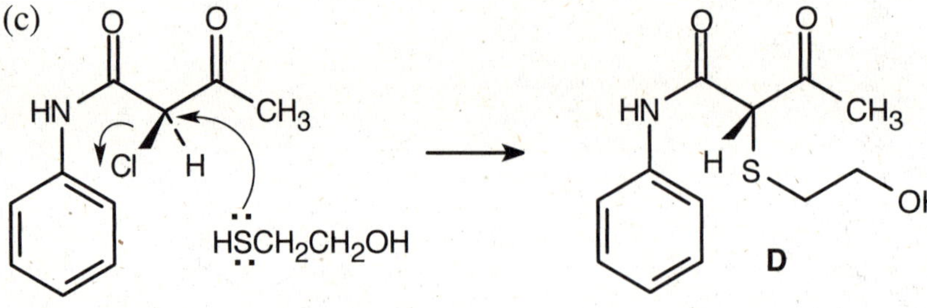


The reaction is an $S_N2$ substitution. The results show that –SH is more nucleophilic than –OH.

(d)

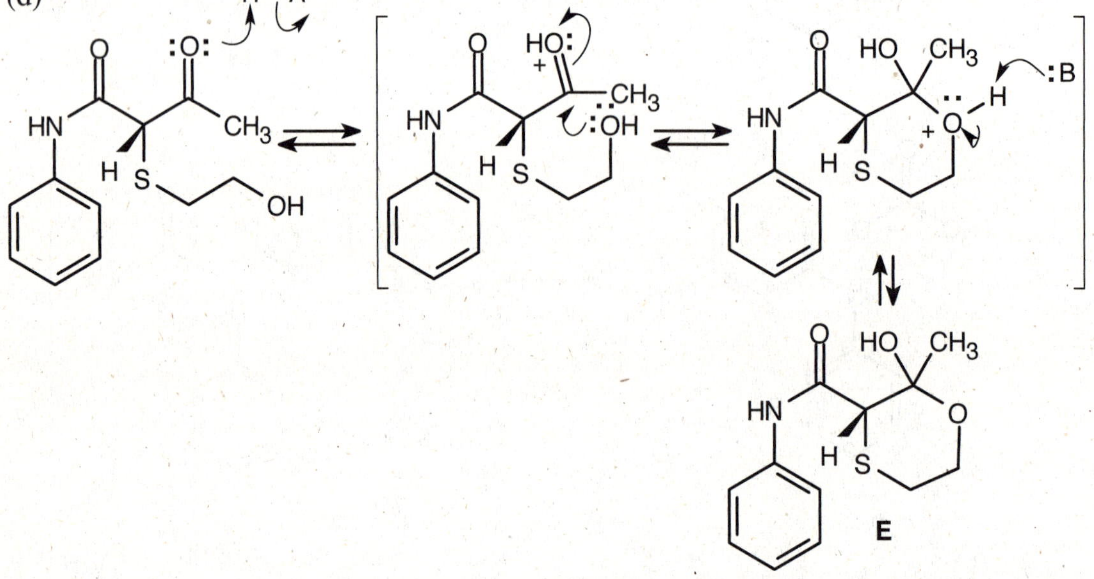


The reaction is an acid-catalyzed hemiacetal formation.

(e)

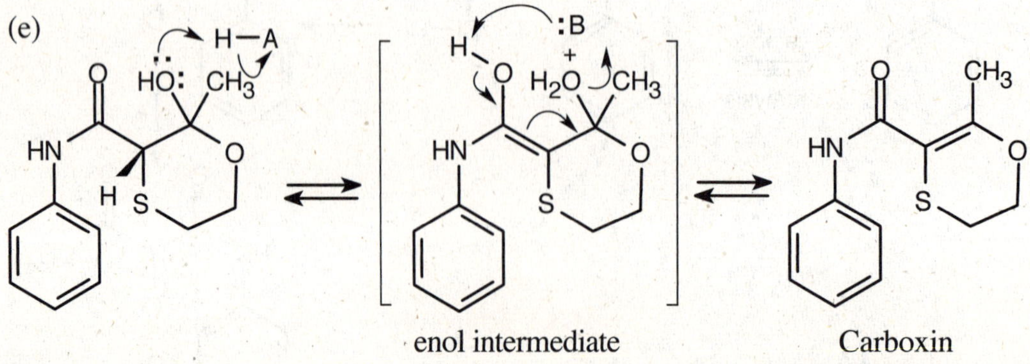


Acid protonates the –OH group, and water is eliminated to form the enone.

# Chapter 12 – Amines

## Chapter Outline

I. Facts about amines (Section 12.1–12.3).
  A. Naming amines (Section 12.1).
    1. Amines are classified as primary ($RNH_2$), secondary ($R_2NH$), tertiary ($R_3N$) or quaternary ammonium salts ($R_4N^+$).
    2. Primary amines are named two ways:
      a. For simple amines, the suffix *-amine* is added to the name of the alkyl substituent.
      b. For more complicated amines, the $-NH_2$ group is an amino substituent on the parent molecule.
    3. Secondary and tertiary amines:
      a. Symmetrical amines are named by using the prefixes *di-* and *tri-* before the name of the alkyl group.
      b. Unsymmetrical amines are named as *N*-substituted primary amines.
        i. The largest group is the parent.
    4. The simplest arylamine is aniline.
    5. Heterocyclic amines (nitrogen is part of a ring) have specific parent names.
      a. Nitrogen receives the lowest possible number.
  B. Structure and properties of amines (Section 12.2).
    1. The three amine bonds and the lone pair occupy the corners of a tetrahedron.
    2. Amines with fewer than 5 carbons are water-soluble.
    3. Amines have higher boiling points than alkanes of similar molecular weight because of hydrogen bonding.
    4. Amines smell nasty.
  C. Amine basicity (Section 12.3).
    1. The lone pair of electrons on nitrogen makes amines both nucleophilic and basic.
    2. The basicity constant $K_b$ is the measure of the equilibrium of an amine with water.
      a. The larger the value of $K_b$ (smaller $pK_b$), the stronger the base.
    3. Base strength.
      a. Alkylamines have similar basicities.
      b. Arylamines and heterocyclic aromatic amines are less basic than alkylamines.
        i. The nitrogen lone pair of an arylamine is delocalized over the ring $\pi$ system and is less available for bonding with an acid.
      c. Amides are nonbasic and are poor nucleophiles.
      d. Amine basicity can be used as a means of separating amines.
        i. An amine can be converted to its salt, extracted from a solution with water, neutralized, and re-extracted with an organic solvent.

II. Synthesis of amines (Section 12.4).
  A. Reduction of amides, nitriles and nitro groups.
    1. $S_N2$ displacement with $^-CN$, followed by reduction, converts a primary alkyl halide into an amine with one more carbon atom.
    2. Amide reduction converts an amide or nitrile into an amine with the same number of carbons.
  B. $S_N2$ reactions of alkyl halides.
    1. It is possible to alkylate ammonia or an amine with RX. Unfortunately, it is difficult to avoid overalkylation.

2. Using azide ion as the nucleophile, followed by reduction of the alkyl azide, produces a primary amine.

C. Reductive amination of aldehydes and ketones.
1. Treatment of an aldehyde or ketone with ammonia or an amine in the presence of a reducing agent yields an alkylated amine.
2. Primary, secondary, and tertiary amines can all be prepared by reductive amination.

D. Reduction of nitrobenzenes.
1. Catalytic hydrogenation can be used if no other interfering groups are present.
2. Fe, Sn, and $SnCl_2$ can also be used.

III. Reactions of amines (Section 12.5).
A. Alkylation and acylation.
1. Alkylation of primary and secondary amines is hard to control.
2. Primary and secondary amines can also be acylated by reaction with acid chlorides or acid anhydrides.

B. 1. Primary amines react with aldehydes or ketones to give imines.

IV. Heterocyclic aromatic amines (Section 12.6).
A. Pyrrole.
1. Although pyrrole is both an amine and a conjugated diene, its behavior is that of an aromatic ring.
2. Pyrrole is nonbasic because all 5 of the nitrogen electrons are part of the aromatic ring.
3. The carbon atoms in pyrrole are electron-rich and are reactive toward electrophiles.
a. Substitution occurs at the position next to nitrogen.

B. Pyridine.
1. Pyridine is the nitrogen-containing analog of benzene.
2. The nitrogen lone pair isn't part of the $\pi$ electron system and is basic.
3. Pyridine is a stronger base than pyrrole but a weaker base than alkylamines.
4. Pyridine is often used as a basic catalyst.

C. Fused-ring heterocycles.

V. Alkaloids (Section 12.7).
A. Alkaloids are naturally-occurring amines.
B. Many alkaloids have biological properties.

## Solutions to Problems

**12.1**

(a)
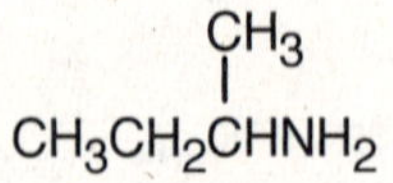
primary amine

(b)
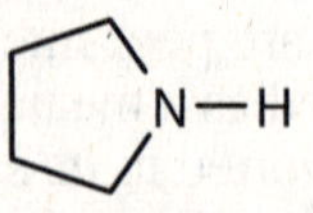
secondary amine

(c)
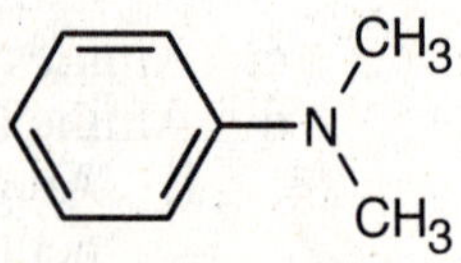
tertiary amine

**12.2** Other structures can be drawn.

(a)
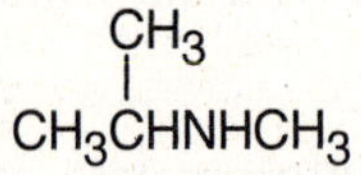

(b)
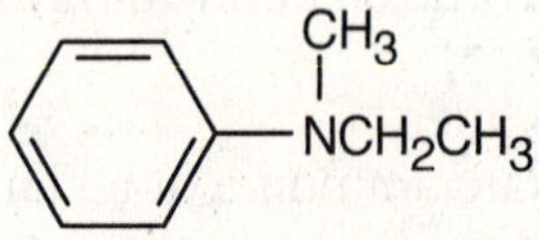

(c)
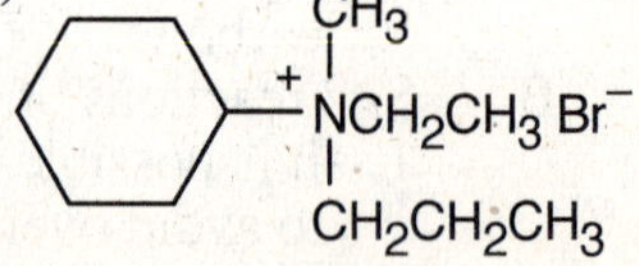

**12.3**

(a)

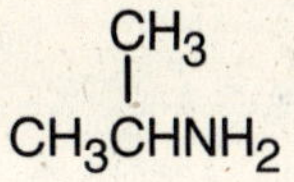

Isopropylamine

(b)

$CH_3CH_2NHCH_2CH_3$

Diethylamine

(c)

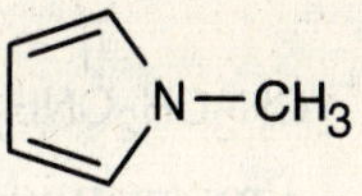

*N*-Methylpyrrole

(d)

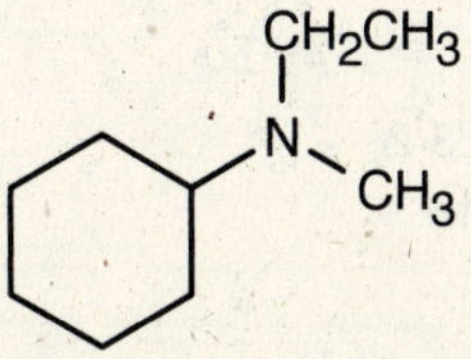

*N*-Ethyl-*N*-methylcyclohexylamine

(e)

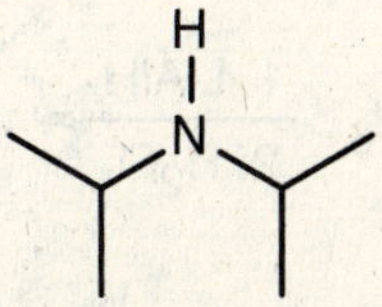

Diisopropylamine

(f)

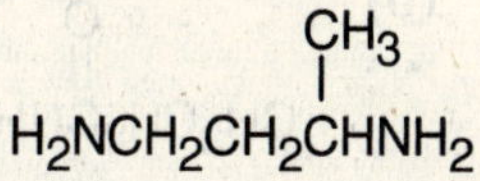

Butane-1,3-diamine

**12.4**

(a)

$(CH_3CH_2)_3N$

Triethylamine

(b)

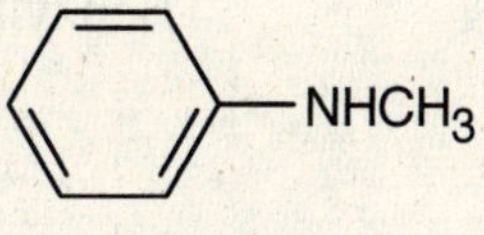

*N*-Methylaniline

(c)

$(CH_3CH_2)_4N^+\ Br^-$

Tetraethylammonium bromide

(d)

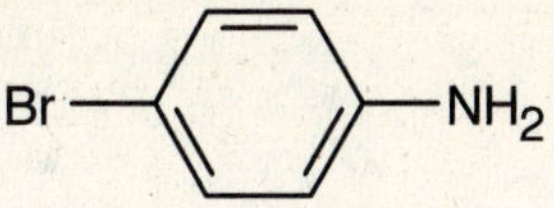

*p*-Bromoaniline

(e)

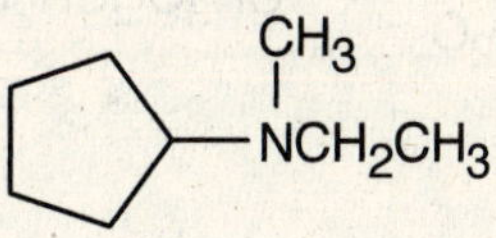

*N*-Ethyl-*N*-methylcyclopentylamine

**12.5**

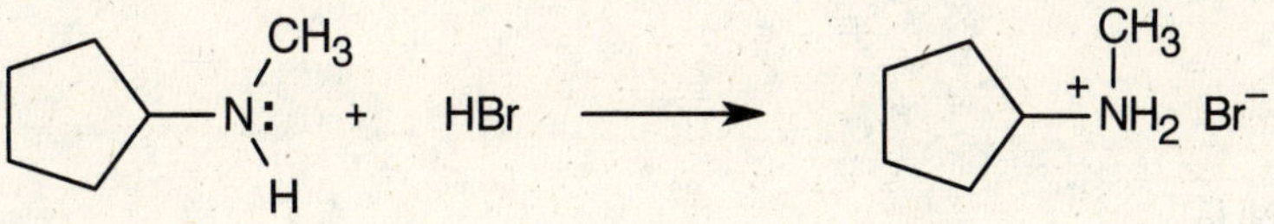

**12.6**

| *More basic* | *Less basic* |
|---|---|
| (a) $CH_3CH_2NH_2$ amine | $CH_3CH_2CONH_2$ amide |
| (b) NaOH hydroxide | $C_6H_5NH_2$ arylamine |
| (c) $CH_3NHCH_3$ alkylamine | $CH_3NHC_6H_5$ arylamine |
| (d) $(CH_3)_3N$ amine | $CH_3OCH_3$ ether |

The basicity order for the above compounds:
hydroxide > alkylamine > arylamine >> amide, ether

**12.7**

(a)

$$CH_3CH_2\overset{O}{\overset{\|}{C}}NH_2 \xrightarrow[2.\ H_2O]{1.\ LiAlH_4} CH_3CH_2CH_2NH_2$$

Propanamide → Propylamine

(b)

$$CH_3CH_2\overset{O}{\overset{\|}{C}}NHCH_2CH_2CH_3 \xrightarrow[2.\ H_2O]{1.\ LiAlH_4} NH(CH_2CH_2CH_3)_2$$

*N*-Propylpropanamide → Dipropylamine

(c)

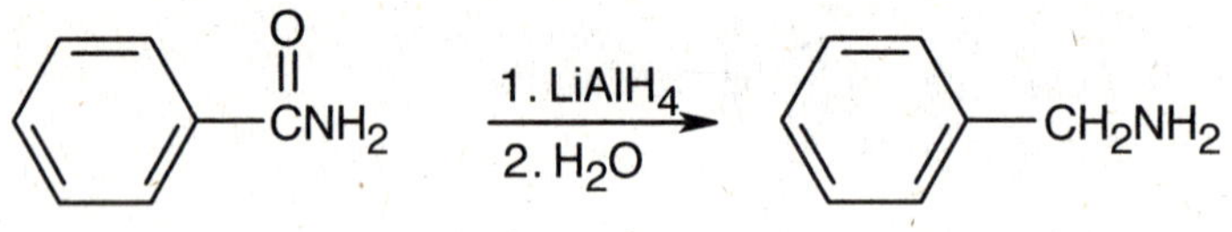

Benzamide → Benzylamine

**12.8**

(a)

$$CH_3\overset{CH_3}{\overset{|}{C}}HCH_2C{\equiv}N \xrightarrow[2.\ H_2O]{1.\ LiAlH_4} CH_3\overset{CH_3}{\overset{|}{C}}HCH_2CH_2NH_2$$

3-Methylbutanenitrile

(b)

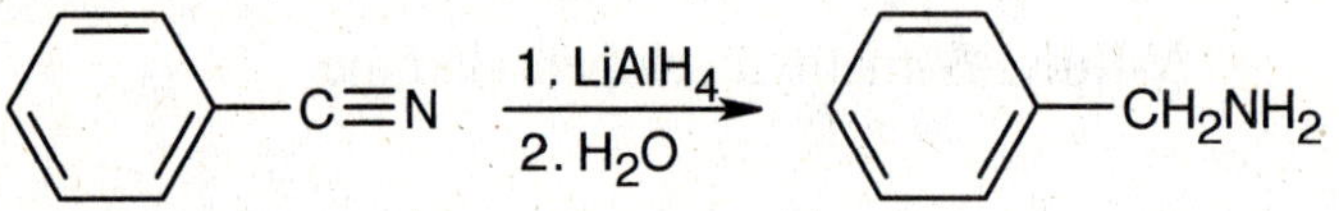

Benzonitrile

**12.9**

(a)

$$3\ CH_3CH_2Br \xrightarrow{NH_3} (CH_3CH_2)_3N$$

Some quaternary ammonium salt is also formed.

(b)

$$4\ CH_3Br \xrightarrow{NH_3} (CH_3)_4N^+\ Br^-$$

**12.10**

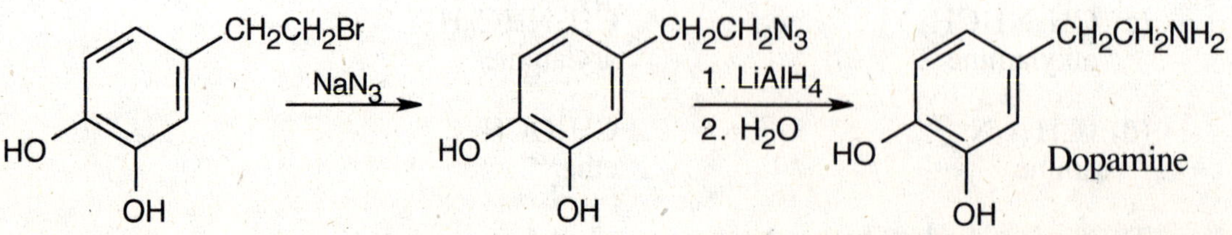

Dopamine

Reaction of an alkyl halide with $NaN_3$ is a good method for synthesizing a primary amine.

**12.11** Look at the target molecule to find the groups bonded to nitrogen. One group comes from the aldehyde/ketone precursor, and the other group comes from the amine precursor. In most cases, two combinations of amine and aldehyde/ketone are possible.

*Amine* *Amine Precursor* *Carbonyl Precursor*

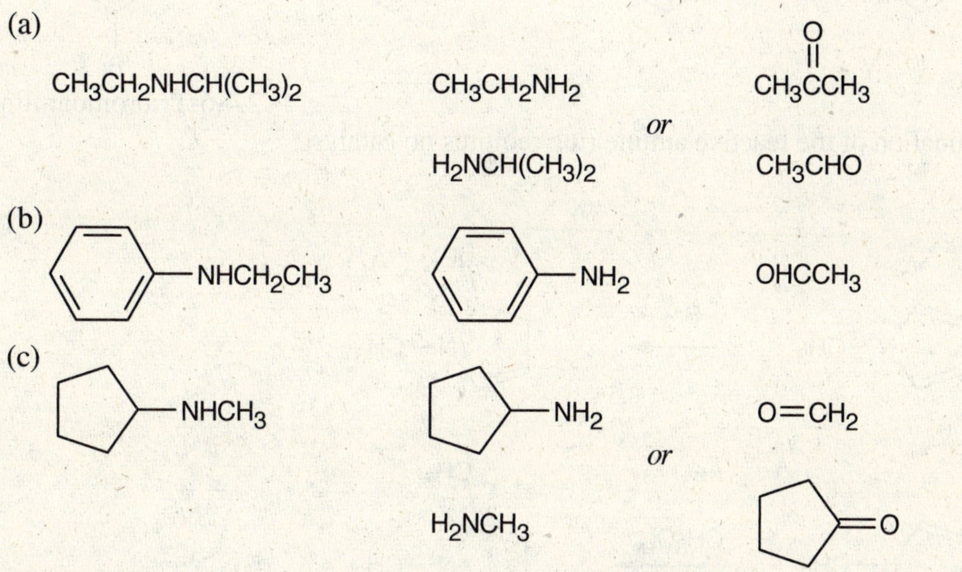


**12.12** The target molecule is a tertiary amine, which can be formed by either of two combinations of aldehyde plus secondary amine. After the nucleophilic addition reaction of the two components, the intermediate imine is reduced by catalytic hydrogenation, using nickel as a catalyst.

*Amine* *Amine Precursor* *Carbonyl Precursor*

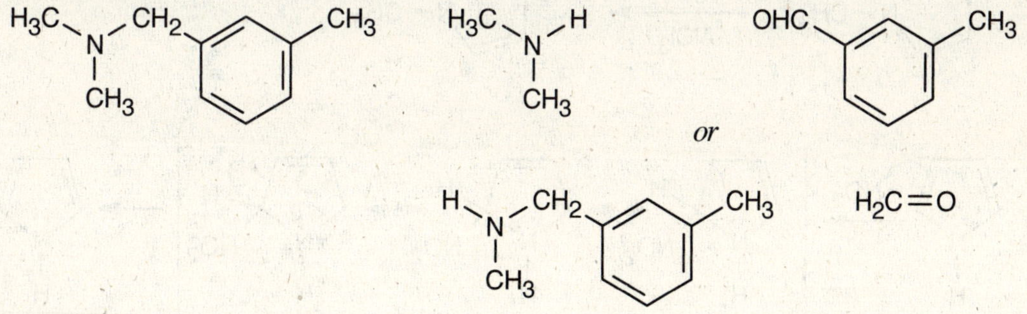


**12.13**

(a)

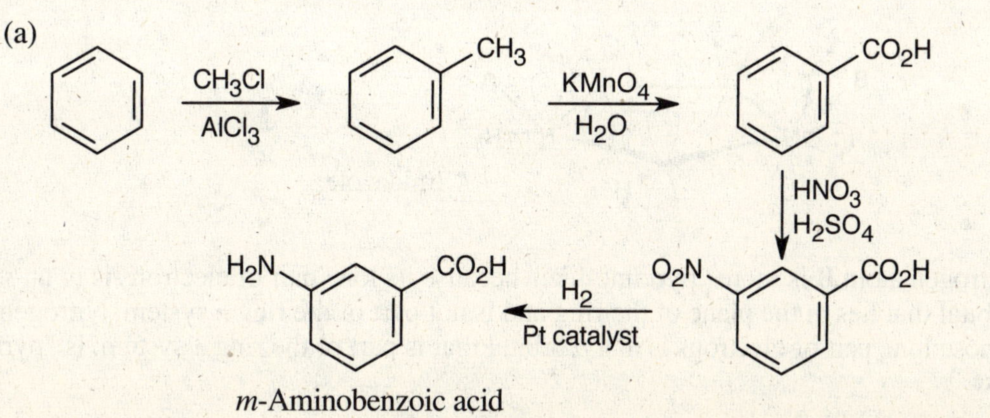


*m*-Aminobenzoic acid

(b)

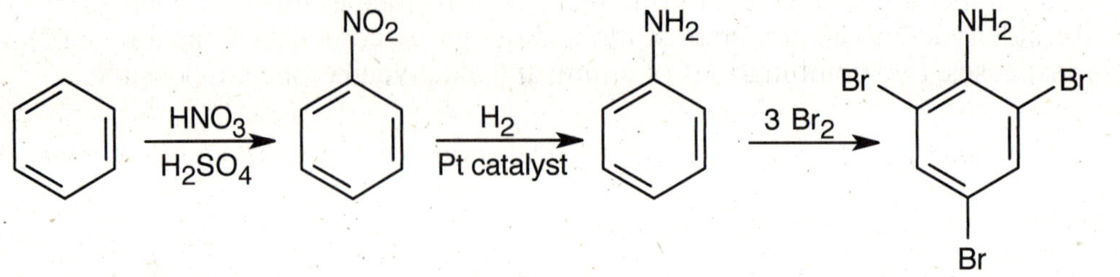


2,4,6-Tribromoaniline

Bromination of the reactive aniline ring requires no catalyst.

**12.14**

(a)

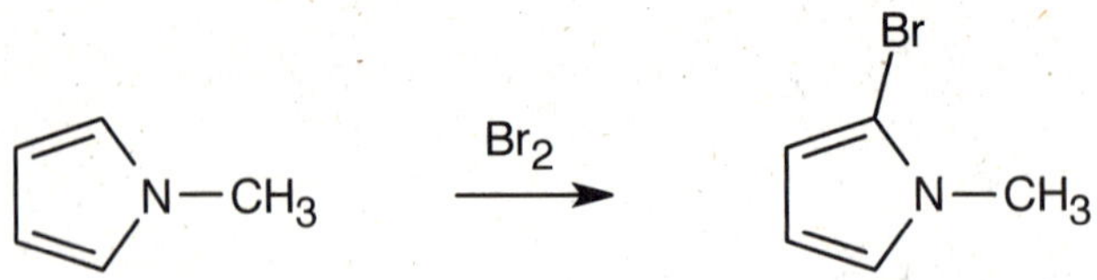


(b)

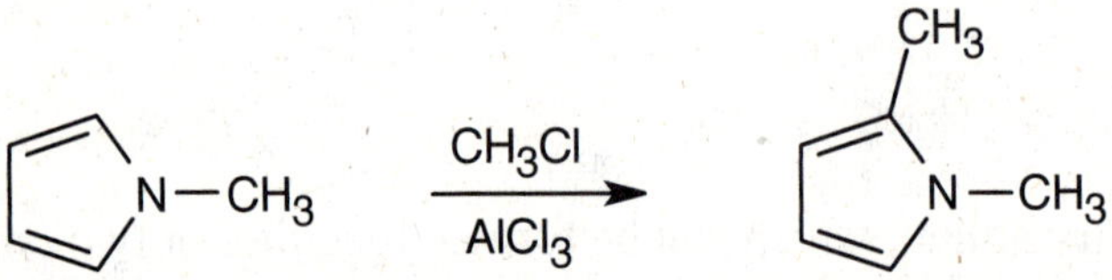


(c)

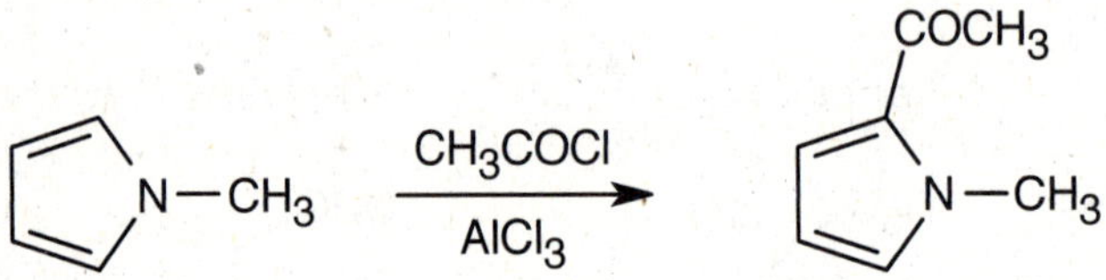


**12.15**

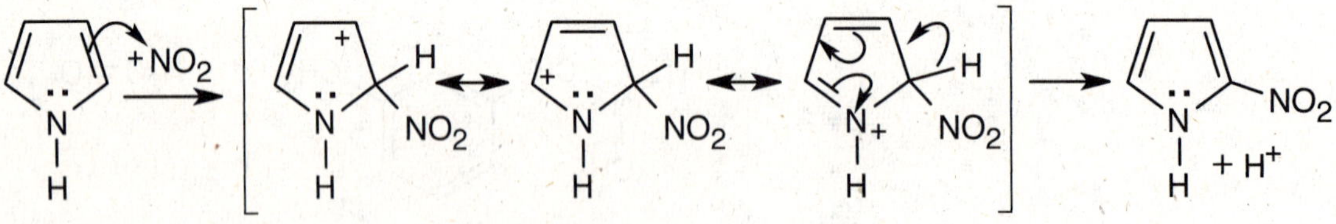


**12.16**

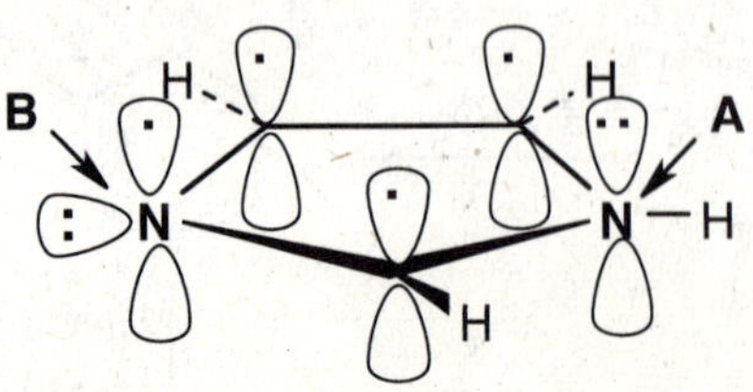


Imidazole

Nitrogen atom **B** is more “pyridine-like” because its lone pair of electrons is in an $sp^2$ orbital that lies in the plane of the ring and is not part of the ring $\pi$ system. Nitrogen **A**, whose lone pair of electrons is in a *p* orbital that is part of the ring $\pi$ system, is "pyrrole-like."

**12.17** The electrostatic potential map shows that the nitrogen on the left (corresponding to **B** in the previous problem) is more basic (red). This makes sense because the lone pair of **A** is part of the ring $\pi$ system and is not available for donation to a Lewis acid.

**12.18**

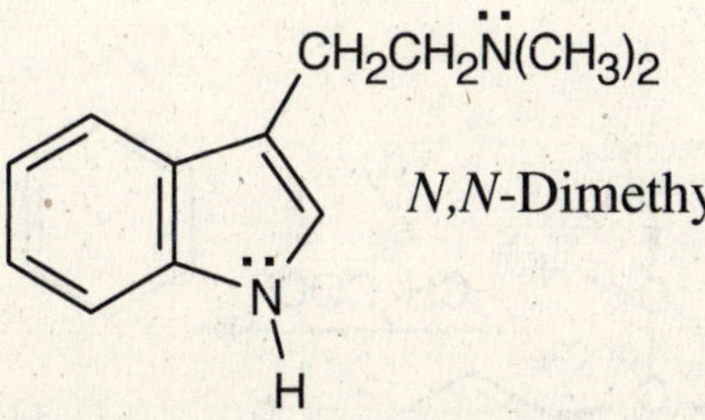

*N,N*-Dimethyltryptamine

The side chain nitrogen atom of *N,N*-dimethyltryptamine is more basic than the ring nitrogen atom because its lone electron pair is more available for donation to a Lewis acid. The aromatic nitrogen electron lone pair is part of the ring $\pi$ electron system.

## Visualizing Chemistry

**12.19**

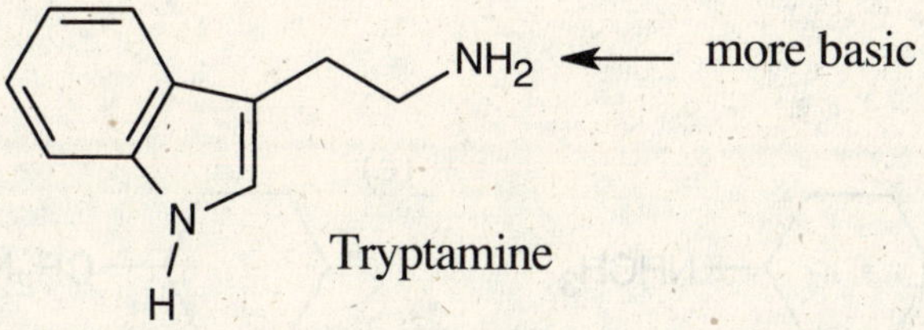


Tryptamine

The electrostatic potential map shows that the indicated nitrogen is more basic because it is more electron-rich. The electrons of the other nitrogen are part of the fused-ring $\pi$ system and are less available for donation to a Lewis acid.

**12.20**

(a)

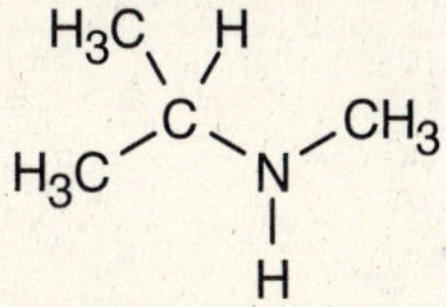

*N*-Methylisopropylamine
secondary amine

(b)

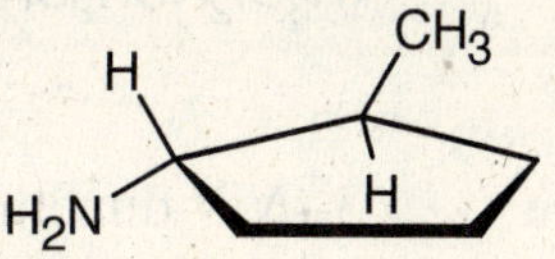

*trans*-(2-Methylcyclopentyl)amine
primary amine

(c)

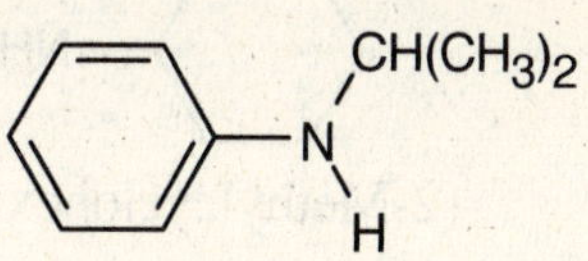

*N*-Isopropylaniline
secondary amine

**12.21**

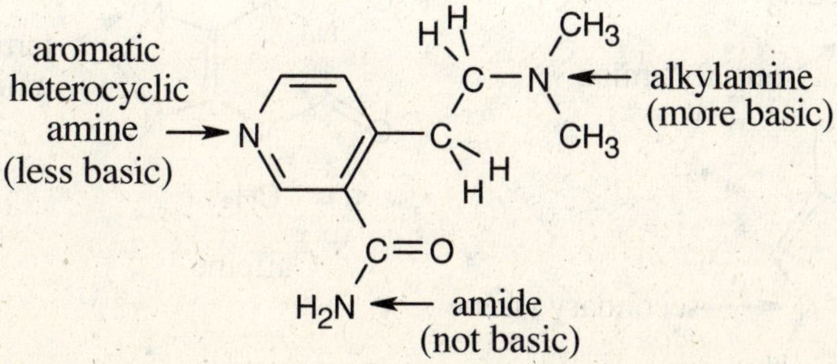

**12.22**

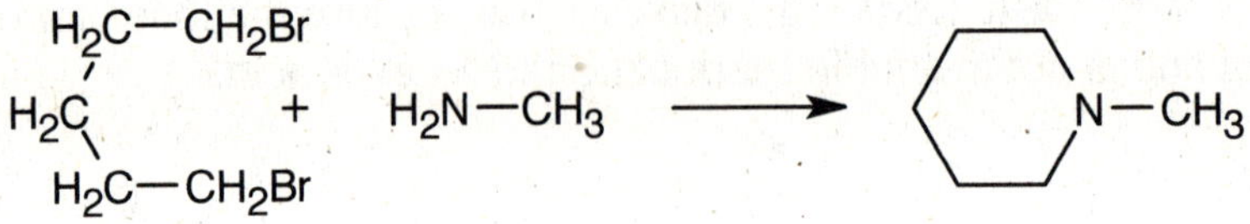

**12.23**

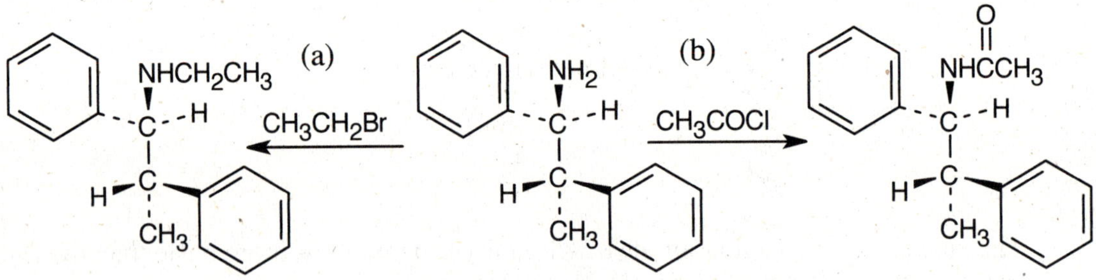

(1*S*,2*S*)-1,2-Diphenylpropylamine

In reaction (a), polyalkylated product will also form.

## Additional Problems

### Naming Amines

**12.24**

(a)

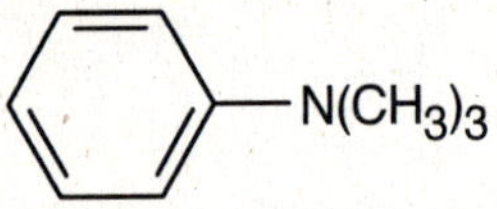

*N*,*N*-Dimethylaniline

(b)

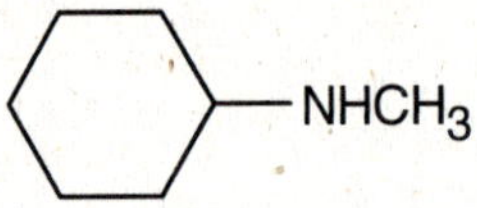

*N*-Methylcyclohexylamine

(c)

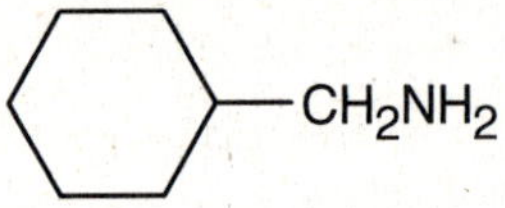

(Cyclohexylmethyl)amine

(d)

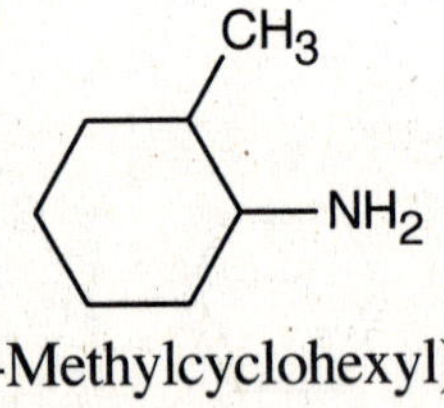

(2-Methylcyclohexyl)amine

(e)

$(H_3C)_2NCH_2CH_2CO_2H$

3-(*N*,*N*-dimethylamino)-propanoic acid

**12.25**

(a)

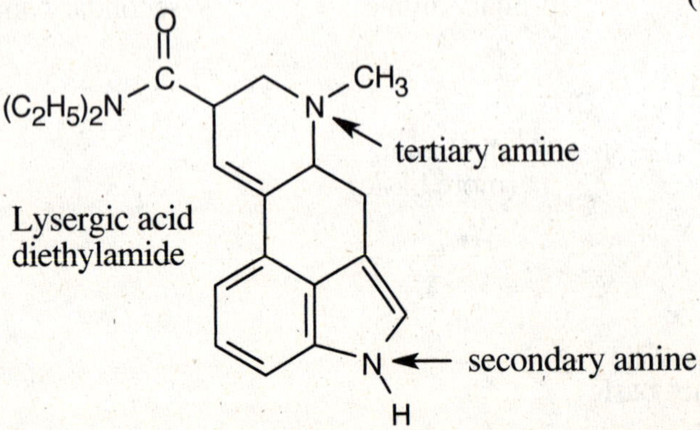


(b)

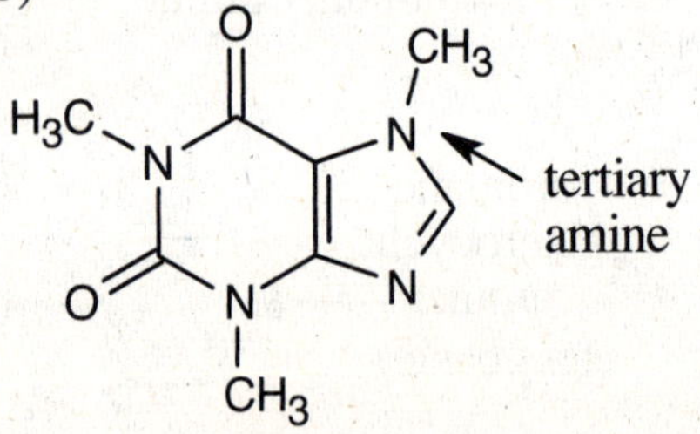


Caffeine

**12.26**

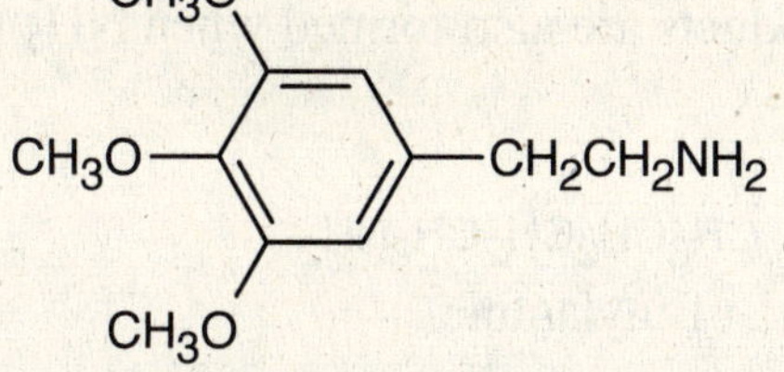

2-(3,4,5-Trimethoxyphenyl)ethylamine

**12.27**

(a)

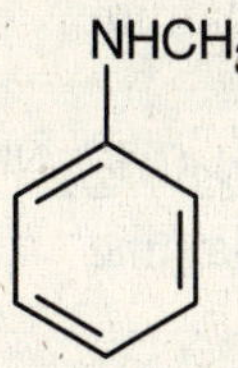

*N*-Methylaniline

(b)

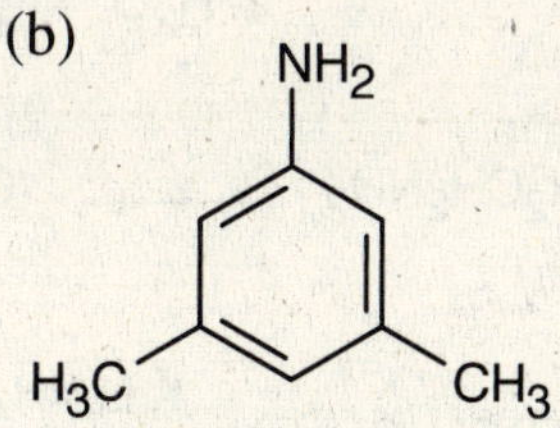

3,5-Dimethylaniline

(c)

$(CH_3)_4N^+\ Br^-$

Tetramethylammonium bromide

(d)

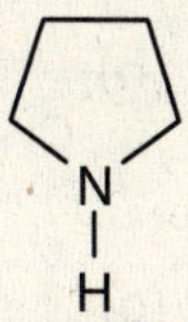

Pyrrolidine

**12.28**

$CH_3CH_2CH_2CH_2NH_2$

Butylamine (primary)

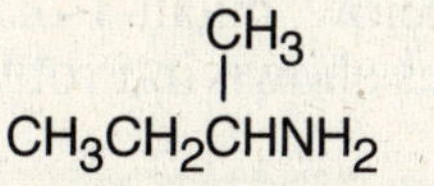

*sec*-Butylamine (primary)

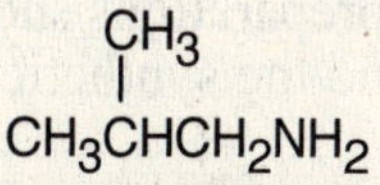

Isobutylamine (primary)

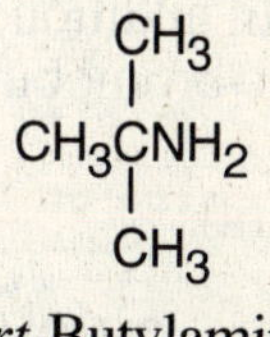

*tert*-Butylamine (primary)

$CH_3CH_2CH_2NHCH_3$

*N*-Methylpropylamine (secondary)

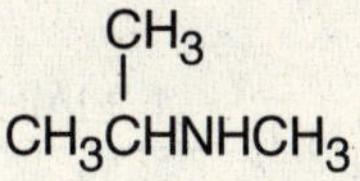

*N*-Methylisopropylamine (secondary)

$CH_3CH_2NHCH_2CH_3$

Diethylamine (secondary)

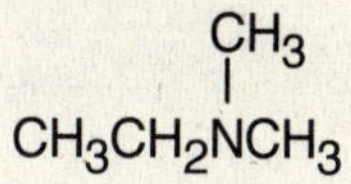

*N,N*-Dimethylethylamine (tertiary)

**12.29**

(a)

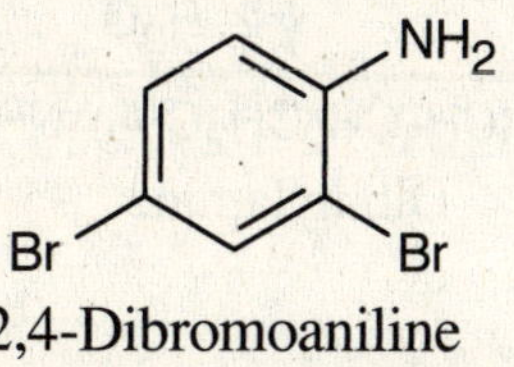

2,4-Dibromoaniline

(b)

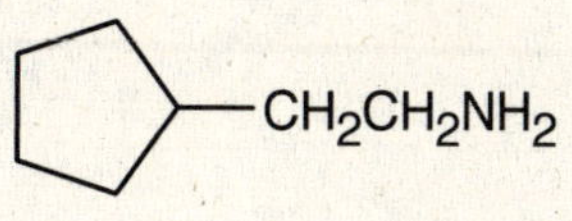

(2-Cyclopentylethyl)amine

(c)

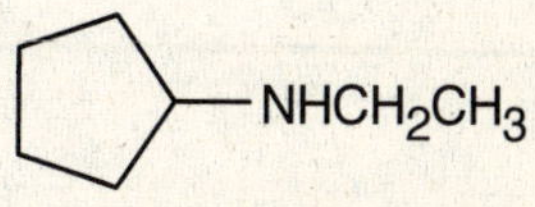

*N*-Ethylcyclopentylamine

(d)

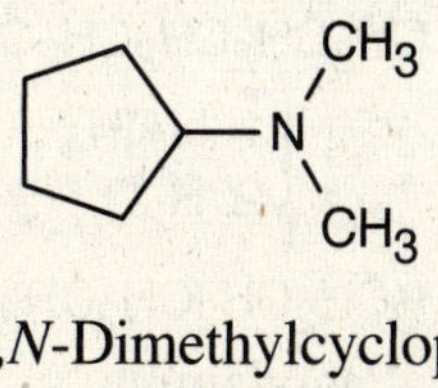

*N,N*-Dimethylcyclopentylamine

(e)

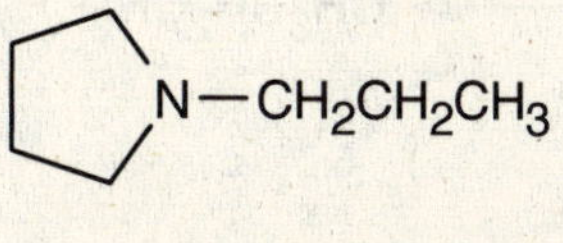

*N*-Propylpyrrolidine

(f)

$H_2NCH_2CH_2CH_2CN$

4-Aminobutanenitrile

## Reactions and Synthesis

**12.30** In parts (a) and (b), overalkylation products are also formed when $NH_3$ is used.

(a)

$$CH_3CH_2CH_2CH_2Br \xrightarrow{\text{excess } NH_3} \underset{\text{Butylamine}}{CH_3CH_2CH_2CH_2NH_2}$$

An alternate (better) route:

$$CH_3CH_2CH_2CH_2Br \xrightarrow{NaN_3} CH_3CH_2CH_2CH_2N_3 \xrightarrow[\text{2. } H_2O]{\text{1. } LiAlH_4} \underset{\text{Butylamine}}{CH_3CH_2CH_2CH_2NH_2}$$

(b)

$$CH_3CH_2CH_2CH_2Br + \underset{\text{from (a)}}{CH_3CH_2CH_2CH_2NH_2} \longrightarrow \underset{\text{Dibutylamine}}{(CH_3CH_2CH_2CH_2)_2NH}$$

(c)

$$CH_3CH_2CH_2CH_2Br \xrightarrow{NaCN} CH_3CH_2CH_2CH_2CN \xrightarrow[\text{2. } H_2O]{\text{1. } LiAlH_4} \underset{\text{Pentylamine}}{CH_3CH_2CH_2CH_2CH_2NH_2}$$

**12.31**

In addition to the preparations shown below, butan-1-ol can be treated with $PBr_3$ to form 1-bromobutane, and the synthetic routes shown in Problem 12.30 can be used.

(a)

$$CH_3CH_2CH_2CH_2OH \xrightarrow[H_3O^+]{CrO_3} CH_3CH_2CH_2\overset{O}{\overset{\|}{C}}OH \xrightarrow{SOCl_2} CH_3CH_2CH_2\overset{O}{\overset{\|}{C}}Cl$$

$$CH_3CH_2CH_2\overset{O}{\overset{\|}{C}}Cl \xrightarrow{2\ NH_3} CH_3CH_2CH_2\overset{O}{\overset{\|}{C}}NH_2$$

$$CH_3CH_2CH_2\overset{O}{\overset{\|}{C}}NH_2 \xrightarrow[\text{2. } H_2O]{\text{1. } LiAlH_4} \underset{\text{Butylamine}}{CH_3CH_2CH_2CH_2NH_2}$$

(b)

$$\underset{\text{from (a)}}{CH_3CH_2CH_2\overset{O}{\overset{\|}{C}}Cl} + \underset{\text{from (a)}}{CH_3CH_2CH_2CH_2NH_2} \longrightarrow CH_2CH_2CH_2\overset{O}{\overset{\|}{C}}NHCH_2CH_2CH_2CH_3$$

$$\xrightarrow[\text{2. } H_2O]{\text{1. } LiAlH_4} \underset{\text{Dibutylamine}}{(CH_3CH_2CH_2CH_2)_2NH}$$

(c)

$$CH_3CH_2CH_2CH_2OH \xrightarrow{PBr_3} CH_3CH_2CH_2CH_2Br \xrightarrow{NaCN} CH_3CH_2CH_2CH_2CN$$

$$CH_3CH_2CH_2CH_2CN \xrightarrow[\text{2. } H_2O]{\text{1. } LiAlH_4} \underset{\text{Pentylamine}}{CH_3CH_2CH_2CH_2CH_2NH_2}$$

**12.32** In these reactions, polyalkylated products are also formed.

(a)

$$CH_3CH_2CH_2CH_2CH_2CH_2Br \xrightarrow{\text{excess } NH_3} CH_3CH_2CH_2CH_2CH_2CH_2NH_2$$

Hexylamine

(b)

$$4\ CH_3I + NH_3 \longrightarrow (CH_3)_4N^+\ I^-$$

Tetramethylammonium iodide

(c)

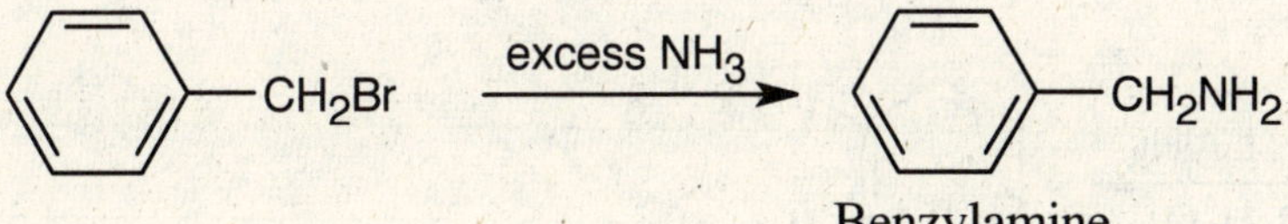

Benzylamine

(d)

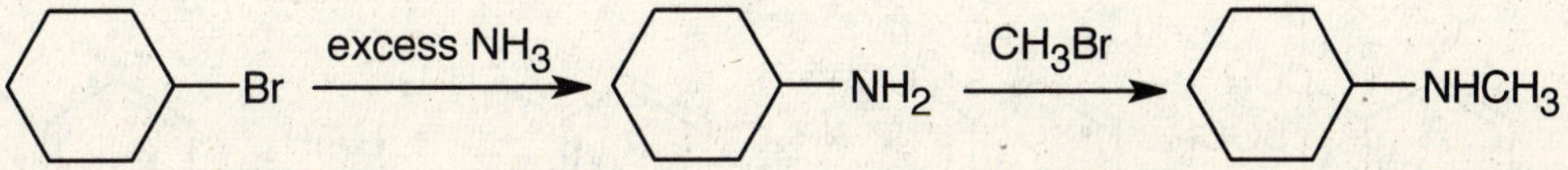

*N*-Methylcyclohexylamine

**12.33**

(a)

$$CH_3CH_2CH_2NH_2 + CH_3Br \longrightarrow CH_3CH_2CH_2NHCH_3$$

Polyalkylated products are also formed.

(b)

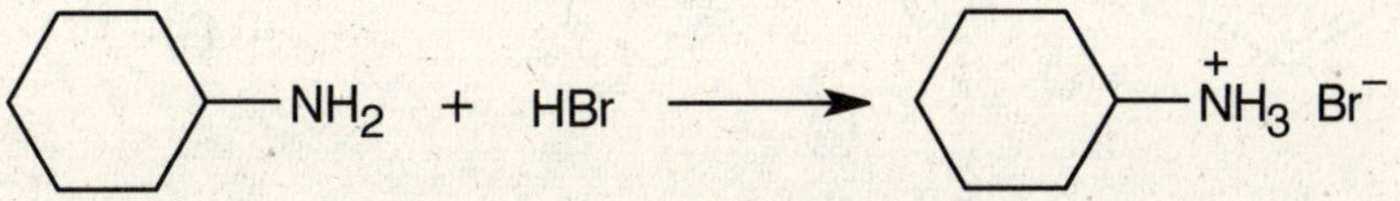

(c)

$$CH_3CH_2C(=O)NH_2 \xrightarrow[2.\ H_2O]{1.\ LiAlH_4} CH_3CH_2CH_2NH_2$$

(d)

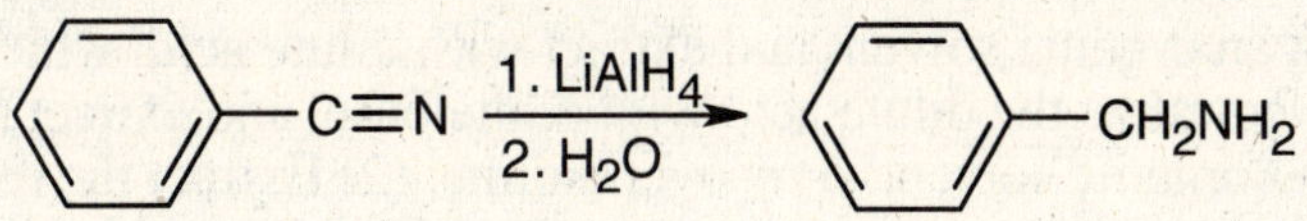

**12.34**

(a)

$$CH_3CH_2CH_2CH_2C(=O)NH_2 \xrightarrow[2.\ H_2O]{1.\ LiAlH_4} CH_3CH_2CH_2CH_2CH_2NH_2$$

(b)

$$CH_3CH_2CH_2CH_2CN \xrightarrow[2.\ H_2O]{1.\ LiAlH_4} CH_3CH_2CH_2CH_2CH_2NH_2$$

(c)

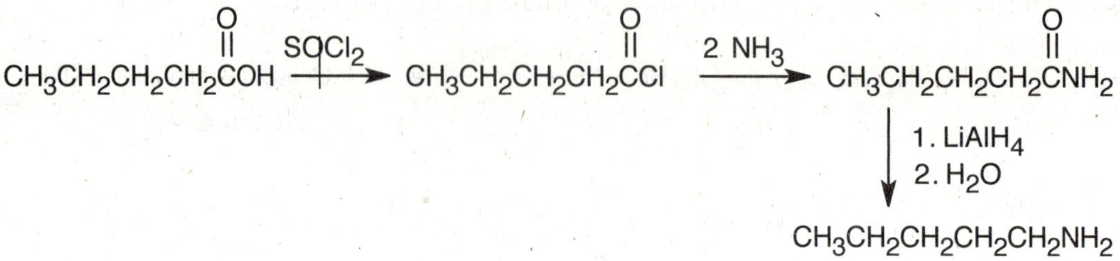

**12.35**

(a)

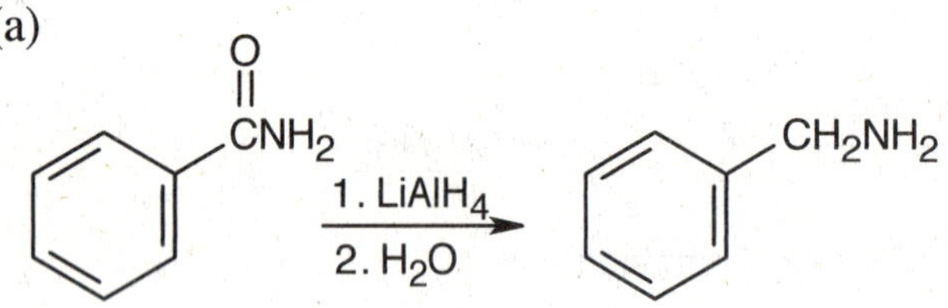

(b)

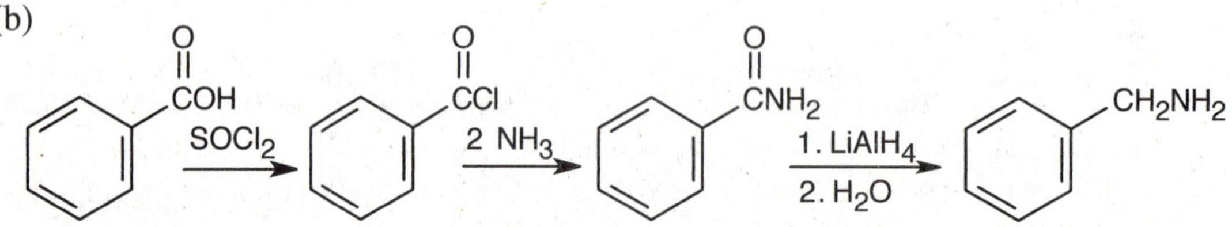

(c)

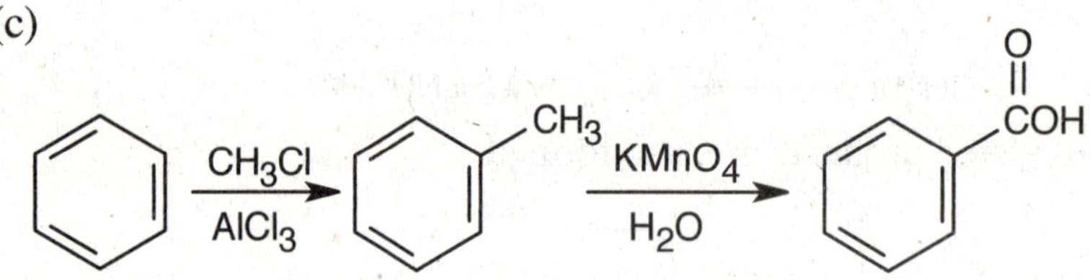

..then proceed as in part (b)

## Basicity

**12.36** Diphenylamine is less basic than aniline. The nitrogen lone-pair electrons of diphenylamine can overlap with the π electron system of both aromatic rings, causing even greater electron delocalization than occurs for aniline.

**12.37** (1) Dissolve the mixture in an organic solvent and extract with dilute acid, which converts aniline to anilinium ion. Neutralize the dilute acid with dilute base, and extract the neutralized solution with an organic solvent to recover aniline. (2) Extract the remaining organic layer from (1) with dilute base, which converts phenol to phenoxide ion. Acidify this solution and extract with an organic solvent to recover phenol. The original organic layer contains toluene.

**12.38** $CH_3CH_2NH_2$ is more basic than $CF_3CH_2NH_2$ because the electron-withdrawing fluorine atoms make the nitrogen of $CF_3CH_2NH_2$ more electron-poor and less basic.

**12.39** The aldehyde group, which is electron-withdrawing, makes *p*-aminobenzaldehyde less basic than aniline.

**12.40** Triethylamine is more basic than aniline. The lone-pair electrons of the aniline nitrogen are delocalized by orbital overlap with the aromatic ring π electrons and are less available for donation to an acid. Thus, the reaction of triethylammonium chloride with aniline does not proceed as written.

## General Problems

**12.41**

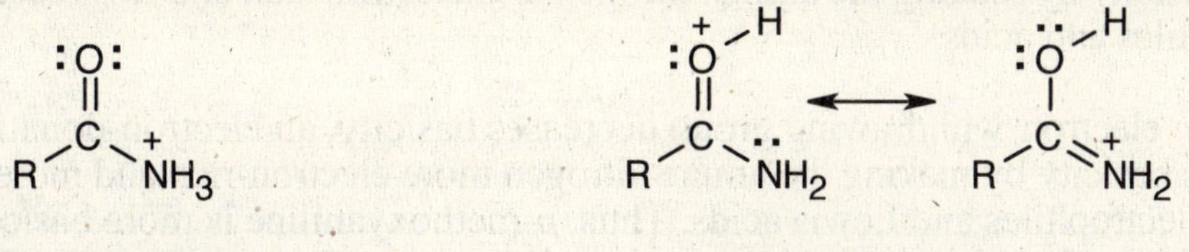

*N*-Protonation (no resonance stabilization) — *O*-Protonation (resonance stabilization)

Protonation occurs on oxygen because an *O*-protonated amide is stabilized by resonance.

**12.42**

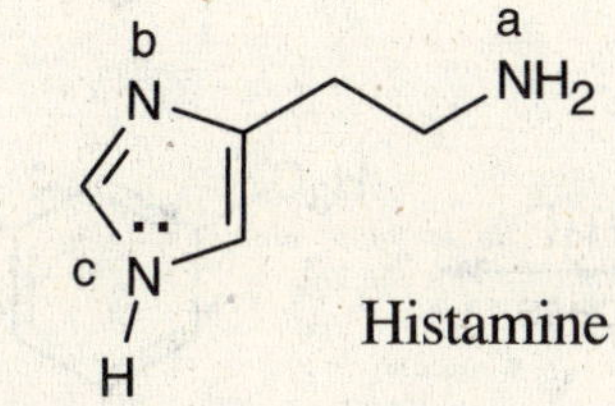

Histamine

The "a" nitrogen is most basic because its electron pair is most available to Lewis acids. The "c" nitrogen is the least basic because the lone-pair electrons of the pyrrole nitrogen are part of the ring $\pi$ electron system.

**12.43** The first step in this sequence is a 1,4 addition to a conjugated diene, as we saw in Chapter 4. An $S_N2$ displacement of chlorine by cyanide yields a dinitrile, which is reduced to the diamine.

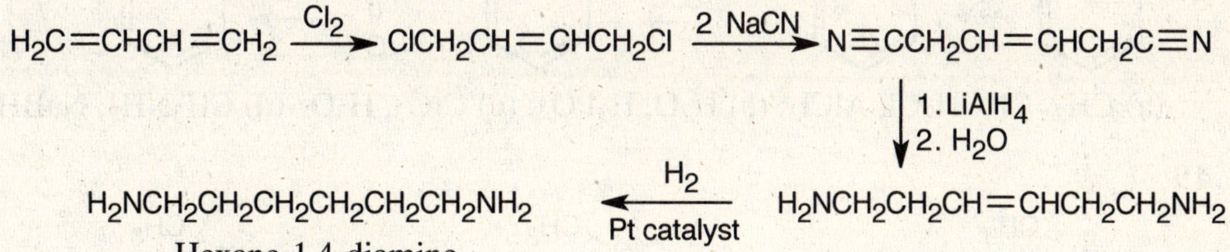

Hexane-1,4-diamine

**12.44**

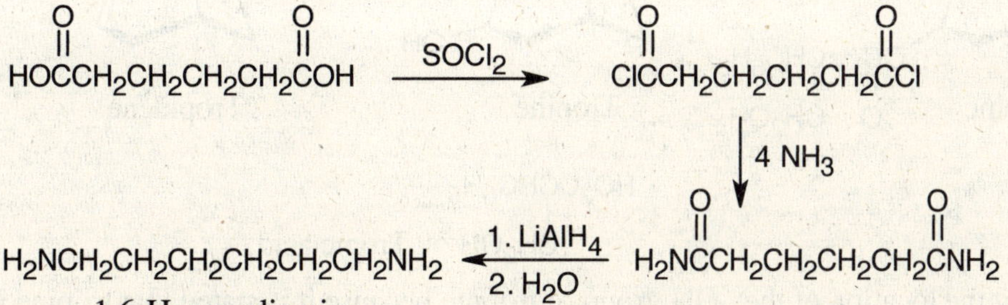

1,6-Hexanediamine

**12.45** In this chapter, we learned that the lone pair electrons of an aromatic amine nitrogen are delocalized over the ring, decreasing basicity. In Chapter 5 we learned that an $-NO_2$ group withdraws electrons from an aromatic ring. Thus, a nitro group further decreases aromatic amine basicity by making the amine nitrogen less electron-rich and less reactive toward electrophiles and acids.

**12.46** Just as an electron-withdrawing group decreases basicity, an electron-donating group increases basicity by making the amine nitrogen more electron-rich and more reactive toward electrophiles and Lewis acids. Thus, *p*-methoxyaniline is more basic than aniline because its methoxyl group is electron-donating.

**12.47**

(a)

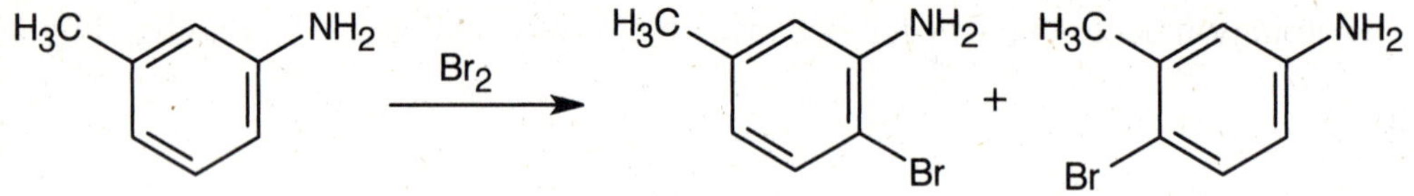


(b)

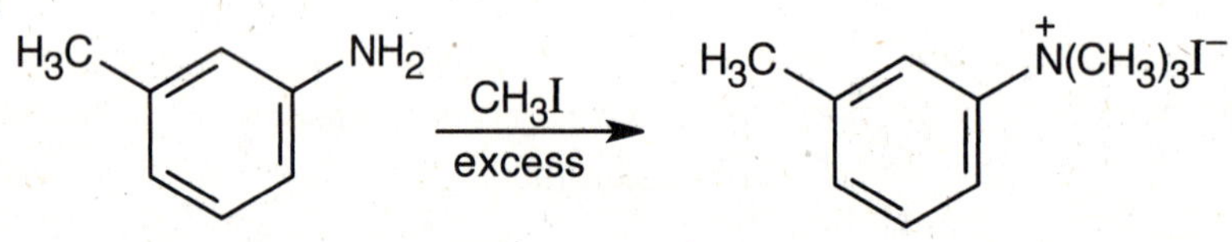


(c)

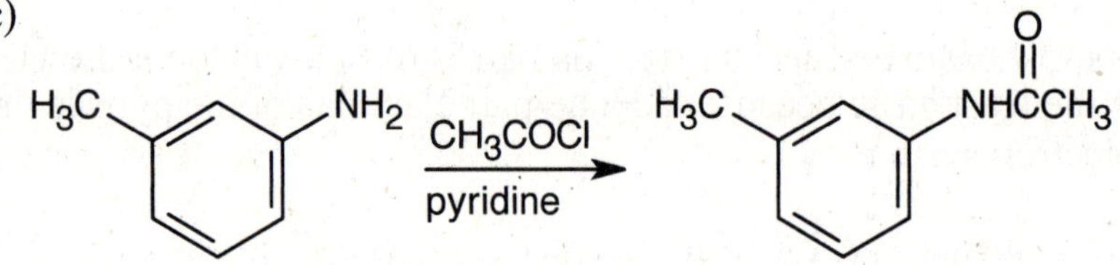


**12.48**

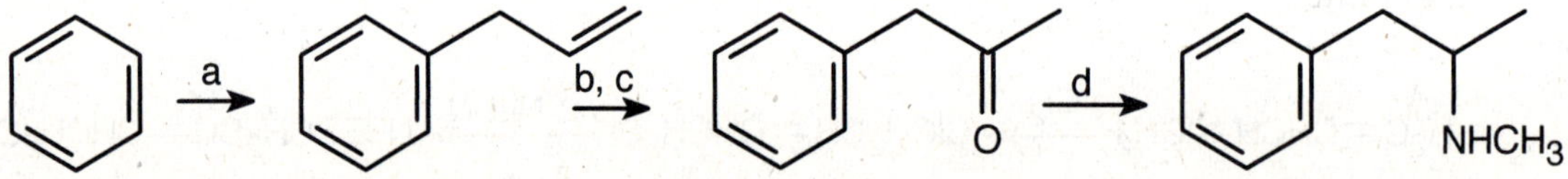


(a) $CH_2{=}CHCH_2Cl$, $AlCl_3$; (b) $H_2O$, $H_3PO_4$; (c) $CrO_3$, $H_3O^+$ (d) $CH_3NH_2$, $NaBH_4$.

**12.49**

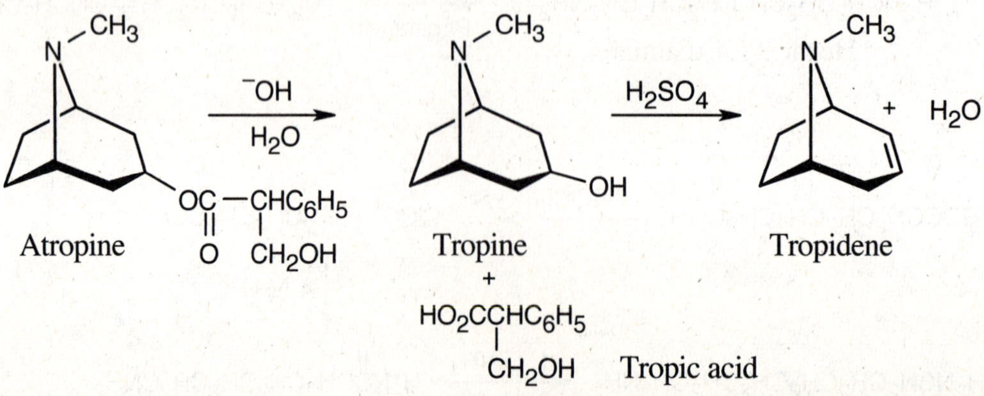


We know the location of the –OH group of tropine because it is stated that tropine is an optically inactive alcohol. This hydroxyl group results from basic hydrolysis of the ester that is composed of tropine and tropic acid.

**12.50**

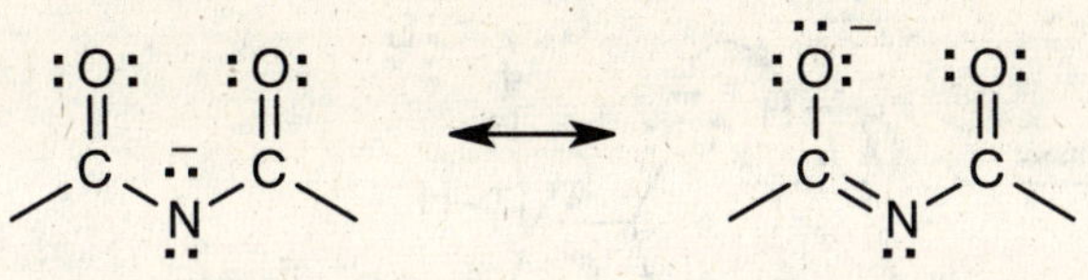

Three resonance forms contribute to the stability of the imide anion.

**12.51** Furan, the oxygen analog of pyrrole, is aromatic because it has 6 $\pi$ electrons in a cyclic, conjugated system. Oxygen contributes two lone-pair electrons from a *p* orbital perpendicular to the plane of the ring. The other oxygen lone pair is in an $sp^2$ orbital that lies in the plane of the furan ring.

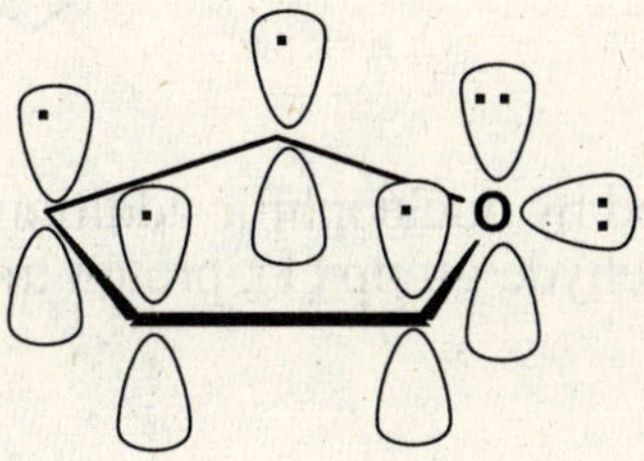

Furan

**12.52**

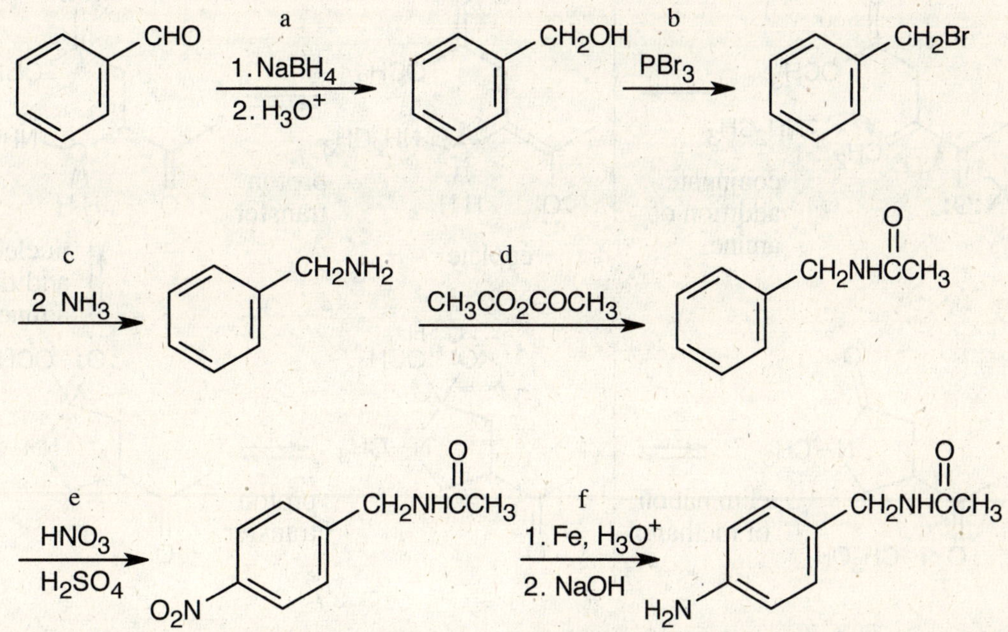

**12.53**

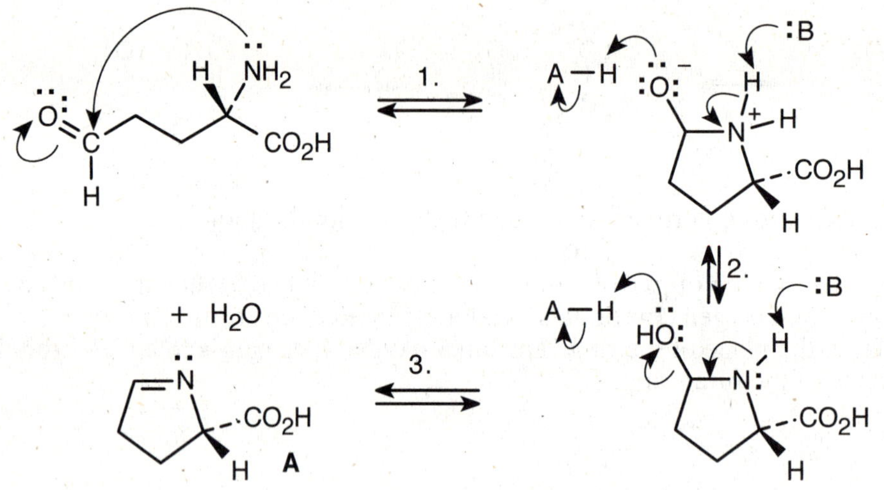

Intermediate **A**, an imine, is formed by nucleophilic addition of the amino group of glutamate semialdehyde to the aldehyde group (1), proton transfer to form a carbinolamine (2), and loss of water (3).

**12.54**

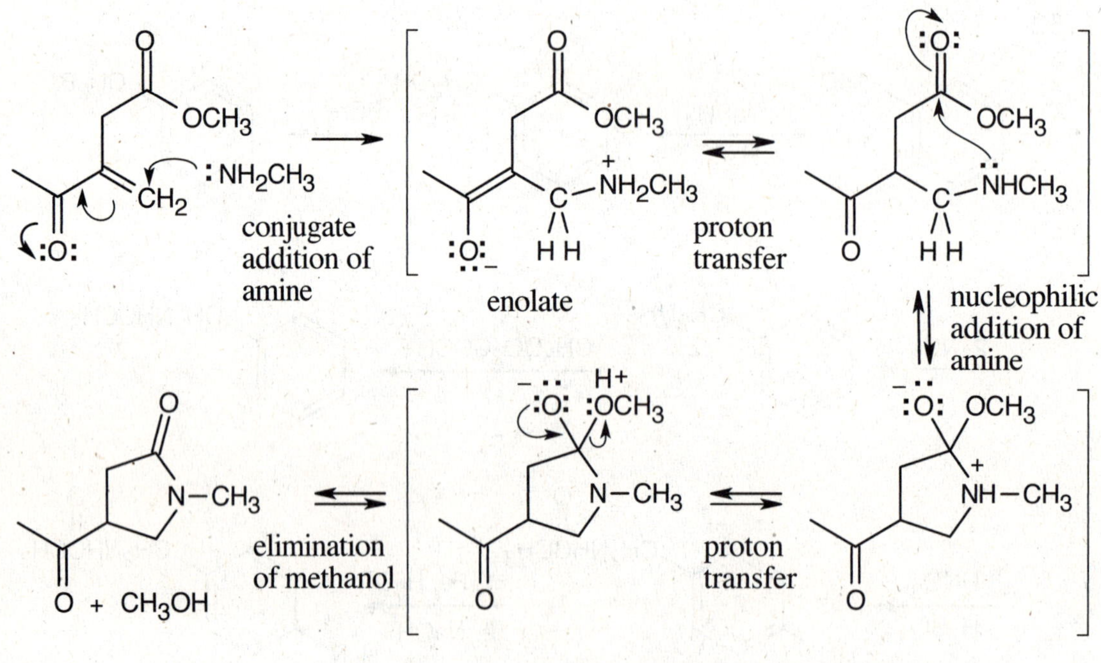

**12.55** The reaction of trimethylamine with ethylene oxide is an $S_N2$ reaction.

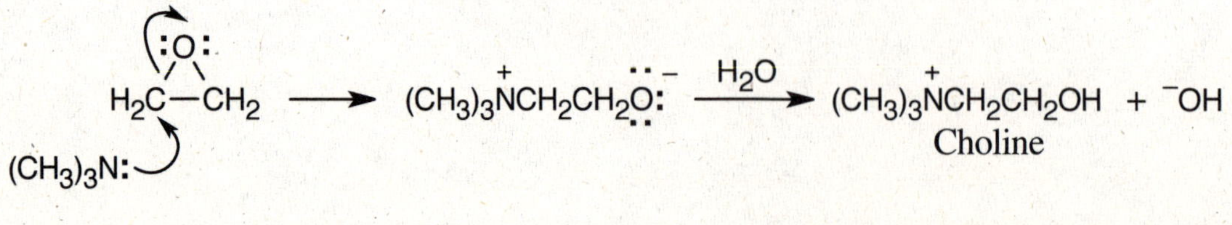

## In the Medicine Cabinet

**12.56** This reaction sequence contains many protonations and deprotonations. To simplify, the two aromatic groups are represented as –Ar.

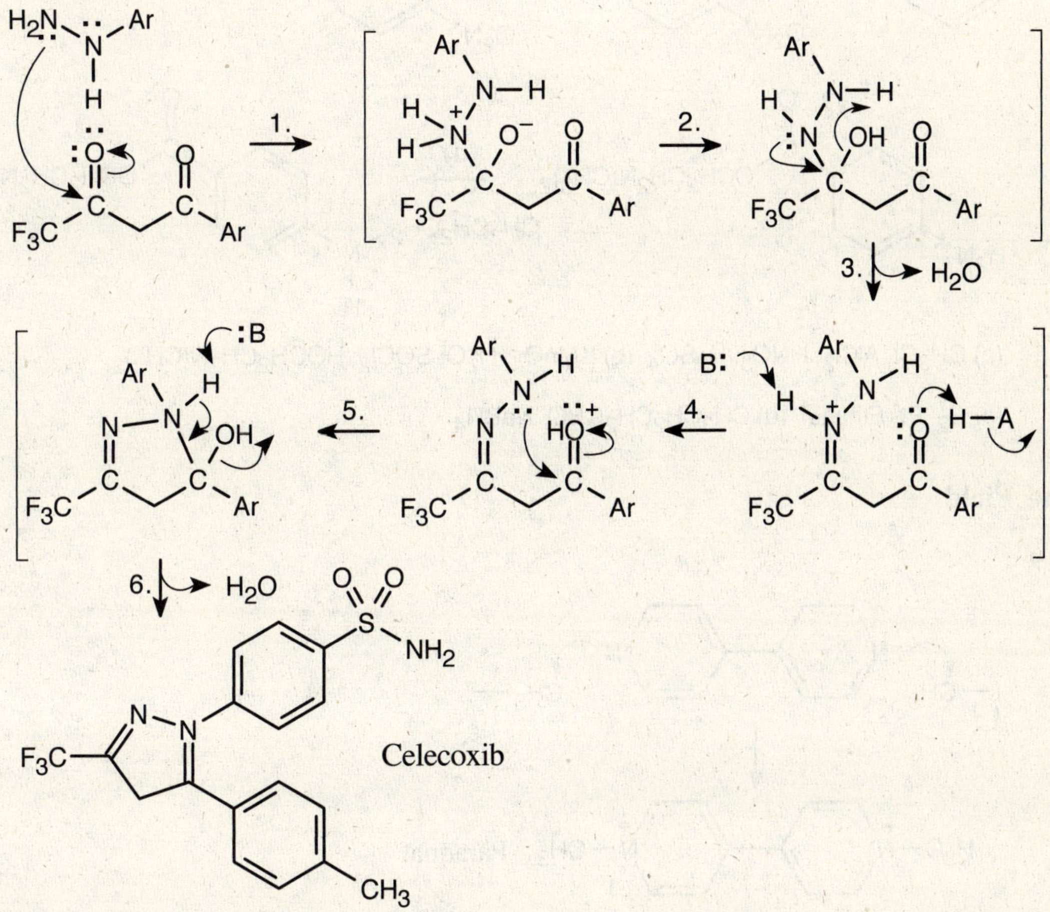

Celecoxib

Steps 1,5: Nucleophilic addition of an amine to a carbonyl group. Steps 2,4: Proton transfers. Steps 3,6: Loss of water.

**12.57**

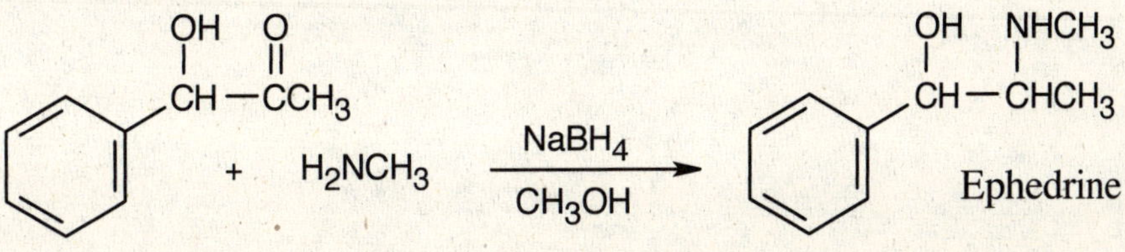

Ephedrine

**12.58**

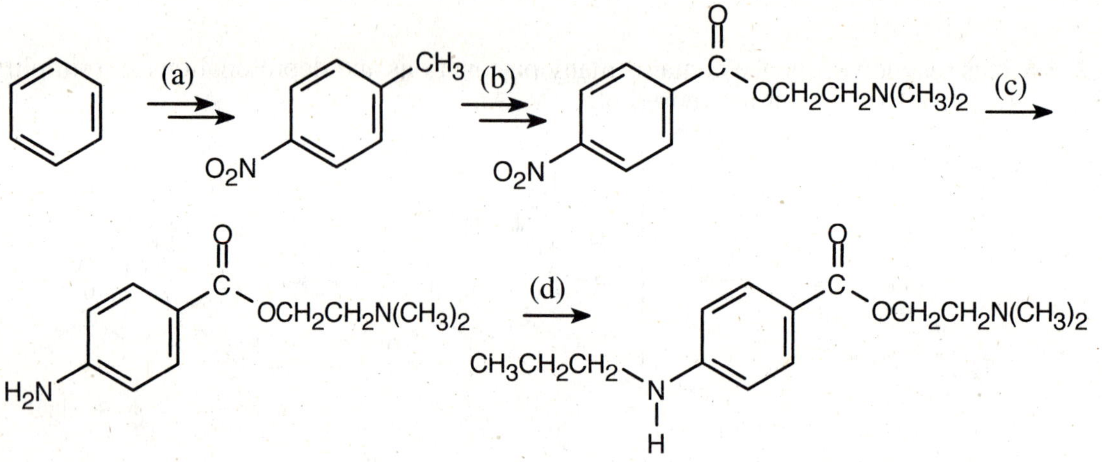

(a) $CH_3Cl$, $AlCl_3$; $HNO_3$, $H_2SO_4$ (b) $KMnO_4$, $H_2O$; $SOCl_2$; $HOCH_2CH_2N(CH_3)_2$

(c) $H_2$, Pt, Ethanol (d) $CH_3CH_2CH_2CHO$, $NaBH_4$

## In the Field

**12.59**

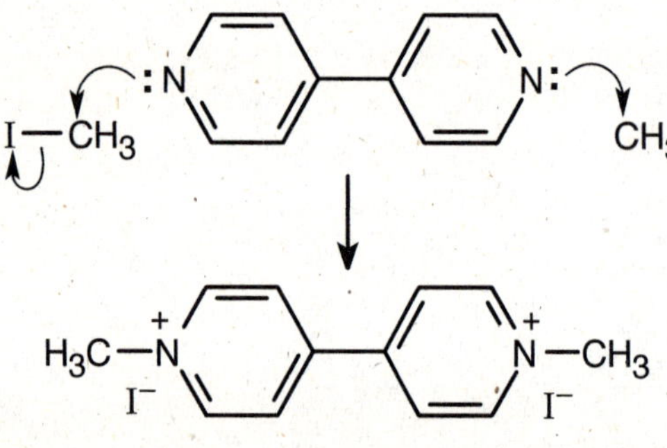

Paraquat

**12.60**

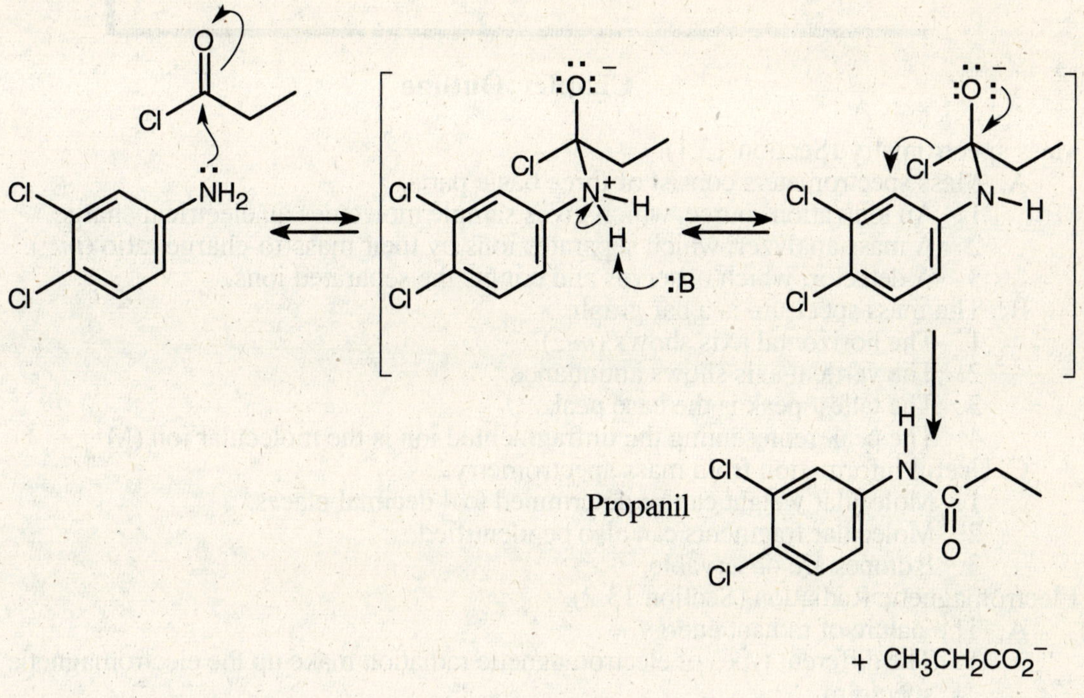


This is a nucleophilic acyl substitution with 3,4 dichloroaniline, the nucleophile, adding to propanoyl chloride.

# Chapter 13 – Structure Determination

## Chapter Outline

I. Mass spectrometry (Section 13.1).
   A. Mass spectrometers consist of three basic parts.
      1. An ionization source, which gives sample molecules an electrical charge.
      2. A mass analyzer, which separates ions by their mass-to-charge ratio ($m/z$).
      3. A detector, which observes and counts the separated ions.
   B. The mass spectrum is a bar graph.
      1. The horizontal axis shows ($m/z$).
      2. The vertical axis shows abundance.
      3. The tallest peak is the base peak.
      4. The peak representing the unfragmented ion is the molecular ion ($M^+$).
   C. Useful information from mass spectrometry.
      1. Molecular weight can be determined to 4 decimal places.
      2. Molecular fragments can also be identified.
      3. Isotopes are observable.

II. Electromagnetic Radiation (Section 13.2).
   A. The nature of radiant energy.
      1. The different types of electromagnetic radiation make up the electromagnetic spectrum.
      2. Electromagnetic radiation behaves both as a particle and as a wave.
      3. Electromagnetic radiation can be characterized by three variables.
         a. The wavelength ($\lambda$) measures the distance from one wave maximum to the next.
         b. The frequency ($\nu$) measures the number of wave maxima that pass a fixed point per unit time.
         c. The amplitude is the wave height measured from the midpoint to the maximum.
      4. Wavelength times frequency equals the speed of light ($\lambda \times \nu = c$).
      5. Electromagnetic energy is transmitted in discrete energy bundles called quanta.
         a. $\varepsilon = h \times \nu$
         b. Energy varies directly with frequency but inversely with wavelength.
   B. Electromagnetic radiation and organic molecules.
      1. When an organic compound is struck by a beam of electromagnetic radiation, it absorbs radiation of certain wavelengths, and transmits radiation of other wavelengths.
      2. If we determine which wavelengths are absorbed and which are transmitted, we can obtain an absorption spectrum of the compound.

III. Infrared spectroscopy and organic molecules (Sections 13.3–13.4).
   A. Infrared radiation (Section 13.3).
      1. The infrared (IR) region of the electromagnetic spectrum extends from $7.8 \times 10^{-7}$ m to $10^{-4}$ m.
         a. Organic chemists use the region from $2.5 \times 10^{-6}$ m to $2.5 \times 10^{-5}$ m.
         b. Wavelengths are usually given in μm, and frequencies are expressed in wavenumbers, which are the reciprocal of wavelength.
         c. The useful range of IR radiation is 4000 $cm^{-1}$ – 400 $cm^{-1}$; this corresponds to energies of 48.0 kJ/mol – 4.80 kJ/mol.
      2. IR radiation causes bonds to stretch and bend and causes other molecular vibrations.

3. Energy is absorbed at a specific frequency that corresponds to the frequency of the vibrational motion of a bond.
4. If we measure the frequencies at which IR energy is absorbed, we can find out the kinds of bonds a compound contains and identify functional groups.

B. Interpreting IR spectra (Section 13.4).
1. Most molecules have very complex IR spectra.
   a. This complexity means that each molecule has a unique fingerprint that allows it to be identified by IR spectroscopy.
   b. Complexity also means that not all absorptions can be identified.
2. Most functional groups have characteristic IR absorption bands that change very little from one compound to another.
3. The significant regions of IR absorptions :
   a. 4000 $cm^{-1}$ – 2500 $cm^{-1}$ corresponds to absorptions by C–H, O–H, and N–H bonds.
   b. 2500 $cm^{-1}$ – 2000 $cm^{-1}$ corresponds to triple-bond stretches.
   c. 2000 $cm^{-1}$ – 1500 $cm^{-1}$ corresponds to double bond stretches.
   d. The region below 1500 $cm^{-1}$ is the fingerprint region, where many complex bond vibrations occur and which is unique for each molecule.

IV. Ultraviolet spectroscopy (Sections 13.5–13.6).

A. Principles of ultraviolet spectroscopy (Section 13.5).
1. The ultraviolet region of interest is between the wavelengths 200 nm and 400 nm.
2. The energy absorbed is used to promote a $\pi$ electron in a conjugated system from one molecular orbital to another.

B. A UV spectrum is a plot of absorbance (A) vs. wavelength in nanometers.

C. UV spectra usually consist of a single broad peak, whose maximum is $\lambda_{max}$.

D. The formula $A = \varepsilon cl$ allows one to measure the concentration of a solution.

E. $\lambda_{max}$ increases with the degree of conjugation of a molecule (Section 13.6).
1. The value of $\lambda_{max}$ indicates the extent of conjugation.

V. Nuclear magnetic resonance spectroscopy (NMR) (Sections 13.7–13.14).

A. General characteristics (Sections 13.7–13.9).
1. Theory of NMR spectroscopy (Section 13.7).
   a. Many nuclei behave as if they were spinning about an axis.
      i. The positively charged nuclei produce a magnetic field that can interact with an externally applied magnetic field.
      ii. The $^{13}C$ nucleus and the $^{1}H$ nucleus behave in this manner.
      iii. In the absence of an external magnetic field the spins of magnetic nuclei are randomly oriented.
   b. When a sample containing these nuclei is placed between the poles of a strong magnet, the nuclei align themselves either with the applied field or against the applied field.
      i. The parallel orientation is slightly lower in energy and is slightly favored.
   c. If the sample is irradiated with radiofrequency energy of the correct frequency, the nuclei of lower energy absorb energy and "spin-flip" to the higher energy state.
      i. The magnetic nuclei are in resonance with the applied radiation.
      ii. The frequency of the rf radiation needed for resonance depends on the magnetic field strength and on the identity of the magnetic nuclei.
      iii. In a strong magnetic field, higher frequency rf energy is needed.
      iv At a magnetic field strength of 4.7 T, rf energy of 200 MHz is needed to bring a $^{1}H$ nucleus into resonance, and energy of 50 MHz is needed for $^{13}C$.

2. The nature of NMR absorptions (Section 13.8).
   a. Not all $^{13}C$ nuclei and not all $^{1}H$ nuclei absorb at the same frequency.
      i. Each magnetic nucleus is surrounded by electrons that set up their own magnetic fields.
      ii. These little fields oppose the applied field and shield the magnetic nuclei. $\boldsymbol{B}_{\text{effective}} = \boldsymbol{B}_{\text{applied}} - \boldsymbol{B}_{\text{local}}$ .This expression shows that the magnetic field felt by a nucleus is less than the applied field.
      iii. These shielded nuclei absorb at slightly different values of magnetic field strength.
      iv. A sensitive NMR spectrometer can detect these small differences.
      v. Thus, NMR spectra can be used to map the carbon–hydrogen framework of a molecule.
   b. NMR spectra.
      i. The horizontal axis shows effective field strength, and the vertical axis shows intensity of absorption.
      ii. Each peak corresponds to a chemically distinct nucleus.
      iii. Zero absorption is at the bottom.
      iv. Absorptions due to both $^{13}C$ and $^{1}H$ can't both be observed at the same time and are displayed on separate spectra.
3. Chemical Shifts (Section 13.9).
   a. Field strength increases from left (downfield) to right (upfield).
      i. Nuclei that absorb downfield require a lower field strength for resonance and are deshielded.
      ii. Nuclei that absorb upfield require a higher field strength and are shielded.
   b. TMS (tetramethylsilane) is used as a reference point in both $^{13}C$ NMR and $^{1}H$ NMR.
      i. The TMS absorption occurs upfield of other absorptions, and is set as the zero point.
   c. The chemical shift is the position along the x-axis where a nucleus absorbs.
   d. NMR charts are calibrated by using an arbitrary scale – the delta scale.
      i. One δ (delta) equals 1 ppm of the spectrometer operating frequency.
      ii. By using this system, all chemical shifts occur at the same value of δ, regardless of the spectrometer operating frequency.

B. $^{1}H$ NMR spectroscopy (Sections 13.10–13.13).
   1. Chemical shifts (Section 13.10).
      a. Most $^{1}H$ NMR chemical shifts are in the range 0–10 δ.
         i. Protons that are $sp^3$-hybridized absorb at higher field strength (upfield).
         ii. Protons that are $sp^2$-hybridized absorb at lower field strength (downfield).
         iii. Protons on carbons that are bonded to electronegative atoms absorb at lower field strength (downfield).
      b. The $^{1}H$ NMR spectrum can be divided into 6 regions:
         i. Saturated (0–1.5 δ).
         ii. Allylic (1.5–2.5 δ).
         iii. H bonded to C next to an electronegative atom (2.5–4.5 δ).
         iv. Vinylic (4.5–6.5 δ).
         v. Aromatic (6.5–8.0 δ).
         vi. Aldehyde and carboxylic acid protons absorb even farther downfield.

2. Integration of $^1H$ NMR signals: proton counting (Section 13.11).
   a. The area of a peak is proportional to the number of protons causing the peak.
   b. Integrated peak areas are superimposed over a spectrum as a stair-step line.
      i. To compare two peaks, measure the relative heights of the steps.
   c. Modern NMR instruments provide a digital readout of relative peak areas.
3. Spin–spin splitting (Section 13.12).
   a. The tiny magnetic field produced by one nucleus can affect the magnetic field felt by other nearby nuclei.
   b. Protons that have $n$ equivalent neighboring protons show a peak in their $^1H$ NMR spectrum that is split into $n + 1$ smaller peaks (a multiplet).
   c. This splitting is caused by the coupling of spins of neighboring nuclei.
   d. The distance between peaks in a multiplet is called the coupling constant ($J$).
      i. The value of $J$ is usually 0–18 Hz.
      ii. The value of $J$ is determined by the geometry of the molecule and is independent of the spectrometer operating frequency.
      iii. The value of $J$ is shared between both groups of hydrogens whose spins are coupled.
      iv. By comparing values of $J$, it is possible to know the atoms whose spins are coupled.
   e. Three rules for spin–spin splitting in $^1H$ NMR:
      i. Chemically identical protons don't show spin–spin splitting.
      ii. The signal of a proton with $n$ equivalent neighboring protons is split into a multiplet of $n + 1$ peaks with coupling constant $J$.
      iii. Two groups of coupled protons have the same value of $J$.
4. Uses of $^1H$ NMR spectroscopy (Section 13.13).
   a. $^1H$ NMR can be used to identify the products of reactions.

C. $^{13}C$ NMR spectroscopy (Section 13.14).
1. Uses of $^{13}C$ NMR spectroscopy.
   a. $^{13}C$ NMR spectroscopy can show the number of nonequivalent carbons in a molecule and can identify symmetry in a molecule.
2. Characteristics of $^{13}C$ NMR spectroscopy.
   a. Each distinct carbon shows a single line.
   b. The chemical shift depends on the electronic environment within a molecule.
      i. Carbons bonded to electronegative atoms absorb downfield.
      ii. Carbons with $sp^3$ hybridization absorb in the range 0–90 $\delta$.
      iii. Carbons with $sp^2$ hybridization absorb in the range 110–220 $\delta$.
         (a). Carbonyl carbons absorb in the range 160–220 $\delta$.
   c. Symmetry reduces the number of absorptions.
   d. Peaks (spectral lines) aren't uniform in size.

## Solutions to Problems

**13.1** The molecular weight of the alkene (56) increases by 254 when $X_2$ is added to the double bond, yielding a product with molecular weight of 310. Since $X_2$ has a molecular weight of 254, X has a molecular weight of 127, corresponding to iodine.

**13.2** The trace impurity, appearing at m/z = 88, has a molecular weight that is 16 units greater than the molecular weight of butanal. Butanoic acid is the likely impurity.

**13.3** At first glance, we know that: (1) energy increases as wavelength decreases, and (2) the wavelength of X-radiation is shorter than the wavelength of infrared radiation. Thus, we estimate that an X ray is of higher energy than an infrared ray.
An exact solution:
$\varepsilon = h\nu = hc/\lambda$; $h = 6.62 \times 10^{-34}$ J·s; $c = 3.00 \times 10^{8}$ m/s
for $\lambda = 1.0 \times 10^{-6}$ m (infrared radiation):

$$\varepsilon = \frac{(6.62 \times 10^{-34}\ \text{J·s})(3.00 \times 10^{8}\ \text{m/s})}{1.0 \times 10^{-6}\ \text{m}} = 2.0 \times 10^{-19}\ \text{J}$$

for $\lambda = 3.0 \times 10^{-9}$ m (X radiation):

$$\varepsilon = \frac{(6.62 \times 10^{-34}\ \text{J·s})(3.00 \times 10^{8}\ \text{m/s})}{3.0 \times 10^{-9}\ \text{m}} = 6.6 \times 10^{-17}\ \text{J}$$

Confirming our estimate, the calculation shows that an X ray is of higher energy than infrared radiation.

**13.4** First, convert radiation in m to radiation in Hz by the equation:

$$\nu = \frac{c}{\lambda} = \frac{3.00 \times 10^{8}\ \text{m/s}}{9.0 \times 10^{-6}\ \text{m}} = 3.3 \times 10^{13}\ \text{Hz}$$

The equation $\varepsilon = h\nu$ says that the greater the value of $\nu$, the greater the energy. Thus, radiation with $\nu = 3 \times 10^{13}$ Hz ($\lambda = 9 \times 10^{-6}$ m) is higher in energy than radiation with $\nu = 4 \times 10^{9}$ Hz.

**13.5** (a) Use the conversion equation shown in the previous problem.

$$\lambda = \frac{c}{\nu} = \frac{3.00 \times 10^{8}\ \text{m/s}}{1.9 \times 10^{9}\ \text{Hz}} = 0.16\ \text{m}$$

(b)

$$E = \frac{1.20 \times 10^{-4}\ \text{kJ/mol}}{0.16} = 7.5 \times 10^{-4}\ \text{kJ/mol for cellphone radiation}$$

The energy of cellphone radiation ($7.5 \times 10^{-4}$ kJ/mol) is less than one ten-thousandth the energy of visible red light (200 kJ/mol).

**13.6** Use Table 13.1 and the descriptions in Section 13.4 to assign these absorptions.

| *IR absorption* | *Due to:* |
|---|---|
| (a) 1715 $cm^{-1}$ | ketone |
| (b) 1540 $cm^{-1}$ | nitro group |
| (c) 2210 $cm^{-1}$ | nitrile group or alkyne |
| (d) 1720 $cm^{-1}$<br>2500-3100 $cm^{-1}$ | carboxylic acid |
| (e) 3500 $cm^{-1}$<br>1735 $cm^{-1}$ | alcohol<br>ester |

**13.7** To use IR to distinguish between isomers, find a strong IR absorption present in one isomer that is absent in the other isomer.

(a)

$CH_3CH_2OH$
Strong hydroxyl band at 3400–3640 $cm^{-1}$

$CH_3OCH_3$
No band in the region 3400–3640 $cm^{-1}$

(b)

$CH_3CH_2CH_2CH_2CH{=}CH_2$
Alkene bands at 3020–3100 $cm^{-1}$ and at 1650–1670 $cm^{-1}$.

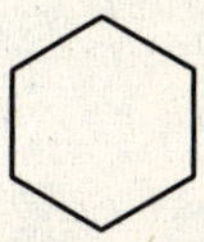

No bands in alkene region.

(c)

$CH_3CH_2CO_2H$
Strong, broad band at 2500–3100 $cm^{-1}$

$HOCH_2CH_2CHO$
Strong band at 3400–3640 $cm^{-1}$

**13.8** (a) An ester next to a double bond absorbs at 1715 $cm^{-1}$. The alkene double bond absorbs at 1640–1680 $cm^{-1}$.

(b) The aldehyde carbonyl group absorbs at 1730 $cm^{-1}$. The alkyne C≡C bond absorbs at 2100–2260 $cm^{-1}$, and the alkyne H–C≡ bond absorbs at 3300 $cm^{-1}$.

(c) The most important absorptions for this compound are due to the alcohol group (a broad, intense band at 3400–3650 $cm^{-1}$) and to the carboxylic acid group, which has a C=O absorption in the range 1710–1760 $cm^{-1}$ and a broad O–H absorption in the range 2500–3100 $cm^{-1}$. Absorptions due to the aromatic ring [3030 $cm^{-1}$ (w) and 1450–1600 $cm^{-1}$ (m)] may also be seen.

**13.9**

$$\varepsilon = \frac{A}{c \times l}$$

Where $\varepsilon$ = molar absorptivity (in $L{\cdot}mol^{-1}{\cdot}cm^{-1}$)
$A$ = absorbance
$l$ = sample pathlength (in cm)
$c$ = concentration (in $mol{\cdot}L^{-1}$)

In this problem:

$\varepsilon = 6{,}100 = 6.1 \times 10^3\ L \cdot mol^{-1}\ cm^{-1}$

$l = 1.0$ cm

$A = 0.20$

$$c = \frac{A}{\varepsilon \times l} = \frac{0.20}{6.1 \times 10^3\ L \cdot mol^{-1} \cdot cm^{-1} \times 1.0\ cm} = 3.3 \times 10^{-5}\ M$$

**13.10** Only conjugated compounds absorb in the region 200 nm–400 nm.

| | *Compound* | *Absorption at 200–400 nm?* | | *Compound* | *Absorption at 200–400 nm?* |
|---|---|---|---|---|---|
| (a) | 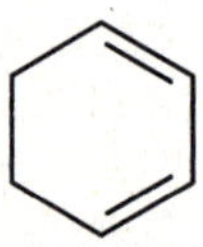 | yes | (b) |  | no |
| (c) | 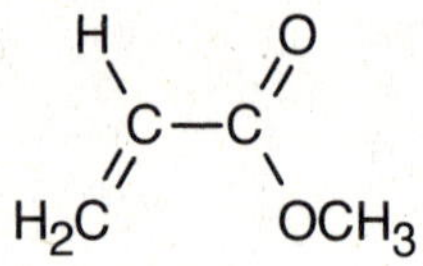 | yes | (d) | 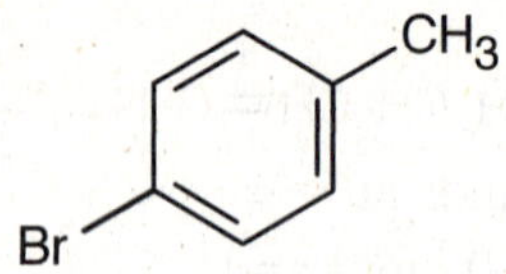 | yes |
| (e) | 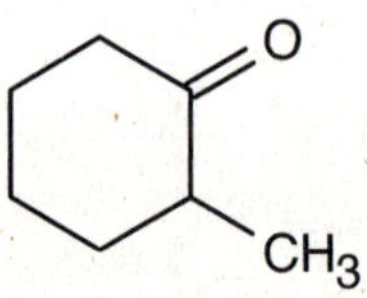 | no | (f) | 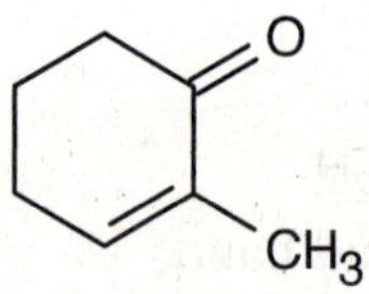 | yes |

**13.11** Both compounds are conjugated and absorb in the ultraviolet region. Hexa-1,3,5-triene has a longer system of conjugated bonds and absorbs at a longer wavelength (lower energy) than hexa-1,3-diene.

**13.12** From Problem 13.1, we find that $\lambda = 1 \times 10^{-6}$ m is a typical value for the wavelength of infrared radiation. Use the equation $\nu = c / \lambda$ to find the frequency of infrared radiation.

$$\nu = \frac{c}{\lambda} = \frac{3 \times 10^{8}\ m/s}{1 \times 10^{-6}\ m} = 3 \times 10^{14}\ Hz$$

Since $\nu$ for NMR radiation ($1 \times 10^8$ Hz) is less than $\nu$ for ultraviolet radiation, the amount of energy used by NMR spectroscopy is less than that used by IR spectroscopy.

**13.13**

| | *Compound* | *Absorptions in ¹H NMR* | *Absorptions in ¹³C NMR* |
|---|---|---|---|
| (a) | $CH_4$ | 1 | 1 |
| (b) | $CH_3CH_3$ | 1 | 1 |
| (c) | $CH_3CH_2CH_3$ | 2 | 2 |
| (d) | 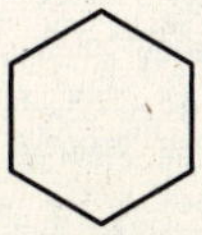 | 1 | 1 |
| (e) | $CH_3OCH_3$ | 1 | 1 |
| (f) | 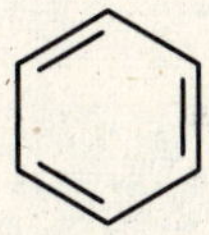 | 1 | 1 |
| (g) | $(CH_3)_3COH$ | 2 | 2 |
| (h) | $CH_3CH_2Cl$ | 2 | 2 |
| (i) | $(CH_3)_2C{=}C(CH_3)_2$ | 1 | 2 |

**13.14**

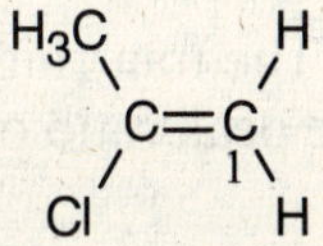


2-Chloropropene

The two protons on C1 are not equivalent: one proton is on the same side of the double bond as chlorine and the other proton is on the opposite side. Since 2-chloropropene has three different types of protons, it shows three signals in its ¹H NMR spectrum.

**13.15**

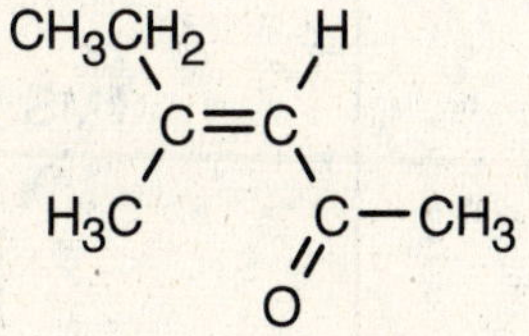


This molecule shows five ¹H NMR signals and seven ¹³C NMR signals.

**13.16** (a) 2.1 ppm x 100 MHz = 210 Hz

(b) The position of absorption in δ units is 2.1 δ for both a 100 MHz and a 220 MHz instrument. A measurement in δ units is independent of the operating frequency of the NMR spectrometer.

(c) 2.1 ppm x 220 MHz = 462 Hz

**13.17**

$$\delta = \frac{\text{Observed chemical shift (in Hz)}}{300 \text{ MHz}}$$

(a) $\delta = \frac{2180 \text{ Hz}}{300 \text{ MHz}} = 7.27\ \delta$ for $CHCl_3$ (b) $\delta = \frac{915 \text{ Hz}}{300 \text{ MHz}} = 3.05\ \delta$ for $CH_3Cl$

(c) $\delta = \frac{1040 \text{ Hz}}{300 \text{ MHz}} = 3.47\ \delta$ for $CH_3OH$ (d) $\delta = \frac{1590 \text{ Hz}}{300 \text{ MHz}} = 5.30\ \delta$ for $CH_2Cl_2$

| **13.18** *Compound* | *$^1H$ Chemical Shift* |
|---|---|
| (a) $CH_3CH_3$ | 0.88 δ |
| (b) $CH_3COCH_3$ | 2.17 δ |
| (c) $C_6H_6$ | 7.17 δ |
| (d) $(CH_3)_3N$ | 2.22 δ |

**13.19**

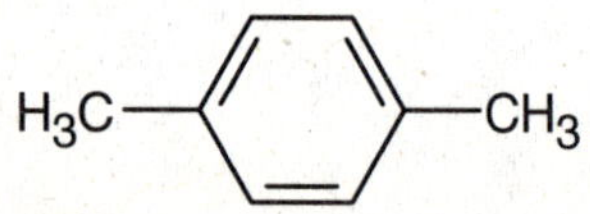


*p*-Xylene

There are two peaks in the $^1H$ NMR spectrum of *p*-xylene. The four ring protons absorb at 7.0 δ, and the six methyl-group protons absorb at 2.3 δ. The peak ratio of methyl protons:ring protons is 3:2.

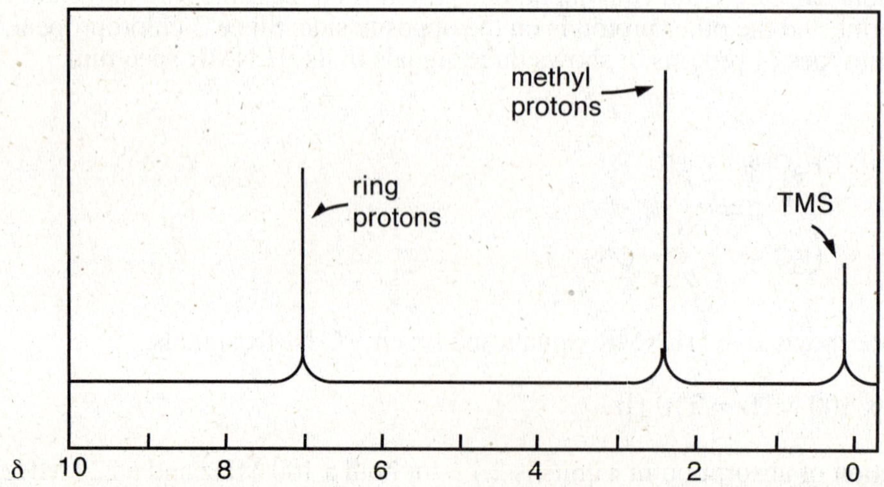

**13.20**

| | Compound | Proton | Number of Adjacent Protons | Splitting |
|---|---|---|---|---|
| (a) | $\overset{1}{CH}Br_2\overset{2}{CH_3}$ | 1 | 3 | quartet |
| | | 2 | 1 | doublet |
| (b) | $\overset{1}{CH_3}O\overset{2}{CH_2}\overset{3}{CH_2}Br$ | 1 | 0 | singlet |
| | | 2 | 2 | triplet |
| | | 3 | 2 | triplet |
| (c) | $Cl\overset{1}{CH_2}\overset{2}{CH_2}\overset{1}{CH_2}Cl$ | 1 | 2 | triplet |
| | | 2 | 4 | quintet |
| (d) | 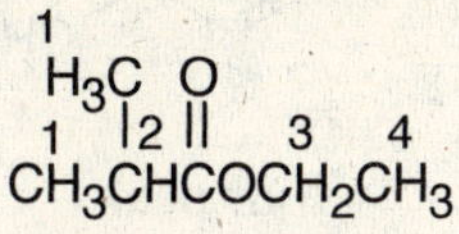 | 1 | 1 | doublet |
| | | 2 | 6 | septet |
| | | 3 | 3 | quartet |
| | | 4 | 2 | triplet |
| (e) | 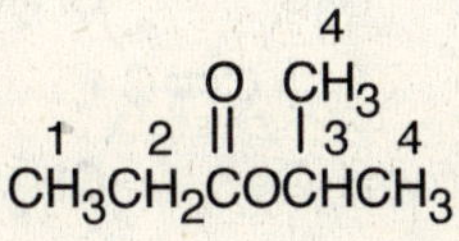 | 1 | 2 | triplet |
| | | 2 | 3 | quartet |
| | | 3 | 6 | septet |
| | | 4 | 1 | doublet |
| (f) | 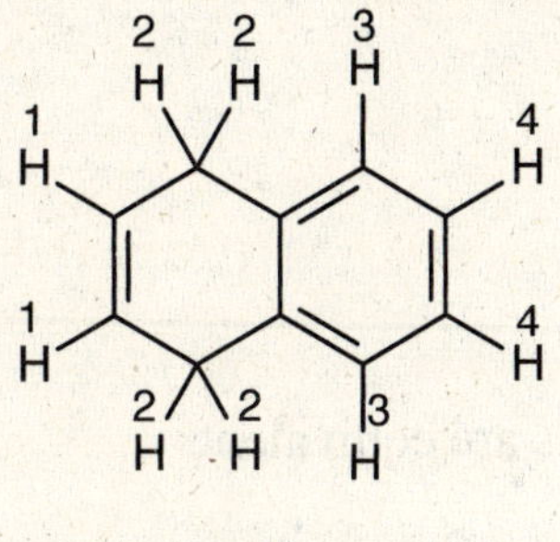 | 1 | 2 | triplet |
| | | 2 | 1 | doublet |
| | | 3 | 1 | multiplet |
| | | 4 | 1 | multiplet |

**13.21**

(a) $C_2H_6O$ has only one type of proton, with no neighbors. Answer: $CH_3OCH_3$

(b) $C_3H_6O_2$ has two types of protons; neither kind has neighbors. Answer: $CH_3CO_2CH_3$

(c) $C_3H_7Cl$ has two types of protons; one kind of proton has six neighbors, and the other kind has one neighbor. Answer: $(CH_3)_2CHCl$

**13.22**

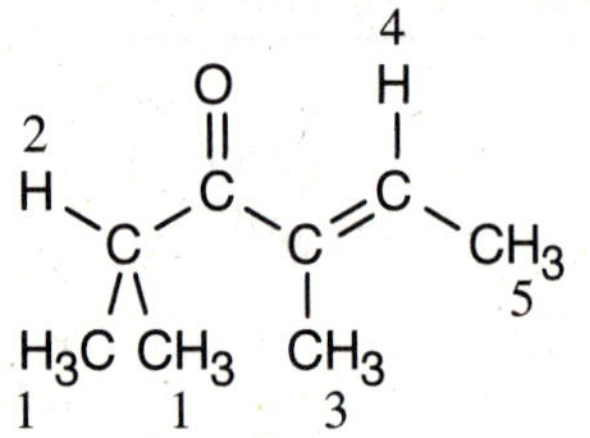

| Proton(s) | Number of Adjacent Protons | Splitting |
|---|---|---|
| 1 | 1 | doublet |
| 2 | 6 | septet |
| 3 | 0 | singlet |
| 4 | 3 | quartet |
| 5 | 1 | doublet |

**13.23** Look for a plane of symmetry that reduces the number of peaks seen in the $^{13}C$ NMR spectrum.

(a)

one peak

(b)

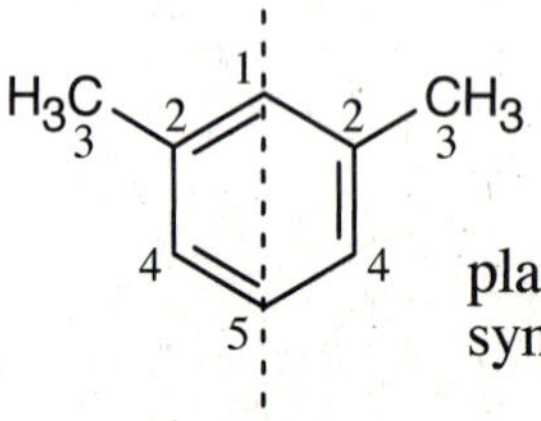


five peaks

(c)

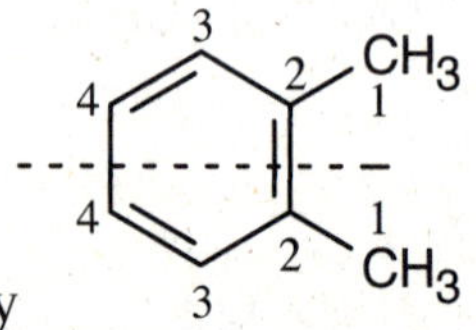

four peaks

(d)

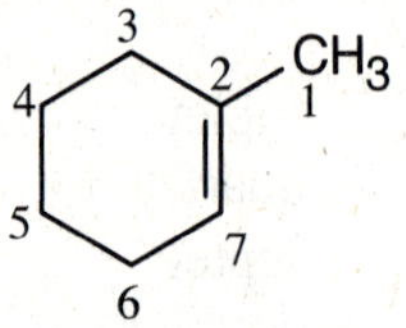

seven peaks

(e)

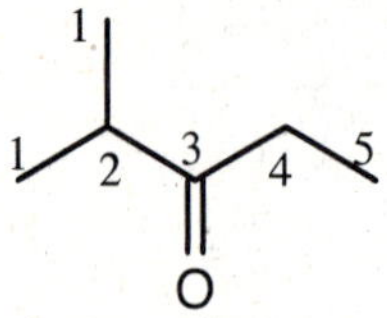

five peaks

(f)

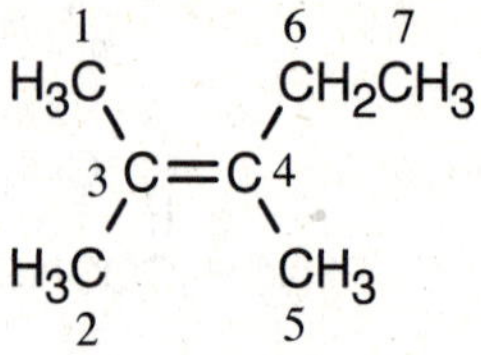

seven peaks

**13.24** (a) $CH_3CH_2CH_2CH_2CH_2CH{=}CH_2$ seven peaks

(b) $(CH_3)_2CHCH_2CH_2CH_3$ five peaks

(c) $(CH_3)_2CHCH_2Cl$ three peaks

Other structures are possible. In (b) and (c), two carbons are equivalent.

## Visualizing Chemistry

**13.25**

(a)

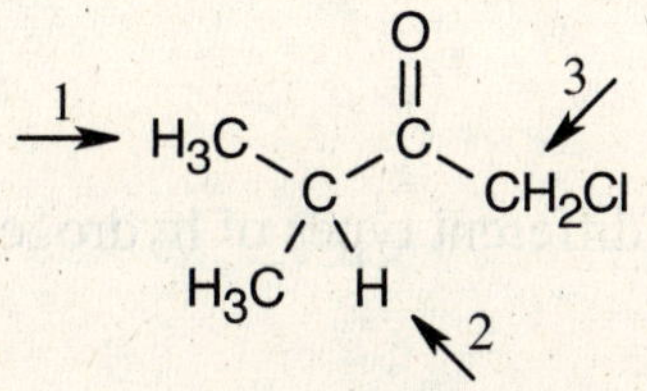

1. doublet
2. septet
3. singlet

(b)

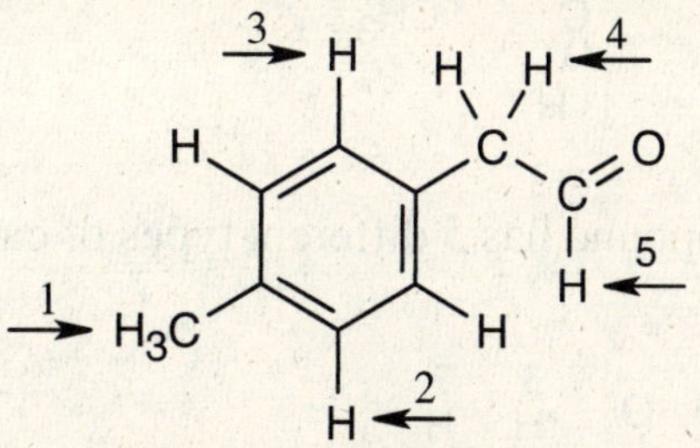

1. singlet
2. doublet
3. doublet
4. doublet
5. triplet

**13.26**

$^{1}H$ NMR: 5 signals

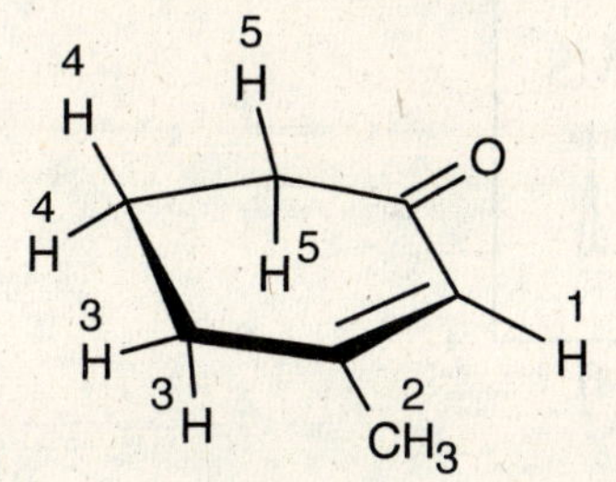

$^{13}C$ NMR: 7 signals

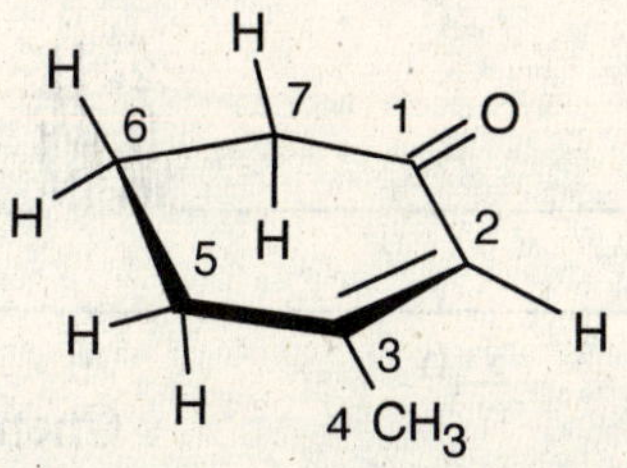

**13.27**

| Compound | Significant IR Absorption | Due to: |
|---|---|---|
| (a) 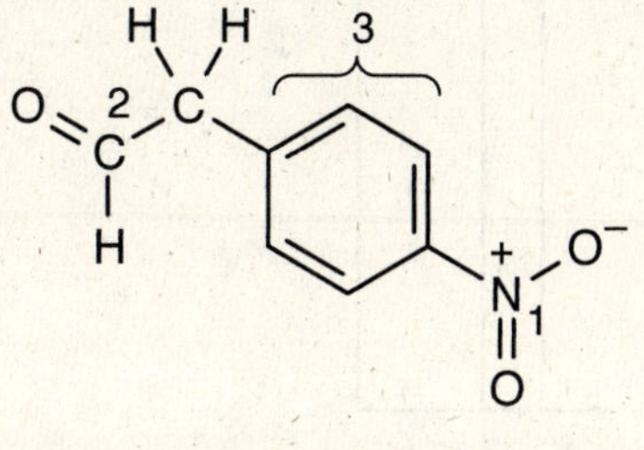 | $1540\ cm^{-1}$ | nitro group (1) |
| | $1725\ cm^{-1}$ | aldehyde (2) |
| | $3030\ cm^{-1}$, | aromatic ring C–H(3) |
| | $1450–1600\ cm^{-1}$ | aromatic ring C=C(3) |
| (b) 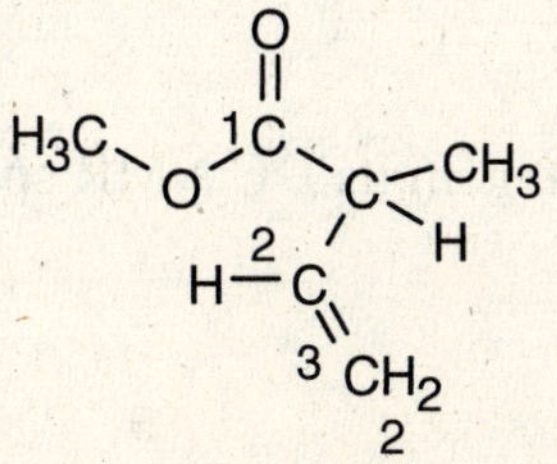 | $1735\ cm^{-1}$ | ester (1) |
| | $3020–3100\ cm^{-1}$ | alkene =C–H(2) |
| | $1640–1680\ cm^{-1}$ | alkene C=C (3) |

(c)

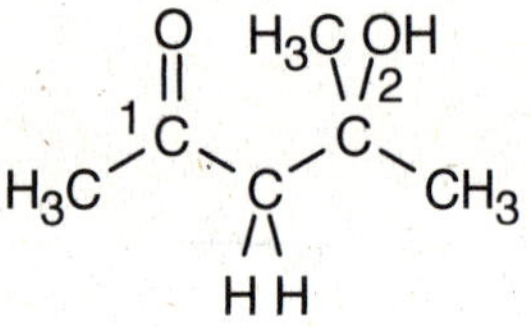

| | |
|---|---|
| 1715 $cm^{-1}$ | ketone (1) |
| 3400–3650 $cm^{-1}$ | alcohol (2) |

**13.28** The compound has 5 different types of carbons and 4 different types of hydrogens.

$^{1}H$

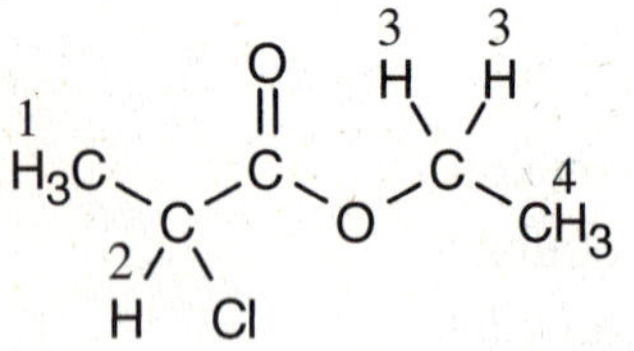

$^{13}C$

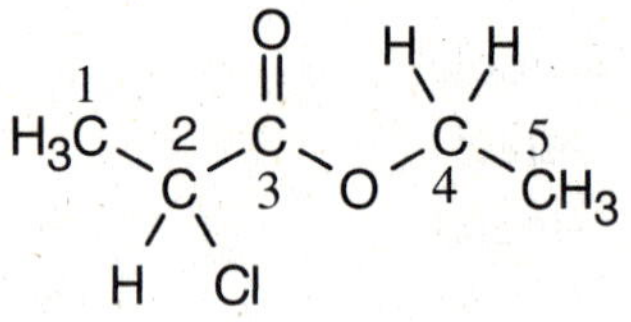

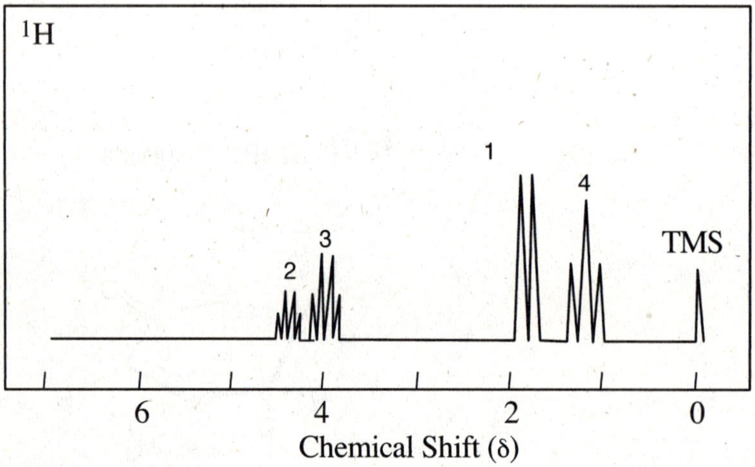


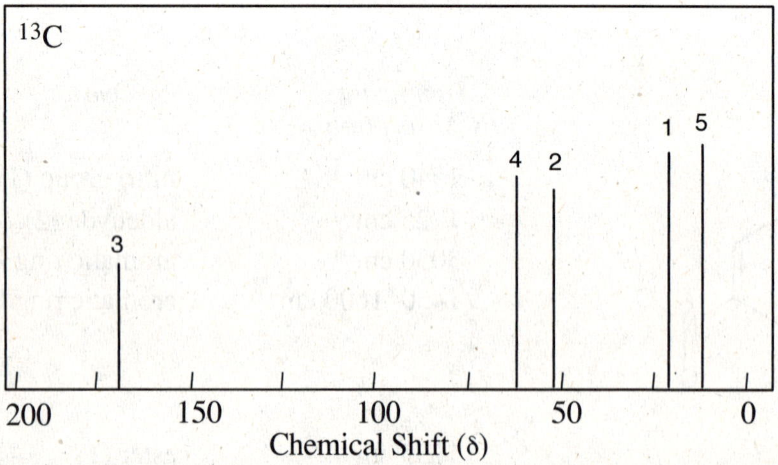


**13.29** Because of symmetry, the compound shows five absorptions in its $^{13}C$ NMR spectrum.

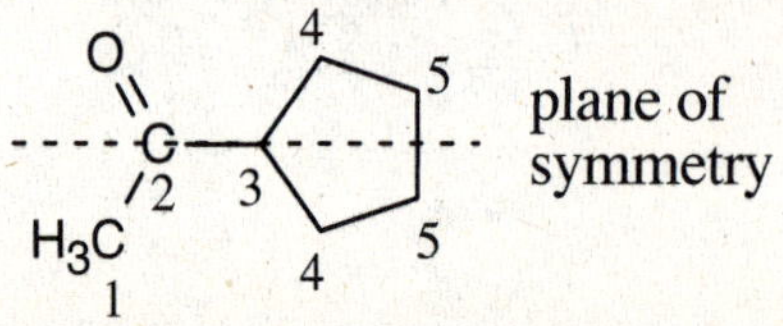

# Additional Problems

## Mass Spectrometry

**13.30** In order to simplify this problem, neglect the $^{13}C$ and $^{2}H$ isotopes in determining the molecular ions of these compounds.

(a) The formula weight of $–CH_3$ is 15, and the atomic masses of the two bromine isotopes are 79 and 81. The two molecular ions of bromoethane occur at $M^+ = 94$ (50.7%) and $M^+ = 96$ (49.3%).

(b) The formula weight of $–C_6H_{13}$ is 85, and the atomic masses of the two chlorine isotopes are 35 and 37. The two molecular ions of 1-chlorohexane occur at $M^+ = 120$ (75.8%) and $M^+ = 122$ (24.2%).

**13.31** If the isotopic masses of the atoms C, H, and O had integral values of 12 amu, 1 amu and 16 amu, many molecular formulas would correspond to a molecular weight of 288 amu. Because isotopic masses are not integral, however, only one molecular formula is associated with a molecular ion at 288.2089 amu.

To reduce the number of possible formulas, assume that the difference in molecular weight between 288 and 288.2089 is due mainly to hydrogen. Divide 0.2089 by 0.00783, the amount by which the atomic weight of one $^{1}H$ atom differs from 1. The answer, 26.67, gives a "ballpark" estimate of the number of hydrogens in testosterone. Then, divide 288 by 12, to determine the maximum number of carbons. Since $288 \div 12 = 24$, we know that testosterone can have no more than 22 carbons if it also includes hydrogen and oxygen. Make a list of *reasonable* molecular formulas containing C, H and O whose mass is 288 and which contain 20–30 hydrogens. Tabulate these, and calculate their exact masses using the values in the text. The only possible formula for testosterone is $C_{19}H_{28}O_2$.

**Isotopic mass**

| Molecular formula | Mass of carbons | Mass of hydrogens | Mass of oxygens | Mass of molecular ion |
|---|---|---|---|---|
| $C_{20}H_{32}O$ | 240.0000 amu | 32.2504 amu | 15.9949 amu | 288.2453 amu |
| $C_{19}H_{28}O_2$ | 228.0000 | 28.2191 | 31.9898 | 288.2089 |
| $C_{18}H_{24}O_3$ | 216.0000 | 24.1879 | 47.9847 | 288.1726 |

**13.32**

| $M^+$ | Molecular Formula | Possible Structure |
|---|---|---|
| (a) 132 | $C_{10}H_{12}$ | 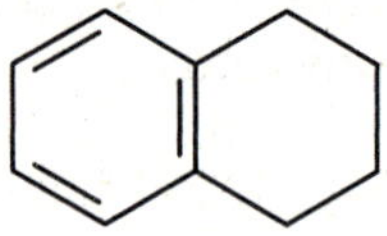 |
| (b) 166 | $C_{13}H_{10}$ | 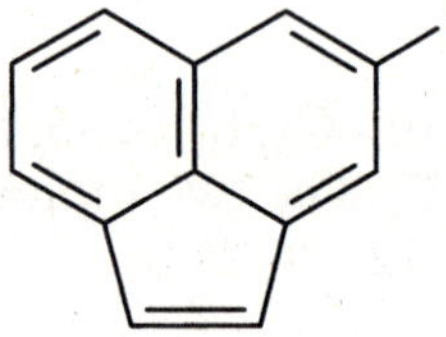 |
| | $C_{12}H_{22}$ | 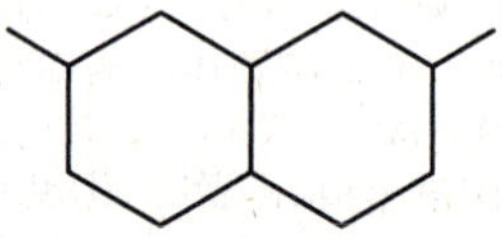 |
| (c) 84 | $C_6H_{12}$ | 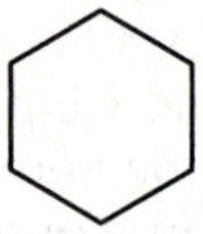 |

## Electromagnetic Radiation and Spectroscopy

**13.33**

$$E = \frac{1.20 \times 10^{-4}\ \text{kJ/mol}}{\lambda\ (\text{in m})} = \frac{1.20 \times 10^{-4}\ \text{kJ/mol}}{6.55 \times 10^{-6}}$$

$$= 0.183 \times 10^{2}\ \text{kJ/mol} = 18.3\ \text{kJ/mol}$$

**13.34**

$$E = \frac{1.20 \times 10^{-4}\ \text{kJ/mol}}{\lambda\ (\text{in m})} \qquad \lambda = 217\ \text{nm} = 2.17 \times 10^{-7}\text{m}$$

$$E = \frac{1.20 \times 10^{-4}\ \text{kJ/mol}}{2.17 \times 10^{-7}} = 5.53 \times 10^{2}\ \text{kJ/mol}$$

Compare this value with the energy required for infrared excitation (Problem 13.33). More energy is required for ultraviolet excitation than for infrared excitation.

**13.35**

$$E = \frac{1.20 \times 10^{-4} \text{ kJ/mol}}{\lambda \text{ (in m)}}$$

To find λ, use the formula

$$\lambda = \frac{c}{\nu} \text{ where } c = 3.0 \times 10^8 \text{ m/s}$$

In this problem, $\nu = 100$ MHz, or $10^8$ Hz

and $\lambda = 3.0$ m

$$E = \frac{1.20 \times 10^{-4} \text{ kJ/mol}}{3.0} = 4 \times 10^{-5} \text{ kJ/mol}$$

for $\nu = 220$ MHz, or $2.2 \times 10^8$ Hz

$\lambda = 1.4$ m

$E = 8.6 \times 10^{-5}$ kJ/mol

Increasing the spectrophotometer frequency from 100 MHz to 220 MHz increases the amount of energy needed for resonance.

**13.36** (a) The *chemical shift* is the exact position at which a nucleus absorbs rf energy in an NMR spectrum.

(b) If the NMR signal of nucleus A is split by the spin of adjacent nucleus B, there is reciprocal splitting of the signal of nucleus B by the spin of nucleus A. The spins of the two nuclei are said to be coupled. The distance between two individual peaks within the multiplet of A is the same as the distance between two individual peaks within the multiplet of B. This distance, measured in Hz, is called the *coupling constant*.

(c) $\lambda_{max}$ is the wavelength in an ultraviolet spectrum at which the absorbance is the greatest.

(d) *Spin-spin splitting* is the splitting of a single NMR resonance into multiple lines. Spin–spin splitting occurs when the effective magnetic field felt by a nucleus is influenced by the small magnetic moments of adjacent nuclei. In $^1$H NMR the signal of a proton with *n* neighboring protons is split into *n+1* peaks. The magnitude of spin–spin splitting is given by the coupling constant *J*.

(e) A *wave number* is the reciprocal of wavelength in centimeters.

(f) The *applied magnetic field* is the magnetic field that is externally applied to a sample by an NMR spectrometer.

## UV Spectroscopy

**13.37** If a concentration of $4.2 \times 10^{-5}$ M ($c_1$) has A = 0.56 ($A_1$) (Problem 13.39), a solution with A = 0.93 ($A_2$) has a concentration of:

$$c_2 = A_2c_1/A_1 = 0.93 \times 4.2 \times 10^{-2} \text{ M} / 0.56 = 6.9 \times 10^{-5} \text{ M}$$

**13.38** The human metabolite has a more extensive conjugated system, and its value of $\lambda_{max}$ should be greater.

**13.39**

$\varepsilon = 13{,}260 \text{ L M}^{-1} \text{ cm}^{-1} = 1.3260 \times 10^4 \text{ L M}^{-1} \text{ cm}^{-1}$

$c = 42\ \mu\text{M} = 4.2 \times 10^{-5}$ M

$l = 1.00$ cm

$A = \varepsilon \times c \times l = 1.3260 \times 10^4 \text{ M}^{-1} \text{ cm}^{-1} \times 4.2 \times 10^{-5} \text{ M} \times 1.00 \text{ cm} = 0.56$

**13.40**

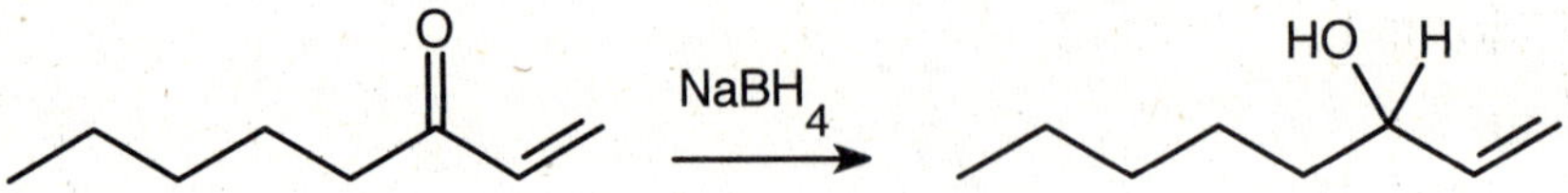


The UV absorption due to the conjugated enone disappears after reduction is complete.

## IR Spectroscopy

**13.41** Reciprocal centimeters ($cm^{-1}$) are units of frequency of radiation. Since frequency is directly proportional to energy, the larger the frequency, the more energy absorbed by a bond, and the stronger the bond. A carbon–oxygen double bond absorbs energy at a higher frequency (1700 $cm^{-1}$) than a carbon–oxygen single bond (1000 $cm^{-1}$) because the C=O double bond is stronger. This answer confirms what we already know about the relative strengths of single and double bonds.

**13.42**

| *IR Absorption* | *Due to:* |
|---|---|
| (a) 1670 $cm^{-1}$ | alkene or carbonyl group |
| (b) 1735 $cm^{-1}$ | ester group |
| (c) 1540 $cm^{-1}$ | nitro group |
| (d) 1715 $cm^{-1}$<br>2500-3100 $cm^{-1}$ | carboxylic acid group |

**13.43** (a) Absorptions at 3300 $cm^{-1}$ and 2150 $cm^{-1}$ are due to a terminal triple bond. Possible structures:

$CH_3CH_2CH_2C{\equiv}CH$ $\quad\quad$ $(CH_3)_2CHC{\equiv}CH$

(b) IR absorption at 3400 $cm^{-1}$ is due to a hydroxyl group. Since no double bond absorption is present, the compound must be a cyclic alcohol.

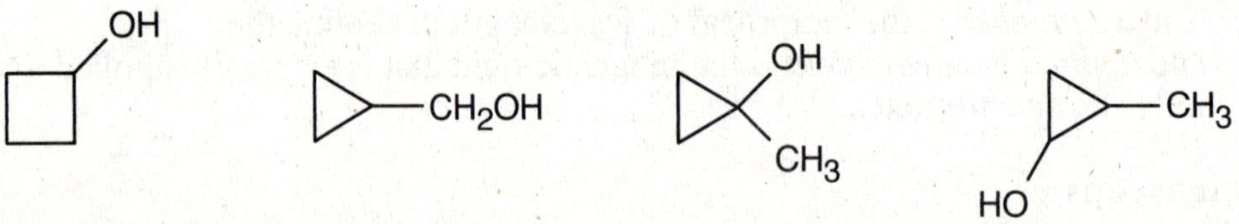


(c) Absorption at 1715 $cm^{-1}$ is due to a ketone. The only possible structure is $CH_3CH_2COCH_3$.

(d) Absorptions at 1600 $cm^{-1}$ and 1500 $cm^{-1}$ are due to an aromatic ring. Possible structures:

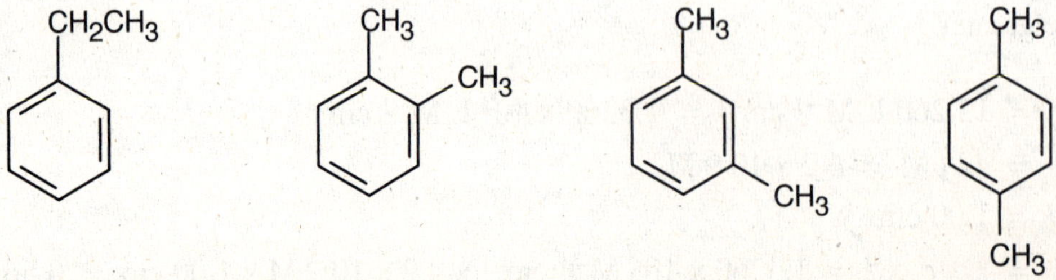

**13.44** Weak absorptions are not listed.

| | *Compound* | *Significant IR Absorption* | *Due to:* |
|---|---|---|---|
| (a) | 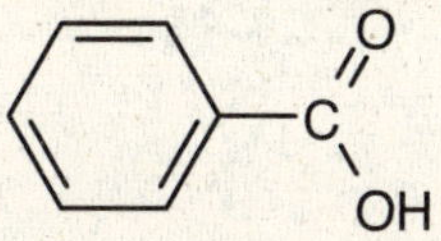 | 2500–3100 $cm^{-1}$<br>1710 $cm^{-1}$<br>1450–1600 $cm^{-1}$ | O–H (carboxylic acid)<br>C=O (carboxylic acid)<br>C=C (aromatic ring) |
| (b) | 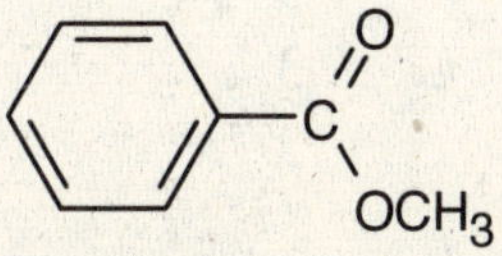 | 1715 $cm^{-1}$<br>1450–1600 $cm^{-1}$ | C=O (conjugated ester)<br>C=C (aromatic ring) |
| (c) | 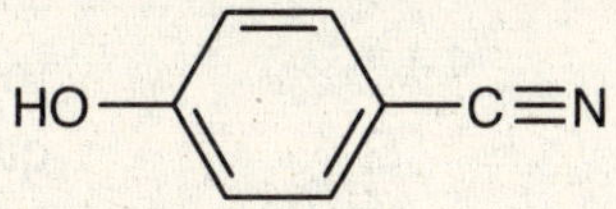 | 2210–2260 $cm^{-1}$<br>3200–3650 $cm^{-1}$<br>1450–1600 $cm^{-1}$ | C≡N (nitrile)<br>O–H (hydroxyl)<br>C=C (aromatic ring) |
| (d) | 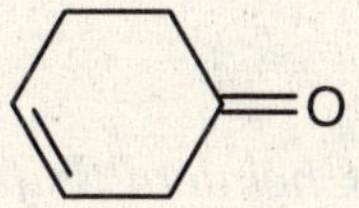 | 1715 $cm^{-1}$<br>1640–1680 $cm^{-1}$ | C=O (ketone)<br>C=C (double bond) |
| (e) | 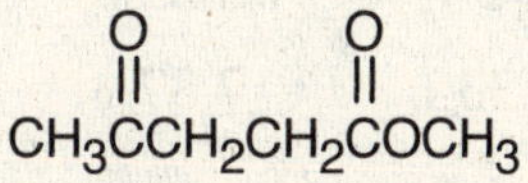 | 1735 $cm^{-1}$<br>1715 $cm^{-1}$ | C=O (ester)<br>C=O (ketone) |

## NMR Spectroscopy

**13.45** (a) 2.1 δ x 300 Hz = 630 Hz (b) 3.45 δ x 300 Hz = 1035 Hz
(c) 6.30 δ x 300 Hz = 1890 Hz

**13.46** (a) Since the symbol "δ" indicates ppm downfield from TMS, chloroform absorbs at 7.3 ppm.

(b) δ x (spectrometer frequency/$10^6$) = observed chemical shift (in Hz)
7.3 x 360 Hz = 2600 Hz

(c) The value of δ is still 7.3 because the chemical shift measured in δ is independent of the operating frequency of the spectrometer.

**13.47** See Problem 13.17 for the method of solution.

(a) $\delta = \dfrac{218 \text{ Hz}}{100 \text{ MHz}} = 2.18\ \delta$ (b) $\delta = \dfrac{478 \text{ Hz}}{100 \text{ MHz}} = 4.78\ \delta$ (c) $\delta = \dfrac{751 \text{ Hz}}{100 \text{ MHz}} = 7.51\ \delta$

**13.48** δ x (spectrometer frequency/$10^6$) = observed chemical shift (in Hz)
Here, spectrometer frequency = 220 MHz

(a) 2.18 x 220 Hz = 480 Hz (b) 4.78 x 220 Hz = 1050 Hz
(c) 7.51 x 220 Hz = 1650 Hz

**13.49**

| | *Compound* | *Protons* | *Chemical Shift* | *Rel. Peak Area* | *Splitting* |
|---|---|---|---|---|---|
| (a) | 1 2<br>$CH_3CHCl_2$ | 1 | 2.0 δ | 3 | doublet |
| | | 2 | 5.9 δ | 1 | quartet |
| (b) | 2 3 1<br>$CH_3CO_2CH_2CH_3$ | 1 | 1.2 δ | 3 | triplet |
| | | 2 | 2.0 δ | 3 | singlet |
| | | 3 | 4.1 δ | 2 | quartet |
| (c) | 1 3 2<br>$(CH_3)_3CCH_2CH_3$ | 1 | 0.9 δ | 9 | singlet |
| | | 2 | 0.9 δ | 3 | triplet |
| | | 3 | 1.2 δ | 2 | quartet |

The peaks due to protons 1 and 2 overlap

**13.50**

*Lowest Chemical Shift* ⟶ *Highest Chemical Shift*

| $CH_4$ | < cyclohexane | < $CH_3COCH_3$ | < $CH_2Cl_2$, | $H_2C{=}CH_2$ | < benzene |
|---|---|---|---|---|---|
| 0.23 δ | 1.43 δ | 2.17 δ | 5.30 δ | 5.33 δ | 7.37 δ |

**13.51**

(a)

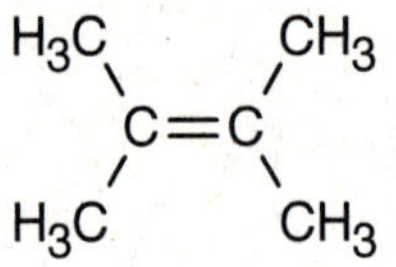

**$^{13}$C:** 2 absorptions
**$^1$H:** 1 absorption

(b)

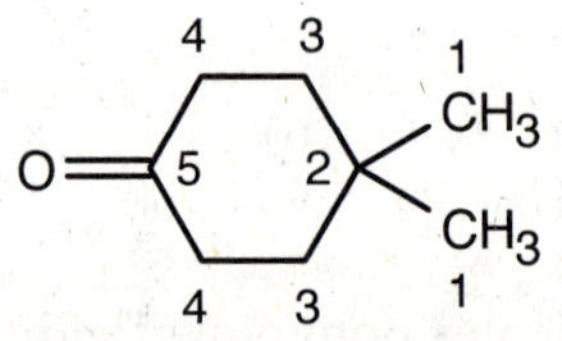

**$^{13}$C:** 5 absorptions (at room temperature)

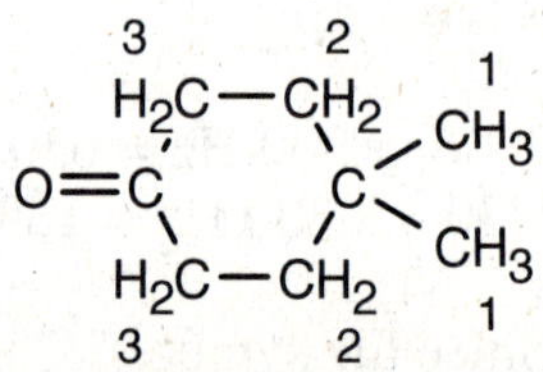

**$^1$H:** 3 absorptions (at room temperature)

(c)

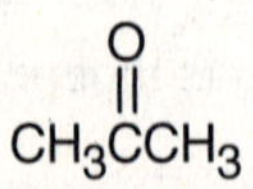

**$^{13}$C:** 2 absorptions
**$^1$H:** 1 absorption

(d)

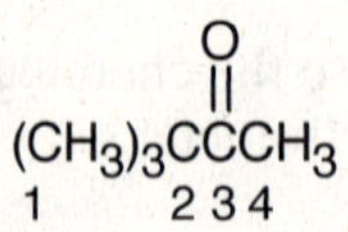

**$^{13}$C:** 4 absorptions

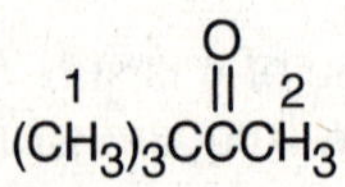

**$^1$H:** 2 absorptions

(e)

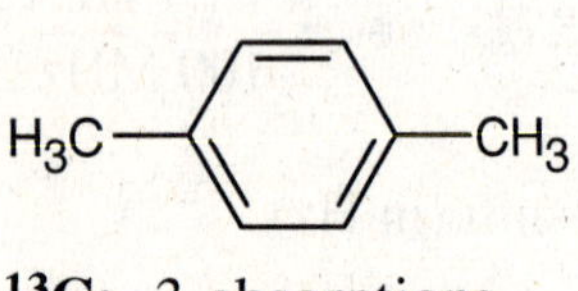

**$^{13}$C:** 3 absorptions
**$^1$H:** 2 absorptions

(f)

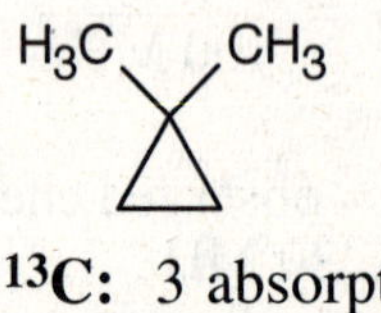

**$^{13}$C:** 3 absorptions
**$^1$H:** 2 absorptions

## General Problems

**13.52**

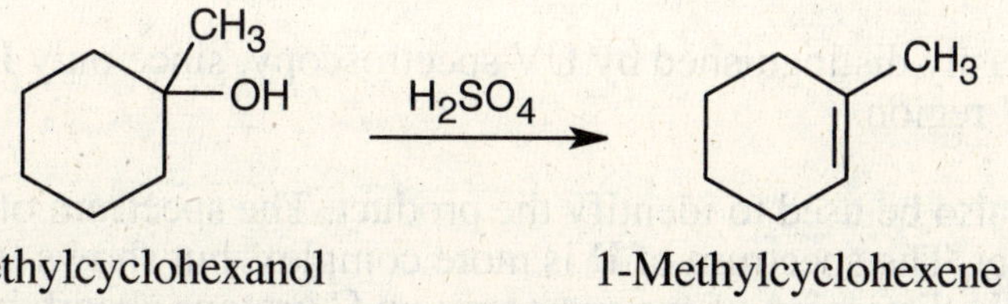

The infrared spectrum of the starting alcohol shows a broad absorption at 3400–3640 $cm^{-1}$, due to an O–H stretch, and another strong absorption at 1050–1100 $cm^{-1}$, due to a C–O stretch. The alkene product exhibits medium intensity absorbances at 1640–1680 $cm^{-1}$ and at 3000–3100 $cm^{-1}$. Monitoring the *disappearance* of one of the alcohol absorptions allows one to decide when the alcohol is totally dehydrated. It is also possible to monitor the *appearance* of one of the alkene absorbances.

**13.53** The mass spectrum of 1-methylcyclohexanol shows a peak at $m/z = 114$. After reaction is complete, this peak disappears and a peak due to 1-methylcyclohexene ($m/z = 96$) appears.

**13.54** The molecular formula shows that the compound is most likely aromatic. The absorption near 1700 $cm^{-1}$ is due to a carbonyl group. The absorptions between 1500–1600 $cm^{-1}$ are due to an aromatic ring. (Peaks at 700–900 $cm^{-1}$ are also characteristic of aromatic rings.) The compound is benzaldehyde.

**13.55**

| | | |
|---|---|---|
| (a) | $CH_3CH_2NHCH_3$<br>N–H absorption at<br>3300–3500 $cm^{-1}$ | $(CH_3)_3N$<br>No absorption<br>at 3300–3500 $cm^{-1}$ |
| (b) | $CH_3COCH_3$<br>Strong ketone absorption<br>at 1715 $cm^{-1}$ | $H_2C{=}CHCH_2OH$<br>Strong alcohol absorption<br>at 3400–3640 $cm^{-1}$ |
| (c) | $CH_3COCH_3$<br>Strong ketone absorption<br>at 1715 $cm^{-1}$ | $CH_3CH_2CHO$<br>Strong aldehyde absorption<br>at 1725 $cm^{-1}$ |

**13.56** One isomer of each pair in Problem 13.55 shows only one peak in its $^1H$ NMR spectrum. In (a), $(CH_3)_3N$ absorbs at 2.12 δ; the other isomer has a more complicated $^1H$ NMR spectrum. In (b) and (c), the acetone absorption occurs at 2.17 δ; the other isomers, again, show more complicated spectra.

**13.57** (a)The $^{13}C$ NMR spectrum of $(CH_3)_3N$ shows only one peak; the spectrum of the other isomer shows 3 peaks.

(b) (c)The spectrum of acetone shows two peaks, one at 30 δ and one at 208 δ. The spectra of the other isomers show three peaks.

**13.58**

$CH_3CH_2C{\equiv}CCH_2CH_3$ — Hex-3-yne (**C**)

$CH_3CH{=}CHCH{=}CHCH_3$ — Hexa-2,4-diene (**D**)

The isomers are easily distinguished by UV spectroscopy, since only **D** is conjugated and absorbs in the UV region.

$^1$H NMR can also be used to identify the product. The spectrum of **C** consists of a quartet and a triplet. The spectrum of **D** is more complex, but shows four protons absorbing in the vinylic region of the spectrum; no **C** protons absorb in this region.

**13.59**

(a)

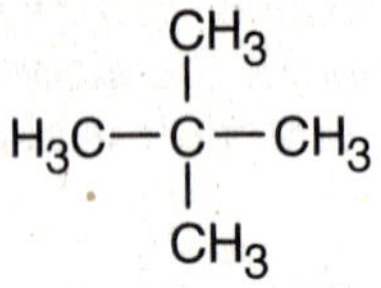

(b)

(c)

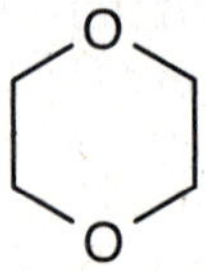

**13.60**

| *Compound* | *Distinguishing Absorption* | *Due to:* |
|---|---|---|
| (a) $CH_3CH_2\overset{O}{\overset{\|}{C}}CH_3$ | 1715 cm$^{-1}$ | C=O (ketone) |
| (b) $(CH_3)_2CHCH_2C{\equiv}CH$ | 2100–2260 cm$^{-1}$<br>3300 cm$^{-1}$ | C≡C<br>C≡C—H |
| (c) $(CH_3)_2CHCH_2CH{=}CH_2$ | 910 cm$^{-1}$, 990 cm$^{-1}$<br>1640–1680 cm$^{-1}$<br>3020–3100 cm$^{-1}$ | $RCH{=}CH_2$<br>C=C<br>=C—H |
| (d) $CH_3CH_2CH_2\overset{O}{\overset{\|}{C}}OCH_3$ | 1735 cm$^{-1}$ | C=O (ester) |
| (e) 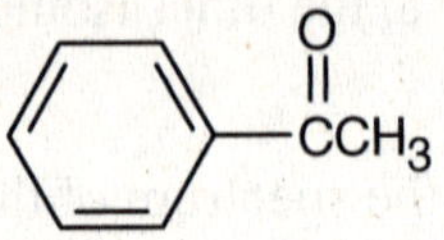 | 1690 cm$^{-1}$<br>1450–1600 cm$^{-1}$ | ketone next to aromatic ring<br>aromatic ring |
| (f) 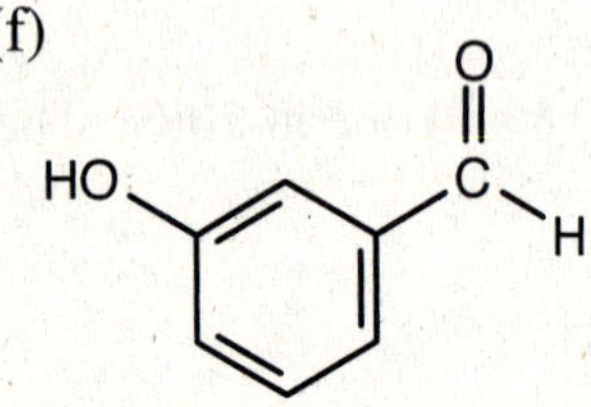 | 1710 cm$^{-1}$<br>3400–3650 cm$^{-1}$<br>1450–1600 cm$^{-1}$ | aldehyde next to aromatic ring<br>alcohol<br>aromatic ring |

**13.61**

| *Compound* | *Kinds of non-equivalent protons* | *Compound* | *Kinds of non-equivalent protons* |
|---|---|---|---|
| (a) 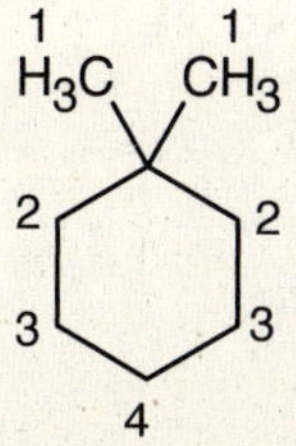 | 4 | (b) $\overset{1}{CH_3}\overset{2}{CH_2}\overset{3}{CH_2}O\overset{4}{CH_3}$ | 4 |

| *Compound* | *Kinds of non-equivalent protons* | *Compound* | *Kinds of non-equivalent protons* |
|---|---|---|---|
| (c) 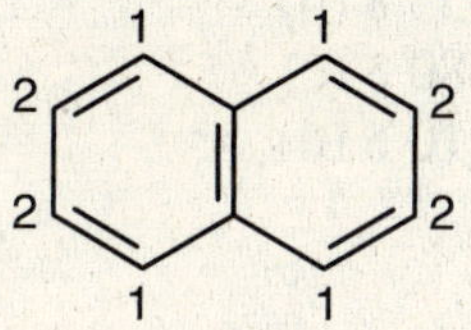 | 2 | (d) 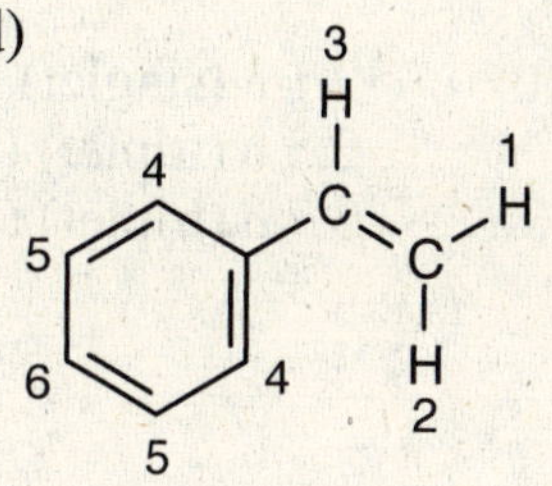 | 6 |
| (e) 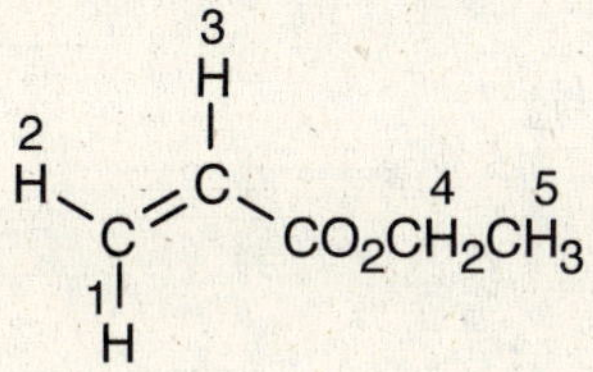 | 5 | | |

**13.62**

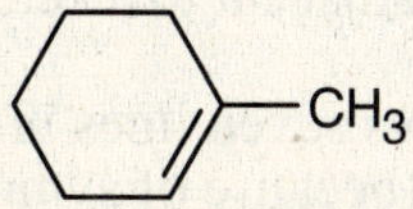

1-Methylcyclohexene (**A**)

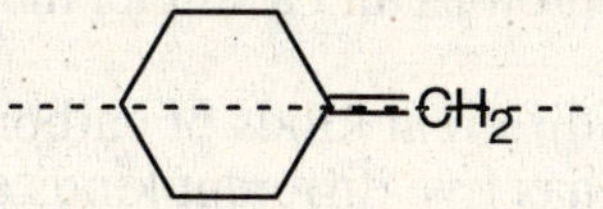


Methylenecyclohexane (**B**)

**$^{13}$C:** Symmetrical methylenecyclohexane (**B**) has only five different kinds of carbons and shows five peaks in its $^{13}$C NMR spectrum. 1-Methylcyclohexene (**A**) has seven different kinds of carbons and shows seven peaks.

**$^{1}$H:** 1-Methylcyclohexene (**A**) has six different kinds of protons; methylenecyclohexane (**B**) has four different kinds of protons. Since several absorptions in each of the spectra overlap (protons on the ring), it is more helpful to focus on specific absorptions in each spectrum. The $^{1}$H NMR spectrum of **A** shows an unsplit methyl group and a vinylic proton signal of relative area 1. The vinylic absorption of **B** has relative area 2.

**13.63**

| | *¹H NMR* | *¹³C NMR* |
|---|---|---|
| (a) | $ClCH_2CH_2CH_2Cl$ (b a b) | $ClCH_2CH_2CH_2Cl$ (b a b) |
| Number of peaks: | 2 | 2 |
| Chemical shift: | 2.2 δ (quintet) (a)<br>3.7 δ (triplet) (b) | 15–55 δ (a)<br>35–80 δ (b) |
| (b) | $CH_3C(=O)CH_2CH_2Cl$ (a b c) | $CH_3C(=O)CH_2CH_2Cl$ (a d b c) |
| Number of peaks: | 3 | 4 |
| Chemical shift: | 2.1 δ (singlet) (a)<br>2.5 δ (triplet) (b)<br>3.7 δ (triplet) (c) | 8–30 δ (a)<br>15–55 δ (b)<br>35–80 δ (c)<br>170–210 δ (d) |

**13.64**

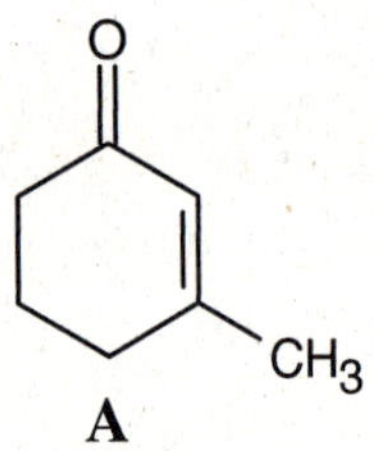


**A**

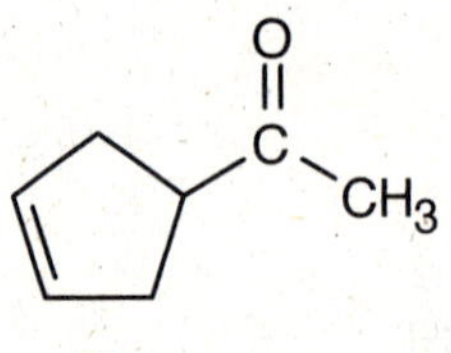


**B**

Distinguishing features of the ¹H NMR spectrum of **A** include one unsplit vinylic proton and a singlet methyl group. For **B**, distinguishing features of the ¹H NMR spectrum include two split vinylic protons and a singlet methyl group adjacent to a ketone.

**13.65** Compound **A** has seven different kinds of carbons and shows seven lines in its ¹³C NMR spectrum. Compound **B** has five different kinds of carbons (because of symmetry) and shows five lines in its ¹³C NMR spectrum.

**13.66** Only **A** is conjugated and shows UV absorption.

**13.67**

| *Compound* | *Proton(s)* | *Chemical Shift* |
|---|---|---|
| (a) 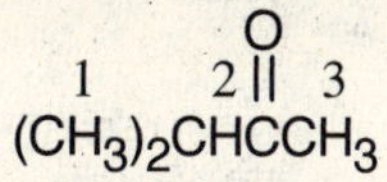 | 1 | 0.95 δ |
| | 2 | 2.43 δ |
| | 3 | 2.10 δ |
| (b) 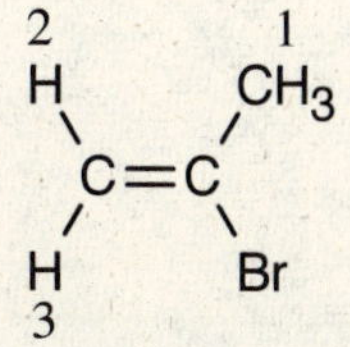 | 1 | 2.32 δ |
| | 2 | 5.25 δ |
| | 3 | 5.54 δ |

**13.68** The four isomers of $C_3H_6Br_2$ are shown below, along with the number of different kinds of protons for each structure.

| *Compound* | *Kinds of protons* | *Compound* | *Kinds of protons* |
|---|---|---|---|
| 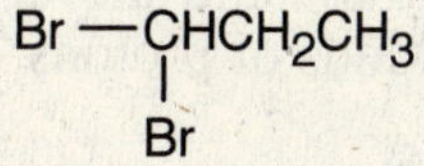 | 3 | 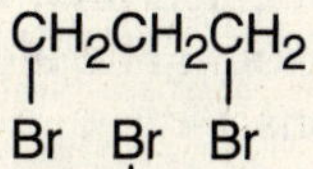 | 2 |
| 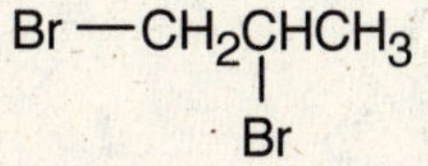 | 3 | 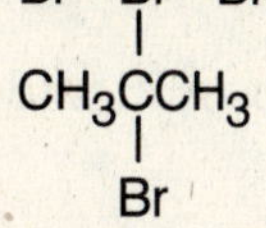 | 1 |

Because the spectrum of $C_3H_6Br_2$ shows two kinds of protons, it must represent 1,3-dibromopropane. The splitting pattern shown in the spectrum (triplet, quintet) is what is expected for 1,3-dibromopropane.

**13.69** Either $^1H$ NMR or $^{13}C$ NMR can be used to distinguish among these isomers. In either case, it is first necessary to find the number of different kinds of protons or carbon atoms.

$^{13}C$ NMR is the preferred method for identifying these compounds since each isomer differs in the number of absorptions in its $^{13}C$ NMR spectrum.

$^1H$ NMR can also be used to distinguish among the isomers. The two isomers that show two $^1H$ NMR peaks differ in their splitting patterns.

| *Compound* | 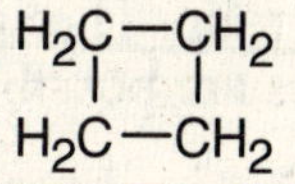 | $H_2C{=}CHCH_2CH_3$ | $CH_3CH{=}CHCH_3$ | $(CH_3)_2C{=}CH_2$ |
|---|---|---|---|---|
| Kinds of protons | 1 | 5 | 2 | 2 |
| Kinds of carbon atoms | 1 | 4 | 2 | 3 |
| Number of $^1H$ NMR peaks | 1 | 5 | 2 | 2 |
| Number of $^{13}C$ NMR peaks | 1 | 4 | 2 | 3 |

**13.70**

(a) Possible structures for $C_3H_6O$:

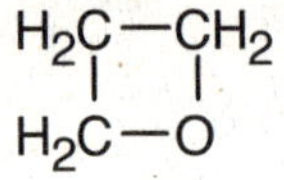

cyclic ether

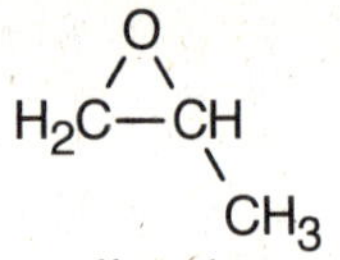

cyclic ether

$H_2C{=}CHOCH_3$

ether, double bond

$H_2C{=}CHCH_2OH$

alcohol, double bond

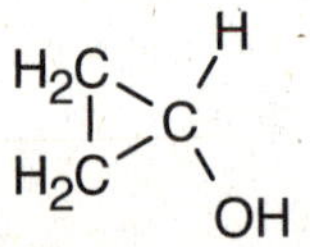

cyclic alcohol

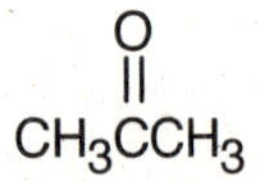

ketone (acetone)

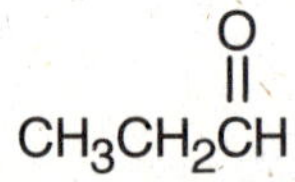

aldehyde

(b) An IR absorption at 1715 $cm^{-1}$ is due to a carbonyl group. Only the last two compounds show an absorption near this region.

(c) The compound must be acetone, which has only one kind of proton and shows only one $^1H$ NMR absorption. (The aldehyde, with three different kinds of protons, would show three absorptions.)

**13.71** Each spectrum shows an absorption due to a methyl ketone.

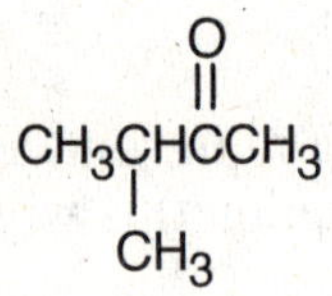

**13.72**

(a)

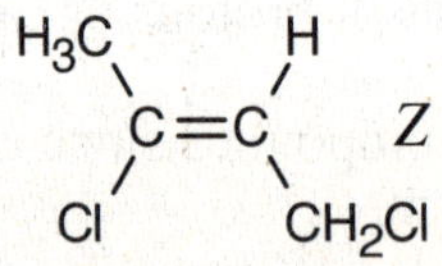

*Z*

The *E* isomer is also an acceptable answer.

(b)

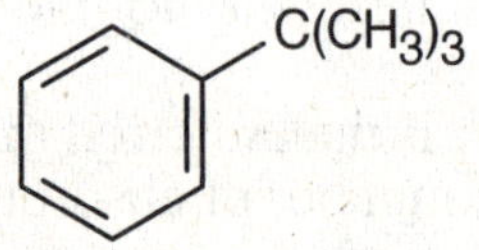

**13.73** The IR absorption at 1740 $cm^{-1}$ is due to an ester group. The splitting pattern (triplet, quartet) is caused by an ethyl group. Two structures are possible at this point.

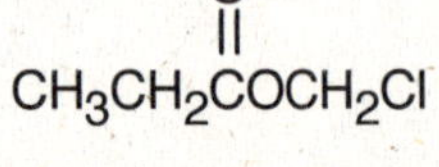

**I**

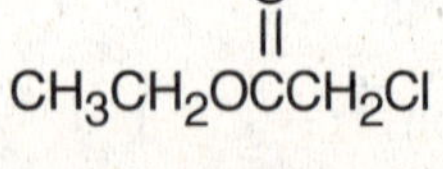

**II**

In **I**, the protons attached to the carbon bonded to both oxygen and chlorine ($-OCH_2Cl$) absorb far downfield (5.0 – 6.0 δ). Because no signal is present in this region of the $^1H$ NMR spectrum given, the unknown must be **II**. In addition, the quartet absorbing at 4.3 δ is typical of a $CH_2$ group next to an electronegative atom and coupled with a methyl group.

**13.74** The spectrum shows two doublets and a multiplet. The peak at 3.31 δ is due to hydrogens bonded to a carbon that is also bonded to an electronegative atom (bromine).

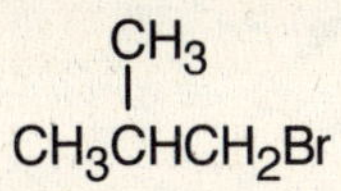

## In the Medicine Cabinet

**13.75**

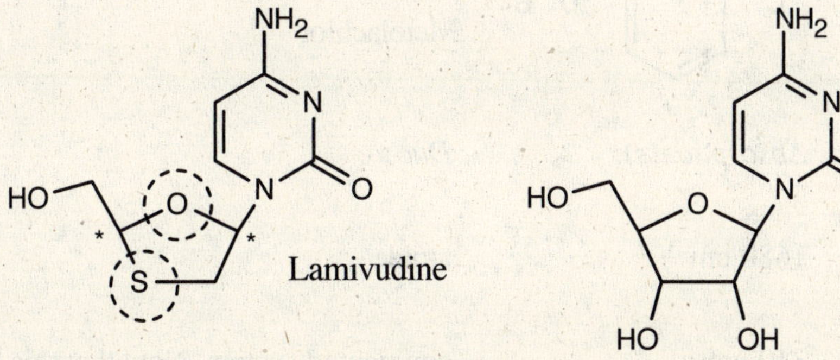

Lamivudine Cytidine

(a) The two chirality centers of lamivudine are starred.

(b) The acid-sensitive acetal-like groups are circled

(c)

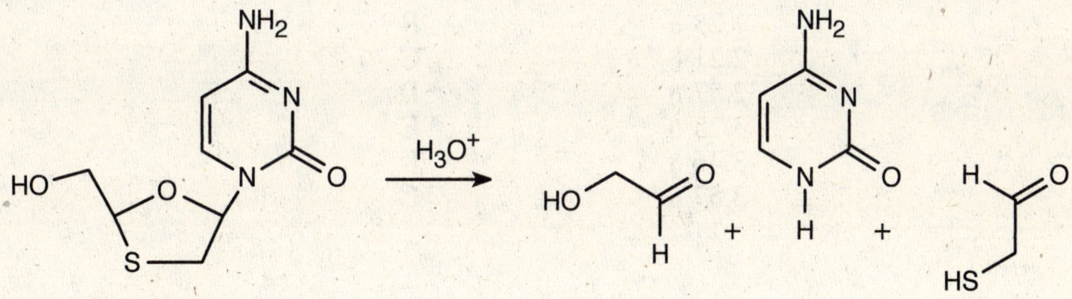

Lamivudine

(d)

$\varepsilon$ = 8,600 $M^{-1}$ $cm^{-1}$ = 8.6 x $10^3$ L $mol^{-1}$ $cm^{-1}$

$l$ = 1.00 cm

$A$ = 0.195

$$c = \frac{A}{\varepsilon \times l} = \frac{0.195}{8.6 \times 10^3\ \text{L·mol}^{-1}\text{·cm}^{-1} \times 1.0\ \text{cm}} = 2.3 \times 10^{-5}\ \text{M}$$

Since the calculated value of $c$ is the concentration after a tenfold dilution, the plasma concentration of lamivudine is 2.3 x $10^{-4}$M, or 0.23 mM.

**In the Field**

**13.76**

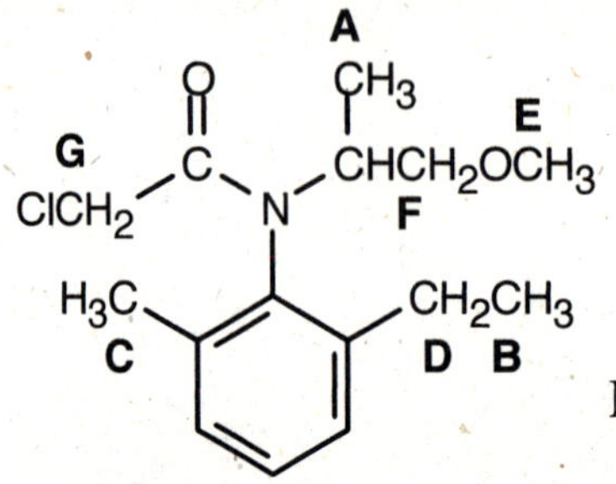


Metolachlor

| *Type of spectroscopy:* | *Absorption(s):* | *Due to:* |
|---|---|---|
| **IR:** | 1680 $cm^{-1}$ | amide |
| **UV/vis:** | 310 nm | conjugated system. Note that this value is larger than the value for a benzene ring because the amide group is also part of the system of conjugation. |
| **$^1$H NMR** | 1.11 δ | **A** |
| | 1.23 δ | **B** |
| | 2.21 δ | **C** |
| | 2.57 δ | **D** |
| | 3.25 δ | **E** |
| | 3.46 δ | **F** |
| | 3.57 δ | **G** |

# Chapter 14 – Biomolecules: Carbohydrates

## Chapter Outline

I. Classification of carbohydrates (Section 14.1).
  A. Simple *vs*. complex:
    1. Simple carbohydrates can't be hydrolyzed to smaller units.
    2. Complex carbohydrates are made up of two or more simple sugars linked together.
      a. A disaccharide is composed of two monosaccharides.
      b. A polysaccharide is composed of three or more monosaccharides.
  B. Aldoses *vs*. ketoses:
    1. A monosaccharide with an aldehyde carbonyl group is an aldose.
    2. A monosaccharide with a ketone carbonyl group is a ketose.
  C. *Tri-*, *tetr-* , *pent-*, etc. indicate the number of carbons in the monosaccharide.

II. Monosaccharides (Sections 14.2–14.8).
  A. Configurations of monosaccharides (Sections 14.2–14.4).
    1. Fischer projections (Section 14.2).
      a. Each chirality center of a monosaccharide is represented by a pair of crossed lines.
        i. Horizontal lines represent bonds coming out of the page.
        ii. Vertical lines represent bonds going into the page.
      b. The carbonyl carbon is placed at or near the top of the Fischer projection.
      c. Fischer projections can be rotated by 180°, but not by 90° or 270°.
      d. Carbohydrates with more than one chirality center are shown by stacking the centers on top of each other.
    2. D,L sugars (Section 14.3).
      a. (*R*)-Glyceraldehyde is also known as D-glyceraldehyde.
      b. In Fischer projections of D sugars, the –OH group farthest from the carbonyl group points to the right.
        i. Most naturally-occurring sugars are D sugars.
      c. In L sugars, the –OH group farthest from the carbonyl group points to the left.
      d. D,L designations refer only to the configuration farthest from the carbonyl carbon and are unrelated to the direction of rotation of plane-polarized light.
    3. Configurations of the aldoses (Section 14.4).
      a. There are 4 aldotetroses – D and L erythrose and threose.
      b. There are 4 D,L pairs of aldopentoses: ribose, arabinose, xylose and lyxose.
      c. There are 8 D,L pairs of aldohexoses : allose, altrose, glucose, mannose, gulose, idose, galactose, and talose.
  B. Cyclic structures of monosaccharides (Sections 14.5–14.6).
    1. Hemiacetal formation (Section 14.5).
      a. Monosaccharides are in equilibrium with their internal hemiacetals.
        i. Glucose exists primarily as a pyranose (six-membered) ring, formed by the –OH group at C5 and the aldehyde.
        ii. Fructose also can exist as a furanose (five-membered) ring.
      b. Structure of pyranose rings.
        i. Pyranose rings have a chair-like geometry.
        ii. The hemiacetal oxygen is at the right rear for D-sugars.
        iii. An –OH group on the right in a Fischer projection is on the bottom in a pyranose ring, and an –OH group on the left is on the top.
        iv. For D sugars, the $–CH_2OH$ group is on the top.

2. Mutarotation (Section 14.6).
   a. When a monosaccharide cyclizes, a new chirality center is generated at C!.
      i. The two diastereomers are anomers.
      ii. The form with the anomeric –OH group trans to the $-CH_2OH$ group is the α anomer (minor anomer).
      iii. The form with the anomeric –OH group cis to the $-CH_2OH$ group is the β anomer (major anomer).
   b. When a solution of either pure anomer is dissolved in water, the optical rotation of the solution reaches a constant value.
      i. This process is called mutarotation.
      ii. Mutarotation is due to the reversible opening and recyclizing of the hemiacetal ring and is catalyzed by both acid and base.

C. Reactions of monosaccharides (Section 14.7).
1. Ester and ether formation.
   a. Esterification occurs by treatment with an acid anhydride or acid chloride.
   b. Ethers are formed by treatment with methyl iodide and $Ag_2O$.
   c. Ester and ether derivatives are crystalline and easy to purify.
2. Glycoside (carbohydrate acetal) formation.
   a. Treatment of a hemiacetal with an alcohol and an acid catalyst yields an acetal.
      i. Acetals aren't in equilibrium with an open-chain form.
      ii. Aqueous acid reconverts the acetal to a monosaccharide.
   b. These acetals, called glycosides, are abundant in nature.
3. Biological ester formation: phosphorylation.
   a. Many carbohydrates are linked to other biomolecules through their anomeric centers to form molecules collectively known as glycoconjugates.
   b. A glycosyl nucleoside diphosphate reacts with the lipid or protein.
   c. Phosphorylation makes the anomeric –OH a better leaving group.
4. Reduction of monosaccharides.
   a. Reaction of a monosaccharide with $NaBH_4$ yields an alditol (polyalcohol).
5. Oxidation of monosaccharides.
   a. Several mild reagents can oxidize the carbonyl group to a carboxylic acid (aldonic acid).
      i. Tollens reagent and Benedict's reagent all serve as tests for reducing sugars.
      ii. All aldoses and some ketoses are reducing sugars, but glycosides are nonreducing.
   b. The more powerful oxidizing agent, dilute $HNO_3$, oxidizes aldoses to dicarboxylic acids (aldaric acids).

D. The eight essential monosaccharides (Section 14.8).
1. D-galactose, D-glucose, D-mannose, D-xylose, *N*-acetyl-D-glucosamine, *N*-acetyl-D-galactosamine, *N*-acetyl-D-neuraminic acid, L-fucose.
2. L-fucose is a deoxy sugar.
3. *N*-acetyl-D-glucosamine, *N*-acetyl-D-galactosamine, and *N*-acetyl-D-neuraminic acid are amino sugars.
4. All essential monosaccharides are biosynthesized from glucose.

III. Other carbohydrates (Sections 14.9–14.11).

A. Disaccharides (Section 14.9).
1. Cellobiose and maltose.
   a. Cellobiose and maltose contain a 1→4 glycosidic acetal bond between two glucose monosaccharide units.
   b. Maltose consists of two glucopyranose units joined by a 1→4-α-glycosidic bond.
   c. Cellobiose consists of two glucopyranose units joined by a 1→4-β-glycosidic bond.

d. Both maltose and cellobiose are reducing sugars and exhibit mutarotation.
e. Humans can't digest cellobiose but can digest maltose.

2. Sucrose.
    a. Sucrose is a disaccharide that yields glucose and fructose on hydrolysis.
        i. Sucrose is called "invert sugar" because the sign of rotation changes when sucrose is hydrolyzed.
        ii. Sucrose is one of the most abundant pure organic chemicals in the world.
    b. The two monosaccharides are joined by a glycosidic link between C1 of glucose and C2 of fructose.
    c. Sucrose isn't a reducing sugar and doesn't exhibit mutarotation.

B. Polysaccharides (Section 14.10).
1. Polysaccharides have a reducing end and undergo mutarotation, but aren't considered to be reducing sugars because of their size.
2. Important polysaccharides.
    a. Cellulose.
        i. Cellulose consists of thousands of D-glucose units linked by 1→4-β-glycosidic bonds.
        ii. In nature, cellulose is used as structural material.
    b. Starch.
        i. Starch consists of thousands of D-glucose units linked by 1→4-α-glycosidic bonds.
        ii. Starch can be separated into amylose and amylopectin fractions.
            (a). Amylose consists of several hundred glucose molecules linked together by 1→4-α-glycosidic bonds.
            (b). Amylopectin is larger and contains 1→6-α-glycosidic branches.
        iii. Starch is digested in the mouth by glycosidase enzymes, which only cleave α-glycosidic bonds.
    c. Glycogen.
        i. Glycogen is an energy-storage polysaccharide.
        ii. Glycogen contains both 1→4 and 1→6 links.

C. Cell surface carbohydrates and carbohydrate vaccines (Section 14.11).
1. Polysaccharides are involved in cell-surface recognition.
    a. Polysaccharide markers on the surface of red blood cells are responsible for blood-group incompatibility, which causes agglutination (clumping).
    b. Red blood cells have two types of markers (antigenic determinants) – A and B.
    c. The monosaccharide components of these markers are among the eight essential sugars.
2. Possible anticancer vaccines have been synthesized from antibodies to cell-surface polysaccharides found on the surface of cancer cells.

## Solutions to Problems

**14.1**

(a)

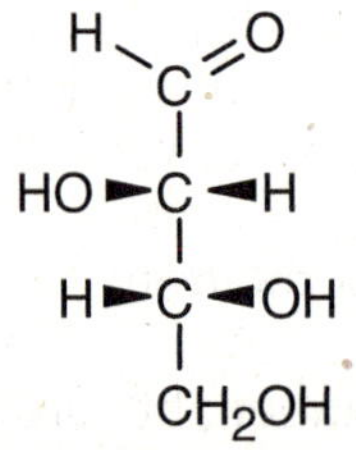

Threose
*an aldotetrose*

(b)

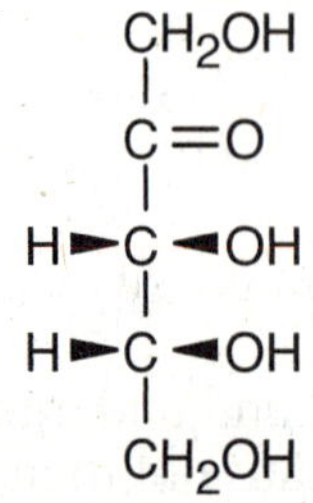

Ribulose
*a ketopentose*

(c)

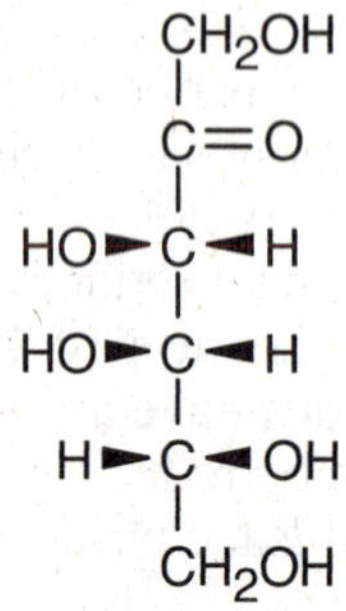

Tagatose
*a ketohexose*

(d)

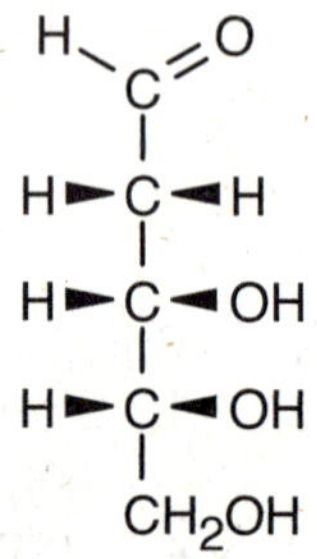

2-Deoxyribose
*an aldopentose*

**14.2** As in Worked Example 14.2, orient the molecule so that two horizontal bonds are pointing out of the page and two vertical bonds are pointing into the page. Then draw two perpendicular lines and arrange the functional groups in the same order as in the second projection.

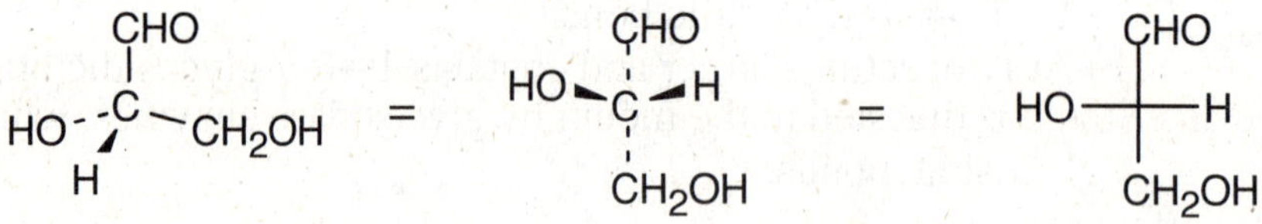

**14.3** To solve this problem, first draw the correct tetrahedral representations of the 2-chlorobutane enantiomers. (If necessary, review Section 6.5 and Worked Example 6.5.) Then, convert the tetrahedral representations to Fischer projections by the method used in Problem 14.2.

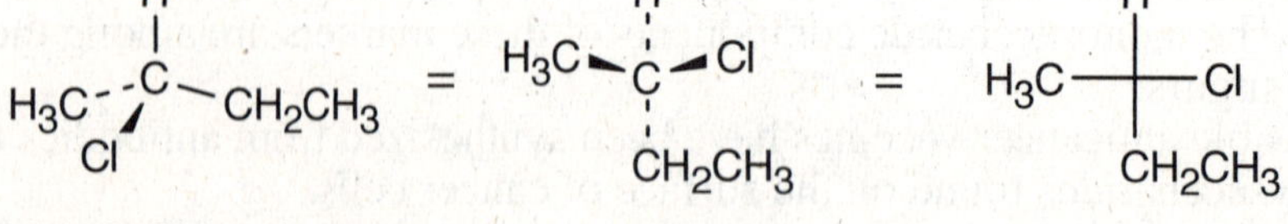

(*R*)-2-Chlorobutane

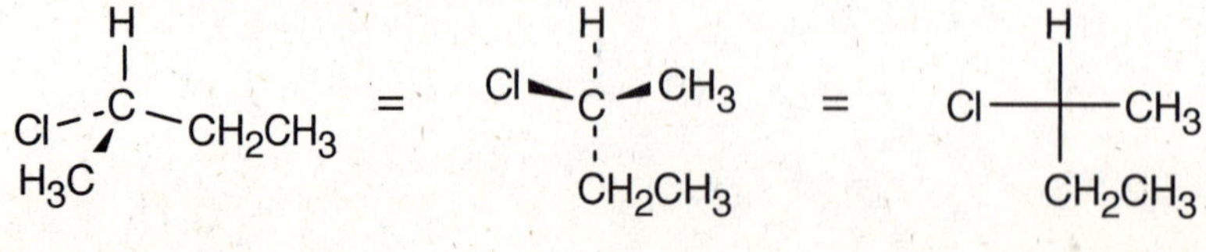

(*S*)-2-Chlorobutane

**14.4** First, convert the Fischer projection to a tetrahedral representation by drawing the horizontal bonds out of the page and the vertical bonds into the page. Then use the sequence rules described in Section 6.5 to assign priorities to the four groups. Rotate the lowest priority group to the rear, and note the direction of rotation of the arrows that go from group 1 to group 2 to group 3. If the arrows indicate clockwise rotation, the stereoisomer has the *R* configuration; if the arrows indicate counterclockwise rotation, the stereoisomer has the *S* configuration.

(a)

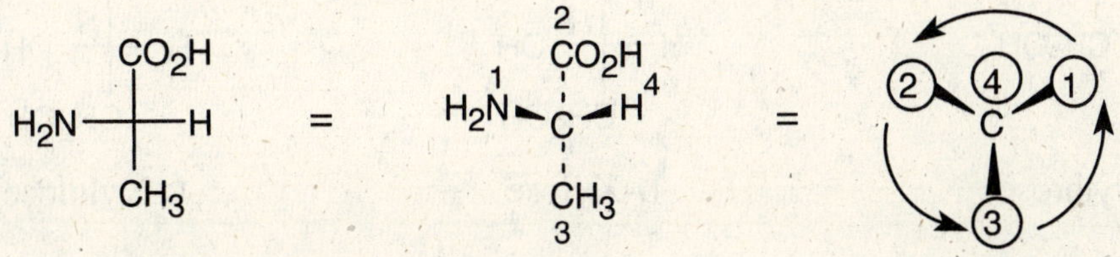

*S*

(b)

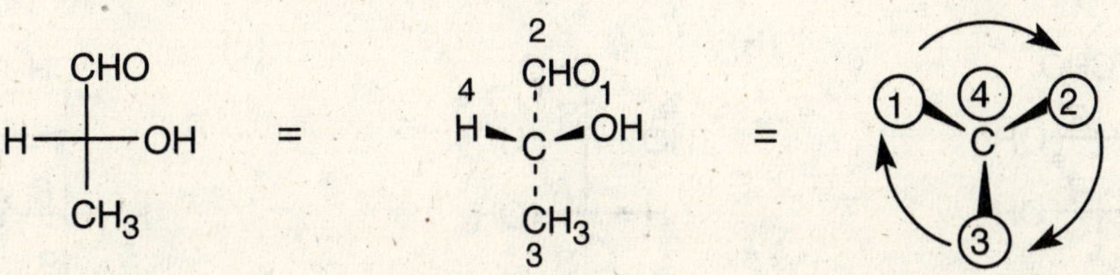

*R*

(c)

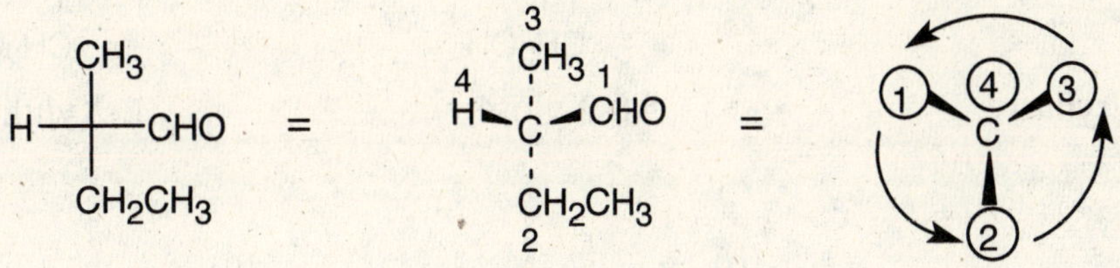

*S*

**14.5** Bring the methyl group slightly forward, so that you have the structure on the left:

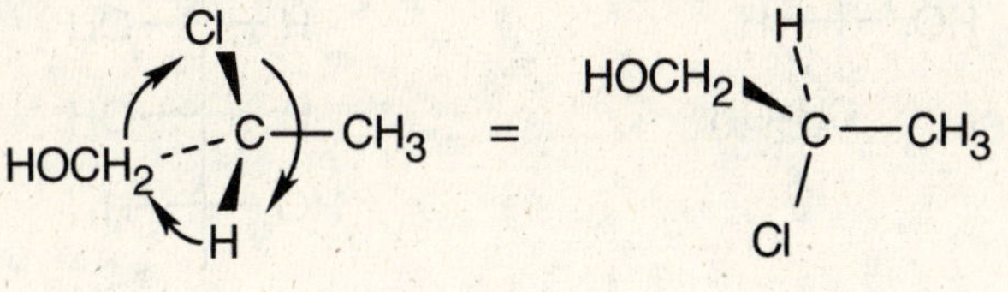

Holding the methyl group steady, perform the indicated rotation to arrive at the second structure shown above. This structure is close in structure to the "steering wheel" structures that we used to assign configurations in Chapter 6. Redraw the second structure above as a "steering wheel" structure, assign priorities to the four substituents, and determine the configuration (*R*).

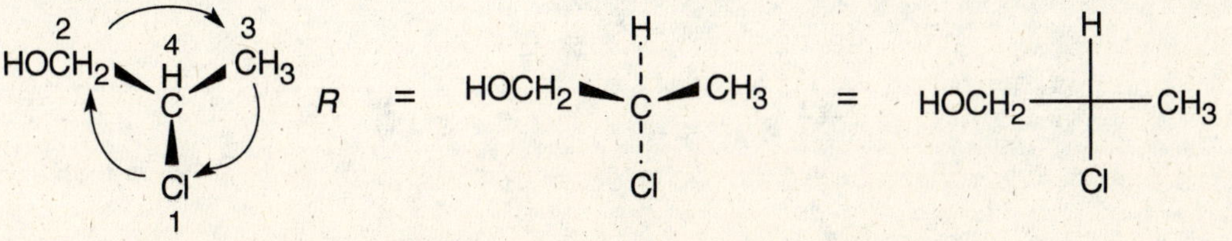

A small tilt of the first structure above produces the second structure, which can easily be drawn as a Fischer projection.

**14.6** For a D sugar, the –OH group on the lowest carbon chirality center is on the right. For an L sugar, it is on the left.

(a)

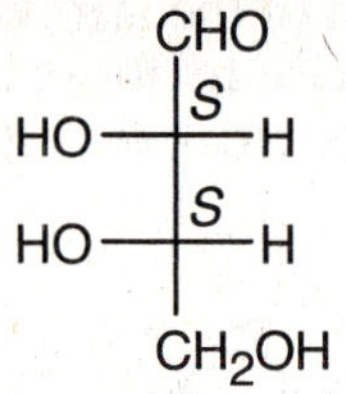


L-Erythrose

(b)

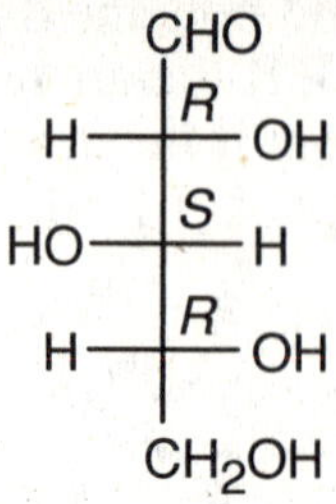


D-Xylose

(c)

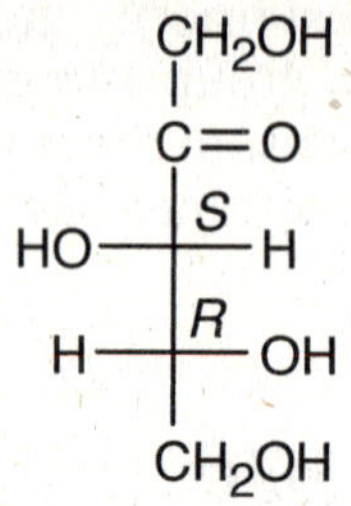


D-Xylulose

**14.7**

(a)

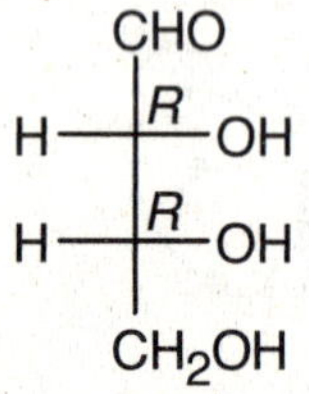


D-Erythrose

(b)

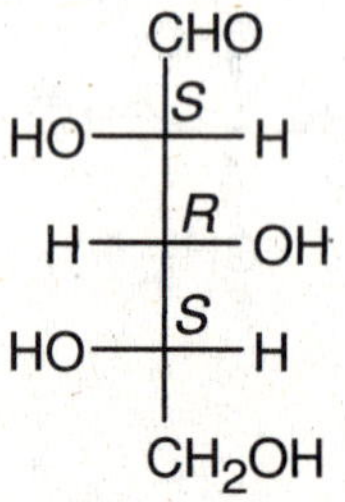


L-Xylose

(c)

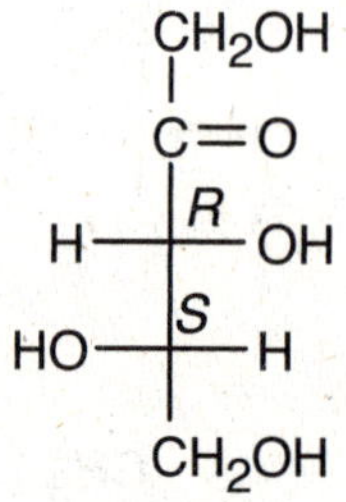


L-Xylulose

**14.8**

(a)

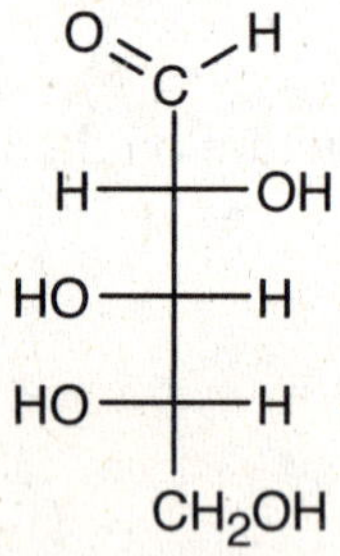


L-Arabinose

(b)

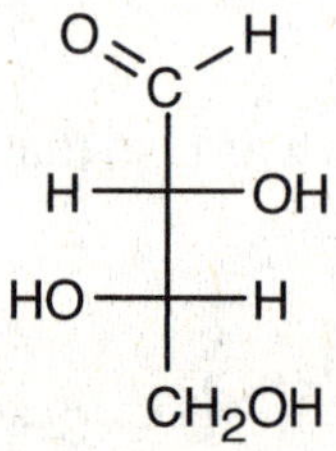


L-Threose

(c)

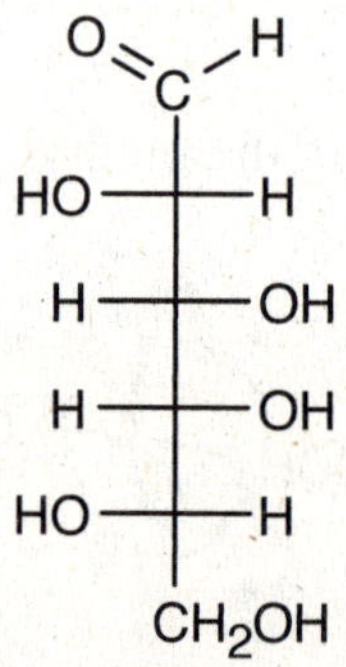


L-Galactose

**14.9** Thirty-two aldoheptoses are possible. Sixteen are D sugars, and sixteen are L sugars.

**14.10**

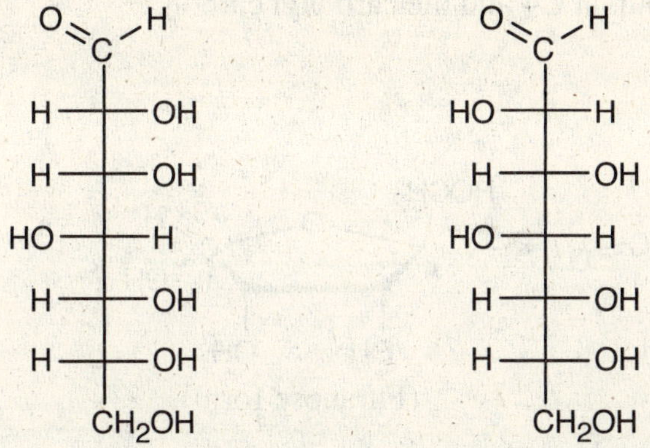

**14.11**

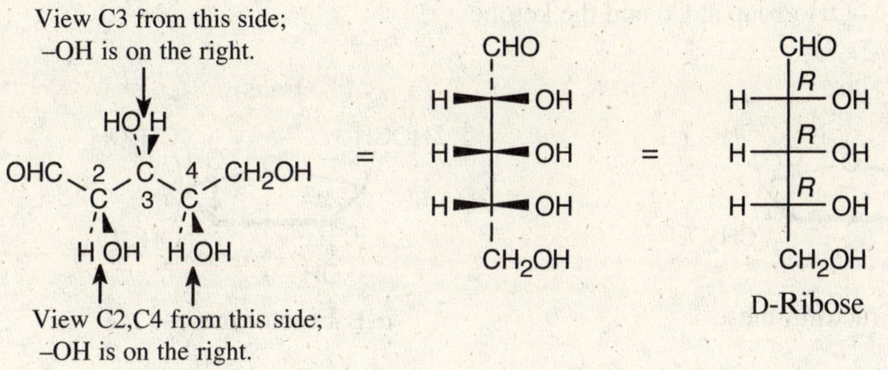


**14.12** Follow the steps in Worked Example 14.5. First, draw the Fischer projection of D-galactose. Lay the projection on its side, and curl it around so that the aldehyde is at the right front, and the $CH_2OH$ is at the left rear. Connect the –OH at C5 to the C1 carbonyl group, raise the left carbon (C4), and lower the right carbon (C1) to form the chair. Hydroxyl groups on the right in a Fischer projection are on the bottom face of the pyranose ring in the middle projection, and –OH groups on the left are on the top face. Note: For simplicity, hydrogens have been left off the ring structures.

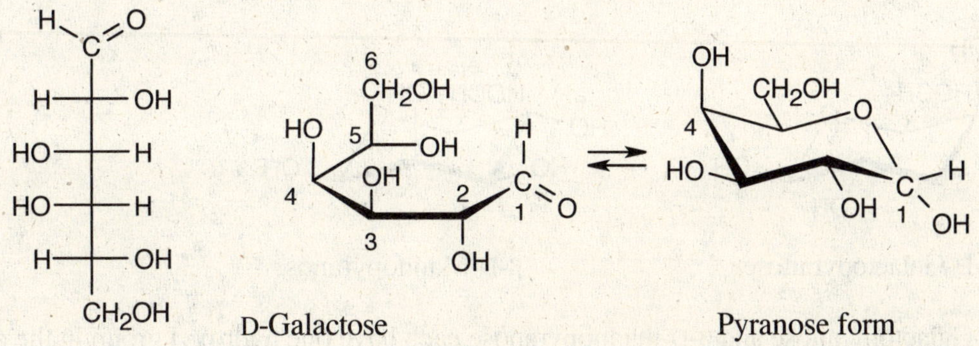

**14.13** The steps for drawing a furanose are similar to the steps for drawing a pyranose. Ring formation occurs between the –OH group at C4 and the carbonyl carbon.

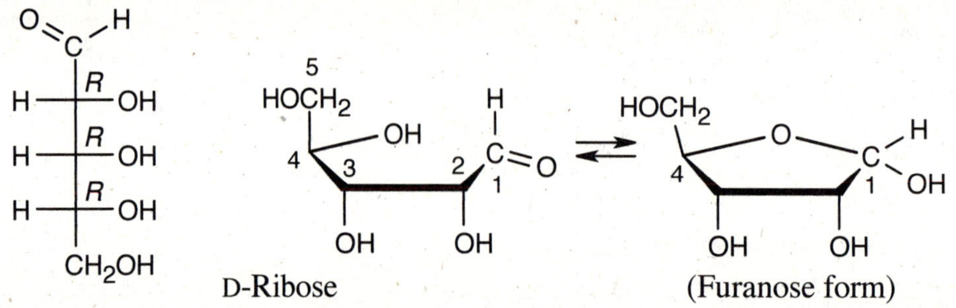

**14.14** The furanose of fructose results from ring formation between the –OH group at C5 and the ketone at C2. In the α anomer, the anomeric –OH group is trans to the C6 –$CH_2OH$ group, and in the β anomer the two groups are cis. In the pyranose form, cyclization occurs between the –OH group at C6 and the ketone.

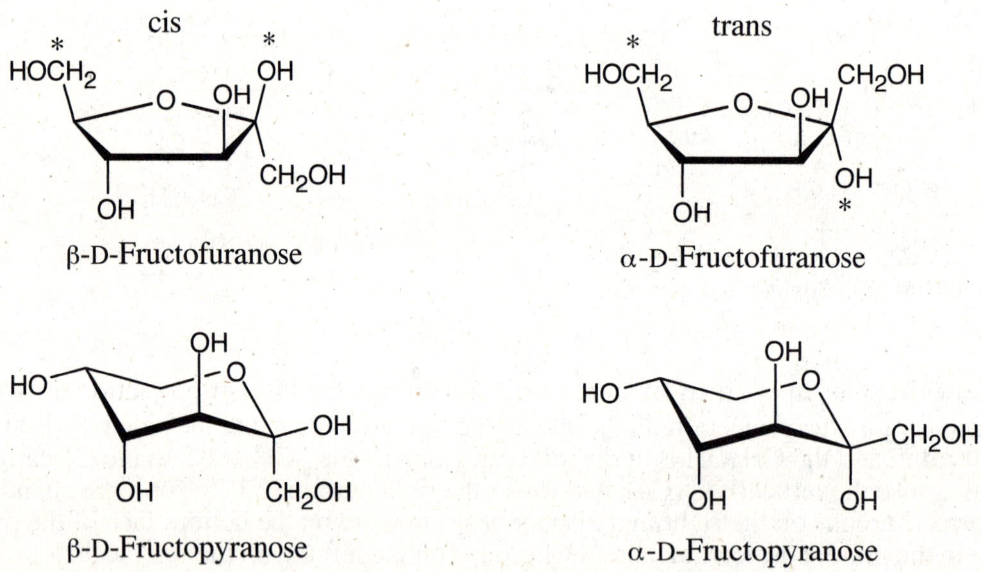

**14.15**

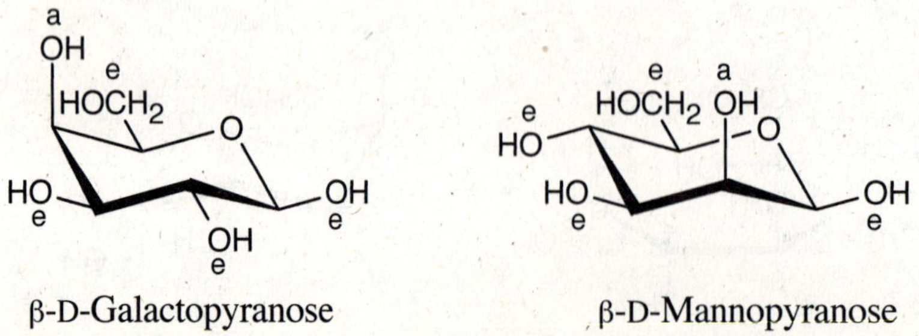

β-D-Galactopyranose and β-D-mannopyranose each have one hydroxyl group in the axial position. Galactose and mannose are of equal stability.

**14.16**

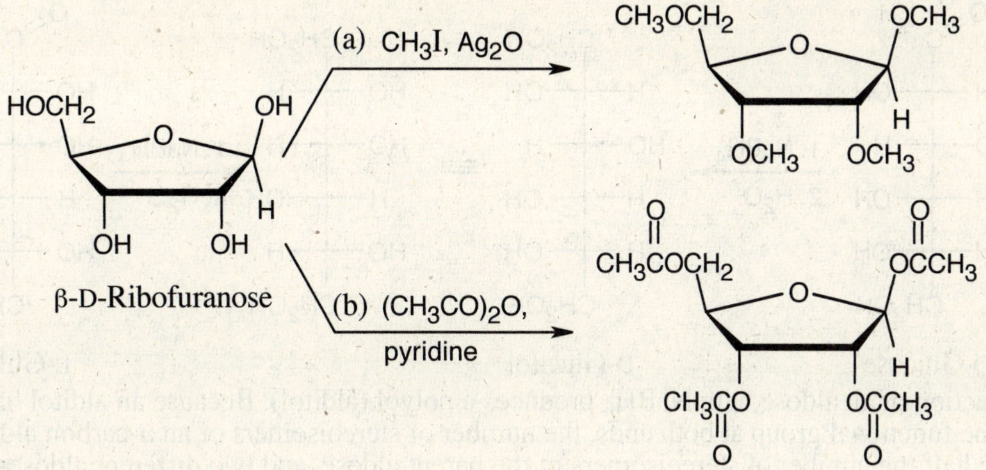

**14.17**

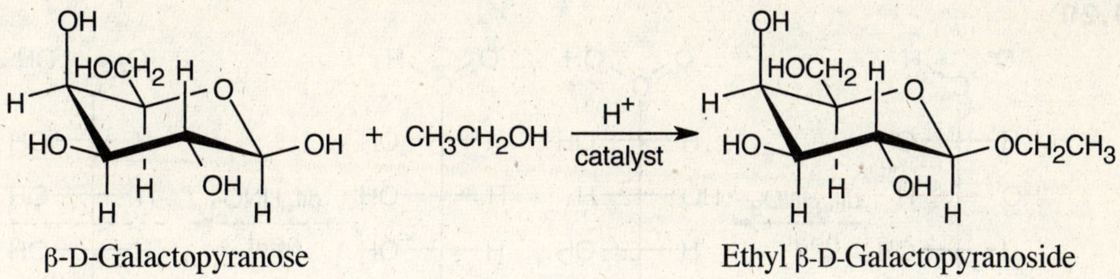

**14.18**

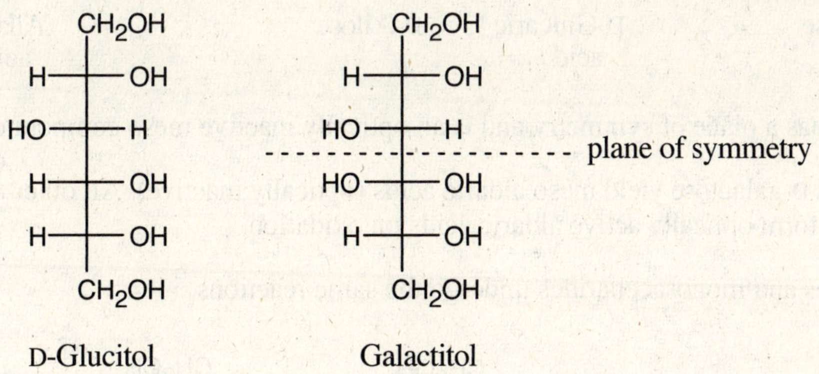

Reduction of the aldehyde group of D-galactose yields an alditol that has a plane of symmetry and is an optically inactive meso compound.

**14.19**

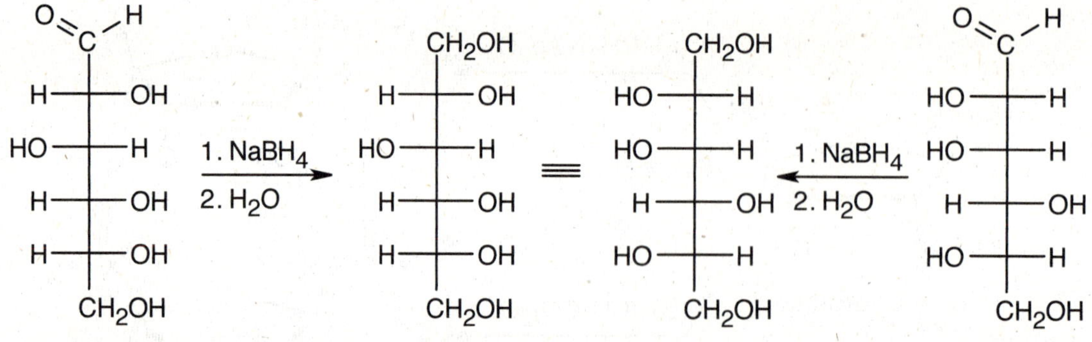


D-Glucose D-Glucitol L-Gulose

Reaction of an aldose with $NaBH_4$ produces a polyol (alditol). Because an alditol has the same functional group at both ends, the number of stereoisomers of an *n*-carbon alditol is one half the number of stereoisomers of the parent aldose, and two different aldoses can yield the same alditol. Here L-gulose and D-glucose form the same alditol (rotate the Fischer projection of L-gulitol 180° to see the identity).

**14.20**

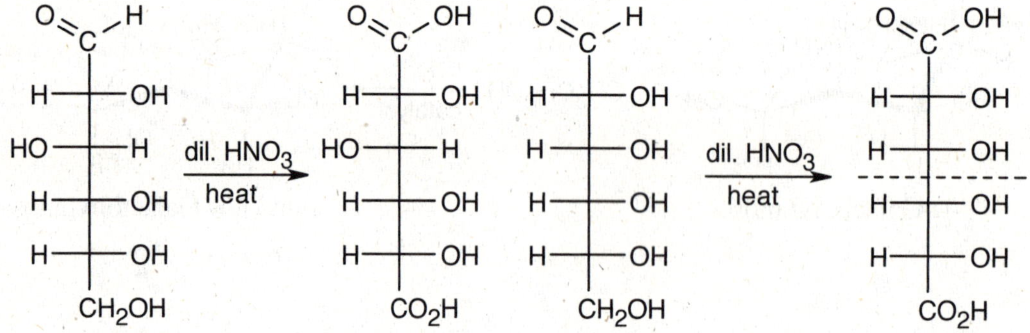


D-Glucose D-Glucaric acid D-Allose Allaric acid

Allaric acid has a plane of symmetry and is an optically inactive meso compound.

**14.21** D-Allose and D-galactose yield meso aldaric acids (optically inactive). All other D-aldohexoses form optically active aldaric acids on oxidation.

**14.22** Disaccharides and monosaccharides undergo the same reactions.

Cellobiose

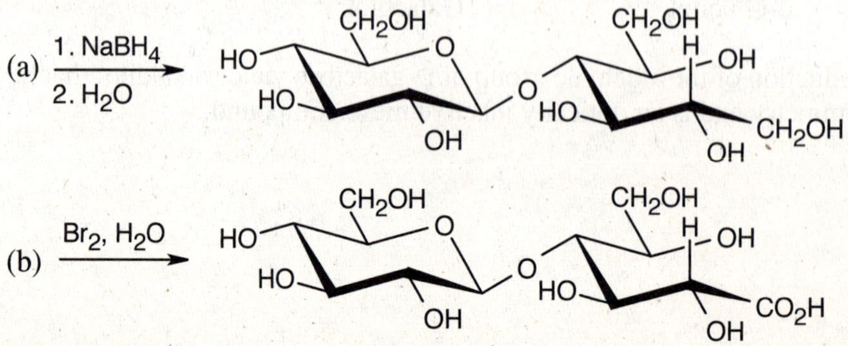

## Visualizing Chemistry

### 14.23

(a) 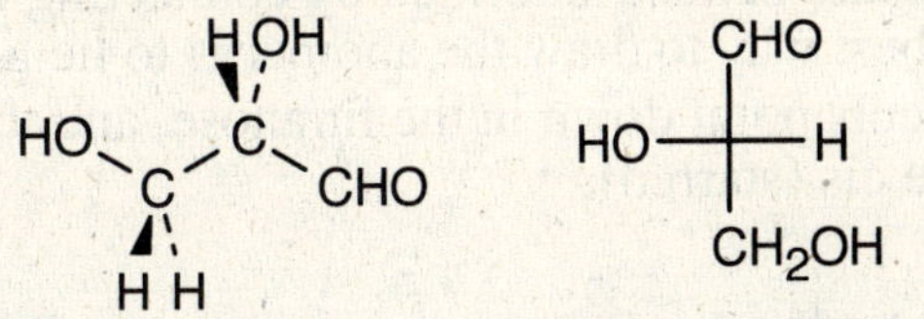

L-Glyceraldehyde

(b) 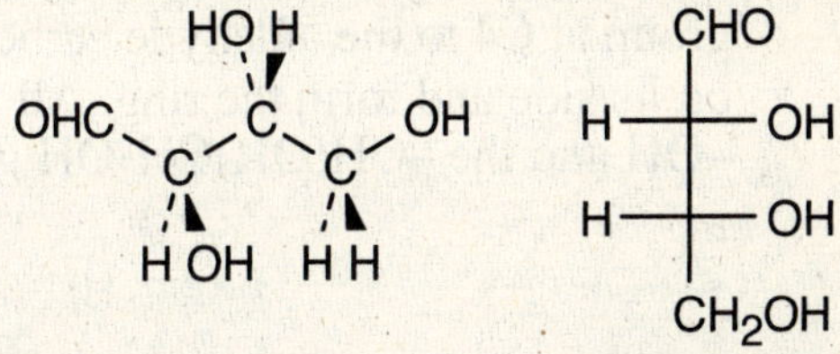

D-Erythrose

**14.24** (a) Convert the model to a Fischer projection, remembering that the aldehyde group is on top, pointing into the page, and that the groups bonded to the carbons below point out of the page. The model represents a D-aldose because the –OH group at the chiral carbon farthest from the aldehyde points to the right.

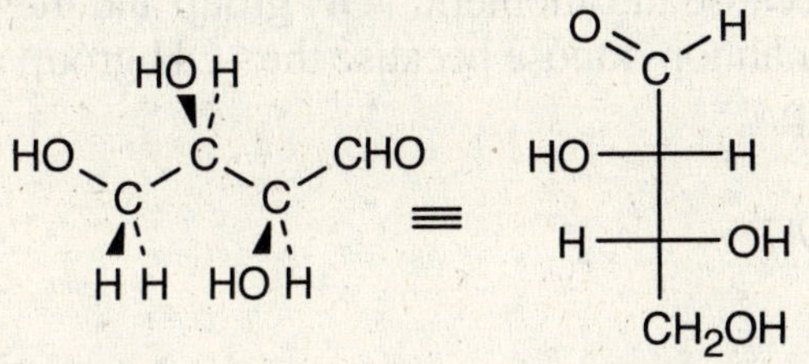

D-Threose

(b) Break the hemiacetal bond and uncoil the aldohexose. Notice that all hydroxyl groups point to the right in the Fischer projection. The model represents the β anomer of D-allopyranose.

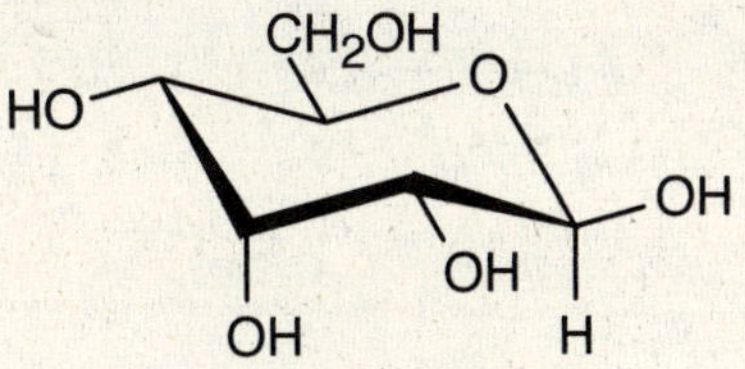

β-D-Allopyranose

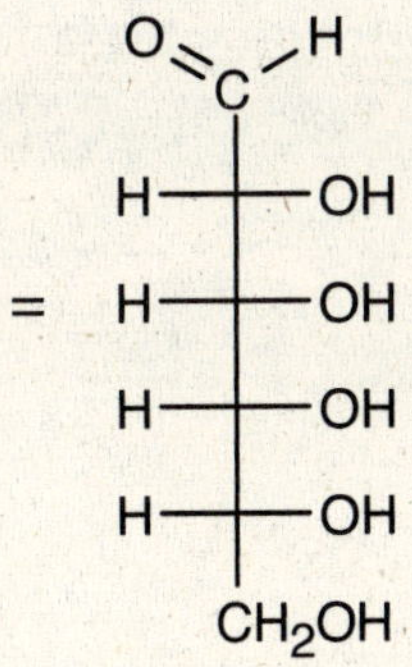

### 14.25

(a) 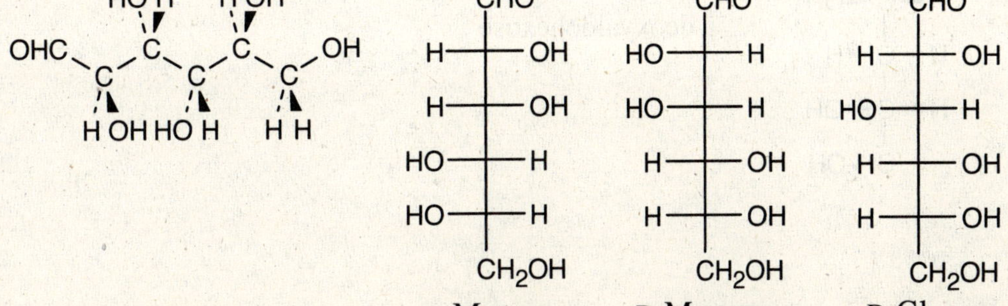

L-Mannose | D-Mannose (enantiomer) | D-Glucose (diastereomer)

(b) The model represents an L-aldohexose because the hydroxyl group on the chiral carbon farthest from the aldehyde group points to the left.

(c) This is tricky! The furanose ring of an aldohexose is formed by connecting the –OH group at C4 to the aldehyde carbon. The best way to draw the anomer is to lie L-mannose on its side and form the ring. All substituents point down in the furanose, and the anomeric –OH and the –CH(OH)$CH_2OH$ group are cis (starred).

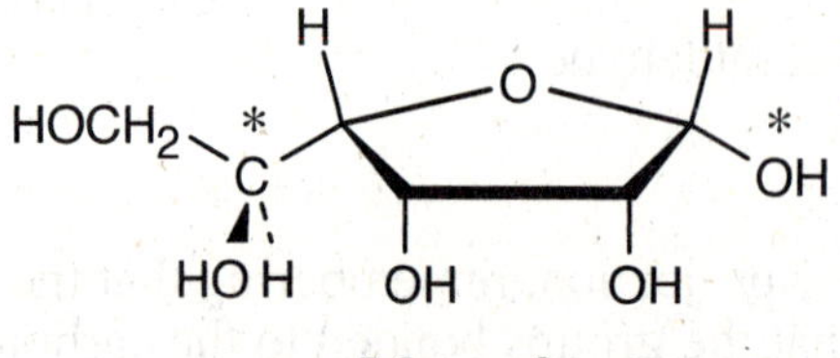

β-L-Mannofuranose

**14.26** The structure represents an α anomer because the anomeric –OH group and the –$CH_2OH$ group are trans. The compound is α-L-mannopyranose because the –OH group at C2 is the only non-anomeric axial hydroxyl group.

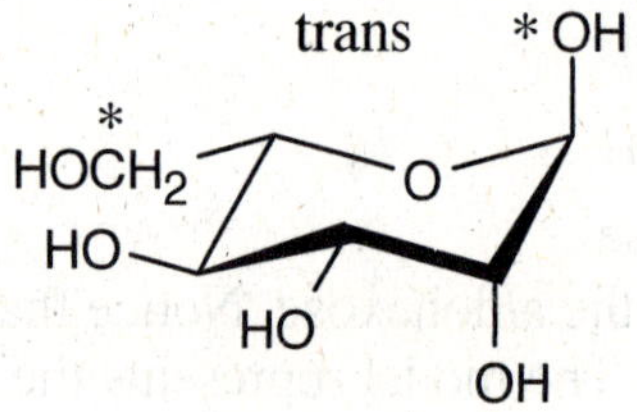

α-L-Mannopyranose

## Additional Problems

### Classifying Monosaccharides

**14.27**

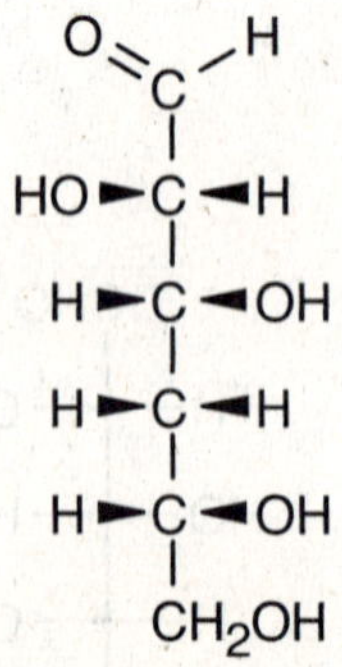

a deoxyaldohexose

**14.28**

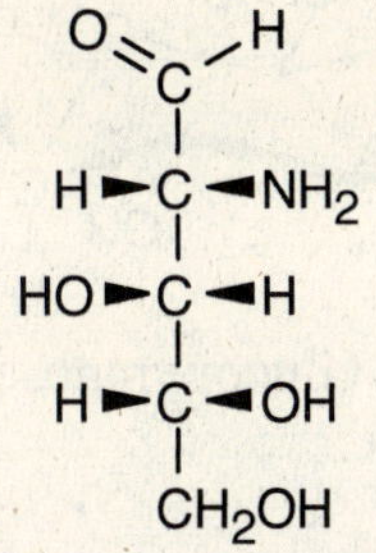

a five-carbon amino sugar

**14.29**

(a)

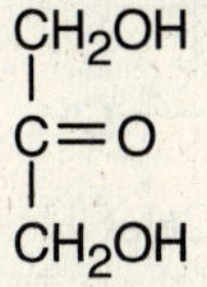

a ketotriose

(b)

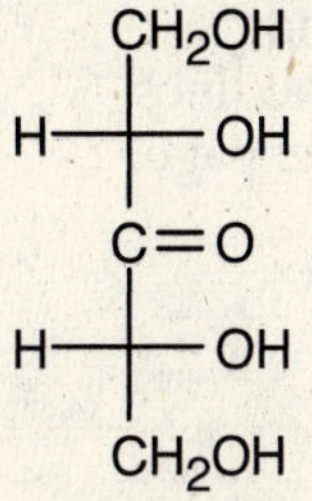

a ketopentose

(c)

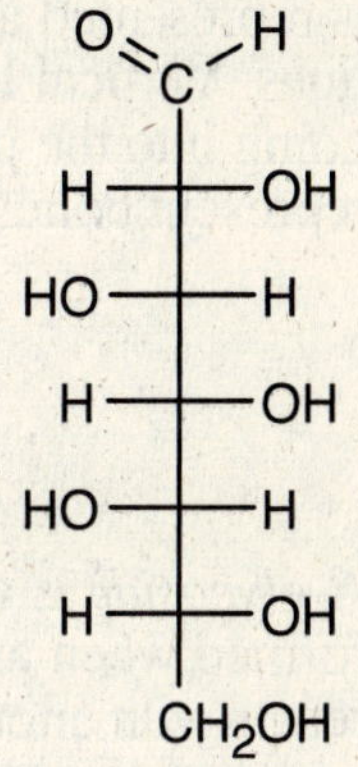

an aldoheptose

**14.30** Other structures can be drawn.

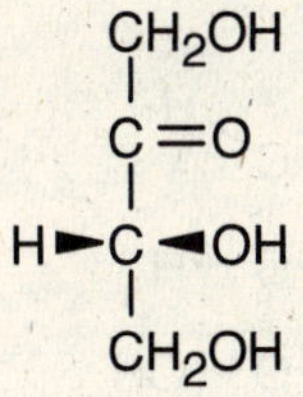

a ketotetrose

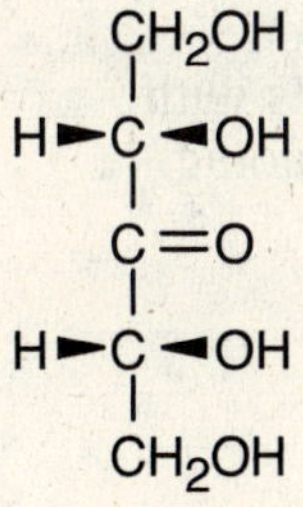

a ketopentose

**14.31**

| *Definition* | *Example* |
|---|---|
| (a) A *monosaccharide* is a carbohydrate that can't be hydrolyzed to smaller units. | 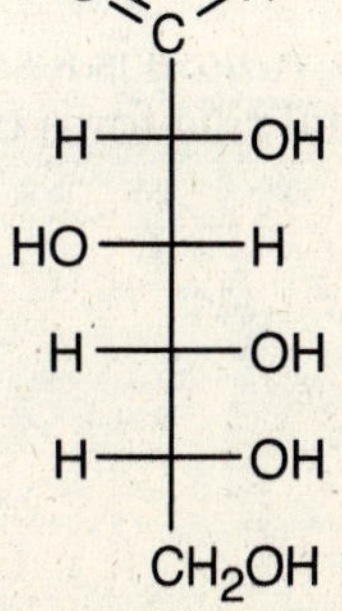 D-Glucose |

(b) An *anomeric center* is a chirality center formed when an open-chain monosaccharide cyclizes to a furanose or a pyranose ring.

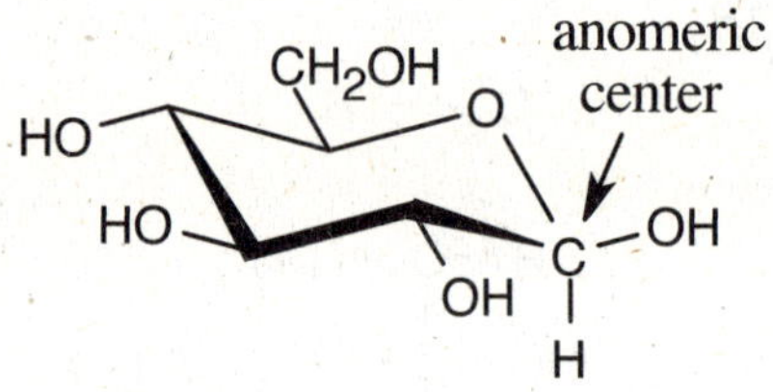

β-D-Glucopyranose

(c) A *Fischer projection* is a drawing of a carbohydrate in which each chirality center is represented as a pair of perpendicular lines. Vertical lines represent bonds going into the page, and horizontal lines represent bonds coming out of the page.

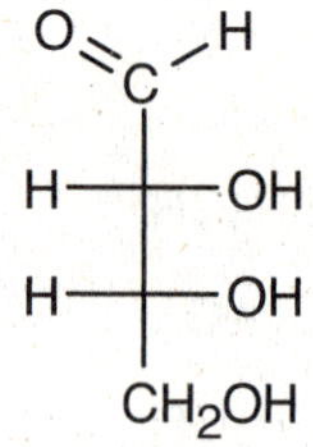

D-Erythrose

(d) A *glycoside* is an acetal of a carbohydrate, formed when an anomeric hydroxyl group reacts with another compound containing a hydroxyl group.

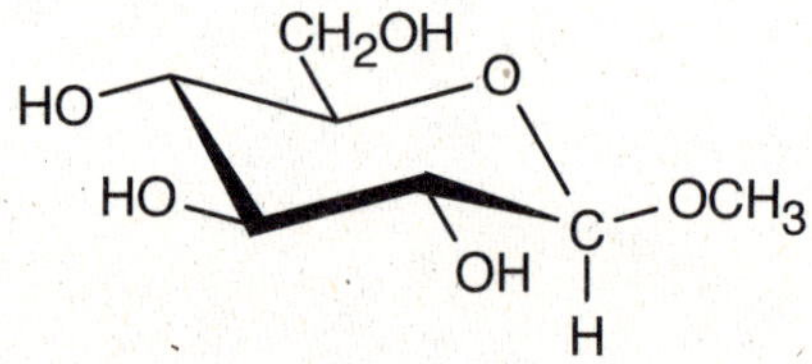

Methyl β-D-glucopyranoside

(e) A *reducing sugar* is a sugar that reacts with any of several reagents to yield an aldonic acid plus reduced reagent.

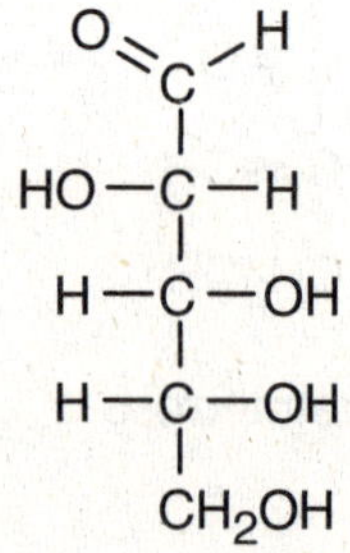

D-Arabinose

(f) A *pyranose* is a six-membered cyclic hemiacetal ring form of a monosaccharide.

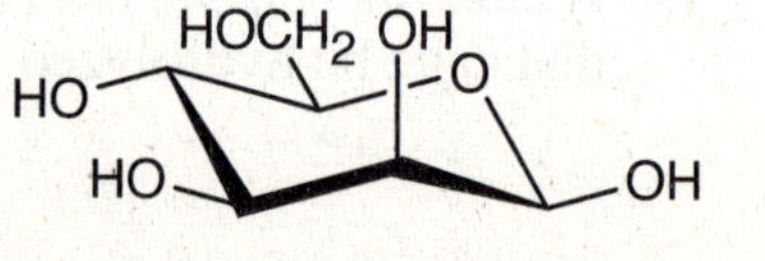

β-D-Mannopyranose

(g) A *1→4 link* occurs when the anomeric hydroxyl group (at carbon 1) of a pyranose or furanose forms a glycosidic bond with the hydroxyl group at carbon 4 of a second monosaccharide.

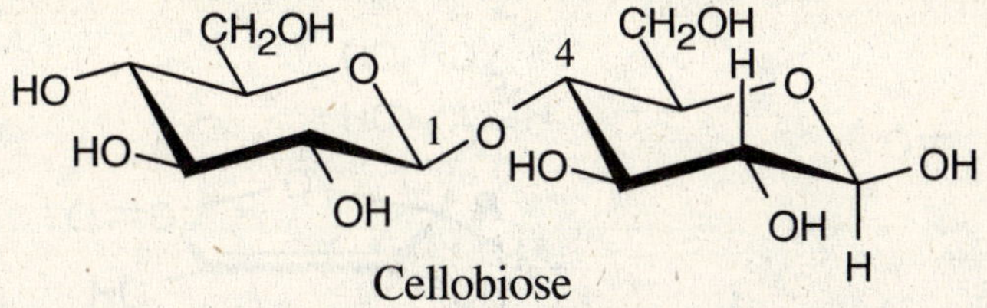

Cellobiose

(h) A D-*sugar* is a sugar in which the hydroxyl group farthest from the carbonyl group points to the right in a Fischer projection.

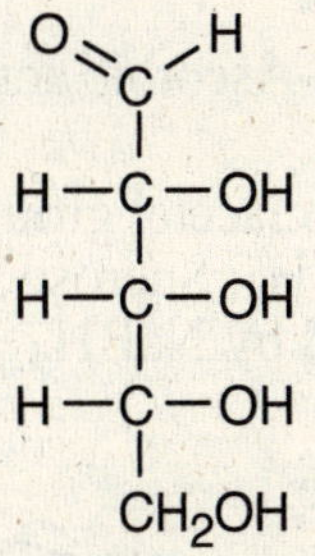

D-Ribose

## Carbohydrate Structures

**14.32**

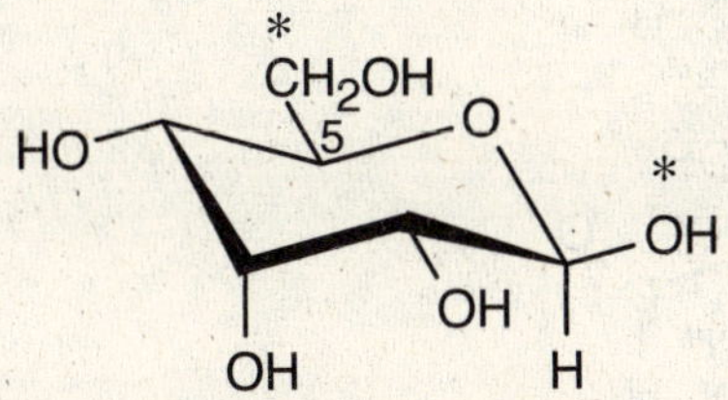

β-D-Allopyranose

This structure is a pyranose (6-membered ring) and is a β anomer ( the C1 hydroxyl group and the $-CH_2OH$ groups are cis). It is a D sugar because the -O- at C5 is on the right in the uncoiled form.

**14.33**

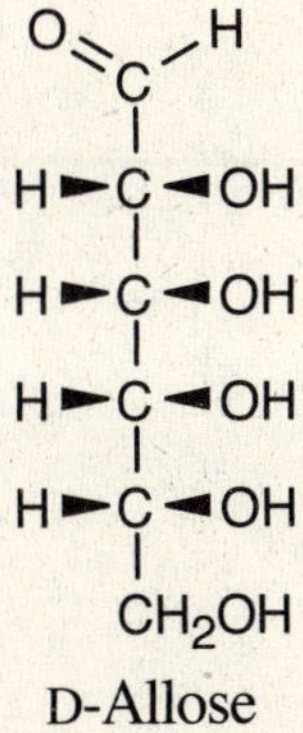

D-Allose

**14.34–14.35**

Ascorbic acid has an L configuration because the hydroxyl group at the lowest chirality center points to the left.

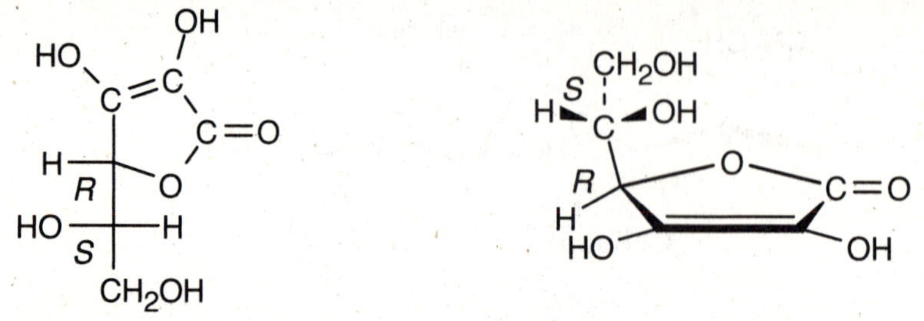

L-Ascorbic acid

**14.36** The hemiacetal group of maltose is in equilibrium with the aldehyde, which can be reduced by $NaBH_4$. Sucrose has no hemiacetal group, and thus no aldehyde is available to be reduced by $NaBH_4$.

Maltose:

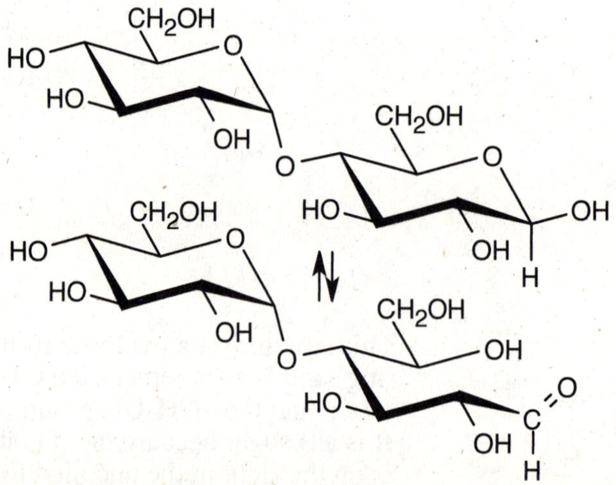

Sucrose:

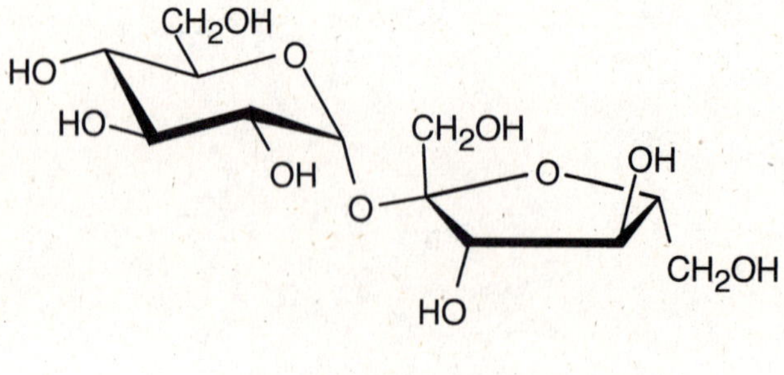

**14.37**

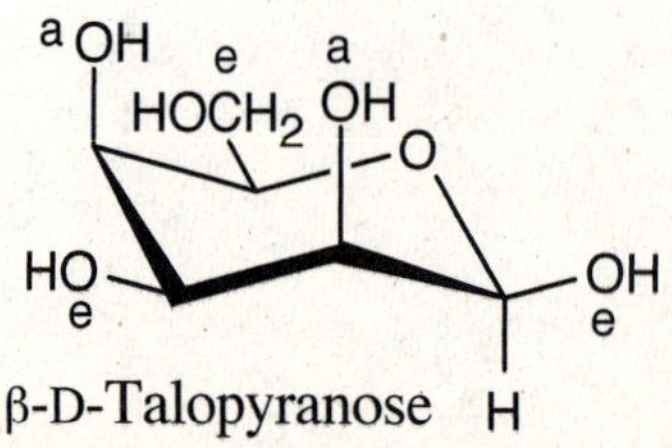

β-D-Talopyranose

**14.38**

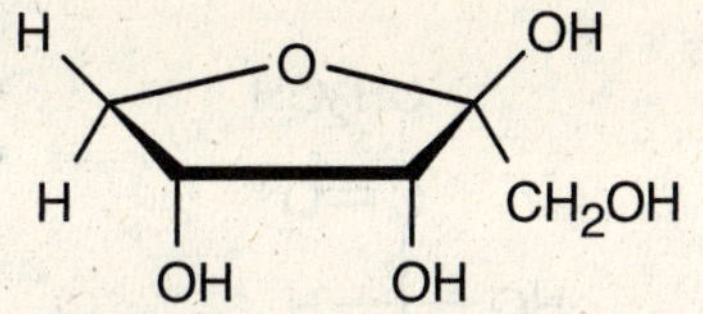

β-D-Ribulofuranose

**14.39**

(a)

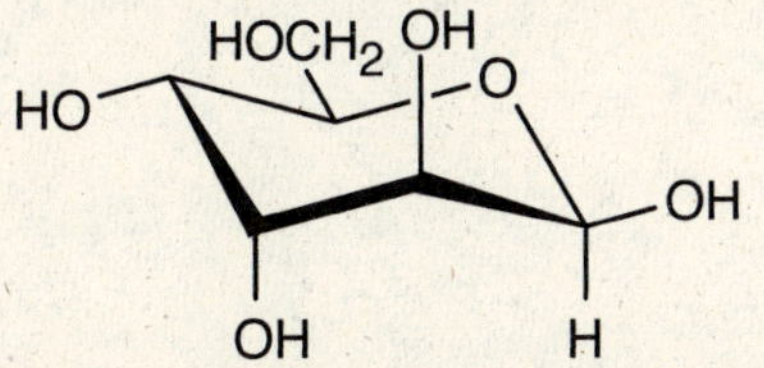

β-D-Altropyranose

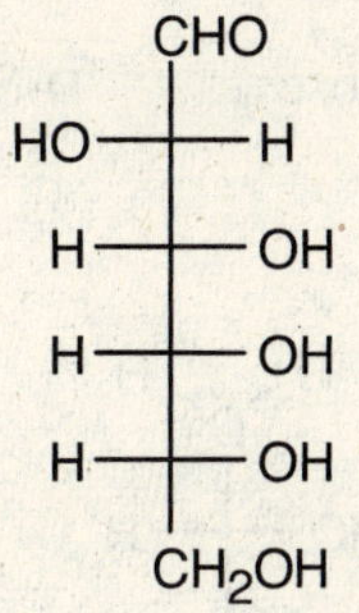

(b)

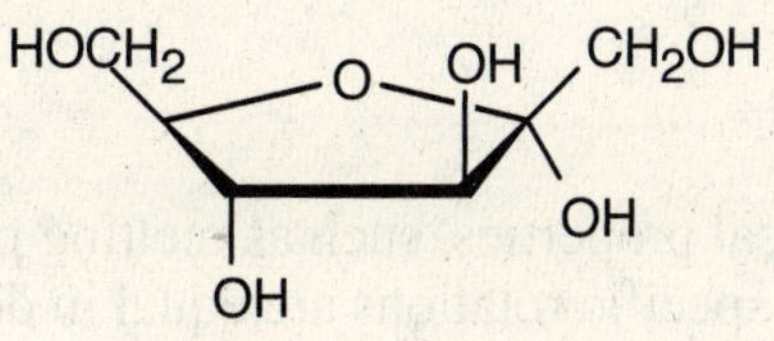

α-D-Fructofuranose

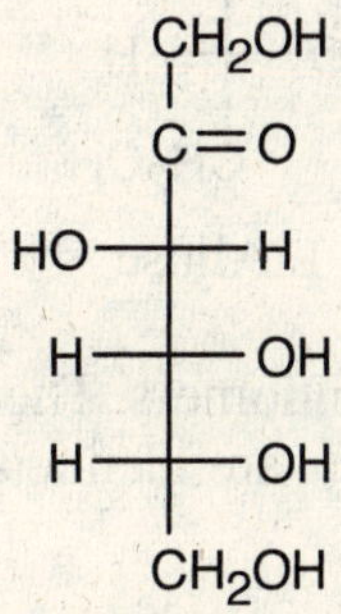

(c)

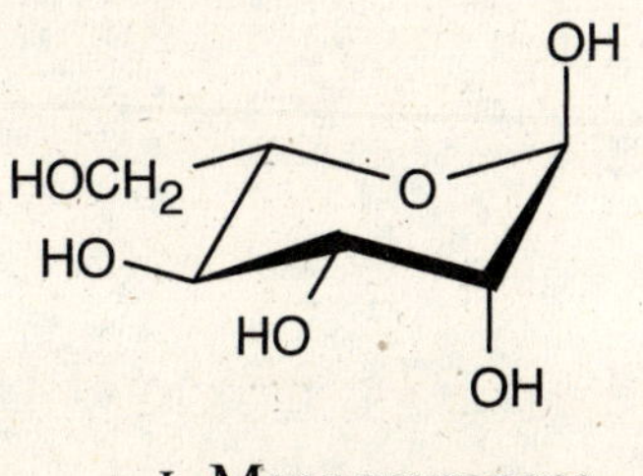

α-L-Mannopyranose

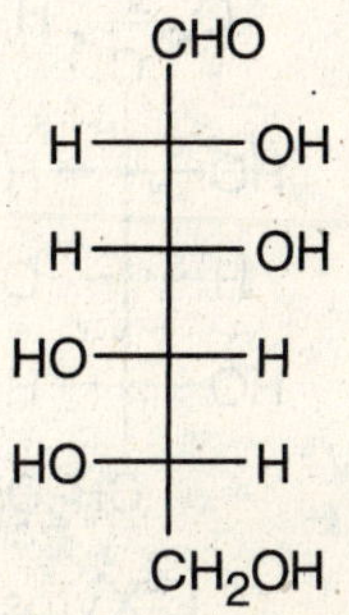

**14.40** There are four D-2-ketohexoses.

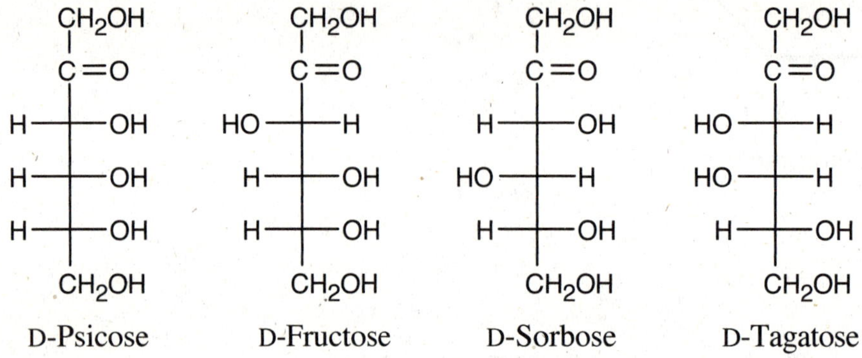

**14.41**

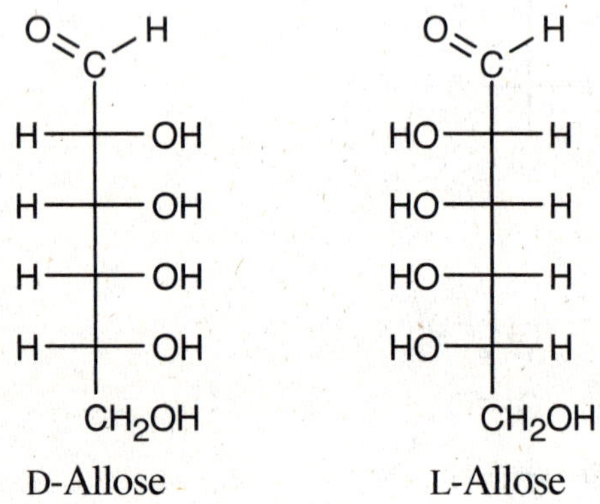

D-Allose and L-allose are enantiomers. Their physical properties, such as melting point, solubility in water, and density, are identical. Their specific rotations are equal in degree but opposite in sign.

**14.42**

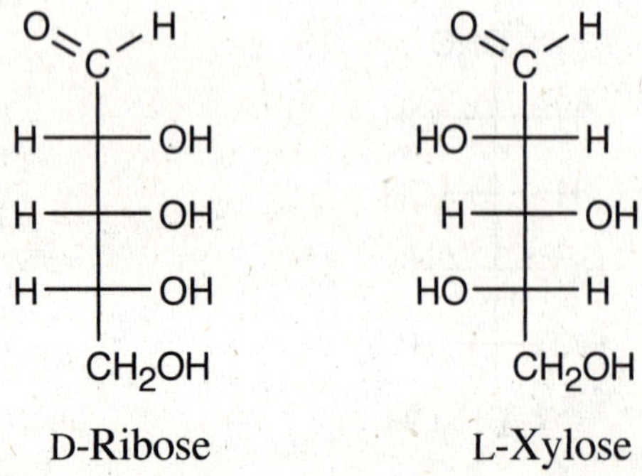

D-Ribose and L-xylose are diastereomers and differ in all physical properties.

## Carbohydrate Reactions

### 14.43–14.44

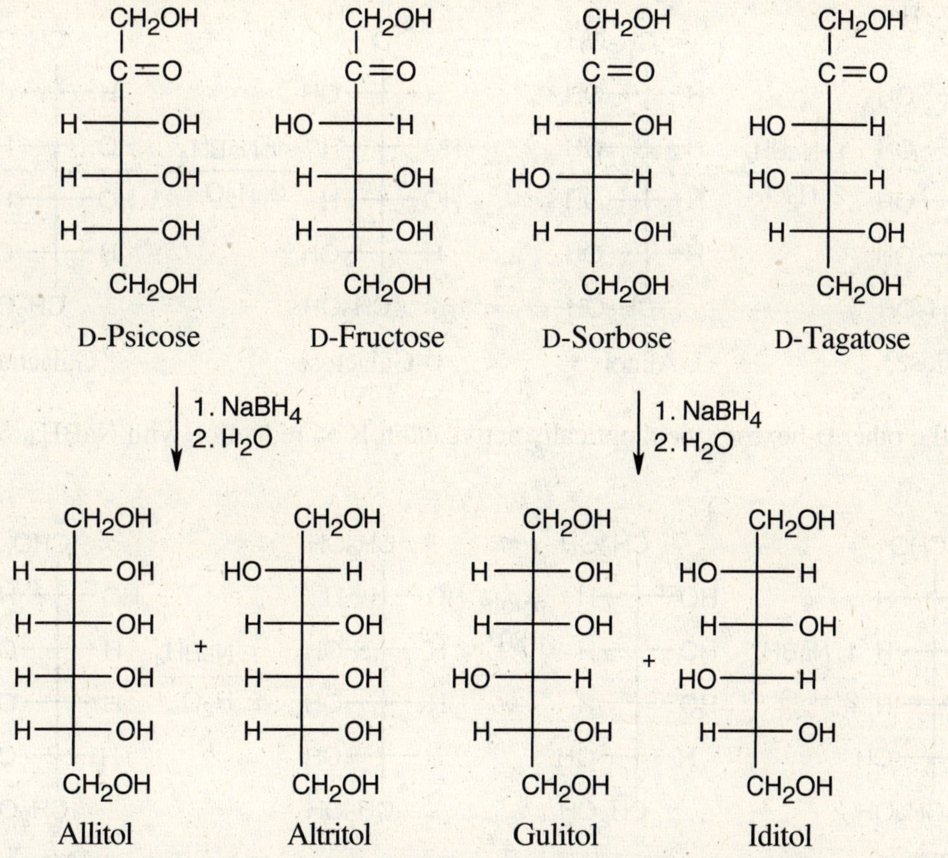


### 14.45

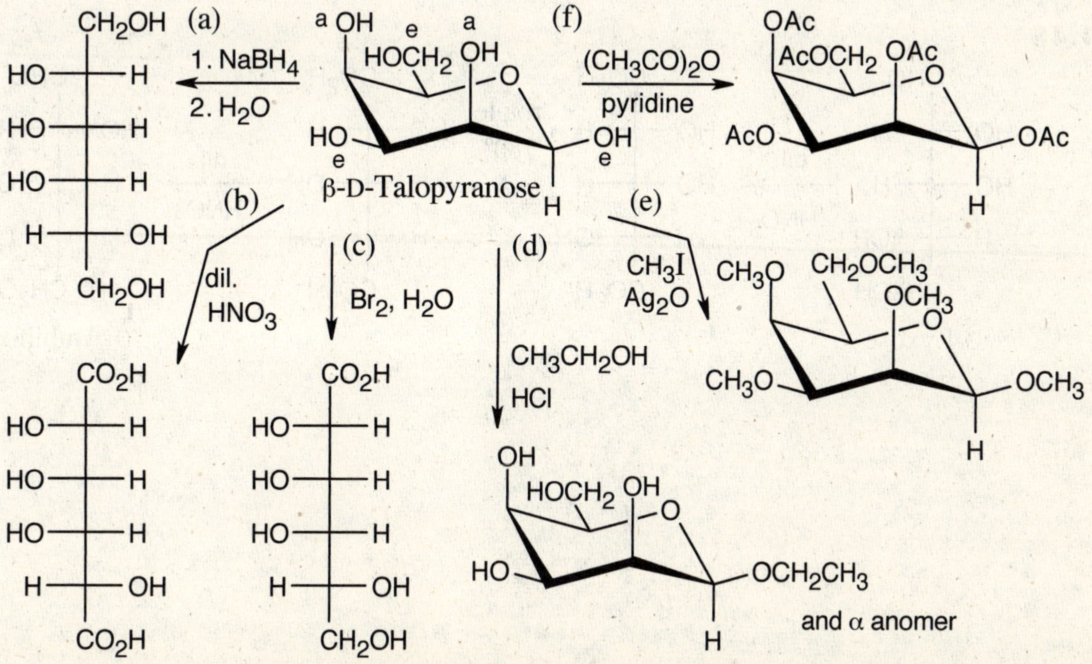

## General Problems

**14.46**

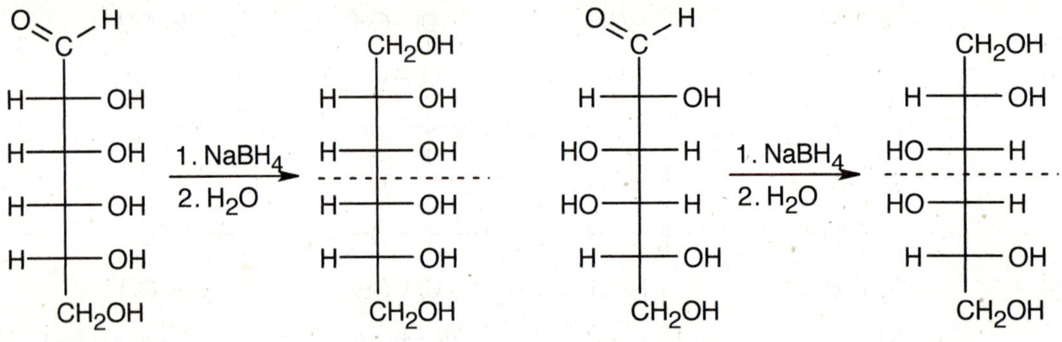

D-Allose Allitol D-Galactose Galactitol

All of the other D-hexoses yield optically active alditols on reduction with $NaBH_4$.

**14.47**

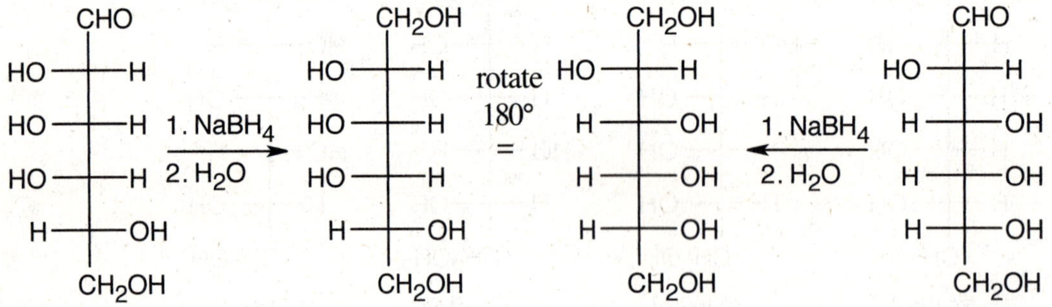

D-Talose D-Altrose

**14.48**

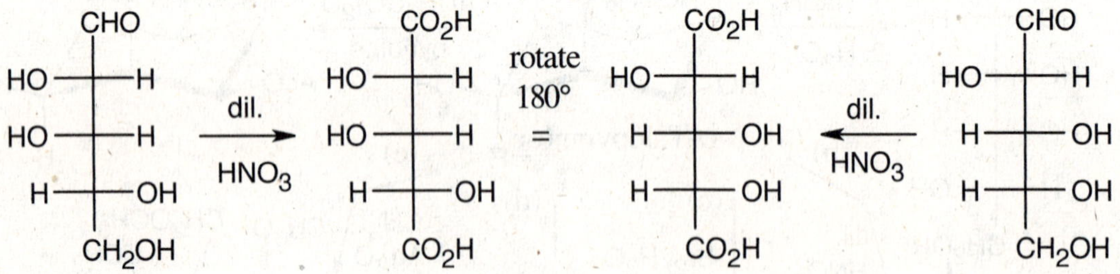

D-Lyxose D-Arabinose

**14.49**

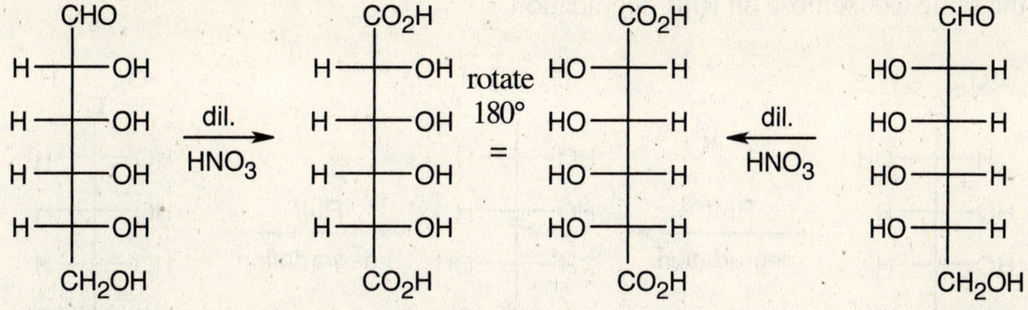

D-Allose L-Allose

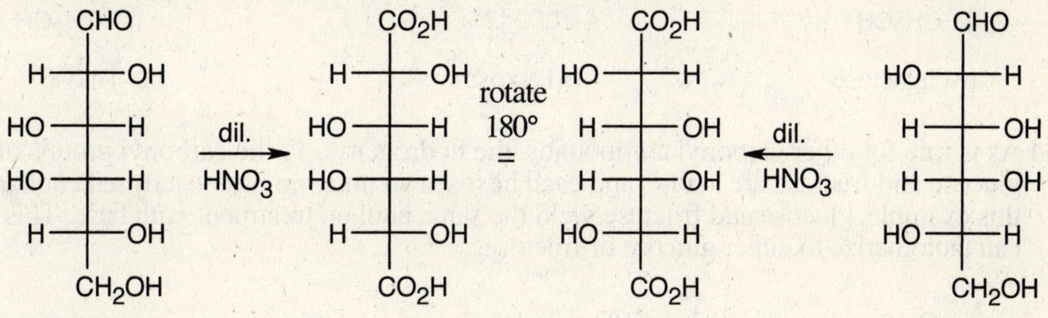

D-Galactose L-Galactose

**14.50**

(a)

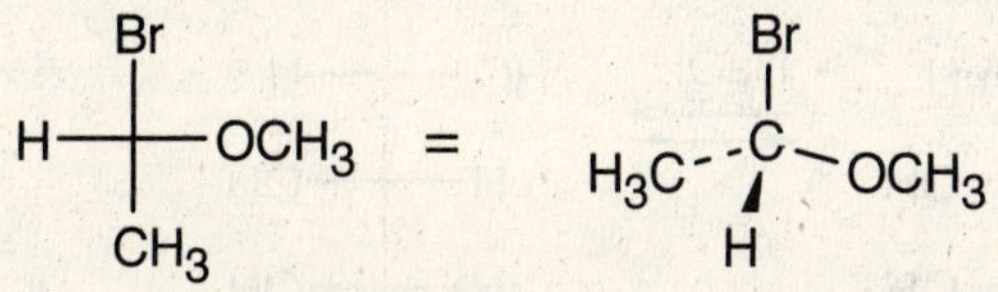

(b)

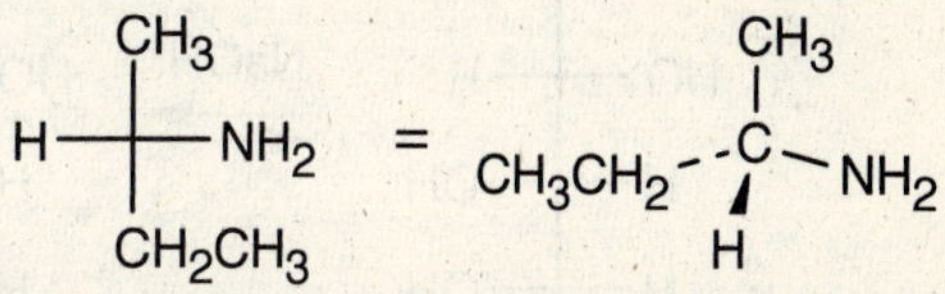

**14.51**

(a)

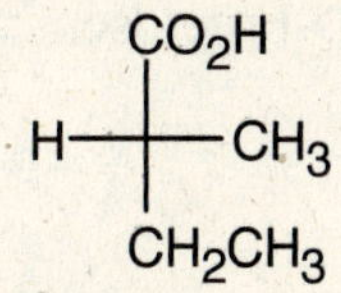

(*R*)-2-Methylbutanoic acid

(b)

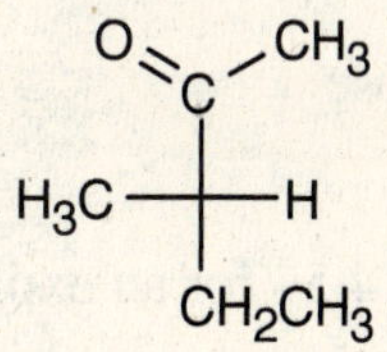

(*S*)-3-Methylpentan-2-one

**14.52**

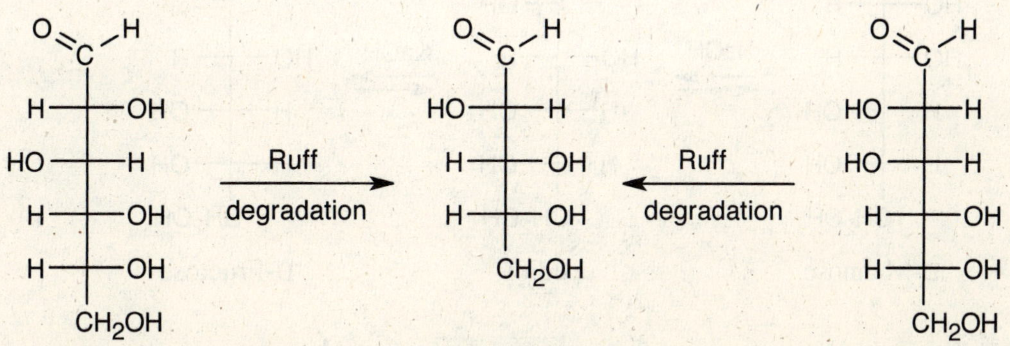

D-Glucose D-Arabinose D-Mannose

**14.53** D-Galactose and D-talose must have the same configuration at C3, C4 and C5 if both yield the same aldopentose on Ruff degradation.

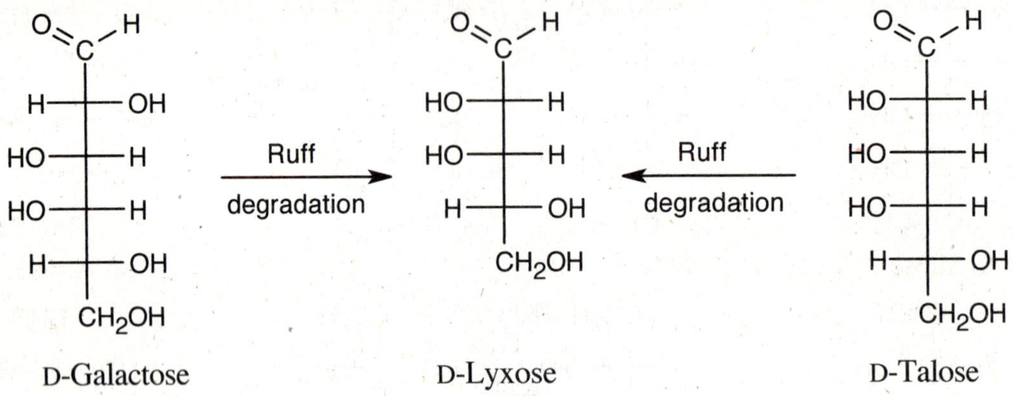


**14.54** As is true for other carbonyl compounds, the hydrogens α to the carbonyl groups of glucose and fructose are acidic, and each hexose can undergo base-catalyzed enolization. In this example, glucose and fructose yield the same enol on treatment with base. This *enediol* can tautomerize to either glucose or fructose.

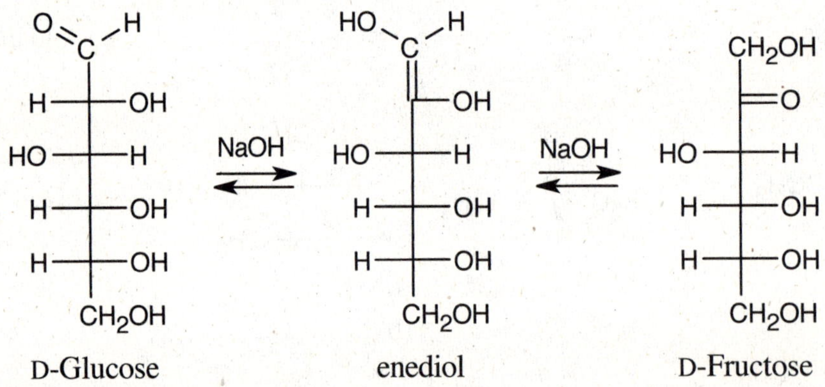


**14.55** See Problem 14.54 for an explanation.

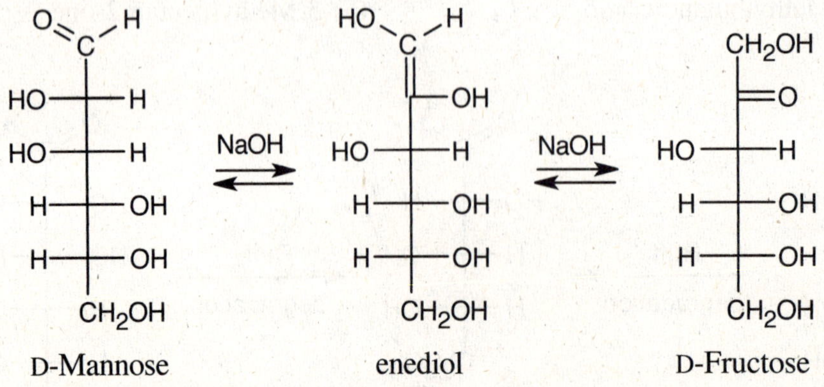

**14.56**

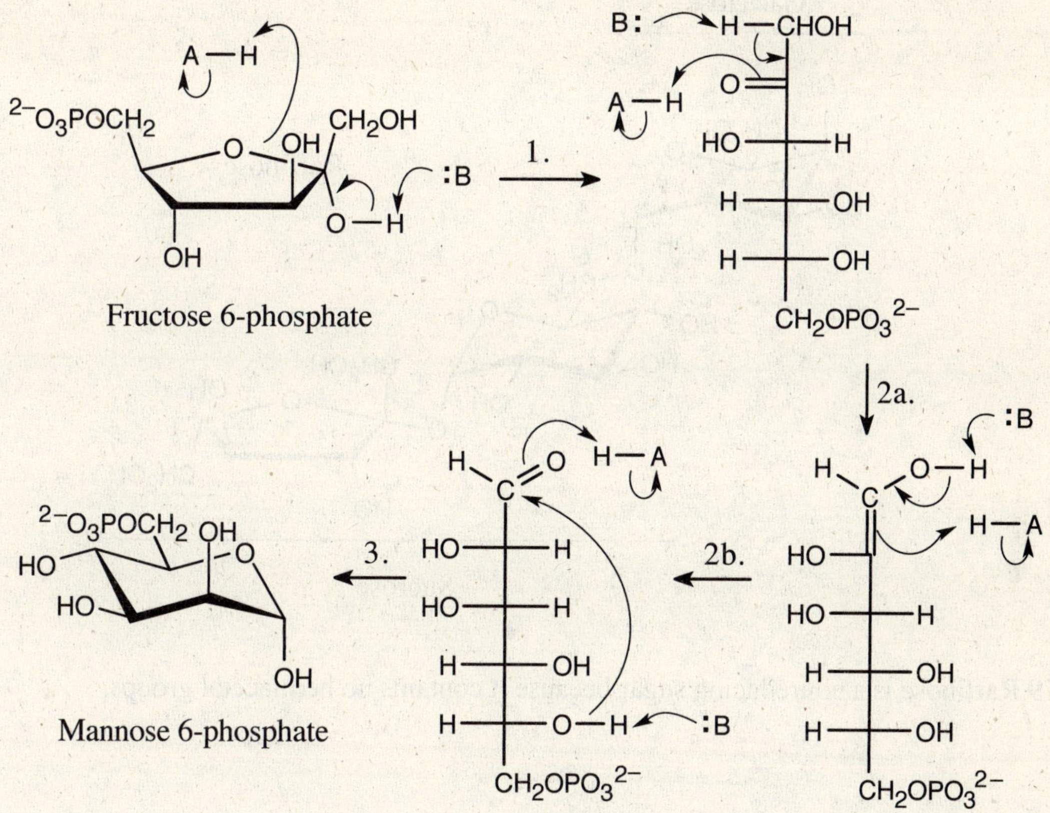

Fructose 6-phosphate

Mannose 6-phosphate

**Step 1:** Hemiacetal ring opening.

**Steps 2a, 2b:** Keto–enol equilibration via an intermediate enediol.

**Step 3:** Hemiacetal ring formation.

**14.57** Two D-aldotetroses are possible. The one with both hydroxyl groups on the right in a Fischer projection yields an optically inactive aldaric acid, and thus must be D-erythrose. The other D-aldotetrose, D-threose, yields an optically active aldaric acid.

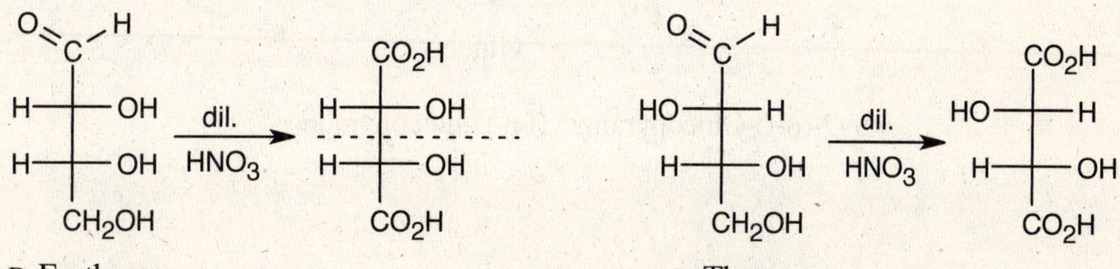

D-Erythrose

D-Threose

**14.58**

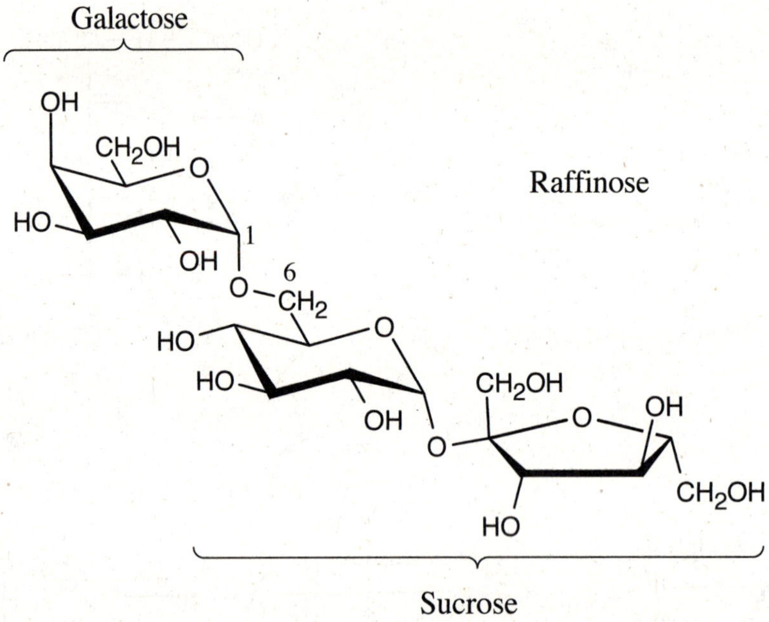


**14.59** Raffinose is a nonreducing sugar because it contains no hemiacetal groups.

**14.60**

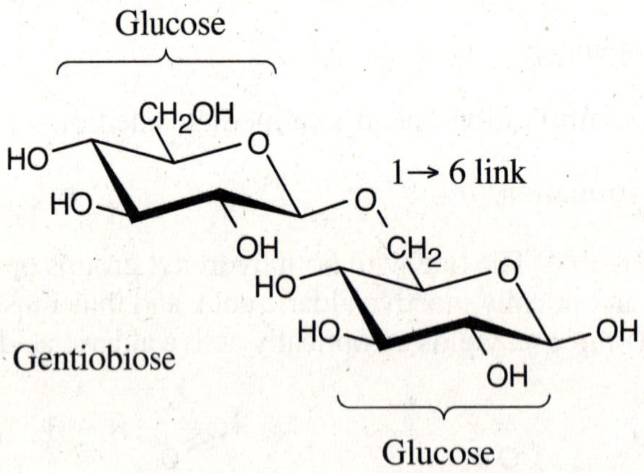


6-*O*-(β-D-Glucopyranosyl)-β-D-glucopyranose

## In the Medicine Cabinet

**14.61**

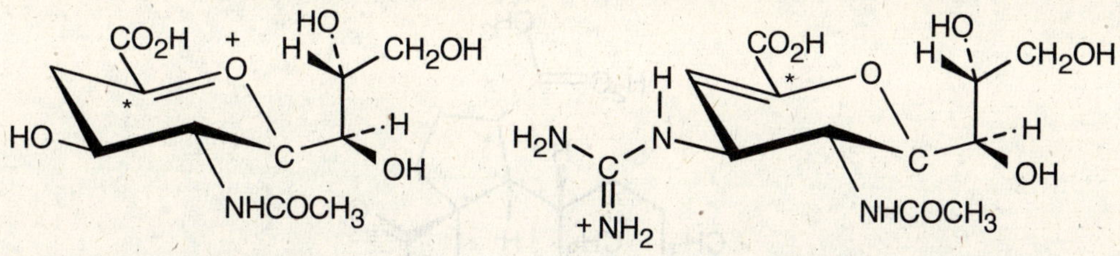


Oxonium ion

Zanamivir (Relenza)

The starred carbon of the oxonium ion has a planar relationship to the other carbons bonded to it. The analogous carbon of Relenza has the same relationship to the carbons bonded to it. Apparently, the viral enzyme that cleaves sialic acid from a cell surface is inhibited by Relenza, which resembles the sialic acid oxonium ion intermediate, and the virus is inactivated.

**14.62**

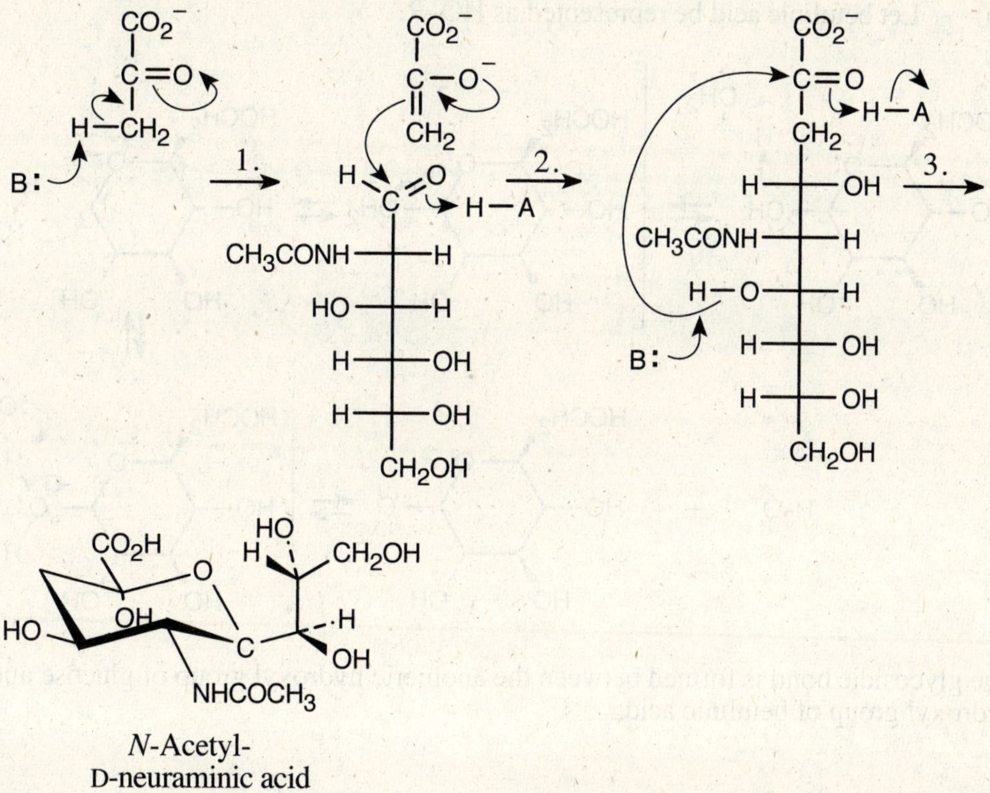


*N*-Acetyl-D-neuraminic acid

**Step 1:** Base-catalyzed formation of pyruvate enolate.
**Step 2:** Aldol condensation between *N*-acetylmannosamine and pyruvate.
**Step 3:** Hemiacetal ring formation.

**In the Field**

**14.63**

(a)

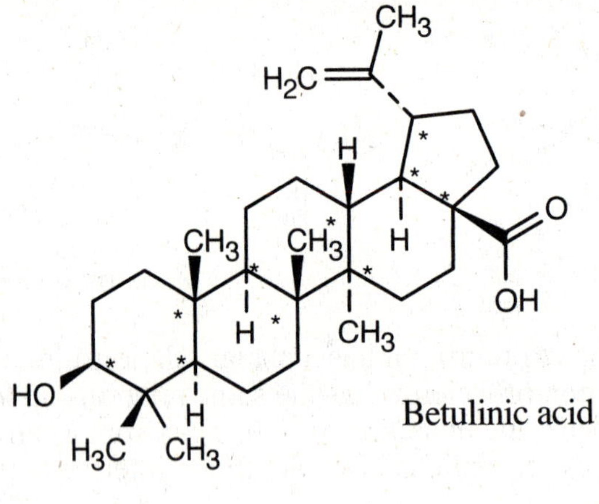


Betulinic acid has ten chirality centers.

(b) Let betulinic acid be represented as HO–R.

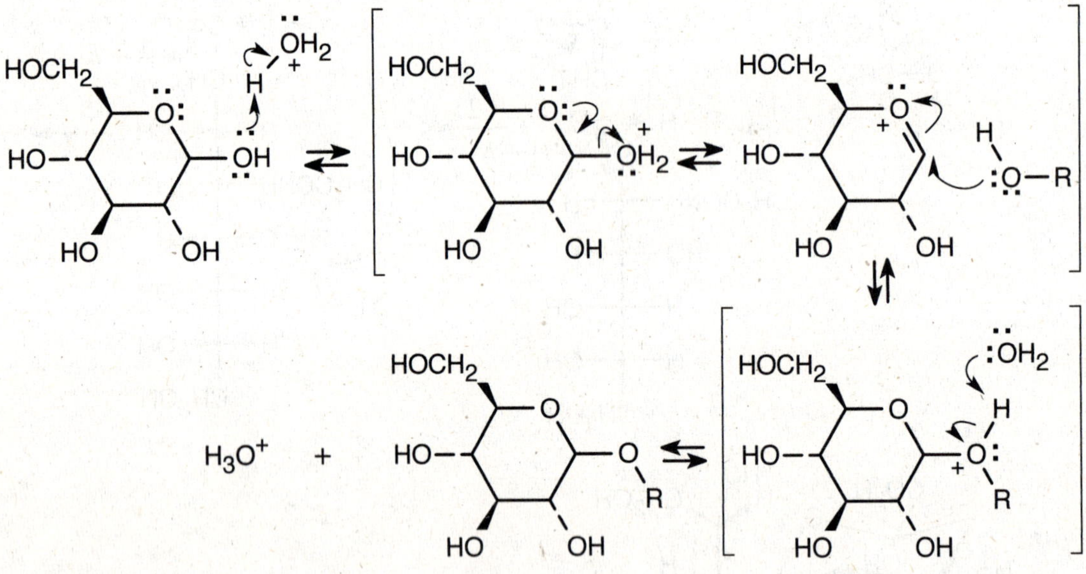


The glycosidic bond is formed between the anomeric hydroxyl group of glucose and the hydroxyl group of betulinic acid.

# Chapter 15 – Biomolecules: Amino Acids, Peptides, and Proteins

## Chapter Outline

I. Amino acids (Sections 15.1–15.2).
   A. Structure of amino acids (Section 15.1).
      1. Amino acids exist in solution as zwitterions.
         a. Zwitterions are internal salts and have many of the properties associated with salts.
            i. They have large dipole moments.
            ii. They are soluble in water.
            iii. They are crystalline and high-melting.
         b. Zwitterions can act either as acids or as bases.
            i. The $-CO_2^-$ group acts as a base.
            ii. The ammonium group ($-NH_3^+$) acts an acid.
      2. All natural amino acids are α-amino acids: the amino group and the carboxylic acid group are bonded to the same carbon.
      3. All but one (proline) of the 20 common amino acids are primary amines.
      4. All of the amino acids are represented by both a three-letter code and a one-letter code.
      5. All amino acids except glycine are chiral.
         a. Only one enantiomer (L) of each pair is naturally-occurring.
         b. In Fischer projections, the carboxylic acid is at the top, and the amino group points to the left.
         c. α-Amino acids are referred to as L-amino acids.
      6. Side chains can be neutral, acidic, or basic.
         a. Fifteen of the amino acids are neutral.
         b. Two (aspartic acid and glutamic acid) are acidic.
            i. At pH = 7.3 (physiological pH), their side chains exist as carboxylate ions.
         c. Three (lysine, arginine and histidine) are basic.
            i. At pH = 7.3, the side chains of lysine and arginine exist as ammonium ions.
            ii. Histidine is not quite basic enough to be protonated at pH = 7.3.
            iii. The double-bonded nitrogen in the histidine ring is basic.
         d. Cysteine and tyrosine are weakly acidic but they are classified as neutral.
   B. Isoelectric points (Section 15.2).
      1. The isoelectric point (p*I*) is the pH at which an amino acid exists as a neutral, dipolar zwitterion.
         a. p*I* is related to side chain structure.
            i. The 13 amino acids that are neutral have p*I* near neutrality.
            ii. The 4 acidic or weakly acidic amino acids have p*I* at a lower pH.
            iii. The 3 basic amino acids have p*I* at a higher pH.
         b. For neutral amino acids, p*I* is the average of the two $pK_a$ values.
            i. For acidic amino acids, p*I* is the average of the two lowest $pK_a$ values.
            ii. For basic amino acids, p*I* is the average of the two highest $pK_a$ values.
         c. The overall p*I* of proteins is determined by the $pK_a$ values of the side chains.
      2. Electrophoresis allows the separation of amino acids by differences in their p*I*.
         a. A buffered solution of amino acids is placed on a paper or gel.
         b. Electrodes are connected to the solution, and current is applied.
         c. Negatively charged amino acids migrate to the positive electrode, and positively charged amino acids migrate to the negative electrode.
         d. Amino acids can be separated because the extent of migration depends on p*I*.

II. Peptides (Sections 15.3–15.7).
  A. Peptide structure (Sections 15.3–15.4).
    1. Peptide bonds (Section 15.3).
      a. A peptide is an amino acid polymer in which the amine group of one amino acid forms an amide bond with the carboxylic acid group of a second amino acid.
      b. This amino acid sequence is known as the backbone of the peptide or protein.
      c. Rotation about the amide bond is restricted.
    2. The N-terminal amino acid of the polypeptide is always drawn on the left.
    3. The C-terminal amino acid of the polypeptide is always drawn on the right.
    4. Peptide structure is described by using the three-letter codes (or the one-letter codes) for the individual amino acids (residues), starting with the N-terminal amino acid.
    5. Disulfide bonds (Section 15.4).
      a. Two cysteines can form a disulfide bond ( –S–S–).
      b. Disulfide bonds can link two polypeptides or introduce a loop within a polypeptide chain.
  B. Structure determination of peptides (Sections 15.5–15.6).
    1. Amino acid analysis (Section 15.5).
      a. Amino acid analysis provides the identity and amount of each amino acid present in a protein or peptide.
      b. First, all disulfide bonds are broken and capped, and all peptide bonds are hydrolyzed.
      c. The mixture is placed on a chromatography column, and the residues are eluted.
      d. As each amino acid elutes, it undergoes reaction with ninhydrin, which produces a purple color that is detected and measured spectrophotometrically.
      e. Amino acid analysis is reproducible on properly maintained equipment; residues always elute at the same time, and only small sample sizes are needed.
    2. Peptide sequencing: The Edman degradation (Section 15.6).
      a. The Edman degradation removes one amino acid at a time from the $-NH_2$ end of a peptide.
        i. The peptide is treated with phenylisothiocyanate (PITC), which reacts with the amino-terminal residue.
        ii. The PITC derivative is split from the peptide.
        iii. The residue undergoes rearrangement to a phenylthiohydantoin (PTH), which is identified chromatographically.
        iv. The shortened chain undergoes another round of Edman degradation.
      b. Since the Edman degradation can only be used on peptides containing fewer than 50 amino acids, a protein must be cleaved into smaller fragments.
        i. Partial acid hydrolysis is unselective and of limited usefulness.
        ii. The enzyme trypsin cleaves proteins at the carboxyl side of Arg and Lys residues.
        iii. The enzyme chymotrypsin cleaves proteins at the carboxyl side of Phe, Tyr and Trp residues.
      c. The complete sequence of a protein results from determining the individual sequences of peptides and overlapping them.
  C. Peptide synthesis (Section 15.7).
    1. Laboratory synthesis.
      a. Groups that are not involved in peptide bond formation are protected.
        i. Carboxyl groups are often protected as methyl or benzyl esters.
        ii. Amino groups are protected as Boc or Fmoc derivatives.
      b. The peptide bond is formed by coupling with dicyclohexylcarbodiimide (DCC).

c. The protecting group are removed.
    i. Boc groups are removed by brief treatment with trifluoroacetic acid.
    ii. Fmoc groups are removed by treatment with base.
    iii. Esters are removed by mild hydrolysis or by hydrogenolysis (benzyl).
2. In solid phase synthesis, the C-terminal amino acid is bound to a support, and the above steps are followed.
    a. Yields are high using this method.

III. Proteins (Sections 15.8–15.10).
  A. Classification of proteins (Section 15.8).
    1. Proteins can be classified by shape.
      a. Fibrous proteins consist of long, filamentous polypeptide chains.
      b. Globular proteins are compact and roughly spherical.
  B. Protein structure.
    1. Primary structure refers to the amino acid sequence of a protein.
    2. Secondary structure refers to the organization of segments of the peptide backbone into a regular pattern, such as an $\alpha$-helix or $\beta$-pleated sheet.
    3. Tertiary structure describes the overall three-dimensional shape of a protein.
      a. Hydrophilic interactions occur between polar side chains and water.
      b. Nonpolar side chains congregate in the middle of a protein, away from water.
      c. Disulfide bonds and salt bridges bring together different parts of a peptide chain.
    4. Quaternary structure describes how polypeptide subunits aggregate into a larger structure.
  C. Enzymes and coenzymes (Sections 15.9–15.10).
    1. Description of enzymes (Section 15.9).
      a. An enzyme is a substance (usually protein) that catalyzes a biochemical reaction.
      b. An enzyme is specific and usually catalyzes the reaction of only one substrate.
        i. Some enzymes, such as papain, can operate on a range of substrates.
      c. The name of an enzyme has two parts, ending with *-ase*.
        i. The first part identifies the substrate.
        ii. The second part identifies the enzyme's class.
      d. Enzymes are grouped into six classes according to the reactions they catalyze.
        i. Hydrolases catalyze hydrolysis reactions.
        ii. Isomerases catalyze isomerizations.
        iii. Ligases catalyze bond formation between two molecules.
        iv. Lyases catalyze the loss of a small molecule from a substrate.
        v. Oxidoreductases catalyze oxidations and reductions.
        vi. Transferases catalyze the transfer of a group from one substrate to another.
      e. Most enzymes are globular proteins, and many consist of a protein portion and a cofactor.
        i. Cofactors may be small organic molecules (coenzymes) or inorganic ions.
        ii. Many coenzymes are derived from vitamins.
    2. How enzymes work – citrate synthase (Section 15.10).
      a. Enzymes work by bringing molecules together, holding them in the correct orientation for reaction, and providing acidic or basic catalysis.
      b. Citrate synthase catalyzes the aldol-like addition of acetyl CoA to oxaloacetate to produce citrate.

## Solutions to Problems

**15.1** Amino acids with aromatic rings:

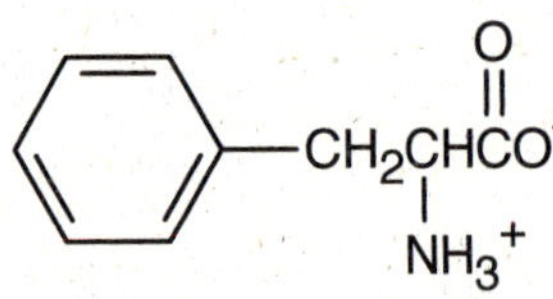

Phenylalanine (Phe) (F)

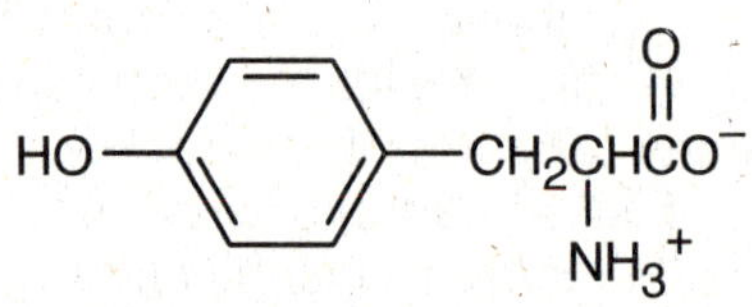

Tyrosine (Tyr) (Y)

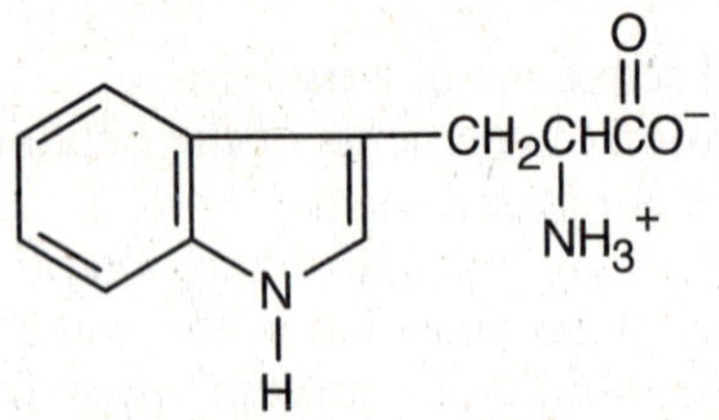

Tryptophan (Trp) (W)

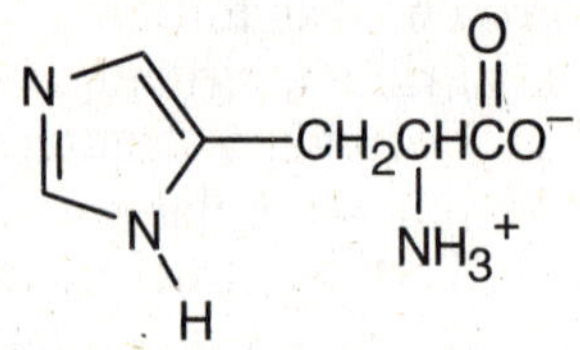

Histidine (His) (H)

Amino acids containing sulfur:

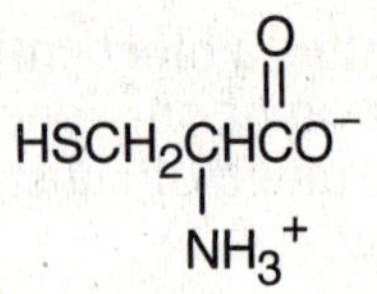

Cysteine (Cys) (C)

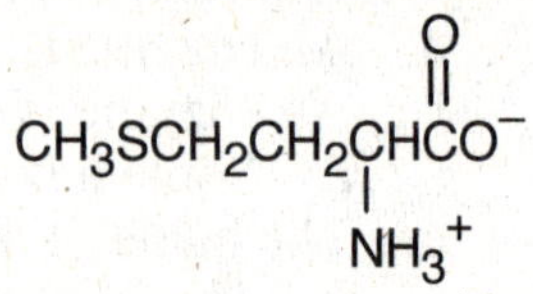

Methionine (Met) (M)

Amino acids that are alcohols:

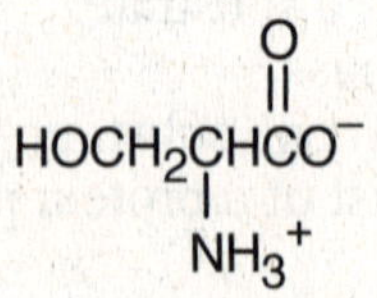

Serine (Ser) (S)

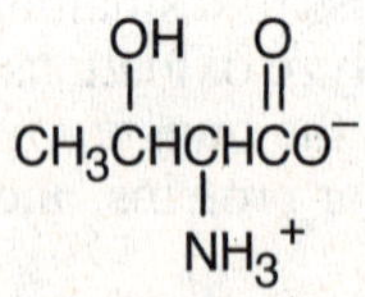

Threonine (Thr) (T)

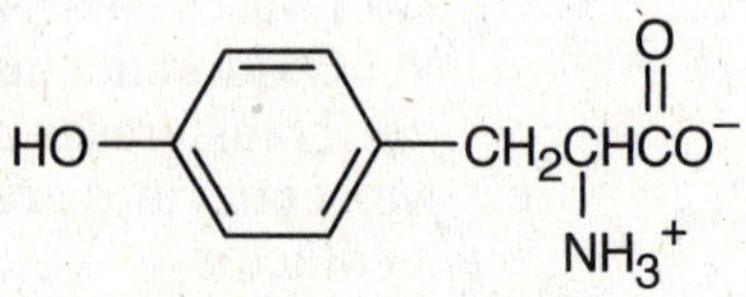

Tyrosine (Tyr) (Y)
a phenol

Amino acids having hydrocarbon side chains:

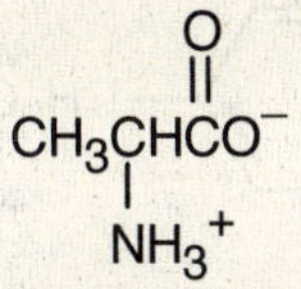

Alanine (Ala) (A)

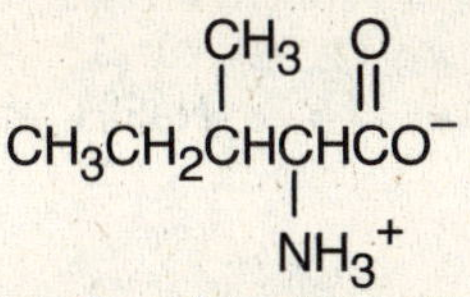

Isoleucine (Ile) (I)

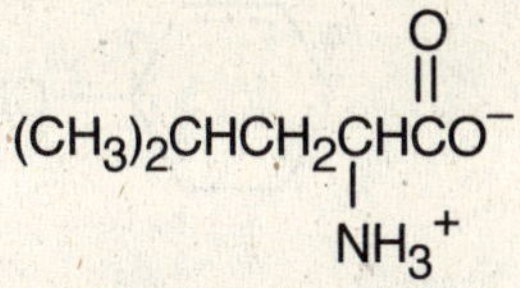

Leucine (Leu) (L)

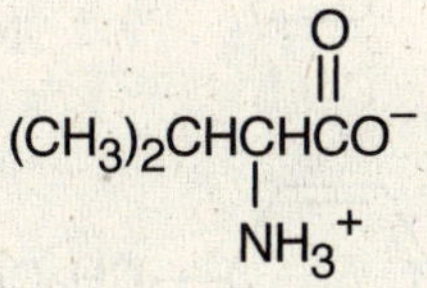

Valine (Val) (V)

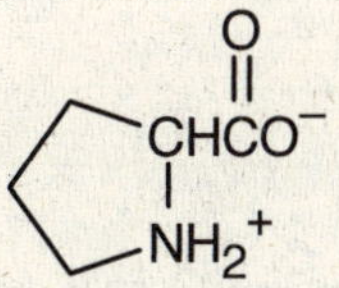

Proline (Pro) (P)

**15.2** A 3-dimensional drawing of the α carbon of an L-amino acid is pictured below.

*For most amino acids:*

| *Group* | *Priority* |
|---|---|
| $-NH_2$ | 1 |
| $-CO_2H$ | 2 |
| –G | 3 |
| –H | 4 |

*For cysteine:*

| *Group* | *Priority* |
|---|---|
| $-NH_2$ | 1 |
| $-CH_2SH$ | 2 |
| $-CO_2H$ | 3 |
| –H | 4 |

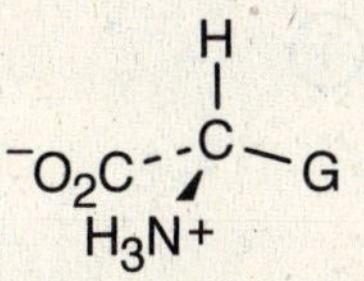

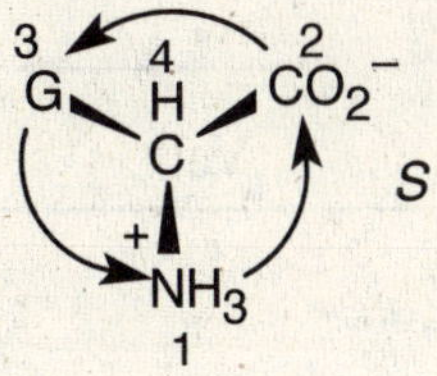

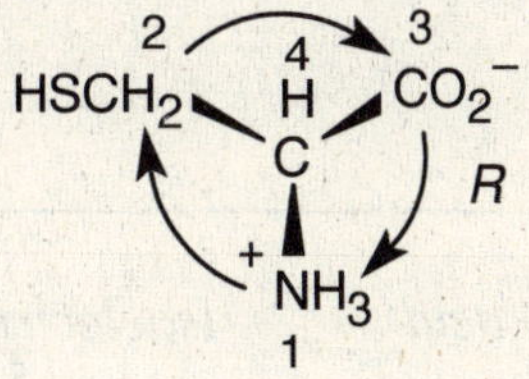

**15.3**

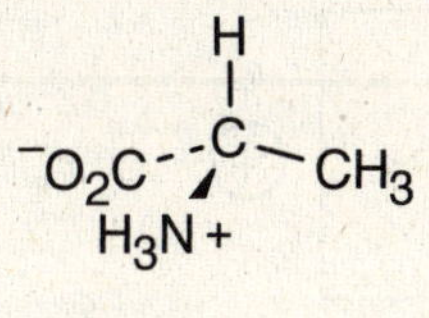

L-Alanine

**15.4**

(a)

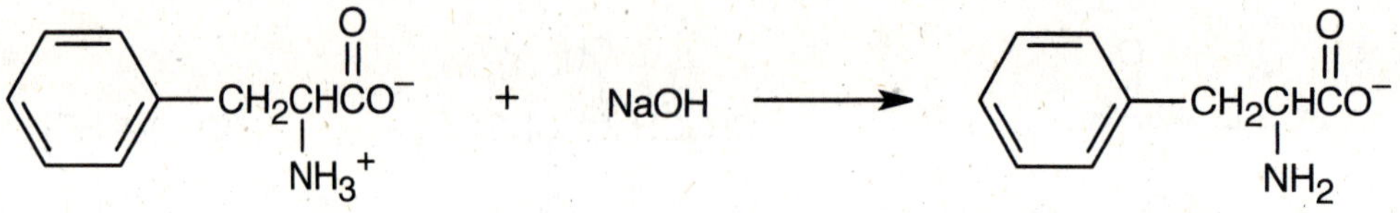


(b)

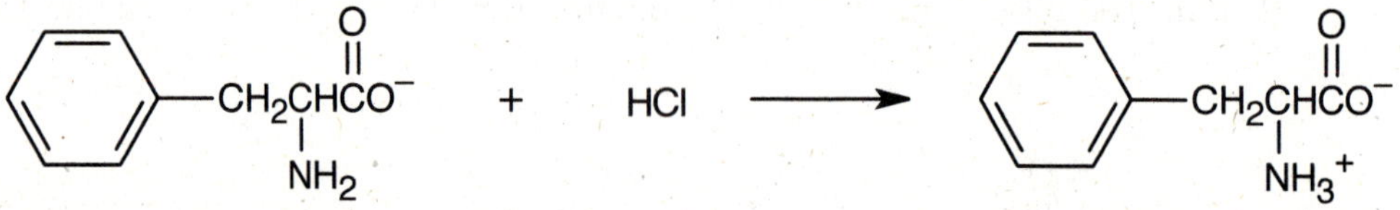


(c)

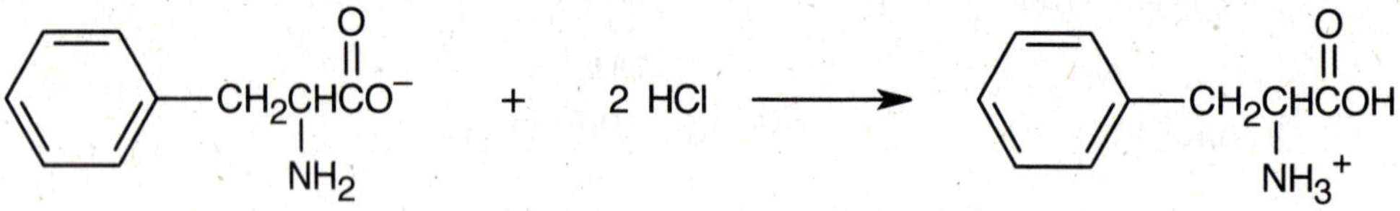


**15.5** Remember that an amino acid is positively charged at a pH lower than its p*I* and migrates to the negative electrode. At a pH higher than p*I*, the amino acid is negatively charged and migrates to the positive electrode. At p*I*, the amino acid is neutral and does not migrate.

(a)

| *Amino acid* | *Isoelectric point* |
|---|---|
| Val | 5.96 |
| Glu | 3.22 |
| His | 7.59 |

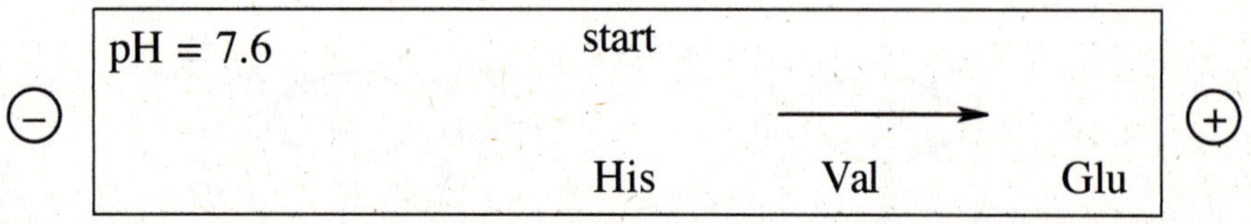


(b)

| *Amino acid* | *Isoelectric point* |
|---|---|
| Gly | 5.97 |
| Phe | 5.48 |
| Ser | 5.68 |

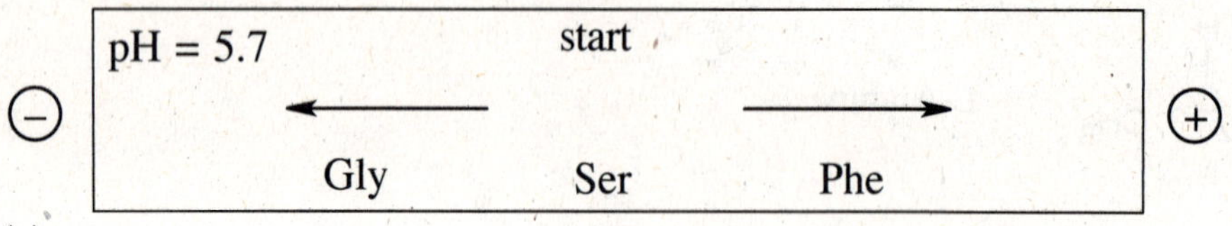


(c)

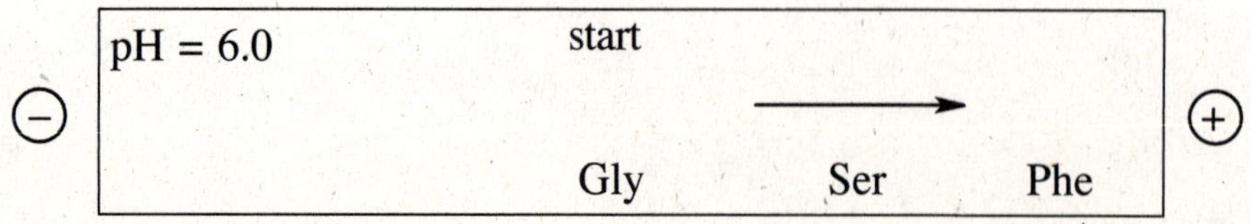

**15.6** On the low pH (acidic) side of p*I*, a protein has a net positive charge, and on the high pH (basic) side of p*I*, a protein has a net negative charge. Thus, hemoglobin (p*I* = 6.8) has a net positive charge at pH = 5.3 and a net negative charge at pH = 7.3.

**15.7**

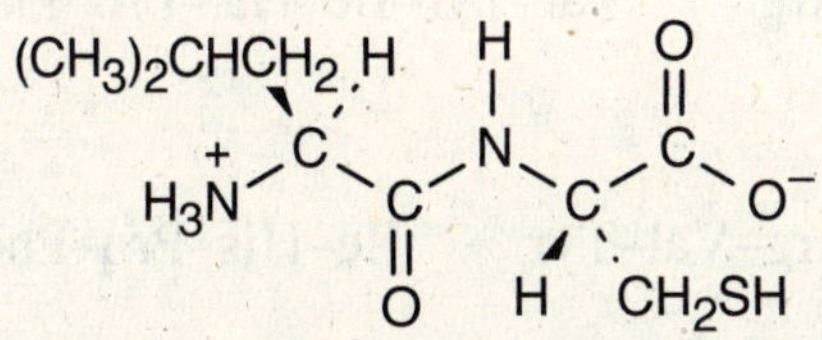

Leu——Cys

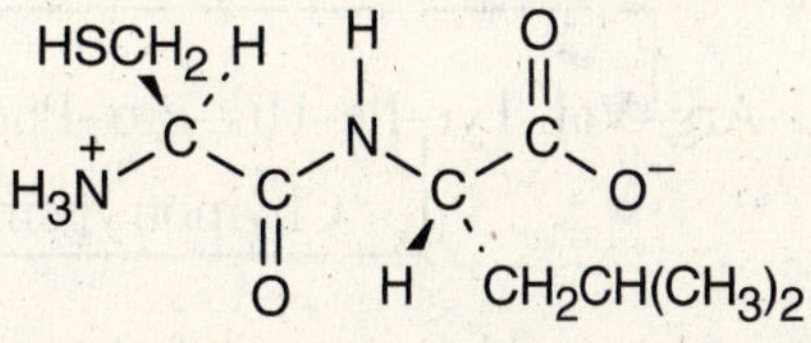

Cys——Leu

**15.8** Val–Tyr–Gly (V–Y–G) Tyr–Gly–Val (Y–G–V) Gly–Val–Tyr (G–V–Y)
Val–Gly–Tyr (V–G–Y) Tyr–Val–Gly (Y–V–G) Gly–Tyr–Val (G–Y–V)

**15.9**

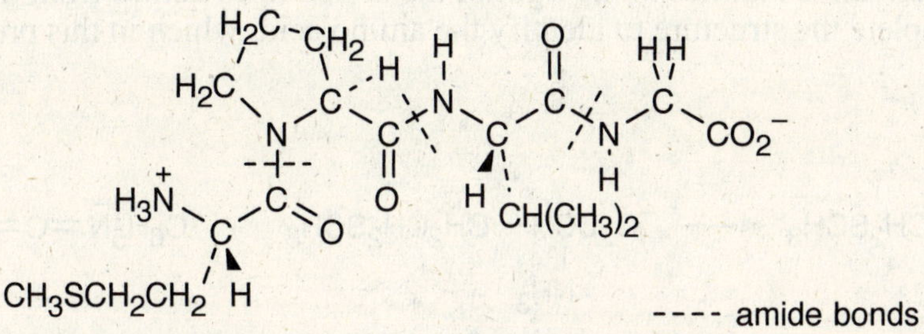


Met —Pro——Val——Gly
M——P——V——G

**15.10**

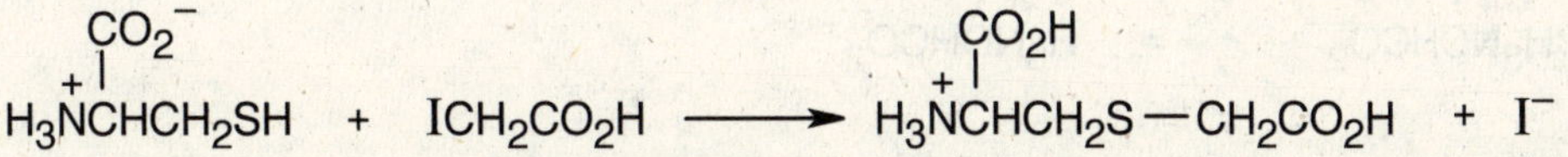

The cysteine sulfur is a good nucleophile, and iodide is a good leaving group.

**15.11** One product of the reaction of an amino acid with ninhydrin is the extensively conjugated purple ninhydrin product. The other major product is the aldehyde that results from reaction of the amino acid with ninhydrin.The identity of the aldehyde is determined by the amino acid side chain. When valine reacts, the resulting aldehyde is 2-methylpropanal. The other products are carbon dioxide and water.

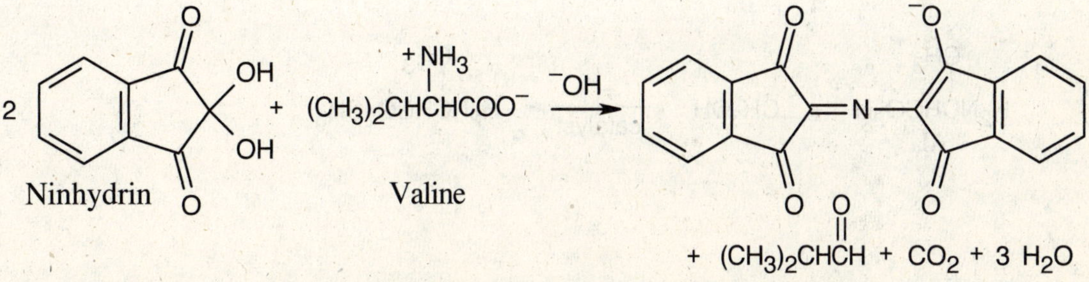

**15.12** Trypsin cleaves peptide bonds at the carboxyl side of *lysine* and *arginine*. Chymotrypsin cleaves peptide bonds at the carboxyl side of *phenylalanine*, *tyrosine* and *tryptophan*. Always remember that the carboxyl side is on the right side of the amino acid.

Asp–Arg–Val–Tyr–Ile–His–Pro–Phe

Trypsin → Asp–Arg + Val–Tyr–Ile–His–Pro–Phe

Chymotrypsin → Asp–Arg–Val–Tyr + Ile–His–Pro–Phe

**15.13** Arg–Pro
Pro–Leu–Gly
Gly–Ile–Val

The complete sequence: Arg–Pro–Leu–Gly–Ile–Val

**15.14** The part of the PTH derivative that lies to the right of the dotted lines comes from the N-terminal residue. Complete the structure to identify the amino acid, which in this problem is methionine.

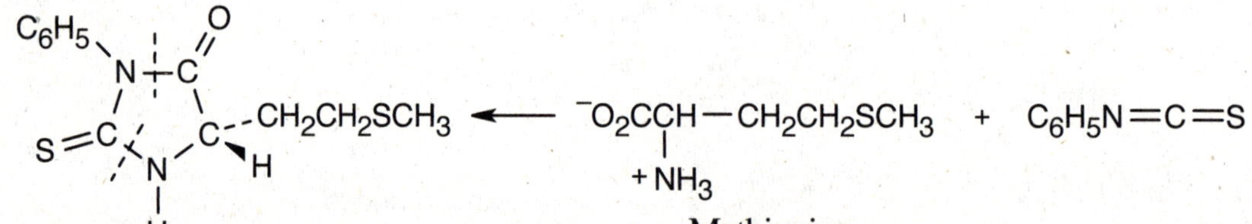


Methionine

**15.15** Refer to Section 15.7.

$$\text{Leu} = \overset{+}{H_3}N\underset{}{C}HCO_2^- \text{ with } CH_2CH(CH_3)_2 \text{ on C} \;=\; H_3\overset{+}{N}CH(R)CO_2^- \qquad R = CH_2CH(CH_3)_2$$

1. Protect the amino group of leucine as the Boc derivative.

$$(CH_3)_3COC(=O)OC(=O)OC(CH_3)_3 + H_2NCH(R)CO_2^- \xrightarrow{Et_3N} (CH_3)_3COC(=O)NHCH(R)CO_2^- + CO_2 + HOC(CH_3)_3$$

(the amino acid reactant is Leu)

2. Protect the carboxylic acid group of alanine as the methyl ester.

$$H_2NCH(CH_3)CO_2H + CH_3OH \xrightarrow[\text{catalyst}]{H^+} H_2NCH(CH_3)CO_2CH_3$$

(the amino acid reactant is Ala)

3. Couple the protected amino acids with DCC.

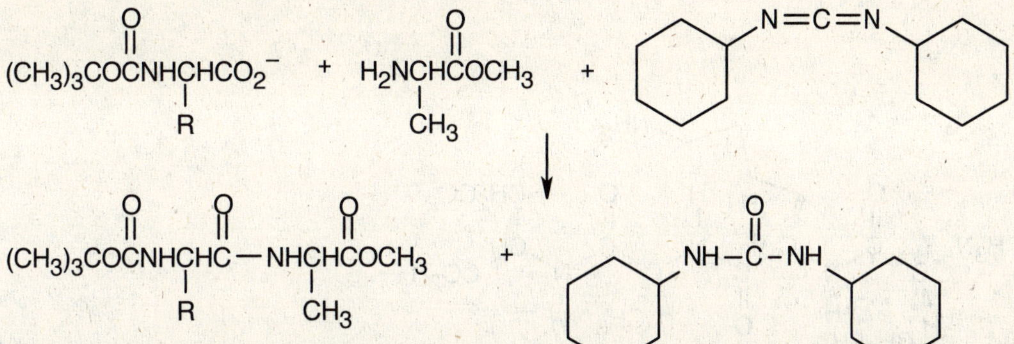

4. Remove the leucine protecting group by treatment with strong acid.

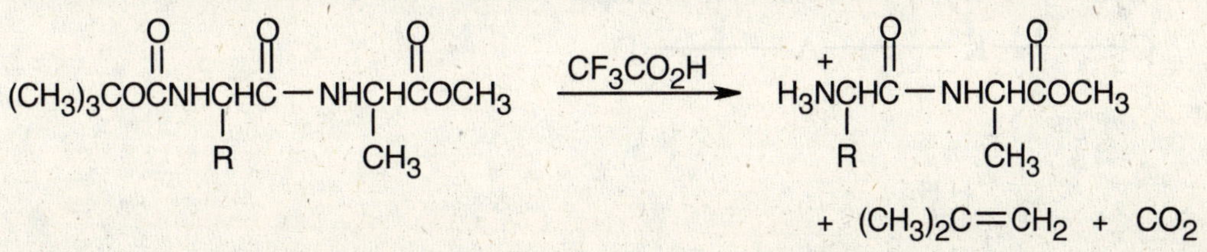

5. Remove the alanine protecting group by base hydrolysis, followed by acidification.

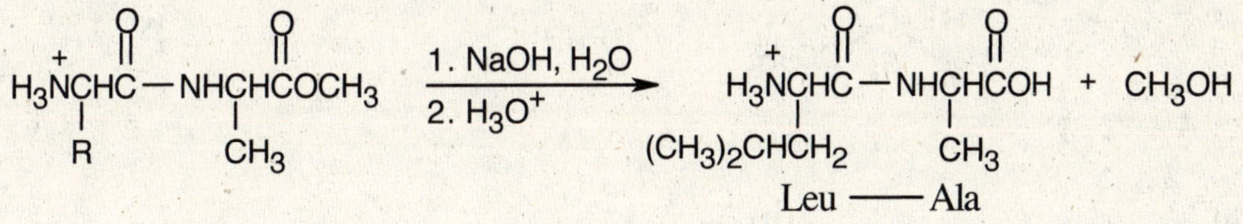

**15.16**

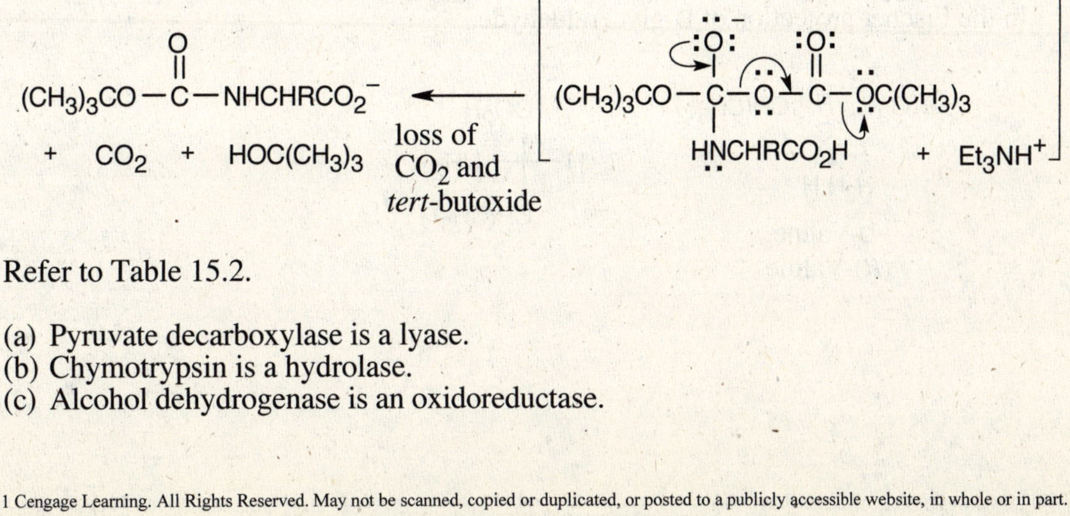

**15.17** Refer to Table 15.2.

(a) Pyruvate decarboxylase is a lyase.
(b) Chymotrypsin is a hydrolase.
(c) Alcohol dehydrogenase is an oxidoreductase.

## Visualizing Chemistry

**15.18**

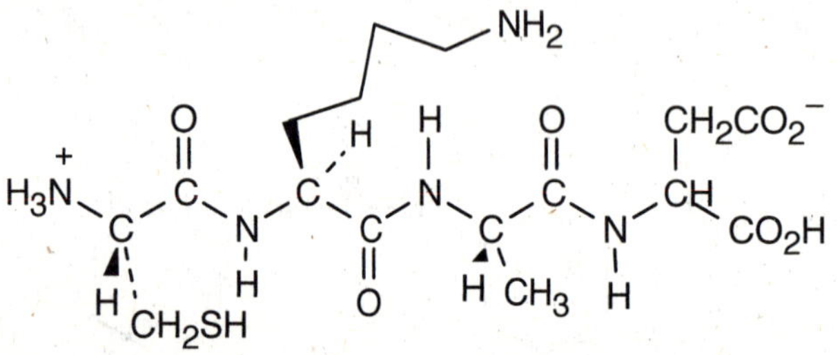

Cys——Lys——Ala——Asp

C——K——A——D

**15.19**

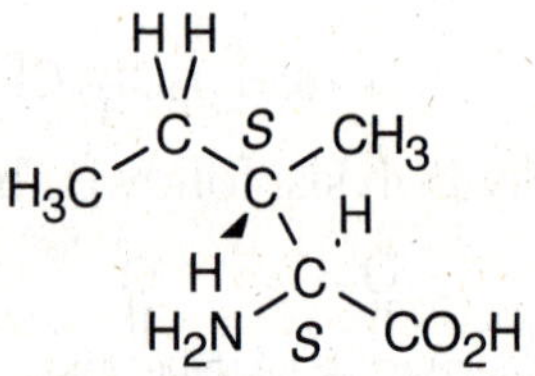

Isoleucine (Ile) (I)

**15.20**

(a)

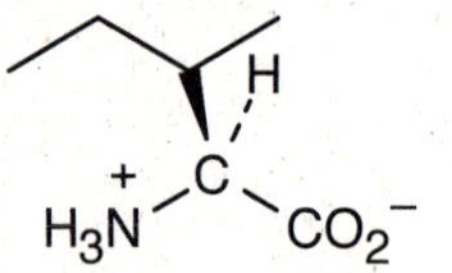

Isoleucine

(b)

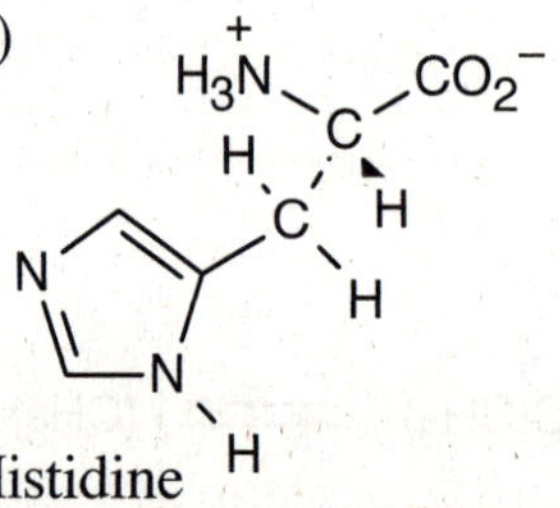

Histidine

(c)

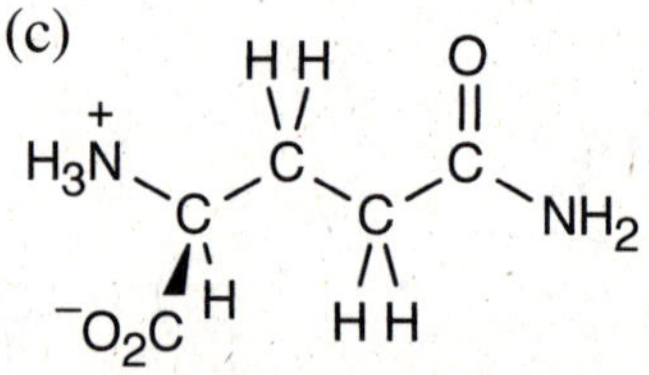

Glutamine

**15.21** It's possible to identify this representation of valine as the D enantiomer by drawing its Fischer projection. The amino group is on the right, just as a hydroxyl group is on the right in the Fischer projection of D-glyceraldehyde.

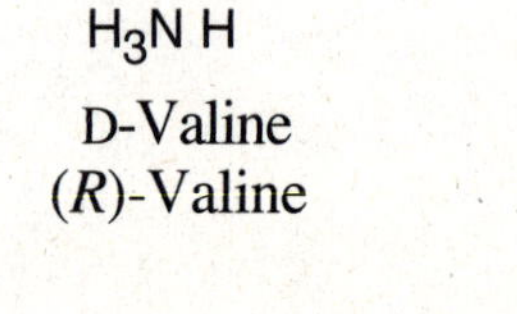

D-Valine

(*R*)-Valine

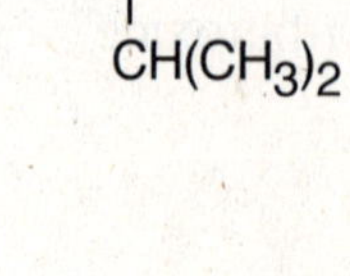

## Additional Problems

### Structures and Chirality

**15.22** The disulfide bridges that cysteine forms help to stabilize a protein's tertiary structure.

**15.23**

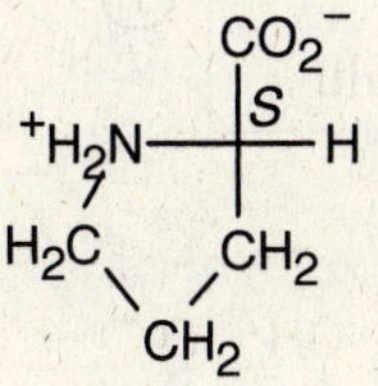

(*S*)-Proline

**15.24–15.25**

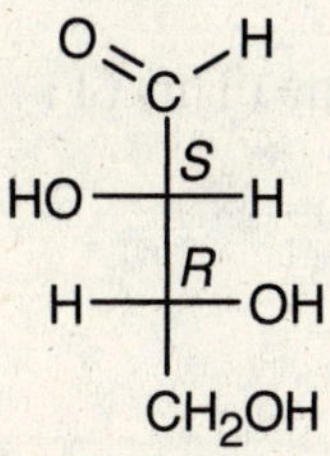

D-Threose

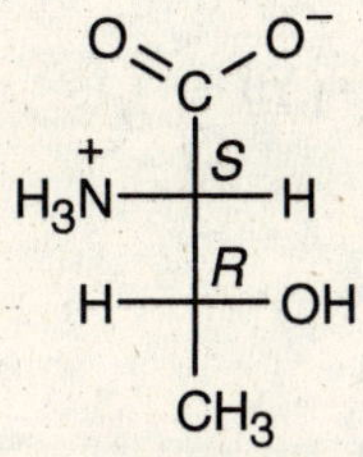

L-Threonine

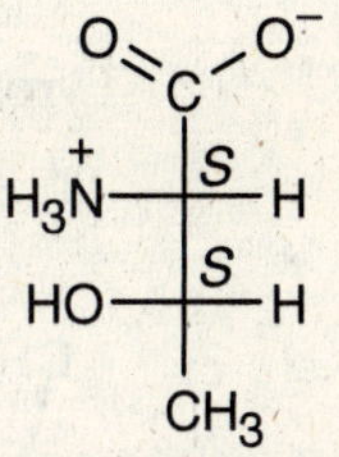

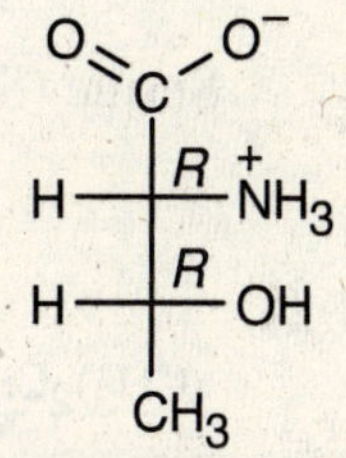

Diastereomers of L-Threonine

**15.26** When referring to an amino acid, the prefix "α" indicates that the amino group is bonded to the carbon atom next to (alpha to) the carbonyl group.

**15.27**

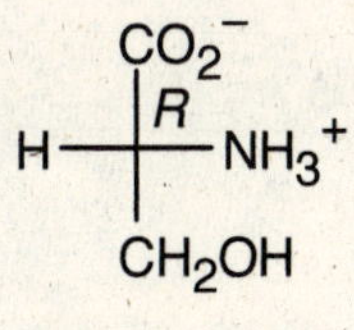

(*R*)-Serine

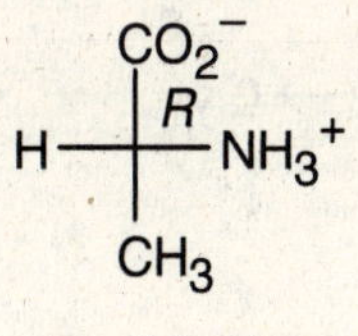

(*R*)-Alanine

Both (*R*)-serine and (*R*)-alanine are D-amino acids.

**15.28** (a) Ser = serine (b) Thr = threonine (c) Pro = proline
(d) F = phenylalanine (e) Q = glutamine (f) D = aspartic acid

**15.29**

| *Amino acid* | *Isoelectric Point* |
|---|---|
| Histidine | 7.59 |
| Serine | 5.68 |
| Glutamic acid | 3.22 |

The optimum pH for the electrophoresis of three amino acids occurs at the isoelectric point of the amino acid intermediate in acidity. At this pH, the least acidic amino acid migrates toward the negative electrode, the most acidic amino acid migrates toward the positive electrode, and the amino acid intermediate in acidity does not migrate. In this example, electrophoresis at pH = 5.7 allows a very good separation of the three amino acids.

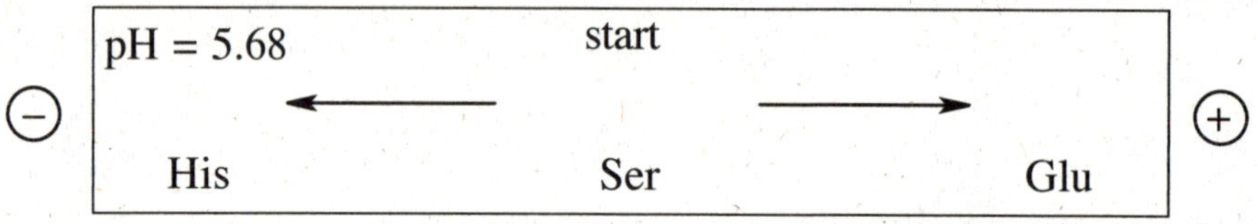


**15.30**

(a)

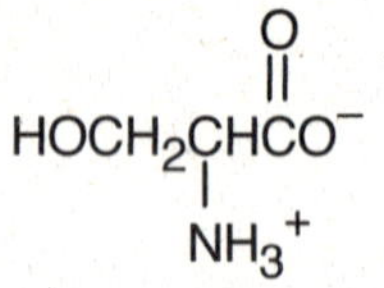

Serine (Ser) (S)

(b)

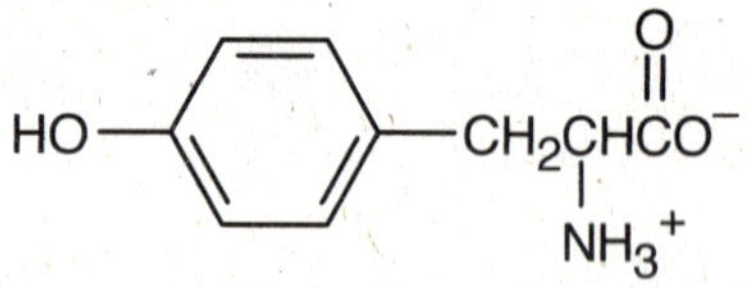

Tyrosine (Tyr) (Y)

(c)

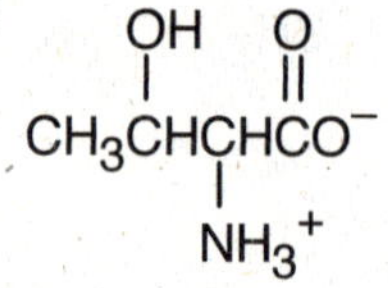

Threonine (Thr) (T)

**15.31**

(a)

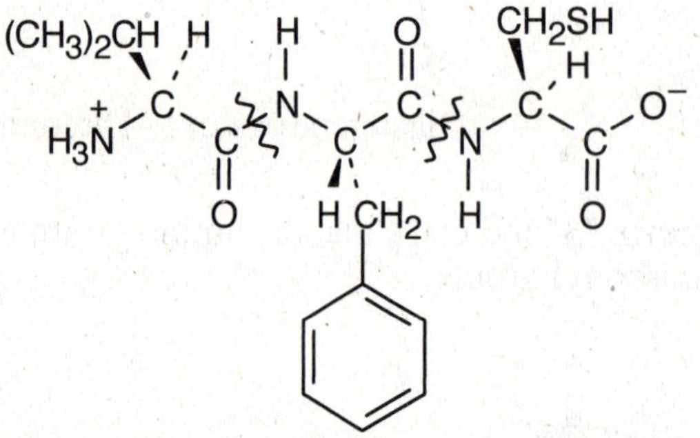

Val —— Phe —— Cys 〰 amide bonds

V —— F —— C

(b)

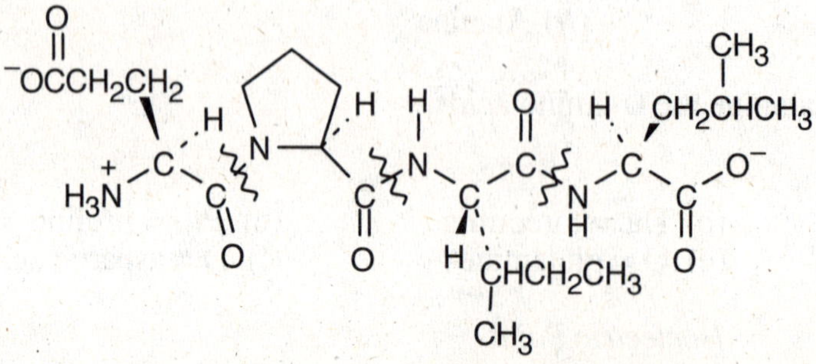

Glu —— Pro —— Ile —— Leu

E —— P —— I —— L

**15.32** Use Table 15.1 as a reference.

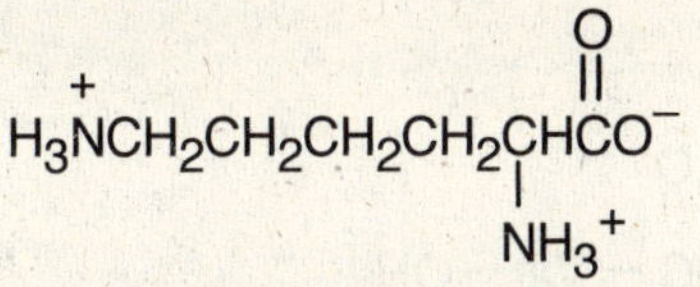

Lysine at pH = 3.0

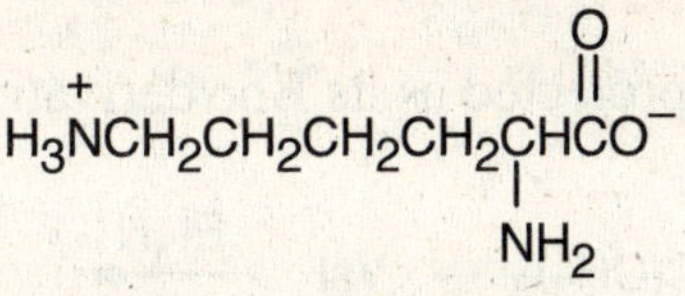

Lysine at pH = 9.7

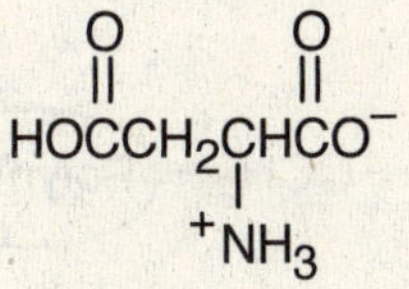

Aspartic acid at pH = 3.0

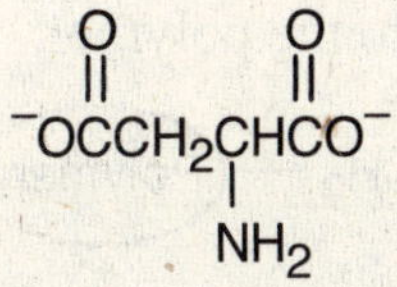

Aspartic acid at pH = 9.7

**15.33** (a) Val–Leu–Ser (V–L–S)
Val–Ser–Leu (V–S–L)
Ser–Leu–Val (S–L–V)
Ser–Val–Leu (S–V–L)
Leu–Val–Ser (L–V–S)
Leu–Ser–Val (L–S–V)

(b) Ser–Leu–Leu–Pro (S–L–L–P)
Ser–Leu–Pro–Leu (S–L–P–L)
Ser–Pro–Leu–Leu (S–P–L–L)
Pro–Leu–Leu–Ser (P–L–L–S)
Pro–Leu–Ser–Leu (P–L–S–L)
Pro–Ser–Leu–Leu (P–S–L–L)
Leu–Leu–Ser–Pro (L–L–S–P)
Leu–Leu–Pro–Ser (L–L–P–S)
Leu–Ser–Leu–Pro (L–S–L–P)
Leu–Ser–Pro–Leu (L–S–P–L)
Leu–Pro–Leu–Ser (L–P–L–S)
Leu–Pro–Ser–Leu (L–P–S–L)

**15.34**

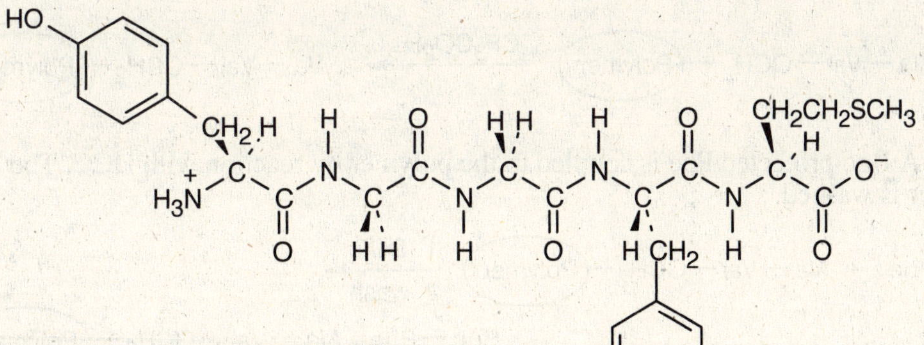

Tyr——Gly——Gly——Phe——Met
Y——G——G——F——M

**15.35** Only primary amines can form the extensively conjugated purple ninhydrin product. A secondary amine such as proline yields a product containing a shorter system of conjugated bonds, which absorbs at a shorter wavelength (440 nm *vs.* 570 nm) and is not purple.

## Reactions

**15.36**

*Step 1*: Valine is protected as its Boc derivative.

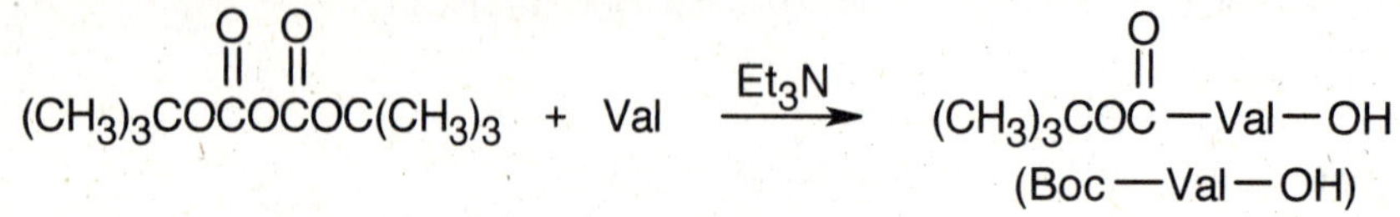


*Step 2*: Boc–Val bonds to the polymer.

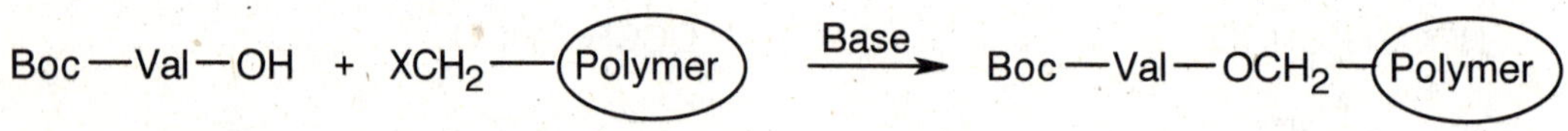


*Step 3*: The polymer is first washed, then is treated with $CF_3CO_2H$ to cleave the Boc group.

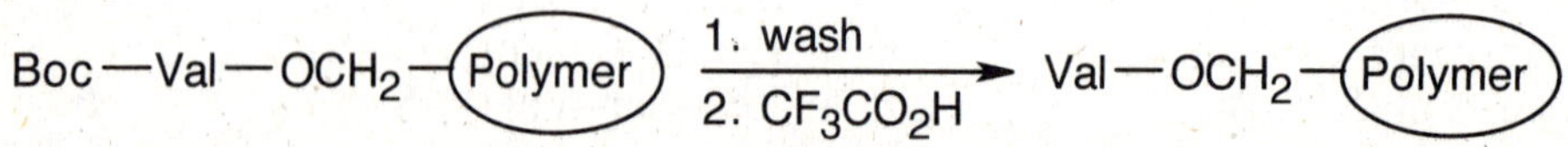


*Step 4*: A Boc-protected Ala is coupled to the polymer-bound valine by reaction with DCC. The polymer is washed.

*Step 5*: The polymer is treated with $CF_3CO_2H$ to remove Boc.

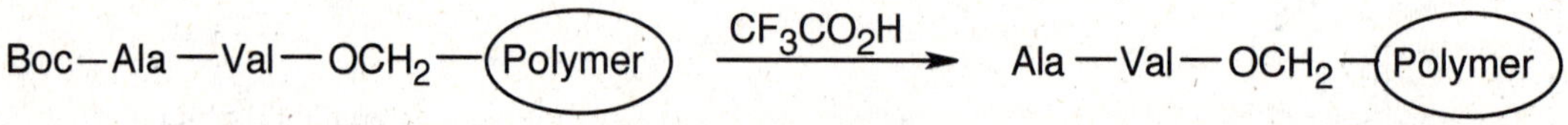


*Step 6*: A Boc-protected Phe is coupled to the polymer by reaction with DCC. The polymer is washed.

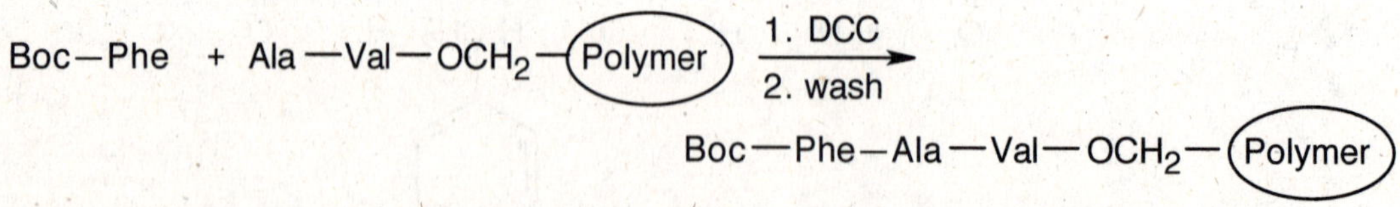


*Step 7*: Treatment with anhydrous HF removes the Boc group and cleaves the ester bond between the peptide and the polymer.

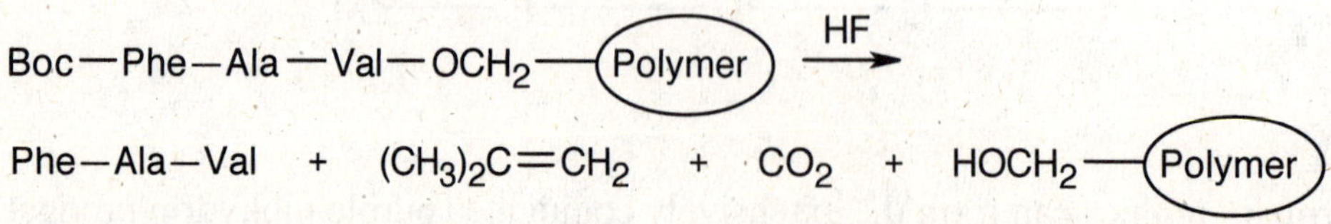


Note: Fmoc protecting groups may also be used but are cleaved by base, rather than acid.

**15.37**

(a)

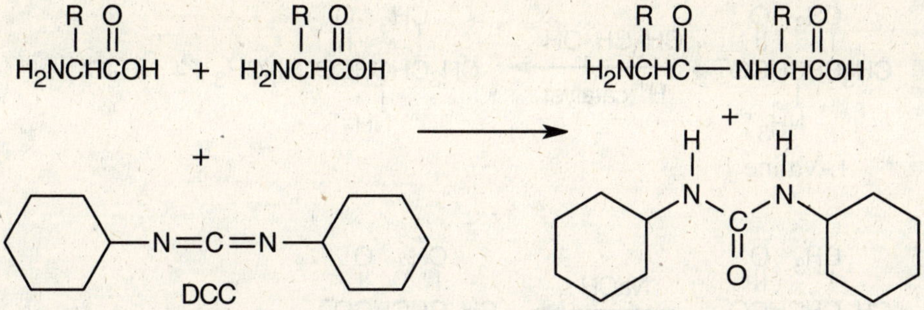


In this step, a dipeptide is formed from two amino acid residues.

(b)

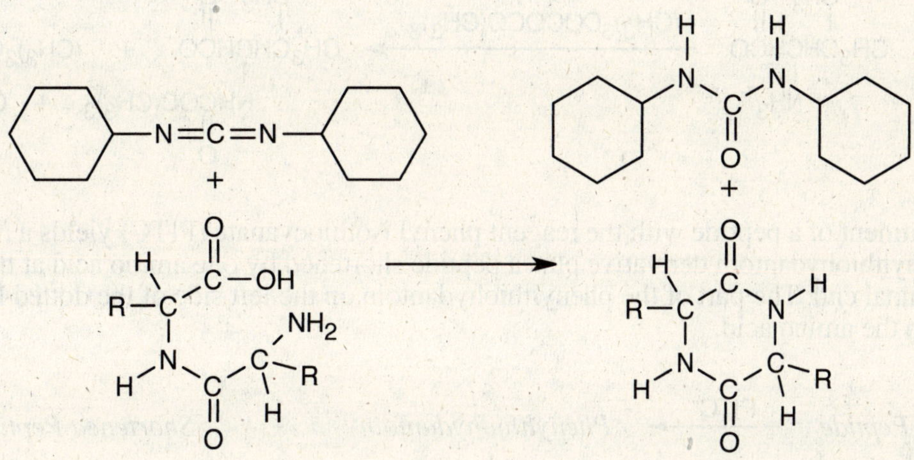

DCC couples the carboxylic acid end of the dipeptide to the amino end to yield the 2,5-diketopiperazine.

**15.38**

(a)

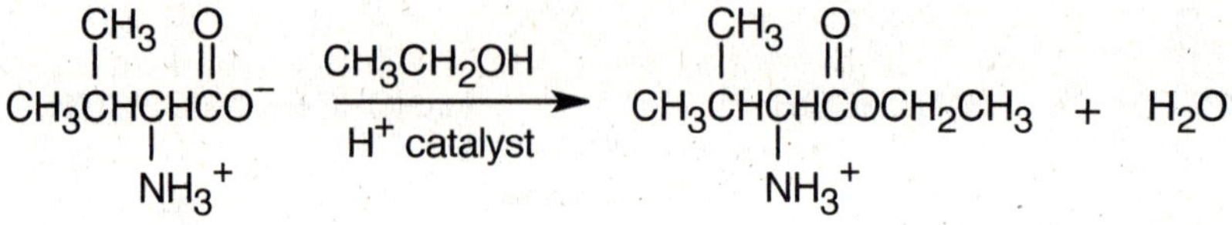

L-Valine

(b)

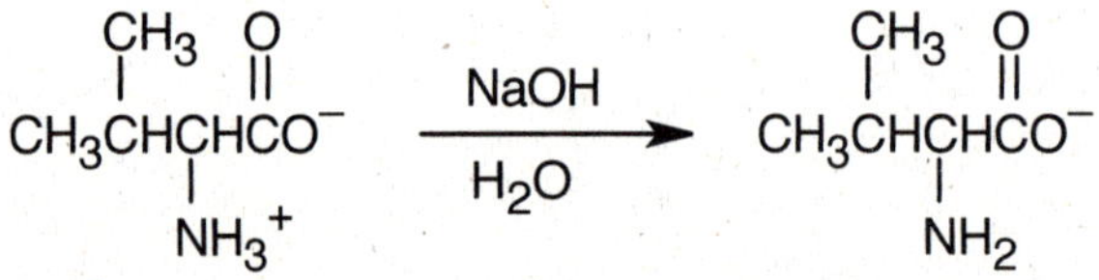

(c)

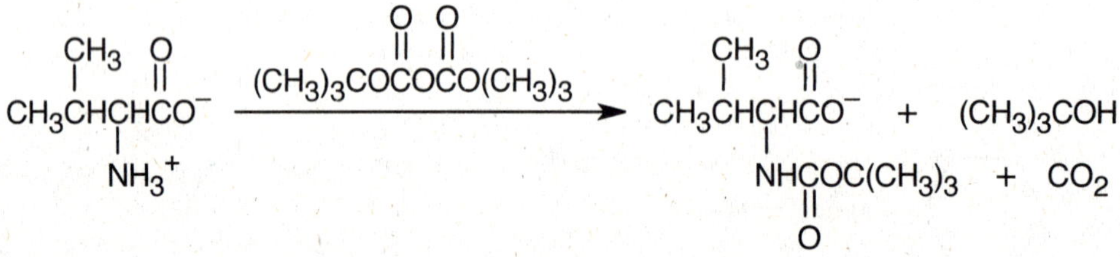

**15.39** Treatment of a peptide with the reagent phenyl isothiocyanate (PITC) yields a *N*-phenylthiohydantoin derivative plus a peptide shortened by one amino acid at the *N*-terminal end. The part of the phenylthiohydantoin on the left side of the dotted line comes from the amino acid.

| *Peptide* $\xrightarrow{\text{PITC}}$ | *Phenylthiohydantoin* | + | *Shortened Peptide* |
|---|---|---|---|
| (a) Val–Leu–Gly | 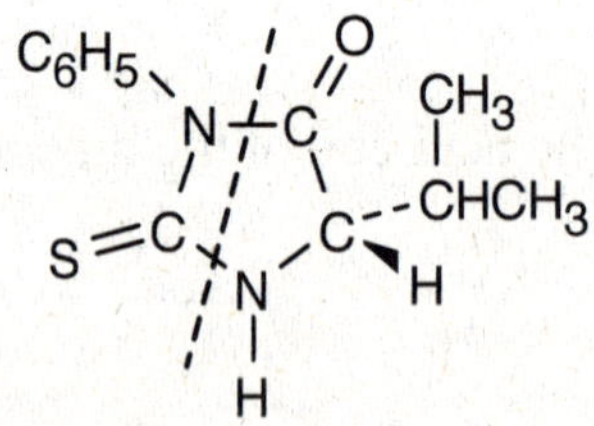 | | Leu–Gly |
| (b) Ala–Pro–Phe | 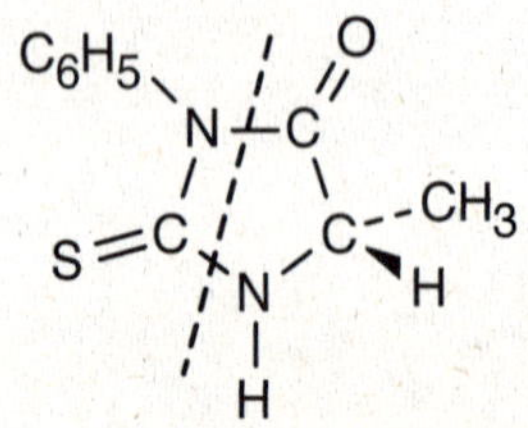 | | Pro–Phe |

## Peptide Sequencing

**15.40** Ser–Ile–Arg–Val–Val–Pro–Tyr–Leu–Arg

**15.41**

Phe ≀ Leu—Met—Lys ┆ Tyr ≀ Asp—Gly—Gly—Arg ┆ Val—Ile—Pro—Tyr

cleaved by trypsin = ------

cleaved by chymotrypsin = ~~~~~

**15.42** Gly

Gly–Asp–Phe–Pro

Phe–Pro–Val

Val–Pro–Leu

The complete sequence: Gly–Gly–Asp–Phe–Pro–Val–Pro–Leu

**15.43** (a) Arg–Pro

Pro–Leu–Gly

Gly–Ile–Val

The complete sequence: Arg–Pro–Leu–Gly–Ile–Val

(b) Val–Met–Trp

Trp–Asp–Val

Val–Leu

The complete sequence: Val–Met–Trp–Asp–Val–Leu

## General Problems

**15.44** (a) A *protease* catalyzes the hydrolysis of an amide group.
(b) A *kinase* catalyzes the transfer of a phosphate group between substrates.
(c) A *carboxylase* catalyzes the addition of $CO_2$ to a substrate.

**15.45** A proline residue in a polypeptide chain interrupts α-helix formation. The amide nitrogen of proline has no hydrogen that can contribute to the hydrogen-bonded structure of an α-helix. In addition, the pyrrolidine ring of proline restricts rotation about the C–N bond and reduces flexibility in the polypeptide chain.

**15.46** (a) *Hydrolases* catalyze the hydrolysis of substrates.
(b) *Lyases* catalyze the removal of a small molecule from a substrate.
(c) *Transferases* catalyze the transfer of a group from one substrate to another.

**15.47**

(a)

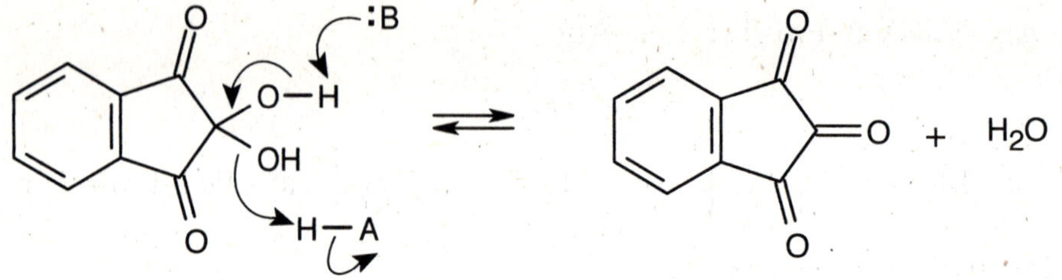

(b)

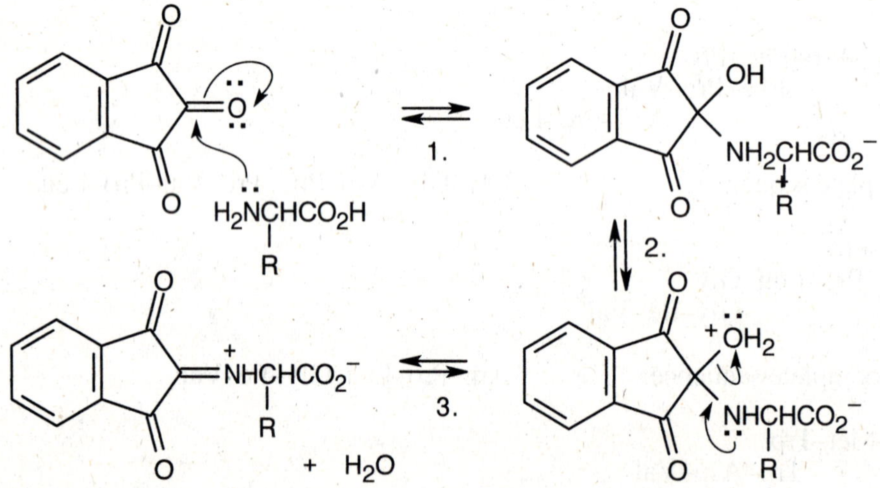

The steps involved in imine formation: (1) nucleophilic addition of the amino group of the amino acid, (2) proton transfer, and (3) loss of water.

(c)

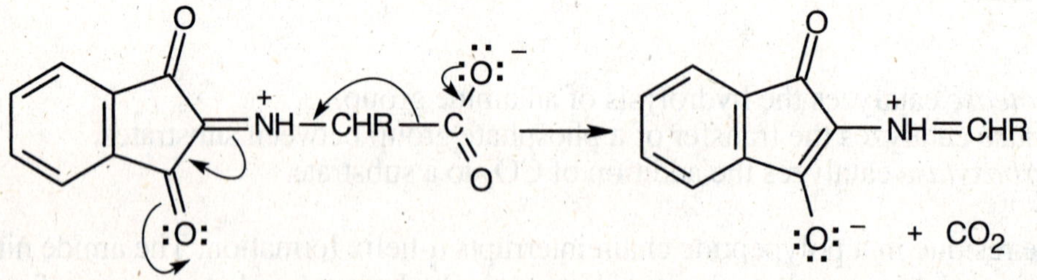

Decarboxylation produces a different imine.

(d)

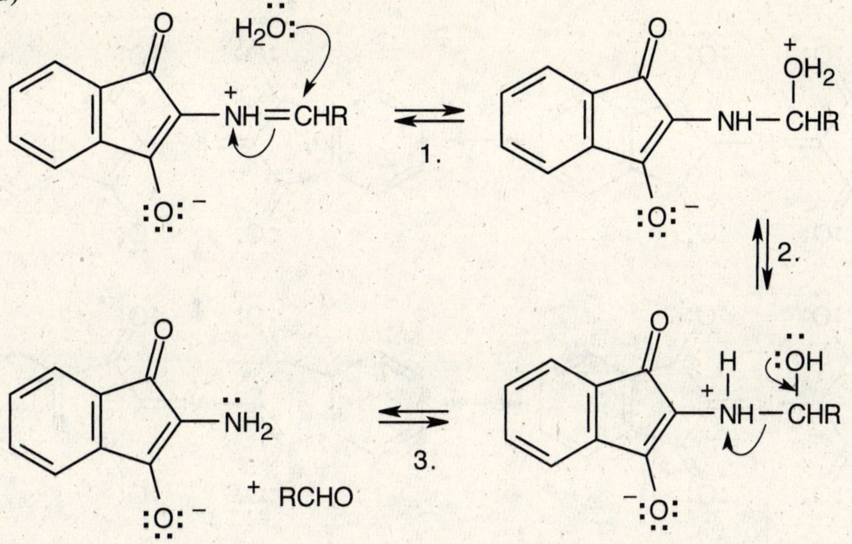


Hydrolysis of this imine occurs by addition of water (1), proton transfer (2), and bond cleavage (3) to yield an aldehyde and an amine.

(e)

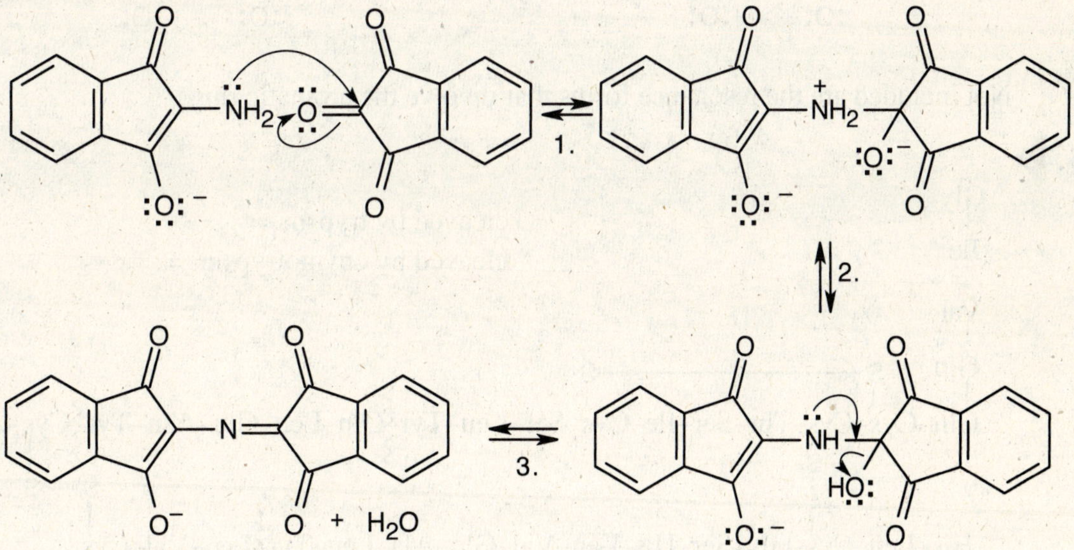


The final series of steps are addition of the amine to a carbonyl carbon of a second ninhydrin molecule (1), a proton shift (2), and loss of water to form the purple anion. Notice that the amine nitrogen is all that remains of the original amino acid.

**15.48**

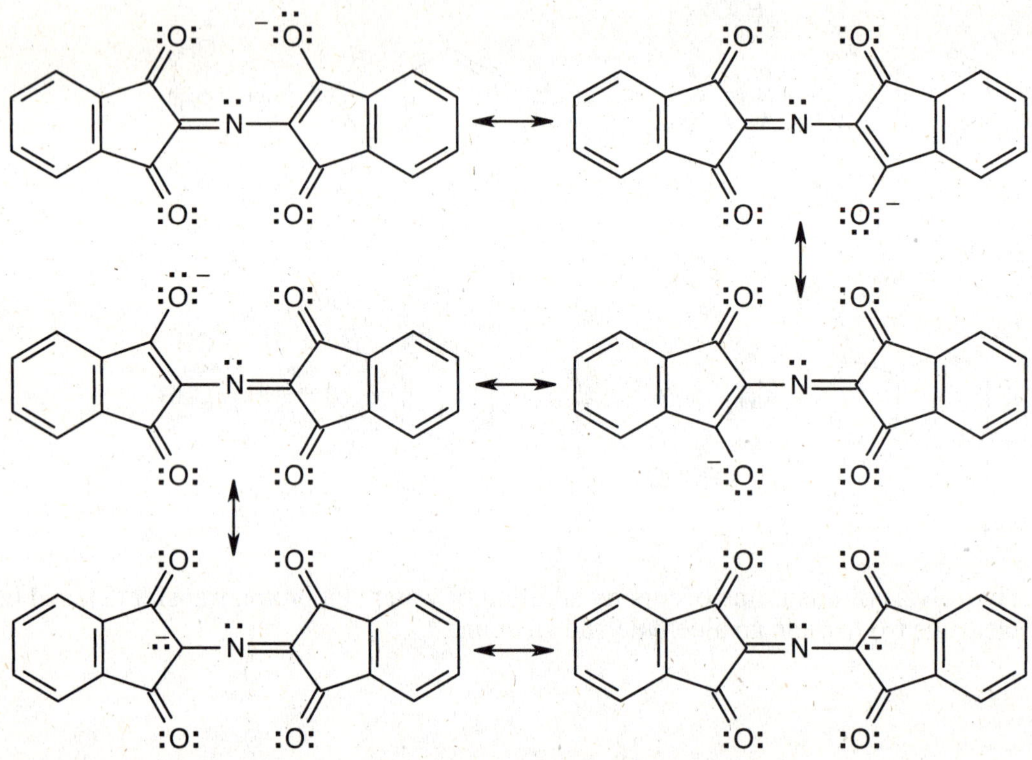

Not included are the resonance forms that involve the aromatic rings.

**15.49**

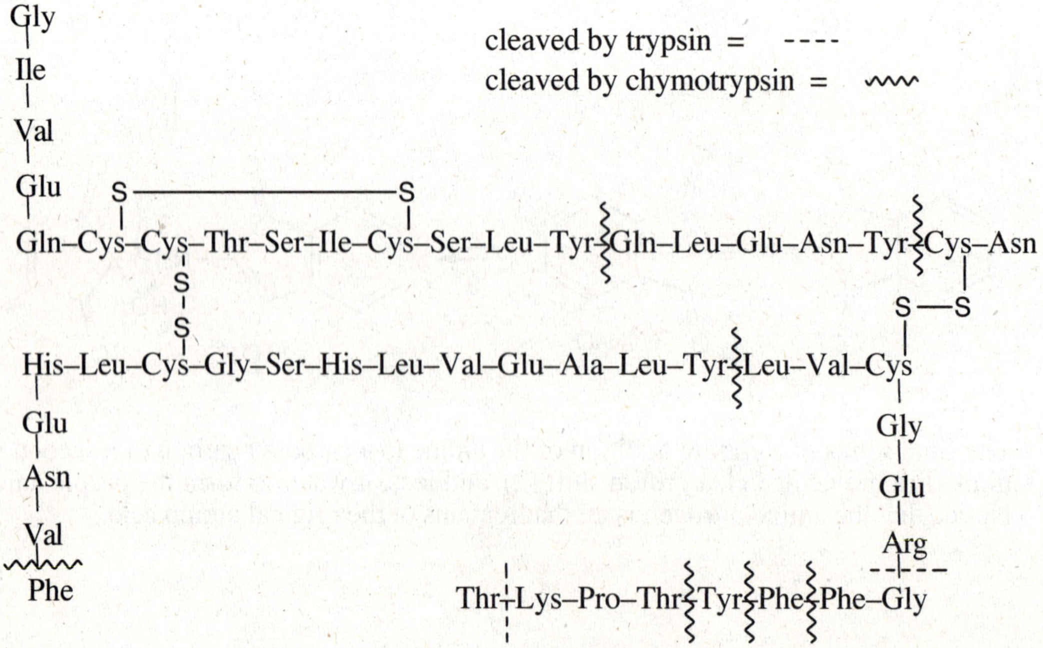

**15.50**

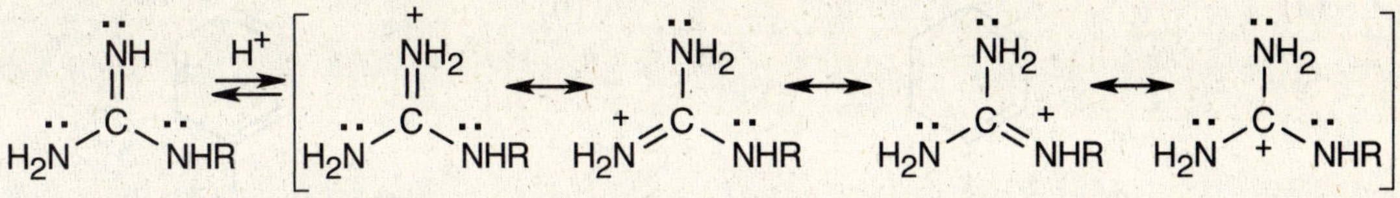

The protonated guanidino group can be stabilized by resonance.

**15.51**

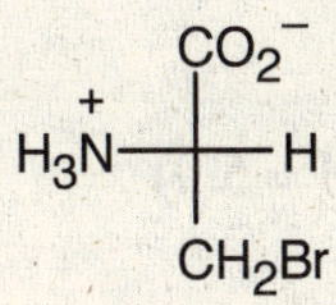

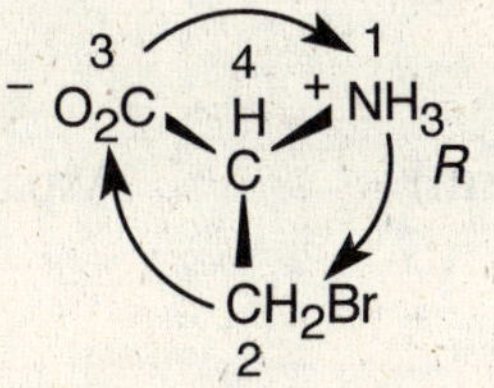

L-Bromoalanine
(*R*)-Bromoalanine

This L "amino acid" also has an *R* configuration because the $–CH_2Br$ side chain is higher in priority than the $–CO_2^-$ group.

**15.52**

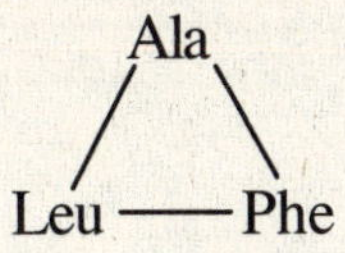

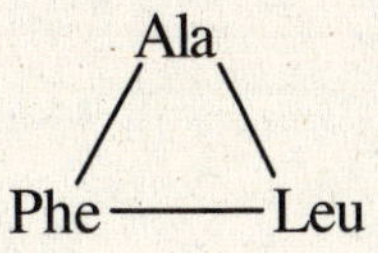

The tripeptide is cyclic and does not have an N-terminal amino acid.

**15.53** Amino acids with hydrocarbon side chains (valine and isoleucine) are more likely to be found on the inside of a globular protein, where they can avoid water. Amino acids with charged side chains (aspartic acid and lysine) are more likely to be found on the outside of a globular protein, where they can hydrogen-bond with water.

**15.54** Pro–Leu–Gly
Gly–Pro–Arg
Arg–Pro

The complete sequence: Pro–Leu–Gly–Pro–Arg–Pro

**15.55**

(a)

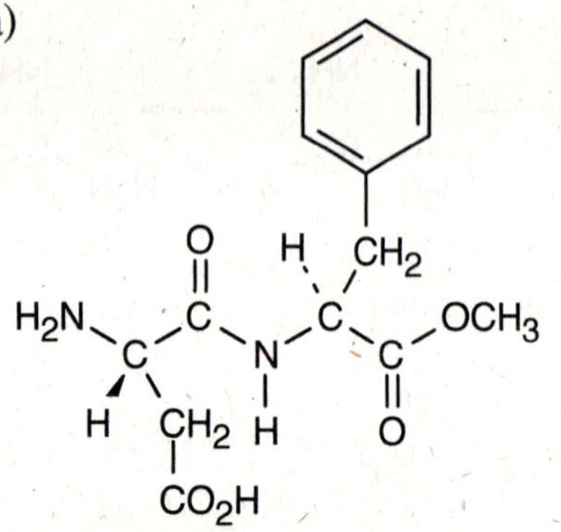

Aspartame (nonzwitterionic form)

(b)

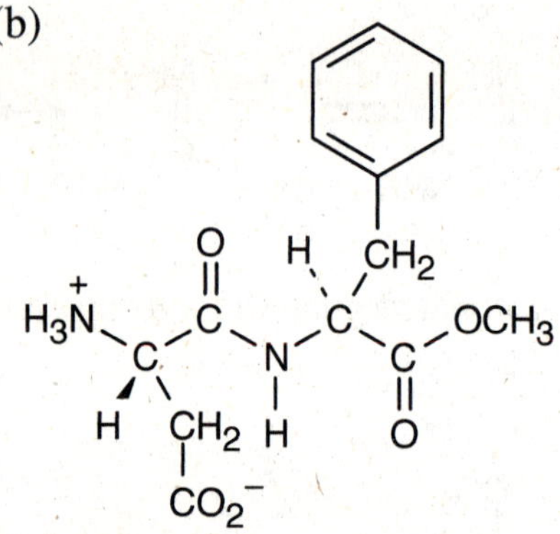

Aspartame at p*I* (pH = 5.9)

(c)

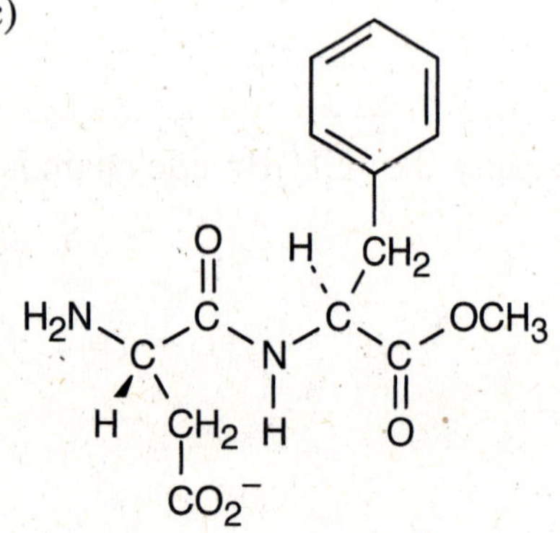

Aspartame at pH = 7.6

(d)

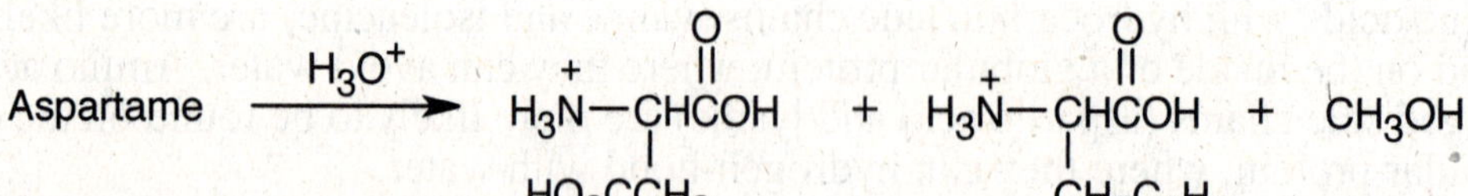

**15.56** 100 g of Cytochrome *c* contains 0.43 g iron.

$$0.43 \text{ g Fe} \times \frac{1 \text{ mol Fe}}{55.8 \text{ g Fe}} = 0.0077 \text{ mol Fe}$$

$$\frac{100 \text{ g Cytochrome } c}{0.0077 \text{ mol Fe}} = \frac{13{,}000 \text{ g Cytochrome } c}{1 \text{ mol Fe}}$$

Cytochrome *c* has a minimum molecular weight of 13,000.

## In the Medicine Cabinet

**15.57**

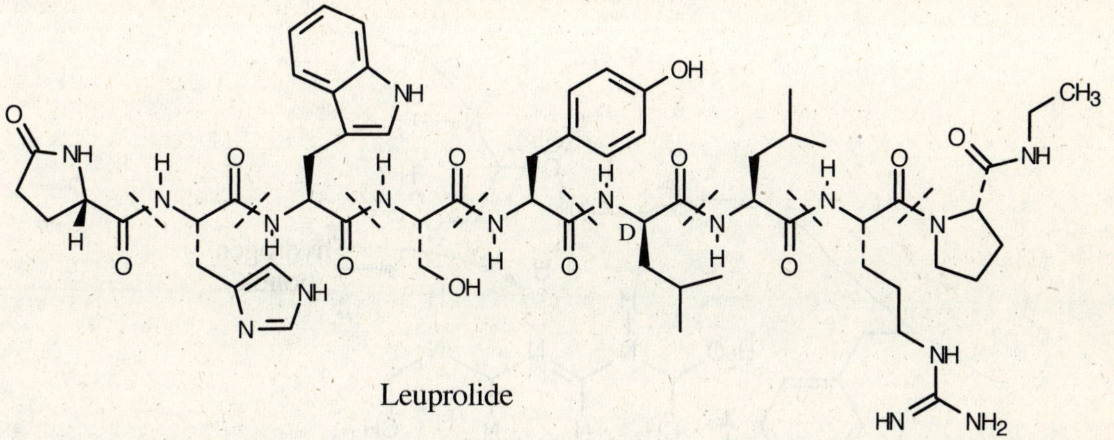

Leuprolide

Glu — His — Trp — Ser — Tyr — Leu — Leu — Arg — Pro — NHEt
E — H — W — S — Y — L — L — R — P — NHEt

(a) The N-terminal glutamic acid is a cyclic lactam. The C-terminal proline is an *N*-ethyl amide.

(b) One of the leucines (indicated above) has D stereochemistry.

(c) See above.

(d) The charge on a peptide is due to the side chains. According to Table 15.1, the only side chain that is charged at neutral pH is arginine. Thus, leuprolide has a charge of +1 at neutral pH.

**15.58** Cys–Tyr
Tyr–Ile–Glu
Ile–Glu
Glu–Asp–Cys
Asp–Cys
Cys–Pro–Leu
Leu–Gly

Reduced oxytocin: Cys–Tyr–Ile–Gln–Asn–Cys–Pro–Leu–Gly–$NH_2$

Oxidized oxytocin: Cys–Tyr–Ile–Gln–Asn–Cys–Pro–Leu–Gly–$NH_2$
S ——————————— S (disulfide bond linking the two Cys residues)

The C–terminal end of oxytocin is actually an amide, but this can't be determined from the information given.

**In the Field**

**15.59**

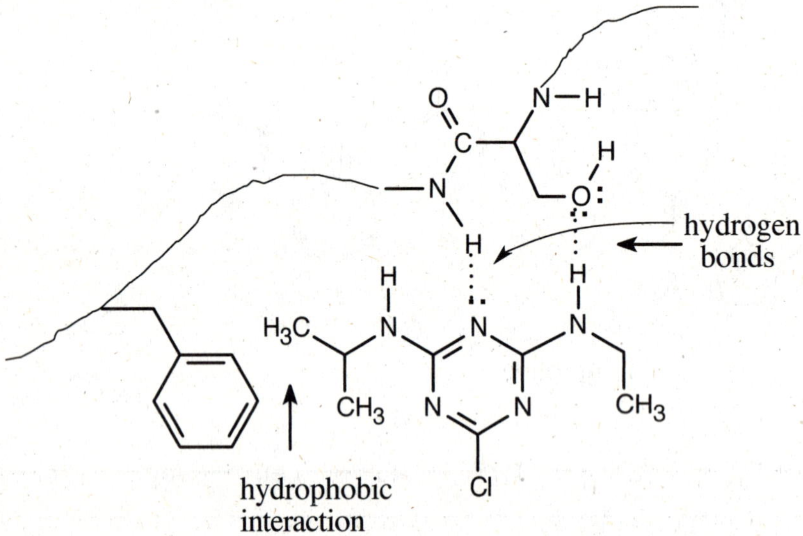

# Chapter 16 – Biomolecules: Lipids and Nucleic Acids

## Chapter Outline

I. Lipids (Sections 16.1–16.4).
  A. Waxes, fats and oils (Section 16.1).
    1. Waxes are esters of long-chain fatty acids with long-chain alcohols.
    2. Fats and oils are triacylglycerols (triglycerides).
      a. Hydrolysis of a fat yields glycerol and three fatty acids.
      b. The fatty acids need not be the same.
    3. Fatty acids.
      a. Fatty acids are even-numbered, unbranched long-chain ($C_{12}$–$C_{20}$) carboxylic acids.
      b. The most abundant saturated fatty acids are palmitic ($C_{16}$) and stearic ($C_{18}$) acids.
      c. The most abundant unsaturated fatty acids are oleic and linoleic acids (both $C_{18}$).
        i. Linoleic and arachidonic acids are polyunsaturated fatty acids.
      d. Unsaturated fatty acids are lower melting than saturated fatty acids because the double bonds keep molecules from packing closely.
      e. The C=C bonds can be catalytically hydrogenated to produce higher-melting fats.
  B. Soaps (Section 16.2).
    1. Soap is a mixture of the sodium and potassium salts of fatty acids produced by hydrolysis (saponification) of animal fat.
    2. Soap acts as a cleanser because the two ends of a soap molecule are different.
      a. The hydrophilic carboxylate end dissolves in water.
      b. The hydrophobic hydrocarbon tails solubilize greasy dirt.
      c. In water, the hydrocarbon tails aggregate into micelles, where greasy dirt can accumulate.
    3. Soaps can form insoluble scum when they encounter $Mg^{2+}$ and $Ca^{2+}$ salts.
      a. This problem is circumvented by synthetic detergents, which don't form insoluble metal salts.
      b. Synthetic detergents are composed of salts of alkylbenzenesulfonic acids.
  C. Phospholipids (Section 16.3).
    1. Glycerophospholipids.
      a. Glycerophospholipids consist of glycerol, two fatty acids (at C1 and C2 of glycerol), and a phosphate group bonded to an amino alcohol at C3 of glycerol.
      b. Glycerophospholipids comprise the major lipids in cell membranes.
        i. The phospholipid molecules are organized into a lipid bilayer, which has polar groups on the inside and outside, and nonpolar tails in the middle.
    2. Sphingomyelins.
      a. Sphingomyelins have sphingosine as their backbone.
      b. They are abundant in brain and nerve tissue.
  D. Steroids (Section 16.4).
    1. Facts about steroids.
      a. Steroids are found in the lipid extracts of plants and animals.
      b. Steroids have a tetracyclic ring structure.
        i. The rings adopt chair conformations but are unable to undergo ring-flips.
      c. Steroids function as hormones in humans.

2. Types of steroid hormones.
   a. Sex hormones.
      i. Androgens (testosterone, androsterone) are male sex hormones.
      ii. Estrogens (estrone, estradiol) and progestins are female sex hormones.
   b. Adrenocortical hormones.
      i. Mineralocorticoids (aldosterone) regulate cellular $Na^+$ and $K^+$ balance.
      ii. Glucocorticoids (hydrocortisone) regulate glucose metabolism and control inflammation.
   c. Synthetic steroids.
      i. Oral contraceptives and anabolic steroids are examples of synthetic steroids.

II. Nucleic acids (Sections 16.5–16.6).

A. Nucleic acids and nucleotides (Section 16.5).

1. The nucleic acids DNA and RNA are the carriers and processors of genetic information.
   a. Both DNA and RNA are biopolymers of nucleotides.
   b. DNA is vastly larger than RNA and is found in the cell nucleus.
2. Nucleosides are composed of a heterocyclic purine or pyrimidine base plus an aldopentose.
   a. In RNA, the purines are adenine and guanine, the pyrimidines are uracil and cytosine, and the sugar is ribose.
   b. In DNA, thymine replaces uracil, and the sugar is 2'-deoxyribose.
   c. Positions on the base receive non-prime superscripts, and positions on the sugar receive prime superscripts.
   d. The heterocyclic base is bonded to C1' of the sugar.
3. In nucleotides, a phosphate group is bonded to C5' of the sugar of a nucleoside.

B. DNA.

1. Structure of DNA.
   a. DNA is composed of nucleotides connected by a phosphodiester bond between a 5' phosphate and the 3' hydroxyl group of a second nucleotide.
      i. One end of the nucleic acid polymer has a free hydroxyl group and is called the 3' end.
      ii. The other end has a free phosphate group and is called the 5' end.
   b. The structure of DNA depends on the order of bases.
   c. The sequence of bases is described by starting at the 5' end and listing the bases in order of occurrence by their one-letter abbreviations.
2. Base-pairing in DNA (Section 16.6).
   a. DNA consists of two polynucleotide strands coiled in a double helix.
      i. Adenine and thymine hydrogen-bond with each other, and cytosine and guanine hydrogen-bond with each other.
   b. Because the two DNA strands are complementary, the amount of A equals the amount of T, and the amount of C equals the amount of G.
   c. The double helix is 2.0 nm wide, there are 10 bases in each turn, and each turn is 3.4 nm in height.
   d. The double helix has a major groove and a minor groove into which polycyclic aromatic molecules can intercalate.
3. The fundamental processes of molecular genetics.
   a. The function of DNA is to store genetic information and to pass it on to RNA, which uses it to make proteins.
   b. Replication, transcription and translation are the three processes that are responsible for carrying out the central dogma.

III. The transfer of genetic information (Sections 16.7–16.9).
  A. Replication of DNA (Section 16.7).
    1. Replication is the enzyme-catalyzed process whereby DNA makes a copy of itself.
    2. Replication is semiconservative: each new strand of DNA consists of one old strand and one newly synthesized strand.
    3. How replication occurs:
      a. The DNA helix partially unwinds.
      b. New nucleotides form base pairs with their complementary partners.
      c. Formation of new bonds is catalyzed by DNA polymerase and takes place in the 5' → 3' direction.
        i. Bond formation occurs by addition of the 3' hydroxyl group of the growing chain to the 5' triphosphate, with loss of a diphosphate leaving group.
  B. Transcription (Section 16.8).
    1. There are 3 major types of RNA:
      a. Messenger RNA (mRNA) carries genetic information to ribosomes when protein synthesis takes place.
      b. Ribosomal RNA (rRNA), complexed with protein, comprises the physical makeup of the ribosomes.
      c. Transfer RNA (tRNA) brings amino acids to the ribosomes, where they are joined to make proteins.
      d. Small RNAs (functional RNAs) have a number of functions within the cell.
    2. mRNA is synthesized in the nucleus by transcription of DNA.
      a. DNA contains "promoter sites", which indicate where mRNA synthesis is to begin, and base sequences that indicate where mRNA synthesis stops.
      b. The DNA partially unwinds, forming a "bubble".
      c. Ribonucleotides form base pairs with their complementary DNA bases.
      d. Bond formation occurs in the 5' → 3' direction and is catalyzed by RNA polymerase.
      e. Only one strand of DNA (the antisense strand ) is transcribed.
      f. Thus, the synthesized mRNA is a copy of the sense strand (with U replacing T).
    3. Much of the DNA chain consists of noncoding sections (introns).
      a. After transcription. the noncoding sections are cut out of the mRNA strand, and the coding sections (exons) are joined together.
  C. Translation (Section 16.9).
    1. Translation is the process in which proteins are synthesized at the ribosomes by using mRNA as a template.
    2. The message delivered by mRNA is contained in "codons" – 3-base groupings that are specific for an amino acid.
      a. Amino acids are coded by 61 of the possible 64 codons.
      b. The other 3 codons are "stop" codons.
    3. Each tRNA is responsible for bringing an amino acid to the growing protein chain.
      a. A tRNA has a cloverleaf-shaped secondary structure and consists of 70–100 ribonucleotides.
      b. Each tRNA contains an anticodon complementary to the mRNA codon.
    4. The protein chain is synthesized by enzyme-catalyzed peptide bond formation.
    5. A 3-base "stop" codon on mRNA signals when synthesis is complete.

IV. DNA technology (Sections 16.10–16.11).
  A. Sanger DNA sequencing (Section 16.10).
    1. The DNA chain is cleaved at specific points by restriction endonucleases.
    2. The following mixture is assembled:
      a. The restriction fragment to be sequenced.
      b. A primer (a small piece of DNA whose sequence is complementary to that on the 3' end of the fragment).
      c. The 4 DNA nucleotide triphosphates.
      d. Small amounts of the four dideoxynucleotide triphosphates, each of which is labeled with a different fluorescent dye.
    3. DNA polymerase is added to the mixture, and a strand begins to grow from the end of the primer.
    4. Whenever a dideoxynucleotide is incorporated, chain growth stops.
    5. When reaction is complete, the fragments are separated by gel electrophoresis.
    6. Because fragments of all possible lengths are represented, the sequence can be read by noting the color of fluorescence of each fragment.
  B. Polymerase chain reaction (Section 16.11).
    1. The polymerase chain reaction (PCR) can produce vast quantities of a DNA fragment.
    2. The key to PCR is *Taq* DNA polymerase, a heat-stable enzyme.
    3. Steps in PCR:
      a. The following mixture is heated to 95 °C (a temperature at which DNA becomes single-stranded);
        i. *Taq* polymerase.
        ii. $Mg^{2+}$ ion.
        iii. The 4 deoxynucleotide triphosphates.
        iv. A large excess of two oligonucleotide primers, each of which is complementary to the ends of the fragment to be synthesized.
        v. Double-strand DNA to be amplified.
      b. The temperature is lowered to 37 °C–50 °C, causing the primers to hydrogen-bond to the single-stranded DNA.
      c. After raising the temperature to 72 °C, *Taq* catalyzes the addition of further nucleotides, yielding two copies of the original DNA.
      d. The process is repeated until the desired quantity of DNA is produced.

## Solutions to Problems

**16.1**

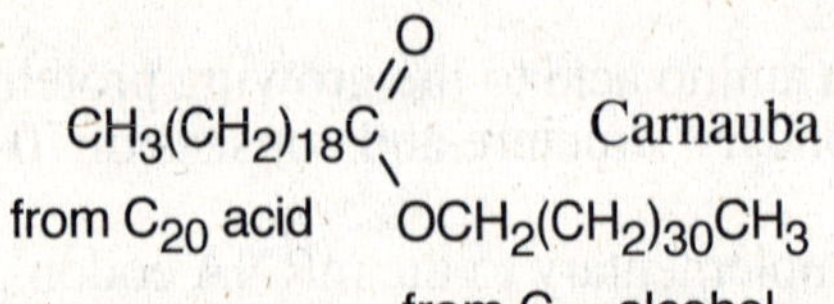


Carnauba wax

**16.2**

(a)

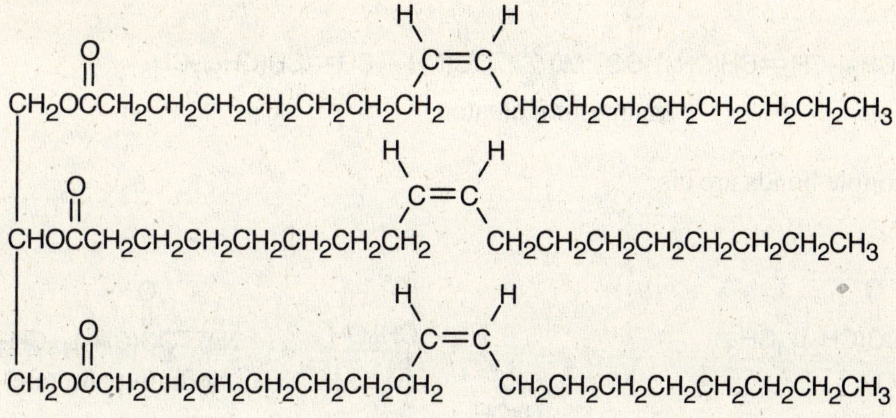

Glyceryl trioleate

(b)

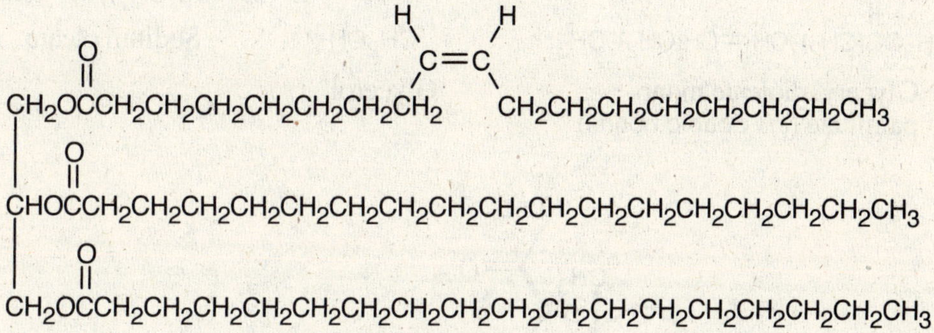

Glyceryl monooleate distearate (one isomer)

Glyceryl monooleate distearate is higher melting because it contains only one unsaturated fatty acid; glyceryl trioleate contains three. Notice that two stereoisomers of fat (b) can be drawn, one with oleic acid bonded to a terminal oxygen of glycerol and one with it bonded to the central oxygen.

**16.3**

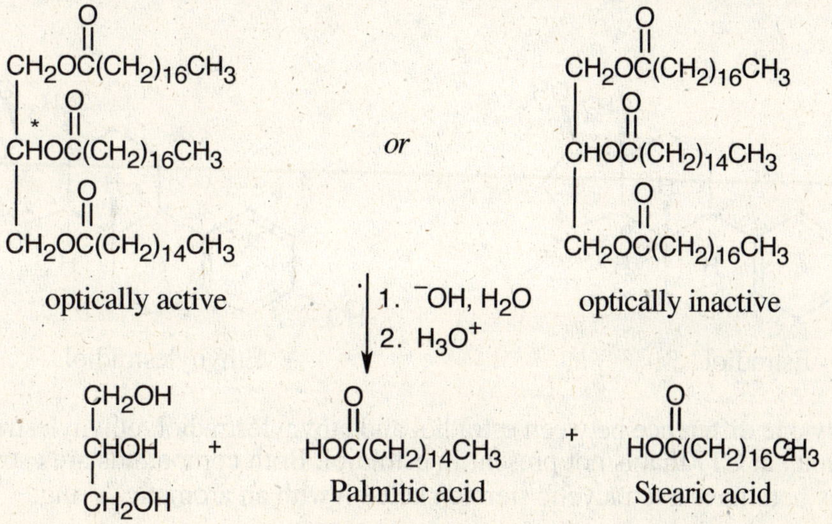


Four different groups are bonded to the central glycerol carbon atom in the optically active fat.

**16.4**

$$CH_3(CH_2)_7CH{=}CH(CH_2)_7\overset{O}{\overset{\|}{C}}O^- \; Mg^{2+} \; {}^-O\overset{O}{\overset{\|}{C}}(CH_2)_7CH{=}CH(CH_2)_7CH_3$$

Magnesium oleate

The double bonds are cis.

**16.5**

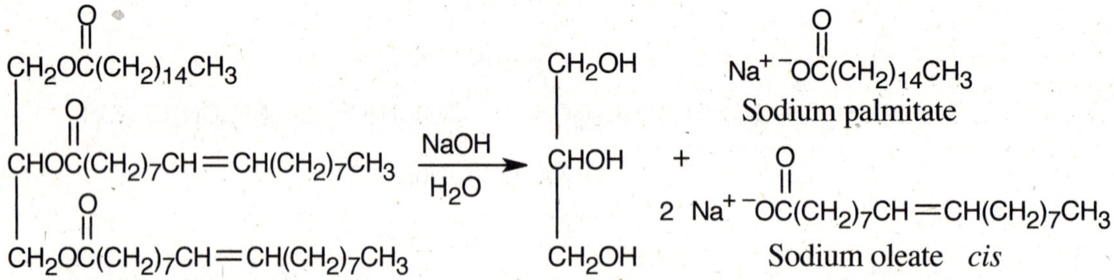


Glyceryl dioleate mono-palmitate (*cis* double bonds)

Glycerol

**16.6**

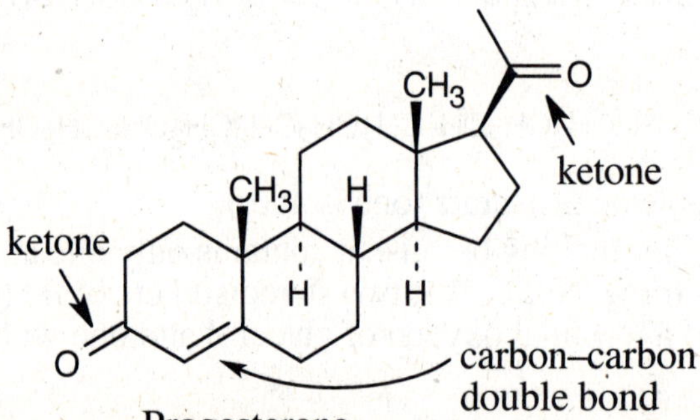


Progesterone

**16.7**

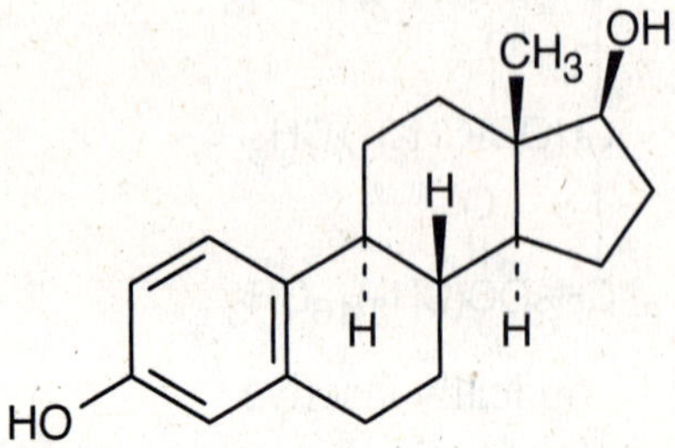

Estradiol

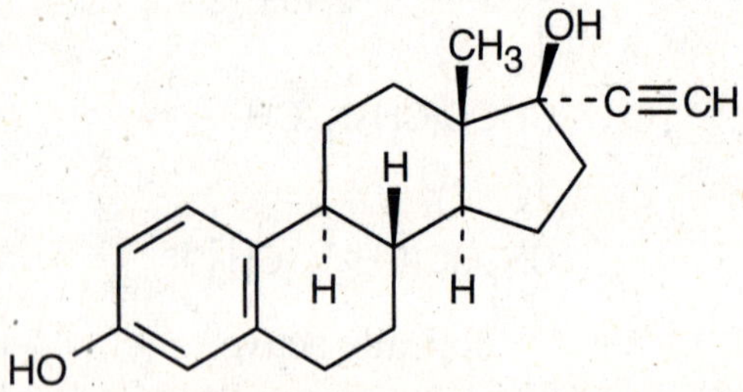

Ethynylestradiol

There is only one difference between estradiol and ethynylestradiol: ethynylestradiol has a –C≡C–H group at C17 that is not present in estradiol. Both compounds are estrogens because they both have a tetracyclic steroid skeleton with an aromatic **A** ring.

**16.8**

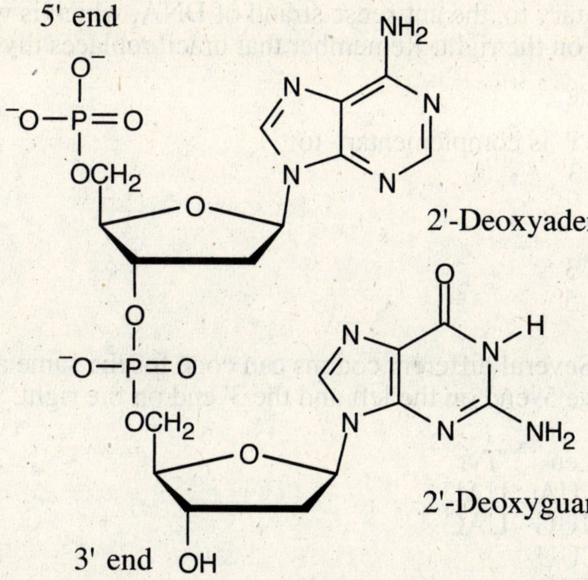


2'-Deoxyadenosine 5'-phosphate (A)

2'-Deoxyguanosine 5'-phosphate (G)

**16.9**

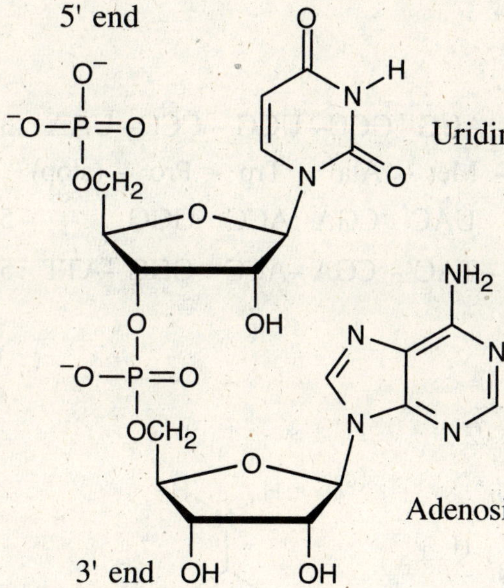


Uridine 5'-phosphate (U)

Adenosine 5'-phosphate (A)

**16.10** Original DNA: 5' GGCTAATCCGT 3'
Complement: 3' CCGATTAGGCA 5'

**16.11**

Uracil 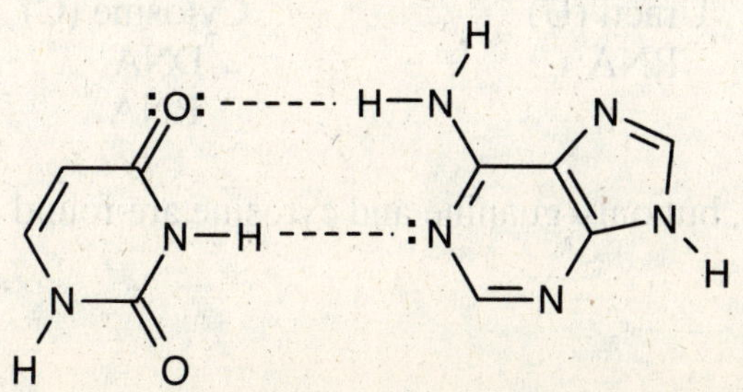 Adenine

**16.12** An mRNA strand is written with the 5' end on the left and the 3' end on the right. It is transcribed from, and is complementary to, the antisense strand of DNA, which is written with the 3' on the left and the 5' end on the right. Remember that uracil replaces thymine in RNA.

| | | | |
|---|---|---|---|
| DNA: | 3' GATTACCGTA | 5' | is complementary to: |
| RNA: | 5' CUAAUGGCAU | 3' | |

**16.13**

| | | |
|---|---|---|
| RNA: | 5' UUCGCAGAGU | 3' |
| DNA: | 3' AAGCGTCTCA | 5' |

**16.14** Use Table 16.3 to find the codons. Several different codons can code for the same amino acid. The codons are written with the 5' end on the left and the 3' end on the right.

| *Amino acid:* | Ala | Phe | Leu | Tyr |
|---|---|---|---|---|
| *Codon sequence:* | GCU | UUU | UUA | UAU |
| | GCC | UUC | UUG | UAC |
| | GCA | | CUU | |
| | GCG | | CUC | |
| | | | CUA | |
| | | | CUG | |

**16.15–16.17**

| | |
|---|---|
| The mRNA base sequence: | 5' CUU– AUG– GCU– UGG – CCC– UAA 3' |
| The amino acid sequence: | Leu – Met – Ala – Trp – Pro – (stop) |
| The tRNA sequences: | 3' GAA UAC CGA ACC GGG 5' |
| The DNA sequence: | 3' GAA – TAC – CGA –ACC – GGG –ATT 5' |

## Visualizing Chemistry

**16.18**

(a)

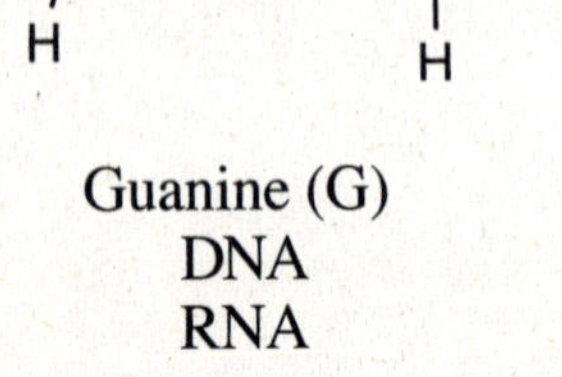

Guanine (G)
DNA
RNA

(b)

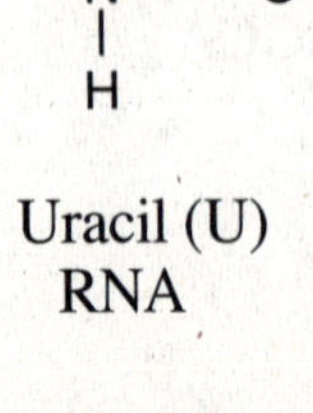

Uracil (U)
RNA

(c)

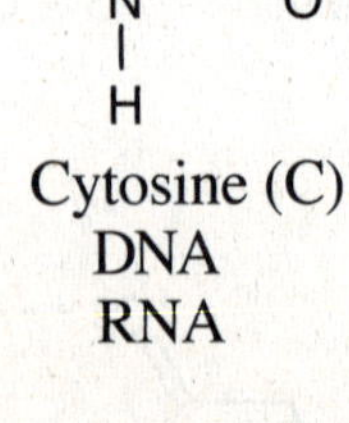

Cytosine (C)
DNA
RNA

All three bases are found in RNA, but only guanine and cytosine are found in DNA.

**16.19** The hydroxyl group in cholesterol is equatorial.

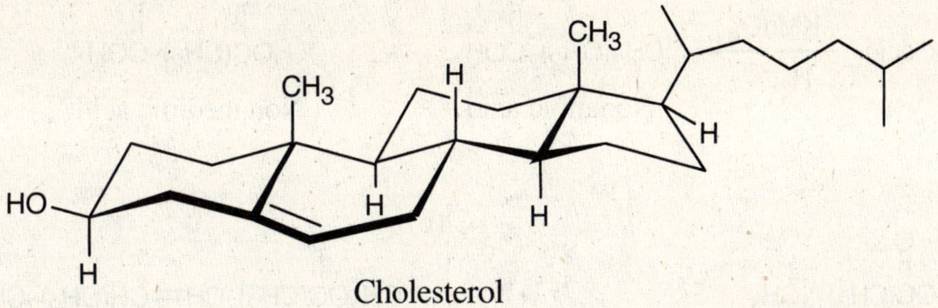

Cholesterol

**16.20**

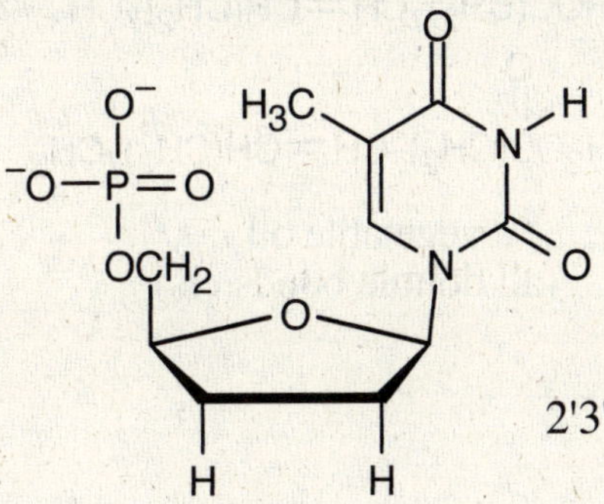

2'3'-Dideoxythymidine 5'-phosphate

The triphosphate made from 2'3'-dideoxythymidine 5'-phosphate is labeled with a fluorescent dye and used in the Sanger method of DNA sequencing. Along with the restriction fragment to be sequenced, a DNA primer, and a mixture of the four dNTPs (2'-deoxyribonucleoside triphosphates), small quantities of the four labeled dideoxyribonucleoside triphosphates are mixed together. DNA polymerase is added, and a strand of DNA complementary to the restriction fragment is synthesized. Whenever a dideoxyribonucleoside is incorporated into the DNA chain, chain growth stops. The fragments are separated by electrophoresis, and each terminal dideoxynucleotide can be identified by the color of its fluorescence. By identifying these terminal dideoxynucleotides, the sequence of the restriction fragment can be read.

## Additional Problems

### Lipids

**16.21**

$$CH_3(CH_2)_7CH{=}CH(CH_2)_7CO_2H \quad \text{(cis)}$$

Oleic acid

(a)

$$\text{Oleic acid} \xrightarrow[\text{H}^+ \text{ catalyst}]{CH_3OH} CH_3(CH_2)_7CH{=}CH(CH_2)_7CO_2CH_3$$

Methyl oleate

(b)

$$\text{Methyl oleate from (a)} \xrightarrow[\text{Pd catalyst}]{H_2} CH_3(CH_2)_{16}CO_2CH_3$$

Methyl stearate

(c)

$$\text{Oleic acid} \xrightarrow[\text{H}_3\text{O}^+]{\text{KMnO}_4} \underset{\text{Nonanoic acid}}{\text{CH}_3(\text{CH}_2)_7\overset{\text{O}}{\overset{\|}{\text{C}}}\text{OH}} + \underset{\text{Nonanedioic acid}}{\text{HO}\overset{\text{O}}{\overset{\|}{\text{C}}}(\text{CH}_2)_7\overset{\text{O}}{\overset{\|}{\text{C}}}\text{OH}}$$

**16.22**

(a)

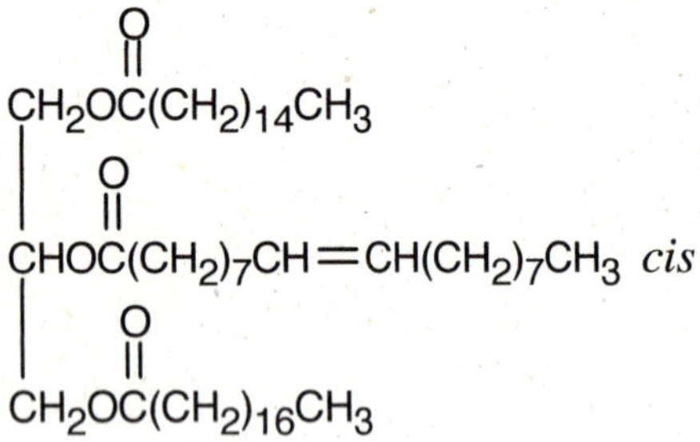

a fat

(b)

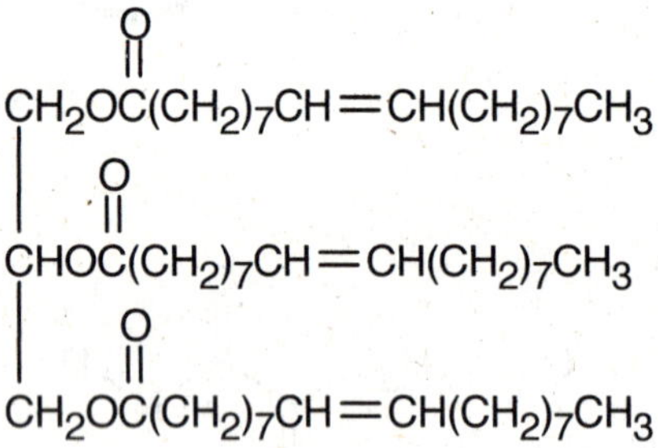

a vegetable oil
(all double bonds cis)

(c)

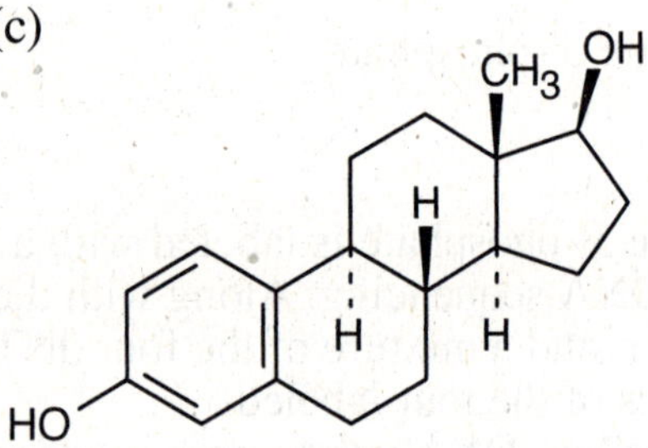

a steroid (Estradiol)

**16.23**

(a)

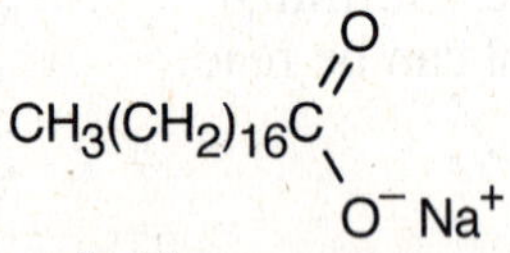

Sodium stearate

(b)

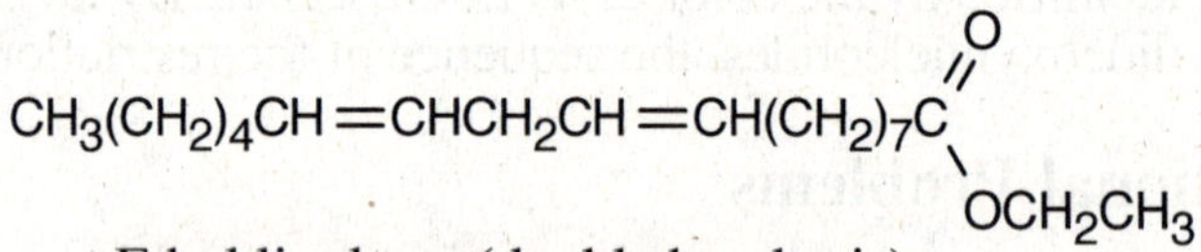

Ethyl linoleate (double bonds cis)

(c)

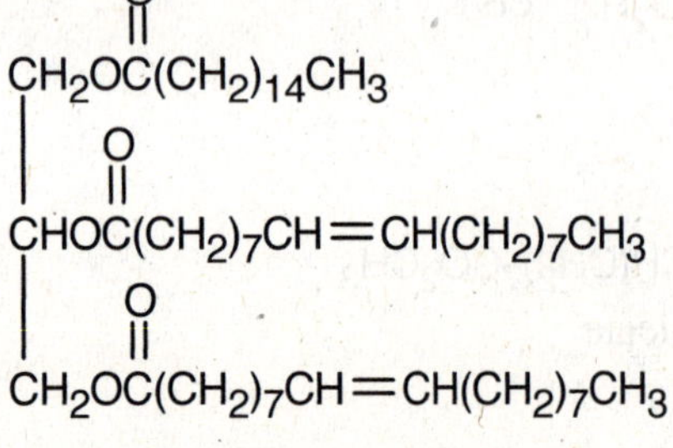

*or*

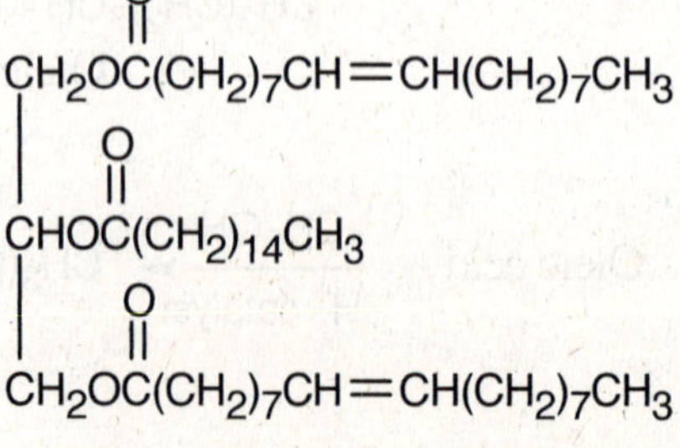

Glyceryl dioleopalmitate

**16.24**

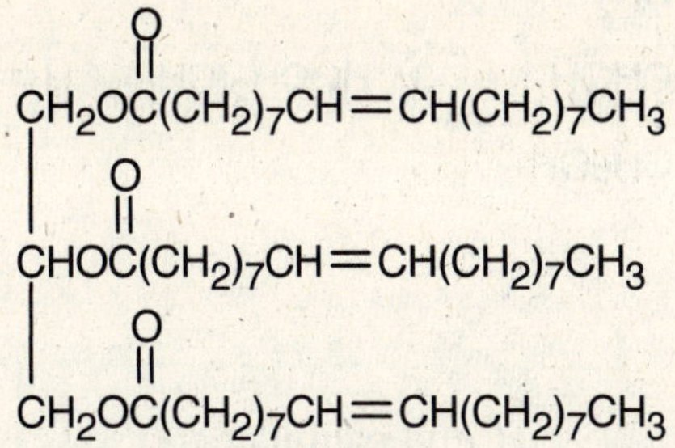


Glyceryl trioleate (double bonds cis)

(a)

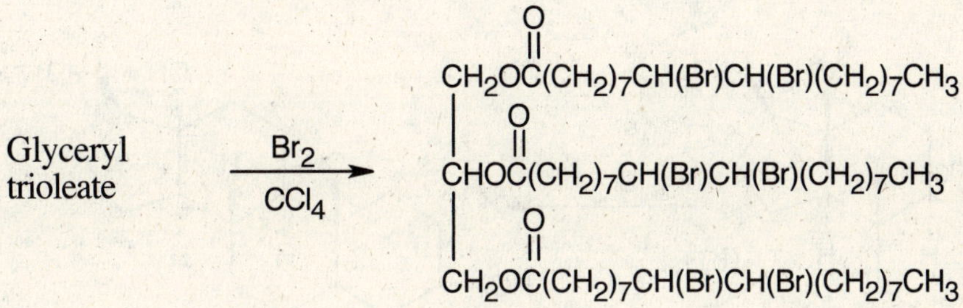


(b)

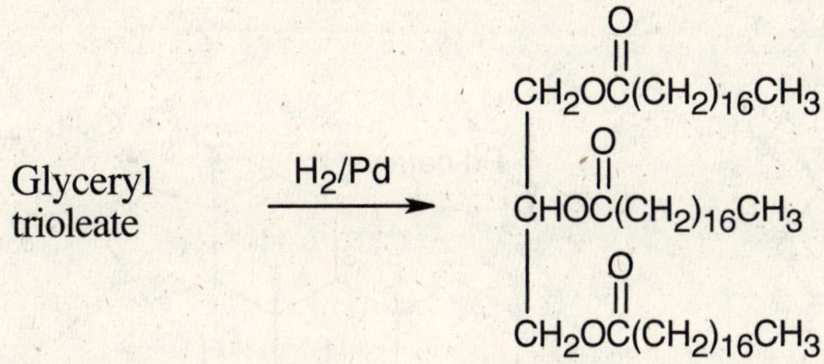


(c)

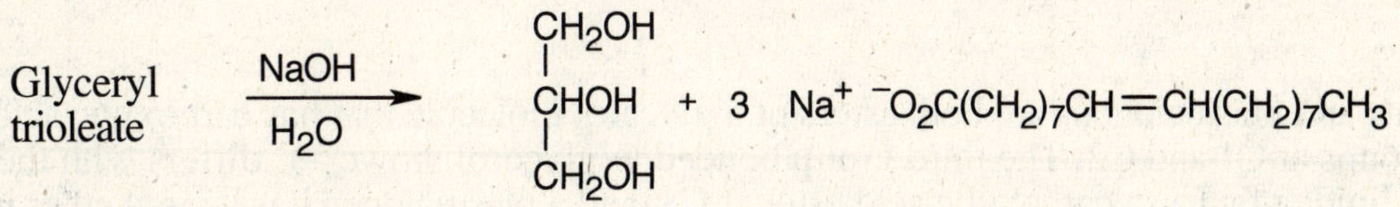


(d)

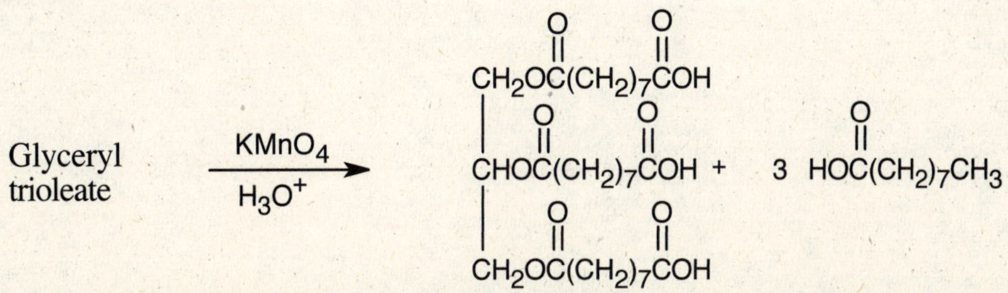

(e)

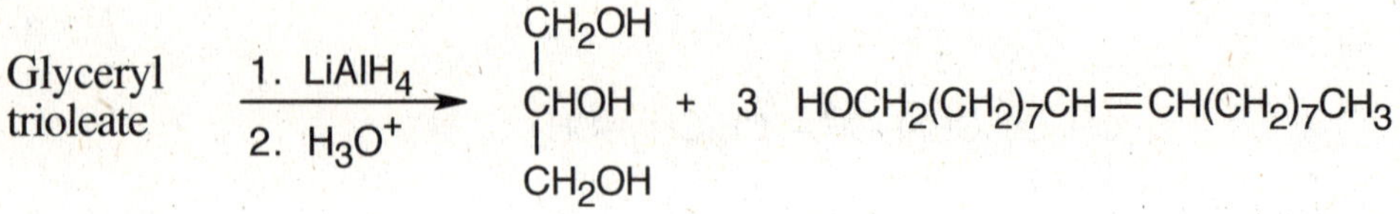


**16.25**

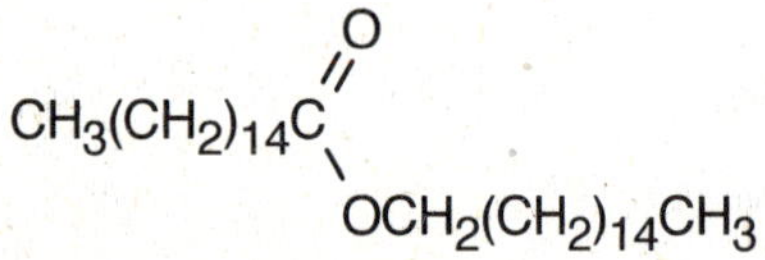


Cetyl palmitate

**16.26**

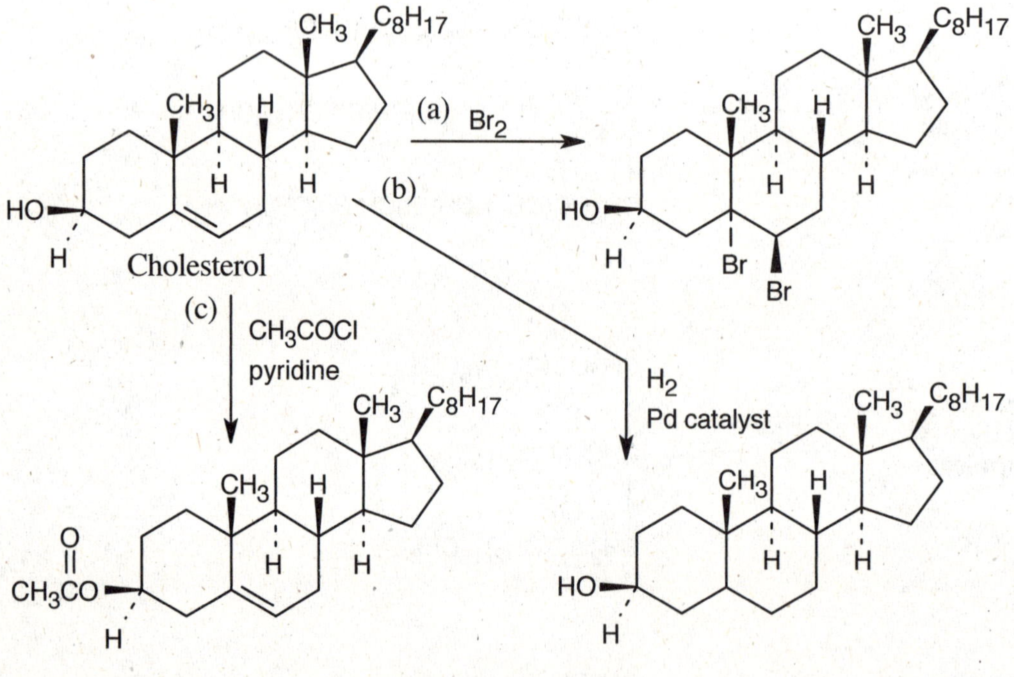


**16.27** Fats and plasmalogens are both esters of a glycerol molecule that has carboxylic acid ester groups at C1 and C2. The third group bonded to glycerol, however, differs with the type of lipid: a fat has a carboxylic acid ester at C3, and a plasmalogen has a vinyl ether in that location.

**16.28**

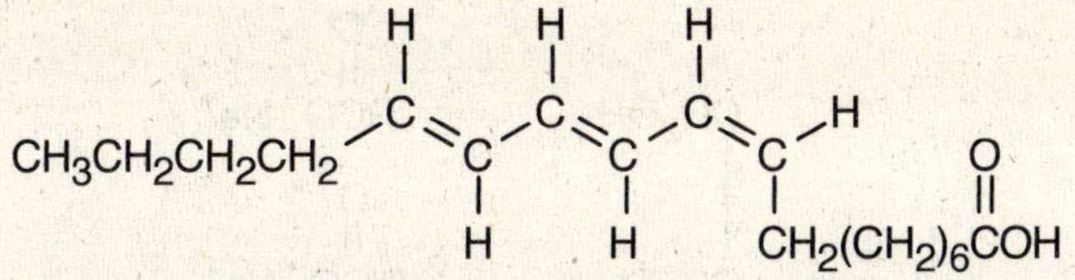

(9*Z*,11*E*,13*E*)-Octadeca-9,11,13-trienoic acid
(Eleostearic acid)

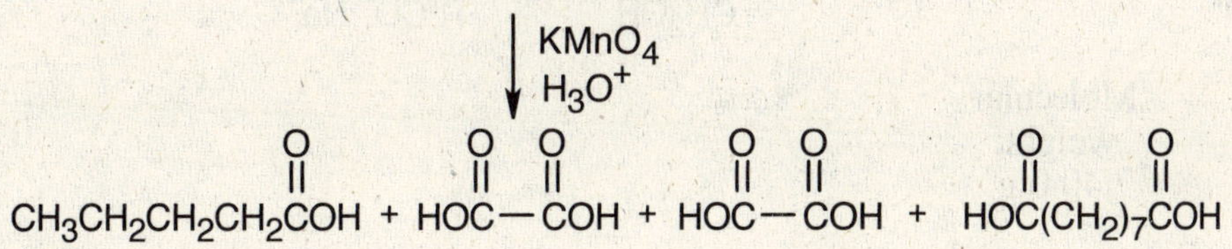

The stereochemistry of the double bonds can't be determined from the information given.

**16.29**

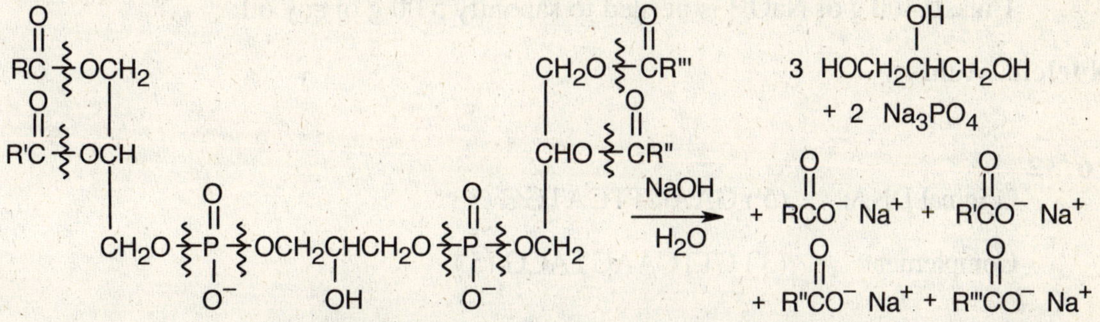

a cardiolipin

Saponification of a cardiolipin yields 4 different fatty acyl carboxylates, 3 equivalents of glycerol and two equivalents of phosphate.

**16.30**

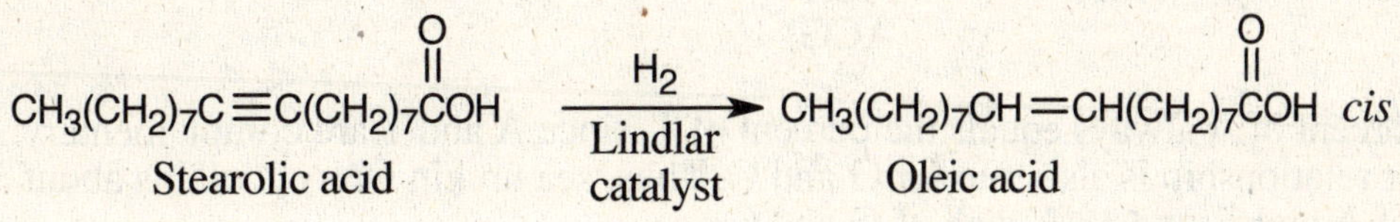

**16.31**

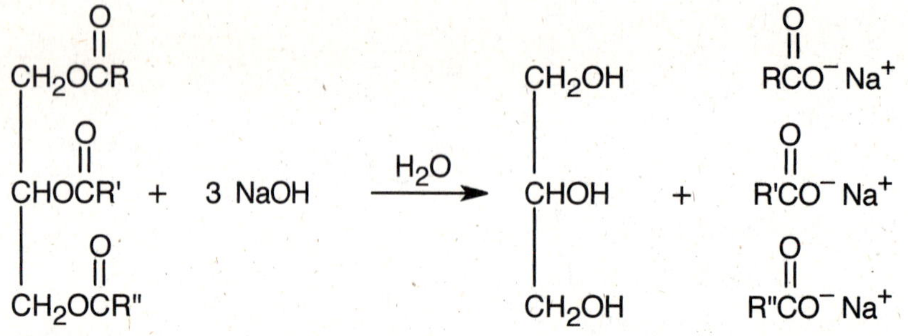

| Molecular weight: 1500 g | Molecular weight: 40.0 g |
|---|---|

$$5.00 \text{ g oil} \times \frac{1 \text{ mol oil}}{1500 \text{ g oil}} \times \frac{3 \text{ mol NaOH}}{1 \text{ mol oil}} \times \frac{40.0 \text{ g NaOH}}{1 \text{ mol NaOH}} = 0.400 \text{ g NaOH}$$

Thus, 0.400 g of NaOH is needed to saponify 5.00 g of soy oil.

## Nucleic Acids

**16.32**

Original DNA: (5') GAAGTTCATGC (3')

Complement: (3') CTTCAAGTACG (5')

**16.33** Table 16.3 gives the codons for each amino acid.

| Amino Acid: | Ile | Asp | Thr |
|---|---|---|---|
| Codon Sequence: (5' –> 3') | AUU<br>AUC<br>AUA | GAU<br>GAC | ACU<br>ACC<br>ACA<br>ACG |

**16.34** The percent of A always equals the percent of T, since A and T are complementary. A similar relationship is also true for G and C. Thus, sea urchin DNA contains about 32% each of A and T, and 18% each of G and C.

**16.35–16.37**

The mRNA codons are written 5' –> 3', and the DNA sequences and tRNA anticodons are written 3' –> 5'.

| mRNAcodon: | a) AAU | b) GAG | c) UCC | d) CAU | e) ACC |
|---|---|---|---|---|---|
| Amino acid | Asn | Glu | Ser | His | Thr |
| DNA sequence: | TTA | CTC | AGG | GTA | TGG |
| tRNA anticodon: | UUA | CUC | AGG | GUA | UGG |

**16.38–16.39** UAC is the codon for tyrosine. It was transcribed from ATG of the DNA chain.

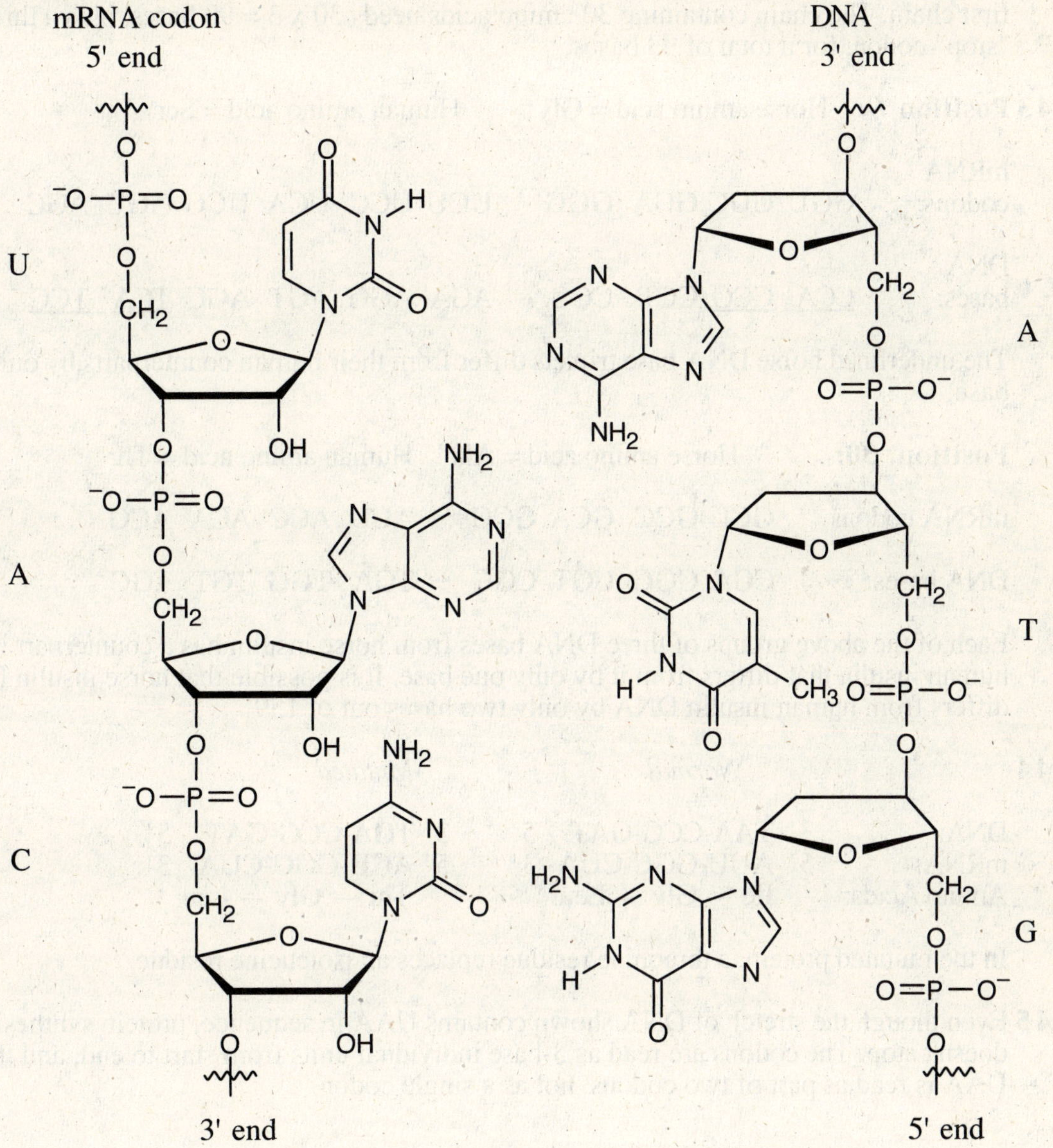


**16.40** Metenkephalin: Tyr — Gly — Gly — Phe — Met is coded by:

| | Tyr | Gly | Gly | Phe | Met | |
|---|---|---|---|---|---|---|
| mRNA: | UAU | GGU | GGU | UUU | AUG | UAA (stop) |
| (5' –> 3') | UAC | GGC | GGC | UUC | | UAG |
| | | GGA | GGA | | | UGA |
| | | GGG | GGG | | | |

**16.41** Use the first set of base pairs to solve this problem.

| | | |
|---|---|---|
| DNA: (antisense strand) | (3' –> 5') | ATA – CCA – CCA – AAA – TAC – ATT |
| DNA: (sense strand = gene) | (5' –> 3') | TAT – GGT – GGT – TTT – ATG – TAA |

**16.42** The chain containing 21 amino acids needs 21 x 3 = 63 bases to code for it. In addition, a three base codon is needed to terminate the chain. Thus, 66 bases are needed to code for the first chain. The chain containing 30 amino acids needs 30 x 3 = 90 bases, plus a three base "stop" codon, for a total of 93 bases.

**16.43 Position 9:** Horse amino acid = Gly Human amino acid = Ser

| | Horse | Human |
|---|---|---|
| mRNA codons: | GGU GGC GGA GGG | UCU UCC UCA UCG AGU AGC |
| DNA bases: | CCA CCG CCT CCC | AGA AGG AGT AGC TCA TCG |

The underlined horse DNA base triplets differ from their human counterparts by only one base.

**Position 30:** Horse amino acid = Ala Human amino acid = Thr

| | Horse | Human |
|---|---|---|
| mRNA codons | GCU GCC GCA GCG | ACU ACC ACA ACG |
| DNA bases: | CGA CGG CGT CGC | TGA TGG TGT TGC |

Each of the above groups of three DNA bases from horse insulin has a counterpart in human insulin that differs from it by only one base. It is possible that horse insulin DNA differs from human insulin DNA by only two bases out of 159!

**16.44**

| | *Normal* | *Mutated* |
|---|---|---|
| DNA: | 3' TAA-CCG-GAT 5' | 3' TGA-CCG-GAT 5' |
| mRNA: | 5' AUU-GGC-CUA 3' | 5' ACU-GGC-CUA 3' |
| Amino Acids: | Ile — Gly — Leu | Thr — Gly — Leu |

In the mutated protein, a threonine residue replaces an isoleucine residue.

**16.45** Even though the stretch of DNA shown contains UAA in sequence, protein synthesis doesn't stop. The codons are read as 3-base individual units from start to end, and the unit UAA is read as part of two codons, not as a single codon.

## General Problems

**16.46**

| Angiotensin II: | Asp | Arg | Val | Tyr | Ile | His | Pro | Phe | (stop) |
|---|---|---|---|---|---|---|---|---|---|
| mRNA sequence: (5'–3') | GAU | CGU | GUU | UAU | AUU | CAU | CCU | UUU | UAA |
| | GAC | CGC | GUC | UAC | AUC | CAC | CCC | UUC | UAG |
| | | CGA | GUA | | AUA | | CCA | | UGA |
| | | CGG | GUG | | | | CCG | | |
| | | AGA | | | | | | | |
| | | AGG | | | | | | | |

**16.47–16.48**

| | |
|---|---|
| mRNA sequence (5'–>3'): | CUA–GAC–CGU–UCC–AAG–UGA |
| Amino Acid: | Leu——Asp——Arg——Ser——Lys (stop) |
| tRNA anticodons (3'–>5'): | GAU CUG GCA AGG UUC |
| DNA sequence (3'–>5'): | GAT–CTG–GCA–AGG– TTC– ACT |
| DNA complement (5'–>3'): | CTA–GAC–CGT– TCC– AAG–TGA |

**16.49**

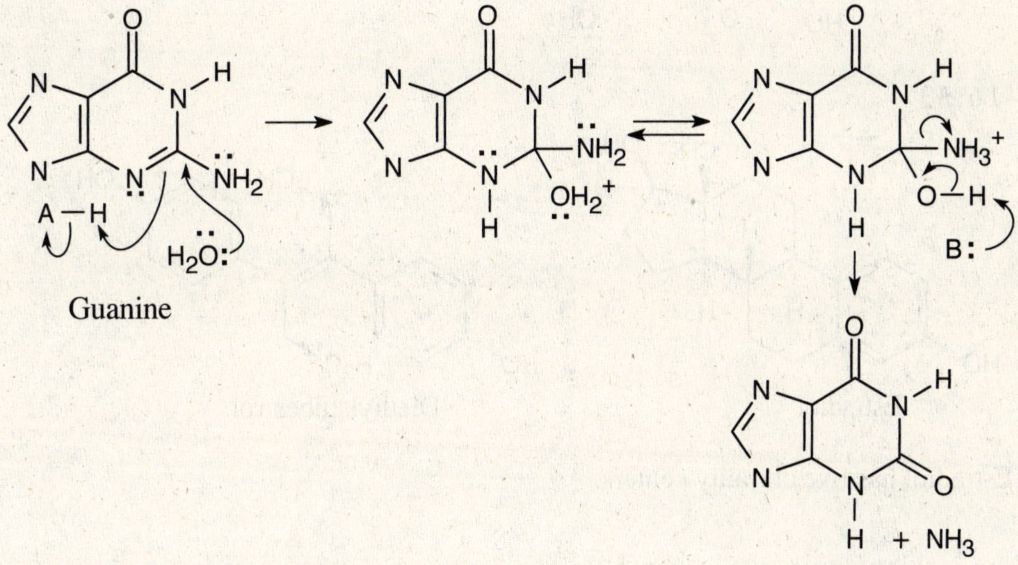


Guanine

Xanthine

**16.50**

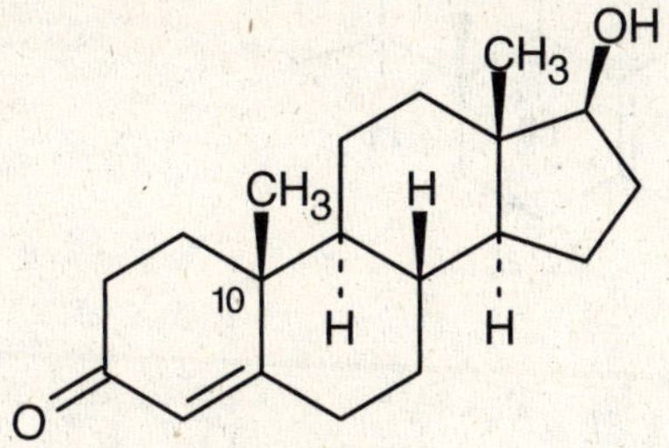


Testosterone

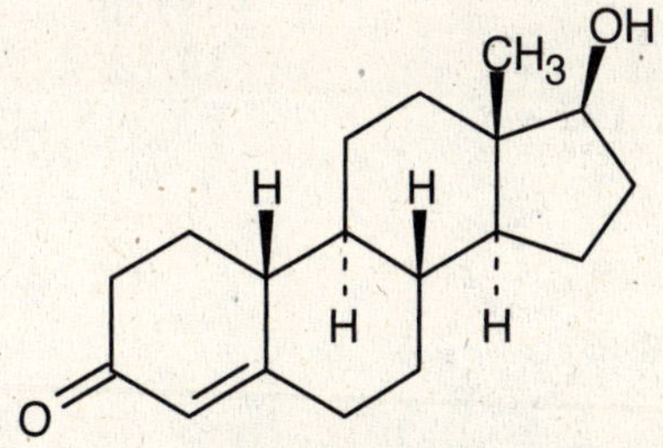


Nandrolone

Testosterone and nandrolone are identical, except for the methyl group at C10 of testosterone, which is replaced by a –H in nandrolone. Both steroids have the same carbon skeleton, both have the same enone group in the A ring, and both have a hydroxyl group at C17 of the D ring.

**16.51**

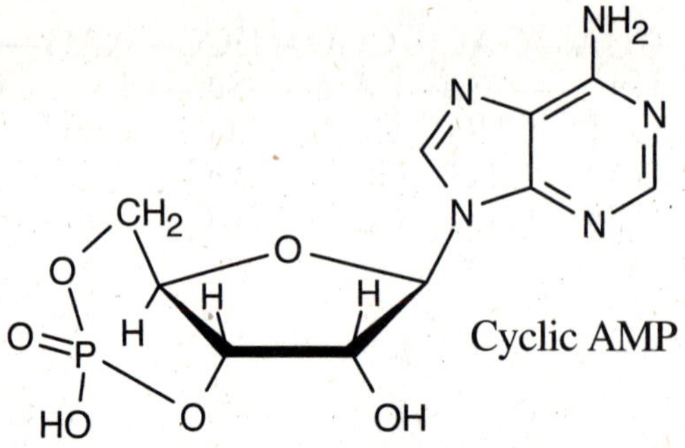


Cyclic AMP

**16.52–16.53**

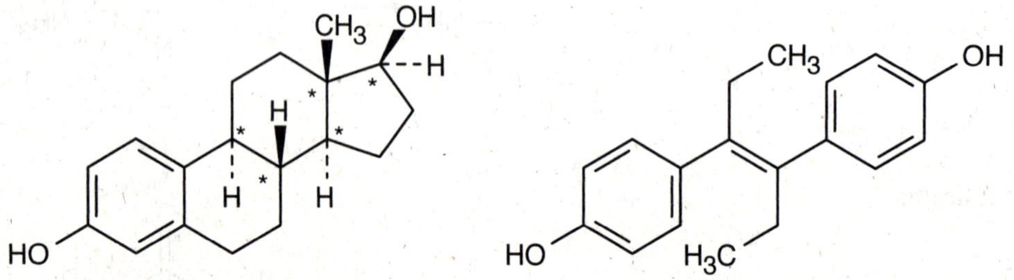


Estradiol

Diethylstilbestrol

Estradiol has five chirality centers.

**16.54**

(a)

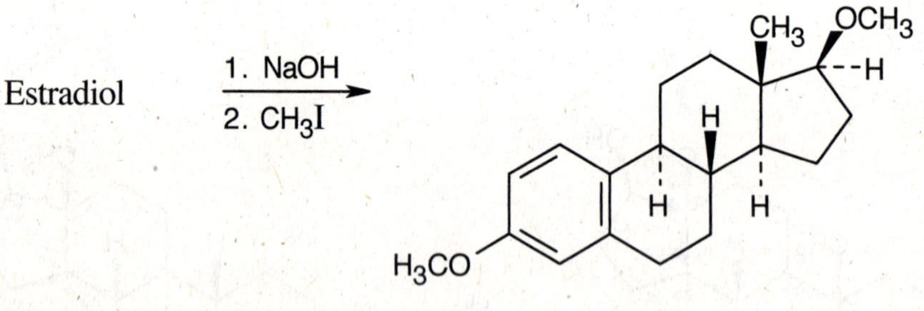


(b)

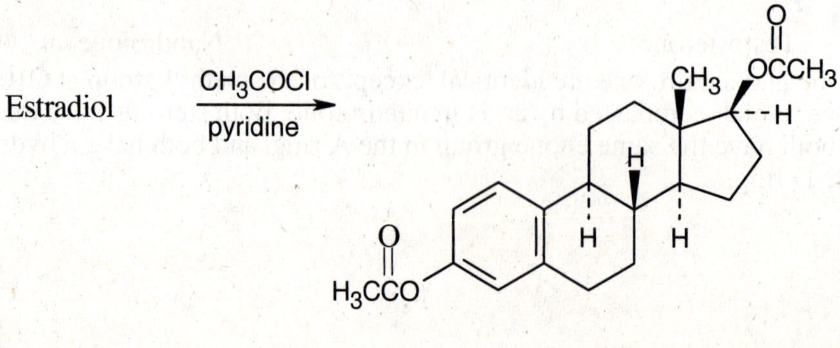

(c)

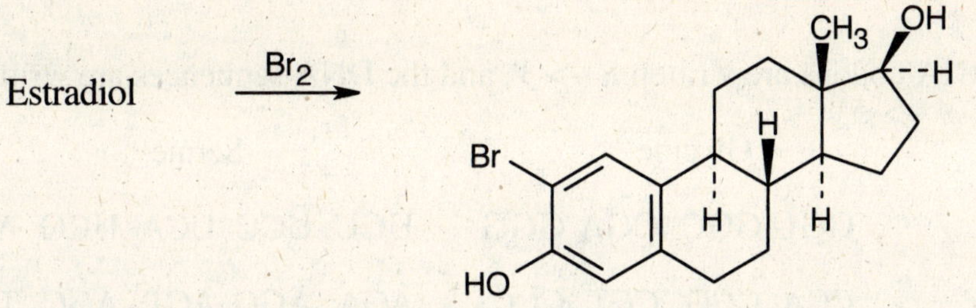


## In the Medicine Cabinet

**16.55**

(a), (b)

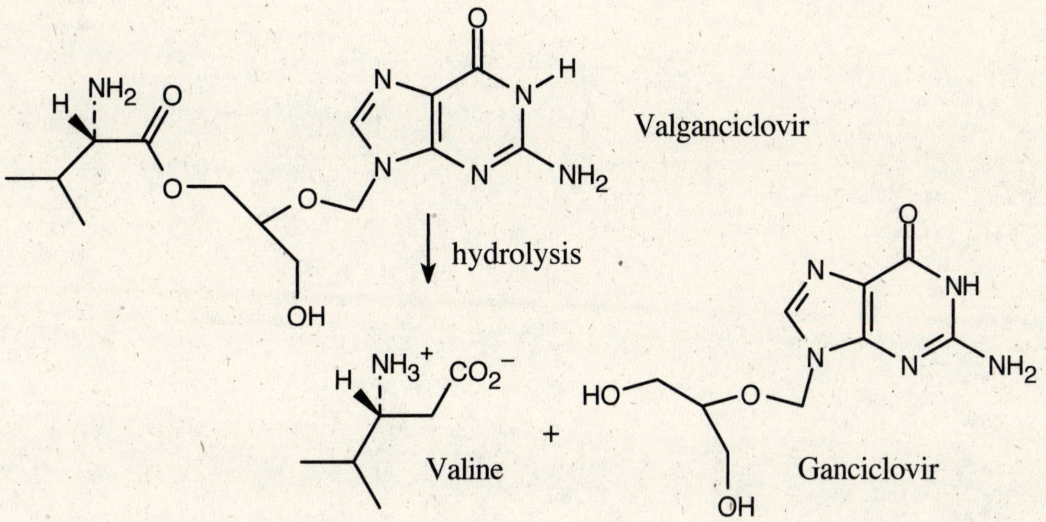


(c)

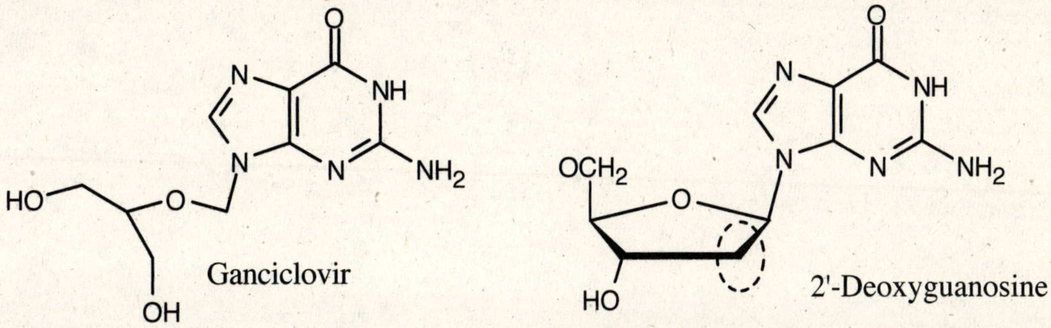


The –$CH_2$– group at C2 of deoxyribose is missing from ganciclovir.

(d) The missing atoms are part of the relatively inflexible deoxyribose ring. Without the ring, the DNA chain is floppy and base pairing to form a double helix can't occur.

(e) As mentioned in (d), the inability to form base pairs stops the replication of DNA.

## In the Field

**16.56** (a) The mRNA codons are written 5' –> 3', and the DNA sequences are written 3' –> 5'.

| | Glycine | Serine |
|---|---|---|
| mRNA codons: | GGU GGC GGA GGG | UCU UCC UCA UCG AGU AGC |
| DNA bases: | <u>CCA</u> <u>CCG</u> CCT CCC | AGA AGG ACT AGC <u>TCA</u> <u>TCG</u> |

(b) The underlined three-base codons for glycine differ from the underlined codons for serine by only one base. Thus, a mutation that converts thymine to cytosine can change an amino acid from serine to glycine.

# Chapter 17 – The Organic Chemistry of Metabolic Pathways

## Chapter Outline

I. Overview of metabolism and biochemical energy (Section 17.1).
  A. Metabolism.
    1. The reactions that take place in the cells of organisms are collectively called metabolism.
      a. The reactions that produce smaller molecules from larger molecules are called catabolism and usually release energy.
      b. The reactions that build larger molecules from smaller molecules are called anabolism and usually absorb energy.
    2. Catabolism can be divided into four stages:
      a. In digestion, bonds in food are hydrolyzed to yield sugars, fatty acids, and amino acids.
      b. These small molecules are degraded to acetyl CoA.
      c. In the citric acid cycle, acetyl CoA is catabolized to $CO_2$, and energy is produced.
      d. Energy from the citric acid cycle enters the electron transport chain, where ATP is synthesized.
  B. Biochemical energy.
    1. ATP, a phosphoric acid anhydride, is the storehouse for biochemical energy.
    2. The breaking of a P–O bond of ATP can be coupled with an energetically unfavorable reaction, so that the overall energy change is favorable.
    3. The resulting phosphates are much more reactive than the original compounds.

II. Catabolism (Sections 17.2–17.5).
  A. Catabolism of fats (Section 17.2).
    1. Triacylglycerols are first hydrolyzed in the stomach and small intestine to yield glycerol plus fatty acids.
      a. Glycerol is phosphorylated and enters glycolysis.
      b. Fatty acids are degraded by β-oxidation, a 4-step spiral that results in the cleavage of an *n*-carbon fatty acid into *n*/2 molecules of acetyl CoA.
      c. Before entering β-oxidation, a fatty acid is first converted to its fatty-acyl CoA.
    2. Steps of β-oxidation.
      a. Introduction of a double bond conjugated with the carbonyl group.
        i. The reaction is catalyzed by acyl CoA dehydrogenase.
        ii. The enzyme cofactor FAD is also involved and is reduced to $FADH_2$.
      b. Conjugate addition of water to form an alcohol.
        i. The reaction is catalyzed by enoyl CoA hydratase.
      c. Alcohol oxidation.
        i. The reaction is catalyzed by L-3-hydroxyacyl-CoA dehydrogenase.
        ii. The coenzyme $NAD^+$ is reduced to $NADH/H^+$ at the same time.
        iii. The mechanism of the reaction involves removal of hydride from the alcohol, followed by conjugate addition of hydride to $NAD^+$ to yield $NADH/H^+$.
      d. Cleavage of acetyl CoA from the chain.
        i. The reaction, which is catalyzed by β-ketoacyl CoA thiolase, is a retro-Claisen reaction.
        ii. Nucleophilic addition of coenzyme A to the keto group is followed by loss of acetyl CoA enolate, leaving behind a chain-shortened fatty-acyl CoA.

3. An *n*-carbon fatty acid yields *n*/2 molecules of acetyl CoA after (*n*/2 – 1) passages of β-oxidation.
   a. Since most fatty acids have an even number of carbons, no carbons are left over after β-oxidation, and all carbons enter the citric acid cycle.
   b. Those with an odd number of carbons require further steps for degradation.

B. Catabolism of carbohydrates (Sections 17.3–17.4).
1. Glycolysis (Section 17.3).
   a. Glycolysis is a 10-step series of reactions that converts glucose to pyruvate.
   b. Steps 1–3: Phosphorylation and isomerization.
      i. Glucose is phosphorylated at the 6 position by reaction with ATP.
         (a). The enzyme hexokinase is involved.
      ii. Glucose 6-P is isomerized to fructose 6-P by glucose-6-P isomerase.
      iii. Fructose 6-P is phosphorylated to yield fructose 1,6-bisphosphate.
         (a). ATP and phosphofructokinase are involved.
   c. Steps 4–5: Cleavage and isomerization.
      i. Fructose 1,6-bisphosphate is cleaved to glyceraldehyde 3-phosphate and dihydroxyacetone phosphate.
         (a). The reaction is a reverse aldol reaction catalyzed by aldolase.
      ii. Dihydroxyacetone phosphate is isomerized to glyceraldehyde 3-phosphate.
      iii. The net result is production of two glyceraldehyde 3-phosphates, both of which pass through the rest of the pathway.
   d. Steps 6–8: Oxidation and phosphorylation.
      i. Glyceraldehyde 3-phosphate is both oxidized and phosphorylated to give 1,3-bisphosphoglycerate.
         (a). Oxidation occurs via a hemithioacetal to yield a product that forms the mixed anhydride.
      ii. The mixed anhydride reacts with ADP to form ATP and 3-phosphoglycerate
         (a). The enzyme phosphoglycerate kinase is involved.
      iii. 3-Phosphoglycerate is isomerized to 2-phosphoglycerate by phosphoglycerate mutase.
   e. Steps 9–10: Dehydration and dephosphorylation.
      i. 2-Phosphoglycerate is dehydrated by enolase to give phosphoenolpyruvate.
      ii. Pyruvate kinase catalyzes the transfer of a phosphate group to ADP, with formation of pyruvate.
   f. Pyruvate can be converted to acetyl CoA or may enter other pathways.
2. The citric acid cycle (Section 17.4).
   a. Characteristics of the citric acid cycle.
      i. The citric acid cycle is a closed loop of eight sequential reactions.
      ii. The intermediates are constantly regenerated.
      iii. The cycle operates as long as $NAD^+$ and FAD are available, which means that oxygen must also be available.
   b. Steps 1–2: Addition to oxaloacetate.
      i. Acetyl CoA adds to oxaloacetate to form citryl CoA, which is hydrolyzed to citrate.
         (a). The reaction is catalyzed by citrate synthase.
      ii. Citrate is isomerized to isocitrate by aconitase.
         (a). The reaction is an E2 dehydration, followed by conjugate addition of water.
   c. Steps 3–4: Oxidation and decarboxylations.
      i. Isocitrate is oxidized by isocitrate dehydrogenase to give a ketone that loses $CO_2$ to give α-ketoglutarate.
      ii. α-Ketoglutarate is transformed to succinyl CoA in a reaction catalyzed by an enzyme complex.

d. Steps 5–6: Hydrolysis and dehydrogenation of succinyl CoA.
   i. Succinyl CoA is converted to an acyl phosphate, which transfers a phosphate group to GDP in a reaction catalyzed by succinyl CoA synthetase.
   ii. Succinate is dehydrogenated by FAD and succinate dehydrogenase to give fumarate.

e. Steps 7–8: Regeneration of oxaloacetate.
   i. Fumarase catalyzes the addition of water to fumarate to produce L-malate.
   ii. L-Malate is oxidized by $NAD^+$ and malate dehydrogenase to complete the cycle.

C. Catabolism of proteins: Transamination (Section 17.5).
1. The pathway to amino acid catabolism:
   a. The amino group is removed by transamination.
   b. What remains is converted to a compound that enters the citric acid cycle.
2. Transamination.
   a. The $-NH_2$ group of an amino acid adds to the aldehyde group of pyridoxal phosphate to form an imine (Schiff base).
   b. The imine rearranges to a different imine.
   c. The second imine is hydrolyzed to give an $\alpha$-keto acid and an amino derivative of pyridoxal phosphate.
   d. The pyridoxal derivative transfers its amino group to $\alpha$-ketoglutarate, to regenerate pyridoxal phosphate and form glutamate.
3. Deamination.
   a. The glutamate from transamination undergoes oxidative deamination to yield ammonium ion and $\alpha$-ketoglutarate.

III. A summary (Section 17.6).
A. The mechanisms of biochemical reactions are almost identical to the mechanisms of laboratory reactions.

## Solutions to Problems

**17.1**

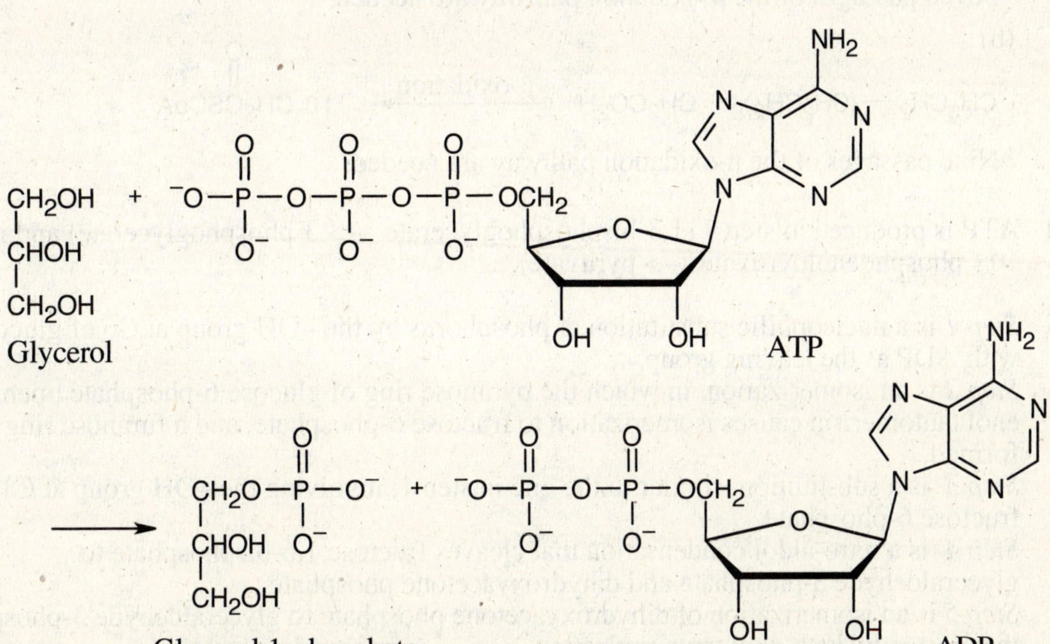

**17.2**

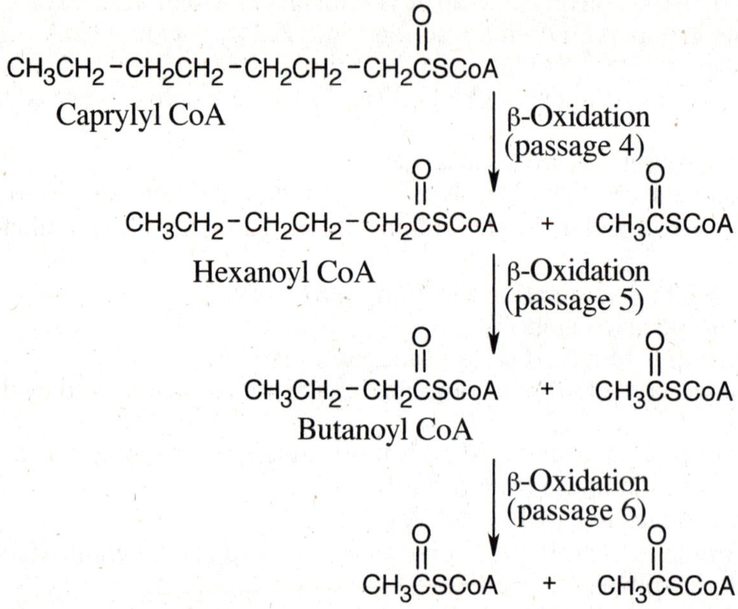

The final step produces two acetyl CoAs.

**17.3**

(a)

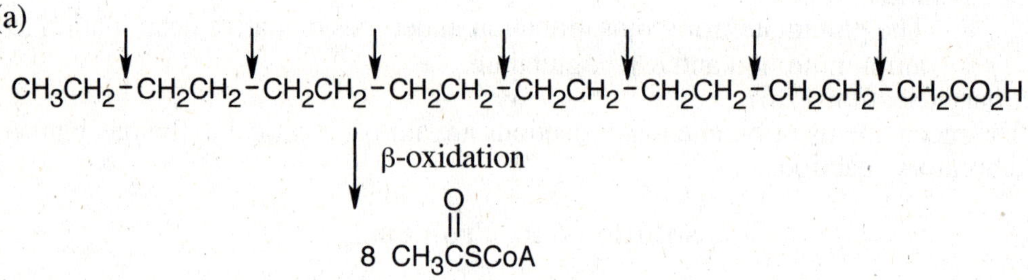

Seven passages of the β-oxidation pathway are needed.

(b)

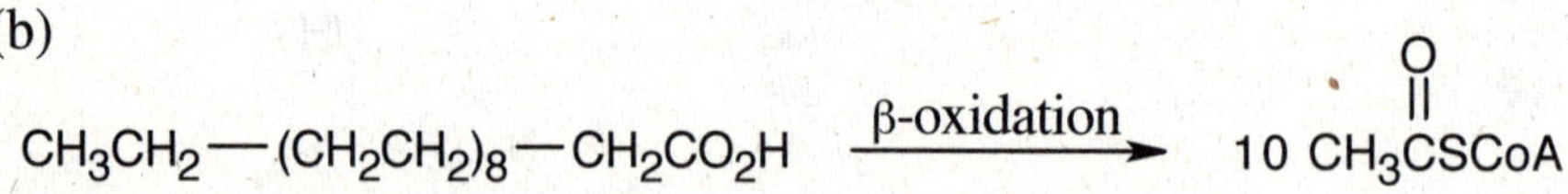

Nine passages of the β-oxidation pathway are needed.

**17.4** ATP is produced in step 7 (1,3-bisphosphoglycerate —> 3-phosphoglycerate) and in step 10 ( phosphoenolpyruvate —> pyruvate).

**17.5** *Step 1* is a nucleophilic substitution at phosphorus by the –OH group at C6 of glucose, with ADP as the leaving group.
*Step 2* is an isomerization, in which the pyranose ring of glucose 6-phosphate opens, keto-enol tautomerism causes isomerization to fructose 6-phosphate, and a furanose ring is formed.
*Step 3* is a substitution, similar to the one in step 1, involving the –OH group at C1 of fructose 6-phosphate.
*Step 4* is a retro-aldol condensation that cleaves fructose 1,6-bisphosphate to glyceraldehyde 3-phosphate and dihydroxyacetone phosphate.
*Step 5* is an isomerization of dihydroxyacetone phosphate to glyceraldehyde 3-phosphate that occurs by keto-enol tautomerization.

In *Step 6*, the aldehyde group of glyceraldehyde 3-phosphate is oxidized by $NAD^+$, followed by a nucleophilic acyl substitution by phosphate, to yield the product 1,3-bisphosphoglycerate.
*Step 7* is a nucleophilic acyl substitution reaction at phosphorus, yielding ATP and 3-phosphoglycerate.
*Step 8* is an isomerization of 3-phosphoglycerate to 2-phosphoglycerate.
*Step 9* is an E1cB elimination of $H_2O$ to form phosphoenolpyruvate.
*Step 10* is a substitution reaction at phosphorus that forms ATP and displaces pyruvate enolate, which tautomerizes to pyruvate.

**17.6** Citrate and isocitrate are tricarboxylic acids.

**17.7**

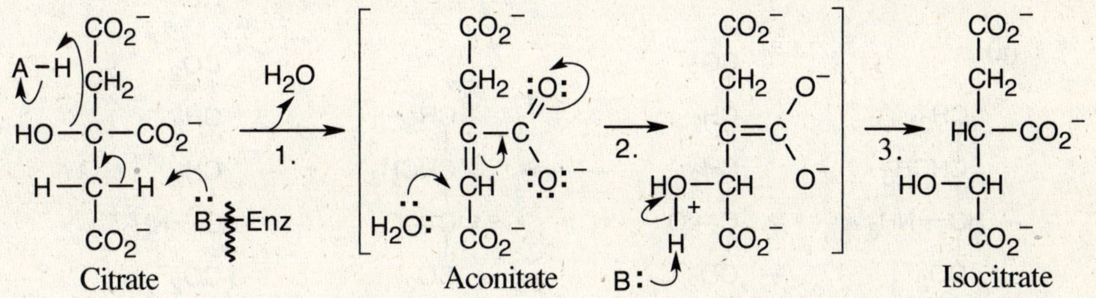

Enzyme-catalyzed E1cB elimination of $H_2O$ (1) is followed by nucleophilic conjugate addition of water (2) to produce an adduct that isomerizes (3) to isocitrate.

**17.8** Position leucine and α-ketoglutarate so that the groups to be exchanged are aligned. This arrangement makes it easy to predict the products of transamination reactions.

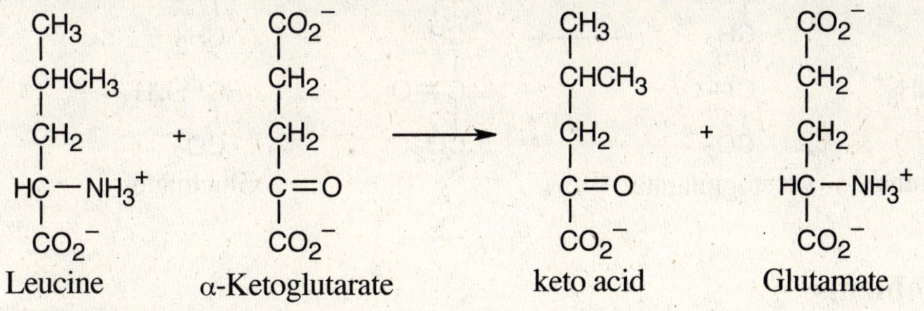

**17.9** Look for the keto group that replaced the amino group. Then try to identify the original amino acid by recognizing the side chain. The α-keto acid was formed from asparagine.

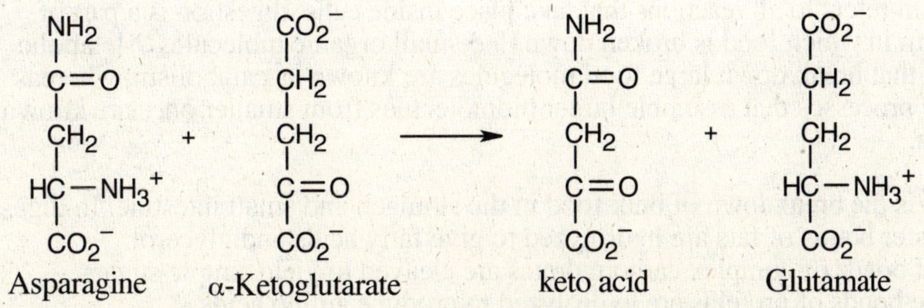

## Visualizing Chemistry

**17.10** The intermediate is *S*-malate.

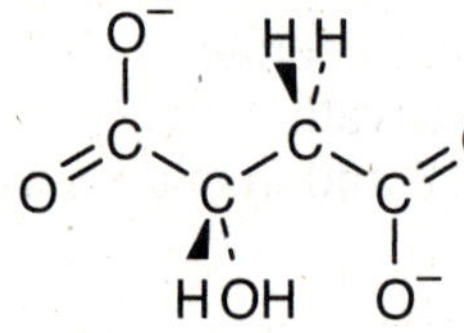

*S*-Malate

**17.11** The amino acid precursors are valine (a) and methionine (b).

(a)

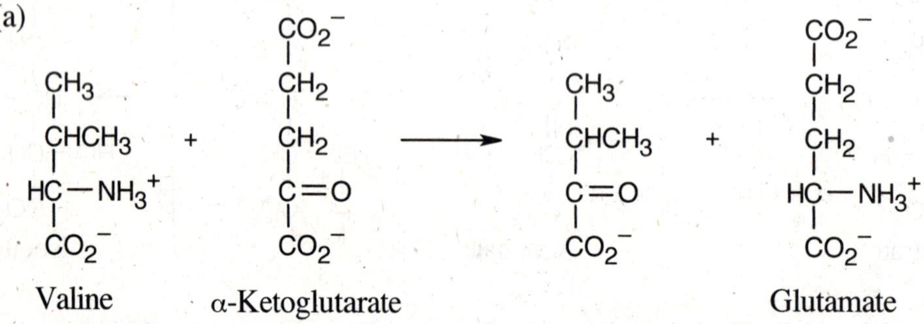

(b)

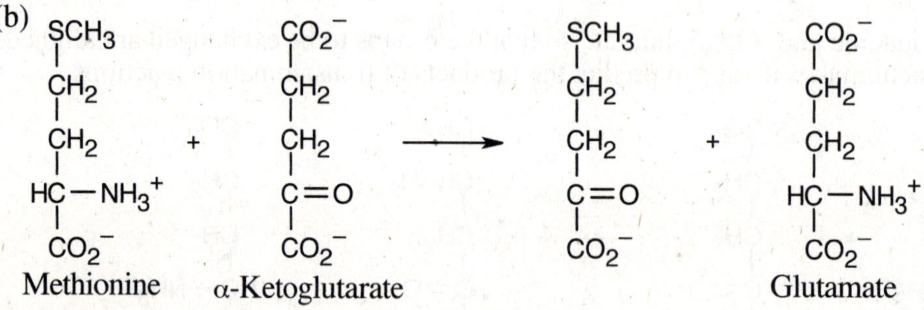

## Additional Problems

### General Metabolism

**17.12** Metabolism refers to all reactions that take place inside cells; digestion is a part of metabolism in which food is broken down into small organic molecules. Metabolic processes that break down large food molecules are known as catabolism, whereas metabolic processes that assemble larger biomolecules from smaller ones are known as anabolism.

**17.13** Digestion is the breakdown of bulk food in the stomach and small intestine. In digestion:
(a) The ester bonds of fats are hydrolyzed to give fatty acids and glycerol.
(b) Acetal bonds of complex carbohydrates are cleaved to yield simple sugars.
(c) Amide bonds of proteins are hydrolyzed to produce amino acids.

**17.14** Problem 17.18 gives the number of moles of acetyl CoA produced for each substance.

| | (a) Glucose | (b) Palmitic acid | (c) Maltose |
|---|---|---|---|
| Molecular weight | 180.2 amu | 256.4 amu | 342.3 amu |
| Moles in 100.0 g | 0.5549 mol | 0.3900 mol | 0.2921 mol |
| Moles of acetyl CoA produced | 2 x 0.5549 mol =1.110 mol | 8 x 0.3900 mol = 3.120 mol | 4 x 0.2921 mol = 1.168 mol |
| Grams acetyl CoA produced | 898.6 grams | 2526 grams | 945.6 grams |

**17.15** Palmitic acid is the most efficient precursor of acetyl CoA on a weight basis.

**17.16**

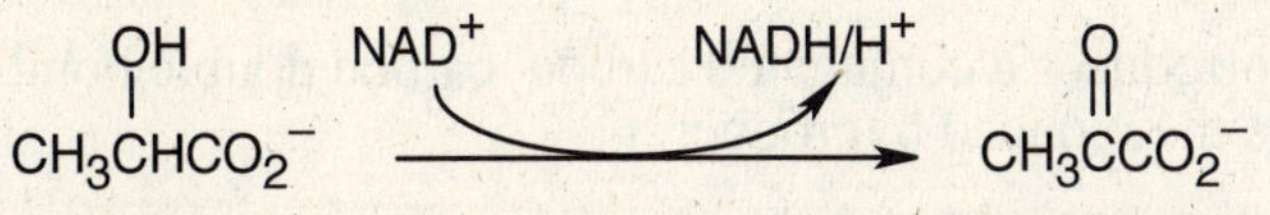


Lactate Pyruvate

$NAD^+$ is needed to convert lactate to pyruvate because the reaction involves the oxidation of an alcohol.

**17.17**

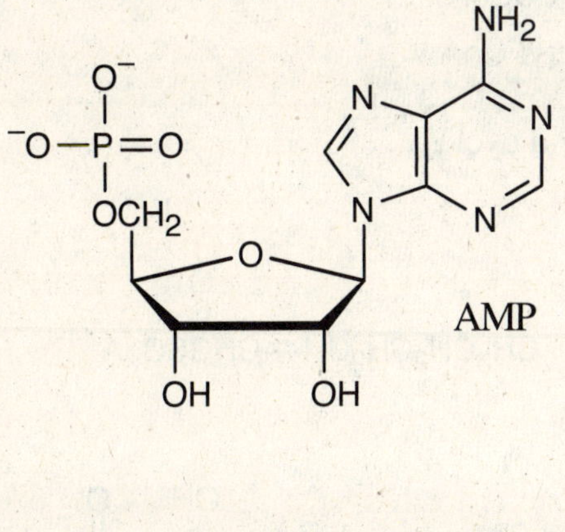


AMP

**17.18** (a) One mole of glucose is catabolized to two moles of pyruvate, each of which yields one mole of acetyl CoA. Thus,

1.0 mol glucose —> 2.0 mol acetyl CoA

(b) A fatty acid with *n* carbons yields *n*/2 moles of acetyl CoA per mole of fatty acid. For palmitic acid ($C_{15}H_{31}CO_2H$),

$$1.0 \text{ mol palmitic acid} \times \frac{8 \text{ mol acetyl CoA}}{1 \text{ mol palmitic acid}} \longrightarrow 8.0 \text{ mol acetyl CoA}$$

(c) Maltose is a disaccharide that yields two moles of glucose on hydrolysis. Since each mole of glucose yields two moles of acetyl CoA,

1.0 mol maltose —> 2.0 mol glucose —> 4.0 mol acetyl CoA

**17.19** ATP is involved with phosphate transfer reactions. ATP is synthesized in catabolic reactions, and transfers a phosphate group to another molecule in anabolic reactions.

**17.21** $NAD^+$ is a biochemical oxidizing agent that converts alcohols to aldehydes or ketones, yielding $NADH/H^+$ as the reduced byproduct.

**17.20** FAD is an oxidizing agent that introduces a conjugated carbon–carbon double bond into a biomolecule, yielding $FADH_2$ as the reduced byproduct.

## Metabolic Pathways

**17.22**

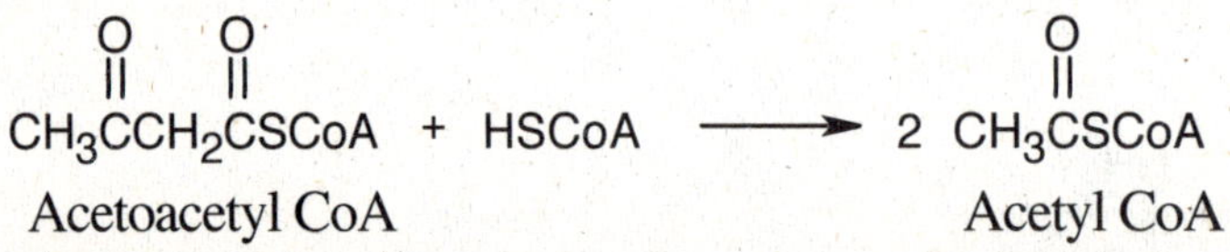


**17.23** Oxaloacetate is the starting point for the citric acid cycle.

**17.24**

(a)

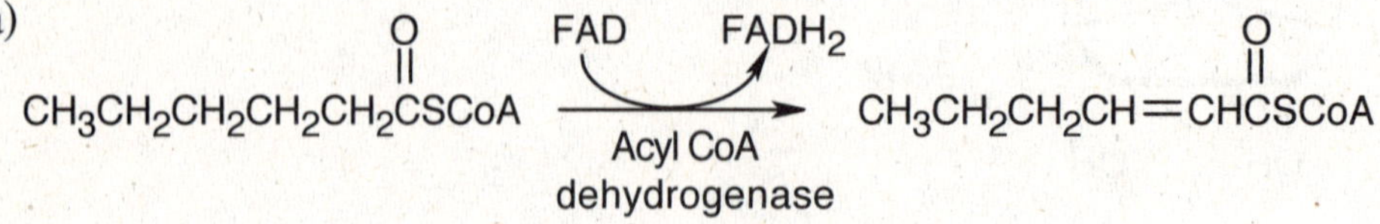


(b)

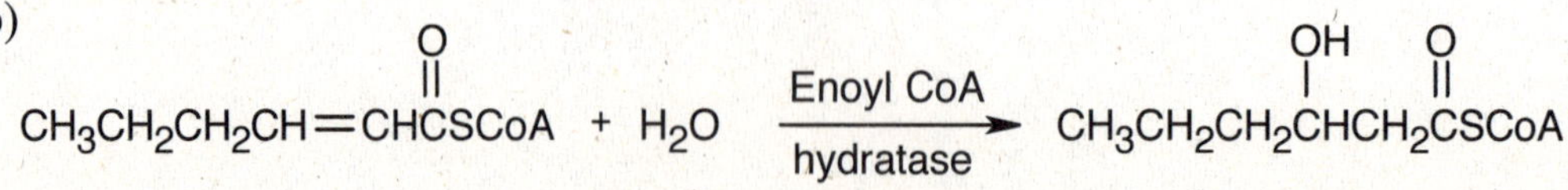


(c)

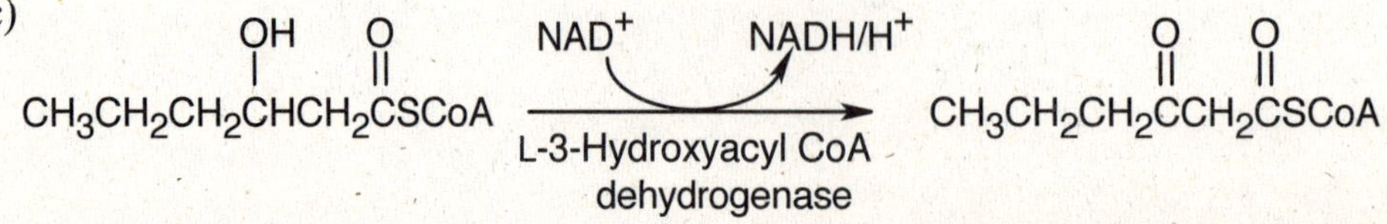

**17.25**

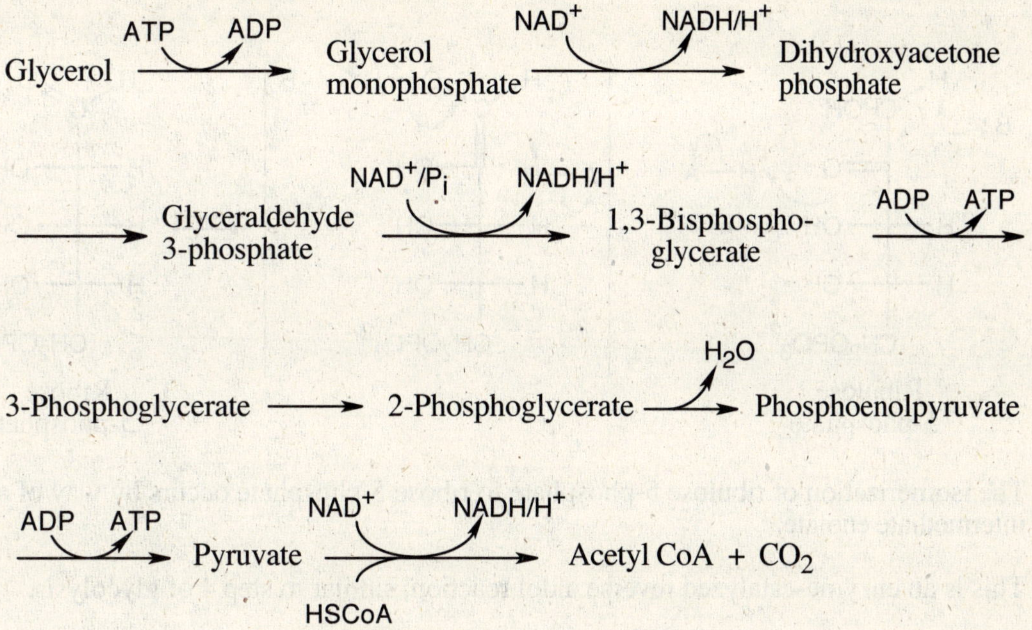


**17.26** Pyridoxal phosphate is the cofactor involved in transamination.

**17.27**

*Amino acid* | α-*Keto acid*

(a)

(b)

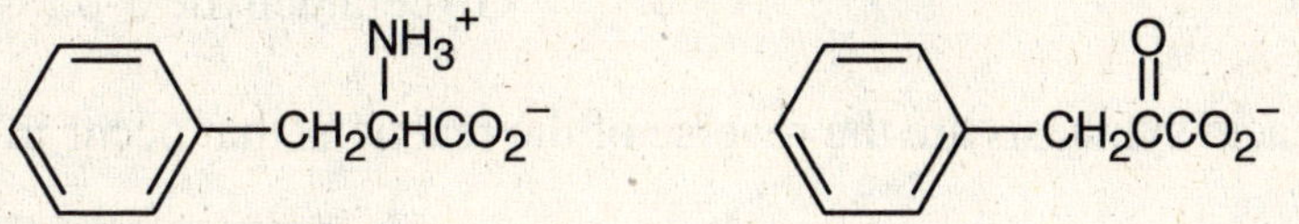


(c)

## General Problems

**17.28**

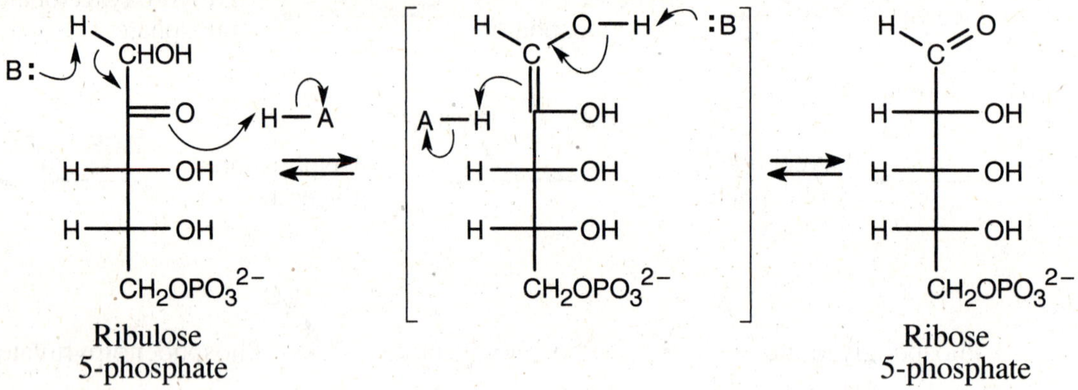


The isomerization of ribulose 5-phosphate to ribose 5-phosphate occurs by way of an intermediate enolate.

**17.29** This is an enzyme-catalyzed reverse aldol reaction, similar to step 4 of glycolysis.

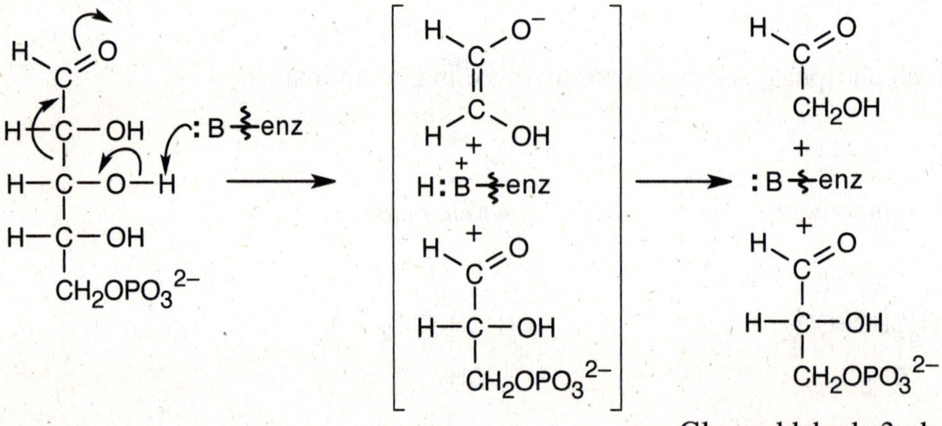

Glyceraldehyde 3-phosphate

**17.30** The reactions of fatty acid synthesis are the reverse of the reactions that occur in the β-oxidation pathway.

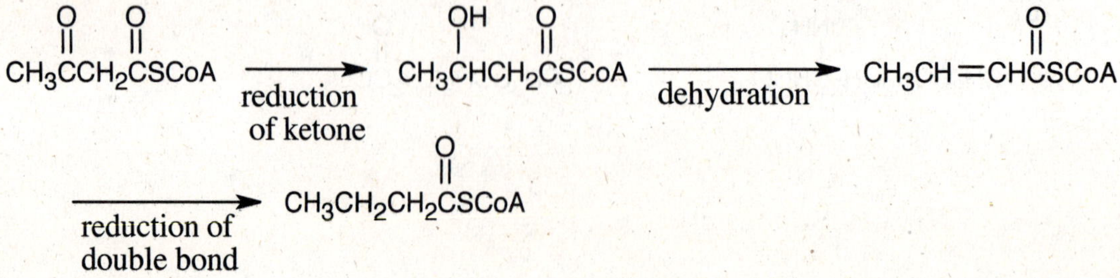

**17.31** The conversion of pyruvate to oxaloacetate is a mixed aldol condensation reaction between $CO_2$ and pyruvate.

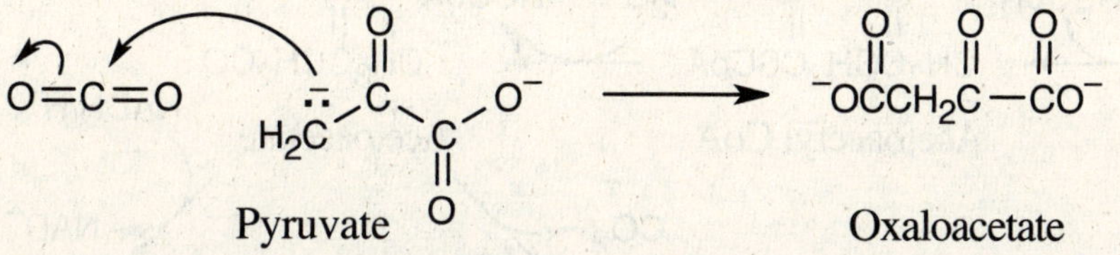


**17.32** Formation of phosphoenolpyruvate occurs by way of a reverse aldol reaction, followed by phosphorylation of the enolate anion.

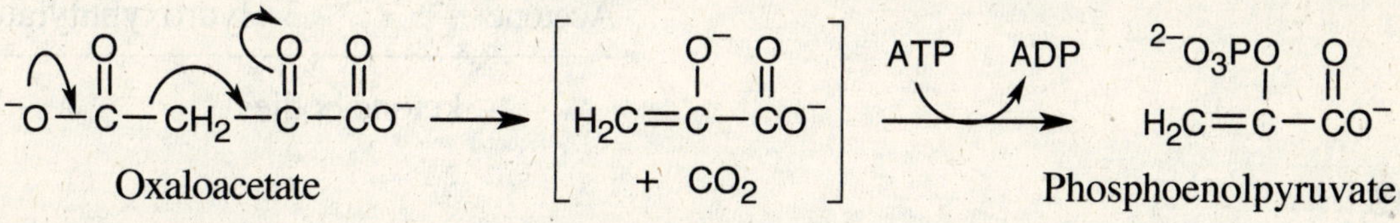


**17.33**

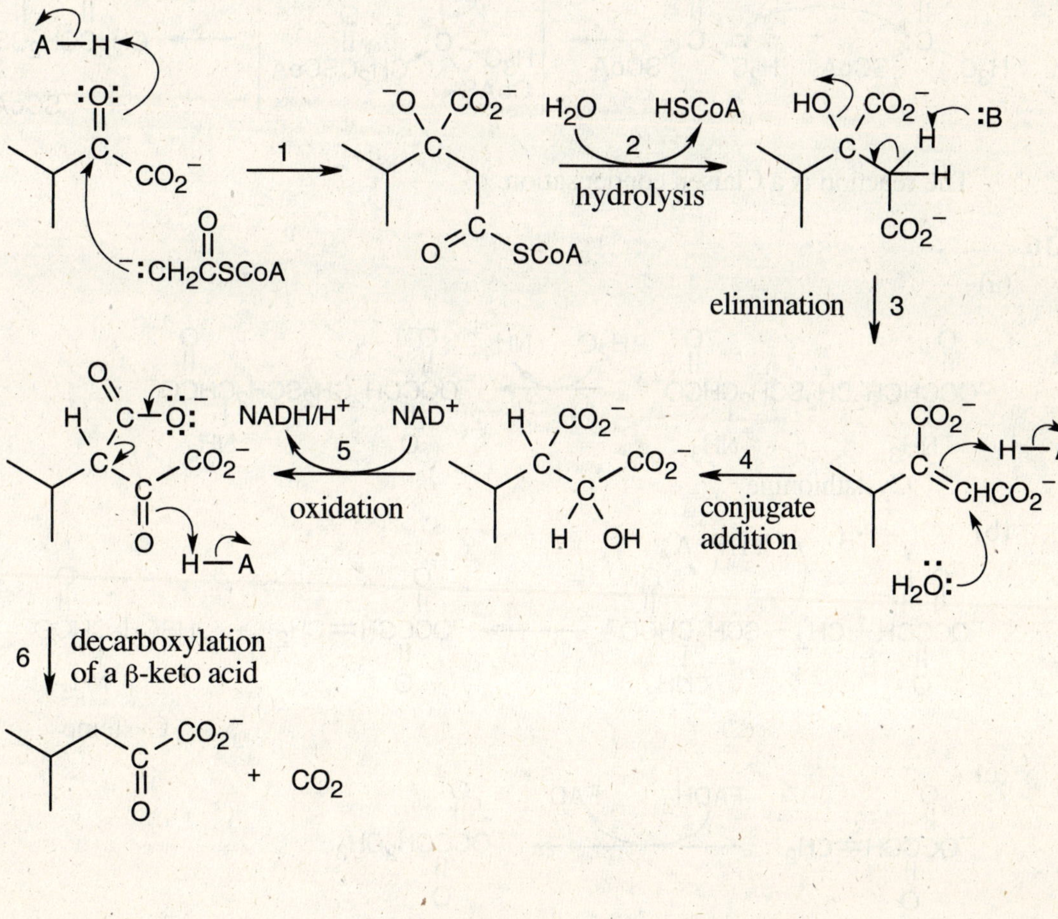

**17.34**

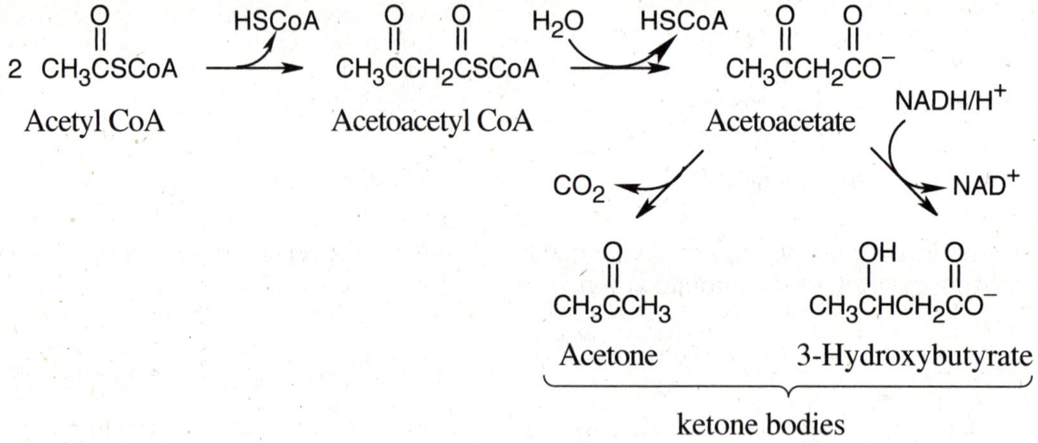


**17.35**

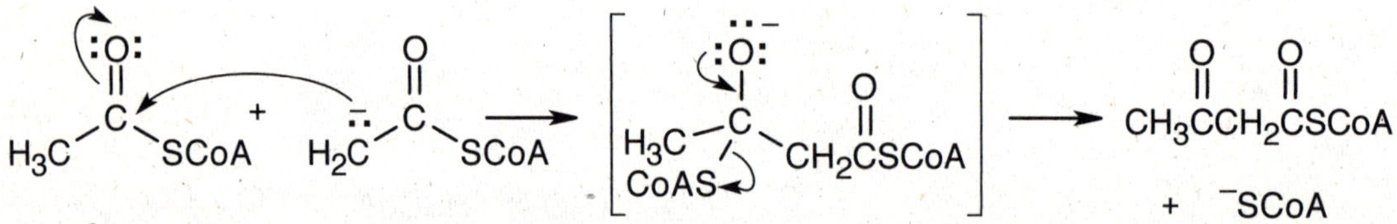


The reaction is a Claisen condensation.

**17.36**

(a)

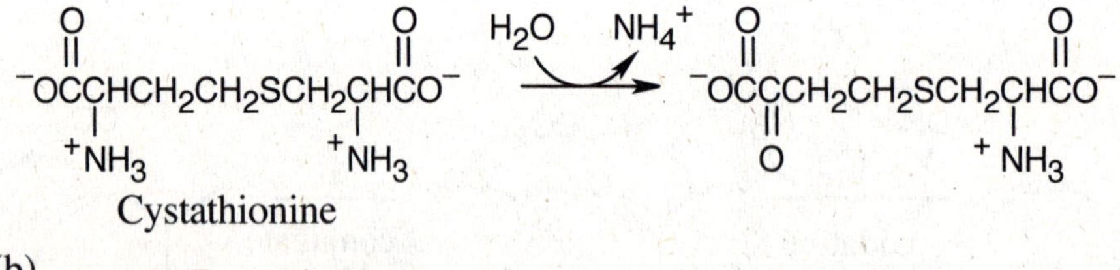


(b)

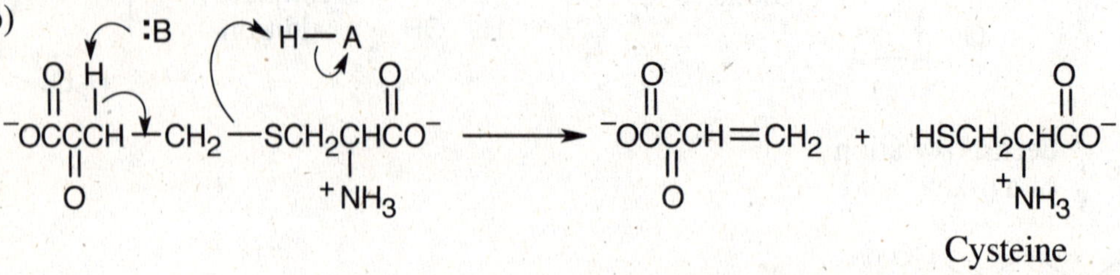


(c)

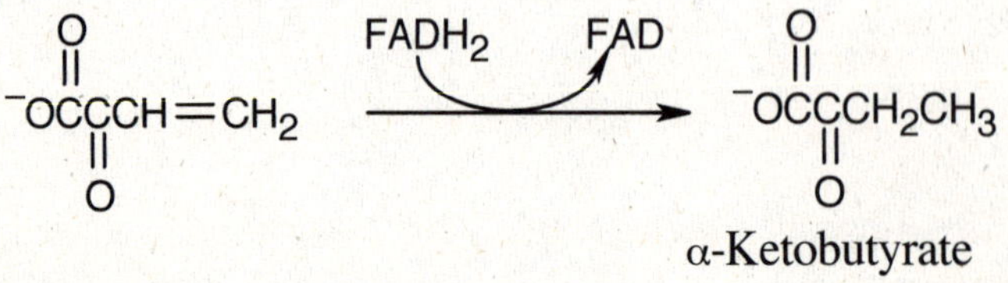


The product of double-bond reduction is α-ketobutyrate. $FADH_2$ is the necessary coenzyme.

**17.37**

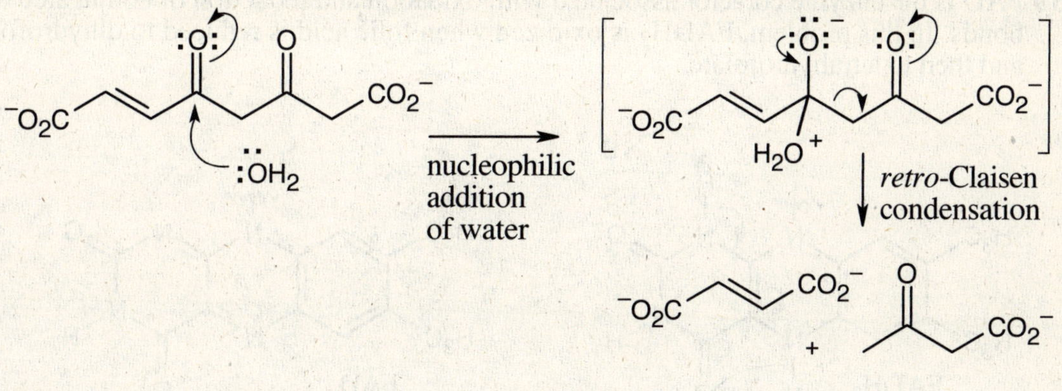


**17.38** The first step in the conversion acetoacetate —> acetyl CoA is the formation of acetoacetyl CoA. This reaction also occurs as the first step in fatty acid degradation.

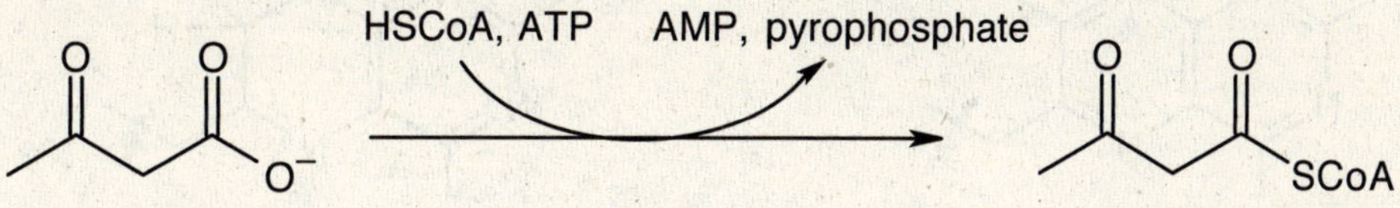


The final step is a retro-Claisen reaction, whose mechanism is pictured in Section 17.2 as Step 4 of β-oxidation of fatty acids.

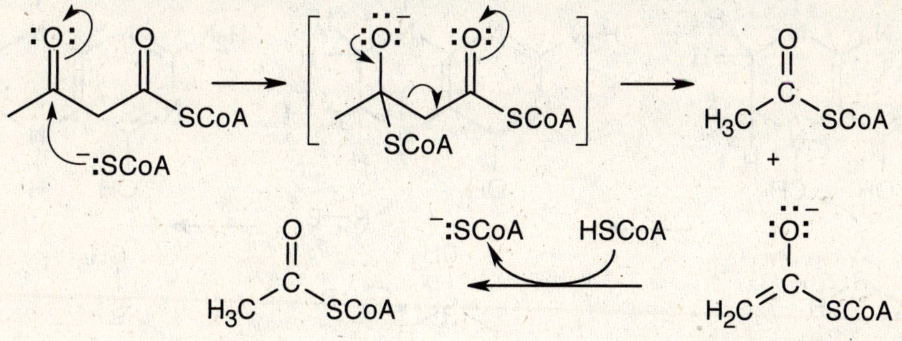

## In the Medicine Cabinet

**17.39** FAD is the enzyme cofactor associated with oxidation and reduction of conjugated double bonds. In this problem, $FADH_2$ is oxidized when folic acid is reduced to dihydrofolate, and then to tetrahydrofolate.

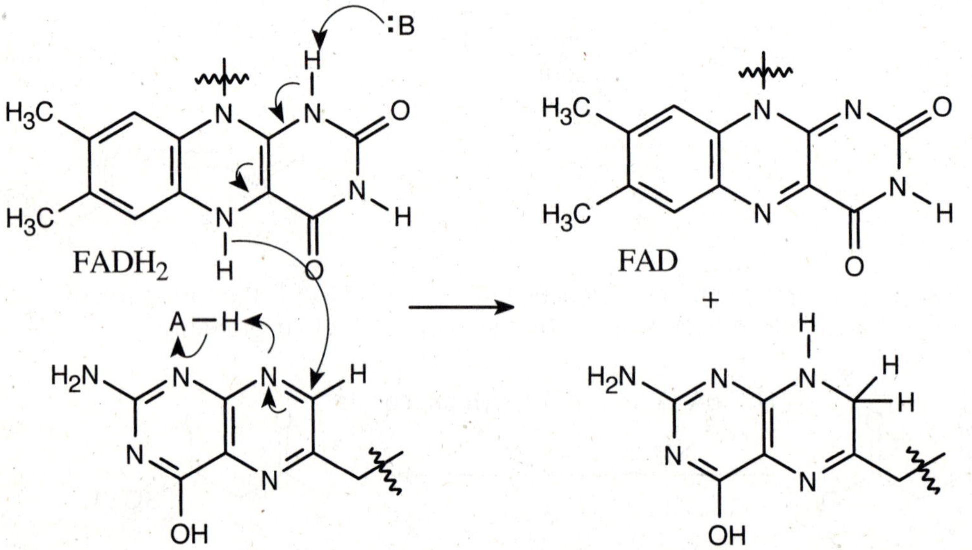


**17.40** The $S_N2$ reaction, in which a methyl group is transferred from the cofactor to homocysteine, produces two charged species that are neutralized in an acid-base reaction.

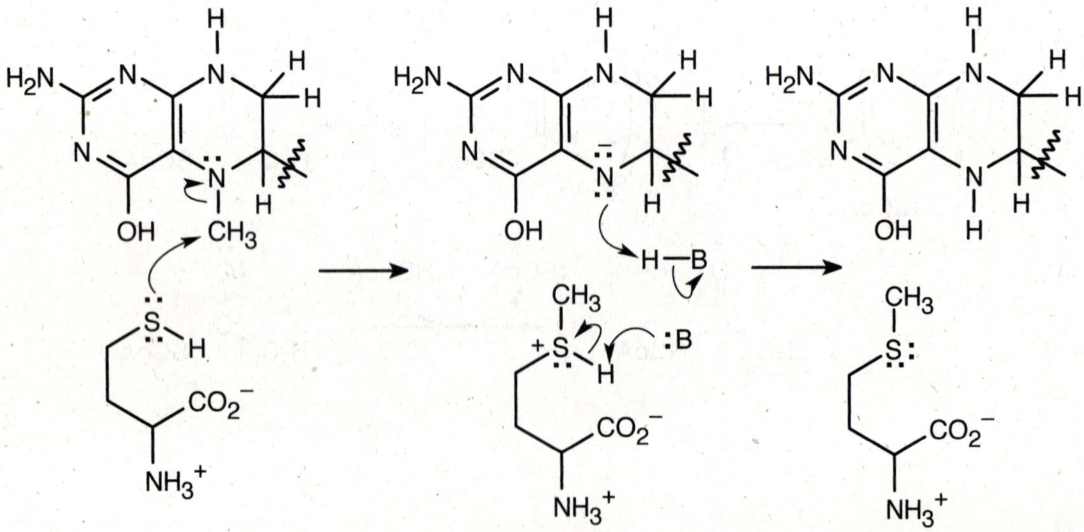

**17.41**

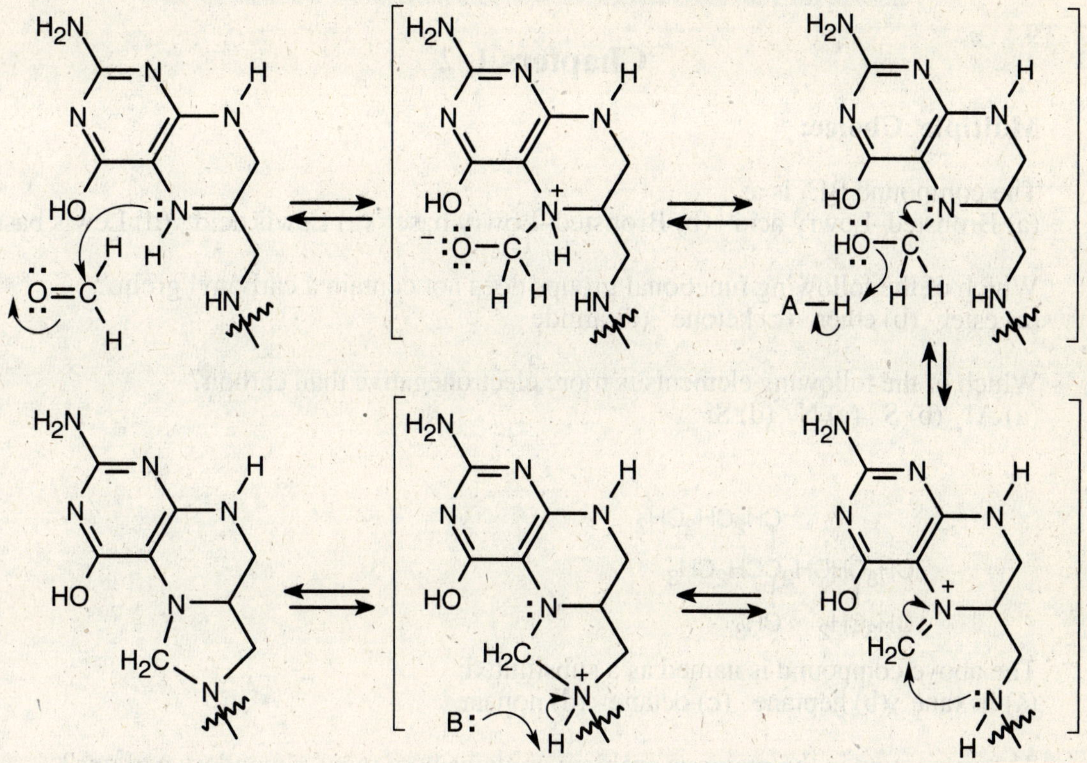

# Review Quizzes

## Chapters 1, 2

**Multiple Choice:**

1. The compound $BF_3$ is a:
(a) Brønsted–Lowry acid (b) Brønsted–Lowry base (c) Lewis acid (d) Lewis base

2. Which of the following functional groups does not contain a carbonyl group?
(a) ester (b) ether (c) ketone (d) amide

3. Which of the following elements is more electronegative than carbon?
(a) Al (b) S (c) N (d) Si

4.

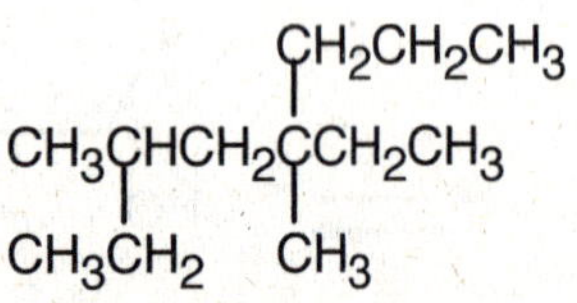

The above compound is named as a substituted:
(a) hexane (b) heptane (c) octane (d) nonane

5. The compound in the previous problem contains how many secondary carbons?
(a) 4 (b) 5 (c) 6 (d) 7

6. The Lewis electron-dot structure for $CS_2$ contains how many electrons?
(a) 40 (b) 16 (c) 12 (d) 8

7. Which of the following elements has no unpaired electrons in its outer shell?
(a) Chlorine (b) Boron (c) Calcium (d) Sulfur

8. The compound H–X has a $pK_a$ of 4, and the compound H–Y has a $pK_a$ of 6. All the following statements are true except:
(a) $[H^+]$ of a 1.0 M solution of HX is approximately 2.
(b) $X^-$ is a stronger base than $Y^-$.
(c) In a solution of 1.0 M HY and 1.0 M NaY the pH is 6.
(d) H–X and H–Y are probably organic acids

9. The formula $C_6H_{14}$ can represent a:
(a) hexane (b) pentane (c) butane (d) all of these

10. In which of the following bonds does carbon have a partial negative charge?
(a) $CH_3$–Na (b) $CH_3$–SH (c) $CH_3$–$OCH_3$ (d) $CH_3$–I

**Tell whether the following statements are true or false:**

1. If HCN has a $pK_a$ of 9.31 and water has a $pK_a$ of 15.74, hydroxide anion is a stronger base than cyanide anion.

2 Both aldehydes and amides can act as Lewis bases.

3. In a graph of energy vs. bond rotation, energy maxima occur when groups are in the staggered conformation.

4. A bond between carbon and sulfur is a polar covalent bond.

5. When a chemical bond is formed, energy is released.

6. A compound with four electrons in its outermost *p* subshell has no unpaired electrons.

7. Pentane is the lowest molecular weight hydrocarbon that is a liquid at room temperature.

8. A compound can be both a Lewis base and a Brønsted base.

9. One isomer of the formula $C_7H_{16}$ is named as a substituted butane.

10. In a bond between carbon and a reactive metal, carbon bears a partial positive charge.

11. A reaction between HCN (p$K_a$ = 9.31) and fluoride (p$K_a$ of HF = 3.45) proceeds as written:

$$HCN + F^- \longrightarrow HF + {}^-CN$$

12. *Trans*–1,4–dimethylcyclohexane is more stable than its cis isomer.

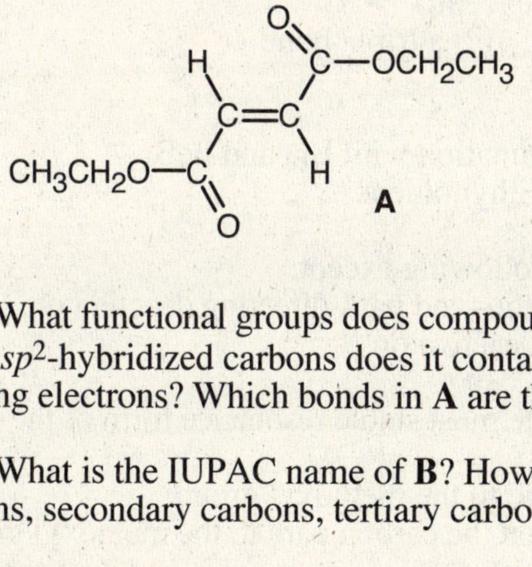
A

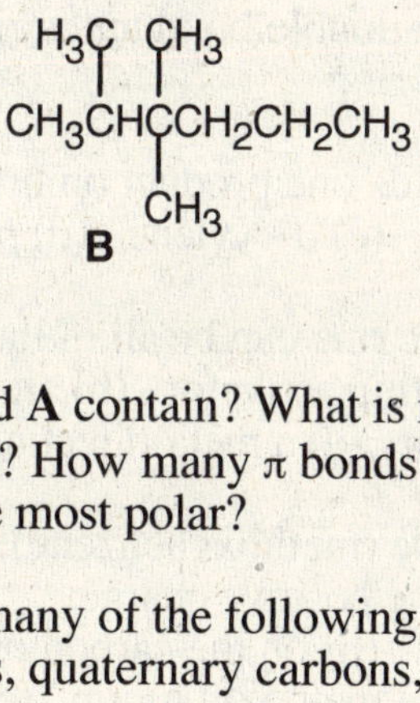
B

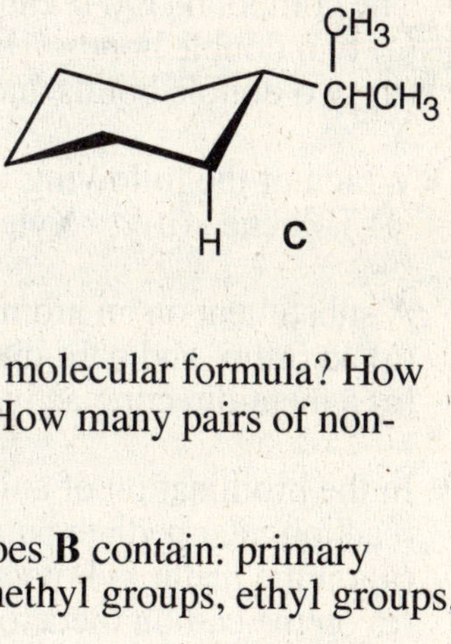
C

What functional groups does compound **A** contain? What is its molecular formula? How many $sp^2$-hybridized carbons does it contain? How many π bonds? How many pairs of non-bonding electrons? Which bonds in **A** are the most polar?

What is the IUPAC name of **B**? How many of the following does **B** contain: primary carbons, secondary carbons, tertiary carbons, quaternary carbons, methyl groups, ethyl groups.

Provide a name for compound **C**. In what conformation is the alkyl group? Is this the more or less stable conformation? How might the two conformations be interconverted? What is the relationship of the indicated hydrogen atom to the alkyl group?

# Chapters 3, 4, 5

## Multiple Choice:

1. Which of the following compounds yields only two identical products on oxidation with acidic $KMnO_4$?
(a) 2,3-Dimethyl-2-butene (b) 1,3-Butadiene (c) 1,3-Cyclohexadiene (d) Cyclohexene

2. The correct priority ranking (from lowest to highest) of the following substituents is:
(a) $-OCH_3$, $-CO_2CH_3$, $-SH$, $-Cl$ (b) $-CO_2CH_3$, $-OCH_3$, $-SH$, $-Cl$
(c) $-Cl$, $-SH$, $-OCH_3$, $-CO_2CH_3$ (d) $-CO_2CH_3$, $-OCH_3$, $-Cl$, $-SH$

3.

$$NaSH + CH_3CH_2CH_2Br \longrightarrow NaBr + CH_3CH_2CH_2SH$$

The preceding reaction is an:
(a) addition reaction (b) elimination reaction (c) substitution reaction
(d) rearrangement reaction

4. All of the following can be oxidized by reaction with acidic or neutral permanganate except:
(a) Toluene (b) 1,3-Pentadiene (c) 1-Hexene (d) Benzene

5. The compound $C_5H_8$ can have all of the following except:
(a) two double bonds (b) a double bond and a ring (c) a triple bond
(d) two double bonds and a ring

6. Which of the following yields one product on bromination with $Br_2$ and $FeBr_3$?
(a) Toluene (b) *o*– Xylene (c) *p*–Xylene (d) *p*–Ethyltoluene

7. A substituent on an aromatic ring can be all of the following except:
(a) an ortho- and para-directing activator (b) an ortho- and para-directing deactivator
(c) a meta-directing activator (d) a meta-directing deactivator

8. In the bromination of anisole (methoxybenzene), the most stable resonance form of the reaction intermediate puts the positive charge:
(a) on the methoxyl oxygen (b) on the carbon ortho to the methoxyl group
(c) on the carbon meta to the methoxyl group (d) on the carbon para to the methoxyl group

9. Which of the following statements about the bromination of benzene is false?
(a) The reaction is two-step. (b) The energy of the products is higher than the energy of the reactants. (c) The second step is faster than the first step. (d) The intermediate can be represented by three resonance forms.

10. Which of the following reactions does not introduce a carbonyl (C=O) group into a hydrocarbon?
(a) Reaction of benzene with acetyl chloride and $FeCl_3$ (b) Treatment of cyclohexene with basic permanganate (c) Hydration of 1–butyne (d) Reaction of toluene with neutral $KMnO_4$

**Tell whether the following statements are true or false:**

1. In ranking a substituent group attached to a double bond, –OH is of higher priority than $-CH_2OCH_3$.

2. Reactions involving symmetrical bond breaking usually occur when the atoms in the bond to be broken differ in electronegativity.

3. A reaction that releases energy also has a negative $E_{act}$.

4. Addition of HBr to 2-methyl-2-pentene yields a single product.

5. A reaction can have an intermediate without having a transition state.

6. Acidic permanganate is used to hydroxylate a double bond.

7. In order for a radical chain reaction to propagate, an odd-electron species must react with an even-electron species.

8. Treatment of 2-pentyne with sodium amide produces an acetylide anion.

9. The substituent $-N(CH_3)_2$ is an ortho, para-directing activator.

10. A benzene ring is reduced with $H_2$ and a Pd catalyst.

11. It is easier to prepare *p*-nitrotoluene than *m* -nitrotoluene.

12 A group that is a meta-director deactivates all positions on the aromatic ring.

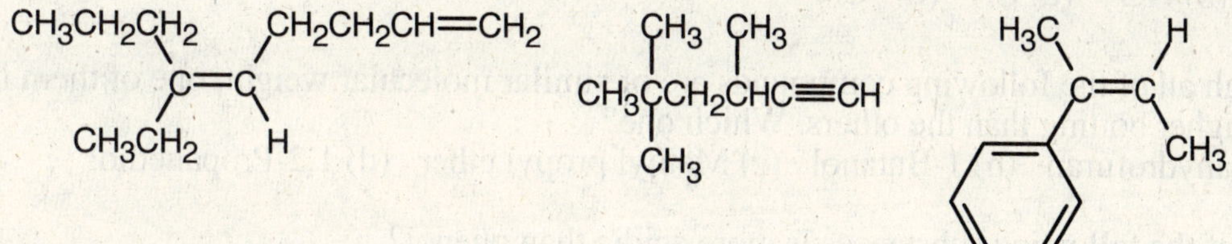

A B C

Name compounds **A** and **B**. Which of the above compounds show *E,Z* isomerism? Assign *E* or *Z* configurations to the isomeric double bonds. Which of the above compounds have conjugated bond systems? Which of the above compounds react with the following reagents? Show the products. (a) $Br_2$, $CH_2Cl_2$ (b) $H_2$, Pd catalyst (c) $Br_2$, $FeBr_3$ (d) $NaNH_2$ (e) HBr, ether

Show the products when compounds **A** and **C** react with : (a) acidic $KMnO_4$ (b) $H_2O$, $H^+$ catalyst. What products result from treatment of **C** with $HNO_3$, $H_2SO_4$? If **C** were to react with acidic $KMnO_4$ and the resulting compound was treated with $HNO_3$, $H_2SO_4$, what product would be formed?

## Chapters 6, 7, 8

**Multiple Choice:**

1. If you were to draw all products of radical bromination of ethane, how many structures would you draw?
(a) 2 (b) 6 (c) 9 (d) 11

2. Which of the following experiments is not useful in helping you decide if a substitution reaction proceeds by an $S_N1$ mechanism or an $S_N2$ mechanism?
(a) changing the nucleophile (b) varying the leaving group (c) varying the concentration of nucleophile (d) measuring the sign of rotation if the product has a stereocenter.

3. A reaction of a tertiary alkyl halide might occur by all of the following routes except:
(a) E1 (b) E2 (c) $S_N1$ (d) $S_N2$

4.

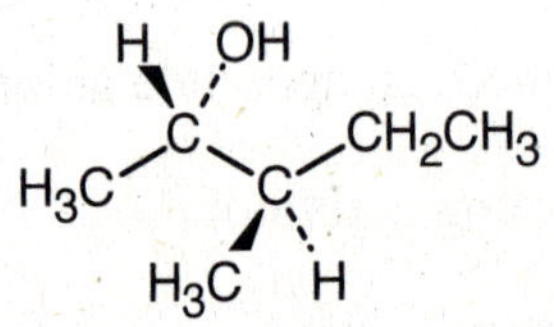

What are the configurations at the carbon stereocenters of the above isomer?
(a) 2*R*,3*R* (b) 2*R*, 3*S* (c) 2*S*,3*R* (d) 2*S*,3*S*

5. Which of the following is both a good nucleophile and a good leaving group?

(a) $I^-$ (b) $HO^-$ (c) $Cl^-$ (d) $^-CN$

6. Although all of the following compounds are of similar molecular weight, one of them is much higher boiling than the others. Which one?
(a) Tetrahydrofuran (b) 1-Butanol (c) Methyl propyl ether (d) 1,2-Propanediol

7. Which of the following substances is more acidic than phenol?
(a) Sodium bicarbonate (b) Water (c) Sodium carbonate (d) *p*–Methylphenol

8. Ethers are unreactive to all of the following except:
(a) nucleophiles (b) halogens (c) strong bases (d) strong acids

9. How many of the alcohol and ether stereoisomers of the formula $C_5H_{12}O$ have chirality centers?
(a) 2 (b) 3 (c) 4 (d) 5

10 Which of the following statements about *p*-nitrophenol is false?
(a) It is more acidic than phenol (b) It is a solid (c) Reaction with $Br_2$, $FeBr_3$ yields 3–bromo–4–nitrophenol (d) It reacts with potassium carbonate to form a phenoxide anion.

**Tell whether the following statements are true or false:**

1. *Cis*-2-pentene and *trans*-2-pentene are diastereomers.
2. The E1 reaction of a tertiary bromide occurs when strong base is present.
3. The $S_N2$ reaction of an *S* enantiomer with a nucleophile yields an *R* enantiomer.
4. A Grignard reagent is formed by reaction of an organohalide with Mg.
5. Two compounds that are diastereomers of a third compound are enantiomers.
6. Acid-catalyzed epoxide ring-opening occurs by a $S_N2$ mechanism.
7. The reaction of a thiolate anion with an alkyl halide yields a disulfide.
8. A peroxyacid is used to form an epoxide from an alkene.
9. The physical properties of a racemic mixture are the same as those of each enantiomer.
10. In a primary alcohol, the hydroxyl group is bonded to a primary carbon.
11. In an $S_N1$ displacement reaction, water is a better leaving group than hydroxide ion.
12. Four different products can result from the cleavage of an unsymmetrical ether with HI.

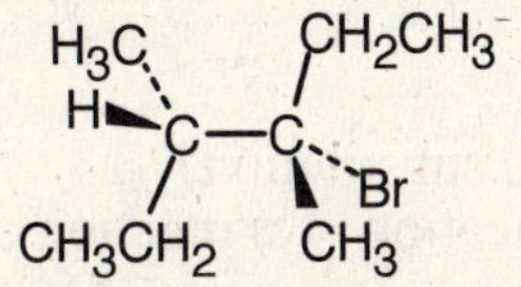


**A**

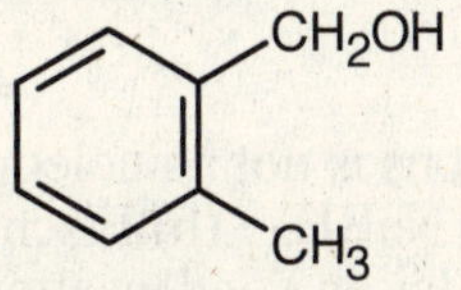


**B**

Give the name of compound **A**. What are the *R,S* configurations at the stereocenters? How many enantiomers of **A** are there? How many diastereomers? What are the configurations of the enantiomers and diastereomers? Is **A** a meso compound? Is **A** a primary, secondary or a tertiary bromide?

A mixture of two isomers of **A** can be prepared from an alcohol and a second reagent. What are the reagents? Why are two isomers formed? What are the configurations of the isomer products at the carbon chirality centers? By what mechanism is the bromide formed? What is the effect of changing the following reaction conditions on the rate of reaction: (a) doubling the concentration of the alcohol (b) tripling the concentration of the other reagent.

Name at least one other product formed as a byproduct of the above reaction to form compound **A**. By what mechanism is it formed? What is the configuration of the carbon atoms in this byproduct?

Give the name of compound **B**. Is **B** an alcohol or a phenol? Show two pathways by which **B** might be synthesized. Is **B** more or less acidic than phenol? What products result from treatment of B with the following reagents: (a) $CrO_3$, $H_3O^+$ (b) $PBr_3$, then $NaSCH_3$ (c) 2 Na (d) product of (c) + $CH_3CH_2I$ (e) product of (d) + HI

# Chapters 9, 10, 11

## Multiple Choice:

1. Both aldehydes and ketones react with all of the following except:
(a) periodinane (b) $NaBH_4$ (c) $CH_3MgBr$ (d) $NH_2OH$

2. Which type of compound is most reactive?
(a) a nitrile (b) an amide (c) an acid chloride (d) an ester

3. Which of the following undergoes aldol self-condensation?
(a) Formaldehyde (b) Acetaldehyde (c) Benzaldehyde (d) 2,2-Dimethylbenzaldehyde

4. What reagent should be used to introduce bromine $\alpha$ to a carbonyl group?
(a) $Br_2$, $FeBr_3$ (b) $Br_2$, $CH_3CO_2H$ (c) $PBr_3$ (d) HBr

5. Which of the following is most acidic?
(a) water (b) a ketone (c) a 1,3-keto ester (d) a 1,3-diketone

6. Which of the following combinations can't be used to produce 2-phenyl-2-butanol?
(a) 2-butanone and phenylmagnesium bromide (b) acetophenone and ethylmagnesium bromide (c) $C_6H_5COCH_2CH_3$ and methylmagnesium bromide (d) butanal and phenylmagnesium bromide

7. What reaction is used to produce $\beta$-keto esters?
(a) Grignard reaction (b) Claisen condensation (c) aldol condensation (d) malonic ester synthesis

8. Which of the following reactions is not a nucleophilic acyl substitution?
(a) reduction of a ketone with $NaBH_4$ (b) Fischer esterification (c) reaction of an acid chloride with an amine (d) Claisen condensation

9. Which of the following has four acidic hydrogens?
(a) 2,4-Pentanedione (b) Cyclohexanone (c) Propanoic acid (d) Acetophenone

10. A Grignard reaction can be used for all of the following except:
(a) converting a halide to an alkane (b) converting an ester to an alcohol
(c) converting a nitrile to a ketone (d) converting an amide to an alcohol

## Tell whether the following statements are true or false:

1. An enolate ion is more reactive than an enol.

2. The Claisen condensation can be viewed as both an $\alpha$-substitution and a nucleophilic acyl substitution.

3. An acetal is stable in aqueous solution.

4. Benzoic acid is a stronger acid than acetic acid.

5. 4-Methyl-2-hexanone is racemized by treatment with dilute acid or base.

6. In aqueous solution, some ketones exist as gem diols.

7. A nitrile can react with a Grignard reagent to yield a ketone.

8. Phenylacetic acid can be synthesized via a malonic ester synthesis.

9. The product of a Claisen condensation is an enone.

10 An amine can be formed from hydrogenation of either an amide or a nitrile.

11. Aldehydes are more reactive than ketones for steric reasons.

12. Enolate formation can be catalyzed by both acid and base.

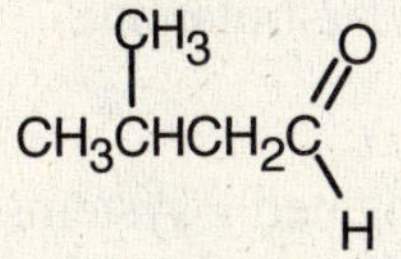


A

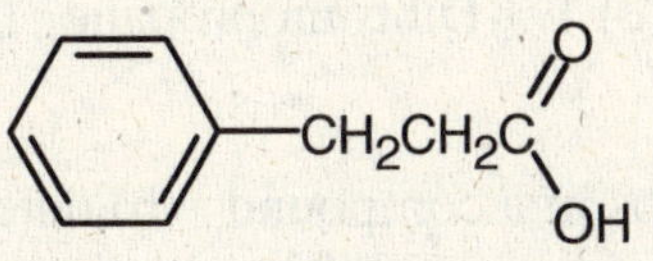


B

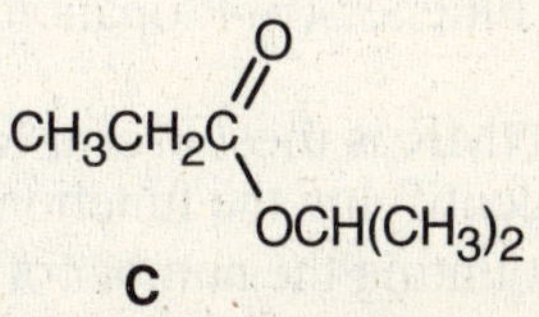


C

What is the name of compound **A**? How might **A** be prepared from the corresponding alcohol? What products might be expected when **A** reacts with (a) $CH_3OH$, $H^+$ catalyst (b) phenylmagnesium bromide (c) $LiAlH_4$ (d) hydroxylamine.

How many acidic hydrogens does **A** have? What products are formed when **A** is treated with: (a) $Br_2$, $CH_3CO_2H$ (b) sodium ethoxide in ethanol (c) product of (b) + heat, acid catalyst.

Give the name of compound **B**. How can **B** be prepared from: (a) $C_6H_5CH_2CH_2Br$ (b) $C_6H_5CH_2CH_2CH_2OH$ (c) $C_6H_5CH_2Br$ (hint: malonic ester synthesis). What is the effect on carboxylic acid acidity if the aromatic ring has a nitro group in the para position?

Predict the products when **B** is treated with: (a) $SOCl_2$ (b) product of (a) + $CH_3CO_2^-$ (c) $CH_3OH$, HCl (d) product of (a) + $CH_3CH_2OH$, pyridine (e) product of (a) + $CH_3NH_2$.

Name Compound **C**. Synthesize **C** from: (a) $CH_3CH_2CO_2H$ (b) $CH_3CH_2COCl$ (c) $CH_3CH_2C{\equiv}N$.

What products are formed when **C** reacts with: (a) $H_2O$, NaOH (b) $LiAlH_4$ (c) $NH_3$, ether (d) $CH_3MgBr$, ether.

How many acidic hydrogens does **C** have? What product is formed when **C** is treated with sodium ethoxide, followed by acidification?

## Chapters 12, 13

### Multiple Choice:

1. Which of the following compounds is the most basic?
(a) 2,4,6–Trimethylaniline (b) Aniline (c) Trimethylamine (d) Acetamide

2. A compound that shows an IR absorption at 1735 $cm^{-1}$ is a:
(a) conjugated diene (b) ester (c) alcohol (d) amine

3. Which of the following is a fused-ring nitrogen heterocycle?
(a) Pyridine (b) Indole (c) Imidazole (d) Pyrrole

4. Which compound shows a 5-peak multiplet in its $^1H$ NMR spectrum?
(a) Toluene (b) Propanoic acid (c) 1,3-Dibromopropane (d) 3-Methyl-2-butanone

5. $^{13}C$ NMR is most useful for :
(a) identifying the functional groups in a compound (b) detecting conjugated $\pi$ systems (c) counting the number of carbon atoms in a molecule (d) calculating the number of hydrogens bonded to an adjacent carbon

6. All of the following statements about alkaloids are true except:
(a) they have pharmaceutical properties (b) the amine group must be part of a ring (c) they are derived from both animal and plant sources (d) they undergo acid–base reactions

7. Which spectroscopic technique is most useful for distinguishing between 1,3,5-octatriene and 1,4,7-octatriene?
(a) UV (b) IR (c) $^{13}C$ NMR (d) $^1H$ NMR

8. The number of peaks seen in the $^{13}C$ NMR spectrum of *cis*-3,4-dimethyl-3-hexene is:
(a) 3 (b) 4 (c) 5 (d) 6

9.

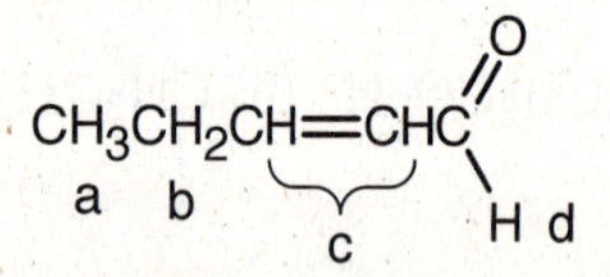


Arrange the $^1H$ NMR absorptions of the indicated hydrogens from downfield to upfield.
(a) a, b, c, d (b) d, c, b, a (c) d, b, c, a (d) b, a, c, d

10. Magnetic field strength is measured in units of:
(a) megahertz (MHz) (b) reciprocal centimeters ($cm^{-1}$) (c) joule (J) (d) tesla (T)

**Tell whether the following statements are true or false:**

1. Quaternary ammonium salts may have nitrogen chirality centers.

2. $^1H$ NMR is a useful technique because each unique hydrogen nucleus in a molecule experiences a slightly different magnetic field on irradiation.

3. A compound with a small value of $pK_b$ gives up a proton more readily than a compound with a large value of $pK_b$.

4. The chemical shift of a carboxylic acid proton occurs at a larger $\delta$ value than does the chemical shift of any other commonly encountered proton.

5. It is difficult to synthesize butylamine from ammonia and 1-bromobutane without also forming dibutylamine and tributylamine.

6. The lone electron pair of the nitrogen atom of pyridine is part of the aromatic ring $\pi$ system.

7. The larger the wave number, the higher the energy of electromagnetic radiation.

8. In the reaction of a primary alkylamine with ammonia, using a large excess of ammonia ensures that only monoalkylation product will be formed.

9. When the NMR spectrometer operating frequency changes from 60 MHz to 200 MHz the value of $\delta$ also changes.

10. Five peaks are visible in the $^{13}C$ NMR spectrum of *p*– nitroethylbenzene.

11. If compound **P** absorbs at a longer UV wavelength than Compound **Q**, the energy required for excitation is more for **P** than for **Q**.

12. $^{13}C$ NMR spectra provide information about the kinds of carbons adjacent to a given carbon.

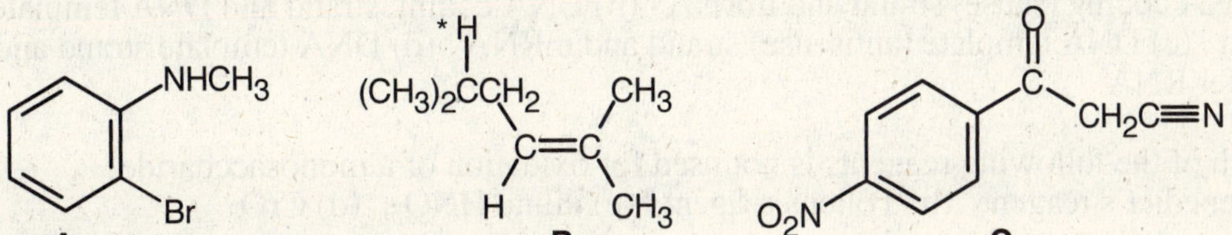


**A** **B** **C**

What is the name of **A**? Is **A** a primary,secondary or tertiary amine? Is **A** more or less basic than methylamine? Does the bromine on the ring make **A** more or less basic than the unbrominated compound? Design a synthetic route to **A** from benzene.

If the $-NHCH_3$ substituent on **A** is changed to $-CH_2NH_2$, what is the name of the new compound(**A′**)? Is **A′** a primary, secondary or tertiary amine? Is it more or less basic than **A**? Synthesize **A′** from benzene. Synthesize **A′** from *o*-bromobenzoic acid.

Name compound **B**. Do either UV or IR spectroscopy provide useful information about the structure of **B**? In the $^{13}C$ NMR spectrum of **B**, how many peaks are visible? What are the approximate chemical shifts of the peaks in the $^{13}C$ NMR spectrum.

How many peaks are visible in the $^1H$ NMR spectrum of **B**? Describe the $^1H$ NMR spectrum of **B**, including approximate chemical shifts and splitting patterns. If the starred atom in **B** is replaced with –Br, describe the resulting changes in the spectrum.

Does Compound **C** show a UV absorption in the range 250-400 nm? Would you expect the nitro group to increase the value of $\lambda_{max}$ relative to the same compound without a nitro group? What significant peaks are visible in the IR spectrum of **C**? Does $^1H$ NMR provide much useful information in this problem?

## Chapters 14, 15, 16, 17

### Multiple Choice:

1. Which of the following amino acids doesn't contain an aromatic ring?
(a) Tyrosine (b) Proline (c) Histidine (d) Tryptophan

2. Secondary structure of a protein refers to:
(a) the sequence of covalent bonds between amino acids (b) the way that several proteins aggregate (c) the regular orientation of segments of the peptide backbone (d) the coiling of the entire molecule into a three-dimensional shape

3. All of the following types of reactions occur during β-oxidation of fatty acids except:
(a) dehydrogenation (b) retro-Claisen reaction (c) oxidation of an alcohol (d) aldol condensation

4. Which of the following is a monounsaturated fatty acid?
(a) Lauric acid (b) Linoleic acid (c) Oleic acid (d) Stearic acid

5. Which of the following monosaccharides has three carbon chirality centers?
(a) an aldohexose (b) a 2-deoxyaldopentose (c) a ketohexose (d) a ketopentose

6. If a strand of a nucleic acid has the sequence –A–C–A–T–G–G–C–T– (strand A) and a strand of a second nucleic acid has the sequence –A–C–A–U–G–G–C–U– (strand B), the relationship of strand A to strand B is:
(a) DNA coding (sense) strand and mRNA (b) DNA coding strand and DNA template strand (c) DNA template (antisense) strand and mRNA (d) DNA template strand and transfer RNA

7. Which of the following reagents is not used for oxidation of a monosaccharide?
(a) Benedict's reagent (b) Tollens reagent (c) dilute $HNO_3$ (d) $CrO_3$

8. A steroid is a lipid because:
(a) it can be extracted from tissues with a nonpolar organic solvent (b) it contains ester linkages (c) it is unsaturated (d) it has a characteristic tetracyclic ring system

9. Which amino acid has only one codon?
(a) Cysteine (b) Tyrosine (c) Lysine (d) Tryptophan

10. Which of the following is characteristic of tertiary structure?
(a) β-pleated sheet (b) hydrophobic interactions (c) superhelix (d) α-helix

### Tell whether the following statements are true or false:

1. All aldoses are reducing sugars.

2. A sugar can be converted to an acetal by treatment with $CH_3I$ and $Ag_2O$.

3. The purpose of the respiratory chain is to convert acetyl groups to $CO_2$ and reduced coenzymes.

4. A dehydrase removes $H_2$ from a substrate.

5. Thymine differs from uracil in having a methyl group bonded to the pyrimidine ring.

6. The Fischer projection shown below represents an *R* enantiomer

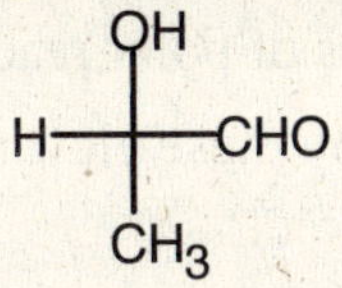

7. In gel electrophoresis of the products of cleavage of a DNA fragment, smaller products migrate farther from the origin than larger products.

8. An enzyme can change the equilibrium constant of a biochemical reaction.

9. Dicyclohexylcarbodiimide is used to protect amino groups during protein synthesis.

10. In the β anomer of an aldohexose, the anomeric hydroxyl group is cis to the hydroxyl group at carbon 6.

11. The hydrolysis of ATP can be coupled with an energetically unfavorable reaction so that the overall reaction has a favorable energy change.

12. Amylose contains 1,4'- and 1,6'-α-glycosidic bonds.

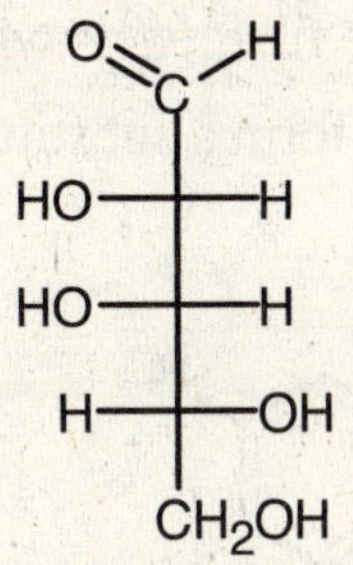

**A**

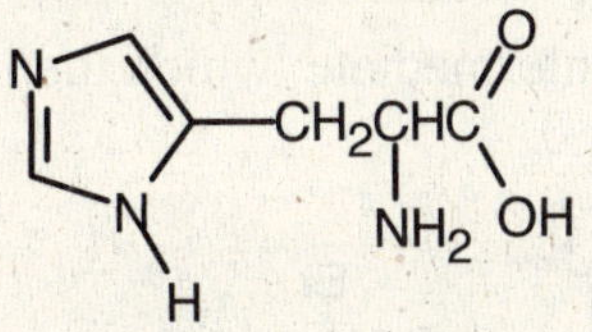

**B**

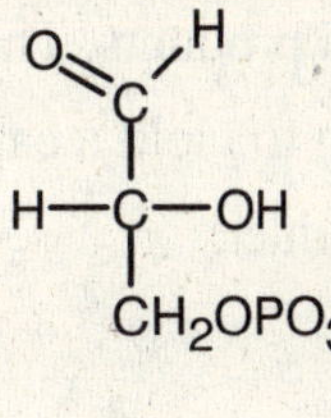

**C**

How is compound **A** classified? Is **A** a D or an L sugar? What is the *R,S* configuration at the carbon next to the carbonyl group? Is **A** a reducing sugar? Draw the β anomer of **A**, and show the products that result from reaction of **A** with: (a) $NaBH_4$ (b) $CH_3OH$, HCl (c) acetic anhydride (d) Tollens reagent (e) dilute $HNO_3$.

Which amino acid is represented by compound **B**? Is **B** acidic, neutral or basic? Draw **B** in its zwitterionic form. Is Compound **B**'s ring aromatic? Give a three-base codon for **B**, and give a three-base tRNA anticodon. What three-base sequence on DNA was used as the template?

Name compound **C**. **C** is an intermediate in which important biochemical pathway? Name the 6-carbon precursor to **C** and the type of reaction that produced **C**. Compound **C** is oxidized and phosphorylated to produce which compound? Which coenzyme is involved in this reaction? What is the end product of this biochemical pathway?

# Summary of Reaction Mechanisms

The following table summarizes the most important types of polar reactions in organic chemistry. Much of organic chemistry is accounted for by these eleven mechanisms.

**1. Electrophilic Additions to Alkenes** (Sections 3.7–3.9)

Alkenes reacts with electrophiles such as HBr to yield saturated addition products. The reaction occurs in two steps: The electrophile first reacts with the alkene double bond to yield a carbocation intermediate that reacts further to yield the addition product.

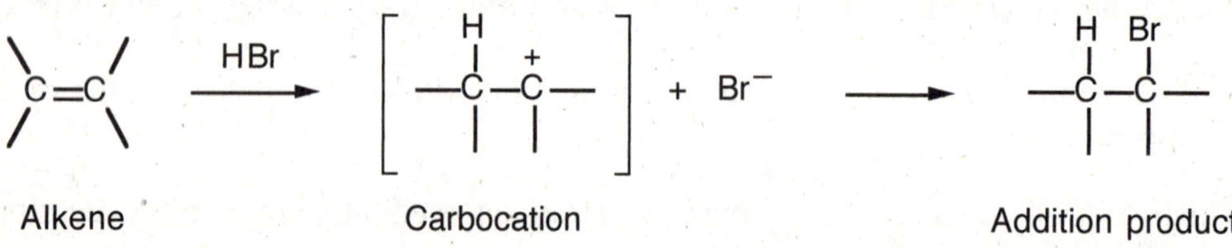


**2. Electrophilic Aromatic Substitutions** (Sections 5.3–5.7)

Aromatic compounds react with electrophiles such as $Br_2$ to yield substitution products rather than addition products. The reaction occurs in two steps: The electrophile first reacts with the aromatic ring to yield a carbocation intermediate, which then loses $H^+$ to form a substituted aromatic product.

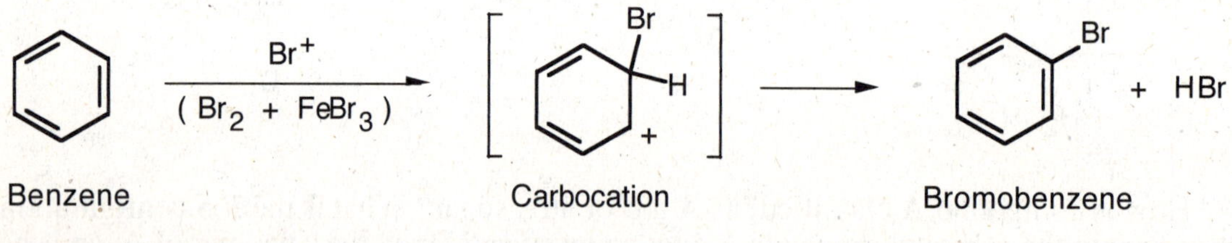


**3. Nucleophilic Substitutions of Alkyl Halides**

**A. $S_N2$ Reaction** (Section 7.5)

Alkyl halides undergo substitution when treated with nucleophiles. Primary and secondary alkyl halides react by the $S_N2$ mechanism, which occurs in a single step when the incoming nucleophile attacks the substrate from a direction 180° away from the leaving group. This results in an umbrella-like inversion of stereochemistry.

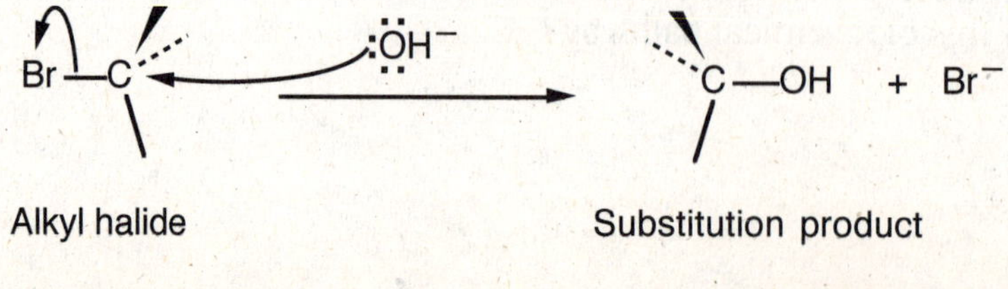

**B. $S_N1$ Reaction** (Section 7.6)

Tertiary alkyl halides undergo nucleophilic substitution by the $S_N1$ mechanism, which occurs in two steps. Spontaneous dissociation of the alkyl halide into an anion and a carbocation intermediate takes place, followed by reaction of the carbocation with a nucleophile. The dissociation step is the slower of the two.

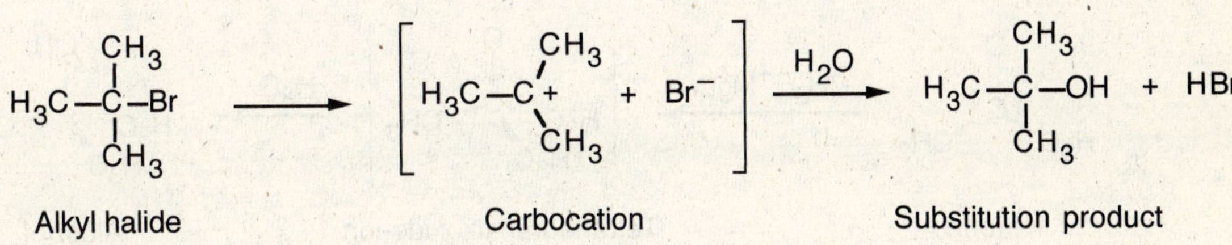


## 4. Elimination Reactions of Alkyl Halides

**A. E2 Reaction** (Section 7.7)

Alkyl halides undergo elimination of HX to yield alkenes on treatment with base. When a strong base such as hydroxide ion ($HO^-$) or alkoxide ion ($RO^-$) is used, alkyl halides react by the E2 mechanism. The E2 reaction occurs in a single step in which base removes a neighboring hydrogen at the same time that the halide ion leaves.

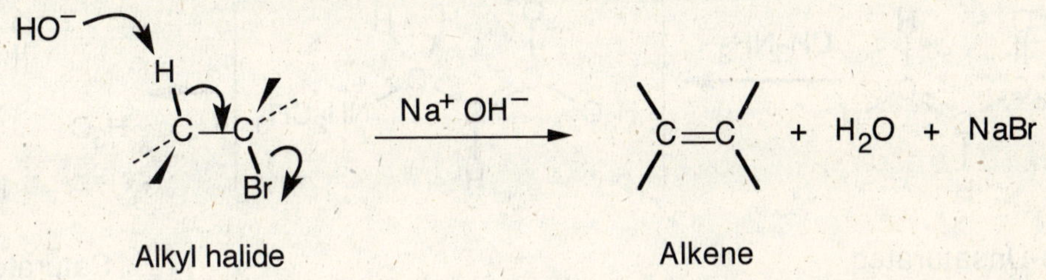


**B. E1 Reaction** (Section 7.8)

Tertiary alkyl halides undergo elimination by the E1 mechanism in competition with $S_N1$ substitution when a nonbasic nucleophile is used in a protic solvent. The reaction takes place in two steps: Spontaneous dissociation of the alkyl halide leads to a carbocation intermediate, which then loses $H^+$.

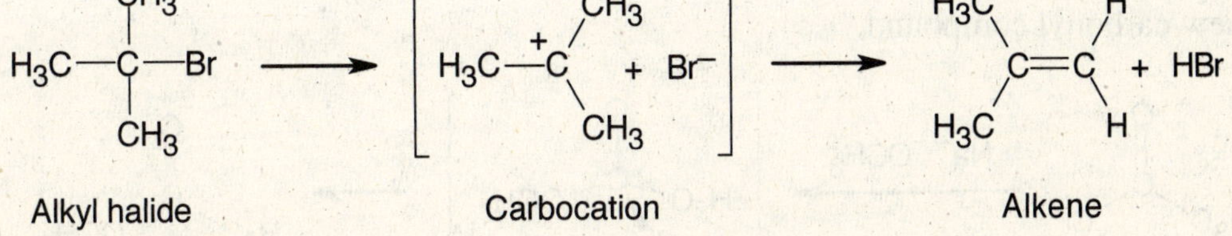

## 5. Nucleophilic Additions to Aldehydes and Ketones

### A. Direct (1,2) Addition (Sections 9.5–9.9)

Aldehydes and ketones react with nucleophiles to yield addition products. The reaction occurs by addition of the nucleophile to the carbonyl group, producing a tetrahedrally hybridized alkoxide-ion intermediate, which is protonated to yield an alcohol.

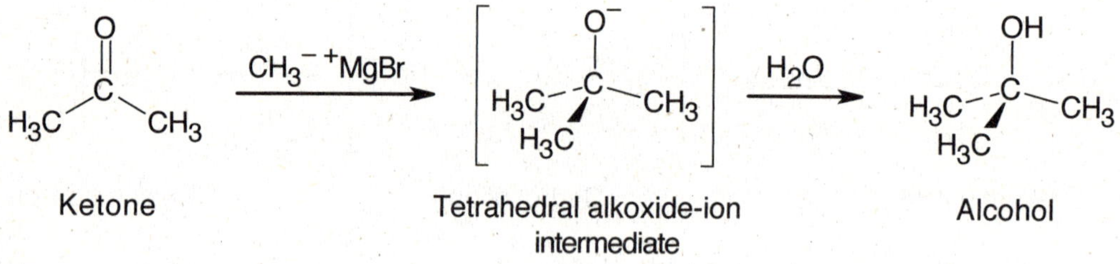


### B. Conjugate (1,4) Addition (Section 9.10)

α,β-Unsaturated aldehydes and ketones react with nucleophiles to yield saturated addition products. The reaction occurs by addition of the nucleophile to the β carbon of the double bond to give an enolate ion, followed by protonation on the α carbon.

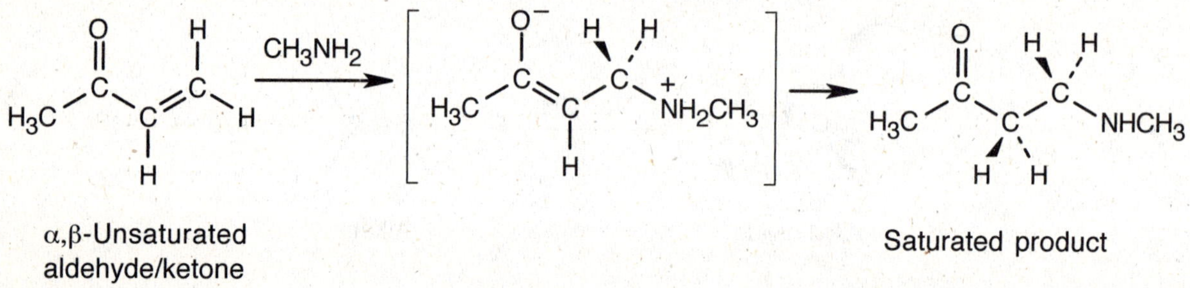


## 6. Nucleophilic Acyl Substitutions (Sections 10.5–10.11)

Carboxylic acid derivatives (acid chlorides, acid anhydrides, esters, amides) undergo a substitution reaction on treatment with nucleophiles. A nucleophilic acyl substitution reaction takes place in two steps: Addition of the nucleophile to the carbonyl group produces a tetrahedrally hybridized alkoxide-ion intermediate, which expels a leaving group to generate a new carbonyl compound.

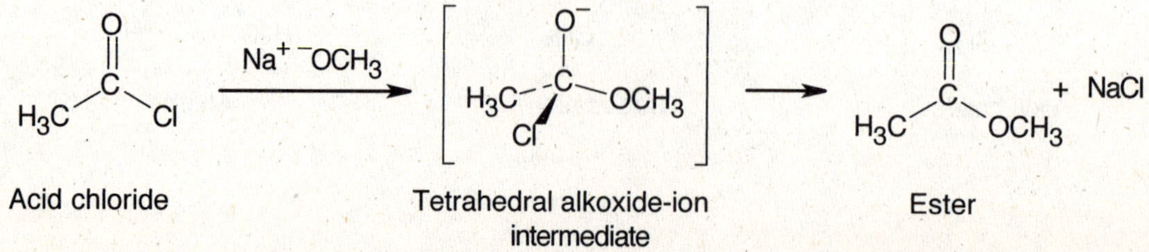

**7. Carbonyl Alpha-Substitution Reactions** (Sections 11.2–11.6)

Carbonyl compounds having alpha hydrogens (hydrogens on carbon next to the carbonyl group) are in equilibrium with their enol tautomers. An alpha-substitution reaction takes place when an enol or enolate ion reacts with an electrophile to produce a substitution product.

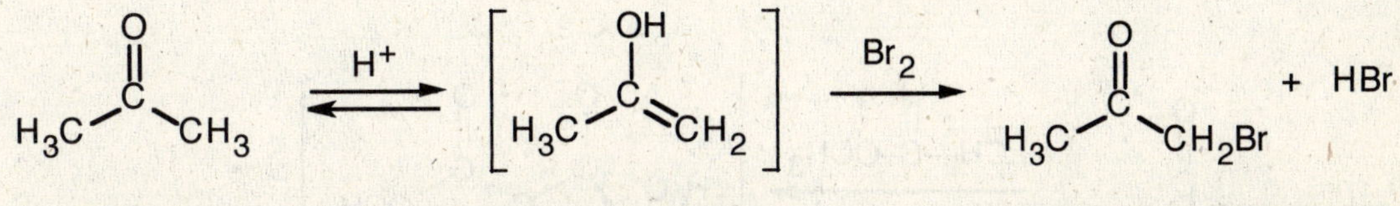


Ketone — Enol intermediate — α–substitution product

**8. Carbonyl Condensation Reactions**

**A. Aldol condensation of aldehydes and ketones** (Sections 11.7–11.9)

Aldehydes and ketones with alpha hydrogens undergo a base-catalyzed dimerization reaction leading to formation of β-hydroxy ketone/aldehyde products. The reaction occurs in two step when one molecule of ketone or aldehyde is converted into its enolate ion, which then does a nucleophilic addition reaction to the carbonyl group of a second molecule.

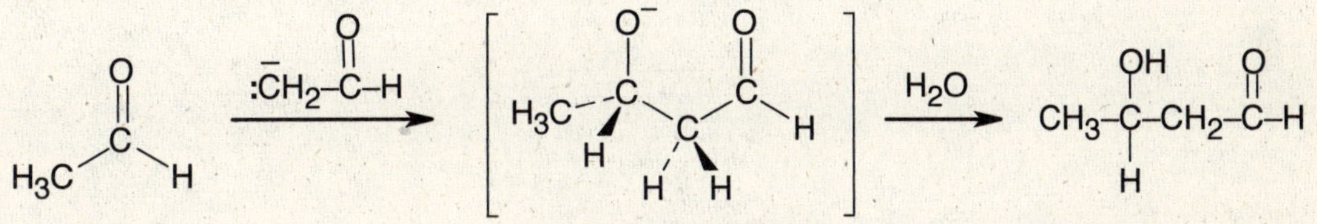


Aldehyde — Tetrahedral alkoxide-ion intermediate — β-Hydroxy aldehyde

**B. Claisen condensation of esters** (Section 11.10)

Esters with alpha hydrogens undergo a base-catalyzed dimerization reaction leading to formation of a β-keto ester product. The reaction occurs in two steps when one molecule of ester is converted into its enolate ion, which does a nucleophilic acyl substitution reaction on a second ester molecule.

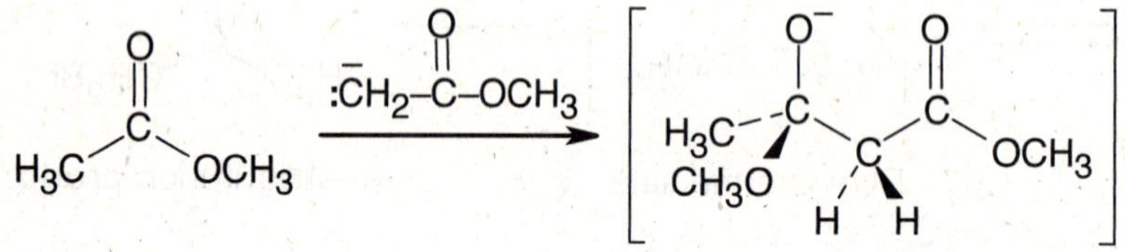

Ester    Tetrahedral alkoxide-ion intermediate

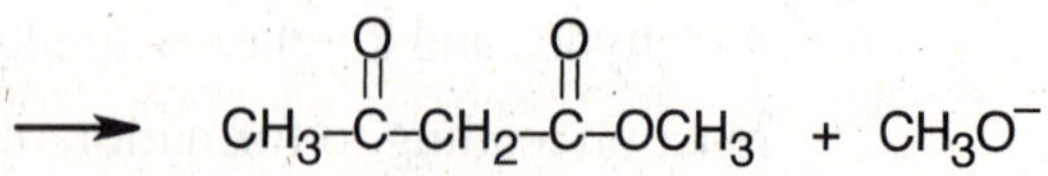

β-Keto ester

## Summary of Functional Group Preparations

The following table summarizes the synthetic methods used to prepare important functional groups. The functional groups are listed alphabetically, followed by reference to the appropriate text section and a brief description of each synthetic method.

**Acetals**

| | |
|---|---|
| (Sec. 9.8) | from ketones and aldehydes by acid-catalyzed reaction with alcohols |

**Acid anhydrides**

| | |
|---|---|
| (Sec. 10.8) | from acid chlorides by reaction with carboxylate ions |

**Acid chlorides**

| | |
|---|---|
| (Sec. 10.7) | from carboxylic acids by reaction with $SOCl_2$ |

**Alcohols**

| | |
|---|---|
| (Sec. 4.6) | from alkenes by hydroxylation with basic $KMnO_4$ |
| (Sec. 4.3) | from alkenes by hydration with aqueous acid |
| (Sec. 7.5) | from alkyl halides by $S_N2$ reaction with hydroxide ion |
| (Sec. 8.6) | from ethers by acid-induced cleavage |
| (Sec. 8.7) | from epoxides by acid-catalyzed ring opening with $H_2O$ |
| (Sec. 8.3) | from ketones and aldehydes by reduction with $NaBH_4$ or $LiAlH_4$ |
| (Sec. 9.6) | from ketones and aldehydes by addition of Grignard reagents |
| (Sec. 8.3) | from carboxylic acids by reduction with $LiAlH_4$ |
| (Secs. 8.4, 10.9) | from esters by reduction with $LiAlH_4$ |
| (Sec. 10.9) | from esters by reaction with Grignard reagents |

**Aldehydes**

| | |
|---|---|
| (Sec. 4.6) | from disubstituted alkenes by reaction with acidic $KMnO_4$ |
| (Secs. 8.4, 9.3) | from primary alcohols by oxidation |

**Alkanes**

| | |
|---|---|
| (Sec. 4.5) | from alkenes by catalytic hydrogenation |
| (Sec. 7.3) | from alkyl halides by protonolysis of Grignard reagents |

**Alkenes**

| | |
|---|---|
| (Sec. 7.7) | from alkyl halides by treatment with strong base (E2 reaction) |
| (Sec. 8.5) | from alcohols by dehydration |
| (Sec. 4.11) | from alkynes by catalytic hydrogenation using the Lindlar catalyst |
| (Sec. 11.3) | from α-bromo ketones by heating with pyridine |

**Amides**

| | |
|---|---|
| (Sec. 10.7) | from acid chlorides by treatment with an amine or ammonia |
| (Sec. 10.8) | from acid anhydrides by treatment with an amine or ammonia |
| (Sec. 10.9) | from esters by treatment with an amine or ammonia |

**Amines**

| | |
|---|---|
| (Secs. 7.5, 12.4) | from primary alkyl halides by treatment with ammonia |
| (Secs. 10.10, 12.4) | from amides by reduction with $LiAlH_4$ |
| (Secs. 10.11, 12.4) | from nitriles by reduction with $LiAlH_4$ |
| (Sec. 12.4) | From aldehydes and ketones by reductive amination |

**Aromatic hydrocarbons (Arenes)**

| | |
|---|---|
| (Sec. 5.5) | from arenes by Friedel-Crafts alkylation with a primary alkyl halide |

**Aromatic sulfonic acids**

| | |
|---|---|
| (Sec. 5.4) | from arenes by electrophilic aromatic substitution with $SO_3/H_2SO_4$ |

**Arylamines**

| | |
|---|---|
| (Sec. 12.4) | from nitroarenes by reduction with either Fe, $SnCl_2$, or $H_2/Pd$ |

**Carboxylic acids**

| | |
|---|---|
| (Sec. 4.6) | from alkenes by reaction with acidic $KMnO_4$ |
| (Sec. 5.8) | from arenes by side-chain oxidation with $Na_2Cr_2O_7$ or $KMnO_4$ |
| (Sec. 9.4) | from aldehydes by oxidation |
| (Sec. 8.4) | from alcohols by oxidation |
| (Secs. 10.4, 10.11) | from nitriles by acid or base hydrolysis |
| (Sec. 10.7) | from acid chlorides by reaction with aqueous base |
| (Sec. 10.8) | from acid anhydrides by reaction with aqueous base |
| (Sec. 10.9) | from esters by hydrolysis with aqueous base |
| (Sec. 10.10) | from amides by hydrolysis with aqueous base |

**Epoxides**

| | |
|---|---|
| (Sec. 8.7) | from alkenes by treatment with a peroxyacid |

**Esters**

| | |
|---|---|
| (Sec. 10.6) | from carboxylic acids by acid-catalyzed reaction with an alcohol (Fischer esterification) |
| (Sec. 10.7) | from acid chlorides by reaction with an alcohol |
| (Sec. 10.8) | from acid anhydrides by reaction with an alcohol |
| (Sec. 11.6) | from alkyl halides by alkylation with diethyl malonate |

**Ethers**

| | |
|---|---|
| (Sec. 8.4) | from primary alkyl halides by $S_N2$ reaction with alkoxide ions (Williamson ether synthesis) |
| (Sec. 8.7) | from alkenes by epoxidation with peroxyacids |
| (Sec. 8.5) | from phenols by reaction of phenoxide ions with primary alkyl halides |

**Halides, alkyl**

| | |
|---|---|
| (Secs. 4.1–4.2) | from alkenes by electrophilic addition of HX |
| (Sec. 4.4) | from alkenes by addition of halogen |
| (Sec. 4.11) | from alkynes by addition of halogen or HX |
| (Secs. 7.2, 8.4) | from alcohols by reaction with HX |
| (Sec. 7.2) | from alcohols by reaction with $SOCl_2$ |
| (Sec. 7.2) | from alcohols by reaction with $PBr_3$ |
| (Sec. 8.6) | from ethers by cleavage with HX |
| (Sec. 11.3) | from ketones by alpha-halogenation with $Br_2$ |

**Halides, aryl**

| | |
|---|---|
| (Secs. 5.3, 5.4) | from arenes by electrophilic aromatic substitution with halogens |

**Imines**

| | |
|---|---|
| (Sec. 9.9) | from ketones or aldehydes by reaction with primary amines |

**Ketones**

| | |
|---|---|
| (Sec. 4.6) | from alkenes by reaction with acidic $KMnO_4$ |
| (Sec. 4.11) | from alkynes by acid-catalyzed hydration |
| (Sec. 5.5) | from arenes by Lewis-acid-catalyzed reaction with an acid chloride (Friedel-Crafts acylation) |
| (Secs. 8.4, 9.3) | from secondary alcohols by oxidation |
| (Sec. 10.11) | from nitriles by reaction with Grignard reagents |

**Nitriles**

| | |
|---|---|
| (Secs. 7.5, 10.11) | from primary alkyl halides by $S_N2$ reaction with $CN^-$ |

**Nitroarenes**

| | |
|---|---|
| (Sec. 5.4) | from arenes by electrophilic aromatic substitution with $HNO_3/H_2SO_4$ |

**Organometallic compounds**

| | |
|---|---|
| (Sec. 7.3) | formation of Grignard reagents from organohalides by reaction with magnesium |

**Quinones**

| | |
|---|---|
| (Sec. 8.5) | from phenols by oxidation with $Na_2Cr_2O_7$ |

**Sulfides**

| | |
|---|---|
| (Sec. 8.8) | from thiols by $S_N2$ reaction of thiolate ions with primary alkyl halides |

**Thiols**

| | |
|---|---|
| (Sec. 8.8) | from primary alkyl halides by $S_N2$ reaction with $HS^-$ |

# Reagents Used in Organic Synthesis

The following table summarizes the uses of some important reagents in organic chemistry. The reagents are listed alphabetically, followed by a brief description of the uses of each and references to the appropriate text sections.

**Acetic anhydride:** Reacts with alcohols to yield acetate esters (Section 10.8).

**Aluminum chloride:** Catalyzes the Friedel-Crafts alkylation and acylation reactions of aromatic compounds (Section 5.5).

**Bromine:** Adds to alkenes yielding 1,2-dibromides (Section 4.4).

- Adds to alkynes yielding either 1,2-dibromoalkenes or 1,1,2,2-tetrabromoalkanes (Section 4.11).
- Reacts with aromatic compounds in the presence of $FeBr_3$ catalyst to yield bromoarenes (Section 5.3).
- Reacts with ketones in acetic acid solvent to yield $\alpha$-bromo ketones (Section 11.3).

**Di-*tert*-butoxy dicarbonate:** Reacts with amino acids to give Boc protected amino acids for use in peptide synthesis (Section 15.7).

**Chlorine:** Adds to alkenes to yield 1,2-dichlorides (Section 4.4).

- Reacts with alkanes in the presence of light to yield chloroalkanes by a radical chain reaction pathway (Section 7.2).
- Reacts with aromatic compounds in the presence of $FeCl_3$ catalyst to yield chloroarenes (Section 5.4).

**Chromium trioxide:** Oxidizes alcohols in aqueous sulfuric acid to yield carbonyl-containing products. Primary alcohols yield carboxylic acids and secondary alcohols yield ketones (Sections 8.4, 9.3, 9.4).

**Dicyclohexylcarbodiimide (DCC):** Couples an amine with a carboxylic acid to yield an amide. DCC is often used in peptide synthesis (Section 15.7).

**Ethylene glycol:** Reacts with ketones or aldehydes in the presence of an acid catalyst to yield acetals, which serve as useful carbonyl protecting groups (Section 9.8).

**Ferric bromide:** Catalyzes the reaction of aromatic compounds with $Br_2$ to yield bromoarenes (Section 5.3).

**Ferric chloride:** Catalyzes the reaction of aromatic compounds with $Cl_2$ to yield chloroarenes (Section 5.4).

**Grignard reagent:** Adds to carbonyl-containing compounds (ketones, aldehydes, esters) to yield alcohols (Sections 9.6, 10.9).

**Hydrogen bromide:** Adds to alkenes to yield alkyl bromides. Markovnikov regiochemistry is observed (Sections 4.1, 4.2).
- Adds to alkynes to yield either bromoalkenes or 1,1-dibromoalkanes (Section 4.11).
- Reacts with alcohols to yield alkyl bromides (Section 7.2).

**Hydrogen chloride:** Adds to alkenes to yield alkyl chlorides. Markovnikov regiochemistry is observed (Section 4.2).
- Adds to alkynes to yield either chloroalkenes or 1,1-dichloroalkanes (Section 4.11).
- Reacts with alcohols to yield alkyl chlorides (Section 7.2).

**Iodomethane:** Reacts with alkoxide anions to yield methyl ethers (Section 8.4).
- Reacts with amines to yield methylated amines (Section 12.4).

**Iron:** Reduces nitroarenes in the presence of acid to yield anilines (Section 12.4).

**Lindlar catalyst:** Catalyzes the hydrogenation of alkynes to yield cis alkenes (Section 4.11).

**Lithium aluminum hydride**: Reduces ketones, aldehydes, esters, and carboxylic acids to yield alcohols (Section 8.3, 10.9).
- Reduces amides to yield amines (Section 10.10).
- Reduces nitriles to yield amines (Section 10.11).

**Magnesium**: Reacts with organohalides to yield Grignard reagents (Section 7.3)

**Mercuric sulfate:** Catalyzes the addition of water to alkynes in the presence of aqueous $H_2SO_4$, yielding ketones (Section 4.11).

**Nitric acid:** Reacts with aromatic compounds in the presence of sulfuric acid to yield nitroarenes (Section 5.4).
- Oxidizes aldoses to yield aldaric acids (Section 14.7).

**Palladium on carbon:** Catalyzes the hydrogenation of carbon–carbon multiple bonds. Alkenes and alkynes are reduced to yield alkanes (Sections 4.5, 4.11).
- Catalyzes the hydrogenation of nitroarenes to yield anilines (Section 12.4).

**Periodinane:** Oxidizes primary alcohols to yield aldehydes and secondary alcohols to yield ketones (Sections 8.4, 9.3).

**Phenyl isothiocyanate:** Used in the Edman degradation of peptides to identify *N*-terminal amino acids (Section 15.6).

**Phosphorus tribromide:** Reacts with alcohols to yield alkyl bromides (Section 7.2).

**Platinum oxide:** Catalyzes the hydrogenation of alkenes and alkynes to yield alkanes (Sections 4.5, 4.11).

**Potassium hydroxide:** Reacts with alkyl halides to yield alkenes by an E2 reaction (Section 7.7).

**Potassium permanganate**: Oxidizes alkenes under basic conditions to yield 1,2-diols (Section 4.6).

- Oxidizes alkenes under neutral or acidic conditions to give carboxylic acids (Sections 4.6).
- Oxidizes aromatic compounds to yield benzoic acids (Section 5.8).

**Pyridine:** Reacts with α-bromo ketones to yield α,β-unsaturated ketones (Section 11.3).

- Catalyzes the reaction of alcohols with acid chlorides to yield esters (Section 10.7).
- Catalyzes the reaction of alcohols with acetic anhydride to yield acetate esters (Section 10.8).

**Silver oxide:** Catalyzes reaction of carbohydrates with iodomethane to yield ethers (Section 14.7).

**Sodium borohydride:** Reduces ketones and aldehydes to yield alcohols (Section 8.3).

- Reduces quinones to yield hydroquinones (Section 8.5).

**Sodium cyanide:** Reacts with alkyl halides to yield nitriles (Sections 7.5, 10.13).

**Sodium dichromate:** Oxidizes primary alcohols to yield carboxylic acids and secondary alcohols to yield ketones (Sections 8.4, 9.3).

- Oxidizes alkylbenzenes to yield benzoic acids (Section 5.8).

**Stannous chloride:** Reduces nitroarenes to yield arylamines (Section 12.4).

- Reduces quinones to yield hydroquinones (Section 8.5).

**Sulfur trioxide:** Reacts with aromatic compounds in $H_2SO_4$ to yield arenesulfonic acids (Section 5.4).

**Sulfuric acid:** Dehydrates alcohols to yield alkenes (Section 8.4).

- Reacts with alkynes in the presence of water and mercuric acetate to yield ketones (Section 4.11).

**Thionyl chloride**: Reacts with primary and secondary alcohols to yield alkyl chlorides (Section 7.2).

- Reacts with carboxylic acids to yield acid chlorides (Section 10.6).

**Trifluoroacetic acid:** Catalyzes cleavage of the Boc protecting group from amino acids in peptide synthesis (Section 15.7).

# Organic Name Reactions

**Aldol condensation reaction** (Section 11.8): The nucleophilic addition of an enol or enolate ion to a ketone or aldehyde, yielding a β-hydroxy ketone.

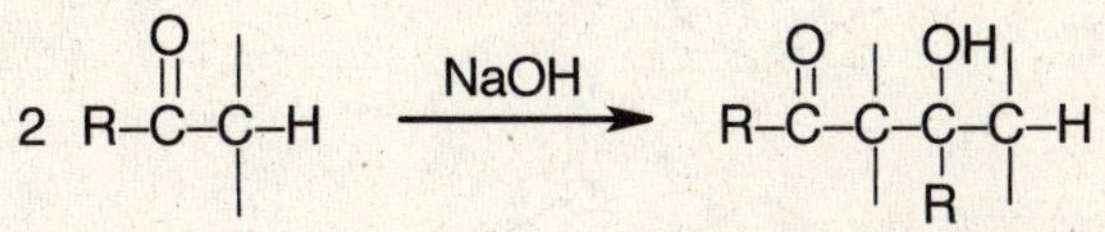

**Claisen condensation reaction** (Section 11.10): A nucleophilic acyl substitution reaction that occurs when an ester enolate ion attacks the carbonyl group of a second ester molecule. The product is a β-keto ester.

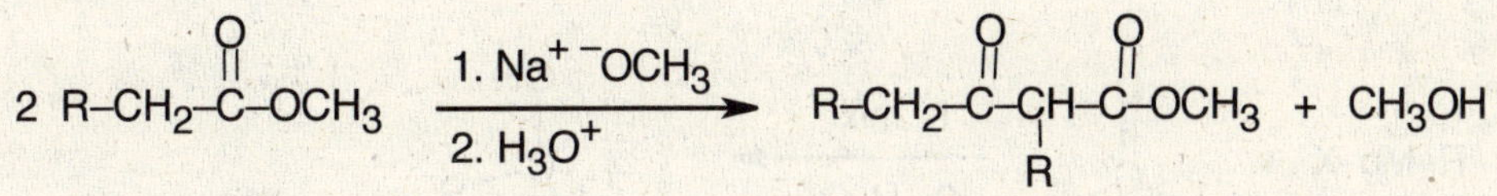

**Edman degradation** (Section 15.7): A method for cleaving the *N*-terminal amino acid from a peptide by treatment of the peptide with *N*-phenylisothiocyanate.

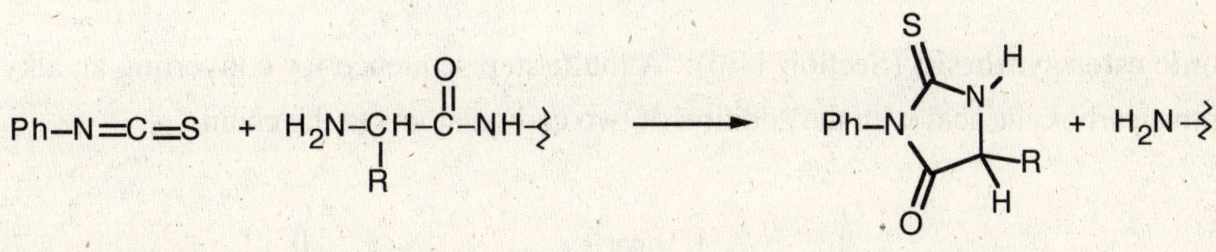

**Fehling's test** (Section 14.7): A chemical test for aldehydes, involving treatment with cupric ion.

**Fischer esterification reaction** (Section 10.6): The acid-catalyzed reaction between a carboxylic acid and an alcohol to yield an ester.

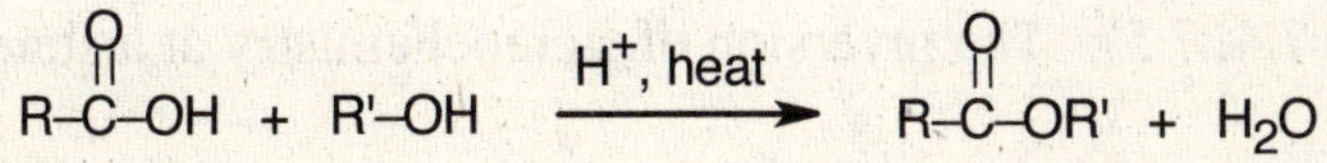

**Friedel-Crafts reaction** (Section 5.5): The alkylation or acylation of an aromatic ring by treatment with an alkyl- or acyl chloride in the presence of a Lewis-acid catalyst.

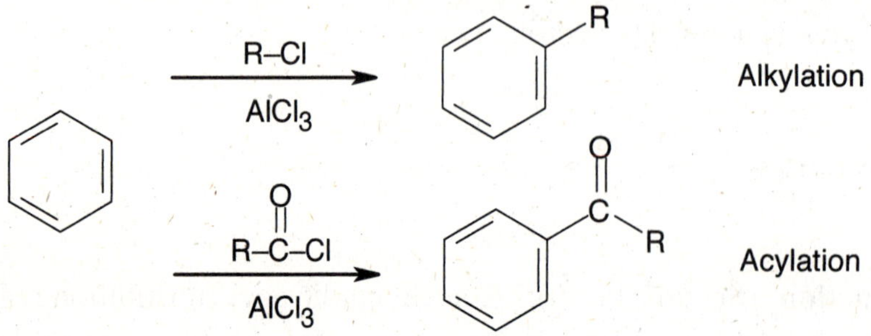


**Grignard reaction** (Section 9.6): The nucleophilic addition reaction of an alkylmagnesium halide to a ketone, aldehyde, or ester carbonyl group.

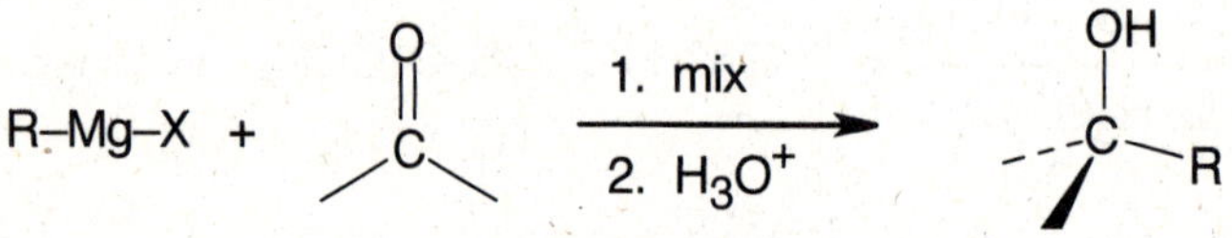


**Grignard reagent** (Section 7.3): An organomagnesium halide, RMgX, prepared by reaction between an organohalide and magnesium metal.

**Malonic ester synthesis** (Section 11.6): A multi-step sequence for converting an alkyl halide into a carboxylic acid with the addition of two carbon atoms to the chain.

$$R{-}CH_2{-}X + Na^{+}\ {}^{-}{:}\underset{\displaystyle COOCH_3}{\underset{|}{CH}}{-}\overset{O}{\overset{\|}{C}}{-}OCH_3 \xrightarrow[\text{2. } H_3O^+,\ \text{heat}]{\text{1. heat}} RCH_2{-}CH_2\overset{O}{\overset{\|}{C}}OCH_3 + CO_2 + CH_3OH$$

**Sanger dideoxy DNA sequencing** (Section 16.10): A rapid and efficient method for sequencing long restriction fragments of DNA by using dye-labeled dideoxynucleotide triphosphates and DNA polymerase.

**Walden inversion** (Sections 7.4, 7.5): The inversion of stereochemistry at a stereocenter that occurs during $S_N2$ reactions.

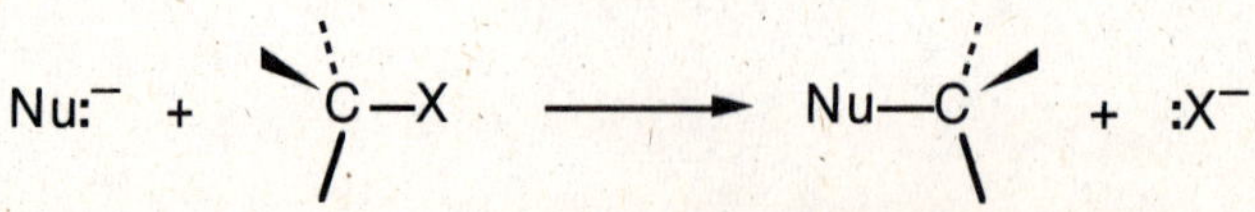

**Williamson ether synthesis** (Section 8.4): A method for preparing ethers by treatment of a primary alkyl halide with an alkoxide ion.

$$R\text{–}O^{-}\ Na^{+} + R'CH_2Br \longrightarrow R\text{–}O\text{–}CH_2R' + NaBr$$

# Abbreviations

| | |
|---|---|
| Å | symbol for Angstrom unit ($10^{-8}$ cm) |
| Ac– | acetyl group, $CH_3C(=O)-$ |
| Ar– | aryl group |
| $[\alpha]_D$ | specific rotation |
| Boc | *tert*-butoxycarbonyl group, $(CH_3)_3COC(=O)-$ |
| bp | boiling point |
| *sec*-Bu | *sec*-butyl group, $CH_3CH_2CH(CH_3)-$ |
| *t* -Bu | *tert*-butyl group, $(CH_3)_3C-$ |
| $cm^{-1}$ | wavenumber or reciprocal centimeter |
| D | stereochemical designation of carbohydrates and amino acids |
| DCC | dicyclohexylcarbodiimide, $C_6H_{11}-N{=}C{=}N-C_6H_{11}$ |
| δ | chemical shift in ppm downfield from TMS |
| Δ | symbol for heat; also symbol for change |
| $\Delta H$ | heat of reaction |
| DNA | deoxyribonucleic acid |
| (*E*) | entgegen, stereochemical designation of double bond geometry |
| $E_{act}$ | activation energy |
| E1 | unimolecular elimination reaction |
| E2 | bimolecular elimination reaction |
| Et | ethyl group, $CH_3CH_2-$ |
| Fmoc | fluorenylmethyloxycarbonyl |
| hν | symbol for light |
| Hz | Hertz |
| *i*- | iso |
| IR | infrared |
| J | Joule |
| *J* | symbol for coupling constant |
| K | Kelvin temperature |
| $K_a$ | symbol for acidity constant |
| $K_b$ | symbol for basicity constant |
| kcal | kilocalories |
| L | stereochemical designation of carbohydrates and amino acids |
| Me | methyl group, $CH_3-$ |
| MHz | megahertz ($10^6$ cycles per second) |
| mol wt. | molecular weight |

| | |
|---|---|
| mp | melting point |
| NMR | nuclear magnetic resonance |
| –OAc | acetate group, 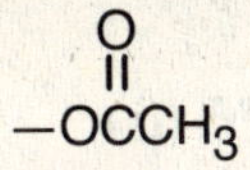 |
| Ph | phenyl group, $-C_6H_5$ |
| pH | measure of acidity of aqueous solution ($= -\log [H_3O^+]$) |
| p$K_a$ | measure of acid strength ($= -\log K_a$) |
| p$K_b$ | measure of base strength ($= -\log K_b$) |
| ppm | parts per million |
| *i*-Pr | isopropyl group, $(CH_3)_2CH-$ |
| R– | symbol for a generalized organic group |
| (*R*) | rectus, stereochemical designation of stereocenters |
| RNA | ribonucleic acid |
| (*S*) | sinister, stereochemical designation of stereocenters |
| *sec-* | secondary |
| $S_N1$ | unimolecular substitution reaction |
| $S_N2$ | bimolecular substitution reaction |
| *tert-* | tertiary |
| THF | tetrahydrofuran |
| TMS | tetramethylsilane, $(CH_3)_4Si$ |
| UV | ultraviolet |
| (*Z*) | zusammen, stereochemical designation of double bond geometry |
| → | chemical reaction in direction indicated |
| ⇌ | reversible chemical reaction |
| ↔ | resonance symbol |
|  | curved arrow indicating direction of electron flow |
| ≡ | is equivalent to |
| > | greater than |
| < | less than |
| ≈ | approximately equal to |
| 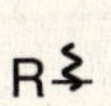 | indicates that the organic fragment shown is a part of a larger molecule |
|  | single bond coming out of the plane of the paper |
| ----- | single bond receding into the plane of the paper |
| ...... | partial bond |
| δ+, δ– | partial charge |
| * | isotopically labeled atom |
| ‡ | denoting the transition state |

# Proton NMR Chemical Shifts

| Type of Proton | | Chemical Shift (δ) |
|---|---|---|
| Alkyl, primary | $R{-}CH_3$ | 0.7 – 1.3 |
| Alkyl, secondary | $R{-}CH_2{-}R$ | 1.2 – 1.4 |
| Alkyl, tertiary | $R_3C{-}H$ | 1.4 – 1.7 |
| Allylic | –C=C–C–H | 1.6 – 1.9 |
| Aliphatic carbonyl | –C(=O)–C–H | 2.0 – 2.3 |
| Benzylic | Ar–C–H | 2.3 – 3.0 |
| Acetylenic | R–C≡C–H | 2.5 – 2.7 |
| Alkyl chloride | Cl–C–H | 3.0 – 4.0 |
| Alkyl bromide | Br–C–H | 2.5 – 4.0 |
| Alkyl iodide | I–C–H | 2.0 – 4.0 |
| Amine | N–C–H | 2.2 – 2.6 |
| Epoxide | C–C–H (O) | 2.5 – 3.5 |
| Alcohol | HO–C–H | 3.5 – 4.5 |
| Ether | RO–C–H | 3.5 – 4.5 |
| Vinylic | –C=C–H | 5.0 – 6.5 |
| Aromatic | Ar–H | 6.5 – 8.0 |
| Aldehyde | R–C(=O)–H | 9.7 – 10.0 |
| Carboxylic acid | R–C(=O)–O–H | 11.0 – 12.0 |
| Alcohol | R–O–H | 3.5 – 4.5 |
| Phenol | Ar–O–H | 2.5 – 6.0 |

# Infrared Absorption Frequencies

| Functional Group Class | | Frequency ($cm^{-1}$) |
|---|---|---|
| Alcohol | –O–H | 3300–3600 (s) |
| | –C–O– | 1050 (s) |
| Aldehyde | –CO–H | 2720, 2820 (m) |
| *aliphatic* | C=O | 1725 (s) |
| *aromatic* | C=O | 1705 (s) |
| Alkane | –C–H | 2850–2960 (s) |
| | –C–C– | 800–1300 (m) |
| Alkene | =C–H | 3020–3100 (m) |
| | C=C | 1650–1670 (m) |
| | $RCH{=}CH_2$ | 910, 990 (m) |
| | $R_2C{=}CH_2$ | 890 (m) |
| Alkyne | ≡C–H | 3300 (s) |
| | –C≡C– | 2100–2260 (m) |
| Alkyl bromide | –C–Br | 500–600 (s) |
| Alkyl chloride | –C–Cl | 600–800 (s) |

| | | |
|---|---|---|
| Amine, *primary* | $-NH_2$ | 3400, 3500 (s) |
| *secondary* | $>N-H$ | 3350 (s) |
| Ammonium salt | $-\overset{+}{N}-H$ | 2200–3000 (broad) |
| Aromatic ring | Ar–H | 3030 (m) |
| *monosubstituted* | Ar–R | 690–710 (s) |
| | | 730–770 (s) |
| *o-disubstituted* | | 735–770 (s) |
| *m-disubstituted* | | 690–710 (s) |
| | | 810–850 (s) |
| |  | |
| *p-disubstituted* | | 810–840 (s) |
| Carboxylic acid | –O–H | 2500–3300 (broad) |
| *associated* | $>C=O$ | 1710 (s) |
| *free* | $>C=O$ | 1760 (s) |
| Acid anhydride | $>C=O$ | 1820, 1760 (s) |
| Acid chloride | | |
| *aliphatic* | $>C=O$ | 1810 (s) |
| *aromatic* | $>C=O$ | 1770 (s) |

| | | |
|---|---|---|
| Amide, *aliphatic* | >C=O | 1690 (s) |
| *aromatic* | >C=O | 1675 (s) |
| *N-substituted* | >C=O | 1680 (s) |
| *N,N-disubstituted* | >C=O | 1650 (s) |
| Ester, *aliphatic* | >C=O | 1735 (s) |
| *aromatic* | >C=O | 1720 (s) |
| Ether | —O—C< (with C bonded to three further groups) | 1050–1150 (s) |
| Ketone, *aliphatic* | >C=O | 1715 (s) |
| *aromatic* | >C=O | 1690 (s) |
| *6-memb. ring* | >C=O | 1715 (s) |
| *5-memb. ring* | >C=O | 1750 (s) |
| Nitrile, *aliphatic* | —C≡N | 2250 (m) |
| *aromatic* | —C≡N | 2230 (m) |
| Phenol | —O—H | 3500 (s) |

(s) = strong; (m) = medium intensity

# Nobel Prizes in Chemistry

1901 **Jacobus H. van't Hoff** (The Netherlands):
"for the discovery of laws of chemical dynamics and of osmotic pressure"

1902 **Emil Fischer** (Germany):
"for syntheses in the groups of sugars and purines"

1903 **Svante A. Arrhenius** (Sweden):
"for his theory of electrolytic dissociation"

1904 **Sir William Ramsey** (Britain):
"for the discovery of gases in different elements in the air and for the determination of their place in the periodic system"

1905 **Adolf von Baeyer** (Germany):
"for his researches on organic dyestuffs and hydroaromatic compounds"

1906 **Henri Moissan** (France):
"for his research on the isolation of the element fluorine and for placing at the service of science the electric furnace that bears his name"

1907 **Eduard Buchner** (Germany):
"for his biochemical researches and his discovery of cell-less formation"

1908 **Ernest Rutherford** (Britain):
"for his investigation into the disintegration of the elements and the chemistry of radioactive substances"

1909 **Wilhelm Ostwald** (Germany):
"for his work on catalysis and on the conditions of chemical equilibrium and velocities of chemical reactions"

1910 **Otto Wallach** (Germany):
"for his services to organic chemistry and the chemical industry by his pioneer work in the field of alicyclic substances"

1911 **Marie Curie** (France):
"for her services to the advancement of chemistry by the discovery of the elements radium and polonium"

1912 **Victor Grignard** (France):
"for the discovery of the so-called Grignard reagent, which has greatly helped in the development of organic chemistry"

**Paul Sabatier** (France):
"for his method of hydrogenating organic compounds in the presence of finely divided metals"

1913 **Alfred Werner** (Switzerland):
"for his work on the linkage of atoms in molecules by which he has thrown new light on earlier investigations and opened up new fields of research especially in inorganic chemistry"

1914 **Theodore W. Richards** (U.S.):
"for his accurate determinations of the atomic weights of a great number of chemical elements"

1915 **Richard M. Willstätter** (Germany):
"for his research on plant pigments, principally on chlorophyll"

1916 No award

1917 No award

1918 **Fritz Haber** (Germany):
"for the synthesis of ammonia from its elements, nitrogen and hydrogen"

1919 No award

1920 **Walther H. Nernst** (Germany):
"for his thermochemical work"

1921 **Frederick Soddy** (Britain):
"for his contributions to the chemistry of radioactive substances and his investigations into the origin and nature of isotopes"

1922 **Francis W. Aston** (Britain):
"for his discovery, by means of his mass spectrograph, of the isotopes of a large number of nonradioactive elements, as well as for his discovery of the whole-number rule"

1923 **Fritz Pregl** (Austria):
"for his invention of the method of microanalysis of organic substances"

1924 No award

1925 **Richard A. Zsigmondy** (Germany):
for his demonstration of the heterogeneous nature of colloid solutions, and for the methods he used, which have since become fundamental in modern colloid chemistry"

1926 **Theodor Svedberg** (Sweden):
"for his work on disperse systems"

1927 **Heinrich O. Wieland** (Germany):
"for his research on bile acids and related substances"

1928 **Adolf O. R. Windaus** (Germany):
"for his studies on the constitution of the sterols and their connection with the vitamins"

1929 **Arthur Harden** (Britain):
**Hans von Euler-Chelpin** (Sweden):
"for their investigation on the fermentation of sugar and of fermentative enzymes"

1930 **Hans Fischer** (Germany):
"for his researches into the constitution of hemin and chlorophyll, and especially for his synthesis of hemin"

1931 **Frederich Bergius** (Germany):
**Carl Bosch** (Germany):
"for their contributions to the invention and development of chemical high-pressure methods"

1932 **Irving Langmuir** (U.S.):
"for his discoveries and investigations in surface chemistry"

1933 No award

1934 **Harold C. Urey** (U.S.):
"for his discovery of heavy hydrogen"

1935 **Frederic Joliot** (France):
**Irene Joliot-Curie** (France):
"for their synthesis of new radioactive elements"

1936 **Peter J. W. Debye** (Netherlands/U.S.):
"for his contributions our knowledge of molecular structure through his investigations on dipole moments and on the diffraction of X rays and electrons in gases"

1937 **Walter N. Haworth** (Britain):
"for his researches into the constitution of carbohydrates and vitamin C"

**Paul Karrer** (Switzerland):
"for his researches into the constitution of carotenoids, flavins, and vitamins A and B"

1938 **Richard Kuhn** (Germany):
"for his work on carotenoids and vitamins"

1939 **Adolf F. J. Butenandt** (Germany):
"for his work on sex hormones"

**Leopold Ruzicka** (Switzerland):
"for his work on polymethylenes and higher terpenes"

1940 No award

1941 No award

1942 No award

1943 **Georg de Hevesy** (Hungary):
"for his work on the use of isotopes as tracer elements in researches on chemical processes"

1944 **Otto Hahn** (Germany):
"for his discovery of the fission of heavy nuclei"

1945 **Artturi I. Virtanen** (Finland):
"for his researches and inventions in agricultural and nutritive chemistry, especially for his fodder preservation method"

1946 **James B. Sumner** (U.S.):
"for his discovery that enzymes can be crystallized"

**John H. Northrop** (U.S.):
**Wendell M. Stanley** (U.S.):
for their preparation of enzymes and virus proteins in a pure form"

1947 **Sir Robert Robinson** (Britain):
"for his investigations on plant products of biological importance, particularly the alkaloids"

1948 **Arne W. K. Tiselius** (Sweden):
"for his researches on electrophoresis and adsorption analysis, especially for his discoveries concerning the complex nature of the serum proteins"

1949 **William F. Giauque** (U.S.):
"for his contributions in the field of chemical thermodynamics, particularly concerning the behavior of substances at extremely low temperatures"

1950 **Kurt Alder** (Germany):
**Otto P. H. Diels** (Germany):
"for their discovery and development of the diene synthesis"

1951 **Edwin M. McMillan** (U.S.):
**Glenn T. Seaborg** (U.S.):
"for their discoveries in the chemistry of the transuranium elements"

1952 **Archer J. P. Martin** (Britain):
**Richard L. M. Synge** (Britain):
"for their development of partition chromatography"

1953 **Hermann Staudinger** (Germany):
"for his discoveries in the field of macromolecular chemistry"

1954 **Linus C. Pauling** (U.S.):
"for his research into the nature of the chemical bond and its application to the elucidation of the structure of complex substances"

1955 **Vincent du Vigneaud** (U.S.):
"for his work on biochemically important sulfur compounds, especially for the first synthesis of a polypeptide hormone"

1956 **Sir Cyril N. Hinshelwood** (Britain):
**Nikolai N. Semenov** (U.S.S.R.):
"for their research in clarifying the mechanisms of chemical reactions in gases"

1957 **Sir Alexander R. Todd** (Britain):
"for his work on nucleotides and nucleotide coenzymes"

1958 **Frederick Sanger** (Britain):
"for his work on the structure of proteins, particularly insulin"

1959 **Jaroslav Heyrovsky** (Czechoslovakia):
"for his discovery and development of the polarographic method of analysis"

1960 **Willard F. Libby** (U.S.):
"for his method to use carbon-14 for age determination in archaeology, geology, geophysics, and other branches of science"

1961 **Melvin Calvin** (U.S.):
"for his research on the carbon dioxide assimilation in plants"

1962 **John C. Kendrew** (Britain):
**Max F. Perutz** (Britain):
"for their studies of the structures of globular proteins"

1963 **Giulio Natta** (Italy):
**Karl Ziegler** (Germany):
"for their work in the controlled polymerization of hydrocarbons through the use of organometallic catalysts"

1964 **Dorothy C. Hodgkin** (Britain):
"for her determinations by X-ray techniques of the structures of important biochemical substances, particularly vitamin $B_{12}$ and penicillin"

1965 **Robert B. Woodward** (U.S.):
"for his outstanding achievements in the 'art' of organic synthesis"

1966 **Robert S. Mulliken** (U.S.):
"for his fundamental work concerning chemical bonds and the electronic structure of molecules by the molecular orbital method"

1967 **Manfred Eigen** (Germany):
**Ronald G. W. Norrish** (Britain):
**George Porter** (Britain):
"for their studies of extremely fast chemical reactions, effected by disturbing the equilibrium with very short pulses of energy"

1968 **Lars Onsager** (U.S.):
"for his discovery of the reciprocal relations bearing his name, which are fundamental for the thermodynamics of irreversible processes"

1969 **Sir Derek H. R. Barton** (Britain):
**Odd Hassel** (Norway):
"for their contributions to the development of the concept of conformation and its application in chemistry"

1970 **Luis F. Leloir** (Argentina):
"for his discovery of sugar nucleotides and their role in the biosynthesis of carbohydrates"

1971 **Gerhard Herzberg** (Canada):
"for his contributions to the knowledge of electronic structure and geometry of molecules, particularly free radicals"

1972 **Christian B. Anfinsen** (U.S.):
"for his work on ribonuclease, especially concerning the connection between the amino acid sequence and the biologically active conformation"

**Stanford Moore** (U.S.):
**William H. Stein** (U.S.):
"for their contribution to the understanding of the connection between chemical structure and catalytic activity of the active center of the ribonuclease molecule"

1973 **Ernst Otto Fischer** (Germany):
**Geoffrey Wilkinson** (Britain):
"for their pioneering work, performed independently, on the chemistry of the organometallic sandwich compounds"

1974 **Paul J. Flory** (U.S.):
"for his fundamental achievements, both theoretical and experimental, in the physical chemistry of macromolecules"

1975 **John Cornforth** (Australia/Britain):
"for his work on the stereochemistry of enzyme-catalyzed reactions"

**Vladimir Prelog** (Yugoslavia/Switzerland):
"for his work on the stereochemistry of organic molecules and reactions"

1976 **William N. Lipscomb** (U.S.):
"for his studies on the structures of boranes illuminating problems of chemical bonding"

1977 **Ilya Pregogine** (Belgium):
"for his contributions to nonequilibrium thermodynamics, particularly the theory of dissipative structures"

1978 **Peter Mitchell** (Britain):
"for his contribution to the understanding of biological energy transfer through the formulation of the chemiosmotic theory"

1979 **Herbert C. Brown** (U.S.):
"for his application of boron compounds to synthetic organic chemistry"

**Georg Wittig** (Germany):
"for developing phosphorus reagents, presently bearing his name"

1980 **Paul Berg** (U.S.):
"for his fundamental studies of the biochemistry of nucleic acids, with particular regard to recombinant DNA"

**Walter Gilbert** (U.S.)
**Frederick Sanger** (Britain):
"for their contributions concerning the determination of base sequences in nucleic acids"

1981 **Kenichi Fukui** (Japan)
**Roald Hoffmann** (U.S.):
for their theories, developed independently, concerning the course of chemical reactions"

1982 **Aaron Klug** (Britain):
"for his development of crystallographic electron microscopy and his structural elucidation of biologically important nucleic acid – protein complexes"

1983 **Henry Taube** (U.S.):
"for his work on the mechanisms of electron transfer reactions, especially in metal complexes"

1984 **R. Bruce Merrifield** (U.S.):
"for his development of methodology for chemical synthesis on a solid matrix"

1985 **Herbert A. Hauptman** (U.S.):
**Jerome Karle** (U.S.):
"for their outstanding achievements in the development of direct methods for the determination of crystal structures"

1986 **John C. Polanyi** (Canada):
"for his pioneering work in the use of infrared chemiluminescence in studying the dynamics of chemical reactions"

**Dudley R. Herschbach** (U.S.):
**Yuan T. Lee** (U.S.):
"for their contributions concerning the dynamics of chemical elementary processes"

1987 **Donald J. Cram** (U.S.):
**Jean-Marie Lehn** (France):
**Charles J. Pedersen** (U.S.):
"for their development and use of molecules with structure-specific interactions of high selectivity"

1988 **Johann Deisenhofer** (Germany):
**Robert Huber** (Germany):
**Hartmut Michel** (Germany):
"for their determination of the structure of the photosynthetic reaction center of bacteria"

1989 **Sidney Altman** (U.S.):
**Thomas R. Cech** (U.S.):
"for their discovery of catalytic properties of RNA"

1990 **Elias J. Corey** (U.S.):
"for his development of the theory and methodology of organic synthesis"

1991 **Richard R. Ernst** (Switzerland):
"for his contributions to the development of the methodology of high resolution NMR spectroscopy"

1992 **Rudolph A. Marcus** (U.S.):
"for his contributions to the theory of electron-transfer reactions in chemical systems"

1993 **Kary B. Mullis** (U.S.):
"for his development of the polymerase chain reaction"

**Michael Smith** (Canada):
"for his fundamental contributions to the establishment of oligonucleotide-based site-directed mutagenesis and its development for protein studies"

1994 **George A Olah** (U.S.):
"for pioneering research on carbocations and their role in the chemical reactions of hydrocarbons"

1995 **F. Sherwood Rowland** (U.S.)
**Mario Molina** (U.S.)
**Paul Crutzen** (Germany)
"for their work in atmospheric chemistry, particularly concerning the formation and decomposition of ozone"

1996 **Robert F. Curl, Jr.** (U.S.)
**Harold W. Kroto** (U.K.)
**Richard E. Smalley** (U.S.)
"for their discovery of carbon atoms bound in the form of a ball (fullerenes)."

1997 **Paul D. Boyer** (U.S.)
**John E. Walker** (U.K.)
"for having elucidated the mechanism by which ATP synthase catalyzes the synthesis of adenosine triphosphate, the energy currency of living cells"

**Jens C. Skou** (Denmark)
"for his discovery of the ion-transporting enzyme $Na^+$–$K^+$ ATPase, the first molecular pump"

1998 **Walter Kohn** (U.S.)
**John A. Pople** (U.S.)
"to Walter Kohn for his development of the density-functional theory and to John Pople for his development of computational methods in quantum chemistry"

1999 **Ahmed H. Zewail** (Egypt, U.S.)
"for his studies of the transition states of chemical reactions using femtosecond spectroscopy."

2000 **Alan J. Heeger** (U.S.)
**Alan G. MacDiarmid** (U.S.)
**Hideki Shirakawa** (Japan)
"for opening and developing the important new field of electrically conductive polymers"

2001 **William S. Knowles** (U.S.)
**Ryoji Noyori** (Japan)
**K. Barry Sharpless** (U.S.)
"for their work on chirally catalysed hydrogenation and oxidation reactions"

2002 **John B. Fenn** (U.S.)
**Koichi Tanaka** (Japan)
"for their development of soft desorption ionisation methods for mass spectrometric analyses of biological macromolecules"

**Kurt Wüthrich** (Switzerland)
"for his development of nuclear magnetic resonance spectroscopy for determining the three-dimensional structure of biological macromolecules in solution"

2003 **Peter Agre** (U.S.)
"for the discovery of water channels in cell membranes"
**Roderick MacKinnon** (U.S.)
"for structural and mechanistic studies of ion channels in cell membranes"

2004 **Aaron Ciechanover** (Israel)
**Avram Hershko** (Israel)
**Irwin Rose** (U.S.)
"for the discovery of ubiquitin-mediated protein degradation"

2005 **Yves Chauvin** (France)
**Robert H. Grubbs** (U.S.)
**Richard R.Schrock** (U.S.)
"for the development of the metathesis method in organic synthesis"

2006 **Roger D. Kornberg** (U.S.)
"for his studies of the molecular basis of eukaryotic transcription"

2007 **Gerhard Ertl** (Germany)
"for his studies of chemical processes on solid surfaces"

2008 **Osamu Shimomura** (Japan)
**Martin Chalfie** (U.S.)
**Roger Y. Tsien** (U.S.)
" for the discovery and development of the green fluorescent protein, GFP"

2009 **Venkatraman Ramakrishnan** (U.K.)
**Thomas A. Steitz** (U.S.)
**Ada E.Yonath** (Israel)
"for studies of the structure and function of the ribosome"